Tehrani's IP Telephony Dictionary

**Published By
Althos Inc.**

404 Wake Chapel Road
Fuquay-Varina, NC 27526 USA
Telephone: 1-800-227-9681
1-919-557-2260
Fax: 1-919-557-2261
email: Success@Althos.com
web: www.Althos.com

Althos

All rights reserved. No part of this book may be reproduced or transmitted in any form or by any means, electronic or mechanical, including photocopying recording or by any information storage and retrieval system without written permission from the authors and publisher, except for the inclusion of brief quotations in a review.

Copyright © 2004 By Althos Publishing

Printed and Bound by Lightning Source Printing, La Vergne, TN USA

Every effort has been made to make this manual as complete and as accurate as possible. However, there may be mistakes both typographical and in content. Therefore, this text should be used only as a general guide and not as the ultimate source of information. Furthermore, this manual contains information on telecommunications accurate only up to the printing date. The purpose of this manual to educate. The authors and Althos Publishing shall have neither liability nor responsibility to any person or entity with respect to any loss or damage caused, or alleged to be caused, directly or indirectly by the information con-

---

International Standard Book Number: 0-9742787-1-8

About the Editor

Mr. Tehrani, President and Group Editor-in-Chief of Technology Marketing Corporation (TMC), is a respected communications technology expert, visionary and evangelist specializing in various leading edge technologies. Tehrani is frequently quoted in newspapers, has served as an expert witness, and is a featured technology editor on Washington, D.C.'s top-rated WMAL radio station. Due to his deep involvement and profound knowledge of various markets, Tehrani is sought out and consulted by companies worldwide.

Starting at TMC in 1982, he has led the company in many capacities, including MIS Director. Mr. Tehrani is the founder of the industry leading Internet Telephony brand which includes a monthly magazine and the biannual INTERNET TELEPHONY® Conference & EXPO, which he is chairman of. Among other responsibilities, he is the Publisher of Customer Inter@ction Solutions magazine. Tehrani frequently speaks at industry events and has been appointed as a speaker at many user group meetings for companies such as Rockwell, Inter-Tel and Empirix. Rich further serves as a keynote/introductory speaker in live and Web seminars throughout the country. Often referred to as the technology guru, Rich has played a pivotal role in steering TMC toward coverage of cutting edge technologies and continues to be a driving force in the creation and implementation of pioneering publications and events. Rich holds a computer hardware and software engineering degree from The University of Connecticut.

Contributing Editors and Advisors

Jon Arnold is the Program Leader for Frost & Sullivan's voice over packet equipment service. His coverage focuses on nextgen infrastructure for service provider networks, which he complements with other Frost analyst coverage in the areas of packet voice services, and enterprise IP telephony. Based in the firm's Toronto office, Jon has been covering the packet voice and IP telephony market continuously since 2001. His current report addresses the standalone enterprise voice gateway market, and will soon be publishing an update on the IP cable telephony market. Central to Jon's coverage is a semi-annual analysis of the global packet voice equipment market, encompassing media gateways, softswitches, media servers, application servers and session border controllers. During 2003 Jon has taken a higher profile in this market on two fronts. First, he has forged a marketing alliance between Frost & Sullivan and the IPCC - the International Packet Communications Consortium. Secondly, he has been actively involved with the VoIP Council since its inception in March 2003. The VoIP Council is an independent industry group serving as an information and educational resource for the media community. Jon is an active participant and speaker at industry conferences, including VON, Supercomm, NGN and ComNet. He is widely quoted on the packet voice industry in both the general business press as well as a variety of industry magazines. Prior to joining Frost, Jon operated a market research consultancy for 15 years, covering many B2B sectors, including telecom and technology.

Mr. Dave Bowler is an independent telecommunications training consultant. He has almost 20 years experience in designing and delivering training in the areas of wireless networks and related technologies, including CDMA, TDMA, GSM and 3G systems. He also has expertise in Wireless Local Loop and microwave radio systems and has designed and delivered a range of training courses on SS7 and other network signaling protocols. Mr. Bowler has worked for a number of telecommunications operators including Cable and Wireless and Mercury Communications and also for Wray Castle a telecommunications training company where he was responsible for the design of training programmes for delivery on a global basis. Mr. Bowler was educated in the United Kingdom and holds a series of specialized maritime electronic engineering certificates.

Kenneth Brown is the President of the Alexis de Tocqueville Institution a non-partisan, non-profit think tank based in Arlington, Virginia. The Alexis de Tocqueville Institution's policy research topics include international trade, immigration and economic policy. Ken oversees all of the Alexis de Tocqueville Institution's policy studies and foundation relationships. Ken is also the Director of AdTI's Technology Program. The AdTI technology policy group focuses on a variety of issues including telecommunications, software licensing, intellectual property and export controls. Kenneth is often quoted for both domestic and international news analysis. His research and editorials have been published in numerous journals including the Wall Street Journal, Journal of Commerce, Washington Times and Internet Regulation Alert. Ken is also Vice-President of Emerging Markets Group, an overseas market investment and advisory firm. Kenneth has a B.A. in English Literature from George Mason University and is married to Erica Brown, an employment attorney at Mintz Levin. Ken and Erica live in Arlington, Virgina.

v

Robert T. Flood has had a distinguished 30-year career in the telecommunications industry. As a renowned speaker at forums around the globe, Robert is a noted authority on Internet Protocol (IP) telephony. Prior to co-founding PingTone Communications, a provider of managed IPT services to corporations and business users worldwide, Robert was the chief technology officer of Cable & Wireless Global, managing a $3 billion capital budget and 1,500 employees worldwide. Robert was previously the chief technology officer and senior vice president of engineering for ICG Communications, driving the Denver-based telecommunications company into 83 markets. At ICG Flood pioneered Voice over Internet Protocol (IP), covering 188 long distance markets within an 8-month period. Prior to that, he worked for CenTel for 19 years in senior engineering positions. Robert serves on the Board of Directors of several technology companies. He earned a bachelor's degree in Economics from the University of Nebraska, completed the executive program at the J.L. Kellogg Graduate School of Management at Northwestern University, and has a master's degree in Economics from the University of Nevada Las Vegas.

Mr. Harte is the president of Althos, an expert information provider covering the communications industry. He has over 29 years of technology analysis, development, implementation, and business management experience. Mr. Harte has worked for leading companies including Ericsson/General Electric, Audiovox/Toshiba and Westinghouse and consulted for hundreds of other companies. Mr. Harte continually researches, analyzes, and tests new communication technologies, applications, and services. He has authored over 30 books on telecommunications technologies on topics including Wireless Mobile, Data Communications, VoIP, Broadband, Prepaid Services, and Communications Billing. Mr. Harte's holds many degrees and certificates include an Executive MBA from Wake Forest University (1995) and a BSET from the University of the State of New York, (1990).

Louis Holder is the Executive Vice President of Vonage Product Development and manages the development of Vonage's technology infrastructure and support services, including systems development and web application development. Before joining Vonage, Mr. Holder was a vice president in the Program Trading Technology Group at PaineWebber, where he performed business and technical analysis and was responsible for the development of a next-generation program trading system. Prior to PaineWebber, Mr. Holder was a senior software developer at Cantor Fitzgerald. Mr. Holder developed several analytical and trading applications for Cantor Fitzgerald's FX Options, Emerging Markets, Government Bond Swaps, Euro Bond Swaps and Interest Rate Swaps desks. Mr. Holder earned a bachelors degree in electrical engineering from Polytechnic University, New York.

Erin Kelly is Associate Director, Next Generation Services at Bell Canada and the team leader for the company's Consumer VoIP unit. For the past 10 years Erin has worked in all aspects of cutting-edge IP-based communications: she developed and helped launch the first streaming video site in the country for Canada's national broadcaster, the CBC, in 1995 and subsequently worked as a product developer for streaming media applications at the former Newbridge startup Televitesse. Erin worked on the launch of Canada's largest high speed Internet service and was Senior Manager of Wireless Applications development at Nortel Networks before joining Bell Canada's Next Generation Services division. At Bell, Canada's premier telephone company and internet and wireless service provider, Erin works on key strategic business initiatives and research projects where she has gained a reputation for her keen insight on the business and marketing issues surrounding the telecommunications industry.

Kelli Lowery is the Vice President of Business Development for Clarisys. Ms Lowery has 20 years experience in both the manufacturing and distribution and fulfillment areas of industry-specific products and solutions to the marketplace. Ms Lowery has been in the communications industry for 12 years. While working to provide the in-house peripheral needs of many of the Fortune 100, she explored product solutions for customers committed to transitioning to converging technologies. During this time she continued her focus to include the emerging ITSP and ISP markets and identified and approved endpoint products that addressed both business-to-business and business-to-consumer VoIP applications. Her extensive work in helping to establish e-commerce and fulfillment business models for these emerging markets resulted in a broad knowledge of application-specific endpoint products developed for the VoIP industry. Ms Lowery currently works with OEMs as well as continuing her work with major VoIP providers.

Ralph Musgrove is vice president of Market Development at Miami-based Venali, Inc., a leader in Enterprise IP fax, providing both desktop solutions and web services. In this capacity, Musgrove is responsible for creating, negotiating and maintaining strategic partnerships with corporations in order to expand Venali's presence within the United States and abroad. Most recently, Musgrove has formed partnerships with Microsoft for the inclusion of Venali's Internet Fax Service in Office 2003 and ACCPAC, a Computer Associates company, to provide fax messaging solutions for the consumer and small to mid-sized business (SMB) market. He is also responsible for negotiating strategic partnerships with telecommunications companies in Europe and Asia. Musgrove has an extensive background in business development, marketing and sales in communications and Internet technologies in the U.S., Canada and Europe. Prior to Venali, he was Director of Product Management for Bazillion, a broadband service provider for consumers and small businesses. Responsible for the company's virtual phones, VoIP and Internet messaging products, Musgrove was instrumental in building third-party relationships with Microsoft and Hewlett Packard.

Avi Ofrane founded the Billing College in 1996, a training company addressing the converging market trends associated with telecommunications Billing and Customer Care. The Billing College is a spin-off company of Mr. Ofrane's technology consulting company, Jupiter Data, Inc., established in 1990. Mr. Ofrane began his career in 1977 as an analyst with the IBM Corporation and has since 1982 concentrated exclusively on the telecommunications industry, in which he is now a recognized expert in Billing and Customer Care. Throughout his extensive career, Mr. Ofrane has been involved in all aspects of the industry, from strategic planning and executive management to vendor evaluation and project implementation. Mr. Ofrane lectures extensively on Billing and Customer Care issues, strategies, methodologies, and practices. He is a frequently requested speaker at major North American and European conferences. Mr. Ofrane is currently President and CEO of the Billing College, as well as a master instructor of the company's courses. Mr. Ofrane is the co-author of the book "Telecom Made Simple" and has written numerous articles for international trade publications. Mr. Ofrane holds a Bachelor of Science in Computer Science from Pennsylvania State University.

Neal Shact is the founder and CEO of CommuniTech (1983) and CommuniTech Services (1997), companies focused on hands-free communications, VoIP, and voice processing. He is also the founder and Chairman of the International VoIP Council. Prior to founding CommuniTech, Neal held management positions at Illinois Bell and General Telecommunications. Mr. Shact has written numerous articles in industry journals such as Internet Week, Telemarketing, and Voice Processing Magazines. He is also a much sought after speaker on the subject of VoIP and telecommunications. A magna cum laude, Phi Beta Kappa alumnus of Clark University, Mr. Shact received his MBA from the Kellogg Graduate School of Management at Northwestern University.

Mr. Eric Stasik has reviewed and assessed thousands of patents. He has negotiated the licensing of patents and found out there are key tools to use when deciding to ignore, acquire, or license IPR. Mr. Stasik has overseen the patent management program and made strategic changes to the system to maximize the value of knowledge within the company. He has developed successful patenting strategies for individual technologies, new products, and open and proprietary standards. Mr. Stasik is a registered US patent agent and has a broad progressive experience in dealing with all aspects of the patent process – from filing and prosecution, to licensing and valuation. Mr. Stasik, is the founder of Patent08, an expert consulting firm that provides patent engineering, mining, and brokering services to high tech firms, research organizations, and individuals who want to profit from their knowledge.

Jeff Stern is Co-Founder and Executive Vice President of GoBeam, Inc. the leading IP Centrex service provider in North America. His responsibilities include management of GoBeam's Wholesale business and the company's business development activities. Mr. Stern's background includes engineering, marketing, sales and investment experience at firms such as: Bell Canada, Nortel; Tymnet-McDonnell Douglas; Independence Technologies which he co-founded and where he was Vice President of Sales and Marketing- the company was sold to BEA Systems; VeriFone, where he co-founded the Client-Server Software business unit; Encanto Networks, where he served as Vice President of Marketing; and Pulsar Venture Group, where he was Vice President of Investment Strategy.

Christian Szpilfogel is the Head of Communication Platforms group within Mitel networks. In this role he is responsible for Enterprise iPBX and PBX products. Prior to this, Christian was responsible for the R&D of Enterprise and SMB Platforms, Desktop phones, and Strategic Technology. Christian joined the Mitel management team in February 2000 to lead the Network Services development team. Previous to that Christian owned and operated a design consulting firm in the Ottawa area and prior to that worked in a variety of R&D roles at Nortel Networks labs in Ottawa. Christian has broad product development experience including real-time software, call processing, OAM&P, mobility, and data communications.
This covered equipment from small embedded systems to large carrier grade networks. Mitel Networks is a leading provider of next-generation IP telephony solutions and a market-leader for voice, video and data convergence. The company creates advanced communication solutions and applications in the area of speech recognition, wireless mobility, unified messaging, and contact center solutions. Mitel Networks is committed to delivering integrated business solutions to the hospitality, education, health care, government, retail and financial markets. The company is headquartered in Ottawa, Canada and supports its customers through offices and partners worldwide.

Craig Walker is the President and CEO of Dialpad Communications, Inc., a leading Internet telephony service provider. Prior to Dialpad, Mr. Walker was a General Partner at Sterling Payot Capital, an early stage telecommunications and Internet focused venture capital fund. Mr. Walker was also a corporate and securities attorney in the Silicon Valley with Gunderson Dettmer and Brobeck Phleger and Harrison where he represented early and late stage venture backed technology companies. Mr. Walker earned a Bachelor of Arts and a Juris Doctorate degree from the University of California at Berkeley and a Masters of Business Administration from Georgetown University. Mr. Walker is a member of the State Bar of California.

Acknowledgements

We thank the many gifted people who gave their technical and emotional support for the creation of this book. In many cases, published sources were not available on this subject area. Experts from manufacturers, Internet Telephone Service Providers (ITSPs), trade associations and other telecommunications related companies gave their personal precious time to help us with the production of this book.

We thank the industry professionals including: Chuck Grothaus from ADC Wireless Systems, Jamie Horton with Alcatel, Mark Burnworth from Alcatel Network Systems, Carol Broniman of Allen Telecom Inc., Ruth Lee with ATX Networks, David Kamm at Avail Networks, Debbie Kline from Avaya Inc., Marguerite Millar with Blonder Tongue Laboratories Inc., Courtney Flaherty and Scott Pollard with Brodeur Worldwide, Genevieve Macias with Sony, Kristin Bagley at Caster Communications, Inc., Dick Wurst from CCC networks, Tim Elliott with Channel Master, Joe Freddoso and Ruth McCullers Lee at Cisco Systems, Kim Sullivan with Convergys, Dana Spangler from Corning Cable Systems, Charles Rothofsky from EMC Technology Inc., Jon Maron with Ericsson , Michelle Maeder from Harris Microwave Corp., Greg Carter of Hartford Concrete Products, Inc., Curtis Bartkowski of Icom America Inc., Robert Selzler from Innomedia Incorporated, David Morris of InCon, Patti Segraves with Iridium Satellite, Tom Sciacca at HVACTool.com, Kim Coxe from Kentrox, Mike Liwanag with Logitech, Steve loudermilk at Lucent, Cathy Harrington with Mack Molding, Brian Sumpter with Mack Molding Southern Operation, Geoffrey Becker of Marconi, Sherri Haupert at Maxon America Inc., Paul Kraska with Multi-Tech Systems Inc., Steve Kopel from NEC America, Inc., Julie Carlson with Next Level Communications (NLC), Joanne Bilger-Guear of Panasonic Telecommunications Systems, Jennifer Righi with Paradyne, Melissa Helms at Plantronics, Evelyn Donis with Hewlett Packard, Michelle Webb of Samsung America, Ann Polanski from Radio Frequency Systems (RFS), Jody Walsh at ROHN, Helena (Hyun-ryung) Lee with Samsung, Paul Sims of Scientific-Atlanta, Suzanne Crow at Siemens,

Tehrani's IP Telephony Dictionary

Agnes Pisarski with Sierra Wireless, Andrea Dyck of Sierra Wireless, Eric Lehmann with Sinclair Technologies Inc., Kristine Freitas with SonicWALL, David Migdal at Sony Electronics, Steve Kuyatt of Tektronix, Sue Frederick from Motorola, Sarah Phillips with Telica, Candy Mizer at Teltronics, Tammy Turner with Thales Communications, Dana Barry from Thorlabs, Stephanie Wnek with Times Microwave Systems, Linda Knapowski at U.S.Robotics Corporation, Patricia Philcox with Xircom (Intel), and Tim Donovan from Zhone Technologies, .Andrea Candler with Actiontech Electronics Inc, Kristen Levy from American Power Conversion Corporation, James Little with Belkin Corporation, Avi Ofrane from The Billing College, Clare Plaisted with Callserve Communications, Neal Shact from Communitech Inc., Michael Scott with D-Link, Jerome Waksman at iConnectHere Powered by Deltathree, Brian Hutchins from Dialpad Communications, Inc., Susan Sparks with Kall8, Kim Coxe from Kentrox, Mike Leshner with MHL Communications, Inc., Bob Schneider at Net2Phone Inc., Doug Hagan with NETGEAR Inc., Suzanne Crow from Siemens, Information and Communication Networks Division, Heather Williams at Symbol Technologies, Claudia Schiepers from Thomson, and Brooke Schulz with Vonage.

There were many definitions provided by and requested from visitors the dictionary's web site. These contributions and requests were essential in assuring this book represents the definitions required by and used by those involved in the industry.

Production of the book was made possible by our gifted staff: Karen Bunn (project manager), Tom Pazderka (illustrator and cover artist), and Katie Jackson (editor).

Foreward

Dear Reader,

It is truly amazing to have witnessed the progression of IP Telephony from its early days as a hobbyist's plaything to its current state as a legitimate alternative to traditional telecom systems. IP Telephony was once a pipe dream and now we're seeing IP PBX deployments become a commonplace occurrence in enterprises around the world.

TMC Research has projected that the service provider VoIP market is showing signs of over 100-percent growth for the period 2003–2006. At that rate the market for calls placed on VoIP networks is scheduled to exceed 100 billion annual minutes by end of 2006.

Any industry that is undergoing such rapid growth needs a resource to allow participants — newcomers and experienced personnel alike — to keep pace with the terms and phrases, the acronyms and jargon that such growth undoubtedly engenders.

This dictionary is intended to be that resource. Featuring Over 10,000 of the latest IP Telephony terms and definitions as well as more than 4,000 of the latest VoIP terms and acronyms the book you are holding in your hands is designed to help you cut through the confusion. The dictionary also features over 400 diagrams and pictures to help explain some of the more complex terms.

I also urge you to stay as up-to-date as possible on the issue and trends shaping the world of IP Telephony by book-marking www.tmcnet.com. This valuable online resource is the Web home of *Internet Telephony*® magazine, and *Internet Telephony*® *Conference and EXPO* and has links to many other areas of interest for you to continue your research into this ever-changing technology.

Thank you for choosing Tehrani's IP Telephony Dictionary as your preferred IP Telephony and VoIP reference manual.

Country Codes

A

Afghanistan - 93
Albania - 355
Algeria - 213
American - Samoa - 684
Andorra - 376
Angola - 244
Anguilla - 809
Antarctica - 672
Antigua - 1 + 268
Argentina - 54
Armenia - 374
Aruba - 297
Ascension - Island - 247
Australia - 61
Austria - 43
Azerbaijan - 994

B

Bahamas - 1 + 242
Bahrain - 973
Bangladesh, - People's - Republic - of - 880
Belarus - 375
Belgium - 32
Belize - 501
Benin - People's - Republic - of - 229
Bermuda - 1+441
Bermuda - 809
Bhutan - 975
Bolivia - 591
Bosnia-Herzegovina - 387
Botswana - 267
Brazil - 55
British - Virgin - Islands - 284
Brunei - 673
Bulgaria - 359
Burkina - Faso - 226
Burundi - 257

C

Cambodia - 855
Cameroon, - United - Republic - of - 237
Canada - 1
Cape - Verde - Islands - 238
Cayman - Islands - 345
Central - African - Republic - 236
Chad - 235
Chile - 56
China, - People's - Republic - of - 86
Christmas - Island - 672
Cocos Island - 672
Colombia - 57
Comoros - 269
Congo - 242
Cook - Islands - 682
Costa - Rica - 506
Croatia, - Republic - of - 385
Cuba - 53
Cyprus - 357
Czech Republic - 42

D

Denmark - 45
Diego - Garcia - 246
Djibouti - 253
Dominica - 767
Dominican Republic - 1 - 809

E

Easter - Island - 56
Ecuador - 593
Egypt - 20
El Salvador - 503
Equatorial Guinea - 240
Eritrea - 291
Estonia - 372
Ethiopia - 251

F

Falkland Islands - 500
Faroe Islands - 298
Finland - 358
France - 33
French - Antilles - 596

www.ITExpo.Com

French - Guiana - 594
French - Polynesia - 689

G

Gabon - Republic - 241
Gambia - 220
Georgia - 995
Germany - 49
Ghana - 233
Gibraltar - 350
Greece - 30
Greenland - 299
Grenada - 473
Guadeloupe - 590
Guam - 671
Guantanamo Bay - 53
Guatemala - 502
Guinea-Bissau - 245
Guinea - 224
Guyana - 592

H

Haiti - 509
Honduras - 504
Hong Kong - 852
Hungary - 36

I

Iceland - 354
India - 91
Indonesia - 62
Iran - 98
Iraq - 964
Ireland - 353
Israel - 972
Italy - 39
Ivory Coast - 225

J

Jamaica - 876
Japan - 81
Jordan - 962

K

Kazakhstan - 7
Kenya - 254
Kiribati - 686
Korea (North) - 850
Korea (South) - 82
Kuwait - 965
Kyrgyzstan - 996

L

Laos - 856
Latvia - 371
Lebanon - 961
Lesotho - 266
Liberia - 231
Libya - 218
Liechtenstein - 423
Lithuania - 370
Luxembourg - 352

M

Macao - 853
Macedonia, Republic of - 389
Madagascar - 261
Malawi - 265
Malaysia - 60
Maldives - 960
Mali Republic - 223
Malta - 356
Marshall Islands - 692
Martinique - 596
Mauritania - 222
Mauritus - 230
Mayotte - 269
Mexico - 52
Micronesia - 691
Molodova - 373
Monaco - 377
Mongolia - 976
Montserrat - 664
Morocco, Kingdom of - 212
Mozambique - 258
Myanmar - 95

Country Codes

N
Namibia - 264
Nepal - 977
Netherlands - 31
Netherlands Antilles - 599
New Caledonia - 687
New Zealand - 64
Nicaragua - 505
Niger - 227
Nigeria - 234
Niue - 683
Norfolk Island 672
Norway - 47

O
Oman - 968

P
Pakistan - 92
Palau - 680
Panama - 507
Paupa New Guinea - 675
Paraguay - 595
Peru - 51
Phillipines - 63
Poland - 48
Portugal - 351

Q
Qatar - 974

R
Reunion Island - 262
Romania - 40
Russia - 7
Rwanda - 250

S
St. Helena - 290
St. Kitts/Nevis - 869
St. Lucia 758
St. Pierre & Miquelon - 508
St. Vincent - 784
Saipan - 670
San Marino - 378

Sao Tome - 239
Saudi Arabia - 966
Senegal - 221
Seychelles Island - 248
Sierra Leone - 232
Singapore - 65
Slovakia - 421
Slovenia - 386
Solomon Islands - 677
Somalia - 252
South Africa - 27
Spain - 34
Sri Lanka - 94
Sudan - 249
Suriname, Republic of - 597
Swaziland - 268
Sweden - 46
Switzerland - 41
Syria - 963

T
Taiwan - 886
Tajikistann - 7
Tanzania - 255
Thailand - 66
Togo - 228
Trinidad & Tobago - 868
Tunisia - 216
Turkey - 90
Turkmenistan - 993
Turks & Caicos Island - 649
Tuvalu - 688

U
Uganda - 256
Ukraine - 380
United Arab Emirates - 971
United Kingdom - 44
United States Of America - 1
Uruguay - 598
Uzbekistan Tashkent - 7

V
Vatican City - 39
Venezuela - 58

www.ITExpo.Com

Tehrani's IP Telephony Dictionary

Vietnam - 84

W

Wallis & Futuna Island - 681
Western Samoa - 685

Y

Yemen - 967
Yugoslavia, Federal States of - 243

Z

Zaire, Republic of - 243
Zambia - 260
Zimbabwe - 263

Table of Contents

ABOUT THE EDITOR III

CONTRIBUTING EDITORS AND ADVISORS V

ACKNOWLEDGEMENTS IX

FOREWORD XI

COUNTRY CODES XIII

DICTIONARY
- Numbers 1
- A 7
- B 57
- C 83
- D 147
- E 185
- F 207
- G 233

H.	245
I.	261
J.	295
K.	299
L.	303
M.	325
N.	361
O.	377
P.	393
Q.	435
R.	439
S.	463
T.	509
U.	539
V.	549
W.	567
X.	579
Y.	581
Z.	583

INDUSTRY MAGAZINES 585

INDUSTRY ASSOCIATIONS 587

INDUSTRY ITSP'S 597

Numbers

 # (push button dial symbol)-This symbol, used on one of the 12 touch-tone keys, is typically used to indicate that the originator has dialed all of the digits in an originating telephone number. It also has other purposes in special systems. In North America this symbol is called several names: pound sign, number sign, tic-tac-toe, or Octothorpe. In other areas of English language use, this sign was historically used as a symbol for the English pound (unit of weight, not money) or as a substitute for the word "number." This symbol also resembles the two vertical parallel lines and the two horizontal parallel lines that one uses when playing the game "tic-tac-toe" (called Naughts and Crosses in British English). In the telephone industry, there is a widely repeated story that the name for this symbol is historically "Octothorpe" but we have found no historical evidence to confirm this. Some people claim that the name "Octothorpe" was promulgated as a joke by a technician named Thorpe! Outside of the English speaking countries, this symbol was not historically used before its appearance on the push button telephone dial, and is typically called "quadrat" since it resembles a square.

*57 (Call Trace)-A dialing feature code that is used by a subscriber (customer) to identify the calling telephone number of an unwanted or harassing telephone call. This call trace feature temporarily stores the dialed digits and alerts the telephone service operator to "tag" the originator's number to allow authorities to investigate the originator of the unwanted or unauthorized call.

.Wav-A program extension code for waveform that is used by Windows for sound files. WAV files are computer digital samples of the analog waveform and they require a relatively large amount of memory compared to compressed sound file types.

.Z-A filename extension that identifies to a Unix system that the file is stored in compressed form by the gzip or the compress utility.

.ZIP-A filename extension that identifies that a file is compressed by the WinZip or PKZip utilities.

µ - Law-Mu

µ, Mu-Greek letter lower case Mu. (1- metric prefix) One millionth or 0.000001. (2- length unit) One micrometer, 0.000001 meter (formerly called one micron). (3- digital code designator) The algorithm for compression or mapping of measured voltage amplitude to binary code value used in conjunction with the DS-1 (T-1) digital PCM encoding of telephone channel waveforms is designated Mu-law or µ-law, in contrast to a similar but distinct algorithm called A-law and used in conjunction with the 2.048 Mbit/s PCM system.
Note that the English letter u is often substituted for µ when the Greek character cannot be produced due to limited typographic capability.
Understandable, but often ambiguous!

0800 Freephone-A service that allows callers to dial and telephone number without being charged for the call. The toll free call is billed to the receiver of the call. In the Americas and other parts of the world, Freephone numbers are sometimes called "Toll Free" and they typically begin with 800, 888, 877, or 866.

1 BASE5-An implementation of the 1 Mbps StarLAN IEEE standard network. The maximum segment length for the 1Base5 system is 500m.

1- Potentiometer-Dead Zone

1.6/2.4 GHz Mobile Satellite Service-A mobile satellite service that operates in the 1610-1626.5 MHz and 2483.5-2500 MHz frequency bands, or in any portion thereof.

10 BASE F-A generic designation for a family of 10 Mb/s baseband Ethernet systems operating over optical fiber.

10 BASE-FB-A baseband Ethernet system operating at 10 Mb/s over two multimode optical fibers using a synchronous active hub.

10 BASE-FL-A Ethernet that transfers data at 10 Mbps over optical fiber. The 10 Base-FL system uses two optical fibers and can include an optional asynchronous active hub.

10 BASE-FP-A Ethernet that transfers data at 10 Mbps over two optical fibers. The 10 Base-FP uses a passive hub to connect data communication devices.

10 Base2-An Ethernet data communication standard that was created by the IEEE for communication over coaxial cable. 10 Base2 has a maximum distance of 185 meters. 10 Base2 is also known as Thin Ethernet.

10 BaseT-A data communications system primarily used for computer networks based on the

Ethernet IEEE standard 802.3. The 10BaseT system media is twisted pair wire at a data rate of 10 Mbps.

10 BROAD36-A broadband Ethernet system operating at 10 Mb/s over three channels (in each direction) of a private CATV system. Segment length is 3600 meters.

10 GE-10 Gigabit Ethernet

10 Gigabit Ethernet (10 GE)-10 Gigabit Ethernet (10 GE) is a data communication system that combines Ethernet technology with fiberoptic cable transmission to provide data communication transmission at 10 Gbps (10,000 Mbps). The specifications for 10 GE are being developed by the Gigabit Ethernet Alliance. The Gigabit Ethernet Alliance is a group of companies that was formed in January 2000.

10 GE is an extension to the IEEE 802.3 protocol that enables data communication at speeds of approximately 10 billion bits per second (Gbps). The IEEE 802.3ae standard supports two Physical Layer protocols: LAN PHY and WAN PHY. Both of these physical layer protocols use the same Media Access Control (MAC) layer and frame formats.

The LAN PHY supports existing Gigabit Ethernet applications at ten times the bandwidth using a cost-effective solution. For compatibility with existing wide area networks based on SONET/SDH, the WAN PHY has been standardized. The WAN PHY includes a simplified SONET/SDH framer that operates at a line-rate of OC-192/STM-64 resulting in a payload rate of approximately 9.29 Gbps while providing compatibility with the existing Ethernet packet format.

10 Gigabit Ethernet Alliance-An alliance between a group of companies that was formed in January 2000 that is assisting in the creation of a 10 Gigabit Ethernet system.

100 BASE-FX-A baseband Ethernet system operating at 100 Mb/s over two multimode optical fibers.

100 BASE-T2-A Ethernet system operating at 100 Mbps over two pairs of Category 3 or higher unshielded twisted pair (UTP) cable.

100 BASE-T4-An Ethernet system operating at 100 Mbps over four pairs of Category 3 or higher using unshielded twisted pair (UTP) cable.

100 BASE-TX-A baseband Ethernet system operating at 100 Mbps over two pairs of STP or Category 5 UTP cable.

100 BASE-X-A generic designation for 100 Base Ethernet systems independent of their underlying physical transmission medium (T- twisted pair or F - fiber).

100 BaseT-A data communications system primarily used for computer networks based on the Ethernet IEEE standard 802.3. This system media is twisted pair wire and its data rate is 100 Mbps.

100 VG-AnyLan-A local area network (LAN) systems that uses a demand priority access method. It is standardized by IEEE 802.12.

1000 BASE-CX-A baseband Ethernet system operating at 1000 Mbps over two pairs of 150 shielded twisted pair (STP) cable.

1000 BASE-LX-A baseband Ethernet system operating at 1000 Mb/s over two multimode or single-mode optical fibers using longwave laser optics.

1000 BASE-SX-An Ethernet system that transmits at 1000 Mbps shortwave laser optics over two multimode optical fibers.

1000 BASE-T-An Ethernet that transmits at 1000 Mbps using four pairs of Category 5 unshielded twisted pair (UTP) cable.

1000 BASE-X-A generic name for an Ethernet systems that transmits at 1000 Mbps.

1024 (K)-Widely used but unofficially, the number 1024 (equal to 2 raised to the 10th power) is represented by the capital letter K. This number normally occurs only when describing file size, memory size or other numbers that are integral powers of 2 because of internal use of the binary number system. Take care to distinguish capital K from small k, which represents 1000.

112-An emergency call number that is used in the European Union.

119-An emergency telephone number that is used in Japan's telecommunication system. It is the equivalent of the 911 emergency number used in the United States.

1996 Telecommunications Act-Legislation designed to spur competition in the telecommunications industry. Resulting deregulation affects local, long distance, and wireless carriers. Direct competition between these suppliers and other 'non-traditional' competitors (cable, utilities, etc.) should result in better prices and more services for the consumer. Signed into law by President Clinton Feb.8, 1996.

1FB-One Flat Business Line

1G-First Generation

1O BASE5-An implementation of the Ethernet IEEE 802.3 standard on Thicket (RG-6/8) coaxial cable, a baseband medium, at 10 Mbps. The maximum segment length is 500m.

1XEVDO-One Channel Evolution Version Data Only

1xRTT-The first phase of cdma2000 technology designed to double voice capacity and support data transmission speeds up to 144 Kbps, or 10 times the data transmission speed that was available in IS-95 CDMA system. 1xRTT is compatible with today's IS-95A and IS-95B cdmaONE.

2 1/2G-Second And A Half Generation

2-Way Trunk-A trunk that can be captured at either end.

2-Wire Circuit-A communication circuit that uses one pair of wires for both directions of transmission.

23B+D-Representation of ISDN's primary rate interface (PRI). A PRI is composed of 23 bearer (B) channels at 64Kbps and 1 data (D) channel at 64Kbps.

24/7-Service that is available 24 hours a day, 7 days a week.

2500 Telephone Set-The term "2500 telephone set" is used in North America to describe a standard analog telephones having a 12-button touch-tone (DTMF) dial. The dial is used both for dialing and for end-to-end signaling via the voice channel. This name was originally a particular model number used by the first manufacturer of this type of telephone set, but has become a generic term today. The design and utility patents covering this set, first manufactured in 1964, have elapsed and many different manufacturers now make virtually indistinguishable replicas of the original design. This device usually provides plain old telephone service (POTS). The 2500 telephone set has replaced the earlier rotary dial 500 set and the briefly manufactured 1500 type set (which had a 10-button touch-tone dial) in many parts of the world.

This figure shows a block diagram of a standard POTS telephone (also known as a 2500 series phone). This telephone continuously monitors the voltage on the telephone line to determine if an incoming ring signal (high voltage tone) is present. When the ring signal is received, the telephone alerts the user through an audio tone (on the ringer). After the customer has picked up the phone, the hook switch is connected. This reduces the line connection resistance (through the hybrid) and this results in a drop in line voltage (typically from 48 VDC to a few volts). This change in voltage is sensed by the telephone switching system and the call is connected. When the customer hangs up the phone, the hook switch is opened increasing the resistance to the line connection. This results in an increase in the line voltage. The increased line voltage is then sensed by the telephone switching system and the call is disconnected.

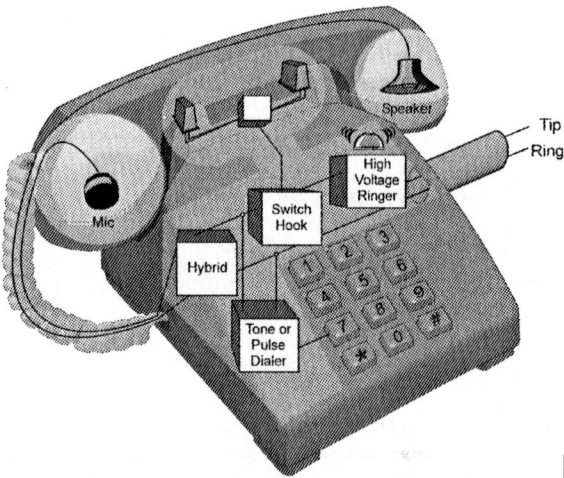

POTS Telephone Block Diagram

2B+D-The combination of two 64 kbps basic rate DS0 (B) and one 16 kbps (D) signaling channels that forms the ISDN basic rate interface (BRI).

2B1Q-Two Binary, One Quarternary

2B4Q Line Coding-Two Binary to Four Quaternary line coding method used for ISDN subscriber lines (the U interface) in North America. In this method of line coding, each two binary bits are mapped into one pulse symbol having one of four distinct voltage levels on the transmission wires. A sequential method of encoding is used to ensure that the same voltage level will not be transmitted in consecutive symbols, so that the waveform has constant alternation and will be accurately transmitted via coupling transformers and coupling capacitors.

2G-Second Generation

3+ Operating System-A network operating system that combines Xerox Network System (XNS) and Microsoft MS-Net file sharing protocols. The 3+ operating system was originally developed by 3Com company.

3.58 MHz-The approximate frequency that is used as a subcarrier in the NTSC video signal to carry color information.

3270-General term used to refer to cluster controller architecture developed by IBM for access to online systems such as CICS. Generically "3270" can refer to controllers, terminals, and printers.

3B2T Line Code-A baseband line code (modulation and signaling structure) that transmits three bits for every two ternary, three level symbols.

3DES-Triple DES Or Three DES

3G-Third Generation

3GPP-3rd Generation Partnership Project

3PCC-Third Party Call Control

3rd Generation-Wideband Cellular Communications

3rd Generation Partnership Project (3GPP)- The 3GPP oversees the creation of industry standards for the 3rd generation of mobile wireless communication systems (W-CDMA). The key members of the 3GPP include standards agencies from Japan, Europe, Korea, China and the United States.

3WC-Three Way Calling

4-Wire Circuit-A communication circuit that uses separate pairs for each direction of transmission. Normally associated with synchronous, dedicated communications where simultaneous two-way, full duplex transmission is required

4.43361875 MHz-The frequency of the subcarrier signal that contains the color information that is used in the PAL video standard.

4:2:2 Digital Video-A CCIR digital video format specification that defines the ratio of luminance samples, to sampling frequencies for each color channel. For every four luminance samples, there are two samples of each color channel.

411-The standard service code for local directory assistance in the United States.

4B/5B Encoding-A data transmission encoding process that is used to transforms each group of four bits to be represented as a five-bit symbol.

4B3T Line Code-A baseband line code (modulation and signaling structure) that transmits four bits for every three levels of symbols.

4G-Fourth Generation

4WL-WDM-Four Wavelength Wave Division Multiplexing

5 M's of Marketing-These are the 5 critical components of any marketing activity.
Market - The group of customers to be targeted
Merchandise - the thing being sold
Media - The type of media being used to contact the prospect (TV, Radio, Phone call etc.)
Message- The message that you are communicating to the prospect about the offer
Margin - the amount of profit that the sale should generate

5 R's of Loyalty -5 major components of customer loyalty to a telephone company:
1. Recourse - providing them with an easy way to fix problems.
2. Recognition - letting the customer know that you think they are valuable
3. Responsiveness - responding quickly to customer inquiries or complaints
4. Respect - treating the customer with respect
5. Reinforcement - re-enforcing the positive messages about your service

511 - North American Traffic Report Number- 511 was designated by the US FCC in the year 2000 as the universal telephone number for vehicular traffic reports.

5250-IBM class of terminals (or printers) for midrange (system 3x and AS/400) environments.

56K Line-A telephone circuit that has a data transmission rate of 56 Kbps. It is sometimes called dataphone digital service (DDS).

5B/6B Encoding-A data transmission encoding process that is used to transforms each group of five bits to be represented as a six bit symbol. This process is used 100BaseVG networks.

611-The universal service code for telephone repair service in the United States.

64 Clear Channel Capability-A channel that allows the end user to transmit 64 kbps without any other constraints such as a maximum ones density or number of-consecutive-zeros restrictions.

64K Line-A digital communication line that allows the end user to transmit 64 kbps.

66 Block-A 25 or 50 pair plastic terminal that serves as a RJ21X network interface or telephone system cross connect and mounted at the demarcation point.

700 Service Access Code-One of several non-geographic numbering plan area (NPA) codes.

711 - TRS Access Number-711 was designated by the US FCC in the year 2000 as the universal access number for state Telecommunications Relay Service (TRS) centers for the deaf. Most TRS centers also continue to use distinct 800 and local numbers as well as 711.

800 Data Base Service-An intelligent network service that facilitates a more efficient provision of the 800 exchange access service through "number portability." The service enables subscribers to change carriers without changing their 800 number. Subscribers can also select any unassigned 800 number and use it with any participating carrier's 800 Service.

800 Number Administration and Service Center-A communication center that allows toll free service providers (800, 888, 877, 866) to access a administrative computer system that provides toll free data base services.

800 Service-A communication services where the receiving party pays for incoming calls. 800 services is often called "Toll Free" In the Americas and "Freephone" in Europe and the dialing digits are 0800. In the United States, additional toll free numbers include 888, 877, and 866.

800 Service Access Code-A non-geographic numbering plan area (NPA) code that indicates that a called party will be charged for a call rather than the calling party being charged.

802.11 Access Point (AP)-A radio access point (wireless data base station) that is used to connect wireless data devices (stations) to a wired local area network (WLAN).

802.11 Association-A process of registering a wireless data device (station) with a specific access point (AP) in an 802.11 specified wireless local area network (WLAN) system.

802.11 Basic Service Set (BSS)-The process used by wireless data devices to directly communicate with other wireless data devices in an 802.11 specified system.

802.11 DeAssociation-A process of un-registering (disassociating) a wireless data device (station) with a specific access point (AP) in an 802.11 specified wireless local area network (WLAN) system. This allows the access point to reuse the logical address that was assigned to the wireless data device.

802.11 Direct Sequence Spread Spectrum (DSSS) Mode-A radio access technology used in a 802.11 specified wireless local area network (WLAN) system that uses a code to spread the data signal direct to a much wider bandwidth than the data signal requires. Because each bit of the signal is spread over a wide frequency band, several spread spectrum signals can exist in the same area at the same time with minimal interference.

802.11 Distributed Coordination Function (DCF) Mode-A mode of operation in a 802.11 specified system that allows the independent operation (distributed access control) of wireless data devices (stations).

802.11 Distribution Service (DS)-The process of distributing packets through access points (APs) and other switching devices in an 802.11 specified wireless local area network system.

802.11 Extended Service Set (ESS)-A set of basic service set (BSS) systems (short range wireless systems) that are capable of receiving and forwarding packets of data to stations in other BSS groups or through access points (AP) to other data networks.

802.11 Frame Structure-The general frame structure for the 802.11 system uses a variable length packet that contains the medium access control (MAC) header, a variable length data body (0-2312 bytes), and a frame check sum (FCS) to check for transmission errors. The MAC header contains a frame control field (type of frame), up to 4 address (used for distribution routing) and a sequence control (to identify a frame in a sequence of frames).

This diagram shows the frame structure for a 802.11 data packet. This diagram shows the packet contains an medium access control (MAC) header and a variable length frame body that holds the data. The data body can vary from 0 to 2312 bytes. The data packet also includes a frame check sum (FCS) to check the packet for transmission errors. This diagram shows that the MAC header is divided into a frame control field (type of frame), contains up to 4 address (used for source, distribution, and destination packet routing) and it contains a sequence control to identify each frame in a sequence of frames.

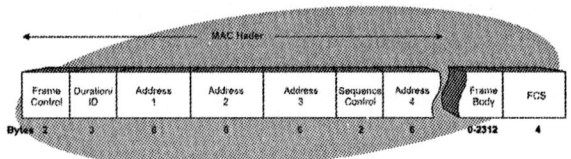

802.11 Frame Structure

802.11 Frequency Hopping Spread Spectrum (FHSS) Mode-A radio access technology used in a 802.11 specified wireless local area network (WLAN) system that uses frequency hopping spread spectrum (FHSS) shared access technology.

802.11 Infrared Mode

FHSS uses a radio transmission process where a message or voice communications is sent on a radio channel that regularly changes frequency (hops) according to a predetermined code. The receiver of the message or voice information must also receive on the same frequencies using the same frequency hopping sequence.

802.11 Infrared Mode-A radio access technology used in a 802.11 specified wireless local area network (WLAN) system that uses infrared communication. The physical specification of the 802.11 system uses carrier (on-off) modulation for the IR signal.

802.11 Point Coordination Function (PCF)-A process used in the 802.11 wireless LAN system to coordinate the transmission of data from other stations in the network that requires each station to receive a token from a serving station (such as an access point) before the station can transmit. This avoids access channel collisions and allows for the coordination of transmission time that is used by each station that is associated (registered) with the network.

802.11 Wireless LAN-The 802.11 wireless local area network (WLAN) is an industry standard developed by the IEEE for wireless network communication. It usually operates in the 2.4 GHz or 5.7 GHz spectrum and permits data transmission speeds from 1 Mbps to 54 Mbps.

802.11A-A version of the 802.11 wireless local area network (WLAN) industry standard that was developed by the IEEE for wireless network communication. It was developed to operate in the 5.7 GHz spectrum and permits data transmission speeds up to 54 Mbps.

802.11B-A wireless local area network (LAN) system that transmits up to 11 Mbps.

802.1P-The IEEE 802.1P signaling technique is an OSI Layer 2 standard for prioritizing network traffic at the data link/Mac sublayer. It can also be defined as best-effort QoS at Layer 2. 802.1P traffic is simply classified and sent to the destination without bandwidth reservations being established. 802.1p is a spin-off of the 802.1Q (VLANS) standard. The prioritization field was never defined in the VLAN standard. The 802.1P implementation defines this prioritization field.

802.1Q-The IEEE 802.1Q standard defines the operation of Virtual LAN (VLAN) Bridges that permit the definition, operation and administration of Virtual LAN topologies within a Bridged LAN infrastructure.

802.3-A network standard defined by the IEEE that specifies carrier sense multiple access with collision detection (CSMA/CD) to coordinate access to the network. This is the access method used by Ethernet networks.

802.4-The IEEE standard that defines the MAC layer for Token Bus networks.

802.5-The IEEE standard that defines the MAC layer for the Token Ring networks.

802.6-A network standard adopted by the Institute of Electrical and Electronics Engineers (IEEE) that defines the architecture of metropolitan area networks (MANs). This standard specifies the protocol that is used between high-speed data networks or network nodes.

811-A standard feature code used in the Americas that permits' the user to reach the business office of a local telephone service provider.

9-Track Tape-A 1/2" magnetic tape storage that stores data from computer systems. Nine track tapes sore data on eight tracks and parity information on the ninth track.

900 Service Access Code-An area code designated for special services, such as information services, tech support lines, polling or other pay-per-use services.

911-The standardized service number assigned for emergency calls in the United States and Canada, and the islands that are included in the North American Numbering Plan.

976 service-A calling party pays (CPP) service that uses the local exchange code (the first 3 digits of the 7 digit local exchange telephone number) to route the calls and identify it is a CPP service.

999-The emergency telephone number that is used by customers to contact public safety officials in countries such as Hong Kong (the countries that are and were part of the British Commonwealth).

A

A-Ampere

A & B Bits-Bits that are designated on a digital transmission line (such as a T1 line) to convey signaling information. These bits generally indicate if loop current is present or if the flash service request (momentary on-hook) has been requested.

A & B Leads-Additional leads that are used in addition to a two-wire E&M interface to provide additional control information. A & B leads are used to connect certain types of private branch exchanges (PBXs).

A & B Signaling-A process of transferring signaling information on a digital communication line that involves the stealing of bits from the digital transmission line (called the A & B bits). This is a type of in-band signaling as it steals some of the channel bandwidth for use as signaling messages. When it is used in T1 transmission line, 1 bit of user data from each of the 24 sub-channels in every sixth frame is stolen (discarded) and replaced with signaling control information. The use of A+B signaling reduces the T1 bandwidth from 1.544Mbps to 1.536 Mbps.

A B Bits-See ABCD Bits

A B C D bits-In associated bit signaling used with 2.048 Mb/s digital multiplexing, four bits in the associated signaling channel are used to convey supervisory or other control information. These bits are designated with the letters A, B, C, and D. In robbed bit signaling used with 1.544 Mb/s digital multiplexing, the robbed bits used in the 6th and 12th frames are designated the A and B bits, respectively. When extended super frame is used, the robbed bits in the 18th and 24th frames are designated the C and D bits, respectively.

A Band-A band of radio frequencies that are used for cellular mobile telephone service. The A band of frequencies are assigned for use by non-incumbent local phone company.

A Interface-The label given to the Radio Data Peripheral Interface for mobile and portable subscriber units.

A Law-The type of non-linear digital voice coding that is commonly used in Europe and other parts of the world, typically with 2.048 Mb/s digital multiplexing. The A Law coding process is used to compress the 13 bit sampling of a digitized audio signal into the equivalent of an 8 bit sample. It does this by assigning a non-uniform voltage difference to each of the consecutive binary code values, except for a small range near zero volts where uniform voltage differences are used between consecutive code values. Another non-linear voice coding system is the Mu Law coding system that is used in North America and Japan with the DS-1 (T-1) digital multiplexing system.

A+ Certification-A certification program that is designed to ensure the competence in basic troubleshooting and repair for computer technicians. The certification is controlled by the Computer Technology Industry Association (CompTIA). A+ certification requires the passing two tests: a core exam and a specialty exam. The core exam tests the general knowledge of PCs technology including installation, configuration, software and hardware upgrading, diagnostics, maintenance, repair, interaction with customers, and safety procedures. The specialty exam tests specific operating system knowledge.

A, B, C, D Shelf-In subscriber loop carrier (SLC), the corresponding multiplexer (mux) and channel bank shelves which breakout 24 X 64 Kb/s channels for subscriber voice or data.

A-B Box-A switching box that allows two or more computers to share a peripheral device such as a printer. It can be switched manually or through software.

A-Band Cellular-A band of frequencies in the 800 MHz range that are assigned to a cellular service provider. In the America's, A band frequencies are assigned to an alternative carrier to the incumbent local telephone service provider.

A-bis Interface-The communication interface that exists inside the base station between the base station controller (BSC) and the base transceiver system (BTS) of a cellular infrastructure system. In contrast to the A-bis interface, the A interface (without the word "bis") exists between the BSC and the mobile-service switching center (MSC) in a cellular communications network. The word "bis" comes from the French language, where it means "additional" or "supplementary" or "again."

A-key-Authentication Key

A-law Encoding-A digital signal companding process that is used for encoding/decoding signals in pulse-code-modulated (PCM) systems. This companding process increases the dynamic range of a binary signal by assigning different weighted values to each bit of information than is defined by the binary system. The A-law encoding system is an international standard. A different companding version is used in the Americas is u-Law.

A-Link-Access Link

A/A1 lines-In an electro-mechanical key telephone system (KTS), an auxiliary pair of wires called the A and A1 wires are provided for each telephone line. When that line is in use for a conversation by one of the telephone sets, these two wires are connected within that telephone set, and they cause the line lamp for that line to remain continuously lit on all telephone sets. When the user momentarily presses the "hold" button, the A and A1 wires are momentarily disconnected, causing a relay in the key service unit (KSU) to bridge ("hold") that telephone line with a 600 Ohm resistor, and start a cyclic interrupter switch to cause the lamps on that line to engage in a flash cadence. When that line is on hold, it can be seized by another telephone set whose corresponding line button is depressed, and/or the first user of that line can use his/her telephone set for calling or answering on another line. In some KTS installations, a single line telephone is permanently connected to one of the telephone lines. To make it compatible with the KSU, spare contacts on the cradle switch are wired to the two spare wires in the line (mounting) cord, such that these two wires are connected to each other when the handset is off hook. These two wires then provide A and A1 busy/idle status indication, and they are connected to the corresponding A and A1 wires in the cable corresponding to that particular telephone line. Thus the line lamp will light on all the multi-line KTS telephone sets to indicate that line is in use, even though the single line telephone is using it. The single line phone cannot put the line on hold, since it does not have a way to momentarily open the A/A1 connection while keeping the telephone off hook.

A/B Switch-Allows the mobile phone to receive signals from both the A and B carrier, A only, B only, priority A, priority B or home only.

A3 Algorithm-A validation process used in a GSM network for authentication of the mobile user that is requesting service. The A3 algorithm is a mathematical process that combines a secret number (called Ki in documents) stored in the SIM chip and a second number that is transferred between the base station and the mobile station. The numerical result of the A3 algorithm calculation is transmitted back to the base station. It is then compared to a previously internally calculated answer also using the A3 algorithm and the same two input quantities, in the base station controller or in the MSC. When the same matching result occurs for both the mobile supplied information and previously stored information, the customer is granted access. See also Authentication.

A4 Paper Size-The ITU T.30 standard for facsimile transmission describes how a sheet of paper is to be scanned. Most facsimile machines are designed to handle the standard "letter" size in North America which is 8.5 by 11 inches (216 by 279 mm). The closest corresponding size in the "A series" of paper sizes described in the German Industrial Standards (Deutsche Industrie Norm - DIN) is size A4, 210 by 297 mm (approximately 8.25 by 11.66 inches). Most facsimile machines can also accommodate smaller paper sizes such as A5 which is 149 by 210 mm, and the A6 size which is 105 by 149 mm.

A5 Algorithm-An encryption algorithm (program) used in the GSM system that is used to provide voice and data privacy.

A8 Algorithm-An encryption program that is stored in the SIM card as part of the GSM network. The A8 algorithm is used to produce a secret key that will be used for voice or data encryption. See also A5 algorithm.

AAA-Authentication, Authorization, Accounting
AAC-Advanced Audio Codec
AAL-ATM Adaptation Layer
AAL 3/4-ATM Adaptation Layer Type 3/4
AAL-2-AAL-CU
AAL-CU (AAL-2)-The method of filling ATM packets with digital voice signals. The AAL-CU composite user (CU) method allows the more efficient packetizing of ATM packets with voice signals.
AAL1-ATM Adaptation Layer 1
AAL2-ATM Adaptation Layer 2
AAL5-ATM Adaptation Layer 5
AAP-Multicast Address Allocation Protocol
AAR-Automatic Alternate Routing
AAT-Above Average Terrain
AB-Abbreviated Burst
AB-Access Burst

Abandoned Attempt-A call attempt that is aborted by the caller before completion.

Abandoned Call-A call setup request (origination request) that is aborted by an originator before the call is answered.

Abandoned in place (AIP)-Abandoned in place (AIP) is cable facilities that are left in place (e.g. in underground conduit or lashed to aerial strand after the old facilities have been deactivated). The new cable facilities are "live" after splicing is complete and may run parallel in the same conduit or in the case of aerial plant may be lashed to the abandoned in place cable.

Abbreviated Burst (AB)-An reduced transmission burst length used in TDMA cellular and PCS systems when accessing the system when the distance from the radio tower to the mobile phone is unknown. This shortened transmission burst is used to protect from overlapping to adjacent bursts due to the relative transmission time delay from near and distant cellular phones within a large cell. After the amount of delay is known, the relative timing of the mobile phone is adjusted and normal transmission bursts may be sent.

ABC-The term that was used for North American Numbering Plan (NANP) to designate the first three digits of a 7-digit telephone number within a specific Numbering Plan Area (NPA). Each ABC code within an NPA was assigned to an end-office switching system. A single switching system could have had more than one ABC assigned to it. The ABC code is now called the NNX/NXX code, office code, or prefix.

ABC digits-The term that was used for North American Numbering Plan (NANP) to designate digits in a telephone number. In the past, the first three digits of a 7-digit telephone number within a specific Numbering Plan Area (NPA) were designated by the letters ABC, or alternatively called the central office (CO) code. Each ABC code within an NPA was assigned to an end-office switching system. A single switching system could have had more than one ABC assigned to it. The ABC code is now also called the NNX or NXX code, office code, or prefix. More recently, the 10 digits in a North American telephone number have been represented in some documents by the first 10 letters of the alphabet thus: ABC-DEF-GHIJ, in which the letters ABC represent the area code and DEF represent the central office code.

ABCD parameters-A matrix of four numeric values that allows computation of the output voltage and current of a two port network based on the input voltage and current, via two linear equations.

ABCD Signaling Bits-These are bits robbed from bytes in each DS-O or T-1 channel in particular sub-frames and used to carry in band all status information such as E&M signaling states. These bits are robbed for signaling with Extended Super Frame (ESF); only A and B are available with Super Frame (SF).

ABD-Average Business Day

Abend-Contraction of abnormal end. A message issued by an operating system when it detects a serious problem, such as a hardware failure or major software damage. It can also mean the abnormal termination of an application.

Abilene-A high speed communication backbone network used in the Internet 2 system.

ABNF-Augmented Backus-Naur Format

Abort-In process used in data transmission to discard and ignore all transmitted bits that have been sent by the sender since the preceding flag sequence. The abort process can be invoked by a primary or secondary sending station.

Above Average Terrain (AAT)-The height of an antenna or signal source that extends above the average terrain level.

ABR-Available Bit Rate

ABS-Average Busy Season

Absolute Address-A data memory address that is used in a computing device to identify a specific storage location. An absolute address usually requires a larger address word than a relative data address reference.

Absolute Addressing-The use of the specific identification code or address number in a LAN or computer system that is permanently assigned to a storage register, location, or device. The antonym for this term is "relative addressing," a method in which the stated relative address must be added to a so-called base number value to obtain the absolute address.

Absolute Novelty-Absolute novelty means that an invention has not been disclosed in a publication, was in public use, offered for sale, or otherwise made known or available to the public prior to the filing date of an application for patent.

With few exceptions, absolute novelty is required in order to obtain a patent in most countries with exceptions such as the United States where an

Absorption

applicant is given a one year "grace period" and required to file an application within one year of such disclosure. Applicants in Japan are allowed a 6 month "grace period."

Absorption-A loss of energy (radio or light) that is the result of absorption within the transmission material or transmission medium. This energy is usually converted to heat.

Abstract Service Primitive (ASP)-A description of an interaction between a service-provider and a service-user at a particular service boundary as defined by Open Systems Interconnection (OSI). The ASP operation is independent of the implementation. ASP is defined as part of ATM systems.

Abstract Syntax Notation number 1 (ASN.1)-A specification that defines how the format of an object (to the bit level) is described. It is defined in recommendation X.680 of ITU-T, ASN.1. As an example, ASN.1 is the language used in creating and defining SNMP MIBs for network management.

Abstract Test Suite (ATS)-A set of procedures that are used to evaluate the testing procedures of products supplied by vendors.

Abuse Of Privilege-An action of a user that was not allowed based upon a law or organizational policy.

AC-Access Customer
AC-Alternating Current
AC-Access Carrier
AC Power-Alternating Current Power

AC Signaling-The insertion of alternating-current signals on a communication circuit to transfer supervisory and address signaling information. Examples of AC signaling include single tone signaling using 2600 Hz or two tone signaling called multifrequency pulsing.

ACA-Automatic Circuit Assurance

Academnet-A communication network within Russia that allows universities to communicate with each other.

ACC-Analog Control Channel

ACC Uplink Message-An analog control channel (ACC) message sent from the wireless phone to the radio base station.

Accelerated Depreciation-A depreciation method or period of time, including the treatment given cost of removal and gross salvage, that is used when calculating depreciation deductions (asset usage) on income tax returns which is different from the depreciation method or period of time prescribed by the Commission for use in calculating depreciation expense recorded in a company's books of account.

Accelerator Board-A assembly or printed circuit board that is added to a computer system that enhances the performance of processor that is installed in the computer.

Acceptable Frame Filter-A process used in a virtual local area network (VLAN) that filters (qualifies) frames as acceptable. This filter may be set to allow all frames or specific frames that have certain tags (control flags) set.

Acceptance Angle-The angle at which light can strike the core of a fiber and still propagate down its length. A large acceptance angle makes alignment of a light source to a fiber less difficult.

Acceptance Test-A test that evaluates the successful operation and/or performance of an electronic assembly or communication system. Acceptance tests usually have specific operation requirements and test measurements. Acceptance tests are often used as a final product approval and may authorize a product for production or purchase.

Accepted Interference Level-A level of interference from other sources or systems that exceeds a pre-defined level (such as a regulatory level) that has been agreed upon between the system owners or operators. Excessive interference levels commonly occur in radio systems due to terrain (such as waterways that transfer signals over long distances) or changing foliage conditions (e.g. leaves fall off trees reducing signal attenuation).

Access-General used in the telecommunications industry to refer ability to connect to a service. Also, a process where a mobile radio competes to gain the services of a wireless system by transmitting an access request message.

Access Attempt-The process of a mobile radio transmitting an access request to a wireless system. If the access attempt is unsuccessful, the mobile radio will typically wait a random amount of time and attempt the access again.

Access Burst (AB)-A transmission burst that is used by a mobile radio to gain the attention of the system. The access burst is shorter than normal bursts to ensure burst transmissions do not overlap. After the system has received the access burst, it responds with an acknowledgment and a time alignment command. After the mobile radio has adjusted its timing, it will begin transmitting normal bursts.

Access Minutes

Access Carrier (AC)-A carrier that provides network access services to an end customer, such as a local exchange company (LEC) or competitive access provider (CAP).

Access Channel-A radio channel in a wireless system that coordinates the random access of mobile radios to the wireless system.

Access Charge-Telecommunications service charges that are approved by the Federal Communications Commission (FCC) that compensate a local exchange carrier (LEC) for connection of local customers to a long distance telephone service company (IXC).

Access Code-Numbers that are dialed by a user to select the access provider for services assigned that access code. The access code usually changes the inter-exchange (long distance) service provider on a per-call basis. When no access code is entered, the default carrier (or no service provider) is selected. At present, North American access codes have the format 1010xx...x and they are dialed as a prefix to the destination number.

Access Control-The actions taken to allow or deny use of the services and features of a communication system to individual users.

Access Control Information (ACI)-A standard directory access control process that is defied by the ITU-T X.500 directory service model where the ACI is used to identify and access a file or directory.

Access Control List (ACL)-A table or list of users, processes, or objects that are authorized to access files or objects. ACL specify access permissions and other parameters such as read, write, execute, and delete.

Access Coupler-A device that allows signals to enter or be extracted (access) from a transmission medium (such as fiber lines).

Access Customer (AC)-A user that purchases end user access services (e.g. leased line) from a communication carrier.

Access Customer Number Abbreviation (ACNA)-A number that is assigned to a Customer for the use in the provisioning and billing of services through an access network.

Access Customer Terminal Location (ACTL)-A code (the CLLI code) that identifies the location of the switch that provides network access to a customer.

Access Discipline-A process that is used by a communication device to gain access to a shared transmission medium. The access discipline can be contention resolution based (all users randomly attempt access) or can be contention free (users must wait for a specific time or token to be given to them.)

Access Domain-The group of communication devices within a network that can directly communicate with each other. These communication devices within a shared system (such as a LAN) operate within the LANs access domain.

Access Failure-The termination (failure) of an attempt to access a system in any manner other than through the initiation of the user.

Access Fee-A fee that is paid for the use of another network to originate, route, or terminate calls. Access fees are commonly paid by a long distance service provider to a local access provider for allowing calls to enter or terminate through the local network.

Access Grant Channel (AGCH)-A logical channel (typically on a wireless cellular system) that is used to assign mobile phone to a channel where it can begin to communicate with the system. An access grant messages is followed by a random access channel (RACH) service request message.

Access Line-The physical link (typically a copper wire or fiber) between a customer and a communications system (typically a central office) that allows a customer to access local and toll switched networks. Access lines may include a subscriber loop, a drop line, inside wiring, and a jack.

Access Link (A-Link)-A communications line (link) that provides access from a service switching point (SSP) and signaling control point (SCP) to signaling transfer points (STPs) in a SS7 network.

Access Method-A set of rules that when followed allow use of a service. The ability and means necessary to store data, retrieve data, or communicate with a system. FDMA, TDMA and CDMA are example..

Access Methods-The process used by a communication device to gain access (obtain services) from a system. Some systems allow communication devices to randomly compete for access (contention based) while other systems assign periods of time or setup events (such as token passing) the precisely controls (non-contention based) the access times and methods.

Access Minutes-The length of time that telecommunications facilities are used for long distance service (interstate or international service). This is for both originating and terminating calls. Access minute timing stops when one of the parties disconnects.

Access Multiplexing-Access multiplexing is a process used by a communications system to coordinate and allow more than one user to access the communication channels within the system. There are four basic access multiplexing technologies used in wireless systems: frequency division multiple access (FDMA), time division multiple access (TDMA), code division multiple access, (CDMA), and space division multiple access (SDMA). Other forms of access multiplexing (such as voice activity multiplexing) use the fundamentals of these access-multiplexing technologies to operate.

This figure shows the common types of access channel-multiplexing technologies used in wireless systems. This diagram shows that FDMA systems have multiple communication channels and each user on the system occupies an entire channel. TDMA systems dynamically assign users to one or more time slots on each radio channel. CDMA systems assign users a unique spreading code to minimize the interference receive and cause with other users. SDMA systems focus radio energy to the geographic area where specific users are operating.

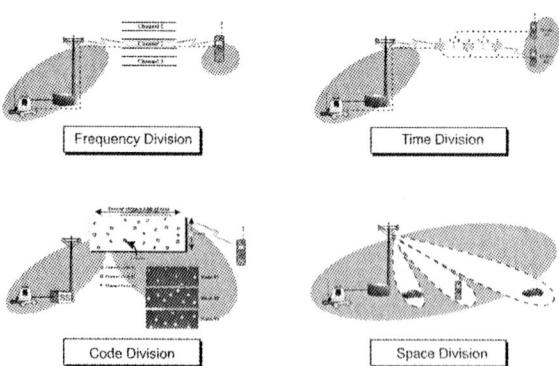

Access Multiplexing Operation

Access Network-A portion of a communication network (such as the public switched telephone network) that allows individual subscribers or devices to connect to the core network.

Access Node-Access nodes are access points or concentration points that allow one or more users to connect to a communication network. Access nodes are usually network devices that allow customers to connect to their communication devices to data networks via wires or wireless connections that use standard transmission protocols such as Ethernet or 802.11 wireless networks.

Access Number-(1-Wireless) A phone number that can be used to directly dial a cellular or PCS customer in the local wireless system when they are roaming. To contact the roaming customer, the access number is dialed first and when a tone is received, the complete mobile phone number (area code and number) of the roaming customer is entered. The local system will then page the roaming customer. Access numbers were very important in the first few years of cellular service because the systems were not automatically connected. Because most cellular and PCS systems can automatically deliver calls to roaming customers, access numbers are rarely used. (2-calling card) A telephone number of a gateway that allows customer's to enter information that permits the connection of a call or access to information services.

Access Overload Class (ACCOLC)-A code assigned to a mobile phone that classifies the type of access it has to a cellular system (overload class). The cellular system continually sends messages that contain the overload classes that enable or disable the access rights to specific mobile phones. For the AMPS cellular system, there are 16 access overload classes. The access overload class for normal customers is typically set to the last digit of the mobile phone's telephone number. Classes higher than 9 are typically reserved for public safety workers. When the cellular system becomes overloaded, the messages sent by the cellular system will begin to restrict groups of mobile phones that attempt to access the system.

Access Permission-A process of assigning permission for a user to gain access to communication facility.

Access Point (AP)-Typically, a point that is readily accessible to customers for access to a wireless or wired system. Also called a Radio Port.

Access Priority-A priority level (user priority level) that is assigned to users or devices within a communication network that is used to coordinate access privileges based on network activity or other factors.

Access Process Parameter (APP)-The parameters used or assigned to a port (logical channel point) in a packet switching, network that coordinate the operation of the port.

Access Protocol-The set of procedures and rules that are used by communication devices to coordinate access to a shared communication media.

Access Provider-The company that provides and controls access to communication devices or to other networks.

Access Request Information System (ARIS)-The software system that is used by a customer service center at a carrier that processes access service orders (service activation).

Access Response Channel-A sub channel (logical channel) used for digital cellular systems to coordinate the random access of mobile phones that want to obtain service from the cellular or PCS system.

Access Response Channel (ARCH)-A communication control channel that is defined in the IS-136 TDMA mobile communication system that is used to coordinate access to the system. The ARCH channel carries wireless system responses from the cell site to the user terminal equipment. ARCH is a logical sub-channel of SPACH Short Message Service (SMS) point-to-point messaging, paging, and access response channel, that is a logical channel of the DCCH (Digital Control Channel), a signaling and control channel that is employed in cellular systems based on TDMA (Time Division Multiple Access).

Access Rights-The properties or access right assignment associated with files, directories, or communication services that define the ability of users to change, interact or access services.

Access Router Card (ARC)-An access router card is a packet data switching device that allows users to connect different geographically located computer systems and local area data networks. The ARC can adapt between different types of traffic and different protocols integrating a data network into a single enterprise-wide network.

Access Server-Access servers are a computer (or computers) that coordinate access for end users who connect to a communication system. The function of access servers can vary from simple access control to advanced call processing services.

Access Service-A service that connects customers that are located within a local access and transport area (LATA) to interexchange carriers (long distance telephone service providers).

Access Service Area-A geographic area established for the provision and administration of communications service. An access service area encompasses one or more exchanges, where an exchange is a unit of the communications network consisting of the distribution facilities within the area served by one or more End Offices, together with the associated facilities used in furnishing communications service within the area.

Access Service Group (ASG)-The group of switching systems associated with the subtending of a tandem switch.

Access Service Request-A request for service that is received from interexchange carrier to allow connection to a customer in a local-exchange telephone network.

Access Service Request (ASR)-A process or form by a local carrier to request services from a local network operator. A form for ASR was developed by the alliance for telecommunications industry solutions (ATIS) in 1984 to help define access requests for the long distance service providers.

Access Software-Software programs that are used to transmit and receive information to and from communication systems.

Access Software Provider-A provider of software that provides or enabling access control using filtering, modification, or redirecting content.

Access Tandem (AT)-A high level switching system that interconnects low level (local exchange) switching systems. An access tandem can also function provide access for nonconforming end offices such as for equal access to other long distance service providers.

Access Tariff-A tariff imposed by federal or state regulators on carriers (typically local exchange carriers) that offer access to telephone exchange services to customers or other companies.

Access Time-The amount of time required to retrieve information from a device or system.

Access Token-An indicator (flag) that is used in a communication network that identifies a process or user that permits access to a system or service. In Windows 2000, an access token contains the combination of security identifiers along with additional information about the user to ensure reliable and authorized access to systems and services.

Accessibility-The ability of network elements, such as servers, trunks, or ports that can be accessed be by users within a group.

ACCH-Associated Control Channel

ACCOLC-Access Overload Class

Account-On LANs or multi-user operating systems, an account is set up for each user. Accounts are usually kept for administrative or security reasons. For communications and online services, accounts identify a subscriber for billing purposes.

Account Average-The average of assets for all telecommunications company assets associated with a particular account that include maintenance, repair experience, and salvage value expectancies.

Account Inquiry Centers (AICs)-A call center that receives and processes calls from customers regarding the status of their accounts.

Account Lockout-In network operating systems (Noose's), a count of the number of invalid logon attempts allowed before a user is locked out.

Account Policy-The set of rules or processes used by networks and multi-user operating systems that define how users are allowed to access to the system based on their predefined account access privilege settings.

Accounting-(1-general) The process of recording, assigning, and tracking the usage of resources on a network to specific accounts or users. (2-FCAPS) Accounting is one of the five functions defined in the FCAPS model for network management. The accounting level might also be called the allocation level, and is devoted to distributing resources optimally and fairly among network subscribers. This makes the most effective use of the systems available, minimizing the cost of operation. This level is also responsible for ensuring that users are billed appropriately.

Accounting Management-The authorizing, recording, and assignment of costs to users and groups based upon their authorization and use of network resources.

Accounting System-A set of accounts, rules, processes, software programs, equipment, and other mechanisms that are used to necessary to operate and evaluate a business system for an operations (business decisions), financial (investor reporting), and regulatory (taxes) perspective.

Accreditation-Recognition from a perceived or defined authority that a product or service meets a specific criteria that is beneficial (e.g. meets UL approval) or required by regulatory laws.

Accredited Standards Development Organization-An organization composed of members that have been accredited by an institution that is responsibility for standards accreditation within an industry.

Accredited Systems Engineer (ASE)-An engineer that is certified Compaq that they can evaluate, install, and administer Compaq workstations. An ASE engineer will have experience running Microsoft Windows 2000 and Novell NetWare network operating systems.

ACCUNET Packet Service (APS)-An AT&T packet service that transmits analog and digital packets of data from 2.4 kbps to 56 kbps.

Accuracy-(1-General) A comparison of an actual signal or measured value as compared to the theoretical or pre-established limits. (2-Network Management) Accuracy is the measurement of interface traffic that does not result in error as used in Network Management. Accuracy can be expressed in terms of a percentage that compares the success rate to the total packet rate over a period of time.

ACD-Automatic Call Distribution
ACELP-Algebraic Code Excited Linear Prediction
Acer Group-One of the top five PC makers in the world, with factories in Malaysia, the Netherlands, the Philippines, Taiwan, and the United States. The Acer Group bought Texas Instrument's notebook computer business in 1997 and has formed business alliances with companies, including 3Com and Hitachi, for the development of advanced digital consumer products such as PC-ready televisions and DVD systems.
For more information on the Acer Group, see www.acer.com.tw.
ACF-Admission Confirm
ACF-Advanced Communications Function
ACI-Access Control Information
ACK-Acknowledgment
Ack-Acknowledgment Message
Ack Message-A message that is used to confirm a person or device is willing to participate in a communication session. The Ack message is defined in session initiation protocol (SIP).

Acknowledge Character-A control character that is transmitted by the receiving station back to the transmitting station to indicate that the last transmission was successfully received.

Acknowledgement Wink-A signaling process where the line voltage is reversed. This indicates that the status of a communications link has changed. Multiple winks indicate various stages of a call. The first wink indicates that a destination carrier has received dialed digits. Subsequent links are used to indicate the status of trunk line availability. Acknowledgement wink is sometimes referred to as a wink acknowledgement.

Acknowledgment (ACK)-Abbreviated ACK. In communications, ACK is a control code, ASCII 06, sent by the receiving computer to indicate that the data has been received without error and that the next part of the transmission may be sent.

Acknowledgment Message (Ack)-A message that is responded by a communications device to confirm a message or portion of information has been successfully received. If a communications device is supposed to send ack messages back to the originator and an acknowledgment message is not received, the system will typically resend the message. See also negative acknowledgment message (Nack).

ACL-Access Control List

ACM-Address Complete Message

ACNA-Access Customer Number Abbreviation

ACONSOLE-A Novell NetWare 3.x workstation utility that controls a modem attached to the workstation. ACONSOLE is used to establish an asynchronous remote console connection to a server. The RS232 NetWare Loadable Module (NLM) must be loaded on the server to which you want to connect. In NetWare 4.x, use RCONSOLE to perform this function.

Acoustic Coupler-A device that is placed against a telephone handset to allows the transmission of audible modem signals over telephone lines.

Acoustic Wave-A wave that is a produced by a mechanical vibration in the of 20 Hz to approximately 20 kHz that can be heard by the human ear. This wave can travel through a variety of mediums including solids, liquids, or gases.

ACPI-Advanced Configuration and Power Interface

Acquire-A process of obtaining a product or service through competition (e.g. acquire a channel) or assignment (token passing).

Acquisition-(1-General) The process of tuning and decoding a signal from a system. (2-Satellite) A process a satellite receiver device of locking on a satellite's global positioning system (GPS) signal. The acquisition includes tuning to the signal frequency and applying automatic gain control (AGC), synchronization (time alignment), and decoding (processing) of the data signal. (3-customer) Acquisition is the process of acquiring new customers from a population of prospects who were not customers in the past.

Acquisition Time-The amount of time it takes for an electronic circuit In a communication system to acquire a specific signal. Acquisition time may include radio signal scanning time in addition to the search for a specific pattern of information (synchronization signal).

ACR-Attenuation to Crosstalk Radio

ACR-Anonymous Call Rejection

ACRE-Authorization & Call Routing Equipment

Across-The-Wire Migration-A method of migrating file-server data, trustee rights, and other information to a Novell NetWare server using the NetWare Migration utility. You can also use across-the-wire migration to upgrade from LAN Manager, LAN Server, and earlier versions of NetWare. A similar process known as BMIGRATE allows users to migrate from Banyan VINES.

ACS-Advanced Communications Service

ACS-Automatic Call Sequencer

Activate-A process or action (such as an operator's service screen entry) that starts or reactivates a service.

Activate/Passive Device-In telecommunications systems where a current loop is used to supply power to communications devices an activate device is capable of providing the power for the loop and a passive device is one that can only consume power from the loop for operation.

Activated Channels-Those channels engineered at the head end of a cable system for the provision of services generally available to residential subscribers of the cable system, regardless of whether such services actually are provided, including any channel designated for public, educational or governmental use.

Activation-The process of inputting specific information into a telephone network database to authorize a service account. For a telephone networks to provide services, the system must be informed of the services and account codes. Activation of services usually requires that certain customer financial criteria must also be met. After this information is input, the service or telecommunications device will become activated and the customer can request and receive services.

Activation Commission-A commission paid to a retailer or other entity for activating a new customer, or adding additional lines of service to an existing customer. The activation commission is typically paid after a contract with the customer is completed.

Activation Fee

Activation Fee-A one-time fee that is charged for the initial setup of communication service. Activation fees are also called "setup fees."

Active-(1-general) A state of a device or circuit where it provides a function. (2-component) A device or circuit that converts a signal through amplification that requires an external source of energy.

Active Directory-In Microsoft Windows 2000, a system of directory services that allows a client to make an IP telephone call via their PC and the Active Directory. The PC asks the Active directory for the phone number of a person within the organization. The Directory find the number, passes it to the PC, who dials the number. The Active Directory is fluid tracking clients around the organization.

Active Hub-A device that amplifies transmission signals in a network, allowing signals to be sent over a much greater distance than is possible with a passive hub. An active hub may have ports for coaxial, twisted-pair, or fiber-optic cable connections, as well as LEDs to show that each port is operating correctly.

Active Matrix Display-A type image display technology that uses an active matrix to display each picture element where each element has its own control transistor. Because of the active elements of the display, the response time of the display can be faster and sharper than passive types of displays. Active matrix display is also called a thin film transistor (TFT) display.

Active Mobile Station-The state of the mobile telephone when it is communicating (transmitting) with the system.

Active Monitor-On a Token Ring network the workstation with the highest medium access control (MAC) address, which participates in the monitor contention process, is chosen as the Active Monitor. The Active Monitor is responsible for maintaining the token and detecting error conditions, such a frame that was not removed by its originator. The Active Monitor is the only station on the ring that provides a clock for ring, either 8 MHz for 4 Mbps or 32 MHz for 16 Mbps operation. The Active Monitor's presence is constantly monitored by the Standby Monitor(s) via the Active Monitor Present process, sometimes referred to as Ring Poll.

Active Monitor Present-A medium access control (MAC) frame that is generated approximately every seven seconds in a token ring network, by a ring's active monitor. The message indicates that an active monitor is operational and alive on the ring. Each standby monitor on the ring monitors for the Active Monitor Present MAC frame. If one is not seen in 14 seconds, the standby monitors will assume the Active Monitor has been removed from the ring and will elect a new active monitor.

Active Sensor-A measuring instrument in the earth exploration-satellite service or in the space research service by means of which information is obtained by transmission and reception of radio waves.(RR)

Active Server Pages (ASP)-An active server page (ASP) is a script interpreter (real time program operation) that allows the development and use of software program modules to operate on a web server (such as a Microsoft Internet Information Server).

Active Signaling Link-A signaling link that is being used in the Signaling System 7 system that has completed the initial alignment procedures and is transporting or ready to transport signaling messages.

Active Termination-A termination used on a SCSI line that results in reduced levels of electrical interference on a string of SCSI devices.

Active Token Monitor-A node (device) in a token ring network that periodically transmits an "active monitor present" control frame to allow other nodes to sense that there is an active monitor present. This allows devices attached to the ring to determine if a token is lost. If a token loss is detected, the active token monitor issues a free token.

ActiveX-A development of Microsoft's COM that adds network capabilities by creating a set of component-based Internet- and intranet-oriented applications.

ActiveX Control-A software control module that requires an ActiveX container (such as a spreadsheet or Web browser) to provide a specific function. ActiveX control allow database access or file accesses that can communicate with another ActiveX containers, ActiveX controls, and to interface with the underlying Windows operating system. ActiveX can directly access files using a security that contain digital certificates that authenticate the source and validity of the control.

Activity Report-A report of call records that contain the call identifier, date, usage amount (e.g. transmission time, kbytes of data transferred), destination address (e.g. telephone number or IP address), and other pertinent information.

ACTL-Access Customer Terminal Location
ACTS-Automated Coin Toll Service
ACTS-Advanced Communications Technologies And Services
Actual Measured Loss-A signal loss of a test carrier wave that is transmitted on a channel or circuit. A common test frequencies are 1004 Hz (old standard) and 1020 Hz (new international standard).
Actuator-A device or assembly that is used to start the operation, calibration, or switching of a circuit or piece of equipment.
ACU-Automatic Calling Unit
ACU-Alarm Controller Unit
Ad Hoc Query-A filtering action or query that is created dynamically or temporarily.
Ad Insertion Module-A process used in cable television or broadcast radio networks that allow the insertion of advertising during pre-determined time segments. This process allows different advertising messages to be inserted in different geographic regions.
Adaptec, Inc.-A leading manufacturer of high-performance networking and connectivity products, including SCSI adapters, RAID products, Fast Ethernet adapters, ATM network interface cards, and server management software. In 1998, Adaptec acquired Ridge Technologies, a manufacturer of RAID and other storage solutions for Microsoft Windows 2000.
Adapter-A printed circuit board that plugs into a computer's expansion bus to provide added capabilities.
Common adapters include video adapters, joy-stick controllers, and I/O adapters, as well as other devices, such as internal modems, CD-ROMs, and network interface cards. One adapter can often support several different devices. Some of today's PC designs incorporate many of the functions previously performed by these individual adapters on the motherboard.
Adapter Board-An assembly or circuit board that can be inserted into system or equipment to allow added functionality or testing.
Adapter Support Interface (ASI)-A communication driver specification that was developed by IBM for networking using IEEE 802.5 token-ring systems.
Adaptive Array Antennas-An antenna system that can dynamically change its signal patterns to optimize the operation of a radio communication link or increase the performance of a system. Adaptive antennas use advanced signal processing algorithms to detect the specific location of a user and to select antenna elements and change signal phase to direct the antenna patterns to specific locations and decrease the antenna patterns to other locations.
Adaptive Channel Allocation-A process of allocating communication channels dynamically based on specific criteria (e.g. system capacity limitations).
Adaptive Differential Pulse Code Modulation (ADPCM)-A process of converting analog voice signals into encoded digital signals through the use of predictive codes that are created by analyzing the previous digital audio signals. ADPCM is derived from the original pulse code modulation (PCM) system that commonly represents an analog signal as 64 kbps (called a DS0). ADPCM systems commonly provide digital signals at 32 kbps and 16 kbps.
Adaptive Equalizer-An signal processing section in a receiver that can analyze the received signal to adjusts the signal reception to compensate for dynamically changing transmission line conditions (such as the elimination or adding of delayed or multipath signals).
Adaptive Jitter Buffer-An adaptive jitter buffer dynamically adjusts the amount of jitter buffering to take into account the variable delay of a network (such as the Internet) over time. When network conditions allow for low delay transfer of audio signals, the jitter buffer is made very small to reduce any perceived audio delays. When network conditions result in long delays in the transfer of audio signals, the jitter buffer is expanded in size to take this delay into account and provide the best possible audio quality under these conditions. also known as a Dynamic Jitter Buffer.
Adaptive Multirate Codec (AMR)-A speech compression process that offers multiple speech coding rates used in third generation (3G) wireless systems. The use of AMR allows a lower bit rate (higher compression rates) coding process to be used when system capacity is limited and more users need to be added to the system.
Adaptive Predictive Coding (APC)-A analog-to-digital conversion process that uses a 1-level or multilevel sampling system for which the system adaptively predicts the future values based on the past values of the quantified signals.

Adaptive Routing-A software mechanism that allows a network to reroute messages dynamically, using the best available path, if a portion of the network fails.

Adaptive Spare-A spare or standby piece of equipment in a communication network that can replace a failed piece of equipment and adaptively change its configuration or performance to match the characteristics of the failed piece of equipment. Use of adaptive spares reduces the number of backup equipment assemblies that are required in a communications network.

Adaptive Speed Leveling-A modem technology that allows a modem to respond to changing line conditions by changing its data rate. As line quality improves, the modem attempts to increase the data rate; as line quality declines, the modem compensates by lowering the data rate. Also known as adaptive equalization.

ADC-Analog to Digital Converter

ADC-Automated Data Collections

ADCCP-Advanced Data Communication Control Procedures

Add On Conference-A conference call feature (usually in a PBX system) that allows additional participants to be added to the conference call. To add-on a conference participant, a participant (or moderator) places the existing call or on hold, obtains system dial tone, and connects to the add-on participant. After the new participant has agreed to participate, the originating participant reactivates (re-connects) to the conference call in progress.

Add-Drop Multiplexer (ADM)-A circuit that inserts (adds) or extracts (drops) signals from a higher speed communication line such as adding a DS0 channel into a DS1 or E1 communication line. In the Synchronous Optical Network (SONET), it is a network element that provides access to all or some subset Synchronous Transport Signal (STS) line signals contained within an Optical Carrier Level N (OC-N).

Add-On-A telephone call center sales process designed to encourage customers to purchase additional products or services secondary to the primary sale. Occasionally, it is incorrectly referred to as up-selling, which means the quantity or quality of the primary product is offered at a premium and higher cost.

Added Channel Framing-A channel frame process that allows signal elements to occupy consecutive time slots. Added channel framing is also called bunched frame alignment.

Added Digit Framing-A channel frame process that allows signal elements to occupy non-consecutive digit time slots. Added channel framing is also called distributed frame alignment signal.

Added Main Line Carrier-A transmission system that uses two analog carrier signals (frequencies) to provide two telephone channels over one telephone line (2 wires).

Additive White Gaussian Noise (AWGN)-A signal that approximates the properties of certain types of "noise" signals normally occurring in nature, that has an even distribution of power density across its frequency range. AWGN is random and unpredictable in theory. It is used for mathematical or laboratory analysis of the performance of systems in the presence of undesired signals such as "noise." The term "noise" is used even when the frequency range of the AWGN is not in the audible range to the human ear, or the signal is electrical rather than audible. Audible white noise is used intentionally in certain situations like offices or libraries as light background noise to soothe and placate such as in an office environment, or to prevent people from hearing and being disturbed by other types of noises.

Address-A grouping of numbers that uniquely identifies a station in a local are network, or a location in computer memory, or a house on a street, etc. For a local area network station or a computer memory location, electronic logic can be arranged to ignore messages not bearing the appropriate address and accept messages that do bear the appropriate address.

Address Bus-The electronic channel, usually from 20 to 64 lines wide, used to transmit the signals that specify locations in memory.

The number of lines in the address bus determines the number of memory locations that the processor can access, because each line carries one bit of the address. A 20-line address bus (used in early Intel 8086/8088 processors) can access 1MB of memory, a 24-line address bus can access 16MB, and a 32-line address bus can access more than 4GB. A 64-line address bus (used in the DEC Alpha APX) can access 16GB.

Address Classes-In a 32-bit IP address the number of bits used to identify the network and the host vary according to the network class of the address, as follows:

Class A is used only for very large networks. The high-order bit in a Class A network is always zero, leaving 7 bits available to define 127 networks. The remaining 24 bits of the address allow each Class A network to hold as many as 16,777,216 hosts. Examples of Class A networks include General Electric, IBM, Hewlett-Packard, Apple Computer, Xerox, Digital Equipment Corporation, and MIT. All the Class A networks are in use, and no more are available.

Class B is used for medium-sized networks. The 2 high-order bits are always 10, and the remaining bits are used to define 16,384 networks, each with as many as 65,535 hosts attached. Examples of Class B networks include Microsoft and Exxon. All Class B networks are in use, and no more are available.

Class C is for smaller networks. The 3 high-order bits are always 110, and the remaining bits are used to define 2,097,152 networks, but each network can have a maximum of only 254 hosts. Class C networks are still available.

Class D is a special multicast address and cannot be used for networks. The 4 high-order bits are always 1110, and the remaining 28 bits allow for more than 268 million possible addresses.

Class E is reserved for experimental purposes. The first four bits in the address are always 1111.

Address Complete Message (ACM)-An ISDN User Part trunk signaling message by which a call's destination SSP acknowledges an initial address message.

Address Field-A section of a message, generally at the beginning, in which the addresses of the message source and destination are found.

Address Resolution-A process that is used to specifically identify differences between computer addressing schemes. The address resolution process may be implemented by mapping channels on layer 3 (network layer) to specific addresses on layer 2 (data link layer).

Address Resolution Protocol (ARP)- Abbreviated ARP. A protocol within TCP/IP (Transmission Control Protocol/Internet Protocol) and AppleTalk networks that allows a host to find the physical address of a node on the same network when it knows only the target's logical or IP address.

Under ARP, a network interface card contains a table (known as the address resolution cache) that maps logical addresses to the hardware addresses of nodes on the network. When a node needs to send a packet, it first checks the address resolution cache to see if the physical address information is already present. If so, that address is used, and network traffic is reduced; otherwise, a normal ARP request is made to determine the address.

Address Signaling-A process of sending the signaling message that includes the dialed telephone number to a exchange carrier. For the public switched telephone network (PSTN), this is performed by dial pulsing (rotary phone) or touch-tone (TM) signals.

Address Signals-Signals that identify call destination information or the digits that are dialed by a caller. The different types of address signaling include Dial Pulse, Multi-Frequency (MF), and Dual-Tone Multi-Frequency (DTMF).

Address Space-A measure of the amount of memory that can be installed on a computer. A computer that utilizes 24 bits to designate a memory location is said to have a 24-bit address space; it can address twice as much memory as a computer having a 23-bit address space.

Address Translation Gateway-A gateway that can convert the address format (field and/or physical layer structure) from one network to another.

Addressability-The ability of communication system to use specific or group addresses to control the distribution of programs from a central location.

Addressed Call Mode-The mode of communication connection that sends the destination address (e.g. telephone number) through a dial up or on-demand connection service.

Addressing Authority-The authority responsible for the unique assignment of Network addresses within a network address domain.

Addressing Domain-The addresses within each level of network hierarchy where every address is part of a group (domain). Domain addresses are hierarchical. If an addressing domain is part of a hierarchically higher addressing domain (which wholly contains it), the authority for the lower domain is authorized by the authority for the higher domain to assign addresses from the lower domain.

Addressing Space-The amount of memory (range of addresses) available to a computer operating system for software control or processing of data.

Adjacency-The relationship between neighboring (adjacent) routers that can exchange (swap) routing information.

Adjacent-Network devices or communication channels that are electrically or physically located next to each other.

Adjacent Channel

Adjacent Channel-A communication channel that has a carrier frequency that is located immediately above or below the frequency of another communication channel.

This diagram shows the relative spacing of adjacent radio communication channels. Additional radio channels are often spaced at bandwidth intervals from the main carrier channel. Radio channels that are located next to a channel are called adjacent channels.

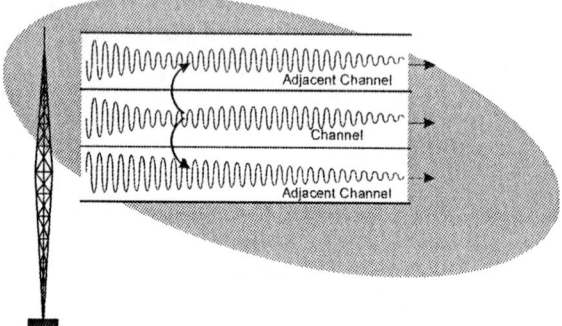

Adjacent Channel System

Adjacent Channel Interference-Adjacent channel interference that occurs when the RF power from an adjacent channel overlaps into the frequency band of the channel that is being used by a radio device.

This example shows that adjacent channel interference occurs when one radio channel interferes with a channel next to it (e.g. channel 412 interferes with 413). Although each radio channel has a limited amount of bandwidth (10 kHz or 30 kHz wide in this example), some radio energy is transmitted at low levels outside this band. A radio transmitter that is operating at full power can produce enough low-level radio energy outside the channel bandwidth to interfere with radios operating on adjacent channels. Because of alternate channel interference, radio channels that are used on the same system cannot usually be spaced adjacent to each other in a single radio tower (e.g. channel 115 and 116). A channel separation of 3 channels is typically sufficient to protect most radio channels from adjacent channel interference in mobile communication systems.

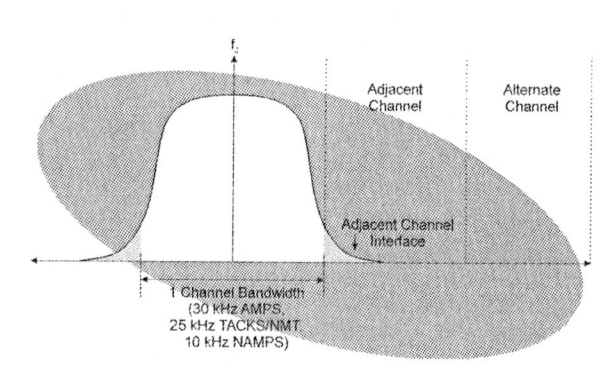

Adjacent Channel Interference Operation

Adjacent Signaling Points-In the Signaling System 7 protocol, any two signaling points that are directly interconnected by one or more signaling links.

Adjudicative Proceeding-Proceedings that involves future rates or practices, initiated upon the Commission's own motion or upon the filing of an application, a petition for special relief or waiver, or a complaint or similar pleading that involves the determination of rights and responsibilities of specific parties.

Adjunct Processor (AP)-Adjunct Processor (AP) is a decentralized form of signaling control point (SCP) databases that are used by switching systems to reduce the requirement of signaling service point (SSP) central office switches from connecting to SCPs.

Adjunct Switch Application Interface (ASAI)-A protocol that is used to interface Lucent DEFINITY PBX switch to computer telephony integration (CTI) applications. The ASAI interface allows provides for call processing control and the providing of event notification between system elements (such as switches and the computer).

ADM-Adaptive Delta Modulation

ADM-Add-Drop Multiplexer

Administrative Alerts-In Windows 2000, informational messages sent to specific accounts, groups, or computers to announce security events, impending shutdown due to loss of server power, performance problems, and printer errors.

When a server generates an administrative alert, the appropriate message is sent to a predefined list of users and computers.

Administrative Distance-A term used by Cisco Systems, Inc., to express the integrity of a routing-information source. Administrative distance is expressed as a value in the range 0 through 255; the higher the value, the lower the quality of the routing information.

Administrative Trunk Groups-Groups of communication trunks that are used to monitor and change the status calls. Administrative trunk group types include announcement, coin supervisory, coin zone, permanent signal, vacant code, and verification. These trunk groups are used to assist in collecting revenues (e.g. for coin collection on payphones), to assist telephone operators, and to protect networks from an overload of messages.

Administrative Unit (AU)-An information element of a communication systems (such as a frame in a SDH system) that contains information that allows the cross-connection and switching of virtual channels (VCs).

Administrator Account-In account on a communications system or software program (such as Microsoft Windows) that is provided with a level of authority and permissions that allows the administrator to assign and remove permissions to users or groups within the system.

Admission-The process of a device requesting and receiving authorization to obtain service from a communication system or network.

Admission Confirm (ACF)-The process or message that confirms a device or services is admitted to a network. An ACF message is defined in H.225.

ADNS-ATM Distributed Network System

ADPCM-Adaptive Differential Pulse Code Modulation

ADSI-Analog Display Services Interface

ADSL-Asymmetric Digital Subscriber Line

ADSL Forum-A forum that was started in 1994. This forum assists manufacturers and service providers with the marketing and development of ADSL products and services. The ADSL forum has been renamed the DSL forum.

ADSL Modem-An electronics assembly device that modulates and demodulates (MoDem) asymmetric digital subscriber line (ADSL) signals. ADSL signals are usually transmitted on a twisted pair of copper wires.

An ADSL modem may be in the form of an internal computer card (e.g. PCI card) or an external device (Ethernet adapter). Most ADSL modems have the ability to change their data transfer rates based on the settings that are programmed by the DSL service provider and as a result of the quality of the communication line (e.g. amount of distortion).

ADSL Router-A packet routing device that is commonly used to interfaces a computer or other residential data communication devices with an ADSL telephone line. The use of a router (as opposed to a hub) allows multiple computers within a home to be assigned different Internet protocol (IP) addresses.

ADSL Transmission Unit - Central Office (ATU-C)-An advanced modem that provides for asynchronous digital subscriber line (ADSL) multi-megabit data rates over unshielded twisted pair (UTP) of copper wires. The ATU-C is usually located at a central switching office or at a remote distribution node (e.g. RTC). The ATU-C is essentially a mirror image of an ATU-R.

ADSL Transmission Unit - Remote (ATU-R)-An advanced modem that provides for asynchronous digital subscriber line (ADSL) multi-megabit data rates over unshielded twisted pair (UTP) of copper wires. The ATU-R is usually located at a customer's premises.

The ATU-R can be in various configurations including and internal computer modem (PCI bus), external modem that connects to the Universal Serial Bus (USB), or a bridge device that converts ADSL signals to an 10BastT or 100BaseT Ethernet form.

ADSL-Lite-A limited version of the standard ADSL transmission system. This limited version of ADSL allows for a simpler filter installation that can often be performed by the end user. The limitation of ADSL-Lite is a reduced data transmission rate of 1 Mbps instead of a maximum rate of 8 Mbps.

This figure shows that an ADSL-lite system is similar to the ADSL network with the primary difference in how the end user equipment is connected to the telephone network. The ADSL-lite system does not require a splitter for the home or business. Instead, the end user can install microfilters between the telephone line and standard telephones. These microfilters block the high speed data signal from interfering with standard telephone equipment. The ADSL-Lite end user modem contains a filter to block out the analog signals.

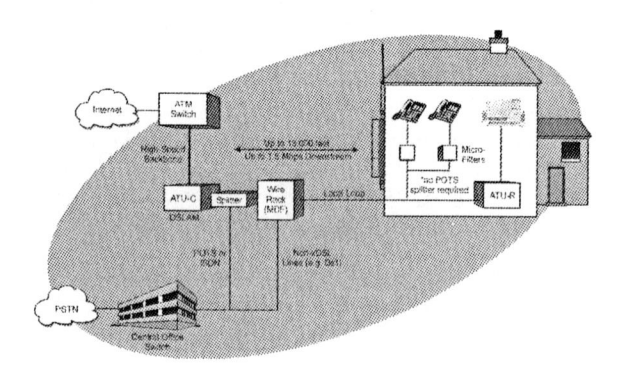

ADSL Lite System

ADSR-Automatic Data Speed Recognition

Advanced 800 Services-A group of toll free (800, 888 number) services that includes call routing that can be based on parameters such as the originating phone number or area code, time of day, or volume of calls. This allows toll free services to be redirected to handle overflow from call centers or redirecting to call centers that are available specific times within a day.

Advanced Audio Codec (AAC)-A lossy audio codec standardized by the ISO/IEC Moving Picture Experts Group (MPEG) committee in 1997 as an improved but non-backward-compatible alternative to MP3. Like MP3, AAC is intended for high-quality audio (like music) and expert listeners have found some AAC-encoded audio to be indistinguishable from the original audio at bit rates around 128 kbps, compared with 192 kbps for MP3.

Advanced Communications Function (ACF)-A group of programs developed by IBM that allows the sharing of computer resources by communications links. This allows the interconnection of two or more domains into a single network that can manage multiple domains.

Advanced Communications Service (ACS)- Abbreviated ACS. A large data-communications network established by AT&T.

Advanced Communications Technologies And Services (ACTS)-A European research program that was started in 1995 to assist with advancements in mobile communication technologies and services.

Advanced Configuration and Power Interface (ACPI)-Abbreviated ACPI. An interface specification developed by Intel, Microsoft, and Toshiba for controlling power use on the PC and all other devices attached to the system. A BIOS-level hardware specification, ACPI depends on specific hardware that allows the operating system to direct power management and system configuration.

Advanced Data Communication Control Procedures (ADCCP)-American version of HDLC. Bit-oriented, link layer protocol established as a standard by ANSI.

Advanced Encryption Standard (AES)-A data encryption standard promoted by the United States government and based on the Rijndael encryption algorithm. The AES standard is supposed to replace the Data Encryption Standard (DES).

Advanced Intelligent Network (AIN)-Advanced intelligent networks (AIN's) are telecommunications networks that are capable of providing advanced services through the use of distributed databases that provide additional information to call processing and routing requests.

In the mid 1980's, Bellcore (now Telcordia) developed a set of software development tools to allow companies to develop advanced services for the telephone network. The advanced intelligent network (AIN) is a combination of the SS7 signaling network, interactive database nodes, and development tools that allow for the processing of signaling messages to provided for advanced telecommunications services.

The AIN system uses a service creation environment (SCE) to created advanced applications. The SCE is a development tool kit that allows the creation of services for an AIN that is used as part of the SS7 network. A service management system (SMS) is the interface between applications and the SS7 telephone network. The SMS is a computer system that administers service between service developers and signal control point databases in the SS7 network. The SMS system supports the development of intelligent database services. The system contains routing instructions and other call processing information.

To enable SCPs to become more interactive, intelligent peripherals (IPs) may be connected to them. IPs are a type of hardware device that can be programmed to perform a intelligent network processing for the SCP database. IPs perform processing services such as interactive voice response (IVR), selected digit capture, feature selection, and account management for prepaid services.

To help reduce the processing requirements of SCP databases in the SS7 network, adjunct processors

(APs) may be used. APs provide some of the database processing services to local switching systems (SSPs).

This diagram shows the basic structure of the AIN. Companies that want to enable information services use the SMS to interface to SCP databases within the SS7 network. This diagram shows how a prepaid calling card company manages a portion of a SCP node using the SCE tool kit. The SCP is connected to an IP that contains an IVR unit that prompts callers to enter the personal identification number (PIN). The IP then reviews the account and determines available credit remains and informs the SCP of the destination number for call routing.

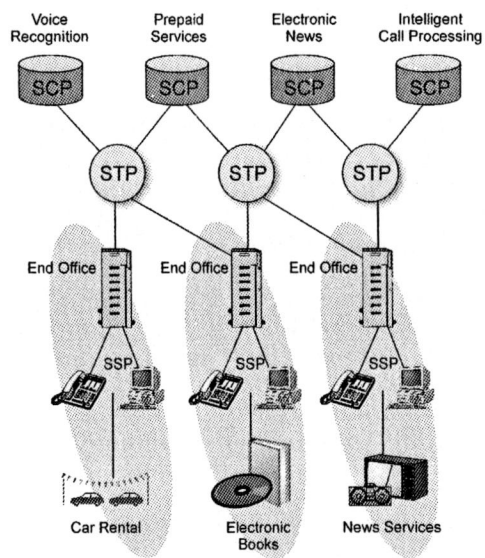

Advanced Intelligent Network (AIN) System

Advanced Interactive Executive (AIX)- Abbreviated AIX. A version of Unix from IBM that runs on its RS/6000 workstations and on minicomputers and mainframes.

Although AIX is based on Unix System V Release 3, it contains many of the features available in System V Release 4, is POSIX-compliant, and meets the Trusted Computer Base (TCB) Level C2 security.

One of the major enhancements of AIX is Visual Systems Management (VSM), a graphical interface into the older Systems Management Interface Tool (SMIT). VSM contains four main elements: Print Manager, Device Manager, Storage Manager, and Users and Groups Manager.

Advanced Micro Devices, Inc. (AMD)- Abbreviated AMD. The fifth largest manufacturer of integrated circuits, flash memory, and microprocessors, specializing in clones of Intel's popular PC chips, including the AMD386, AMD486, AMDK5, and the AMDK6.

Advanced Mobile Phone System (AMPS)-An analog cellular communications system that uses frequency-division multiple Access (FDMA) for control and frequency division dupiex (FDD) for two transmission. The AMPS radio channel types include 30 kHz FSK control channels and 30 kHz voice channels. It operates in the 825 MHz to 890 MHz range.

Advanced Peer-To-Peer Networking (APPN)- Abbreviated APPN. IBM's SNA (Systems Network Architecture) protocol, based on APPC (Advanced Program-to-Program Communications). APPN allows nodes on the network to interact without a mainframe host computer and implements dynamic network directories and dynamic routing in a SNA network.

APPN can run over a variety of network media, including Ethernet, token ring, FDDI, ISDN, X.25, SDLC, and higher-speed links such as B-ISDN or ATM.

Advanced Power Management (APM)- Abbreviated APM. An API specification from Microsoft and Intel intended to monitor and extend battery life on a laptop computer by shutting down certain system components after a period of inactivity.

Advanced Program-To-Program Communications (APPC)-Abbreviated APPC. A set of protocols developed by IBM as a part of its SNA (Systems Network Architecture), designed to allow applications running on PCs and mid-range hosts to exchange data easily and directly with mainframes. APPC can be used over a SNA, Ethernet, X.25, or Token Ring network and is an open, published communications protocol.

APPC/PC is a PC-based version of APPC used over a Token Ring network.

Advanced Radio Data Information Service (ARDIS)-A two-way wireless data network. The ARDIS system allows simulcast radio transmission to increase the penetration and reliability of wireless data signals into buildings. The data rate used by an ARDIS device is approximately 9.6 kbps and the frequency range commonly used for ARDIS systems are in the 800-900 MHz range.

Advanced Radio Research Projects Agency Network (ARPAnet)- A computer network that was developed by the Advanced Research Projects Agency of the U.S. Department of Defense. ARPAnet was the predecessor to the Internet. The objective of ARPAnet was to allow continuous communications in the event portions of the network were disabled (possibly due to military or nuclear weapon attack).

Advanced Radio Technology Subcommittee (ARTS)- Sponsored by the CTIA to study the industry needs, technology available, and manufacturers support to develop new cellular technology.

Advanced Replacement- A process that allows the end user or repair facility to obtain a replacement component (transmitter, receiver, software, etc.) before returning the defective equipment or software. This process usually requires the requesting person or company to obtain a reference number for the advanced replacement component that is requested. The person or company often uses the box from the replacement equipment to return the defective equipment or software.

Advanced SCSI Programming Interface (ASPI)- This is the standardized interface for the management and control of SCSI2 devices within a computer system. It is specified by the ANSI standard X3.131-1994.

Advanced Technology Attachment (ATA)- A interface system for disk drive interfaces based on ANSI standard x3T10. It is usually called Integrated Drive Electronics (IDE).

Advanced Television Enhancement Forum (ATVEF)- A group of companies and industry expert that work towards the creation of industry standards for combining Internet content broadcast television.

Advanced Television Services- Television services that are provided using alternative technologies (such as digital television). Advanced television services are defined the report: "Advanced Television Systems and Their Impact Upon the Existing Television Broadcast Service", MM Docket 87-268.

Advanced Television Systems Committee (ATSC)- A government-appointed committee that assists with the development of advanced television technologies systems.

Advertising- The process by which services on a network inform other devices on the network of their availability. Novell NetWare uses the Service Advertising Protocol (SAP) for this purpose. Such a process is used in packet networks to inform switches and router of new routes and route changes.

Advice Of Charge (AOC)- The ability of a telecommunications system to advise of the actual costs of telephone calls either prior or after the calls are made. For some systems, (such as a mobile phone system) the AOC feature is delivered by short message service.

Advisory Tones- Audio tones that are provided from the telephone system to inform the customer of the call status or change in call status. Advisory tones include busy, dialtone, ringing, fast-busy, call-waiting, and other tones.

AEB- Analog Expansion Bus

Aerial Lift- A bucket or ladder truck that is used by telephone company (telco) outside plant personnel to access overhead aerial plant, such as terminals, splices, cabling repeaters, dropwire, etc. for new construction or maintenance activities.

Aerial Telephone Cable- Telephone distribution cable that is mounted above ground. The majority of aerial cable is 50-600 pair unshielded twisted pair copper conductor, poly outer sheath, and lashed onto metal twisted strand strung from utility poles

Aeronautical Advisory Station (unicom)- An aeronautical information broadcast station that is use to provide advisories and civil defense information that is primarily related to private aircraft.

Aeronautical Fixed Service- A wireless communication service that is used for air navigation safety and to coordinate efficient operation of air transportation. These communication services are between fixed transceivers (stations).

Aeronautical Fixed Station- A transceiver that is used for fixed aeronautical communication service.

Aeronautical Mobile Satellite Service- A satellite communication service that allows aircraft to directly communicate with or through orbiting satellites. For this service, mobile Earth stations are loaded on board aircraft. Other mobile satellite stations such as emergency position-indicating radio-beacon stations may be part of this service.

Aeronautical Mobile Service- Mobile communication service between aeronautical transceivers (stations) and aircraft stations or directly between aircraft stations.

Aeronautical Radionavigation Satellite Service- A radio communication service that provides position information for aircraft navigation through the use of satellite radio transmission.

Aeronautical Radionavigation Service-A radio communication service that provides position information for aircraft navigation.

Aeronautical Station-A land station in the aeronautical mobile service. In certain instances an aeronautical station may be located, for example, on board ship or on a platform at sea.

AES-Aircraft Earth Station

AES-Advanced Encryption Standard

AFACTS-Automatic Facilities Test System

Affiliate-A company or person that owns, controls or is owned or controlled by another company or person. Regulations may specify the definition of "affiliate" based on an equity percentage of ownership. Regulations may limit the maximum amount of ownership (e.g. 10%) by an affiliate that is involved in related industries (such as television and radio system ownership) or the amount of ownership by a foreign person or company.

Affiliated Companies-The companies that directly or indirectly own or control a company or resource. Regulations may specify the definition of an "affiliated company" based on an equity percentage of ownership. Regulations may limit the types of companies that may participate as an affiliated company based on an amount of ownership (e.g. 10%) and the actions these companies may perform or may be required to perform because they are considered an affiliate company.

Affiliation Through Common Facilities-An affiliation that results when a company or person shares resources (such as a communication system or office space) with other companies or people. This affiliation is usually visible when more than one person or company who uses the shared resource can change the shared resources that have potential control of others involved in the use of the resource.

Affiliation Through Common Management-Affiliation between entities (corporation or individuals) that arises where agents of the company serve as a controlling element of the management or board of directors of another entity.

Affiliation Through Contractual Relationships-Affiliation between entities (corporation or individuals) that arises where the control of agents of an entity is dependent upon contractual terms of another entity.

Affiliation Through Stock Ownership-Affiliation between entities (corporation or individuals) that arises where equity owners (shareholders) own or have the power to control (e.g. vote proxies) of more than 50 percent of the voting stock.

AFP-AppleTalk Filing Protocol

AFS-Andrews File System

Aftermarket-The market for related hardware, software, and peripheral devices created by the sale of a large number of computers of a specific type.

AGC-Automatic Gain Control

AGCH-Access Grant Channel

Agency-(1-general regulatory) A Commission, board of Commissioners, Committee, or other group of commissioners who are authorized to act on behalf of the commission or regulatory department. (2-communications) The Federal Communications Commission (FCC) agency of the U.S. Government as defined by section 105 of title 5 U.S.C., the U.S. Postal Service, the U.S. Postal Rate Commission, a military department as defined by section 102 of title 5 U.S.C., an agency or court of the judicial branch, or and an agency of the legislative branch, including the U.S. Senate and the House of Representatives.

Agency Head-The head (chairman) of the Federal Communications Commission agency.

Agent-A program that performs a task in the background and informs the user when the task reaches a certain milestone or is complete.

A program that searches through archives looking for information specified by the user. A good example is a spider that searches Usenet articles. Sometimes called an intelligent agent.

In SNMP (Simple Network Management Protocol), a program that monitors network traffic.

In client-server applications, a program that mediates between the client and the server.

Agent Channel-Distribution channel that makes use of a third party sales force (individual and corporate) to deliver products and services directly to customers.

Aggregate Customer Information-Collective data that relate to a group or category of services or customers, from which individual customer identities and characteristics have been removed.

Aggregate Route-Based IP Switching (ARIS)-A process that is used to establish virtual circuits though switched networks through a network without the need to make switching (routing) decisions at each node through the use of tags (on each packet) that are used to guide the packets through the virtual circuits. Some of the protocols used in the ARIS system include Open Shortest Path First (OSPF) and Border Gateway Protocol (BGP).

Aggregated Link-A process of combining two or more physical links to provide a single high-speed interface to higher layer protocol layers. This process is also called inverse multiplexing.

Aggregator-The entity that performs the operations required to make multiple physical links function as an aggregated link. Also a seller of discounted network services based the aggregator's ability to contract for bulk services for significantly reduced rates.

Aging Process-Within the Spanning Tree Protocol, the process that removes dynamic entries from the filtering database when the stations with which those entities are associated have been inactive for a specific time (the aging time).

Aging Time-Within the Spanning Tree Protocol, the time after which a dynamic filtering database entry will be removed if its associated station has been continuously inactive.

AGRAS-Air-Ground Radiotelephone Automated Service

AH-Authentication Header Protocol

AIC-Automatic Intercept Center

AICs-Account Inquiry Centers

AID-Attention Identification

AIFF-Audio Interchange File Format

AIN-Advanced Intelligent Network

AINTCC-Automated Intercept Call Completion

AIOD-Automatic Identification Of Outward Dialing

AIP-Application Infrastructure Service Provider

AIP-Abandoned in place

AIR-Allowed Information Rate

Air Carrier Aircraft Station-A mobile station on board an aircraft which is engaged in, or essential to, the transportation of passengers or cargo for hire.

Air Interface7-The specification that devices the access control and channel sharing on radio and other free space transmission systems. Some of the common air interface specifications include global system for mobile communications (GSM) and wideband code division multiple access (WCDMA)..

Air Pipe-A plastic pipe with an aluminum inner sheath used in cable air pressurization delivery systems. Common air pipes have a 5/8" inner dimension (ID) and they originates at the central office (CO) and are placed in parallel with underground telephone cabling. Air pipes are used to distribute and provide dry air throughout distribution cabling.

Air Vent-Air vents are tubes that are used to equalize atmospheric pressure within a cable. Air vent sizes include 2.5 " vents that are commonly used in land based (inland) regions and larger 4.5 " vents are often used along the seacoast region.

Air-Ground Radiotelephone Automated Service (AGRAS)-A telephone service that is used for aircraft.

Air-Ground Radiotelephone Service-A wireless communications service for subscribers in aircraft.

Air-Ground Service-A wireless service between aircraft and a telephone or data network. There are two basic types of air-ground services; general aviation and passenger or commercial service.

Air-To-Ground (ATG)-Communications services that is provided between aircraft and ground stations. Initially, these services have involved voice communication from aircraft to ground systems (such as the public telephone network). It has evolved to include digital Air-to-Ground (ATG) systems and two-way (ground-to-air) service was introduced in 1994.

Airborne Station-A mobile radio that is authorized to provide Air-Ground Radiotelephone Service.

Aircraft Earth Station (AES)-A mobile radio that is located on an aircraft to provide mobile-satellite service. AES stations communicate from the aircraft through satellites to reach their ultimate destination.

Aircraft Station-A mobile communication transceiver that is located on an aircraft that provides for communication in aeronautical mobile service. An aircraft station does not include other radio devices such as a survival craft transceiver.

This figure shows a public aircraft telephone system. This diagram shows that aircraft may be served by terrestrial (land-based) systems or satellite communication systems. In either case, the aircraft communicates with a gateway that links the radio system to the public telephone system

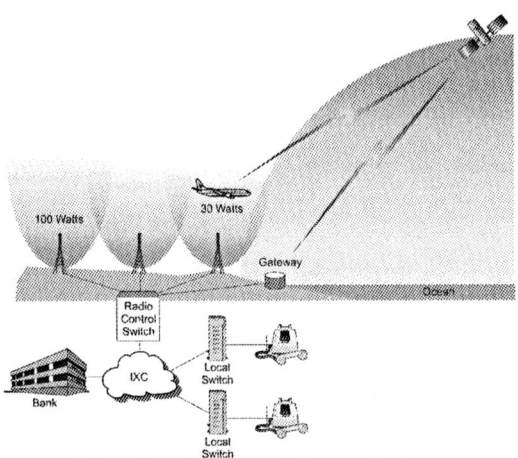

Public Aircraft Telephone System

Airport Control Tower Station-An aeronautical transceiver that is used to provide communication between aircraft and a airport control tower.
AirTime-The length of time radio communication service (radio transmission time) is used.
AIX-Advanced Interactive Executive
AK-Authorization Key
Alarm-An audible or visible indication of a trouble condition. Alarms are classified as minor, major, or critical, depending on the degree of service degradation or disruption.
Alarm Controller Unit (ACU)-A control unit that coordinates the reporting of remote alarm information. A common use of an ACU is on a T1-D4 channel bank that used in subscriber loop carrier (SLC) system that interfaces both central office (CO) and remote terminals.
Alarm Indication Signal-An indication signal (such as a code or voltage) that indicate that a failure has been detected on a communication assembly or the failure of a process/ program operation.
Alarm Monitoring Service-A service or company that detects and responds to equipment failures, alarm triggers, or incidents.
Alarm Suppressor Unit (ASU)-An equipment plug-in module that is used in conjunction with T-carrier systems when fiberoptic lines are used for the transmission line. The use of an ASU eliminates false alarms on a T-carrier span.
ALC-Automatic Level Control
ALC-Automatic Light Control
ALE-Automatic Link Establishment

Alert Tones-The types of alert tones that are available to indicate a particular status of a telecommunications event. An example of an alert tone is a sound that alerts the user that a new short message has been received.
Alert With Info-An alert message that is sent to a communications device (e.g. start ringing) that also contains additional information (e.g. calling name or calling number).
Alert With Info Ack-A message that is sent back to a sender that confirms the alert with information message was successfully received.
Alerter Service-A messaging service that is integrated with an application (e.g. such as Microsoft Windows 2000) that warns a list of users of an administrative alert.
Alerting Sequence-A group of symbols or sequences of data that is followed by a control or data message.
Alerting Signal-A prompt sent on a subscriber line to indicate an incoming call.
Alerting Signals-Signals that are used to alert a user of a service request (e.g. incoming call) or status change (message waiting).
ALG-Application Layer Gateway
Algorithm-A set of well defined steps or rules that allows for the solution of a problem or processing of information. Commonly the name for a portion of a software program.
Algorithm A3-(1- GSM) An authentication algorithm that is used to validate a mobile station. The Algorithm uses a random number sent by the system (RAND) and a secret key (Ki) to computer a product (SRES) by its cryptographic algorithm.
Algorithm A5-(1- GSM) A cryptographic authentication algorithm to produces a product using a secret key (Kc).
Algorithm A8-(1- GSM) An authentication algorithm that is uses a random number sent by the GSM system (RAND) and a secret key (Ki) to produce another secret key (Kc).
Algorithm Specific Signal Processor (ASSP)-An integrated circuit (IC) that is specifically designed for a unique application or series of related applications (e.g. radio signal filtering).
ALI-Automatic Location Identifier
Aliasing-(1-network addressing) Providing temporary or alternative identification codes or names to identify a channel or service. Aliasing allows devices or services to be addressed using a short-

Alignment

ened code or allows the user of a name that hides the underlying addressing information. (2-sampled data waveform processing) Signals appearing in the wrong part of the frequency spectrum due to insufficiently frequent sampling of the original waveform.

Alignment-The process of adjusting a circuit or the status of system components that interact with each other and can be changed so the performance of the system can be enhanced. An example of alignment is the tuning of multiple tuned circuits in a transmitter or receiver assembly so the signal can be amplified in each stage. When all the amplifiers are tuned to the correct frequencies, the system is said to be aligned.

Alignment Error Rate Monitoring-A procedure for measuring the error rate of a signaling link during an initial alignment in the SS7 system.

ALIT-Automatic Line Insulation Testing

All Call Paging-A process of sending an audio broadcast announcement to someone (a verbal page) through the speakers of all the telephones or loudspeakers in a building or area.

All Trunks Busy (ATB)-A measurement in a communication network of the amount of time or number of times that all trunks in a group are busy (unavailable for service because they are in-use with other channels).

All-Call Voice Page-A telephone call-handling feature that allows simultaneous voice messages to all paging areas.

All-Dielectric Cable-A cable that has the separating materials between the conductors that is made entirely of dielectric (insulating) materials without any metal conductors, armor, or strength members.

All-Routes Explorer Packet-A packet that is sent in a bridging network to hunt for another device that is attached somewhere in the network.

Alliance for Telecommunications Industry Solutions (ATIS)-A North American organization established by Bellcore to develop standards and guidelines for the methods and procedures needed in the telecommunications industry. ATIS has four committees, T1, SONET, Internet Work Interoperability Test Coordination, and Order and Billing Form.

Alligator Clip-A spring-loaded connector that allows the temporary connection of a test lead to a wire or electronic circuit component. The alligator clip has a long section that contains teeth keep the clip from sliding off the connection (looks like an alligator's mouth).

This photo image shows how an alligator clip can be mounted on an extension assembly to protect a user from an electric shock. A rubber isolator is used to isolate the electrical signal from reaching the user.

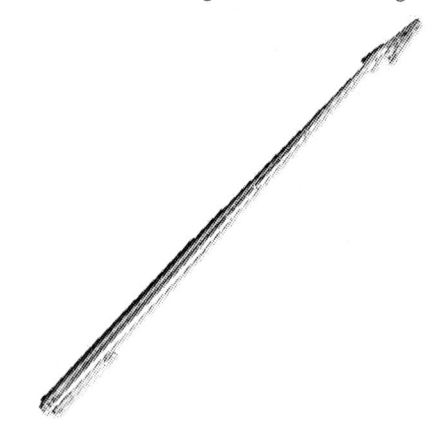

Alligator Clip
Source: HVACtool.com

Allocated Call Reference-A call reference identifier that is assigned to a call in the SS7 and Integrated Services Digital Network (ISDN) Q.931 protocols.

Allocated Channel-A channel that is either connected or reserved for a call. Allocated channels are defined in signaling system 7 (SS7) and Integrated Services Digital Network (ISDN) Q.931.

Allocated Circuit-A circuit designed and reserved for the use of a particular customer.

Allocation Group-In automated facility planning, any message trunk group, specially defined special-service circuit group, or specially defined carrier system group that creates a demand for facilities and equipment.

Allocation Of A Frequency Band-Entry in the Table of Frequency Allocations of a given frequency band for the purpose of its use by one or more terrestrial or space radio-communication service or the radio astronomy service under specified conditions. This term shall also be applied to the frequency band concerned.

Allotment of A Radio Frequency Or Radio Frequency Channel-An entry of a designated radio frequency channel in an agreed upon plan, adopted for use by one or more administrations for a terrestrial or space radio-communication service.

Allowed Information Rate (AIR)-In a frame-relay Data Link Connection (DLC) the maximum data rate is given by the sum of the Committed

Information Rate (CIR) and the Excess Information Rate (EIR) of the connection.

ALOHAnet-A wireless radio network developed at the University of Hawaii to connect their dispersed campus for data networking. ALOHAnet used what would later be refereed to as carrier sense multiple access with collision detection (CSMA/CD). This later evolved into the IEEE 802.3 standard for Ethernet. In this scheme, a user wishing to transmit does so at will. Collisions are resolved by retransmitting after a random period of time.

Alpha Paging-Text messages which are sent via operator or computer to an Alpha Pager.

This diagram shows how an alpha paging system receives voice, text, or data messages from callers and forwards these messages in text form to an alphanumeric pager. In this example, a sender can access the system by voice or by sending email messages via the Internet. When accessing the system by voice, a caller dials a paging access number and is either connected to an interactive voice response (IVR) unit or to an operator. When connected to an IVR, the user may be given options for specific messages (canned messages) or their voice may be converted to text messages. When connected to an operator, the operator converts (keys in) their messages to text form. When messages are sent via the Internet, their format is changed to a form suitable for the alpha paging system. In any of these cases, the message are placed in a message queue that holds the message until the system is available (not other messages waiting) before it. When the message reaches the top of the queue (available time to send), it will be encoded (formatted) to a form suitable for transmission on a radio channel. In this example, the message is sent as part of group 4. Sending the messages in groups allows the pager to sleep during transmission of pages from other groups that are not intended to reach the alphanumeric pager. The text message includes the pager address along with the text message in digital form. During the reception of the message, it is stored into the message paging memory area so the pager can display the message after it is received.

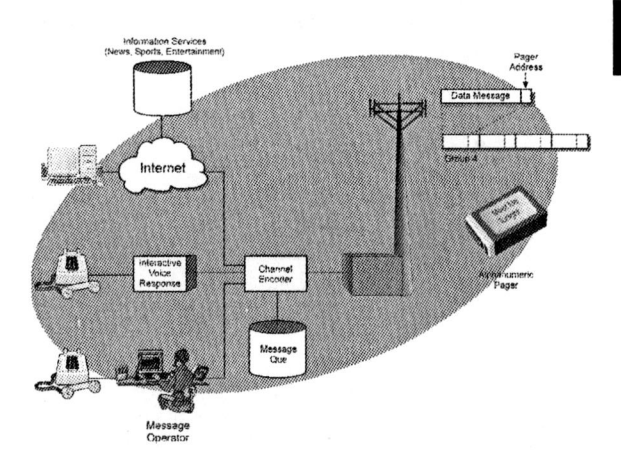

Alpha Paging Operation

Alpha Testing-The first stage in testing a new hardware or software product, usually performed by the in-house developers or programmers.

Alphanumeric-A generic term for alphabetic letters, numerical digits, and special characters that can be processed and displayed by a machine. Alphanumeric displays provide a character set that includes letters, numbers, and punctuation marks.

ALS-Alternate Line Service

Alt Newsgroups-A set of Usenet newsgroups containing articles on controversial subjects often considered outside the mainstream. Alt is an abbreviation for alternative.

These newsgroups were originally created to avoid the rigorous process required to create an ordinary newsgroup. Some alt newsgroups contain valuable discussions on subjects ranging from agriculture to wolves, others contain sexually explicit material, and others are just for fun. Not all ISPs and online services give access to the complete set of alt newsgroups.

Alternate Billing Services-Calling-card, collect, and third-number-billed calls whose originating party does not pay for the call.

Alternate Buffer-In a data communications device, a section of memory set aside for the transmission or reception of data when the primary buffer is filled. This allows for uninterrupted flow of data in either transmission or reception mode.

Alternate Channel-A communication channel that has a carrier frequency that is located one channel bandwidth above or below the frequency of another communication channel.

Alternate Channel Interference

This diagram shows the relative spacing of alternate radio communication channels. Alternate radio channels are often spaced at two bandwidth intervals from the main carrier channel.

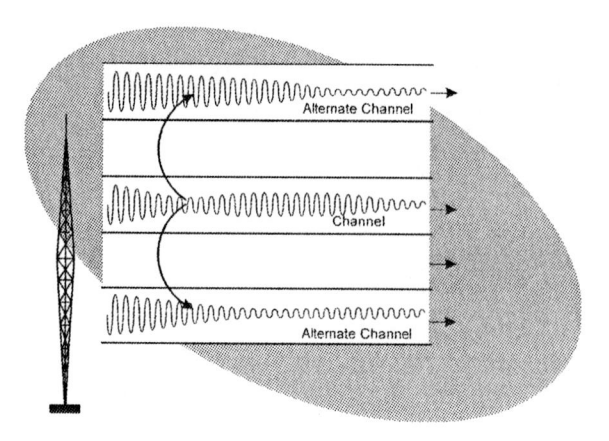

Alternate Channel System

Alternate Channel Interference-Alternate channel interference occurs when the RF power from an alternate channel (2 channels away from desired channel) overlaps into the frequency band of the channel being used by a radio device.

Alternate Gatekeeper-A gatekeeper that is assigned to provide backup connections in the event that a primary gatekeeper fails or becomes unavailable.

Alternate Line Service (ALS)-A multiple telephone number service where a customer can have two or more telephone numbers with a single telephone device. ALS is used to allow a business phone number and personal phone number to share the same telephone. Many telephones allow for the selection of different ring tones for different ALS numbers. ALS is also very important for telephone devices that may receive voice, data or fax information. For example, one ALS phone number could be used for a dedicated fax number the other for voice. If a call is incoming on a fax number, the telephone device knows to answer as a fax call.

Alternate Mark Inversion (AMI)-A line coding method used in T-1 and other digital wire transmission systems. AMI in T-1 systems uses return-to-zero (RZ) in an alternating bipolar pulse string, with logical binary zero corresponding to zero volts and logical binary one corresponding to alternating serial positive or negative 3 volt pulses. This process allows self-synchronization when a limited number of zeros is transmitted. There are therefore limitations (called zeros density limitations) on the number of consecutive zero volt pulses in such systems (a limit of 15 consecutive zero volt pulses is the maximum allowed for DS-1 systems, for example).

This diagram shows how alternate mark inversion (AMI) uses an alternating bipolar scheme to represent digital information. The AMI system represents information by alternating between a positive (e.g +3 Volts) and negative (e.g. -3 Volts) levels. This AMI process allows for self-synchronization by forcing during each bit sequence transmission

Alternate Mark Inversion (AMI) Operation

Alternate Operator Services (AOS)-Operator services that are provided by another company (third-party operator). Alternative operator services are often used by interexchange carriers or small companies that do not have their own operators. Examples of common businesses that commonly use alternate operator services include small retail businesses, hotels, hospitals, and independent pay-phone companies.

Alternate Operator Services Provider (AOSP)- A company that provides operator (verbal call assisted) services other than the company providing that is providing the physical communication (transmission) service.

Alternate Power Supply-A backup power supply which converts commercial ac power to dc to maintain batteries at appropriate charging levels. Alternate power supplies usually include an inverter and regulator for dc/ac conversion to maintain remote site terminals and central office equipment in the event of a commercial power outage.

Alternate Recovery Facility (ARF)-This is a backup facility that is equipped and configured to perform the command, control and data communication operations necessary to maintain the orbital parameters of a particular satellite.

Alternate Route-A alternative path or connection of circuits between two points that is used as second or next-choice in the event a primary route is disconnected.

Alternate Routing-A network switching feature that enables alternate routing of trunk or path assignments. Alternate routing may be enabled if a failure occurs in the primary route (path) or for least-cost routing service.

Alternate Voice Data Circuits-Circuits that are electrically conditioned to handle both voice and data signals.

Alternating Current (AC)-Refers to the type electricity that is characterized as a cyclic wave of energy where the voltage varies continuously from positive to negative to positive. When graphed the wave form is sinusoidal. The opposite of AC is DC (direct current).

Alternating Current Power (AC Power)-Electrical power that is supplied in a form of alternating current (AC). AC power is usually supplied from an external source such as Utility or generation as the main supply to equipment or assemblies.

Alternative Access Provider-A telecommunications service that provides an access connection between the end customer and a telecommunications network. This provider is a different company than the established LEC or PTT company.

Alternative Route-A secondary communications path to a specific destination. An alternative route is used when the primary path is not available.

Alternative Routing Of Signaling-The routing of signaling messages (such as in a Signaling System 7 system) through alternative paths as a result in the failure of a primary routing path.

ALU-Arithmetic Logic Unit

Always On-A connection to a communications network (such as the Internet) that appears always on to the customer. Although always on connections appear as a dedicated connection to the end user (no need to initiate a dial up sequence), the connection may be temporary and automatically re-established each time the user accesses the network.

AM-Amplitude Modulation

AM Broadcast Band-The band of frequencies extending from 535 to 1705 kHz.

AM Broadcast Channel-The band of frequencies occupied by the carrier and the upper and lower sidebands of an AM broadcast signal with the carrier frequency at the center. Channels are designated by their assigned carrier frequencies. The 117 carrier frequencies assigned to AM broadcast stations begin at 540 kHz and progress in 10 kHz steps to 1700 kHz.

AM Broadcast Station-A broadcast station licensed for the dissemination of radio communications intended to be received by the public and operated on a channel in the AM broadcast band.

AMA-Automatic Message Accounting

Amateur Operator-A person that has authorization from a regulatory agency that allows the operation of an amateur radio station. Amateur operators are people that use radio communication for their own personal goals without financial or commercial service provider interest.

Amateur Radio Services-Radio service authorized for use by amateur operators that include satellite and civil emergency service.

Amateur Satellite Service-A band of frequencies allocated on some communications satellites for use by amateur radio operators in carrying out voice and data communications.

Amateur Service-A radio-communication service for the purpose of self-training, intercommunication and technical investigations carried out by amateurs. Amateur operators are people that use radio communication for their own personal goals without financial or commercial service provider interest.

Amateur Station-A radio station (transceiver) that is operated by a amateur operator. Amateur operators are people that use radio communication for their own personal goals without financial or commercial service provider interest.

Ambient Current-A level of electrical current that results from the voltages that are created by random movement of electrons in a circuit without the addition of power.

Ambient Voltage-Voltages that are created by the random vibration (movement) of electrons in a circuit without the supply of power.

AMD-Advanced Micro Devices, Inc.

American Mobile Satellite Corporation (AMSC)

American Mobile Satellite Corporation (AMSC)-A consortium formed to provide a mobile satellite system.

American National Standards Institute (ANSI)-The US organization that sets the rules and procedures for, and also authorizes specific standards setting organizations. ATIS and EIA/TIA are two ANSI authorized standards setting organizations in the US in the subject area of telecommunications.

American Standard Code for Information Interchange (ASCII)-A widely accepted standard for data communications that uses a 7-bit digital character code to represent text and numeric characters. When companies use ASCII as a standard, they are able to transfer text messages between computers and display devices regardless of the device manufacturer.

This table shows the symbols that can be represented by the 7 bit ASCII code. The columns indicate the upper 3 bits of the ASCII code (b5 - 67) and the rows indicate the lower 4 bits of the ASCII code (b1-b4). For example, to represent the letter A, the ASCII code would be 100 (upper) + 0001 (lower) resulting in ASCII code 1000001. Computers that receive the code 1000001 would display the capital letter A.

	000 (0)	001 (1)	010 (2)	011 (3)	100 (4)	101 (5)	110 (6)	111 (7)
0000 (0)	NUL	DLE	SP	0	@	P	`	p
0001 (1)	SOH	DC1	!	1	A	Q	a	q
0010 (2)	STX	DC2	"	2	B	R	b	r
0011 (3)	ETX	DC3	#	3	C	S	c	s
0100 (4)	EOT	DC4	$	4	D	T	d	t
0101 (5)	ENQ	NAK	%	5	E	U	e	u
0110 (6)	ACK	SYN	&	6	F	V	f	v
0111 (7)	BEL	ETB	'	7	G	W	g	w
1000 (8)	BS	CAN	(	8	H	X	h	x
1001 (9)	HT	EM	)	9	I	Y	i	y
1010 (A)	LF	SUB	*	:	J	Z	j	z
1011 (B)	VT	ESC	+	;	K	[	k	{
1100 (C)	FF	FS	,	<	L	\	l		
1101 (D)	CR	GS	-	=	M	]	m	}
1110 (E)	SOH	RS	.	>	N	^	n	~
1111 (F)	SI	US	/	?	O	_	o	DEL

ACK - Acknowledge
BEL - Bell
BS - Backspace
CAN - Cancel
CR - Carriage Return
DC - Direct Control
DEL - Delete Idle
DLH - Data Link Escape
EM - End of Medium
ENQ - Enquiry
EOT - End of Transmission
ESC - Escape
ETB - End of Transmission Block
ETX - End of Text
FF - Form Feed
FS - Form Separator
GS - Group Separator
HT - Horizontal Tab
LF - Line Feed
NAK - Negative Acknowledge
NUL - Null
RS - Record Separator
SI - Shift In
SO - Shift Out
SOH - Start of Heading
STX - Start of Text
SUB - Substitute
SYN - Synchronous Idle
US - Unit Separator
VT - Vertical Tab

American Standard Code for Information Interchange (ASCII)

American Standards Association (ASA)-Predecessor organization to ANSI.

American Wire Gauge (AWG)-A measurement system that provides the diameter of conductors (typically copper wire). The larger the thickness of the wire (higher the gauge), the lower the AWG number and the better of the ability of the line to carry electrical signals. Many local telephone access loops use 24 AWG or 26 AWG copper lines.

AMI-Alternate Mark Inversion

AMIS-Audio Messaging Interface Specification

AMIS-Audio Messaging Interchange Specification

Amortization-The assignment of cost of an asset or acquisition for distribution over the time period that the asset or acquisition will be used.

Ampere (A)-Unit of electric current, equal to a flow rate of approximately 6 250 000 000 000 000 000 electrons per second. The precise definition of an ampere is based on the force between two current-carrying wires. If two straight parallel conductors of infinite length and negligible circular cross-section diameter, are placed 1 meter apart in vacuum, a force equal to 0.000 000 2 newton per meter of length is produced between them when they each carry 1 ampere. Named for André M. Ampère, 19th century French physicist.

Amplified Handset-A handset (microphone and speaker) that contains an integrated (built-in) amplifier. Amplified handsets may be used by the hearing impaired.

Amplifier-An amplifier converts an input signal (usually low level) into a larger version of itself. An amplifier device provides this conversion process. Amplifiers increase both the desired signal and unwanted noise signals. Noise signals are any random disturbance or unwanted signal in a communication system that tends to obscure the clarity of a signal in relation to its intended use.

This figure shows the basic process of amplification. An amplifier uses an input signal to control the current or voltage of a device that has an external power supply. The ability of the input signal to control the current or voltage allows the replication of the original signal. The ratio of input signal to the output signal level is called the gain (amount of amplification).

Analog

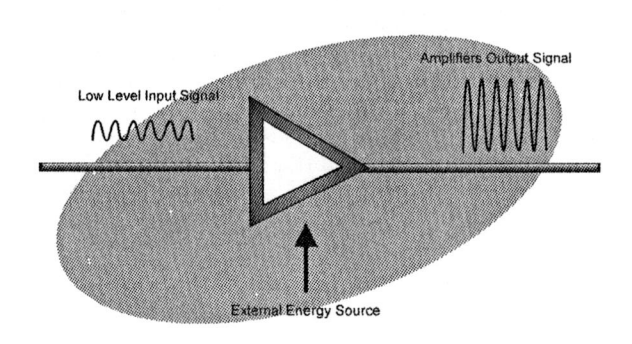

Signal Amplification Operation

Amplifier Nonliearity-An indicator of distortion that is caused by inconsistent amplification (higher or lower gain to different levels or frequencies of input signals) of a device or system.

Amplitude-A number proportional to the signal voltage or current of a waveform. One measure of amplitude is the peak-to-peak voltage (a measurement of the difference between the largest positive voltage and the most negative voltage of the waveform). Another measure is the maximum positive voltage. Yet another measure is the time-average voltage. The amplitude of a waveform is a signed (positive or negative) quantity.

Amplitude Compandored Single Sideband Modulation (ASSB)-A transmission technique where only one sideband is used for transmission with its amplitude compressed. The amplitude of the sideband signal is expanded at the receiver. No carrier is transmitted and the other sideband is suppressed.

Amplitude Distortion-A variance between the desired or expected amplitude and a signal and the actual received signal. Amplitude distortion can be measured by subtracting the expected test signal from a distorted signals to allow the measurement of the amount of distortion signal.

Amplitude Equalizer-A corrective network that is designed to modify the amplitude characteristics of a circuit or system over a desired frequency range. Such devices may be fixed, manually adjustable, or automatic.

Amplitude Modulation (AM)-Amplitude modulation is the transferring of information onto a radio wave by varying the amplitude (intensity) of the radio carrier signal.

This figure shows that amplitude modulation involves the transferring of information onto a carrier signal by varying the amplitude (intensity) of the carrier signal. This diagram shows an example of an AM modulated radio signal (on bottom) where the high of the radio carrier signal is change by using the signal amplitude or voltage of the audio signal (on top).

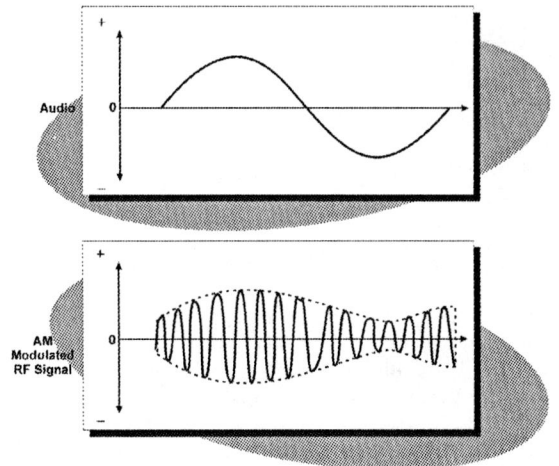

Amplitude Modulation (AM) Operation

Amplitude Modulator Stage-The last amplifier stage in a transmitter where the assembly (stage) modulates a radio-frequency signal.
AMPS-EIA-553
AMPS-Advanced Mobile Phone System
AMR-Adaptive Multirate Codec
AMR-Automatic Meter Reading
AMS-Audio/visual Multimedia Services
AMSC-American Mobile Satellite Corporation
AMTS-Automated Maritime Telecommunications Systems
AMVER-Automated Mutual-Assistance Vessel Rescue System
AMX-Analog Matrix Switch
Analog-Information that can be represented by a continuous and smoothly varying amplitude or frequency over a certain range such as voice or music. Analog lines allows the representation of information to closely resembles the original information.

www.ITExpo.Com

33

Analog Access Channel

This figure shows a sample analog signal created by sound pressure waves. In this example, as the sound pressure from a person's voice is detected by a microphone, it is converted to its equivalent electrical signal. Also, the analog audio signal continuously varies in amplitude (height, loudness, or energy) as time progresses.

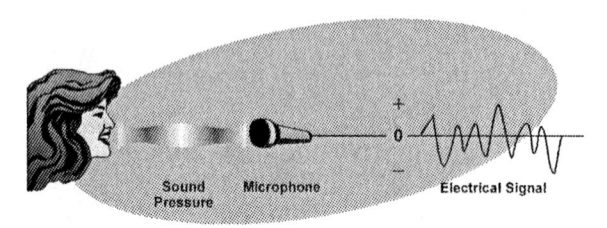

Analog Audio Signal Processing

Analog Access Channel-A control channel on an analog cellular system that uses frequency shift keying (FSK) modulation to pass digital messaging signals between the mobile phone and cell site. See: access channel.

Analog Bridge-A circuit or system that allows an analog communication session to be connected (to include) another users.

Analog Carrier System-A transmission system that uses the modulation (amplitude, frequency, or phase) of carrier signal to transport an analog information signal.

Analog Cellular-An industry term given to cellular systems that transmit voice information using a form of analog modulation (e.g. FM). Analog cellular systems may have digital control channels.

This figure shows a basic analog cellular system. This diagram shows that there are two types of radio channels; control channels and voice channels. Control channels typically use frequency shift keying (FSK) to send control messages (data) between the mobile phone and the base station. Voice channels typically use FM modulation with brief bursts of digital information to allow control messages (such as handoff) during conversation. Base stations typically have two antennas for receiving and one for transmitting. Dual receiver antennas increases the ability to receive the radio signal from mobile telephones which typically have a much lower transmitter power level than the transmitters in the base station. Base stations are connected to a mobile switching center (MSC) typically by a high speed telephone line or microwave radio system. This interconnection must allow both voice and control information to be exchanged between the switching system and the base station. The MSC is connected to the telephone network to allow mobile telephones to be connected to standard landline telephones.

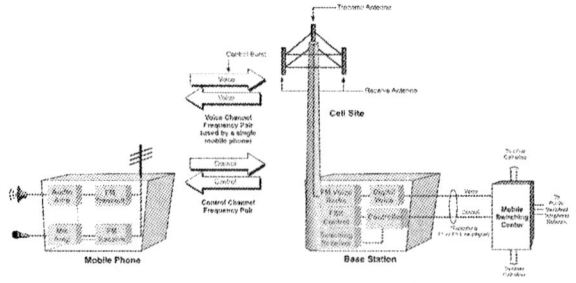

Analog Cellular System (1st Generation)

Analog Control Channel (ACC)-An analog channel used for the transmission of control signaling information from a base station to a cellular phone, or from a cellular phone to a base station.

Analog Display Services Interface (ADSI)-A Bellcore/Telcordia industry standard that defines the flow of information between a network element (e.g. network server, switch, voice mail system) and a customer's telephone, PC or other terminal device that contains a screen display.

Analog Expansion Bus (AEB)-The electrical connection between network interface modules and analog resource modules that are used in computer telephony systems. The AEB standard was originally developed by Dialogic corporation.

Analog Facsimile-A facsimile that transmits images on an analog communication line through the conversion of images or shades of images to analog signals (tones).

Analog Matrix Switch (AMX)-A switching system that can interconnect analog telephone lines or analog telephone devices.

Analog Microwave-A microwave transmission system that transfers information through the modulation of a microwave carrier signal. The type of modulation used may be amplitude, frequency or phase shift, but the analog input signal is used as the source of modulation information.

Analog Mobile-A cellular telephone that is limited to utilizing only analog FM radio transmission for voice communications.

Analog Multiline-A device that uses frequency division to multiplex two telephone lines on a single 2-wire facility (copper pair). Analog multiline is used by telephone companies (telcos) in service areas where access to wire facilities are limited.

Analog Signal-An analog signal is a direct representation of a physical process. For instance, an analog electromagnetic signal representing your voice on a telephone line is represented by continuous variations in voltage. Loud sounds are represented by large voltages, soft ones by small. High voices are represented by high frequency variations in the voltage, low ones by low frequencies. Analog signals provide the most nuanced and precise record of a physical process because of this exact representation. However, they are more easily distorted by noise and other factors than are digital signals.

Analog Signal Processing-Analog signal processing involves the conversion of analog signals into another form using analog (continuous) circuits or systems. Analog signal processing includes filters, shaping circuits, combiners, and amplifiers to change their shape and modify the content of analog signals.

Analog Simultaneous Voice and Data (AVSD)- A the combining of analog voice and digital data signals on the same signal carrier.

Analog Subscriber Loop Carrier (ASLC)-A high efficiently analog transmission system that uses existing distribution cabling systems to transfer analog information between the telephone system (central office) and a users telephone. An analog double sideband carrier system is sometimes used in facilities-starved areas as a permanent or temporary engineering solution. It employs bi-directional transmission over a single exchange grade cable pair used for control. One control pair derives 8 channels using FDM over a 8 to 144 kHz spectrum. These systems are designed for 140 dB with repeater spacing at 35 dB intervals and are successful in low-density suburban applications (e.g. summer home areas).

Analog Switching System-A switching system in which the connection route uses analog transmission.

Analog Telephone Adapter (ATA)-Analog Telephone Adapter (ATA) is a device that converts analog telephone signals into another format (such as digital Internet protocol). These adapter boxes may provide a single function such as providing Internet telephone service or they may convert digital signals into several different forms such as audio, data, and video. When adapter boxes convert into multiple information forms, they may be called multimedia terminal adapters (MTAs) or integrated access devices (IADs).

Analog telephone adapters (ATA) must convert both the audio signals (voice) and control signals (such as touch tone or hold requests) into forms that can be sent and received via the Internet.

Analog Terminal Adapter (ATA)-A communications adapter that allows analog telephone devices (e.g. a computer modem) to interconnect to digital telephone systems.

Analog to Digital Converter (ADC)-A signal converter that changes a continuously varying signal (analog) into a digital value. A typical conversion process includes an initial filtering process to remove extremely high and low frequencies that could confuse the digital converter. A periodic sampling section that at fixed intervals locks in the instantaneous analog signal voltage, and a converter that changes the sampled voltage into its equivalent digital number or pulses.

This diagram shows how an analog signal is converted to a digital signal. This diagram shows that an acoustic (sound) signal is converted to an audio electrical signal (continuously varying signal) by a microphone. This signal is sent through an audio band-pass filter that only allows frequency ranges within the desired audio band (removes unwanted noise and other non-audio frequency components). The audio signal is then sampled every 125 microseconds (8,000 times per second) and converted into 8 digital bits. The digital bits represent the amplitude of the input analog signal.

Analog Transmission

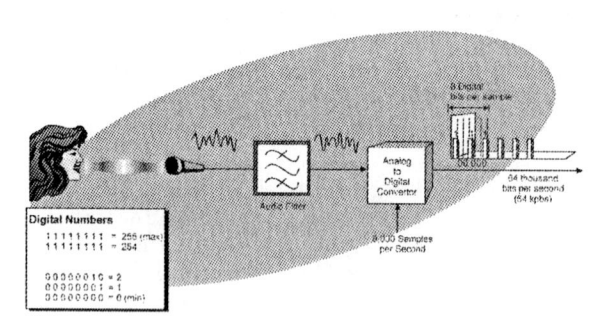

Signal Digitization Operation

Analog Transmission-A system that is capable of transferring continuously varying signals (analog signals) between points. The system may directly transfer the analog signal or the analog signal may modify another carrier (such as a radio carrier).

Analog Video-Analog video is the representation of a series of multiple images (video) through the us of a rapidly changing signal (analog). This analog signal indicates the luminance and color information within the video signal.

Sending a video picture involves the creation and transfer of a sequence of individual still pictures called frames. Each frame is divided into horizontal and vertical lines. To create a single frame picture on a television set, the frame is drawn line by line. The process of drawing these lines on the screen is called scanning. The frames are drawn to the screen in two separate scans. The first scan draws half of the picture and the second scan draws between the lines of the first scan. This scanning method is called interlacing. Each line is divided into pixels that are the smallest possible parts of the picture. The number of pixels that can be displayed determines the resolution (quality) of the video signal. The video signal television picture into three parts: the picture brightness (luminance), the color (chrominance), and the audio.

Analog Voice Channel (AVC)-A radio channel on a cellular system that typically uses FM modulation to transfer voice (audio) signals. The analog voice channel also sends digital messages during brief periods by muting the voice signal.

Anchor MSC-The interfacing (e.g. gateway) mobile switching center (MSC) that is responsible for connecting the call to all other MSCs as a mobile telephone moves to different system areas during a call.

Ancillary Services-Additional features or services provided, including call features, detail billing, voice mail, etc.

Andrews File System (AFS)-A protocol that allows communication devices to access file servers in remote locations using TCP/IP protocol. AFS protocol was developed at Carnegie Mellon University. AFS includes a memory cache that helps to the administration of the file transfer process.

Angle Modulation-Modulation of the phase angle of a carrier, as in some forms of phase and frequency modulation.

Angstrom-A measure of the length of light wavelengths. A single angstrom is 0.1nm or 10-10m.

ANI-Automatic Number Identification

ANI Identification-A software processing function at an end office (local) switch that forwards the billing number to the termination point. ANI is needed for the caller identification service offered by many service providers.

ANM-Answer Message

Announcement Service-A service that provides a caller with a predefined message when an incoming call is received.

Announcement Trunk Group-A trunk group used to inform customers or operators about call status or to access announcement services.

Anonymous Call Rejection (ACR)-A service that allows a telephone customer (wireless or wired) to reject calls from callers who have selected a privacy feature that disables the display of their calling telephone number.

Anonymous FTP-A security process that allows general users to limited have access to file servers without the need for registering for account identification and passwords. The user enters anonymous as an unregistered user and the password is usually your e-mail address.

ANSI-American National Standards Institute

Answer Message (ANM)-An ISDN User Part trunk signaling message that indicates the called party has answered the call.

Answer Mode-A call processing function that determines how a telephone device (e.g. a fax or a modem) will respond to incoming calls. The answer mode may be set to no answer, number of rings, or immediate answer.

Answer Signal-A control message (signal) that indicates in incoming call (or service request) has been answered or accepted. The answer signal can be a current flow (loop start) or a digital message that is sent on a separate signaling channel.

Answer Supervision-The sending of an off-hook supervisory signal back to an exchange carrier's point of termination to show that the called party has answered.

Antenna-A device used to convert signals between electrical and electromagnetic form. Antennas are usually designed to operate over a specific frequency range. Directional antennas are designed to focus (concentrate) the transmitted energy in a particular direction to achieve antenna gain.

This diagram shows how an antenna system is used to connect a transmitter and receiver to each other along with the key characteristics of the transmission. The transmitter provides an electrical signal in the RF signal range. This signal is transmitted through a cable or transmission line to the antenna. As the signal travels through the transmission line, some of the energy leaks from the cable and some of the energy is converted to heat in the cable. The antenna converts the electrical energy to electromagnetic energy that transfers through air and other mediums (e.g. buildings and trees). The amount of energy that is converted depends on the impedance match between the transmission line and the antenna. When there is an impedance mismatch, some of the energy is reflected back through the transmission line and is not converted. Some antennas focus energy in a particular direction. Depending on the amount of directional focus, the antenna can appear to have gain in a particular direction (at the expense of reduced gain in other directions). The electromagnetic signal is then transferred between antennas at a loss of approximately 20 dB per decade in free space transmission (through air). This means that for every 10 times the distance (e.g. 10 meters to 100 meters), the signal energy will drop by 99% (decrease to 1%). The receiving antenna converts electromagnetic energy into electrical energy. Receiving antennas can sense energy in a particular direction. Depending on the amount of directional focus, the receiving antenna can appear to have gain in a particular direction (at the expense of reduced gain from signals that are received from other directions). This signal is then transferred through a cable or transmission line to the receiver. As the signal travels through the transmission line, some of the energy leaks from the cable and some of the energy is converted to heat in the cable. The amount of energy that is received from the cable depends on the impedance match between the transmission line and the receiver. When there is an impedance mismatch, some of the energy is reflected back through the transmission line and is not received.

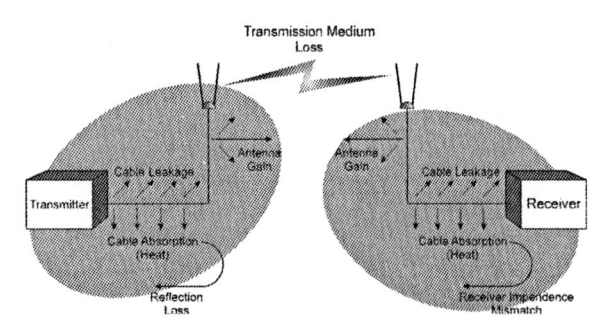

Antenna System

Antenna Array-A group of several antennas linked together to facilitate a specified degree of directivity.

Antenna Bandwidth-The frequency range (bandwidth) that an antenna can effectively convert electrical signals into electromagnetic waves.

Antenna Combiner-A group filters (usually tuned cavity filters) that allow the outputs of multiple transmitters to be connected to the same antenna assembly. Because each tuned cavity in the antenna combiner is tuned to the specific frequency of its transmitter, high power radio energy from other transmitters at a different frequency is blocked (highly attenuated) from entering into the transmitter.

This diagram shows how tuned cavity filters are used to connect the output of four high power transmitters to the same antenna assembly. In this example, each transmitter is tuned to a different frequency along with their associated tuned cavity. Each tuned cavity allows a specific narrow band of frequencies to pass through it. This diagram shows that when other high power signals are applied to the tuned cavity with a different frequency, the signal is highly attenuated (blocked) from entering back into other transmitters.

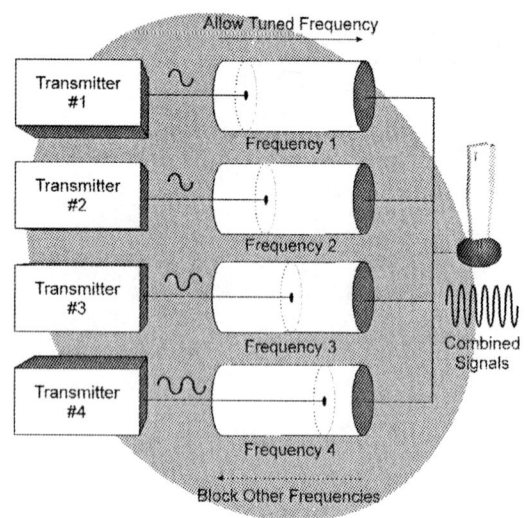

Antenna Electrical Beam Tilt

Antenna Combiner Operation

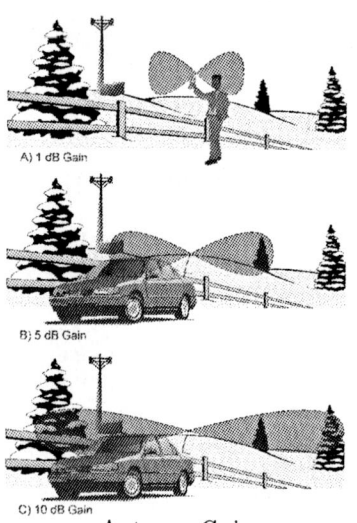

Antenna Gain

Antenna Electrical Beam Tilt-The vertical directing of an antenna radiation pattern through the use of electrical means (phase shifting) so that the radiation pattern can be adjusted along the horizontal plane.

Antenna Farm-A geographic location (area) that is used to group antennas that may have a common impact on aviation as designated by the Federal Communications Commission.

Antenna Frequency-The center of a frequency band that an antenna is capable of transmitting. The band of frequencies (antenna bandwidth) the antenna may be capable of transmitting ranges from less than one percent (e.g. waveguide antenna) to over 10%, dependant on the design and type of the antenna.

Antenna Gain-Antenna gain is the ratio of energy that is supplied to an antenna to the amount of energy transmitted from the antenna in a specific direction. The antenna gain is usually referenced to the direction of maximum radiation.

This diagram shows antenna systems that have different amounts of gain (directivity). In this example, a handheld telephone has a small amount of gain so they can transmit equally in most directions. The antenna that is mounted on a car can have more gain (directivity) as it will not be tilted (changed angle) as much as the portable telephone.

Antenna Height Above Average Terrain (HAAT)-The average height of an antenna (the center of the transmitted signal) above the average terrain.

Antenna Height Above Sea Level-The height of the topmost point of the antenna above mean sea level.

Antenna Input Power-The product of the square of the antenna current and the antenna resistance at the point where the current is measured.

Antenna Lightning Protection-A system that protects radio system electronic equipment from a lightning strikes to an attached antenna system.

Antenna Mechanical Beam Tilt-The vertical directing of an antenna radiation pattern through the use of mechanical means (physical positioning of the antenna) so that the radiation pattern can be adjusted along the horizontal plane.

Antenna Power Input-The amount of power (RMS or radio signal peak power) that is supplied to the antenna from the antenna transmission line.

Antenna Resistance-The total resistance of the transmitting antenna system at the operating frequency and at the point at which the antenna current is measured.

Antenna Structure-The supporting structures along with the transmitting and/or receiving antenna system that are mounted to the structure.

Anti Aliasing-(1-imaging) An imaging term that refers to a process that smoothes the sharp contrasts between two image areas of different colors. Properly done, this eliminates the jagged edges of text or colored objects. (2-audio) Used in voice processing, anti-alias usually refers to the process of over sampling or frequency limiting of analog signals so that the digital signal representation cannot be interpreted as different analog signal frequencies (alias).

Anti Reflection Coating-A coating used to reduce the amount of light reflected from a surface and increase the amount transmitted through it. The coating may be of one or several thin films, typically one quarter wavelength thick each. The material for the coating is selected primarily based on its index of refraction. The combination of film thickness and appropriate index supports constructive interference of the light to be reflected and destructive interference of the light to be transmitted. This technology has application in many optical components where reflected light represents lost signal intensity.

Antitrust Laws-Has the meaning given it in subsection (a) of the first section of the Clayton Act, except that such term includes section 5 of the Federal Trade Commission Act to the extent that such section 5 applies to unfair methods of competition.

Antivirus Program-A program that can detect the presence of and possibly eliminate a computer virus. Anti-virus program usually locate and identify viruses by looking for previously identified patterns or suspicious activity in the system.

Any Person Aggrieved-Includes any person with proprietary rights that has their communication intercepted by wire or radio, including wholesale or retail distributors of satellite cable programming and may include any person engaged in the lawful manufacture, distribution, or sale of equipment necessary to authorize or receive satellite cable programming.

AOC-Advice Of Charge
AOS-Alternate Operator Services
AOSP-Alternate Operator Services Provider
AP-802.11 Access Point
AP-Adjunct Processor
AP-Access Point

Apache HTTP Server-The Apache Web Server is the outgrowth of the original National Super Computing Applications (NSCA) http server. The NSCA released the source code under a freeware licensing scheme that allowed it to be extended by researchers and other computer programmers as long as the modified software was given back into the public domain for free use.

In its infancy the software was far from perfect and required a lot of software updates to fix bugs and to add needed functionality. These software updates came as small new sections of code referred to as patches. Just as Unix was a play on the Multics operating system the unix community named the NCSA web server Apache since it required so many patches to fix it.

Today the Apache Web Server is the most widely used http server of the world wide web and is supported on all of the major operating systems. The best part; it's still free.ost, excellent performance, good scalability, and great flexibility. Don't expect easy graphical configuration programs and hypertext help; you'll get the command line and the man pages instead, so it certainly helps to have staff with Unix experience.

Apache Server is available as part of the Red Hat Software Linux distribution, which also provides developers with full support for CGI, Perl, Tcl, a C or C++ compiler, an Apache server API, and a SQL database.

APC-Adaptive Predictive Coding
APCO-Associated Public Safety Communication Officials
Apco 16-Associated Public Safety Communication Officials 16 Standard
Apco 25-A trunked radio system specification that is primarily used in public safety communication systems.
Apco 25-Associated Public Safety Communication Officials Project 25 Standard
APCO Project 16A-A suite of operational requirements developed by APCO for Public Safety trunked radio systems. It is titled "900 MHz Trunked Communications System Functional Requirements Development, Dated March 1979."
APD-Avalanche Photodiode
APDU-Application Protocol Data Units
API-Application Programming Interface
API-Application Program Interface
APL-Automatic Program Load
APL-Average Picture Level
APM-Advanced Power Management
APM Factor-Attempts Per Message
Apogee-The point in a satellite's orbit around the Earth that has the largest distance from the center of the Earth. This distance can be used to calculate the worst case transmission loss and delay time.
APON-ATM Passive Optical Network
APP-Access Process Parameter

APPC-Advanced Program-To-Program Communications

Applet-A small program that uses the Java programming language.

AppleTalk-A protocol suite developed by Apple Computer, used in Macintosh computers and other compatible devices.

AppleTalk Filing Protocol (AFP)-A protocol that is used in the presentation and application layers of the protocol stack in the AppleTalk systems. AFP provides access for remote users and includes file access security control.

AppleTalk Transaction Protocol (ATP)-A transport protocol that provides reliable (guaranteed) transaction service between communication ports (sockets). in an AppleTalk system. ATP coordinates information exchanges between two socket connections where one socket acts as a client, requesting the other socket to perform a particular function or task. ATP relates (binds) the request and response with each other to guarantee the exchange of a request with the specific response to that request.

Applicant-The entity that submits a form or application to participate in an event (such as a communication license auction) that may include all holders of partnership and other ownership interests and a percent of stock (equity) interest or outstanding stock in the entity (e.g. more than 5%) and officers and directors of that entity.

Application-A software program that is designed to perform operations using commands or information from other sources (such as a user at a keyboard). Popular applications that involve human interface include electronic mail programs, word processing programs, and spreadsheets. Some applications (such as embedded program applications) do not involve regular human interaction such as automotive ignition control systems.

Application Based Call Routing-A process that controls call routing of an incoming call based on the type of selected application such as sales, customer service, or order tracking.

Application Class-A group of client applications that perform similar services, such as voice messaging or fax-back services.

Application Decomposition-The ability to divide an application into functional parts. This allows for applications to more easily evolve into more advanced programs by the upgrading of the individual functional parts.

Application Flow-A stream of frames or packets among communicating processes within a set of end stations.

Application for Patent-A document satisfying certain minimum formal requirements which describes and claims an invention for which patent protection is requested.

In the United States, for example, an application for patent includes:

-Utility Patent Application Transmittal Form or Transmittal Letter
-Fee Transmittal Form and Appropriate Fee
-Application Data Sheet (see 37 CFR § 1.76)
-Specification (with at least one claim)
-Drawings (when necessary)
-Oath or Declaration

Many countries, including the United States, allow applications for patent to be filed electronically.

Application Infrastructure Service Provider (AIP)-Companies that provide communication application services such as email, web hosting, and voice communication.

Application Layer-The application layer protocol coordinates the information interface between the communication device and the end user. The application layer receives data from the underlying protocols and processes this information into a form required or requested by the user or endpoint device. The application layer usually requests or responds to requests for a communication session. The location of the application layer is at the top of protocol stacks. The application layer is layer 7 in the open system interconnection (OSI) protocol layer model.

Application Management-The management of the installation, operation, and resource allocation for software and communication applications.

Application Note-Descriptions or instructions that are provide to assist in the application or design of a device, product, or service into a system. Manufacturers commonly provide application notes to help their customers to use their products or services in their systems.

Application Policy Server (APS)-A communications server (computer with a software application) that coordinates the allocation of network resources and the priority for service for clients who use application services (such as email and voice communication).

Application Program Interface (API)-Defined and documented entry points into a software application where other programs may interact with the application in order to provide customized extensions or perform special processing functions. Typically an API is a public function call that then itself calls on the services of the application. In this way the API hides the underlying details and complexities of the application software making it easier for programmers to add custom functionality.

Application Programming Interface (API)-A software program that has defined information entry and exit points in that allows other programs or devices to communicate with the software program. API programs often allow end users to define the operation or sequence of tasks for complicated systems. Examples of API include Telephony Services API (TSAPI) and Java Telephony API (JTAPI).

Application Protocol Data Units (APDU)-A packet structure that is used in the H.450 (packet voice services) specification series.

Application Protocols-Application protocols are commands and procedures used by software programs to perform operations using information or messages that are received from or sent to other sources (such as a user at a keyboard). Application protocols are independent of the underlying technologies and communication protocols. The use of well-defined application protocols (agreed commands and processes) allows the software applications to interoperate with other programs that use the application protocol independent from the underlying technologies that link them together (such as wires or wireless connections.)

Application Server (AS)-A computer and associated software that is connected to a communication network and provides information services (applications) for clients (users). Application servers are usually optimized to provide specific applications such as database information access or sales contact management.

Application servers are a key component of nextgen networks, and an enabler of IP-based enhanced services. These enhanced services will generate much-needed new revenue streams for service providers. Examples include all forms of conferencing, voice mail and unified messaging. Being softswitch-based, application servers have the flexibility to easily offer services that go beyond the feature set of legacy switched telephony. In terms of network configuration, the application server works in tandem with the media server, providing it with business logic and instructions for delivering enhanced services.

Application Service Element (ASE)-A software program or portion of a communication protocol that is part of an application layer of a protocol stack. Several ASEs may be combined to form a complete application protocol.

Application Service Provider (ASP)-A company that provides an end user with an information service. An ASP owns or leases computer hardware and software system that allows one or more users to access information services on or through that computer systems.

Application Specific Integrated Circuit (ASIC)-An integrated circuit (IC) that is designed to provide specific signal processing functions. This is in contrast to general purpose IC's that perform more general signal processing functions. ASICs are often created from a gate array (batches of logical gates) through the use of a custom mask that interconnection the gates so they can perform specific functional operations.

Applications-Applications are the software programs that require voice over data communication technology. Many of the communications applications and services that were available in the year 2000 were designed for the narrow-band applications (below 56 kbps). These applications included limited graphic web browsing, text based on-line shopping, email and word processor file transfer. Low cost broadband services such as xDSL systems provide a tremendous opportunity for the development of richer, more enhanced applications. These applications require services such as streaming video, rapid image file transfer or high speed data file transfer services.

Applications Generator (Apps Gen)-A software capability that enables users to develop programmatic coding using high-level input to save time and reduce the programming tasks.

Applications Processor-A computer system or information service provider that is dedicated to processing applications such as voice mail or billing systems.

APPN-Advanced Peer-To-Peer Networking
Apps Gen-Applications Generator
APS-Application Policy Server
APS-ACCUNET Packet Service
ARAM-Audio grade DRAM
Arbitrated Loop-Arbitrated loop is a shared data network that can allow up to 126 devices and one switch to be connected, Arbitrated loops operate at 200 Mbps (full duplex).

Arbitration-(1-general) A process or set of rules that is used to manage conflicts. (2-computers) The process of competing for computer resources such as memory or peripheral devices, made by multiple processes or users.

ARC-Access Router Card

ARCH-Access Response Channel

Archie-A program that locates (searches) for files or information on Internet sites that have file transfer protocol (FTP) capability using a particular string.

Architecture-The functional design of a network, computer or telecommunications system elements and the relationships between them. The architecture usually includes hardware and software components.

Architecture for Voice, Video and Integrated Data (AVVID)-AVVID is a network structure standard that defines the types of devices used in a voice over data (multimedia) network and how they are interconnected and used within the network. The AVVID structure allows for system expansion, efficient feature deployment, security, and increased reliability. AVVID was developed by Cisco.

Archival-A process or media storage device that has a long-term minimum life-spans over which the information will not become corrupted.

Archiving Files-This is a process where the information contained in an active computer file is made ready for storing in a non-active file, perhaps in off-line or near-line storage. Typically when files are archived, they are compressed to reduce their size. To restore the file to its original size requires a process know as unarchiving.

ARCnet-Attached Resource Computer Network

ARD-Automatic Ring Down

ARDIS-Advanced Radio Data Information Service

Area Code-A 3-digit number that generally identifies a geographic area of a switch that provides service to a telephone device. In North America, the Numbering Plan Area (NPA) is the area code. In countries other than North America, the area code may have any number of digits depending on the regulation of telecommunication in that country.

Area Exchange-Geopolitical areas that are defined as region authorized to provide local telephone services. Small metropolitan areas or a collection of towns often share a single area exchange.

Area Wide Centrex-A service that extends the advanced services offered by Centrex system through the use of an Intelligent Network (IN) to allow Centrex features to operate throughout a large geographic area without the need for dedicated facilities for each Centrex telephone.

ARF-Alternate Recovery Facility

Argument Separator-A code or structure that separates information elements for records or data within programs such as databases, program languages, and spreadsheets. These separators delimit the information elements to allow the software programs to identify and process each element according to their program's requirements. Examples of an argument separator include a comma, tab, or semicolon.

ARI-Automatic Room Identification

ARIB-Association for Radio Industries and Business

ARIS-Aggregate Route-Based IP Switching

ARIS-Access Request Information System

Arithmetic Coding-A data compression technique for a sequence of data. Arithmetic coding can compress with more efficiency than Huffman coding. This is because arithmetic coding allows compressed bits to be shared between two (or more) encoded symbols, unlike Huffman coding in which each compressed bit belongs exclusively to a single encoded symbol. However arithmetic coding has requirements on arithmetic accuracy, buffer memory, and computation which Huffman coding can avoid.

Arithmetic Logic Unit (ALU)-A part of a central processing unit (CPU) that performs arithmetic and logical operations on data.

Arithmetic Overflow-A condition that occurs in a digital computing device or system when the result of a calculation that is greater than the computing device can store or display.

Arithmetic Register-A short-term memory storage location (register) that holds the operands or the results of an arithmetic operation.

Arithmetic Unit-A processing portion of a computing system that performs a majority of the arithmetic operations processed by the computer.

Armor-A protective layer or mechanical shield that is located around a cable or assembly.

ARN-Authentication Random Number

ARP-Address Resolution Protocol

ARP Cache-A data structure that provides the current mapping of 32-bit IP addresses to 48-bit MAC addresses.

ARPAnet-Advanced Radio Research Projects Agency Network

ARPM-Average Rate Per Minute

ARPU-Average Revenue Per User

ARR-Automatic Alternate Routing

Arrears-The assessment of charges for services which have been used (e.g. usage).

Arrival Rate-As realted to call centers, the arrival rate is the pattern at which incoming calls are received at the call center. Arrival rates are classified as steady, random or peaked. This term is also used in queuing theory to describe the rate at which entities enter into an ordered list to be processed.

ARRL-American Radio Relay League

ARS-Automatic Route Selection

ARSG-Australian Radiocommunications Study Group

Artifacts-In general, results, effects or modifications of the natural environment produced by people. In the processing or transmission of audio or video signals, a distortion or modification produced due to the actions of people or due to a process designed by people. Unintended, unwanted visual aberrations in a video image. In all kinds of computer graphics, including any display on a monitor, artifacts are things you don't want to see. They fall into many categories (such as speckles, in scanned pictures) but they all have one thing in common: they are chunks of stray pixels that don't belong in the image.

Artificial Intelligence-A deductive reasoning process that can be applied on an automated basis by computer processing. Artificial intelligence simulates the reasoning capabilities much like the human mind through user input. The Artificial intelligence term was introduced in the 1950's by British mathematician Alan Turning.

Artificial Line Interface-The ability of a piece of transmission equipment to attenuate its output level to meet the required transmission level.

ARTS-Advanced Radio Technology Subcommittee

ARU-Audio Response Unit

AS-Application Server

ASA-American Standards Association

ASA-Average Speed of Answer

ASAI-Adjunct Switch Application Interface

ASBM-Set Asynchronous Balanced Mode

ASCII-American Standard Code for Information Interchange

ASCII Extended Character Set-The character codes with values from 128 to 255 in the ASCII Character Set. These codes are used to represent special characters and non-printing control commands to computer hardware. The codes in this range are not standardized and may be used by computer hardware manufactures and software developers for special purposes and therefore may not be compatible across different systems. Examples of use are for alternate language symbols, mathematical symbols and line drawing symbols.

ASCII File-A text file that only contains text characters from the ASCII character set. An ASCII character set only includes letters, numbers, and punctuation symbols. The extended ASCII code contains non-standard graphics and characters that can be used as text-formatting codes.

ASCII Standard Character Set-The American Standard Code for Information Interchange (ASCII) that uses an 8 bit binary code to represent the letters of the alphabet. The numerical values from 0 to 127 are known as the Standard ASCII set and contain the most commonly used punctuation symbols and non-printing control codes. The values from 128 to 255 are known as the Extended ASCII character set.

ASE-Accredited Systems Engineer

ASE-Application Service Element

ASG-Access Service Group

ASI-Adapter Support Interface

ASIC-Application Specific Integrated Circuit

ASLC-Analog Subscriber Loop Carrier

ASN-Autonomous System Number

ASN.1-Abstract Syntax Notation number 1

ASP-Active Server Pages

ASP-Application Server Process

ASP-Abstract Service Primitive

ASP-Application Service Provider

Aspect Ratio-The ratio of the number of items (such as pixels on a screen) as compared to their height and width.

ASPI-Advanced SCSI Programming Interface

ASR-Access Service Request

ASR-Average Service Rate

ASR-Authorized Sales Representative

ASSB-Amplitude Compandored Single Sideband Modulation

Assembler-An executable program that translates a set of statements written in assembly level programming language into the machine specific code that executes the instructions specified in the statements. Assembly language is very closely related to the actual instructions and architecture of the microprocessor being targeted and therefore assembly language instructions typically map one-for-one to machine level instructions. In a high-level language such as C, Fortran or Pascal; one statement may create dozens of machine level instructions.

Assembly Language-A low-level programming language that each statement corresponds to a single machine language instruction that a processor can execute. Assembly language is very closely related to the actual instructions and architecture of the microprocessor being targeted and therefore assembly language instructions typically map one-for-one to machine level instructions. Assembly languages are specific to a given microprocessor.

Assembly Line-A group of machines or assembly areas that form a sequence of operations that are used to assemble circuit boards or products.

Assigned Frequency Band-The frequency band where emission of a mobile station (transmitter /receiver) is authorized at specific levels.

Assigned Pairs-Communication lines (wire pairs) that are assigned for customer service. These lines may be working or idle.

Assignment Of A Radio Frequency Or Radio Frequency Channel-The sending of a channel assignment or authorization message that instructs a mobile station (mobile radio) to use a specific radio frequency, time slot, and/or channel code for a communication session (voice or data).

Assignment Of Authorization-A transfer of authorization to provide communication services from one party to another.

Associate Session-A communication session that is associated with another session. These sessions may be synchronized (such as video and audio).

Associated Bit Signaling-In 2.048 Mb/s digital multiplexing systems, one of the 32 distinct 64 kb/s channels (the time slot designated as channel 16) is set aside for associated or common channel signaling. In older systems, associated bit signaling is carried in this channel. First a multi-frame sequence comprising 16 frames is established based on a synchronizing bit sequence that occurs in the synchronizing channel (the physically first channel in the 2.048 Mb/s system, designated channel zero). In each of the 16 frames in this signaling multi-frame sequence, the 8 bits in the associated signaling channel are logically viewed as two separate 4-bit signals, thus giving 32 distinct 4-bit signals during each 16 frames. Each one of the 4-bit signals is logically associated with one of the 32 distinct physical channels. The bit groups associated with channels zero and 16 are not used for conveying supervisory information, since these two channels do not carry traffic. The four bits in each group are designated A, B, C, and D in documentation. In most public telephone network installations, only the A bit is used to represent the supervisory (idle vs. busy) status of that channel. In tie line or other applications, additional bits (B, C or D) are also used and the bits may have other meanings. This associated bit signaling is historically associated with various types of R2 signaling. Associated bit signaling has the advantage of not altering the bits in the traffic channel (as North American Robbed Bit signaling does) so it is convenient for ISDN and other applications where the traffic channel is used for 64 kb/s subscriber data. However, it does not have the versatility of common channel No. 7 signaling. At present, associated bit signaling has been replaced, using the same physical channel, by common channel No. 7 signaling in many European, African, Asian and Latin American telephone systems.

Associated Broadcasting Station-A broadcasting station or stations that are part of a system or network that is licensed as an auxiliary broadcast location for which it is principally used.

Associated Control Channel (ACCH)-A logical control channel that is associated with another channel. An example of an ACCH is the slow associated control channel (SACCH). The SACCH is associated with the dedicated control channels (DCCH).

Associated Mode Of Signaling-In Signaling System 7 and some other common-channel protocols, the mode in which messages for a signaling relation between two adjacent signaling points are conveyed over an interconnecting signaling link.

Associated Public Safety Communication Officials (APCO)-An association that represents public safety communications officials for the development and implementation of communications systems and training.

Associated Public Safety Communication Officials 16 Standard (Apco 16)-The Association of Public Safety Officers and Communications Officials (APCO) Project 16 (APCO 16) created an industry standard for LMR systems in the late 1970s. This standard was updated to revision A (APCO 16A) in the early 1980s. This standard was primarily a User Performance Requirements wish list. No specific technical standards resulted. Each manufacturer was free to develop a solution to meet the basic requirements of the Project 16 requirements.

Project 16 systems is an analog system that transmits using 25 kHz channels. Some of these systems have migrated to digital with voice and data encryption supported. Various TIA/EIA documents (e.g., TIA/EIA-603) recommended specific performance measurements for LMR.

Associated Public Safety Communication Officials Project 25 Standard (Apco 25)- Associated public safety communication officials (APCO) project 25 is a digital trunked land mobile radio standard that is primarily used in the Americas. The standard was accepted in 1993 and APCO Project 25 compliant systems are primarily used for public safety applications.

One of the key objectives for the Project 25 standard was backward compatibility with standard analog FM radios. The Project 25 standard allows for a simple migration from analog to digital Project 25 systems. This permits users to gradually replace analog radios and infrastructure equipment with digital Project 25 systems.

There are two radio channel bandwidths that are used in the Project 25 system; 12.5 kHz and 6.25 kHz. The 12.5 kHz radio channel uses (C4FM) modulation and the 6.25 kHz radio channel uses (CQPSK) modulation to achieve a more efficient data transmission rate. Because of the modulation types selected, the typical receiver is capable of demodulating either the C4FM and the CQPSK signals. The digitized voice uses a 4400 bps improved multiband excitation (IMBE) voice coder. This is the same voice coder type selected by INMARSAT, for use in satellite communication devices.

This photograph shows the Racal 25 digital land mobile radio that is used by police, firefighters, and other public safety professionals.

Digital Land Mobile Radio

Associated Television Broadcast Station-A television broadcast station licensed to the licensee of the television auxiliary broadcast station and with which the television auxiliary station is licensed as an auxiliary facility.

Association for Radio Industries and Business (ARIB)-An association in Japan that oversees the creation of telecommunications standards.

ASSP-Algorithm Specific Signal Processor

Assurance-The process of making sure that customers receive the levels of service that they have purchased or agreed to purchase. Assurance may include commitments to a high-level of overall customer satisfaction of quality of service.

Assurance Level-A level of probability that a service or product will meet a specific criteria or range of limits. Assurance level is often expressed as a percent. For example, there is 99% assurance (probability) that a dialtone will be available in a subscriber loop.

Assured Delivery-A protocol is said to provide assured delivery if each packet is guaranteed to be delivered. This is accomplished via receiver acknowledgement and retransmission by the sender when packets are not acknowledged. Examples of protocols which provide assured delivery are TCP/IP and IEEE 802.2 LLC connection oriented services.

AST-Automatic Scheduled Testing

Asterisk-The symbol denoted with *. This symbol is often used as a wildcard within a computer operating and file system. It is typically used to match one or more characters within a specific filename. For example a file system with the files Dog.txt, Hog.txt and Fog.txt would return all three files when asked for *og.txt.

ASU-Alarm Suppressor Unit

Asymmetric-Two-way communication that allows for transmission rates that can vary by direction. For example, the downlink broadcast channel may be a high-speed channel (e.g. 6 Mbps) and an uplink (reverse direction) channel may only be 640 kbps.

Asymmetric Compression-Techniques where the computational complexity of the decoding process is not the same as that of the compression process. Many compression techniques are asymmetric because only the encoding process requires a search over possible choices of encodings or requires analysis of the input. There are many compression scenarios where asymmetric compression is tolerable, for example: decoding will be done many times for

Asymmetric Digital Subscriber Line (ADSL)

each encoding; decoding needs to be done on a device with much less computational power than the encoding device; encoding and decoding will be done by each device so that the total computing power required on each device is equal.

Asymmetric Digital Subscriber Line (ADSL)-A communication system that transfers both analog and digital information on a copper wire pair. The analog information can be a standard POTS or ISDN signal. The maximum downstream digital transmission rate (data rate to the end user) can vary from 1.5 Mbps to 9 Mbps downstream and the maximum upstream digital transmission rate (from the customer to the network) varies from 16 kbps to approximately 800 kbps. The data transmission rate varies depending on distance, line distortion and settings from the ADSL service provider.

This figure shows that a typical ADSL system can allow a single copper access line (twisted pair) to be connected to different networks. These include the public switched telephone network (PSTN) and the data communications network (usually the Internet or media server). The ability of ADSL systems to combine and separate low frequency signal (POTS or IDSN) is made possible through the use of a splitter. The splitter is composed of two frequency filters; one for low pass and one for high pass. The DSL modems are ADSL transceiver unit at the central office (ATU-C) and the ADSL transceiver unit at the remote home or business (ATU-R). The digital subscriber line access module (DSLAM) is connected to the access line via the main distribution frame (MDF). The MDF is the termination point of copper access lines that connect end users to the central office.

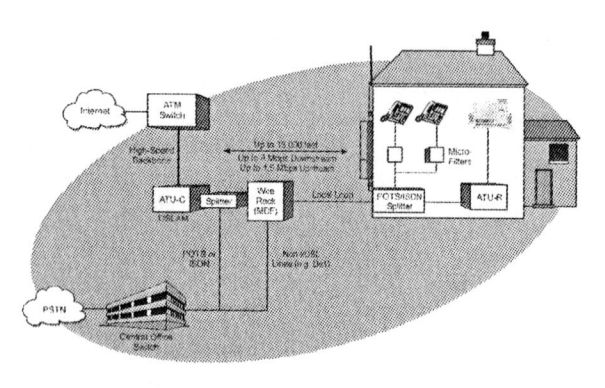

Asymmetrical-A type of Asynchronous Connectionless (ACL) link that operates at two different speeds in the upstream and downstream directions. An example of an asymmetrical connection is the Bluetooth ACL link. The Bluetooth specification specifies a maximum data rate of up to 723.2 Kbps in the downstream direction, while permitting 57.6 Kbps in the upstream direction. See also symmetrical.

Asymmetrical Private Virtual Circuit (Asymmetrical PVC)-A virtual circuit that permits uneven (asymmetrical) data transmission rates for each direction of transmission.

Asymmetrical PVC-Asymmetrical Private Virtual Circuit

Asymptotic Coding Gain-A processing gain of a coding system that can be obtained when the signal to noise ratio (SNR) approaches infinity.

Asynchronous-A signal which does not have synchronization with some other reference signal. The communications on an asynchronous channel is not sequential and may appear random in nature.

This diagram shows the process of data transmission using asynchronous (unscheduled) transmission. In this example, each message or block of data that is transmitted in an asynchronous data communication system must include indicators (delimiters) that identify the start and stop of a block of data. The blocks of data usually include some bits of information that are dedicated for flow control (e.g. routing and/or error protection).

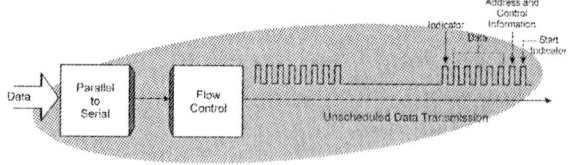

Asymetric Digital Subscriber Line (ADSL) System

Asynchronous Data Transmission

Asynchronous Communication System-A data communication system that dynamically can send and receive data blocks where data blocks are individually synchronized.

Asynchronous Devices-Devices that transmit communication signals at irregular time intervals. Examples of asynchronous devices include mobile telephones and local area network data communication devices.

Asynchronous Interface Protocol-A protocol that enables network signaling messages to be sent to a data terminal as an ASCII message. This protocol was developed by the Consultative Committee for International Telephony And Telegraphy (CCITT) for access to public packet-switched service.

Asynchronous Time-Division Multiplexing-An asynchronous transmission system that uses time slots to transfer uncoordinated data transfer (non-synchronized).

Asynchronous Transfer Mode (ATM)-A packet data and switching technique that transfers information by using fixed length 53 byte cells. The ATM system uses high-speed transmission (155 Mbps) and is a connection-based system. When an ATM circuit is established, a patch through multiple switches is setup and remains in place until the connection is completed. ATM service was developed to allow one communication medium (high speed packet data) to provide for voice, data and video service.

As of the 1990's, ATM has become a standard for high-speed digital backbone networks. ATM networks are widely used by large telecommunications service providers to interconnect their network parts (e.g. DSLAMs and Routers). ATM aggregators operate networks that consolidate data traffic from multiple feeders (such as DSL lines and ISP links) to transport different types of media (voice, data and video).

This figure shows a functional diagram of an ATM packet switching system. This diagram shows that there are three signal sources going through an ATM network to different destinations. The audio signal source (signal 1) is a 64 kbps voice circuit. The data from the voice circuit is divided into short packets and sent to the ATM switch 1. ATM switch 1 looks in its routing table and determines the packet is destined for ATM switch 4 and ATM switch 4 adapts (slows down the transmission speed) and routes it to it destination voice circuit. The routing from ATM switch 1 to ATM switch 4 is accomplished by assigning the ATM packet a virtual circuit identifier (VCI) that ATM switch can understand (the packet routing address). This VCI code remains for the duration of the communication. The second signal source is a 384 kbps Internet session. ATM switch 1 determines the destination of these packets is ATM switch 4 through ATM switch 3. The third signal source is a 1 Mbps digital video signal from a digital video camera. ATM switch 1 determines this signal is destined for ATM switch 4 for a digital television. In this case, the communication path is through ATM switches 1, 2, and 4.

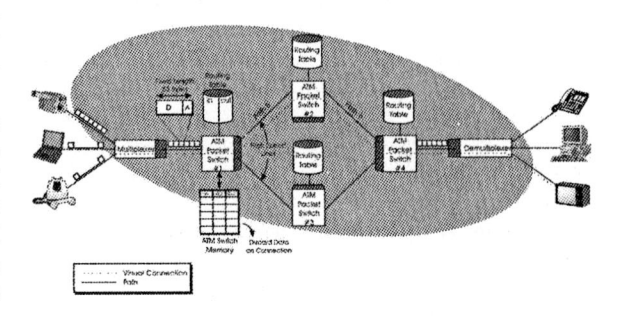

Asynchronous Transfer Mode (ATM) System

Asynchronous Transfer Mode 25 Mbps (ATM25)-A 25 Mbps version of ATM. The ATM25 standard was developed primarily for corporate networks. However, the QoS advantages of ATM and customer needs for switched services for digital video and Internet access has stimulated interest in ATM for the DSL industry.

AT-Access Tandem

AT-Prefix For Dialing Using A Modem

AT Command Set-A simple communication language (list of commands) that is used to setup and control modulator/demodulator (modem) devices. The AT command set was developed by Hayes Microcomputer Products to control their modems. Some of the commands include telephone dialing (pulse or tone), adjustment the audio volumes, data transmission rates, and programming modem on how to answer incoming calls.

AT&T Consent Decree-The order for AT&T to divest parts (the break-up) of its business and the restriction not to allow AT&T to provide local telephone service.

ATA-Advanced Technology Attachment

ATA-Analog Telephone Adapter

ATA-Analog Terminal Adapter

ATAPI

ATAPI-Attachment Packet Interface
ATB-All Trunks Busy
ATC-Autotune Combiner
ATC-Adaptive Transform Coding
ATD-Attention Dial The Phone
ATDT-Attention Dial the Phone in Touchtone Mode
ATG-Air-To-Ground
ATHD-Average 10-High Day
ATIS-Alliance for Telecommunications Industry Solutions
Atlas/Centaur-A fueled-propulsion rocket that is used to launch communications satellites into their geosynchronous Earth orbits. This design can accommodate satellites that are up to a 900 kg.
ATM-Asynchronous Transfer Mode
ATM-Automatic Teller Machine
ATM Adaptation Layer (AAL)-A set of standard protocols that translate user traffic into a size and format that can be contained in the small payload of an ATM cell. User traffic is returned to its original form at the destination. This process is called segmentation and reassembly. All AAL functions occur at the ATM end-station rather than at the switch. These protocol layers are designed to allow for constant bit rate (CBR), variable bit rate (VBR), unspecified bit rate (UBR), and other types of services.
ATM Adaptation Layer 1 (AAL1)-The layer within the ATM protocol that converts the 53 byte packets from the network into the form used by the customer for constant bit rate (CBR) services.
ATM Adaptation Layer 2 (AAL2)-The layer within the ATM protocol that converts the 53 byte packets from the network into the form used by the customer for variable bit rate (VBR) services. AAL2 Supports connection oriented traffic such as compressed voice and data.
ATM Adaptation Layer 5 (AAL5)-A more efficient class of service for ATM than AAL1. AAL5 was previously called Simple and Efficient AAL (SEAL).
ATM Address-A 20 byte address used in asynchronous transfer mode (ATM) systems that identify the country, area, and end-system identifiers. ATM address formats are defined in the user network interface (UNI) specification.
ATM Backbone Switch-A switch that receives and forward standard 53 byte asynchronous transfer mode (ATM) packets and is located in the interconnection backbone networks of carriers to interconnect slower switches or edge (network interface) switches. Because the ATM backbone switch can prioritize packets for different types of services based with varying bandwidth and quality of service (QoS) requirements, ATM multi-service switches are well suited as backbone switches. The processing of standard size packets allow ATM backbone switches to provide bandwidth over 200 Gbps.
ATM Cell-A 53 byte packet of data (called a "cell") that is used in an ATM network. An ATM cell is usually divided into a 5 byte header and 48 byte payload. The ATM header is primarily used for local connection routing information to the next switching point.
ATM DXI-ATM Data Exchange Interface
ATM ELSU-ATM Ethernet LAN Service Unit
ATM Forum-A forum that was started to assist manufacturers and service providers with marketing and development of ATM products and services. The ATM forum was started in 1991.
ATM header-The 5 byte portion of the 53 byte ATM cell that contains the addressing information for the ATM cell. There are different ATM cell header structures; user network interface (UNI) and network to network interface (NNI). The header structure for UNI provides for end-to-end addressing and the NNI structure only provides for inter-network addressing.
ATM Packet-A 53 byte packet that contains an 5 byte header and 48 byte payload.
This diagram shows that ATM packets have a fixed length of 53 bytes. Of these, 5 bytes are used for address and control information and 48 bytes are dedicated for data. This diagram also shows that the ATM packet structure varies dependent on its use. When it is used for user to network interface (UNI), it contains additional generic flow control (GFC) bits. When it is used for network to network interface (NNI) communication, it uses the additional bits for additional virtual path indicator (VPI) channels.

Attendant Switchboard

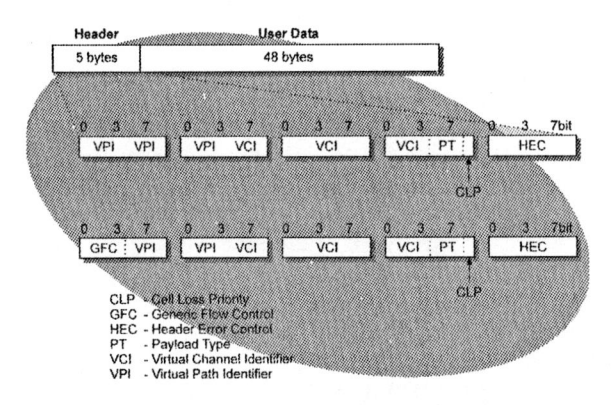

ATM Packet Structure

ATM Passive Optical Network (APON)-An ATM passive optical network (APON) combines, routes, and separates optical signals through the use of passive optical filters and ATM protocol. The APON distributes and routes signals without the need to convert them to electrical signals for routing through switches.

ATM Switch-A packet switch in an ATM network that receives and forwards fixed length 53 byte cells. The ATM switch is connection oriented so the paths for packets are pre-established at the beginning of a communication session. The ATM switch can prioritize the routing of packets based on the routing address or content of the packet.

ATM TLSU-ATM Token Ring LAN Service Unit

ATM25-Asynchronous Transfer Mode 25 Mbps

ATMARP-ATM Address Resolution Protocol

Atomicity-An attribute of a transaction that guarantees that all of the constituent steps are completed without any intervening actions or interruptions that would compromise the state of the system. Any failures during the transaction initiate a roll-back to the previous state of the system.

ATOP-Automatic Traffic Overload Protection

ATP-AppleTalk Transaction Protocol

ATS-Abstract Test Suite

ATSC-Australian Telecommunication Standardization Committee

ATSC-Advanced Television Systems Committee

ATT-Automatic Toll Ticketing

ATT-Average Talk Time

Attach-The process of establishing a connection (a physical and/or logical connection) between a client (e.g. workstation) and a network server (e.g. a file server).

Attach Terminal-The process that is used by a software application program that assigns a computer terminal for exclusive use by the application.

Attachment-An attachment to an electronic mail is non-text data included in an e-mail using Multipurpose Internet Mail Extensions (MIME). An e-mail message may contain any number of attachments. Each attachment has a "MIME type" property that suggests to the user's e-mail application the data type of the attachment, for example HTML text or a JPEG image. Depending on the software, in some cases the user's e-mail application will display attachments within the e-mail application (like JPEG images) and in other cases an attempt will be made to open an external application, sometimes requiring assistance or permission from the user, or the user may save the attachment as a file.

Attack-In network and computer security parlance and attack is an attempt to disable or gain unauthorized access to a computer system or network by exploiting weaknesses within the operating system or implemented security measures.

Attempt-Any process of requesting or demanding service from a communications system.

Attendant-An attendant is a person who answers, screens, or directs calls in a communication system.

Attendant Call Waiting Indication-The indication light or message on the attendant console that indicates that one or several calls are in queue to be answered. The indication may change (e.g. flash or ring) when additional thresholds (e.g. maximum number of waiting calls) are reached.

Attendant Conference-A PBX system feature that permits an attendant to establish a conference call between the public telephone lines and PBX extensions.

Attendant Forced Release-An attendant or operator activated release of a call that will "disconnect" all parties on that circuit entered by the attendant.

Attendant Switchboard-An attendant switchboard is a communication device (phone and display) that provides the ability for a receptionist (attendant) to identify and answer incoming calls, interact with callers, and redirect (transfer) calls to the proper extension. Attendant switchboard consoles in an IPBX system are software programs that typically operate on a standard multimedia computer. Attendant switchboards display incoming Caller ID information, have graphical call status

Attention Identification (AID)

indications (hold, in-use), allow quick access to company directories, and permit the simple transfer of calls through the use double-clicks.

Attention Identification (AID)-An access control attention identifier that is used for initiating communication between a data terminal and a host system.

Attentuator-A device that reduces the signal energy (e.g. power) that is transferred through it. Attenuator are normally rated as a ratio in decibels (dB) as compared between the input and output signal levels.

Attenuation-The amount of signal energy (power or level) caused by a transmission channel. Attenuation can be caused by absorption (conversion of energy to heat), mismatching of transmission lines (causing signal energy to be reflected), or other forms of distortion that reduces the received signal level energy.

Attenuation to Crosstalk Radio (ACR)-A comparison between signal attenuation and the amount of crosstalk. ACR is usually measured in decibels.

Attenuator-A device that reduces the power or amplitude level of a signal as it passes through the device. Attenuators can be fixed (often used to isolate electrical assemblies) or variable (such as a audio volume level control).

This diagram shows a chip attenuator, model 33-1006-03.00, that is inserted on a printed circuit card to reduce (attenuate) the signal level.

Attenuator

ATU-C-ADSL Transmission Unit - Central Office
ATU-R-ADSL Transmission Unit - Remote
ATUG-Australian Telecommunications Users Group
ATVEF-Advanced Television Enhancement Forum
ATX-A specification used in the personal computer industry that specifies the form factor and placement of connectors on the motherboard. This allows manufactures of motherboards and cases to interoperate and to reduce the cost of manufacturing computers.

AU-Audio File Format
AU-Administrative Unit
AuC-Authentication Center
Audibility-The quality of being able to be heard. For most humans, the frequency range of audibility extends roughly from 20 Hz to 20 kHz.

Audible-The range of sound that can be heard by a normal human.

Audible Sound-Audible sound spans a range of frequencies from around 20 hertz (cycles per second) to 20 kilohertz. To be audible, the power intensity of the sound must be at least 1.0E-12 watt/square meter. This acoustic power intensity level is also denoted as 0 acoustic decibels. (Note that 0 dB in electric power measurement is 1 milliwatt of power, a different base unit that does not involve an area but only a power measurement.)

Audio-A signal that is composed of frequencies that can be created and heard by humans. A typical human can hear sounds from about 15 to 20,000 hertz.

Audio Bridge-In telecommunications, a device that mixes multiple audio inputs, then provides the composite audio back to each communication device, less that devices audio input. An example of an audio bridge is a conference call.

Audio Chat Room-Audio chat rooms are real-time communication services that allow several participants (typically 10 to 20) to interact act much like an audio conference session. Audio conference chat rooms may be public (allow anyone to participate) or private (restricted to those with invitations or access codes.)

Audio Conference-Audio conferencing (also called teleconferencing) is a process of conducting a meeting between two or more people through the use of telecommunications circuits and equipment.

Audio Frequency-The frequency in Hertz (Hz) of an audio signal. The typical adult can hear audio frequencies between 20 Hertz and 20000 Hertz. Audio compact discs can play frequencies up to 22050 Hertz.

Audio Messaging Interchange Specification (AMIS)-An industry specification that defines how messages are exchanged on a network between voice mail systems.

Audio Response Unit (ARU)-A device or system that can translate data files (usually stored on a computer) into audio voice messages.

Audio Streaming-A real-time system for delivering audio, typically over the internet. Upon request, a server system will deliver a stream of audio (usually compressed) to a client. The client will receive the data stream and (after a short buffering delay) decode the audio and play it to a user. Internet audio streaming systems are used for delivering audio from 2 kbps (for telephone-quality speech) up to hundreds of kbps (for audiophile-quality music).

Audio Video Transport (AVT)-An IETF working group that is responsible for audio and video transport protocols.

Audio/Video Interleave (AVI)-A type of Windows file format that can produce multimedia (video and audio sound) signals. The elements are of multimedia signals are stored in alternate, interleaved blocks with the data file.

Augmented Backus-Naur Format (ABNF)-A meta-language syntax used in defining and utilizing the session initiation protocol (SIP) as defined in RFC2543. An example ABNF construct used to describe a SIP message is as follows: SIP-message = Request | Response.

AUI-Autonomous Unit Interface

AUI-Attachment Unit Interface

AUI Cable-Attachment Unit Interface Cable

Authenticated SMTP-Authenticated Simple Message Transfer Protocol

Authentication-Authentication is a process of exchanging information between a communications device (typically a user device such as a mobile phone) and a communications network that allows the carrier or network operator to confirm the true identity of the user (or device). This validation of the authenticity of the user or device allows a service provider to deny service to users that cannot be identified. Thus, authentication inhibits fraudulent use of a communication device that does not contain the proper identification information.

This figure shows the operation of a basic authentication process used in a radio communication system. As part of a typical authentication process, a random number that changes periodically (RAND) is sent from the base station. This number is regularly received and temporarily stored by the mobile radio. The random number is then processed with the shared secret data that has been previously stored in the mobile radio along with other information in the subscriber to create an authentication response (AUTHR). The authentication response is sent back to the system to validate the mobile radio. The system processes the same information to create its own authentication response. If both the authentication responses match, service may be provided. This process avoids sending any secret information over the radio communication channel.

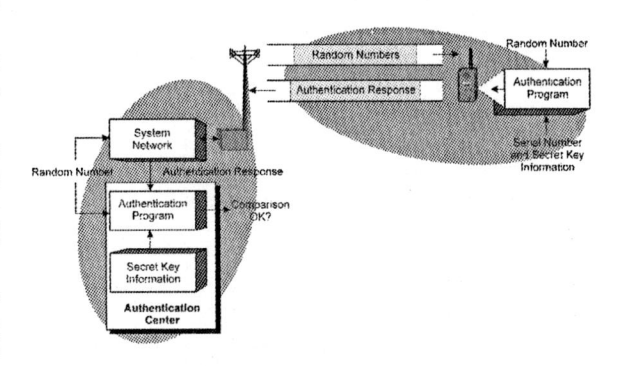

Basic Authentication Process

Authentication Center (AuC)-A part of a network that manages the encryption keys that validate the identity of customers and enable voice privacy services. A single authentication center may process validation requests using different keys, random numbers and encryption algorithms.

Authentication Key (A-key)-A secret key that is stored in the cellular or PCS system and entered into a mobile phone to create the shared secret data (SSD) that is used to validate the identity of the subscriber. See also shared secret data (SSD).

Authentication Procedure-The sequence of steps carried out by two end points of a communication system to exchange the information necessary to insure that some aspect of the communication session are valid. This may include user validation, data validation or service validation.

Authentication URL-A URL or address for a device (such as an IP telephone) where the authentication information is maintained to allow the proper security for incoming information.

Authentication via Challenge-Response -A method of authentication used by digital cellular telephones, military communication systems, Microsoft Windows 2000 and other systems. This authentication process responds with a "challenge" number value (different for each instance of authentication) when a user contacts a server, upon which the user then performs a cryptographic operation, and then returns the result to the server. The user equipment and the server both contain a copy of an internal secret number, used in the cryptographic operation. The server also performs the same cryptographic operation, and if the two results are the same, the user is considered authentic. Although the transactions can be intercepted and observed by an eavesdropper, they cannot be imitated or used to produce a false authentication, because different numbers are exchanged in each instance of the authentication process.

Authentication, Authorization, Accounting (AAA)-In network management or remote dial access networks AAA stands for Authentication, Authorization, and Accounting. Authentication is the process of validating the claimed identity of an end user or a device, such as a host, server, switch, or router. Authorization is the act of granting access rights to a user, groups of users, system, or a process. Accounting is the method to establish who, or what, performed a certain action, such as tracking user connection and logging system users.

Authoring Tools-Software and/or hardware tools to allow creation of a digital media presentation from multiple source media. For example, a video authoring tool may allow editing of video clips into a single video presentation, while a web authoring tool may allow creation of graphical and text content for pages on a web site. Authoring tools may also function as encoding tools.

Authorization-Authorization is the enabling of services to a device or customer that requests services. Authorization is often part of the billing and customer care (BCC) system and is maintained in a customer database service profile.

Services are initially enabled in a network as a result of provisioning. Provisioning is a process within a company that allows for establishment of new accounts, activation, termination of features, and coordinating and dispatching the resources necessary to fill those service orders. Provisioning is usually part of customer care systems.

Auto Attendant-The auto attendant feature is used to route incoming telephone calls based on selections or information provided by the incoming caller. The auto attendant feature may use interactive voice response (IVR) to prompt the caller to select the call routing based on category choices or it may use the calling number identification to determine the destination (e.g. a telephone number for a specific sales group).

Auto Dial Number in Short Message-A feature that allows a telephone (typically a mobile telephone) to automatically call a telephone number that is part of a short message.

Auto Discovery-Automatic discovery is a process where a network manager automatically searches through a range of network addresses and discovers specific types or all types of devices present in that range of addresses. The auto discovery process may be manually initiated by the network administrator or it may be initiated after a new device automatically registers after it is connected to the network.

Auto Redial-The ability for a telephone to repeatedly dial a telephone number in the event the first dialing attempt is unsuccessful. In a wireless system, the maximum number of attempts for redialing may be limited to ensure the system does not become overloaded with requests for service.

Auto Sensing-Process during which a network device automatically senses the speed of another device. For example, Ethernet hardware with auto-sensing 10 base-T / 100 base-T ports will automatically determine which of those two network speeds is supported by a device connected to the port, then use the faster choice if possible.

Auto-Discovery-Auto-discovery is a method of discovering devices or services that may be attached or disconnected from a system or network. Auto-discovery is typically performed via layer 3 network connectivity and there are many vendors who have written network auto-discovery tools. Cisco's Cisco discovery protocol (CDP) is an example of a discovery tool Cisco uses to identify it's equipment connections and parameters.

AUTODIN-Automatic Digital Network

Automated Attendant System-A processor control system that performs telephone console attendant functions such as answering a call, transferring callers to specific user stations, directing callers to voice mail, or performing other related call-routing functions without the assistance of a live attendant. The caller's activation's of these features occurs through pressing keys that activate DTMF signaling.

Automatic Callback

This diagram shows how computer telephony system can be used to create virtual (simulated) call attendants. In this diagram, a call is received to the main telephone number of the company to the computer telephony board. The automated telephony call processing software detects a ring signal, answers the phone (creates an off-hook signal) and plays a pre-recorded messaging informing the caller of options they may choose to direct the call to a specific extension. In this example, the automated call attendant software decodes DTMF tones or limited list of voice commands to determine the routing of the call. The automated call attendant software then determines if the destination choice is within the option list and if the extension is available. If the extension is available, the automated attendant will send a command to the computer telephony board switching the call to the selected extension. If the extension is not valid or not available, the automated attendant will provide a new voice prompt with updated information and additional options.

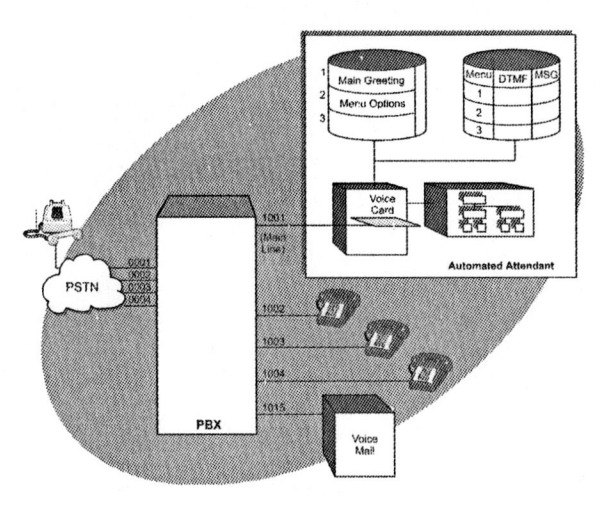

Automated Attendant Operation

Automated Voice Response Systems (AVRS)-A system that will automatically answer an incoming telephone call and provide voice instructions or information to the caller. The caller's response to these instructions may be keypad tones or even spoken words and will be used by the system to route the call to the appropriate extension or to other sources of additional information.

Automatic Busy Redial-A feature for automatically redialing a busy number at certain intervals.

Automatic Call Distribution (ACD)-ACD is a system that automatically distributes incoming telephone to specific telephone sets or stations calls based on the characteristics of the call. These characteristics can include an incoming phone number or options selected by a caller using an interactive voice response (IVR) system. ACD is the process of management and control of incoming calls so that the calls are distributed evenly to attendant positions. Calls are served in the approximate order of their arrival and are routed to service positions as positions become available for handling calls.

This figure shows a sample automatic call distribution (ACD) system that uses an interactive voice response (IVR) system to determine call routing. When an incoming is initially received, the ACD system coordinates with the IVR system to determine the customer's selection. The ACD system then looks into the databases to retrieve the customers' account or other relevant information and transfer the call through the PBX to a qualified customer service representative (CSR). This diagram also shows that the ACD system may also transfer customer or related product information to the CSR.

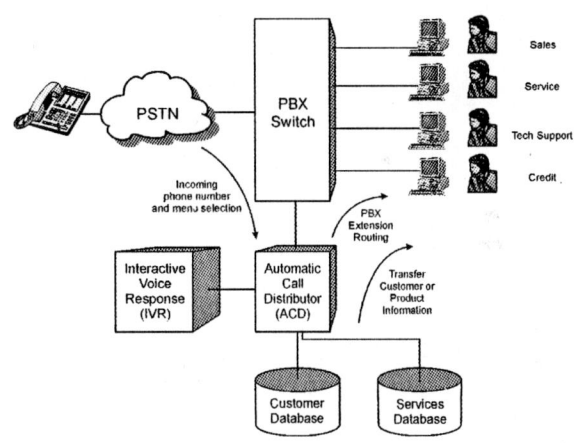

Automatic Call Distribution (ACD) Operation

Automatic Callback-A CLASS service feature that allows a caller to complete a call to a busy station by dialing an activation code (usually a single

Automatic Digital Network (AUTODIN)

digit) and hanging up. The system automatically rings both parties when the lines are available.

This figure shows the basic operation of automatic callback. To activate automatic callback service, after a call has dialed a number that is busy, the customer dials an automatic callback feature code and hangs up the telephone. The local switch (caller's phone carrier) informs the remote (distant) switch of automatic callback request. This reserves (blocks) the called number from receiving additional calls until the automatic callback service is completed. When the called number becomes available, the remote switch sends a message to the local switch and this rings the original caller's number (possibly with distinctive ring feature.)

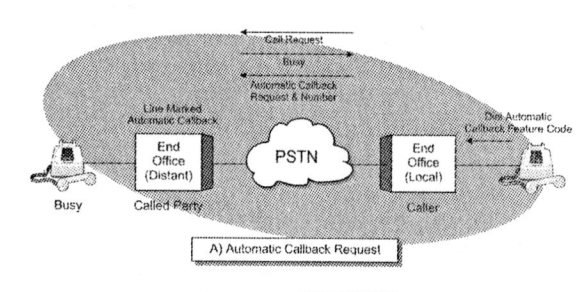

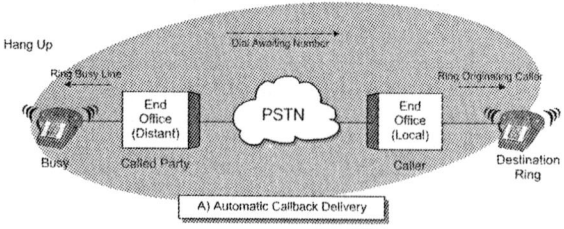

Automatic Callback

Automatic Digital Network (AUTODIN)-The worldwide circuit switched data communications network of the U.S. Department of Defense.

Automatic Gain Control (AGC)-An assembly or circuit that is part of a communications receiver that automatically adjusts the received signal level so that its level is approximately the same regardless of the received radio signal level. AGC is often used to supply a constant level signal to a demodulator assembly.

This diagram shows how a varying level signal that is supplied to a communication receiver can be adjusted to a fairly constant by an automatic gain control (AGC) system. This diagram shows that a varying radio signal is supplied to a signal level detector (diode) and a variable gain amplifier. The output of the detector is used to inversely vary the gain of the amplifier. As a result, the amplifier produces a near constant signal level that can be provided to the demodulator assembly.

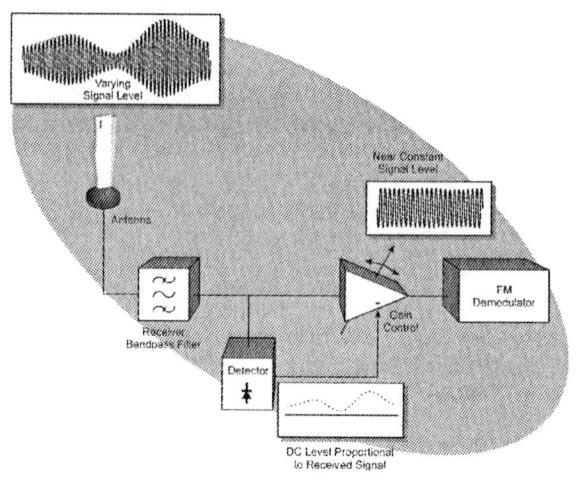

Automatic Gain Control (AGC) Operation

Automatic Identification Of Outward Dialing (AIOD)-The capability of some PBX or Centrex switches to provide an itemized breakdown of calls and charges, including individual charges for toll calls, for calls made from each telephone extension station. See also ANI.

Automatic Location Identifier (ALI)-A number (such as a telephone number or MAC address) that identifies a location of a device or assembly.

Automatic Message Accounting (AMA)-An automatic system for recording data describing the origination time of day, dialed number and time duration of a call for purposes of billing. The earliest systems used punched paper tape, later replaced by magnetic computer tape and then later magnetic computer disk. AMA is a term mostly used in the public network, and similar terms, some used in private, PBX, or inter-carrier systems are Call Detail Recording (CDR), Station Detail Message Recording (SMDR), and Automatic (calling) Number Identification (ACNI or ANI).

Automatic Meter Reading (AMR)-The process of reading meters (generally electric or water utility meters) via a communication systems such as wireless or wired technology.

Automatic Number Identification (ANI)-The providing of the originating telephone number, including an extension number in a Centrex system or PBX system. The ANI is an administrative number provided by the telephone system and may not be the actual originating number.

Automatic Roaming-The ability of a cellular telephone customer to make and receive calls automatically outside of the customer's home area.

Automatic Secure Voice Communications Network (AUTOSEVOCOM)-This is a worldwide secure communications network developed by the Department of Defense (DOD) and used by the armed forces and other governmental institutions engaged in national security and defense. This network is comprised of both voice and data encryption units, dedicated switched circuits, packet networks, wireless terrestrial and satellite communications systems.

Automatic Vehicle Location (AVL)-A system that determines the location of vehicle as it moves within a given geographic area. The position of the vehicle can be determined by system sensing (such as the Teletrac system) or by the vehicle reporting its location using position locating systems (such as the Global Positioning System).

Automatic Voice Network (AUTOVON)-The principal long-haul voice communications network within the Defense Communications System. It is notable, among other reasons, because originators can indicate from the dial the priority level of each call, and higher priority calls can force pre-existing lower priority calls to disconnect in order to free a channel.

Autonomous Cells (system)-Cell sits in a wireless network that are not directly controlled by the mobile switching center for the normal assignment of calls. These autonomous cells may be used as private systems and are not listed as a potential candidate for call transfer for neighboring cells that are part of the public system.

Autonomous Registration-A process where a mobile radio independently transmits information to a wireless system that informs it that it is available and operating in the system. This allows the system to send paging alerts and command messages to the mobile radio. The mobile radio may be stimulated to register with the system when it detections it has entered into a new radio coverage area or it detects a registration request message.

Autonomous System Number (ASN)-A unique number identifies autonomous systems connected to the Internet. ASNs are assigned by the InterNIC and are used by routing protocols such as Border Gateway Protocol (BGP) to identify the autonomous system.

AUTOSEVOCOM-Automatic Secure Voice Communications Network

Autotune Combiner (ATC)-A group of tuned cavity RF filters that can be automatically re-tuned to different radio frequencies. The autotune combiner is used in mobile radio communication systems to allow multiple transmitters to share a single antenna system and to allow the frequencies of the transmitters to change without the need to dispatch a technician to retune the tuned cavity RF filters.

AUTOVON-Automatic Voice Network

AUX-Auxiliary

Auxiliary Joint-A lead sleeve installed over polysheath telco cable that will provide a soldering surface to facilitate the subsequent soldering of a main sleeve as a splice closure.

Availability-A measurement that indicates the connection status or a commitment to provide a minimum amount of connection status of a network during a period of time. Availability may be measured by a connection time or by a minimum performance measurement (e.g. at a minimum data transfer rate). Availability is often tied to reliability.

Available Bit Rate (ABR)-Available bit rate (ABR) is a communications service category that provides the user with a data transmission rate that varies dependent on the availability of the network resources. ABR service may provide the user with feedback as to the changed data transfer rate and may have established minimum and maximum levels of data transmission rates.

Avalanche Multiplication-A current-multiplying phenomenon that occurs in a semiconductor diode or photodiode that is reverse-biased just below its breakdown voltage. Under such a condition, electrons are swept across the junction with sufficient energy to raise the energy of additional valence electrons via collisions, creating additional conduction electrons in a regenerative action similar to a chemical chain reaction. Avalanche current multiplication occurs in avalanche photo diode detectors

Avalanche Photodiode (APD)

used with fiber optics, and also occurs in the breakdown voltage regions of Zener avalanche diodes. A similar electron emission chain reaction occurs in photo-multiplier vacuum tubes and in the Farnsworth image dissector television camera tube.

Avalanche Photodiode (APD)-A semiconductor diode device that has a barrier region that is very sensitive to photons and the reception of photons causes an avalanche of electrons across p/n junction. This causes a rapid increase in current with the detection of light (photon) energy. APDs are used in optical networks to detect light, typically at the receiver. It is a semiconductor pn junction diode that is heavily biased so that each time a photon is absorbed by the device, the resulting conduction electron has so much energy that it can knock loose many other electrons and each of those electrons can generate others, resulting in an avalanche of electrons and therefore a large current. This causes a rapid increase in current with the detection of light (photon) energy. APDs can detect faint optical signals. However, APDs require higher voltages than other types of semiconductor devices.

AVC-Automatic Volume Control
AVC-Analog Voice Channel
Average Busy Season-Busy Hour
Average Power-The measure of power or energy as sampled over a period of time.
Average Revenue Per User (ARPU)-One Indicator of a wireless business's operating performance. ARPU measures the average monthly revenue generated for each customer unit, such as a cellular phone or pager, that a carrier has in operation. Severely declining ARPU typically is a negative sign that may indicate a carrier is adding too many low-revenue generating customers to its roll.
AVI-Audio/Video Interleave
AVK-Audio Video Kernel
AVL-Automatic Vehicle Location
AVRS-Automated Voice Response Systems
AVSD-Analog Simultaneous Voice and Data
AVSS-Audio-Video Support System
AVT-Audio Video Transport
AVVID-Architecture for Voice, Video and Integrated Data

AWG-American Wire Gauge
AWGN-Additive White Gaussian Noise
AWOS-Automatic Weather Observation Station
Azimuth-An angle measured (usually in degrees) in the horizontal plane between a reference direction (such as geographic north, or the strongest beam direction of an antenna) and a specific direction.

B

B-Byte
b-Bit
B 911-Basic 911
B Band-A band of radio frequencies that are used for cellular mobile telephone service. The B band of frequencies are assigned for use by the incumbent (local) phone company.
B Battery-Part of a telephone system power supply that supplies direct current for operating relays and other components. The B battery is typically 20 volts.
B Channel-A logical channel in an Integrated Services Digital Network (ISDN) that provides a 64 kbps connection to a switching system or to the non-switched portion of a network. This channel is used for transferring voice or data information between customer premises equipment and an end-office switching system.
B Connector-A connector that is used for splicing twisted wire pairs. B connectors are sometimes called beans. B connectors are shaped like a plastic tube that is approximately 2.5 cm (1 inch) long. They include metal teeth inside to help ensure a good connection is made when the B connector is crimped (compressed) and to ensure that the connector will not slide off the splice. Some B connectors contain a water-retardant jelly inside to reduce the effects of corrosion on the wire splice.
B Frame-Bi-Directional Frame
B Side-The wireline carrier who had landline service in place prior to building out the cellular network.
B Wire-European term corresponds to North American Ring wire.
B-Band Cellular-A band of frequencies in the 800 MHz range that are assigned to a cellular service provider. In the America's, B band frequencies are assigned to an incumbent local telephone service provider that have decided to offer cellular telephone service.
B-CDMA-Broadband Code Division Multiple Access
B-DCS-Broadband Digital Cross-connect System
B-frame-Bi-directional Predicted Frame
B-ICI-Broadband Intercarrier Interface
B-ICI SAAL-Broadband Inter-Carrier Interface Signaling ATM Adaptation Layer
B-ISDN-Broadband Integrated Services Digital Network
B-ISDN ICI-Broadband Integrated Services Digital Network Inter-Carrier Interface
B-ISUP-Broadband Integrated Services Digital Network User Part
B-Link-Bridge Link
B-LLI-Broadband Lower Layer Information
B-TA-Broadband Terminal Adapter
B-TE-Broadband Terminal Equipment
B2B-Business-To-Business
B2BUA-Back to Back User Agent
B2C-Business-To-Consumer
B3ZS-Binary 3-Zero Substitution
B3ZS-Bipolar with 3-Zero Substitution
B8ZS-Bipolar with 8 Zero Substitution
B8ZS Line Code-Bipower With 8-Zero Substitution
BAA-Blanket Authorization Agreement
Babble-The aggregate crosstalk from many interfering channels.
Babbling Tributary-A communication device (station) that continuously transmits unnecessary messages.
BABT-British Approvals Board for Telecommunications
Back Door-In an otherwise secure system, an intentional way for a trusted party to circumvent the security with secret knowledge such as the nature of a designed-in security flaw. Since back doors might be found and exploited by malicious parties, they weaken the security of what may appear to be a well-protected system.
Back End-A system or database server function that processes data via a network connection.
Back Haul-Back haul is the process of extending the use of communication facilities or more efficient circuits by using communicating routing lines that are longer than would be typical for a specific type of service. Back haul allows for cost effective sharing of facilities by either sharing network facilities (such as a switching system) or communication circuits (sharing long haul lines with many more users). An example of using back haul to reduce cost is the use of long distance lines to connect cellular radio towers to distant mobile cellular switching facilities. The use of back haul lines reduces or eliminates the need to install mobile switching equipment in a local system.

This diagram shows how long distance communication lines can be used to back haul communication circuits to a distant switching system to eliminate the need to install a local switching system. In this example, the distant mobile radio communication towers (cell sites) operate as they would if they were connected to a local switching system. The long distance back haul lines carry the control information and the voice communication channels.

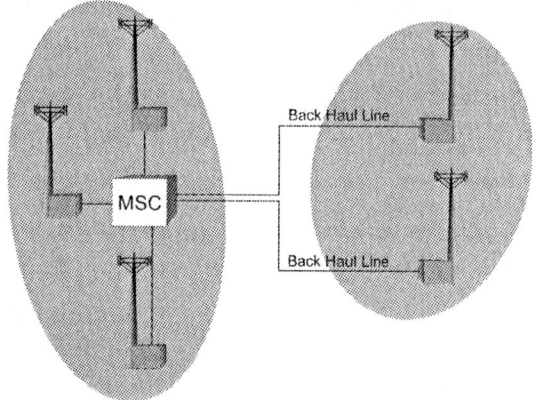

Back Haul System

Back Hoe Fade-A reduction in the ability of a communication system to route calls due to the cutting of a buried communication cable (e.g. fiber optic cable). The reduction in capacity comes from the automatic re-routing of communication circuits through other systems that have a lower capacity than the original communication circuit.

Back Porch-The short time interval in an analog television video waveform immediately following the horizontal synchronizing pulse. The color synchronizing burst is transmitted during the back porch time interval. See the figure appearing with the term Horizontal Scan.

Back to Back User Agent (B2BUA)-A user agent (UA) that is part of a SIP system that provides some advanced call processing features.

Back-End Processor-This is another computer system or dedicated microprocessor that is optimized to perform a specialized task in order to offload work from the main processing resources.

Back-To-Back Connection-A connection between the output of a receiver and the input of a transmitter. Also may refer to private line circuits that are connect via modems or CSU connected back to back using the appropriate roll-over cable.

Backbone-The core infrastructure of a network that connects several major network components together. A backbone system is usually a high-speed communications network such as ATM or FDDI.

Backbone Network-A communications network that connects the primary switches or nodes within the network.

Backbone Route-A high-capacity transmission line or combination of circuits that interconnects various multiplexing or distribution points. Backbone routs are efficient transmission facilities to allow communication traffic aggregation at selected points to maintain the circuit loading (percentage of user of the backbone transmission line).

Backcharging-A process of charging for usage of a service when the service request is initiated rather than when the service is connected. An example of backcharging is the process of starting the billing time on a mobile telephone when the user initiates the call, not when the call is actually connected.

Background Communication-Communication that is simultaneously performed on another channel than the communication channel that is in use. An example of background communication is the operation of a white board during a video conference call.

Background Noise-Electrical signal energy in a channel or frequency bandwidth that is not part of the desired received signals.

Background Noise Regeneration-Speech codecs like G.729 may use a silence suppression technique to avoid sending unnecessary bits during periods of no speech, however the lack of any sound at the decoder during these times sounds odd to a listener. For a more natural sound during these moments, a background noise regeneration technique may be used to play audio that sounds like the normal background noise.

Background Processing-The process by which lower priority programs are scheduled to use computer system resources only when no other higher priority tasks are available for execution. This allows the computing resources to be made available to higher priority programs that must perform an immediate action.

Background Program-A separate program or processing thread that runs at a lower priority and therefore only uses the resources of the computer system when no other higher priority tasks (foreground tasks) are available for processing.

Background Task-A processing job (task) that is performed concurrently with a primary job where the background processing steps are usually performed during periods of low activity or idle periods during the processing of the primary task.

Backhaul-The act of moving packets or communication signals over relatively long distances to a separate location for processing.

Backhauling-The linking of cell sites to a mobile switching center (MSC) through telephone lines. The term backhauling is often used when referring to cellular or PCS systems that provide service by backhauling their cell sites to a MSC that is located in a different geographic region. In some rural areas, it is not economically viable to install and connect a switching center to PSTN to service the cell sites. Cellular or PCS carriers in adjacent markets sometimes enter into relationships to provide the switching services via backhauling.

Backing Out-The process of returning the state of a system to its prior known state after a transaction fails to complete. This process guarantees the integrity of the information contained within the system. Used primarily for database and distributed systems.

Backoff-A process that is used in a data communications network to prevent repeated collisions during data access attempts. Using the backoff procedure, when a failed access attempt occurs, the communication device will delay its next access transmission request (backoff) before a second attempt is made to access the network.

Backplane Bus-A collection of electrical wiring that interconnects the slots where modules in a computer system are inserted. The backplane bus wiring connects the same signal to identically numbered pins of each module in the computer system. For example, the "power on" signal might be connected to pin 3 of each module via the backplane bus wiring.

Backscattering-The scattering of light in a direction that is opposite to the direction that it was originally traveling.

Backup-(1-data) A copy of data information, usually on a storage device that is stable over relatively long periods of time. (2-process) The process of transferring files or information to a storage device or media.

Backup Link-A communication link that is used for communication in the event a primary or specified link becomes inactive or disabled.

Backup Power Supply-A redundant power supply that takes over if the primary power supply fails. This may or may not be automatic although when supplied as part of the overall system it usually is.

Backup Program-This is an application program that is used to make archival copies of the contents of a computer file system. Most backup programs provide methods for both full and incremental backups. Incremental backups only archive files that have been modified since the last full backup. Most commercial backup programs provide a means of also restoring the archived information if the need arises.

Backup Server-A program that administers the copying of users' files so that at least two up-to-date copies exist.

Backward Channel-In channel in a data transmission system that is opposite that of the primary channel. The backward channel may be a low capacity channel that is used for transmission control or supervisory signals.

Backward Compatibility-Ability of new hardware or software to operate effectively with older versions of the same.

Backward Congestion Notification-Backward congestion notification indicates to sending (downstream) switching devices in a data communication network that congestion is occurring and packets that are received may be discarded. The sending switch can then change the priority of packet discarding and send and indication to other switches indicating network congestion. This should eventually reduce the amount of data end users are sending into the network.

Backward Error Correction (BEC)-Error correction techniques in which the receiver detects any errors and requests retransmission of the erroneous data. When error rates are low or zero, BEC can be very efficient. However if the same error occurs repeatedly BEC techniques can never transmit the data properly. Compare to forward error correction (FEC).

Backward Explicit Congestion Notification (BECN)-A control bit carried within the overhead of a data packet in a frame-relay network that indicates network congestion exists in the backward (opposite) direction of its flow. This control bit allows higher-level protocols within the data communications equipment (DCE) and data terminal equipment (DTE) to take appropriate bandwidth allocation action if necessary.

Backward Indicator Bit (BIB)-A bit in an SS7 signaling message, by its status change at the remote end, requests retransmission because of message received out of sequence.

Backward Sequence Number (BSN)-A field in an SS7 signal unit sent that contains the forward sequence number of a correctly received signal unit being acknowledged.

BACP-Bandwidth Allocation Control Protocol

BACR-Billing Account Cross Reference

Bad Block-Within a mass storage device the storage medium is divided up into blocks that each contain a fixed amount of data, typically 512 or 1024 bytes. If a particular block on the medium is defective it is marked as a bad block so that the operating system will not use it to store information.

Bad Track Table-A list of the unusable areas on a storage disk, usually defective tracks.

Badged-An English term that represents that a product manufactured by one firm that is sold by another. The Original Equipment Manufacturer (OEM) produces the product with identification and/or brand of the selling firm.

BAF-Bellcore Automatic Message Accounting Format

Bag Phone-A mobile telephone that is contained in a bag or soft case. Bag phones are often high powered mobile telephones and they may or may not contain batteries.

BAIC-Barring of All Incoming Calls

Balance condition-The condition in which two signals are effectively separated or equal levels.

Balance Test Line-Equipment in a central office that provides a proper point to terminate a line for echo-balance and noise testing. Access to such equipment through a dial-up telephone line facilitates testing of trunks by measurement systems and by field service personnel.

Balanced Line-A transmission line that uses two conductors where each of the conductors have equal voltage but opposite polarity levels to form a balanced line.

Balanced link access procedure (LAPB)-The data link layer protocol that connects terminals and computers to a packet switched network. It can be viewed as the same as the asynchronous balanced mode of HDLC.

Balanced Modulator-A modulator that combines the information signal and the carrier so that the output contains the two sidebands without the carrier.

Ballistics Test-An outside plant cable test that measures the charging and discharging capacity in a pair of conductors. This meter deflection allows a technician to approximate the length of the conductors.

Balloon Antenna System-A communication system that uses balloons to elevate communication equipment. Balloon based communication systems use wireless repeaters (combined receiver/transmitters) to efficiently provide radio coverage to relatively large geographic areas.

This photograph shows the Space Data Corporation's SkySitesTM balloon based antenna system that is used to form a network of wireless repeaters. These repeaters can provide efficiently provide wireless communication service to wide geographic areas. Each repeater is carried by small, biodegradable, latex weather balloons operating at 100,000 feet. The SkySitesTM Space Data's advanced transceivers can be launched from weather monitoring sites. Weather balloons have been reliably launched worldwide for more than half a century gathering meteorological data from the surface of the earth to 100,000 feet. In 2002, the United States had 70 sites located in 48 states that launched balloons twice a day. Space Data's network supports terrestrial wireless networks and is compatible with existing subscriber equipment.

Balloon Antenna System
Source: Space Data Corporation

BAN-Billing Account Number

Band-A range of frequencies defined specified upper and lower frequency limits.

Band Elimination Filter (BEF)-A filter that has a an attenuation band for a specific range of frequencies. See band reject filter.

Band Reject Filter-A filter that rejects a specific band of frequencies. A filter that theoretically produces exactly zero output for one specific sine wave frequency (having a mathematical zero of transmission precisely on the imaginary frequency axis in the complex frequency plane) is called a notch filter.

Band Splitter-A band multiplexer or filter assembly that is used split a frequency band into several smaller frequency bands.

Band Stop Filter (BSF)- A device which blocks (stops) a specific range of frequencies and allows all other frequencies to pass through with minimum attenuation. See band reject filter.

Bandpass-A range of frequencies (between upper and lower limit frequencies) that allows signals to pass through without low signal loss (low attenuation).

Bandpass Filter-A filter designed to reject or block all frequencies not within a given bandwidth. Such filters may be used to reject noise or other signal bands close in frequency to that of the desired signal. Filters typically require a trade-off among how much signal they pass (the amount of loss), how strongly they reject the undesirable signal, and how sharp the dividing line between passed and unpassed signals is.

This diagram shows a banpass filter that is used to block high and low frequency component parts of an input signal. In this example, both the high frequency noise and low frequency noise are attenuated by the bandpass filter. This example shows the desired frequency is allowed to pass through the filter with minimal attenuation.

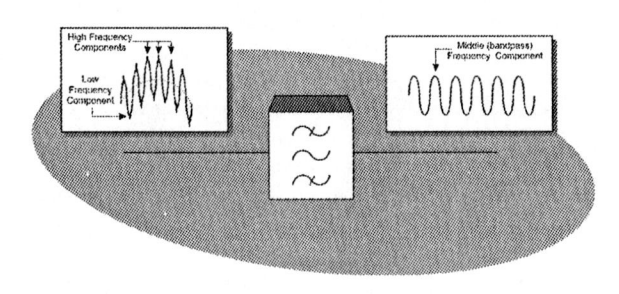

Bandpass Filter Operation

Bandwidth-A term that defines the signals that occupy a portion of a frequency spectrum, particularly a radio system or their data transmission rate. Analysis or measurement of the signals or signal waveforms of such a system will show that most (or substantially all) of the power contained in that signal can be found in a designated portion of the frequency spectrum. The difference between the highest and lowest frequency describing that portion of the spectrum is the bandwidth of the signal. Frequency (radio spectrum) bandwidth is measured in units of hertz or cycles per second and data transmission bandwidth is measured in bits per second.

Bandwidth Confirm (BCF)-The process or message that confirms the bandwidth assigned to a communication connection. An BCF message is defined in H.225.

Bandwidth On Demand (BoD)-A system that allows different data transmission rates based on requests from the customer, their application (e.g. voice or video), and the data transmission capability of the system.

BAOC-Barring of All Outgoing Calls

Bar Code-A standardized sequence of typically black vertical bars separated by white spaces that may be read with an optical decoder. The decoder reads the sequence of bars and interprets them as alphanumeric characters. The pattern of adjacent thick and thin bars is located on a contrasting background. Each decimal digit of an identification number is represented by a binary bit group in a bar code, and parity check digits are typically appended. A bar code can be scanned by an optical scanner device for input to a computer system for purposes of inventory control or the like.

Bare Board-A particular form factor of a circuit board with the necessary system connectors, power and ground conductors and standardized parts layout but without components. These boards are typically used for building prototypes and specialized functions that are not commercially available.

Bare Wire-An electrical conductor that does not contain any type of shielding over its surface area.

BARG-Billing and Accounting for Roaming Group

Barge In-A call processing feature that allows another caller to be added to a communication line that is already in use. The barge in feature is sometimes called executive override.

Barge Out-A call processing feature that allows a user to leaving a call in progress without any notice (usually after a call barge in).

Barrel Connector-A small cylindrical connector made with two female ends so that it may be used to connect between two cables both terminated with male connectors.

Barring of All Incoming Calls (BAIC)-A system feature that restricts the delivery of all incoming calls to a phone.

Barring of All Outgoing Calls (BAOC)-A system feature that restricts the delivery of all outgoing calls from a phone.

Base Address-A numeric value used as a reference in the calculation of addresses during the execution of a computer program. Usually defined by the beginning real address of a partition or other segment of computer memory.

Base Rate-A fixed amount charged for a service per a specific time period. Normally this term is associated with a service that has a fixed charge (for a basic amount or type of service) and variable charge (applied only when more than the basic services are used).

Base Station (BS)-The radio part of a cellular radio transmission site (cell site). A single base station usually contains several radio transmitters, receivers, control sections and power supplies. Base stations are sometimes called a land station or cell site.

A base station contains amplifiers, radio transceivers, RF combiners, control sections, communications links, a scanning receiver, backup power supplies, and an antenna assembly. The transceiver sections are similar to the mobile telephone transceiver as they convert audio to RF signals and RF to audio signals. The transmitter output side of these radio transceivers is supplied to a high power RF amplifier (typically 10 to 50 Watts). The RF combiner allows separate radio channels to be combined onto one or several antenna assemblies without interfering with each other. This combined RF signal is routed to the transmitter antenna on top of the radio tower via low energy-loss coaxial cable.

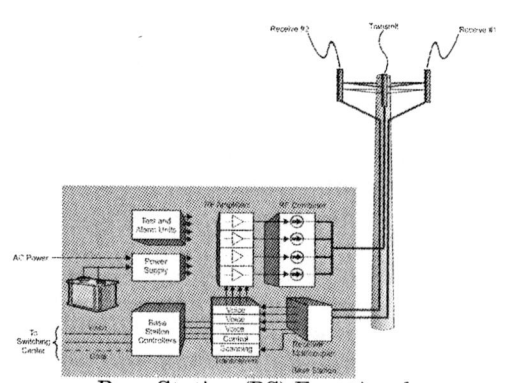

Base Station (BS) Functional

Base Station Controller (BSC)-A automatic coordinator (controller) that allows one or more base transceiver station (BTS) in a wireless network to communicate with a mobile switching center.

Base Station Subsystem (BSS)-The radio networks parts of a GSM network. This includes the base transceiver sections (BTS) including their radio towers and the base station controllers (BSCs).

Base Station Transceiver-A combined transmitter and receiver assembly (transceiver) that is used as the radio processing part of a cellular system radio transmission site (cell site). The base station transceiver consists of all radio transmission and reception equipment, it provides coverage to a geographic area, and is controlled by a base station controller.

Base Traffic Load-In trunk forecasting, the average load offered on the first route available between two identified areas. Base load is found by averaging the traffic measured during the same 1-hour period each day over a period of several days. Base loads often are used to forecast future loads.

Base Transceiver Station (BTS)-The radio part of a wireless network (typically cellular or PCS) that includes the transmitters and receivers, antennas and tower that is used to communicate with mobile radios. A BTS is connected to a base station controller (BSC).

Base Year-Any 12 consecutive months for which data is collected for determining base loads.

BASE64-A binary-to-text encoding mechanism specified in RFC 1521, used to convert arbitrary sequences of binary data into a strictly limited subset of printable ASCII characters. Each set of 3 bytes (that is, 24 bits) is converted to a set of 4 characters from the ASCII subset. One application is for converting binary attachments into printable text for MIME-formatted e-mail attachments.

Baseband-A digital signal that occupies the entire bandwidth of a channel. Used in short range networks such a LAN's.

baseband channel-A information content (channel) that is used to modulate or encode a transmission medium. When used with radio signals, the high frequency component is called the broadband channel.

Baseband LAN-A local area network (LAN) that uses a single carrier frequency on the communication channel that represents the actual information that is transferred (no conversion required). Ethernet, Token Ring and ArcNet LAN's are examples of baseband transmission systems.

Baseband Signal-The original form of a information signal. Baseband signals can be applied to a modulator to impose it's information on a carrier signal (radio wave or optical) to produce a broadband signal.

Baseband Signaling-The direct transmission of a signal information without any conversion to another transmission medium.

Baseline-The process of determining and documenting network throughput and other performance information when the network is operating under what is considered a normal load. Measured performance characteristics might include error-rate and data-transfer information, along with information about the most active users and their applications.

BASIC-Beginners All-Purpose Symbolic Instruction Code

Basic 911 (B 911)-Basic 911 is an emergency service call code that allows users to contact a emergency service operators. B 911 is the early version of emergency call service as it does not provide location information of the caller to the emergency operator.

Basic Exchange Telecommunications Radio Service (BETRS)-An wireless extension to a local telephone system to allow telephone service in rural areas. The BETRS system is a digital radio system that operates at frequencies near 150, 450, and 850 MHz. The system is used in primarily in areas where wired telephone service is not economically viable. BETRS is sometimes referred to as Basic Exchange Radio (BEXR) Service.

Basic Input/Output System (BIOS)-The software or processes that controls the input and output of information to devices connected to a computer. The BIOS contains the set of instructions that sets up and tests the hardware when the computer is first turned on (booted). It starts the loading of the operating system and coordinates the operation of computing devices, such as floppy drives, hard disks, CD ROMs, video cards, mouse, and keyboards. The BIOS program is stored in non-volatile memory and is rarely changed (if ever). It is pronounced "bye + Ose."

Basic Input/Output System Enumerator (BIOS Enumerator)-The BIOS enumerator identifies all hardware devices on a computer, obtains their communication and configuration parameters, and assigns them temporary identification numbers during the computer's operation.

Basic Rate Interface (BRI)-The standard basic interface that is used in the integrated services digital network (ISDN) system. The BRI interface provides up to 144 kbps of information that is divided into two 64 kbps channels (voice or data) and one 16 kbps control channel (data). The 64 kbps channels are referred to as the B channels and the 16 kbps channel is called the D channel.

Basic Trading Area (BTA)-A geographic region in the United States where area residents do most of their commerce activities. The United States has been divided into 493 BTAs. Some PCS licenses issued in the United States in the mid 1990's (bands C through F) were awarded based on BTA values.

Batch Processing-A process that does not occur in a real time mode. In a batch process, events are collected and then forwarded in batches to the processor which is "idle" until required to begin.

Battery Recharging-The process of recharging batteries. Recharging is typically performed for a time period or until an indication from the battery indicates the battery is fully charged. See also: memory effect and knee voltage.

Battery, Overvoltage, Ringing, Supervision, Coding, Hybrid, Test (BORSCHT)-A group of functions provided in Subscriber Line Circuits (LCs). It stands for:

B: Battery supply to subscriber line.

O: Over voltage protection (prevents damage to equipment or hazard to subscriber in case of lightning pulses or a "cross" connection with a high voltage power line).

R: Ringing current supply.

S: Supervision of subscriber set (determining busy/idle or off hook vs. on hook status).

C: Coder and decoder for analog-digital conversion when analog telephone is used with a digital switch or transmission system.

H: Hybrid coils or "induction" coil (2 wire to 4 wire conversion) when audio frequency voltage waveforms flow in both directions on the subscriber lines but are separated in the telephone set for connections to the microphone and earphone. Also, a similar separation/combination of two opposite unidirectional signal flows in the switch or trunk transmission system vis-à-vis the two-wire analog subscriber lines.

T: Test.

Borscht is a group of functions provided to an analog line from a line circuit of a digital central office switch. An analog electronic switch can omit C and possibly H. A line circuit on a switch with metallic matrix (SXS, Xbar, 1,2,3ESS) only detects call origination and disconnects itself. The term BORSCHT is attributed to John W. Iwerson of AT&T Bell Laboratories in the 1960s.

Baud

Baud-Name for the unit of data symbols per second. This name, abbreviated Bd, is taken from the name of the 19th century French teletypewriter machine innovator Emiel Baudot. For a method of modulation or encoding in which there is a choice of only two symbol values per symbol interval, or one bit per symbol (such as two-level pulse voltages) the baud rate is equal to the bit rate (bits per second). For a method of modulation or encoding in which there are more than two symbol values per symbol interval (and thus 2 or more bits per symbol) the bit rate is higher than the baud rate. For example, QPSK phase modulation and 2B4Q pulse coding both have 4 symbol values per symbol interval and thus the bit rate (bits per second) is twice the symbol (baud) rate. (Please do not make the error of writing "baud per second.")

Baud Rate-The number of the signaling elements (symbols) per second on a transmission medium. For some line codes, such as bipolar, baud rate is the same as bit rate. However, in many applications, the baud rate is below the bit rate. For example, in 2B1Q coding, each quaternary signaling element conveys 2 bits of information, so the baud rate is one-half the bit rate. The spectral characteristics of a line signal depend on the baud rate, not the bit rate. For high-speed digital communications systems, one state change can be made to represent more than one data bit.

This diagram shows that the baud rate is not always the same as the bit rate as each baud (symbol) can have several states that represent multiple binary bits.

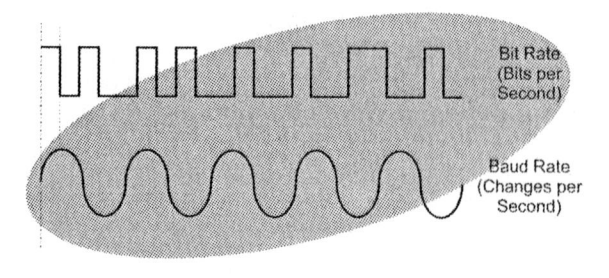

Baud Rate

Baud Rate Equalizers-A signal processing circuit or system that removes unwanted components (delayed repeated signals) from the received signal. Baud rate equalizers operate at the baud rate (symbol rate) of the communications circuit.

Baudot-Murray Code-Baudot-Murray (ITU alphabet No. 2) is a 5-bit character code used by older teletypewriters and telex machines, and also by most teletypewriters for the deaf. Frequently called Baudot code. It is typically transmitted with an added start bit and a lengthened stop bit (7 bits total) in an asynchronous sequence.

Bay-An equipment casing that can hold electronic assemblies such as transmission or call processing assemblies. Commonly called an "Equipment Bay."

BBS-Bulletin Board System

Bc-Committed Burst

Bc-Committed Burst Size

BCC-Blind Carbon Copy

BCC-Billing and Customer Care

BCCH-Broadcast Channel

BCCs-Quired By Bellcore Client Companies

BCD-Binary Coded Decimal

BCE-Bandwidth-Control Elements

BCF-Bandwidth Confirm

BCH Code-Bose, Chaudhuri, And Hocquenhem

BCI-Billing Correlation Identifier

BCL-Base General Premises Cabling License

BCOB-Broadband Connection Oriented Bearer

BCOB-A-Bearer Class A

BCOB-C-Bearer Class C

BCOB-X-Bearer Class X

BCP-Best Current Practice

BCSM-Basic Call State Model

BD-Building Distributor

BDFB-Breaker Distribution Fuse Bay

BDN-Bell Data Network

Be-Excess Burst

Be, or Burst excess-Excess Burst Size

Beacon Channel-A channel used in the Bluetooth system to support parked slaves where the master establishes a beacon channel when one or more slaves are parked. The beacon channel consists of one beacon slot or a train of equidistant beacon slots, and is transmitted periodically with a constant time interval. When parked, the slave will receive the beacon parameters through a Link Management Protocol (LMP) command. See also Park Mode.

Beaconing-In a token-ring network, a process used by all nodes on a faulting ring to isolate the failure. When a node determines that it cannot transmit and receive reliably, the device will "beacon". This involves transmitting a special Beacon

MAC frame every few milliseconds. Any station receiving a Beacon MAC frame must repeat the frame. Within a few seconds, the exact location of a cable fault can be isolated and the MAC addresses of the network interface cards between the fault discovered. When a ring is beaconing, user communication cannot take place.

Beamsplitter-A assembly or device that receives an input light wave and divides it into two or more separate beams.

Bearer Services (BS)-Bearer services are telecommunication services that are used to transfer user data and control signals between two pieces of equipment. Bearer services can range from the transfer of low speed messages (300 bps) to very high-speed data signals (10+ Gbps).

Bearer services are typically categorized by their information transfer characteristics, methods of accessing the service, inter-working requirements (to other networks) and other general attributes. Information characteristics include data transfer rate, direction(s) of data flow, type of data transfer (circuit or packet) and other physical characteristics. The access methods determine what parts of the system control could be affected by the bearer service. Some bearer services must cross different types of networks (e.g. wireless and wired) and the data and control information may need to be adjusted depending on the type of network. Other general attributes might specify a minimum quality level for the service or special conditional procedures such as automatic re-establishment of a bearer service after the service has been disconnected due to interference. Some categories of bearer services available via the telephone system include synchronous and asynchronous data, packet data and alternate speech and data.

This figure shows a data transmission bearer service. In this diagram, a customer decides to send a data file to a computer that is connected to a public telephone network (at the office). In this example, the bearer service is 9.6 kbps circuit-switched data. The customer uses a modem to adapt their portable computer to the telephone network. The portable dials the office computer telephone number via the modem. The telephone system routes this call to a modem that connects the office computer to the telephone network. When the office computer modem accepts the call, the customer's modem begins to send data directly on the telephone line at channel at 28.8 kbps. Because the telephone system bearer service only provides 9.6 kbps data communication, the end-to-end transmission will be adjusted to 9.6 kbps.

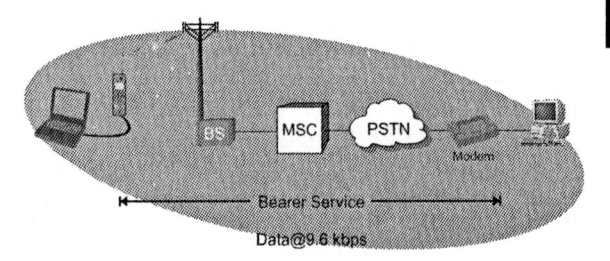

Bearer Service Operation

BEC-Backward Error Correction
BECN-Backward Explicit Congestion Notification
BEF-Band Elimination Filter
Behavioral Score-A measure of a customer's credit worthiness that is based on past payment history. Sometimes termed "Internal Score", or "Payment Behavior Score".
Bel-A unit of power measurement where each positive unit represents a tenfold (10x) increase in power and each negative unit represents a tenfold decrease in power ratio (1/10th). This unit of measurement is named in honor of Alexander Graham Bell.

The commonly used version of the Bel unit is one tenth of a bell, or a decibel (dB). One bell equals 10 db. The Bel unit is a logarithmic value. If an amplifier increases the power of a signal by 100 times, the power gain of the amplifier is equal to 2 Bel, or 20 dB.

Bellcore-A telecommunications research and development consortium that was developed by the seven regional companies that resulted from AT&T's divestiture of the Bell Operating Companies. These were Ameritech, Bell Atlantic Corporation, BellSouth, NYNEX, Pacific Telesis, Southwestern Bell Corporation, and US West. Bellcore was sold in the late 1990s to Science Associates, and is now called Telcordia.

Bent Pipe-Bent pipe satellite transmission is a process of transferring radio signals through a satellite by redirecting the signal (bending) to a new direction rather than receiving, processing, and retransmitting the signal. The bent pipe method reduces the complexity of the transferring device as it is only retransmits the original signal in a new direction without analyzing or altering its information content.

BER-Basic Encoding Rules
BER-Bit Error Rate
BER Threshold-Bit Error Rate
Berkeley Internet Name Domain (BIND)-A Domain Name Server (DNS) implementation available for all major server operating systems. BIND is by far the most widely deployed DNS implementation on the internet.
BERT-Bit Error Rate Test
Best Effort-A level of service in a communications system that doesn't have a guaranteed level of quality of service (QoS).
Best Mode-In accordance with United States law, an application for patent must include a description of the best mode contemplated by the inventor, as of the time of filing the application for patent, of practicing or implementing the invention which is claimed.
The purpose of the best mode requirement is to force the inventor to not only disclose his or her invention, but also the best way in which the invention can be used. The best mode requirement is intended to prevent an inventor from obtaining a patent on an invention while keeping secret the best way to exploit the invention.
Best Path-The selection of an optimal route between source and destination and stations through a wide area network. The best path can be determined through routing protocols such as RIP and OSPF, best path can be based on lowest delay, cost or other criteria. As a result, the path chosen in one direction is not necessarily the same as the opposite direction.
Beta-(1- general) Second letter of the Greek alphabet. Used as a mathematical symbol for the current amplification factor of a junction transistor. (2- testing) A second stage of product testing (following alpha testing), typically allowing selected potential consumers use the product in the field. (3- video) A now-obsolescent videotape format for consumer magnetic tape video recording, and a shortened form of a broadcast quality video recording system, named Betacam, that uses metallic particle video recording tape in a cassette having the same dimensions as the consumer Beta product. Betacam is used by several television network news broadcasters.
Beta-Betacam
BETRS-Basic Exchange Telecommunications Radio Service
Between Half-Power Points-Spectral Bandwidth
BFSK-Binary Frequency Shift Keying

BFT-Binary File Transfer
BFV-Bipolar Violations
BG-Border Gateway
BGMP-Border Gateway Multicast Protocol
BGP-Border Gateway Protocol
BGP4-Border Gateway Protocol Version 4
BHANG-Broadband High Layer Information
BHCA-Busy Hour Call Attempts
BHCC-Busy Hour Call Completion
BHM-Busy Hour Minutes
BHMC-Busy Hour Minutes of Capacity
Bi-Directional Frame (B Frame)-A type of frame within the MPEG digital video compression process that are used reference frames.
Bi-directional Predicted Frame (B-frame)-In video compression, a frame encoded with reference to one past reference frame and one future reference frame which are P-frames and/or I-frames. Using motion compensation, the difference between the pixels of the B-frame and the reference frames is encoded in a way similar to JPEG for still images. Also called B-pictures. B frames typically produce the most compression.
Bi-Polar Violation (BPV)-In digital telephone multiplexing systems, two level logic digital signal pulses in the electronic multiplexer are converted into three level pulses for purposes of transmitting the pulses via transmission paths that incorporate transformers or coupling capacitors. These devices cause waveform distortions if the transmitted waveform does not have a zero average voltage over the time interval corresponding to several pulses. Before transmitting, a binary 1 signal is converted into a zero volt pulse (an interval with no pulse voltage). A binary zero signal is converted into either a positive pulse (of about 3 volts amplitude) or a negative pulse (with negative 3 volts amplitude). A simple memory device in the coder causes each zero value pulse output to be set to the opposite polarity from the previous zero value pulse output. This line coding method is called Alternate Mark Inversion (AMI), and is sometimes called "pseudo-ternary" since it has three voltage pulse levels (+3, 0, -3 volts) but the three levels do not represent three independent code values but are instead dependent on two underlying code values. At the receiving location, the three different pulse levels are converted back into just two levels. Because of the design, no two distinct consecutive 3 volt pulses should ever have the same polarity. If they appear to do so, we call this a bi-polar violation. It indicates that either a pulse has been "lost" in the transmission, or that an

extra pulse has been produced (perhaps due to lightning or some other source of electrical interference. Certain other types of line coding such as HDB3 or B8ZS intentionally produce BPVs for certain all-zero binary data strings.

BIA-Burned In Address

BIB-Backward Indicator Bit

BICSI-Building Industry Consulting Service International

Bidirectional-The transmission of information that occurs in both forward and reverse directions.

Bidirectional Carrier-A telephone company (telco) carrier system that utilizes 1 cable runs. This cable run uses 100 pair separation between transmit and receive pairs.

Bidirectional Line Switched Ring (BLSR)-A form of synchronous optical network (SONET) transmission that uses 2 rings for communication. The data flow on these two rings is opposite: clockwise on one and counterclockwise on the other. The use of BLSR transmission allows half of the data communication devices to be served by one ring and half to be served by the other.

Bidirectional Repeater-A repeater that regenerates a signal in both directions of transmission. It can be used in single pair cable operation. Bi-directional repeaters are placed within T1 spans at 3,000-6,000 feet intervals.

Bill Cycle Code-An identifier assigned to all customer who are to be billed on the same run of the billing process.

Bill Of Materials (BOM)-A list of parts (components and other assemblies) that are the materials that are used to produce a product, quantity of product, or an assembly. The listing of parts in a BOM is often assigned a cost to estimate the construction cost of the product or assembly.

Bill Period-The period for which the recurring charges apply. In monthly billing for instance, the Bill Period typically corresponds to the period between the current bill and the next bill for charges billed in advance, and between the previous bill and the current bill for charges billed in arrears (e.g. April 5th to May 4th).

Bill Pool-A group of call records that have been updated by the call processing stage in a billing system to include charging rate information. The bill pool usually contains records that are ready for the final stage of bill processing.

Bill Posting-The process of recording an invoice payment into a billing system.

Bill Run-The process of triggering the start of the billing process that gathers and processes billing charges associated with specific customers or accounts. Also sometimes referred to as "Bill Round", or "Billing", or "Month End", etc. Typically, one or more cycles are assigned to a Bill Run; i.e. all customers who have been assigned those cycles will be processed during this Bill Run.

Billed Minutes-The actual time, in reported minutes, for messages of a duration equal to or greater than an initial period. When the reported minutes are less than the initial period, the initial period minutes are shown as billed minutes. For example, a one-minute conversation on a connection having an initial period of three minutes would be billed as three minutes.

Billing-The process of grouping billing information for specific accounts or customers, producing and sending invoices, and recording (posting) payments made to customer accounts.

Billing Aggregator-A billing aggregator is a company that gathers billing records from one or more companies and posts them to another billing system. An example of a billing aggregator is the gathering of billing of information from many companies that provide information or value added services (e.g. news or messaging services) and transferring these to another basic services carrier for direct billing to a customer. The customer pays the carrier, the carrier pays the aggregator, and the aggregator pays the value added service provider company. The billing aggregator performs services similar to a clearing house.

Billing and Accounting for Roaming Group (BARG)-A subgroup of the GSM Association that focuses on the financial, administrative, and procedural issues related to roaming (visiting) customers.

Billing and Customer Care (BCC)-A set of functions and processes related to generating customer invoices. These generally include Event (network usage) Rating, Invoicing, and tools to establish and maintain the customer profile.

Billing and custom care systems convert the transfer of bits and bytes of digital information within the network into the money that will be received by the service provider. To accomplish this, billing and customer care systems provide account activation and tracking, service feature selection, selection of billing rates for specific calls, invoice creation, payment entry and management of communication with the customer.

Billing Company

Billing and customer care systems are the link between end users and the telecommunications network equipment. Telecommunications service providers manage networks, setup the networks to allow customers to transfer information (provisioning), and bill end users for their use of the system. Customers who need telecommunication services select carriers by evaluating service and equipment costs, reviewing the reliability of the network, and comparing how specific services (features) match their communication needs. Because most network operations have access to systems with the same technology, the billing and customer care system is one of the key methods used to differentiate one service provider from another.

This figure shows an overview of a billing and customer care system that is used for communication services. This diagram shows the key steps for billing systems. First, the network records events that contain usage information (for example, connection time) that is related to a specific call. Next, these events are combined and reformatted into a single call detail record (CDR). Because these events only contain network usage information, the identity of the user must be matched (guided) to the call detail record and the charging rate for the call must be determined. After the total charge for the call is calculated using the charging rate, the billing record is updated and is sent to a bill pool (list of ready-to-bill call records). Periodically, a bill is produced for the customer and as payments are received, they are recorded (posted) to the customer's account.

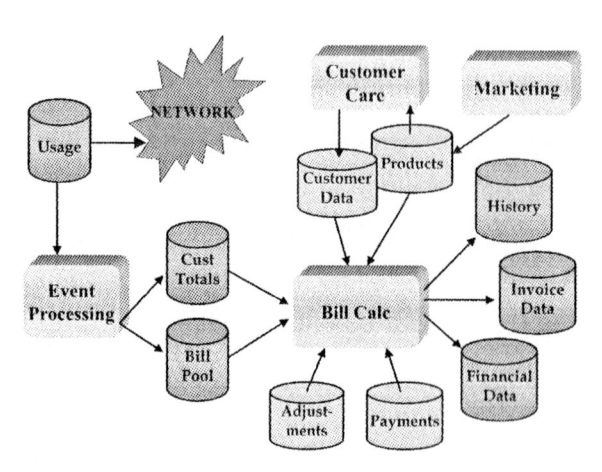

Billing and Customer Care (BCC) System

Billing Company-A company that bills customers for services such as collect calls or long distance changes. The billing company may or may not be the same as the company that provides the service.

Billing Correlation Identifier (BCI)-A code (16-octets) that is used to identify a particular call that is made within the PacketCable system.

Billing Cycle-To distribute the bill processing requirements, billing systems divide customers into groups that allows their bills to be process in specific cycles (or "billing cycles.") The billing cycles are different for groups of customers. This allows the billing system to only batch a portion of the billing records each time. These billing records must be forwarded for delivery (to a bill printer or for electronic distribution).

This diagram shows how a billing system can divide the invoicing process into billing cycles. In this example, each month is divided into 4 groups of billing cycles. At the end of each billing cycle, the billing records for each customer in a particular billing group are gathered, processed, and converted into invoices.

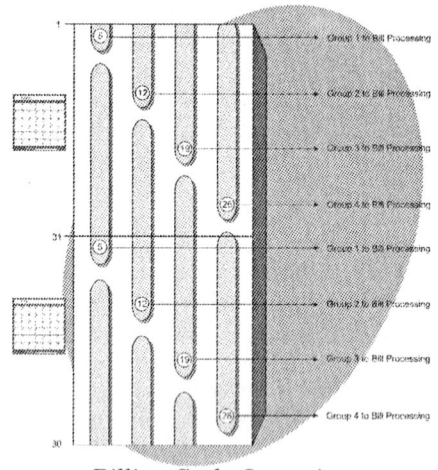

Billing Cycle Operation

Billing Data-All billing data collected during a telephone call. Billing data usually includes call time, originating port number, calling telephone number, call class, dialed number, connection indicator, home/roam indicator (wireless), answer time, disconnect time, timeout, and other billing related information. Billing data is usually transferred on an AMA tape or transferred electronically.

Billing Event-A measurable condition in a network that represents the usage of a network resource by a customer. Several billing events usually occur for each communications session (e.g. telephone call). Billing events are often supplied to a mediation device that combines the billing events into a single call detail record (CDR).

Billing Increment-The smallest amount of time or resource that is can be billed or charged. Billing increments include units of time (e.g. 1 minute or 6 second increments), number of messages, thousand bytes of data (kBytes), or other amount that can be calculated into a charge or criteria for a billing system.

Billing Interconnection Percentage (BIP)-The percentage of interconnection charge that is calculated based on the percentage of a communication route or amount of resources used when there are multiple carriers that are providing a service.

Billing Media Converter (BMC)-A device that transfers automatic message accounting (AMA) data from end offices to regional accounting offices using magnetic tape.

Billing Name and Address-The name and address provided to carrier (such as a local exchange company) by each of its customers to which the company directs bills for its services.

Billing Service Bureau-A billing service bureau is a company that provides billing services to other companies. The services provided can range from billing consolidation to complete billing operations that include gathering billing records, processing invoices, mailing or issuing the invoices, and posting payments.

Billing System-A system (usually a combination of software and hardware) that receives call detail and service usage information, grouping this information for specific accounts or customers, produces invoices, creating reports for management, and recording (posting) payments made to customer accounts.

This figure shows a standard billing process. In this diagram, the customer calls customer care or works with an activation agent to establish a new wireless account. The agent (customer care) enters the customer's service preferences into the system, checks for credit worthiness, and provides the customer with a phone number so that the customer may make and receive calls through the telephone network. As the customer makes calls, the connections made by the network (such as switches) create records of their activities. These records include the identification of the customer and other relevant information that are passed onto the billing system.

The billing system also receives records from other carriers (such as a long distance service provider, or a roaming partner). The billing system now guides and updates these call detail records (CDRs) to their correct customer and rating information. As information about the customer is discovered (e.g. rate plan), the updated billing records are placed in a billing pool so that they may be combined into a single invoice that is sent to the customer. The customer then sends his payment to the telecom service provider. Payments are recorded in the billing system. History files are then updated for the use of customer service representatives (CSRs) and auditing managers.

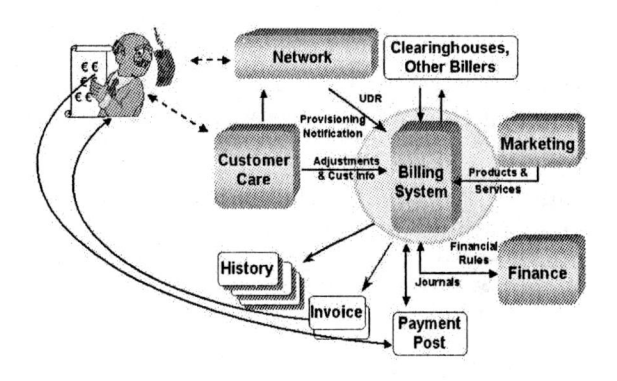

Standard Billing Process Operation

Billing Tape-Monthly call detail records which are generated by the carrier. Resellers often use the billing tape to generate their own bills. (see Call Report).

Billing Telephone Number (BTN)-A number recorded by the switch on a Call Detail Record (CDR) identifying the party to be billed for the call. This number is not always necessarily a telephone number, it could be a Calling Card Number or any other number used to identify the party responsible for payment.

Binary-A numbering system based upon the powers of 2 used to represent data in digital computer systems. Binary data consists of a positional sequence of 1s and 0s. Each position is used to represent a specific power of 2. For example, the decimal number 19 is written in binary as 10011. This corresponds to (2 to the 4th power) + (2 to the 1st) + 2 to the 0th) which is 16 + 2 + 1 = 19.

Binary Digit-Bit

Binary Frequency Shift Keying (BFSK)

Binary Frequency Shift Keying (BFSK)-A digital modulation process where each binary digit is represented a different (unique) frequency.

Binary Message-A string of binary digits that are defined using a message code table to determine the meaning of the message. The use of binary messages results in many different messages being defined in a limited number of data bits.

Binary Phase-Shift Keying (BPSK)-A method of transmitting binary bits in a form of phase shift, without substantially changing the frequency of the carrier waveform. The phase of a carrier is the relative time of the peaks and valleys of the sine wave relative to the time of an unmodulated "clock" sine wave of the same frequency. BPSK uses only two phase angles, corresponding to a phase shift of zero or a half cycle (that is, zero or 180 degrees of angle). BPSK and FM (FSK) are shown in the diagram. If you look closely, you can see the changes in the waveforms that represent the switch from a one to a zero value.

BIND-Berkeley Internet Name Domain

Binding-The relating of an identifier (such as a telephone number) to another identifier or resource (such as an IP address or service).

BIOS-Basic Input/Output System

BIOS Enumerator-Basic Input/Output System Enumerator

BIP-Billing Interconnection Percentage

BIP-Bit Interleaved Parity

BIP-Borderline Interconnection Percent

BIP-N-Bit Interleaved Parity-N

Bipolar with 8 Zero Substitution (B8ZS)-A technique specified in the ITU-T G.703 recommendation, B8ZS is used in some North American DS-1 (T-1) links to permit transmission of unrestricted binary data strings, including consecutive zero binary values, and thus provide clear channel digital transmission. B8ZS substitutes one of two pre-defined special bi-polar violation (BPV) pulse sequences for strings of eight consecutive zero voltage values, with the intentional bipolar violation codes being inserted in bit positions 4 and 7 of the eight bit group. The bipolar violation codes are of opposite polarity, thus producing a necessary average voltage of zero over the entire eight bit group. The specific bipolar violation code of the two is chosen depending on the polarity of the last preceding pulse. B8ZS, which generally is used on newer DS-1 (T-1) and ISDN PRI circuits, offers clear channel communications of 64 kb/s on each channel. AMI (Alternate Mark Inversion) the older technique, suffers from a loss of timing recovery when 8 consecutive zeros are transmitted. The 2.048 Mb/s E-1 primary rate multiplexing standards utilize a similar (but different in detail) pulse substitution method, HDB3 (High Density Bipolar 3) line coding, which also is specified in G.703.

Birds of Feathers (BOF)-A group of interested people and companies who form a temporary BOF group to assess the interest and need for a new Internet protocol specification.

Biscuit-A term sometimes used for external wall mounted RJ-45/RJ-48 jack.

BISYNC-Binary Synchronous Protocol

BISYNC-BISYNChronous Transmission

Bit (b)-In digital signals, a bit is a single unit of data. Generally, a bit has a value of either one or zero corresponding to on or off for an optical or electrical signal. Bits are typically clustered into groups of eight, called bytes or octads. Different coding schemes, such as ASCII, have been developed to assign meaning to the bytes.

Bit Duration-The time it takes to transmit a single encoded bit on the physical transmission medium.

Bit Error-A bit error occurs when one or more encoded bits received by the receiver were corrupted at some point during transmission.

Bit Error-Random

Bit Error Rate (BER)-BER is calculated by dividing the number of bits received in error by the total number of bits transmitted. It is generally used to denote the quality of a digital transmission channel.

Bit Error Rate (BER) Threshold-The maximum rate of bit errors (usually in percentage form) for which the number of bits received in error exceeds the error correction capability of the digital communication system.

Bit Interleaved Parity (BIP)-A simple parity check mechanism. SDH implements two BIP parity checks, BIP-8 and BIP-24. BIP-8 is used in the RSOH, for regenerator section error control, and in the VC overheads, of any level. BIP-24 is used in the MSOH, for regenerator section error control.

Bit Oriented-Term used to denote that fields consisting of one or more bits are used to convey control information within the communications protocol.

BIT Rate-A measurement of the transfer rate of digital signals through a channel. The bit rate is the number of bits transmitted in a specified amount of time. Bit rate is usually expressed as bits per second (bps).

Bit Slice Processor-A microprocessor that has its control circuits and arithmetic and logic units on separate integrated circuits. For example, a 16-bit microprocessor function might be sliced into four 4-bit functions.

Bit Slippage-A phenomenon that occurs in a parallel digital data bus when one or more bits of the bus become mistimed in relation to the rest. The result is that when the data on the parallel bus is next latched, erroneous data is generated. The most common cause for bit slippage is differential cable length for one or more bits of the parallel bus.

Bit Stream-(1-data flow) The sequence of encoded bits that are flowing over a digital communications channel. (2-humor) A place where engineers go to find bitty fish that byte in channels. The engineers often throw back the bitty ones.

Bit Stream Transmission-The transmission of characters at fixed time intervals without stop and start elements. The bits that make up the characters follow each other in sequence without interruption.

Bit String-A linear sequence of bits.

BIT Stuffing-In digital multiplexing, the insertion of additional bits in a Plesiochronous data stream to correct for frequency differences between the clock of unsynchronized inputs. It is a method for changing the rate of a digital signal in a controlled manner so that it can match a rate different from its own inherent rate, usually without loss of information. Bit stuffing is also called pulse stuffing.

BITBLT-Bit Block Transfer

BITE-Built-In Test Equipment

BITNET-Because It's Time Network

BITS-Building Integrated Timing Supply

Bits Per Inch (BPI)-The number of bits that a magnetic tape or computer tape cartridge can store per linear inch of length.

Bits Per Second (BPS)-The common measurement for data transmission that indicates the number of bits that can be transferred to or from a communications device in one second.

Bitstream-The stream of compressed data that is the input to a decoder, especially for compressed audio or video.

BIU-Bus Interface Unit

BLA-Bridged Line Appearance

BLAM-Binary Logarithmic Access Method

Blank And Burst Signaling-A form of signaling between telecommunications devices where control data temporarily mutes voice or user data and replaces it with a control message. This is also known as in-band signaling.

Blast-Blocked Asynchronous Transmission

BLEC-Broadband Local Exchange Carrier

BLEC-Building Local Exchange Carrier

BLER-Block Error Ratio

BLERT-Block Error Rate Test

BLES-Broadband Loop Emulation Service

BLF-Busy Lamp Field

Blind Transfer (BT)-The process of transferring a call to another extension or phone number without telling the person who's calling that they are being transferred. Blind transfers are sometimes called cold transfer or unsupervised transfer.

BLOB-Binary Large Objects

Block-(1-data) An information unit or group of data bits that usually include a header, information element, and an ending error-checking code. (2-image) In encoding of an image or one frame of video, the image is divided into blocks of pixels like a checkerboard, for example each block is a set of 8 by 8 pixels. Each block is then processed separately.

Block Call-A feature that allows a telephone user to block incoming calls from specific telephone numbers.

Block Check Character-A character added at the end of a message or transmission block to facilitate error detection. Generally block check characters are derived via an algorithm that is applied to the block of information being transmitted.

Block Code-A series of bits or a number that is appended to a group of bits or batch of information that allows for the detecting and/or correcting of information that has been transmitted. Block codes use mathematical formulas that perform an operation on the data that will be transmitted. This produces a resulting number that is related to the transmitted data. Depending on how complex the mathematical formula is and how many bits the result may be, the bock code can be used to detect and correct one or more bits of information.

Block Convertor-A device or assembly that shifts a band of frequencies to another band of frequencies.

This diagram shows that a frequency block converter shifts an entire band or group of frequency channels to a new frequency band. This diagram shows that the incoming frequency band A is shifted up to a new frequency band B. All the channels that are received from frequency band A are shifted up by the same amount.

Block Diagram

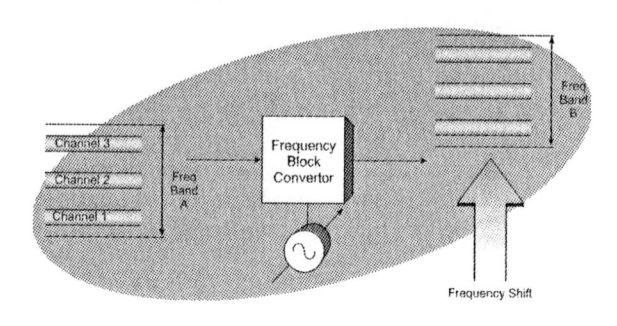

Block Convertor Operation

Block Diagram-A high level abstraction of the various functions or components of a network, system or circuit depicted using simple geometric figures. Lines and arrows between the geometric figures denote an interaction of some type between the connected blocks. Block diagrams are typically used to simplify the understanding of a complex system.

Block Error Probability-The ratio of incorrect blocks to the total number of successful block deliveries during a measurement period.

Blockage-The temporary lack of access because of high traffic in a switching system or in a subscriber line concentrator.

Blocked Attempt-An attempted call that cannot be further advanced toward its destination because of an equipment shortage or failure in a network.

Blockiness-In video and image compression, a common artifact of low-quality compression when too few bits are used. Blockiness is an obviously perceptible contrast of color at the boundaries of the encoding blocks with a codec like JPEG or MPEG video. Blockiness can be partially concealed by post processing the image to blur the block boundaries.

Blocking Capacitor-A capacitor included in a circuit to stop the passage of direct current.

Blocking Factor-The likelihood of no free channels in a cellular telephone system being available when dialed.

Blocking IPR-An intellectual property right (IPR), patent, copyright or other right proprietary to an individual, group or company, which precludes someone else from making, using or selling that invention.

Blocking Objective-The design objective for the maximum ratio of allowed unsuccessful access compared to total access attempts (average blocking ratio) for which a group of switches, servers or networks is engineered or administered.

Blocking Probability-The percentage of calls that cannot be completed within a one hour period due to capacity limitations. For example, if within one hour 100 users attempt accessing the system, and 10 attempts fail, the blocking probability is ten percent.

Blocking Ratio-For a group of servers, the ratio of unsuccessful access attempts compared to total access attempts within a specified time interval.

Blown Fiber-A process of using compressed air is used to install a fiber cable in a tube or conduit. The compressed air pushes a float at the front of the fiber cable so it travels through the tube or conduit.

BLSR-Bidirectional Line Switched Ring

Blue Signal-In telecommunications applications, an alarm signal composed of ones and zeros (101010, etc.) substituted for loss of a valid input signal to indicate loss of the signal to downstream equipment.

Bluetooth-Bluetooth is a wireless personal area network (WPAN) communication system standard that allows for wireless data connections to be dynamically added and removed between nearby devices. Each Bluetooth wireless network can contain up to 8 active devices and is called a Piconet. Piconets can be linked to form Scatternets. Information about Bluetooth technology and wireless data can be found at www.Bluetooth.com.

The system control for Bluetooth requires one device to operate as the coordinating device (a master) and all the other devices are slaves. This is very similar to the structure of a universal serial bus (USB) system that is commonly used in personal computers and devices such as digital cameras. However, unlike USB connections, most Bluetooth devices can operate as either a master (coordinator) or slave and Bluetooth devices can reverse their roles if necessary.

This diagram shows the basic radio transmission process used in the Bluetooth system. This diagram shows that the frequency range of the Bluetooth system ranges from 2.4 GHz to 2.483 GHz and that the basic radio transmission packet time slot is 625 usec. It also shows that one device in a Bluetooth piconet is the master (controller) and other devices are slaves to the master. Each radio packet contains a local area piconet ID, device ID, and logical channel identifier. This diagram also shows that the hopping sequence is normally determined by the master's Bluetooth device address. However, when a device is not under control of the master, it does not know what hopping sequence to use to it listens for inquiries on a standard hopping sequence and then listens for pages using its own Bluetooth device address.

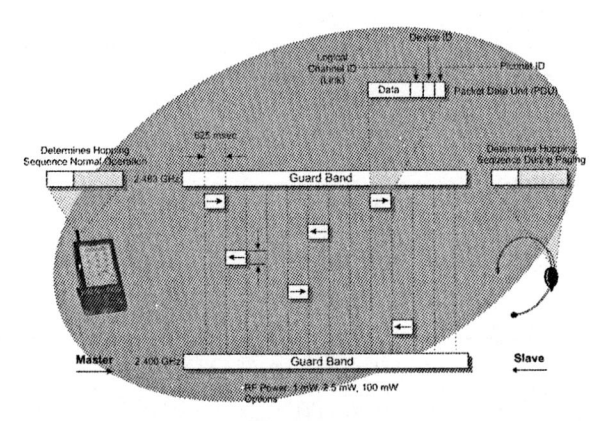

Bluetooth Radio

This diagram shows how a Bluetooth system can connect multiple devices on a single radio link. This diagram shows that a laptop computer has requested a data file from a desktop computer. When this laptop computer first requests the data file, it accessed the Bluetooth radio through a serial data communication port. The serial data port was adapted to Bluetooth protocol (RFComm) and a physical radio channel was requested from the local device (master) to the remote computing device (slave). The link manager of the master Bluetooth device requests a physical link to the remote Bluetooth radio. After the physical link is created, the logical link controller sends a message to the remote device requesting a logical channel be connected between the laptop computer and the remote computer. The logical link continually transmits data between the devices. In this diagram, the user then requests that a CD Player send digital audio to a headset at the remote computer. Because a physical channel is already established, the logical link controller only needs to setup a 2nd logical link between the master Bluetooth device and the remote Bluetooth device. Now data from the CD ROM will be routed over the same physical link between the two Bluetooth devices.

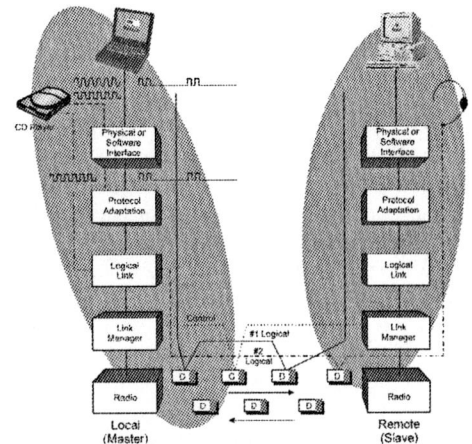

Bluetooth System Operation

This photograph shows a wireless headset that uses the Bluetooth wireless standard communications protocol. This device conforms to the Bluetooth headset profile requirements that allows it to be used with devices from any manufacturer that have tested and certify that they conform to the Bluetooth technology and headset profile.

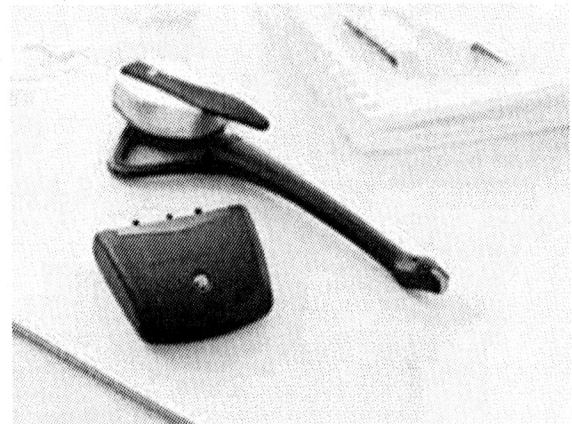

Bluetooth Device
Source: Sony Ericsson

Bluetooth Host-A computing device such as a mobile telephone or data access point that communicates and provides services with other Bluetooth communication devices.

Bluetooth Packet Data Unit-Bluetooth packet data units (PDUs) are transmitted between master and slave devices within a Bluetooth Piconet. Each PDU contains the address code of the Piconet, device identifier, logical channel identifiers, and a

Bluetooth Qualification Body (BQB)

payload of data. The length of the PDU can vary to fit within 1, 3 or 5 time slot period (625 usec per time slot). Control message PDUs (e.g. link control) always fit within 1 time slot.

-Bluetooth Qualification Body (BQB)-A specific PERSON who is authorized by the Bluetooth qualification review board (BQRB) to be responsible for the checking of declarations and documents against requirements, reviewing product test reports, and listing conforming products in the official database of Bluetooth qualified products.

Bluetooth Radio Packet-Bluetooth radio systems use radio 1 MHz wide radio channels that hop over 79 channels. A Bluetooth radio transceiver transmits and receives modulated electrical signals wirelessly between peer Bluetooth devices.

This diagram shows the basic Bluetooth radio packet structure. The first part of the Bluetooth radio packet is the address code. The address code usually indicates the identification of the Piconet. This is followed by the header that defines the device number in the Piconet for this packets destination. Next, a link channel (L_CH) flag is used to identify if this packet is a link control packet or if it contains data for the designated device. The data payload will usually contain a logical channel identifier that allows the Bluetooth devices to route packets to the specific application port for which the data is designated for. The payload may also contain a protocol service multiplexer (PSM) flag that indicates which protocol is to be used (e.g. serial port or telephony control) to coordinate the logical channel.

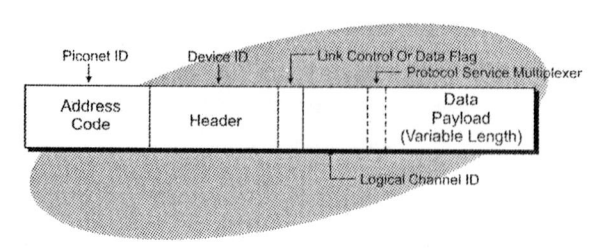

Bluetooth Packet Structure

BMC-Broadcast/Multicast Control Protocol
BMC-Billing Media Converter
BMC-Business Management Computer

BMP-Basic Multilingual Plane
BMTI-Block Mode Terminal Interface
BNAP-Broadband Network Access Point
BNC-British Naval Connector
BNR-Bell-Northern Research
Board-To-Board Test-A test performed before a cutover from an existing switching system to a replacement switching system. The test ensures the proper connection and assignment of lines on the new system by physical comparison to the lines on the existing switching system.
BOB-Breakout Box
BOC-Bell Operating Company
BOC-Business Office Code
BoD-Bandwidth On Demand
BOF-Birds of Feathers
BOF-Business Operations Framework
BOIC-Barring of Outgoing International Calls
BOM-Bill Of Materials
BOM-Beginning of Message
BONDING-Bandwidth On Demand Interoperability Group
Bong-A unique tone that telephone carriers and to an audio signal while a call is in progress to alert the user that another action on the users part is required (such as to dial an access code number for a calling card).
BooP-Bootstrap Protocol
Boot Loader-A small program, typically in ROM, that executes after the central processing unit is reset or powered on that takes care of loading the operating system and necessary system drivers from disk.
Boot PROM-Boot-Programmable Read-Only Memory
BOOTP-Boot Protocol
BOOTP-Bootstrap Protocol
Bootstrap Protocol (BooP)-BooP is a protocol that is used to acquire initializing information from a data communications network that a device is connected to. The use of BooP allows a communication device to obtain a network address and to initially determine where to initially start transmitting packets of data.
BOP-Bit Oriented Protocol
Border-A perimeter of a system or network. Borders can be physical or logical boundaries that can be crossed at locations known as border gateways.
Border Gateway (BG)-A router or data processing device that connects one network (such as the Internet) to another network (such as a private

LAN). Because the border gateway is single point of entry, it filters all data going between the networks.

Border Gateway Multicast Protocol (BGMP)-A protocol that is used to communicate between routers to help determine the selection of routes for multicast packets over multiple network domains.

Border Gateway Protocol (BGP)-A routing protocol used by routers that are located between different networks to evaluate each of the possible routes for the best one before choosing the routing path for a packet it has received. BGP is designed to replace external gateway protocol (EGP).

Border Gateway Protocol Version 4 (BGP4)-An inter-domain routing protocol used that is used to span (link) autonomous systems on the Internet on the Internet. BGP4 is capable of aggregating routes listed within a router's memory and the use of BGP4 results in a reduction of the size of routing tables.

BORSCHT-Battery, Overvoltage, Ringing, Supervision, Coding, Hybrid, Test

BOS-Best Effort Service

BOS-Bill Output Specifications

Bose, Chaudhuri, And Hocquenhem (BCH) Code-A class of cyclic redundancy codes used to provide error detection and correction capabilities to a communication link.

Bot-Robot

Bottleneck-A point in a data communications path or computer processing flow that limits overall throughput or performance.

BP-Block Pair

BP 14-Body Part 14

BPAD-Bisynchronous Packet Assembler Dissasembler

BPDU-Bridge Protocol Data Unit

BPI-Bytes Per Inch

BPI-Bits Per Inch

BPI+-Baseline Privacy Interface Plus

BPKM-Baseline Privacy Key Mangement

BPP-Brokered Private Peering

BPP-Bits Per Pixel

BPS-Bits Per Second

BPSK-Binary Phase-Shift Keying

BPV-Bi-Polar Violation

BQB-Bluetooth Qualification Body

BQM-Business Quality Messaging

BQRB-Bluetooth Qualification Review Board

BRADS-Bell Rating Administrative Data System

Bragg Grating-A parallel array of reflecting planes with a precise, periodic spacing. Light with wavelengths approximately twice the separation of the planes will be reflected due to interference while light of other wavelengths will pass through undisturbed for all practical purposes. This phenomenon is widely exploited in optical network components, such as fiber Bragg gratings.

Branch Splice-A cable splicing method whereby cable feeder or distribution counts are separated within a splice closure. The total incoming cable count splits into 2 or more separate paths for final distribution via drop-off at various fixed or ready access terminals along the way. For example cable 10 pairs 1-600 separates into cable 101 pairs 1-300 and cable 101 pairs 301-600 in a Y type of arrangement.

Brand Equity -Brand equity is the term used to describe the extra value that a company or product can command in the marketplace because of the branding activity associated with it.

BRB-Be Right Back

BRCS-Business and Residence Customer Services

Breadboard-A board used for prototype assembly. The term came from the use of low cost wooden breadboards that allowed inventors to easily attach sockets and other electrical or electronic components so they could be interconnected by temporary wires. A breadboard is often used to describe circuits or systems that are assembled to test if a design or technology will work.

Break Interval-In a dial telephone, that portion of a pulse cycle during which the dial contacts are open.

Break Permanent Connection-A facility modification that requires a dedicated permanent connection (working or idle) to be broken for a service order or related line and station transfer. It includes instances when a dedicated feeder pair (working or idle) is disconnected at a serving area interface, or a dedicated distribution pair (working or idle) is broken at a serving terminal beyond the interface. Break permanent connection includes the subcategories break access point, break dedicated address, and break control-point pair.

Break-In-Busy Override

Breakage-A final balance on a prepaid service that is never used.

Breakdown Test-A troubleshooting method of using high DC (approximately 500 volts) voltage to apply sealing current to a cable pair. The sealing current is applied at the central office MDF and sent over the two conductors to locate and either weld a high-resistance open closed or to make a leaky trouble fault solid for accurate testing, isolation and repair.

Breakpoint-A point, measured as a distance from a central office, beyond which service is not provided without a mileage charge. A technology breakpoint is one beyond which service cannot be providing by an existing transmission system. An economic breakpoint is one at which it is more costly to provide service on one technology than on another.

BRI-Basic Rate ISDN

BRI-Busy-Reserved-Idle Field

BRI-Basic Rate Interface

Brickwall Filter-An ideal type of security filter that is used in a data communications network. A brick wall filter has rapid response with no loss to passband data and infinite loss stopband.

Bridge-A bridge is a data communication device that connects two or more segments of data communication networks by forwarding packets between them. Bridges extend the reach of the LAN from one segment to another.

Bridge devices operate at layer 2 of the OSI reference model, used to connect two or more LAN segments. Bridges provide more intelligence and fault isolation than repeater, which operate at Layer 1. However, unlike routers, which operate at Layer 3, bridges do not provide broadcast domains.

Bridges operate by inspecting each packets MAC header and forwarding the packet based only on this information. Each bridge keeps a table which enables it to determine the egress port to which each packet should be forwarded.

In general, there are two classes of bridges: Transparent bridges are commonly found in Ethernet-like networks and make use of only MAC addresses. Source Route bridges are commonly found in Token-Ring networks, and utilize a Routing Information Field in the MAC header to forward packets.

The terms Routing Information Field and Source Routing are slightly misleading. While these terms have the word "route" in them, these are purely Layer 2 protocols and should not be confused with Layer 3 routing protocols such as RIP or OSPF.

This figure shows the basic operation of a bridge that is connecting 3 segments of a LAN network. Segment 1 of the LAN has addresses 101 through 103, segment 2 of the LAN has addresses 201 through 203, and segment 3 of the LAN has addresses 301 through 303. The table contained in the bridge indicates the address ranges that should be forwarded to specific ports. This diagram shows a packet that is received from LAN segment 3 that contains the address 102 will be forwarded to LAN segment 1. When a data packet from computer 303 contains the address 301, the bridge will receive the packet but the bridge will ignore (not forward) the packet.

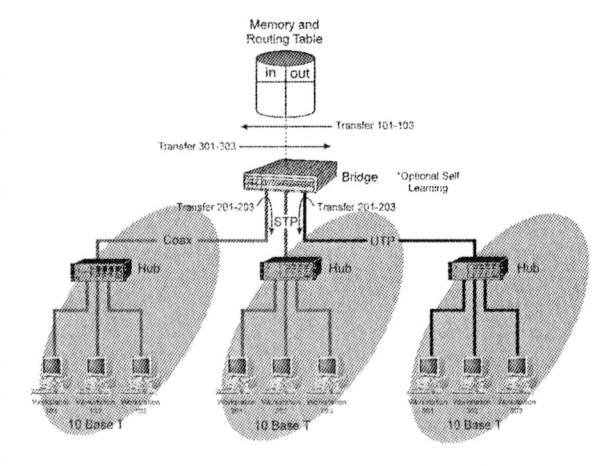

Data Bridge Operation

Bridge Identifier-In the Spanning Tree Protocol, a catenation of a Bridge Priority value and the bridge's MAC address. The bridge with the numerically-lowest bridge identifier value will become the Root Bridge in the spanning tree.

Bridge Lifter-A device that isolates idle pairs from active pairs in order to reduce transmission loss when two or more subscriber pairs are combined into multiparty telephone service. The lifter usually is installed in a central office, but may be field-mounted near a customer's premises. One type of bridge lifter uses an inductor to approximately cancel out the capacitive shunt or parallel impedance of the idle pair over a range of audio frequencies. Another type of bridge lifter uses a diode to prevent current flow into the idle pair when the appropriate voltage polarity is applied. See Bridged Tap.

Bridge Link (B-Link)-A "B" (bridge) link connects a signal transfer point (STP) to another STP in an SS7 network. Typically, a quad of "B" links interconnect peer (or primary) STPs (e.g., the STPs from one network to the STPs of another network). The distinction between a "B" link and a "D" link is rather arbitrary. For this reason, such links may be referred to as "B/D" links.

Bridge Number-A locally-assigned, ring unique number identifying each bridge in a source routed catenet.

Bridge Port-A network interface on a bridge.

Bridge Protocol Data Unit (BPDU)-A type of packet data that is used in an ATM system to exchange management and control information between other systems (bridge).

Bridge Tap-A bridge tap is an extension to a communication line that is used to attach two (or more) end points (user access lines) to a central office. Bridge taps provide connection options to the telephone company on connecting different communication lines to a central office without having to install new pairs of wires each time a customer requests a new telephone line.

This figure shows how reflections from bridge line tap can cause distortion. This signal shows that some of the energy from the bridge tap is reflected back to the communications line. This reflected signal is a delayed representation of the original signal. Typically, bridge taps must be removed from communications lines that use DSL technology.

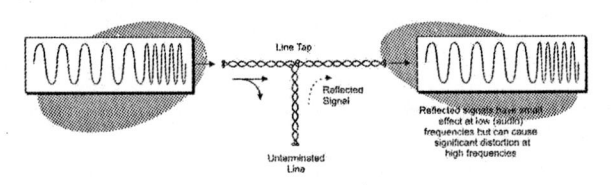

Bridge Tap Reflections

Bridge Transit Delay-The delay between the receipt of a frame on one port and the forwarding of that frame onto another port of a bridge.

Bridge-Router (Brouter)-When the functionality of a router is combined with a bridge function, it is called a BRouter. The BRouter can route one or more specific protocols, such as TCP/IP, and bridge other protocols. This allows a BRouter to operate at either the data-link layer or the network layer of the OSI Reference Model.

Bridged LAN-A network that is formed my connecting two or more LAN segments with a bridge. See bridge.

Bridged Line Appearance (BLA)-A communication line that is designed to appear as a bridged communication line.

Bridged Tap-A communication line (tap) that is connected to another communication line between a receiver and a transmitter. The bridged tap appears as a stub or side branch on the main line. Bridged taps allow communication lines to have other possible termination points (possibly to allow connection to different customers in the future). At normal audio signal frequencies (300Hz to 3300Hz), a bridged tap does not significantly affect the electrical transmission characteristics. However, at high frequencies (such as those used in DSL technologies), bridged taps can distort the transmission of electrical signals (commonly called a shunting affect).

Bridger-Bridger Amplifier

Bridging-The shunting or paralleling of one circuit with another.

Bridging Connection-A parallel connection that draws some of the signal energy from a circuit, often with imperceptible effect on the normal operation of the circuit.

Bridging Loss-The loss at a given frequency resulting from connecting an impedance across a transmission line.

Bridging Repeater-A specially designed central-office repeater used for patching pulse code modulation (PCM) lines so that service will not be interrupted when a patch is removed.

British Naval Connector (BNC)-A twist-locking bayonet connector that is used to connect coaxial cable. This connector was originally developed by the British Navy.

This diagram shows several types of British Naval Connectors (BNC). A BNC connector is a twist-locking bayonet connector that is used to connect coaxial cable. This connector was originally developed by the British Navy. BNC connectors are common connectors in a network environment because of their versatility and universal acceptance. Most network peripherals will have a port that allows a cable fitted with a BNC connector to be attached.

Broadband

British Naval Connector (BNC)

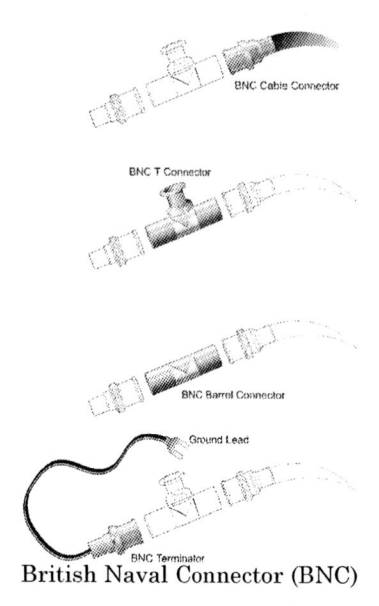

Broadband-(1-data transfer) A term that is commonly associated with high-speed data transfer connections. When applied to consumer access networks, broadband often refers to data transmission rates of 1 Mbps or higher. When referred to LANs, MANs, or WANs, broadband data transmission rates are 45 Mbps or higher. (2-radio bandwidth) A frequency bandwidth that is much larger than the required bandwidth to transfer the information signal. For example, using a 1 MHz wide radio channel to transmit a 4 kHz limited audio signal.

Broadband Communications-Voice, data, and/or video communications at rates greater than wideband communications rates (1.544 Mbps).

Broadband Integrated Services Digital Network (B-ISDN)-Usually refers to the portions of a digital network operating at data transfer rates in excess of 1.544 or 2.048 Mbps. The B-ISDN network often uses ATM to enables transport and switching of voice, data, image, and video over the same network equipment.

Broadband Integrated Services Digital Network User Part (B-ISUP)-An SS7 protocol that defines the signaling messages that are used to control ATM broadband connections and services.

Broadband Local Exchange Carrier (BLEC)-A service carrier (provider) that offers broadband services locally. Broadband services is usually defined as the ability to transmit a large amount of information, including voice, data, and video, that has a combined data transmission rate that exceeds 1 Mbps.

Broadband Network-A network that is capable of transmitting a large amount of information, including voice, data, and video that have a combined data transmission rate that exceeds 1 Mbps. Sometimes called wideband transmission, it is based on the same technology used by cable television.

Broadband Signal-A term that is used to describe high-speed data communication signal. For consumer data communication, broadband signals have a data transmission rate above 1 Mbps are usually considered Broadband signals. For telecommunication or data networks, data transmission rates above 45 Mbps (OC1) are considered broadband.

Broadband Telephony-The use of a broadband data connection (such as high-speed DSL or cable modem Internet connection) to make telephone calls.

Broadcast-Transmission of an information signal to a specified geographic area or network. This allows the same information to be received by all customers in that geographic area that can successfully receive (demodulate) and decode the information.

This diagram shows two broadcast examples: radio broadcast and network broadcast. Part (a) shows a radio broadcast tower that is sending an audio broadcast to all radios that are within its radio signal coverage area. Part (b) shows a network broadcast system that sends a data message that is coded to indicate the message is a broadcast message. This message contains an address that indicates it is a broadcast message. When routers or other data distribution devices receive this message, each distribution device forwards the data broadcast message to the other network parts for which it is connected to. All communication devices that are connected to the network can receive the broadcast message.

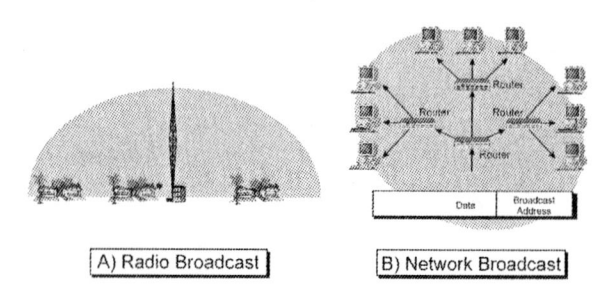

Broadcast Communication Operation

Broadcast Address-A well-known multicast address signifying the set of all stations.

Broadcast and Unknown Server (BUS)-A process used for LAN Emulation (LANE) in an asynchronous transfer mode (ATM) system to handle broadcast and multicast data. The BUS communicates with the LAN emulation server (LES) to register and resolve addresses between a 48-bit Ethernet address and ATM addresses. The BUS system labels each device transmission with both addresses.

Broadcast Channel (BCCH)-The broadcast channel contains several logical channels that are multiplexed onto one communications channel that is continuously broadcast from a cell site. The broadcast channel typically provides the mobile phone with system information, lists of neighboring radio channels and other system configuration information.

Broadcast Fax-A process or service that broadcasts a fax message to a list of pre-defined recipients.

Broadcast Messaging-A service that broadcasts the same message to a group of recipients that are connected to a network or who can receive the broadcast message via a communications medium.

This diagram illustrates how broadcast SMS messages are delivered to users in a cellular (wireless) system. Like the point-to-point messaging service, the message first goes to a message center (step 1) where the message center determines that this message is designated for all communication devices. The message is stored with a broadcast message code identifier (step 2). The cellular system then transmits the broadcast message on designated message channels (step 3). These broadcast channels may be part of a control channel in every cell site where the broadcast message is designated to be received or may be on other types of channels (e.g. voice/traffic channels). Unlike point-to-point and point-to-multipoint messages, communication devices do not acknowledge receipt of broadcast messages. If a communication device is off or is not tuned to the message channel, it misses the broadcast message. To address this limitation, messages may be broadcast several times, and mobile telephones that have already received message may ignore repeats (steps 4-5).

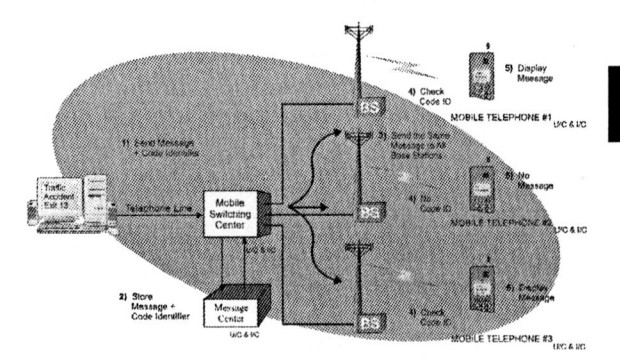

Broadcast Messaging Operation

Broadcast Radio-Broadcast radio technology allows a single transmitted signal to be heard by many radio receivers. The first radio broadcast transmission systems used amplitude modulated (AM) radio transmission to provide audio service. Radio broadcast technologies have evolved to allow for enhanced audio (stereo) and data signals (such as radio broadcast data system).

This figure shows how a broadcast radio system can operate. In this diagram, one high power transmitter is located at a high point above the city. The transmitter may be broadcasting at over 50,000 watts to allow a radio signal to reach many homes and businesses within a 50-mile radius. Typically, any receiver that is capable of tuning into the broadcast signal can receive and listen to it. However, some systems (such as movie channels) are scrambled so only receivers with a signal decoding capability can listen or view the signal.

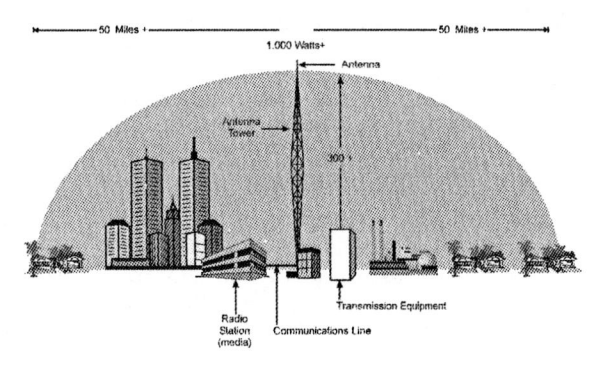

Radio Broadcast System

Broadcast Television-The technology that is used for television broadcast was developed in the 1940s. The success of the television marketplace is due to standardized, reliable, and relatively inexpensive television receivers and a large selection of media sources. The first television transmission standards used analog radio transmission to provide black and white video service. These initial television technologies have evolved to allow for both black and white and color television signals, along with advanced services such as stereo audio and closed caption text. This was a very important evolution as new television services (such as color television) can be on the same radio channel as black and white television services.

Broadcast Transmissions-Sending the same signal to many different places, like television broadcasting station. Broadcast transmission can be over optical fibers if the same signal is sent to many subscribers. For LAN's broadcast messages are usually associated with Ethernet that uses CSMA/CD at its MAC layer.

Broadcast/Multicast Control Protocol (BMC)- The data protocol that is used in the universal mobile telephone service (UMTS) system for cell and system broadcast messaging.

Broadcasting-Broadcasting is a process that sends voice, data, or video signals simultaneously to group of people or companies in a specific geographic area or who are connected to the broadcast network system (e.g. satellite or cable television system). Broadcasting is typically associated with radio or television radio transmission systems that send the same radio signal to many receivers in a geographic area. Broadcasting can also be applied to distribution or point to point networks where all users that are connected to the network can receive the same information signal.

Brouter-Bridge-Router

Brownout-A reduction in the servicing of customers as a result of the demand for service exceeding the service processing capability of the service provider's equipment or staff. Brownout usually occurs during a peak period. Because service access attempts by customers increase during a brownout period (customers repeatedly attempt to get service), service providers may discontinue services to groups of customers during brownout.

Browser-A software program or module (called a client) that is used to convert information that is available on the Web portion of the Internet into forms usable by a person (text, graphics and sound). Also called a web browser.

BS-Base Station
BS-Bearer Services
BSA-Basic Serving Arrangement
BSC-Base Station Controller
BSCS-Business Support And Control System
BSD-Berkeley Software Distribution
BSD Unix-Berkeley Software Distribution Unix
BSE-Basic Service Element
BSF-Band Stop Filter
BSGL-Branch Systems General Licence
BSI-British Standards Institution
BSIC-Base Station Identity Code
BSN-Backward Sequence Number
BSPs-Bell System Practices
BSS-Base Station Subsystem
BSS-Business Support System
BSS-802.11 Basic Service Set
BSS-Broadband Switching System
BT-Burst Tolerance
BT-Blind Transfer
BTA-Broadband Telecommunications Architecture
BTA-Business Technology Association
BTA-Basic Trading Area
Btag-Beginning Tag
BTAM-Basic Telecommunications Access Method
BTFD-Blind Transport Format Detection
BTN-Billing Telephone Number
BTN-Billed Telephone Number
BTO-Built To Order
BTS-Base Transceiver Station
BTSC-Broadcast Television Systems Committee
BTSM-Base Transceiver Station Management
BTU-British Thermal Unit
BTV-Business Television
BTW-By The Way

Budget Management Discipline -A process utilized by telephone company (telco) management personnel to get departments and project managers to adhere more tightly to standardized proposal, funding and follow-up procedures for budgeted projects and organizational units.

Buffer-(1-memory) An allocation of memory storage that set aside for temporary storage of data. The use of memory buffers allow data transfer rates to vary so that differences in communication speeds between devices do not interrupt normal processing operations. (2-optical) A layer placed between a fiber and its jacket to provide additional protection to the fiber. The layer is often made of a thermoplastic material.

Bug-(1-problem) A problem that occurs in software or hardware that can be fixed by changes to the software or design changes to the hardware. The term seems to have started from a problem that occurred with an early model computer system when a moth got caught inside the machinery resulting in a bug problem. (2-audio) A microphone or listening device that is concealed for audio surveillance.

Building Industry Consulting Service International (BICSI)-BCSI is the overseeing company for Registered Communications Distribution Designer (RCDD) certification.

Building Local Exchange Carrier (BLEC)-A building owner, real estate owner, or corporation that provides local exchange carrier services to the tenants or customer's of their building(s).

Bulk Basis-Billing system that is based on the billing name and address information for all the local exchange service subscribers of local exchange carrier.

Bulk Billing-A method of billing telephone customers in which the charges for all messages of a given type, chargeable to a particular account for a billing period, are combined and billed as a single amount. Message-unit billing is the most frequent application of the bulk-billing procedure.

Bundespatentgericht-The Federal German Patent Court

Bundling-The process of combining different products and services into a "package" offer, and then offering it to a customer at a separate, combined price.

Bundling Services-The combining of different services into one service offering so the customer can communicate with one company for several different services. An example of bundling is the combination of cellular, PCS, local and long distance services as one service package.

Burst-(1- general) A short transmission of information (data). (2- GSM) A time slot of information that lasts 577 usec. There are several burst types including normal, synchronization, frequency correction, access and dummy. (3-CDMA) A time slot of information that lasts 1.25 msec.

Burst Collisions-The overlapping of transmission bursts between time slots at a receiver. This may occur from varying amounts of propagation time in a mobile communication time where some mobile radios are close and others are far away. Dynamic time alignment (variable transmission time periods) was created to solve this challenge. Dynamic time alignment allows mobile radios at a distance to start the transmission early so their transmission bursts will be received in the correct time slot period.

Burst Switching-The techniques of switching packets in bursts through a network.

Bursty Data-An attribute of a communications channel where the data comes in unpredictable bursts instead of a continuous stream. Typical voice communications is an example of bursty data because there are long periods of silence followed by periods of speech.

BUS-Broadcast and Unknown Server

Bus (Signal Path)-A central conductor for the primary signal path. The term bus also may refer to a signal path to which a number of inputs may be connected for feed to one or more outputs.

Bus Network-A network where each communication device (typically computers) are connected to a common bus. Each computer that is connected to the bus network uses a transmission control process to sense availability and contention (simultaneous access) on the network.

Bus Segment-A section or piece of bus that is electrically continuous, with no intervening components such as repeaters. Electrical continuity may be maintained using fixtures such as coaxial cable connectors, but no signal boosting equipment (repeaters) are used, the piece of cable on the other side of such equipment is defined as another bus segment.

Bus Topology-In a local area network (LAN), a topology consisting of a single shared medium, typically coaxial cable, to which multiple devices are connected. These devices monitor signals on the medium and selectively copy the data addressed to them. Modern day Ethernet hubs implement a logical bus topology while providing a star-wired network structure. (See also: bus topology, ring topology, tree topology.)

This figure shows a bus network. This diagram shows that data communication devices can hear each other on the common (shared) bus.

Business Group

Bus Network (Shared Medium)

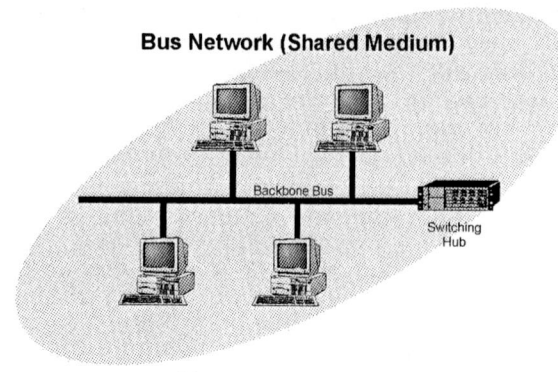

Physical Bus Topology

Business Group-A collection of lines that serve a single business location, share a common number dialing plan, and are assigned other business features. It is commonly called a Centrex group.

Business Services Data Base-An application residing in a service control point (SCP) that provides call-handling instructions to a service switching point (SSP).

Business Support And Control System (BSCS)-The billing system used in the GSM network.

Busy Hour-The hour in a day when the total usage of the network, trunk connection, or the switching system is greater than at any other hour during the day. Telephone systems and networks are typically designed to meet a specific quality level that can be provided during the busy hour (e.g. maximum number of blocked calls).

Busy Hour Minutes (BHM)-The number of minutes during "busy hours" ordered by a customer. Other terms: BHMOT represents originating usage, BHMTT represents terminating usage.

Busy Hour Minutes of Capacity (BHMC)-The maximum amount of access minutes that an interconnecting carrier (such as an IXC) expects to route through end office switch at peak activity.

Busy Override (Break-In)-A feature that allows access to a busy number for an emergency.

Busy Tone-An audible tone or signal that indicates to a caller that a their call cannot be received because a called line is unavailable (in use).

Busy/Idle Flag-A busy/idle flag used in cellular radio transmission is an indicator that is transmitted by the Mobile Data Base Station (MDBS) periodically to indicate whether the reverse channel is currently in the busy state or the idle state.

BVA-Billing Validation Application

Bye Message-A message that is used to inform a person or device that a communication session is ending. The Bye message is defined in session initiation protocol (SIP).

Bypass-The routing of calls or communication sessions around any other networks facilities to avoid toll charges. Bypass is commonly discussed with the ability to bypass local exchange carriers (LEC) to save on interconnection charges.

Bypass LAN-The capability of a station to be electronically or optically isolated from a network while maintaining the integrity of the network ring. This is found in the newer hub-based token ring networks.

Bypass Relays-Relays in a ring network that permits message traffic to travel between two nodes that are not normally adjacent. Usually, such relays are arranged so that any node can be removed from the ring for servicing and the two nodes on either side of the removed node are now connected via the bypass relay.

Byte (B)-A group of bits, typically eight, used to represent data values from 0 to 255. These values may represent alphabetic or control characters or numbers less than 255. A single byte is typically the smallest data value manipulated by a computer. In order to represent larger values, multiple bytes are grouped together into words that are 16, 32, 64 or 128 bits in length.

Byte Multiplexing-A form of time-division multiplexing (TDM) in which the whole of a byte from one subchannel is sent as a unit, interlacing in successive time slots with complete bytes from other subchannels.

Bytes-Unassigned X-Bits

BZT-Bundesamt fur Zulassungen in der Telekommunikation

BZW-Blizzard Warning

C

C-coulomb
C Wire-European term corresponds to North American Sleeve wire.
C&C-Computers and Communications
C- Band-A band of frequencies from 4.0 to 8.0 GHz used for both satellite and terrestrial microwave transmission.
C-DTE-Character Mode Data Terminal Equipment
C-Link-Cross Link
C-Message-The weighting equivalent of a conventional 500-type telephone set. A filter, having this response, is inserted into a test instrument to measure circuit noise. This is a standard for noise on an idle voice channel.
C/A Code-Clear/Acquisition Code
C/A Interference Ratio-Carrier to Adjacent Channel
C/I-Carrier To Interference Signal Ratio
C/I-Cochannel Interference
C/I-Carrier-To-Interference Ratio
C/N-Carrier to Noise Ratio
C/R-Command Response
C2-Command and Control
C3-Command, Control, and Communications
CA-Certification Authority
CA-Certificate Authority
CA-Communications Assistant
CA-Call Agent
Cabinet-In communication systems, an enclosure that is used to hold equipment or electronic assemblies.
Cable Binder- A device that holds multiple cable pairs.
Cable Converter-Cable converters, commonly called a "set top" box are electronic devices that convert an incoming cable television signal into a form that can be displayed on a video device typically a television or computer. The set-top box is typically located in a customer's home to enable the reception and/or interaction with services on the customer's television or computer.
Cable Entrance Facility (CEF)-The steel, wooden, tile, or plastic duct work or pathway that allows for cabling to enter a building.
Cable Head End-The portion of a cable television communication system that receives, formats, and transmits the carrier signals to the distribution system. Because cable television companies have started to offer two-way telecommunication services such as Internet and telephone service, the head end equipment has been expanded to include voice and data gateways.

This figure shows a diagram of a simple head-end system. This diagram shows that the head-end allows the selection of multiple video sources. Some of these video sources are scrambled to prevent unauthorized viewing before being sent to the cable distribution system. The video signals are supplied to video modulators the convert the low frequency video signals into their radio frequency television channel. The output of each modulator is combined and connected to the distribution trunk.

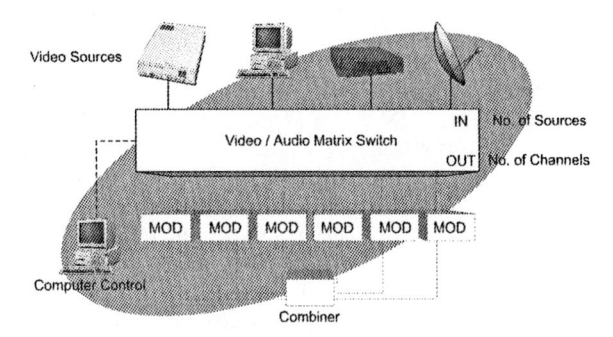

Head End System

Cable Modem (CM)-A communication device that Modulates and demodulates (MoDem) data signals to and from a cable television system. Cable modems select and decode high data-rate signals on the cable television system (CATV) into digital signals that are designated for a specific user.
There are two generations of cable modems; First Generation one-way cable modems transmit high speed data to all the users into a portion of a cable network and return low speed data through telephone lines or via a shared channel on the CATV system. First generation cable modems used asymmetrical data transmission where the data transfer rate in the downstream direction was typically much higher than the data transfer in the upstream

Cable Modem Termination System (CMTS)

direction. The typical gross (system) downstream data rates ranged up to 30 Mbps and gross upstream data rates typically range up to 2 Mbps. Because 500 to 2000 users typically share the gross data transfer rate on a cable system, cable modems also have the requirement to divide the high-speed digital signals into low speed connections for each user. The average data rates for a first generation cable modem user rage up to 720 kbps. Until the late 1990's, most cable modems used first generation technology.

Second generation cable modems offered much data transmission rates in both downstream and upstream directions. Second generation cable television systems use two-way fiber optic cable for the head end and feeder distribution systems. This allows a much higher data transmission rate and many more channels available for each cable modem. As of the year 2000, approximately 35% of the total cable lines in the United States had already been converted to HFC technology.

This figure shows a block diagram of a cable modem. This diagram shows that a cable modem has a tuner to convert an incoming 6 MHz RF channel to a low frequency baseband signal. This signal is demodulated to a digital format, demultiplexed (separated) from other digital channels, and is decompressed to a single data signal. This data signal is connected to a computer typically in Ethernet (e.g. 10 Base T Ethernet) or universal serial bus (USB) data format. Data that is sent to the modem is converted to either audio signals for transfer via a telephone line (hybrid system) or converted to an RF signal for transmission back through the cable network.

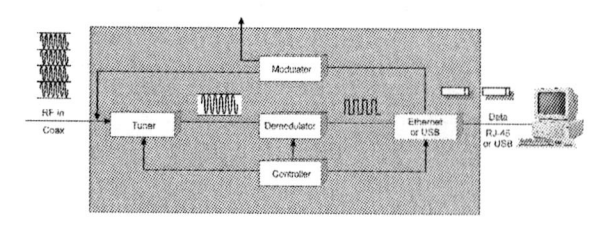

Cable Modem Termination System (CMTS)- The device in the cable television headend that provides network control for a cable modem network. The CMTS not only provides a data path to the connected Internet, but also provides cable modem authentication, IP address assignment, billing functions and is responsible for the majority of Media Access Control (MAC) functionality in a cable modem network. A single CMTS typically controls hundreds or even thousands of end-user cable modems.

Cable Stub-A short length of cable, usually less than 25 feet long, that is terminated and shipped with telephone company (telco) terminal hardware, left slack and readied for splicing at the time of activating the terminal with "live" working facilities.

Cable Telephony-A process of providing telecommunications services through the use of community access television (CATV) systems. Cable telephony services usually combine voice telephone, Internet access, digital cable television (TV), and analog cable TV.

This diagram shows a CATV system that offers cable telephony services. This diagram shows that a two-way digital CATV system can be enhanced to offer cable telephony services by adding voice gateways to the cable network's head-end CMTS system and media terminal adapters (MTAs) at the residence or business. The voice gateway connects and converts signals from the public telephone network into data signals that can be transported on the cable modem system. The CMTS system uses a portion of the cable modem signal (data channel) to communicate with the MTA. The MTA converts the telephony data signal to its analog audio component for connection to standard telephones. MTAs are sometimes called integrated access devices (IADs).

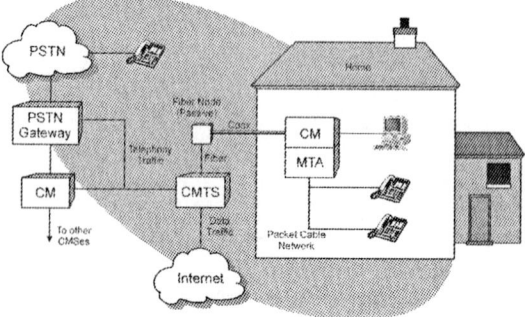

Cable Modem Operation

Cable Telephony System

www.ITMag.Com

This picture shows a cable telephone Network Interface Unit (NIU) that allows a cable television line to provide service for up to 4 analog telephones in addition to standard telephone services.

Cable Telephony NIU
Source: ARRIS

Cable Television Network

Cable Television (CATV)-A television distribution system that uses a network of cables to deliver multiple video and audio channels. CATV systems typically have 50 or more video channels. In the late 1990's, many cable systems started converting to digital transmission using fiber optic cable and digital signal compression.

This figure shows a cable television network that has both television distribution and cable modem transmission capability. This diagram shows that the head-end of a cable television (CATV) system is the initial distribution center for the CATV system. The head end is where incoming video and television signal sources (e.g., video tape, satellites, local studios) are received, amplified, and modulated onto TV carrier channels for transmission on the CATV cabling system. The cable distribution system is a cable (fiber or coax) that is used to transfer signals from the head end to the end-users. The cable is attached to the television through a set-top box. The set-top box is an electronic device that adapts a communications medium to a format that is accessible by the end-user.

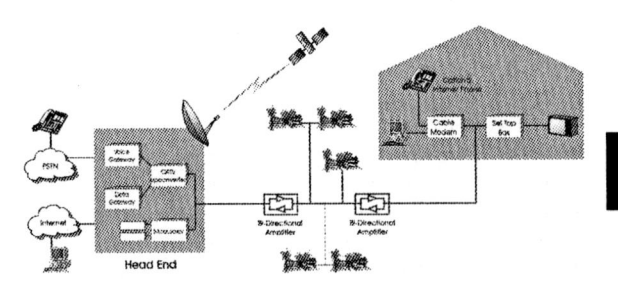

Cable Transfer-Cable transfer occurs when working cables are spliced and the cable count is changed per engineering documents.

Cable Transmission Loss-The amount of signal or radio frequency (RF) energy that is lost while it travels on or through a cable. Signal loss during cable transmission may result from the cable's electrical characteristics (e.g. shape), the type of conductors, and materials. High frequency signal energy transmitted through twisted pair cable usually radiates (leaks) energy from the conductors. For coaxial cable, much of the high frequency energy is kept within the coaxial cable. Coaxial cables generally have greater signal loss at higher frequencies than lower frequencies due to signal absorption (conversion to heat). As a result, cable losses are usually calculated for the highest frequency that the cable is rated for.

Cable Vault-A concrete enclosure that are designed for routing, splicing and connecting communication cable lines. Cable vaults are also known as a manholes.

CableLabs-Cable Laboratories, Inc.
CABS-Carrier Access Billing System
CAC-Card Authorization Center
CAC-Call Admission Control
CAC-Carrier Access Code
CAC-Circuit Access Code
CAC-Customer Administration Center
CAC-Connection Admission Control

Cache-A memory storage area that temporarily stores information that is repeatedly accessed several times. The cache memory can be access more quickly than other memory storage areas such as a CD ROM or hard disk. This allows computers to process information faster. A cache is also used to temporarily store files from other location (such as

Internet Web pages) so it is possible to return to this information without having to transfer the information again.

CAD-Computer Aided Dispatch

CAD-Computer Aided Design

CAFC-Court of Appeals of the Federal Circuit

CAI-Computer Associates International

CAI-Computer Assisted Instruction

CAI-Common Air Interface

CALC-Carrier Access Line Charge

CALC-Customer Access Line Charge

CALEA-Communications Assistance For Law Enforcement

Call Accounting-A the processing call routing information that tracks the services used and their amount of use associated with the call.

Call Admission Control (CAC)-(1-ATM) A traffic management feature of ATM networks that ensures that virtual channel connections are not offered unless enough bandwidth is available on its network. (2-voice) Call Admission Control (CAC) is a concept that applies to voice traffic. CAC is a deterministic and informed decision that is made before a voice call is established and is based on whether the required network resources are available to provide suitable QoS for the new call.

Call Agent (CA)-In a communication network, the call agent is responsible for call processing functions related to setup, maintain, and teardown communication sessions (e.g. voice calls).

Call Barring-A telecommunications network feature which restricts the origination or delivery of calls to a device operating in its system. Call barring can restrict some or all parts of outgoing or incoming calls. When all calls are restricted (may be activated when the customer does not pay their bill), no calls may be made from or answered by the telephone device. Outgoing call restriction may limit the calls to emergency only, no international calls or local calls only. Restrictions on incoming calls may allow calls from specific people in a closed user group (CUG) or non-toll calls (such as collect calls).

Call Block-Block the Blocker

Call Blocking-(1-system unavailable) The limitation on the ability to originate or deliver of a call. Call blocking can occur on origination if telephone system resources are not available (such as when all the channels are occupied on a cellular system). Call blocking can also be a system feature where the delivery of calls can be blocked. (2-service) A service that permits a customer to restrict the delivery of incoming calls from numbers on a pre-defined (blocked number) list.

Call Center-A call center is a place where calls are answered and originated, typically between a company and a customer. Call centers assist customers with requests for new service activation and help with product features and services. A call center usually has many stations for call center agents that communicate with customers. When call agents assist customers, they are typically called customer service representatives (CSRs).

Call centers use telephone systems that usually include sophisticated automatic call distribution (ACD) systems and computer telephone integration (CTI) systems. ACD systems route the incoming calls to the correct (qualified) customer service representative (CSR). CTI systems link the telephone calls to the accounting databases to allow the CSR to see the account history (usually producing a "screen-pop" of information).

This figure shows a sample call center. This diagram shows that calls may be received or originated from the call center. The customer traditionally communicates with the call center by telephone. When a call is received by a call center, the user is typically provides with a list of options by an automated interactive voice response (IVR) unit. As the user selects from the list of options, an ACD system routes the call to a CSR station that is qualified to assist the customer (e.g. sales agent or technician). When the CSR agent answers the call, some of the customer's account information may become available on the CSR's computer screen ("screen pop"). The CSR will communicate with the customers and should make notes in the customer's account regarding the activity that progressed.

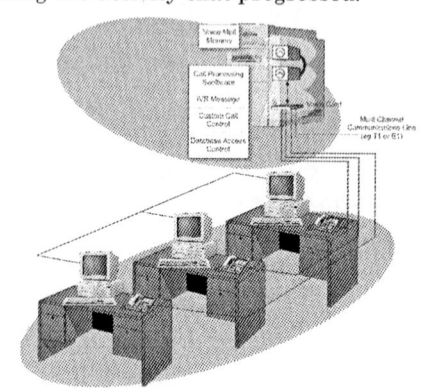

Call Center Operation

Call Clearing-The process that are used (e.g. signaling messages) to clear a call connection. Call clearing involves informing the network elements of the release of connection ports and call processing (e.g. echo canceling) activities.

Call Congestion-The ratio of calls lost due to a lack of system resources to the total number of calls over a long interval of time.

Call Control (CC)-The processes associated with establishing, maintaining, and releasing a voice or data calls.

Call Delay-The time delay that is experienced when a call arrives at a switching device or system and resources are not immediately available.

Call Detail Record (CDR)-A data record that holds information related to a telephone call. This information usually contains the origination and destination address of the call, time of day the call was connected, added toll charges through other networks, and duration of the call.

This figure shows the basic structure of a call detail record (CDR). A CDR contains the identification number (unique identifier) of who originated the call, the called number, the time the call started and completed. This diagram shows that CDRs can grow in size and they can hold other types of information. In this example, an additional charge for operator assistance is added.f

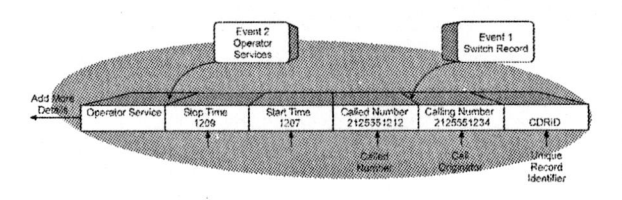

Call Detail Record (CDR) Structure

Call Detail Record (CDR) Processing-The process of creating a call billing record through the gathering and manipulating of billing event and service record information related to the call. Call record processing involves receiving the call record, guiding the record to the specific billing account, rating the record, and routing the record to the appropriate database for collections or account settlement.

This figure shows the general process that is used to identify and rate (bill) a call. This diagram shows that a call detail record evolves as it passes through the rating process. In the first step, a rate band is determined. Then, the identification information on the CDR is used to identify a specific customer's account (guide the record). The customer's rate plan is discovered and the unit (usage) and fixed (per event) charging rates are gathered and calculated. The new information (rate band, call charge amount) is added to the call detail record and it is moved to the bill pool, as it is ready to be billed.

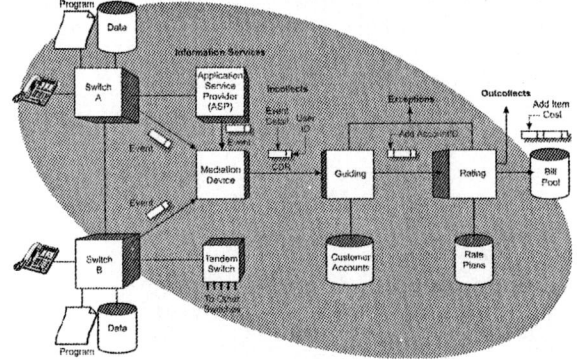

Call Detail Record (CDR) Processing Operation

Call Detail Record Server (CDR Server)-A server that collects and processes call usage information for the creation of call detail record (CDR) accounting and billing information.

Call Diversion-The automatic forwarding of telephone call to a programmable number if the called telephone number is non-operational. This number is chosen and changed by the user.

Call For Votes (CFV)-A prompt for users within a group (such as a Usenet newsgroup) to submit responses to one or more questions within a voting period. Usually one or more email addresses are included to allow the users to respond with their vote.

Call Forward Busy-The process of forwarding a call to another extension or telephone number when the selected extension or telephone number is in use (busy).

Call Forwarding-A call processing feature allows a user to have telephones calls automatically redirected to another telephone number or device (such

Call Forwarding-Busy (CFB)

as a voice mail system). There can be conditional or unconditional reasons for call forwarding. If the user selects that all calls are forwarded to another telephone device (such as a telephone number or voice mailbox), this is unconditional. Conditional reasons for call forwarding include if the user is busy, does not answer or is not reachable (such as when a mobile phone is out of service area).

This diagram shows how call forwarding can be used to automatically redirect telephone calls based on specific conditions. This example shows that a call may be redirected by the switching system to one extension if the user is busy and to a different extension if the user has programmed the extension as "Do Not Disturb" or if it is busy. When the call is received by the system destined for extension 1001, the call processing system uses the indication of busy along with a redirection table to determine the call must be automatically transferred to extension 1003.

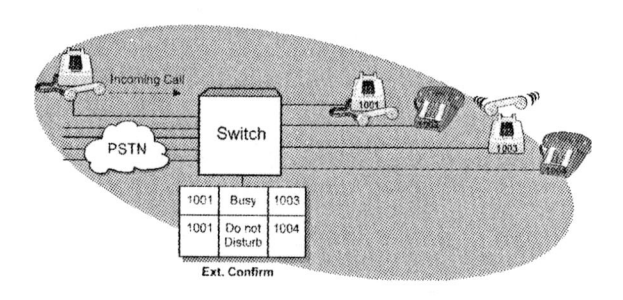

Call Hold-A feature that allows a user to temporarily hold and incoming call, typically to use other features such as transfer or to originate a 3rd party call. During the call hold period, the caller may hear silence or music depending on the network or telephone feature.

This diagram shows how a call can be temporarily placed on hold so the call can stay connected without the user having to continue conversation with the caller. During hold, the audio from the user is muted. For an analog line, the call hold feature involves placing a load (connection) across the line so that current may continue to flow through the circuit. For digital systems, the call hold feature may send a call hold message back to the system (such as a signaling message on the signaling channel) so the system can know that the status of the telephone station has changed to "hold."

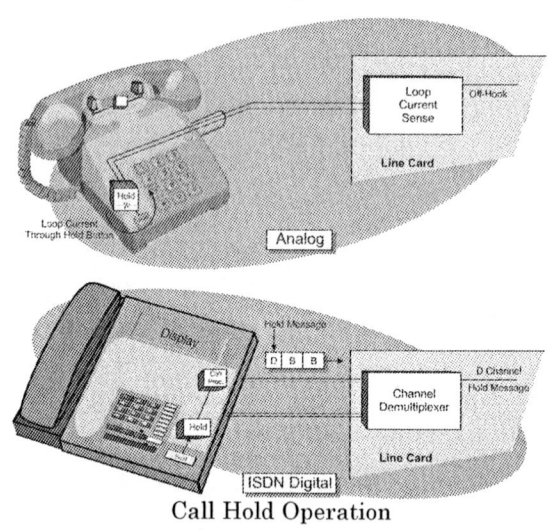

Call Forwarding Operation

Call Hold Operation

Call Forwarding-Busy (CFB)-A call routing process that redirects incoming calls addressed to a called subscriber's telephone number to another telephone number (forwarded to number) when the telephone number is busy processing a call. CFB is commonly used to redirect an incoming call to a voice mailbox.

Call Gapping-A network management control that limits the rate of traffic flow to a specific destination code or station address.

Call Hand Off-The process of transferring a mobile communication call from one radio channel (transceiver) to another radio channel, usually from one cellular site (or cellular transmitter) to another.

Call Intrusion-The unannounced or un-requested connection of another caller to a call that is in progress.

Call Log-The recording of information that is associated with a communication device or service.

Call Management Record (CMR)-Records relating to a call that was processed by a call manager in an Enterprise communication system (such as an iPBX system that uses AVVID protocols).

Call Origination-Call origination on a wireless system is the initiation of a call or communication session by a mobile or fixed radio. The call origination process begins with a communication session request (such as the mobile radio user dialing a number and pressing send). This is followed by an attempt to access the communication system on an access channel. After attempting to access the system, the mobile radio listens to see if a response to its' communication request has been processed (called "contention resolution). If it does not hear a response in a pre-determined maximum time, it will wait a random amount of time before attempting access again to prevent repeated collisions with other users. If the system is not busy, it will send out an acknowledgment message and assign the mobile radio to a communication channel.

Call Paging-Call paging is a process of delivering a call alert (command for a ring signal) via a communications system or radio signal to a person whose exact whereabouts are unknown by the sender of the message. When the mobile radio receives a page message, it begins to alert the user that an incoming call is waiting to be received. If the user responds (usually by pressing SEND or GO,) the mobile radio will respond to the system with an answer message and a communication channel can be assigned.

Call Park-A call hold feature that allows a telephone station (usually a PBX telephone) to initiate other calls or services while a call is parked.

Call Pickup-A telephone call processing (switch control) feature that enables a telephone user to answer a telephone from another telephone station. When a call is received, a key sequence is entered from specific groups or any telephone (dependent on how the system is setup) and the call is redirected to the extension or line that has picked up (entered the code) the line.

This diagram shows the operation of a telephone system that has a call pickup group feature. In this example, an incoming call is received and is directed toward the main extension 1001. When the call is received for extension 1001, the system uses the call pickup group list to determine that extensions 1001, 1002, 1003, and 1004 are programmed for call pickup. This allows any of these extensions to answer the call by pressing a key sequence (key "3" in this example).

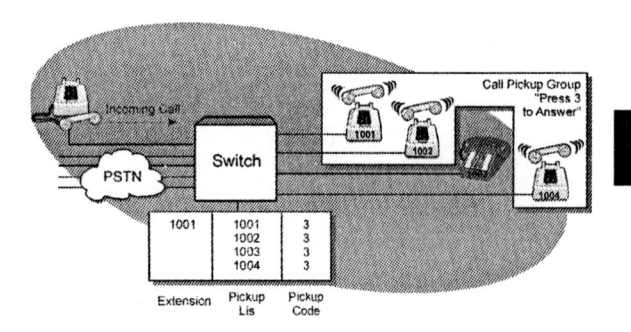

Call Pickup Operation

Call Proc-In the Integrated Services Digital Network (ISDN) Q.931 protocol, a Layer 3 call-proceeding message that indicates that all information has been received for a call and that a call to the next entity has been started. When received from customer premises equipment, Call Proc is a request for additional time.

Call Processing-Steps that occur during the duration of a call. These steps are typically associated with the routing and control of the call. When used as part of a billing system, call process involves gathering messages from various sources (event records), reformatting and edit the messages, calculate a charge for the message, assign a customer to the message, and getting the message ready for billing.

This figure shows the basic functions of the call processing section of a billing system. This diagram shows how different event sources are received by the call processing system. These event sources may be from the network elements or from other companies that have provided services to your customers. These records are reformatted to a common CDR format and duplicate CDRs are eliminated. Identification information in each call detail record is used to guide (match) the record to an account in the customer database. The customer's information determines the rate plan to use in charge calculation. The rating database uses rate tables, the customers selected rate plan, and possibly other information (e.g. distance, time of day) to calculate the actual charge for each call. All of the information is added to the CDR and it is either placed in the bill pool (ready for billing), or it is sent to another company to be billed if the customer identification is not

Call Processing Language (CPL)

part of this network's customer database. If there are any problems with call processing, the call detail records are sent to message investigation for further analysis.

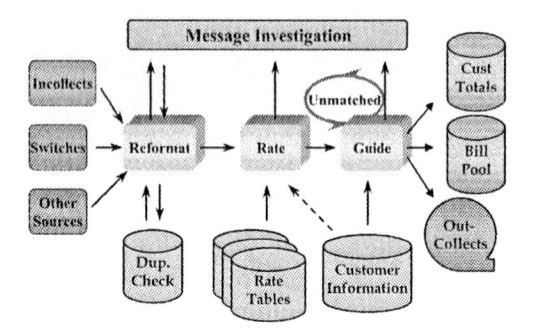

Billing Record Call Processing Operation

Call Processing Language (CPL)-A language that is based on extended markup language (XML) that is used to describe and control Internet (VoIP) telephony services. CPL is used by end users to create telephone services that integrate VoIP with existing Email, WEB, and other applications. It can be used with other VoIP protocols such as session initiation protocol (SIP).

Call Processing Redundancy-The ability of a call processing system to operate when some or all of the call processing network parts experience operational difficulties or communication failures.

Call Progress Signals-Voice band tones or tone combinations and announcements used to inform an originator or operator of the progress or disposition of a call. Examples include busy tone, voice announcement, special information tones, and audible ringing. The tones or announcements can be used after a number has been dialed and until the called telephone is answered or the attempt is abandoned. Because these signals are intended for a human listener and due to variations in precise timing and loudness, it is difficult for an electronic device to use these tones to determine the status of a call. Also called Advisory Tones or signals or messages.

Call Progress Tones-Signaling control tones that are sent to inform the device or system of the progress of a call. Examples of call progress tones include dial tone, ringback, and busy tone.

Call Rate (CR)-The number of calls that are received over a defined period of time (such as during an hour or day.) The call rate may be characterized by a specific period of time such as the busy hour (BH) so its usage often include enough qualifying information to assure that it will be properly understood.

Call Rating-A process of assigning a value or cost to a telephone call or communication session.

This figure shows the basic call rating process used in a billing system. This diagram shows that a call detail record evolves as it passes through the rating process. In the first step, a rate band is determined. Then, the identification information on the CDR is used to identify a specific customer's account (guide the record). The customers rate plan is discovered and the unit (usage) and fixed (per event) charging rates is gathered and calculated. The new information (rate band, call charge amount) is added to the call detail record and it is moved to the bill pool as it is ready to be billed.

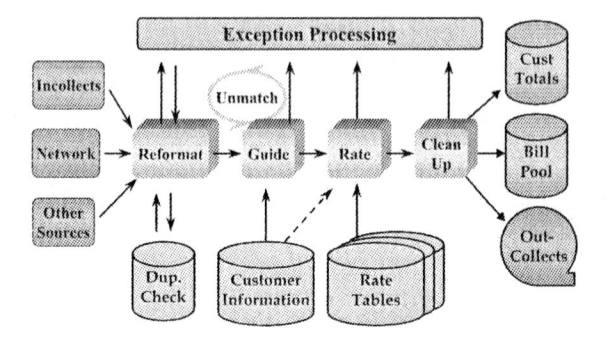

Billing System Call Rating

Call Reference-In the Integrated Services Digital Network (ISDN) Q.931 protocol, an information element that identifies the call that corresponds to a Layer 3 message.

Call Reference Flag-In the Integrated Services Digital Network (ISDN) Q.931 protocol, a flag that indicates which side of an interface allocates the call reference value.

Call Reference Value-In the Integrated Services Digital Network (ISDN) Q.931 protocol, the numerical value of a call reference.

Call Report-A cellular telephone record stored on a data acquisition system (DAS) tape containing the overall timing information, mobile number, dialed digits, and appropriate indicators to ticket a call for every call completed or attempted through the system. (See billing Tape).

Call Restrictions-In call-processing feature that restricts telephone call origination to specific groups of authorized numbers (typically local phone numbers or no-international calls).

Call Routing-(1-circuit switching) The process of determining the path of a call from point of origination to point of destination. (2-packet switching) The steps taken to ensure a connection can be made from an origination point to its destination point. This packets that transfer through this connection may actually take different paths.

Call Routing Table-A table that holds the dial plans associated with how call routing will be processed based on the calling number and the dialed digits received from the caller. Call routing tables determine if the call is allowed from the caller (international calls may be restricted) and which network will handle the call (local telephone line or long-distance network).

Call Routing Tree-A graphic display (looks like a tree) of the different call routing paths that can be taken by call routing decision logic.

Call Screening-A process that allows a call recipient to review the incoming telephone number (and name if available) to determine if they desire to answer the incoming call or to transfer the call to another person.

Call Server-A call server is a particular form of application server that manages the setup or connection of telephone calls. The call server will receive call setup request messages, determine the status of destination devices, check the authorization of users to originate and/or receive calls, and create and send the necessary messages to process the call requests.

Call Signaling-The process of sending control information during a call. Call signaling may be in band (muting the audio while sending control information) or out of band (on a separate signaling channel (such as SS7) during the call.

Call Stateful Proxy-A proxy server that remembers the call states during a communication session. All the call state messages associated with a communication must pass through the call stateful proxy. The call stateful proxy is defined in the session initiation protocol (SIP) standard.

Call Supervision-The process of monitoring a communication line or trunk for changes in call status. These changes can include start-dial (off-hook), feature change requests, or call termination (on-hook).

Call Termination-The process of terminating a call or communication session. Call termination can be initiated by any of the users of the call or the system that is connecting the call.

Call Timers-Clocks timers that are used count time or events that occur with a telecommunications device. This may be the airtime or usage of a telecommunications service or how many calls were made during a particular time period.

Call Topology-The interconnection relationships between switches or routers in a communication network. Interconnection types can vary from circuit switched TDM to packet switched VoIP.

Call Trace-Call trace allows a subscriber to initiate a call trace request message that allows the dialed digits of a caller to be stored for investigation. The activation of the call trace service alerts the telephone service operator to "tag" the originator's number to allow authorities to investigate the originator of the unwanted or unauthorized call. Some of the call trace activation codes include *57 on a touchtone phone or the dialing of 1157 on a rotary (pulse) phone. If the call trace of the last call was completed successfully, an announcement should be heard. The service operator will usually release the call trace information to law enforcement agencies and a signed authorization from the subscriber may be required.

Call Trace-*57

This diagram shows the call trace operation. The call trace operation starts with the reception of an unwanted call. When the call is received, an event record is created in the switch that records the connection through the switch (resources used) and this record usually contains the calling party number associated with the incoming call. This example shows that the customer dials a call trace feature code to inform the carrier to trace the last call. The local switch (recipients phone carrier) reviews the call detail information stored within the switch (or billing system) to determine if a calling number was provided. If the number was provided and is available, the information will be stored in the carriers system. The customer then calls the carrier and requests that the number be sent to local authori-

Call Transfer

ties for further action. This example shows that the customer completes a call trace release form that authorizes the transfer of the number to authorities.

between 1001 and 1011 is disconnected and a connection is made between incoming line 1001 and 1011.

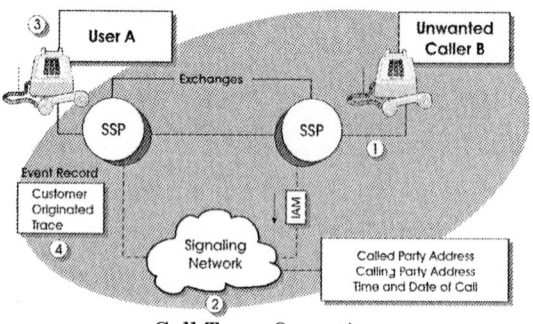

Call Trace Operation

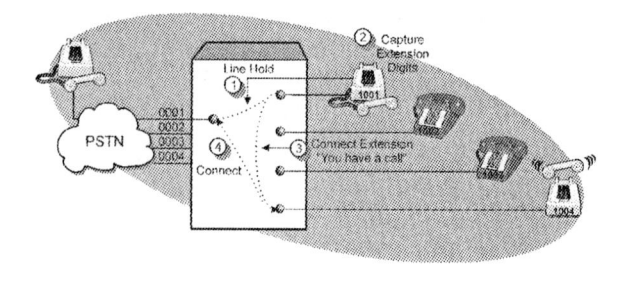

Call Transfer Operation

Call Transfer-A call connection routing feature that transfers a call from one telephone or extension to another. Call can occur at various stages of the call conversation through the use of a system call request feature. The system special service request feature is often called a "Flash" feature. The flash feature is created to indicate a desire to recall a service function or to activate a custom calling feature (such as a call transfer request).

A flash feature service request can be created when the user initiates a short on-hook interval or through the sending of a special service request message. The short on-hook interval is created by a momentary operation of the telephone switch hook, during a prolonged off-hook period. The special service request message can be sent by a button on a telephone (such as a PBX telephone) or by pressing the SEND key on a mobile telephone.

This diagram shows a typical call transfer process used in a private telephone system. In this example, a caller is connected from the public telephone network to extension 1001 via incoming line 0001 on a telephone system (PBX). The user on extension 1001 desires to transfer the call to extension 1011. The first step involves extension 1001 sending a hold signal to the telephone system that allows it to place incoming line 0001 on hold. This is followed by the user at extension 1001 sending the digits indicating the destination of the call transfer (1011). The system then connects extension 1001 with 1011 and the user at 1001 may inform the user at 1011 that a call is being transferred to them. When the user at 1001 hangs up, the connection

Call Type - Billing-The Call Type is a parameter used in rating a call. It is typically derived from a set of network elements recorded by the network (Ingress, Egress, Service Type, Routing, Special Features, etc.) Examples of a Call Type that are rated differently: Operator Assisted, International, Roaming, etc.

Call Waiting (CW)-A telephone call processing feature that notifies a telephone user that a another incoming call is waiting to be answered. This is typically provided by a brief tone that is not heard by the other callers. Some advanced telephones (such as digital mobile telephones) are capable of displaying the incoming phone number of the waiting call.

After the service provides the subscriber with the notification of an incoming call while the subscriber's call, controlling subscriber can either answer or ignore the incoming call. If the controlling subscriber answers the second call, it may alternate between the two calls.

This diagram shows how call waiting service may be provided on an analog telephone line. In this example, a call is in process with caller 1. During the call, a second caller (caller 2) dials the telephone number of the user that has call waiting. The system discovers that the line or extension is busy on another call. The system also determines that this user has the call waiting service processing feature available so it sends a call waiting message tone to the user (only heard by the user). If the user desires to answer the call, the user sends a flash message (a momentary open on the line) that indicates to the telephone system to place the current call in progress on hold and switch to the other incoming call (caller 2). Each time a flash message is sent, the line alternates between each incoming caller.

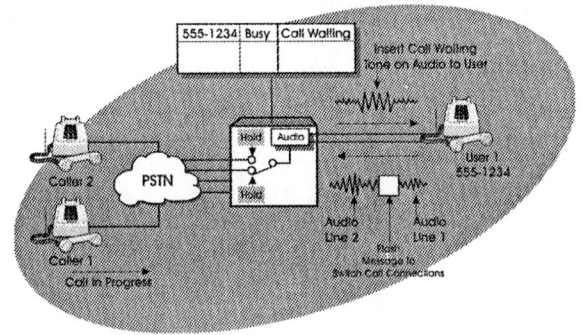

Call Waiting Operation

Call Waiting Identification (CWID)-A telephone service that provides a receiving telephone device with the phone number of a new originating caller (call to be received) while the callee is on the line with another caller.

Call-Back Unit-A device that calls a user at a pre-programmed telephone number when a connection is requested to prevent unauthorized access from unknown location, or to establish low cost international calls between a nation with low originating costs, automatically directing call-backs into a target nation with higher originating costs.

Callback-A call processing service that reverses the connection of calls. This process is divided into the call setup (dial-in) and callback stages. The caller dials a number that provides access to the callback service. The callback gateway receives the call and prompts the caller to say or enter (e.g. by touch tone) the number they desire to be connected to and the number they want the callback service to connect to. The callback center then originates calls to both numbers and connects the two individuals to each other.

Called Line Identification-The identification information carried by an SS7 packet that provides the destination receiver to identify the source of the call.

Caller ID-An optional telephone service that provides a receiving telephone device with the phone number of the originating caller, which can be displayed to the destination person prior to receiving the call. The caller ID is transmitted as a data parameter in the SS7 Initial Address Message from the originating end switch to the destination end switch in the process of setting up the call. Some caller ID services can also provide directory name listing information, derived separately from the LIDB data base. Caller ID information is typically transferred as a type-202-modem-compatible data signal between the first two ringing cadence cycles of the alerting tone.

Caller ID Type II-The ability of the system to announce the phone number of the caller who is calling in while the line is in use (call waiting).

Calling Card-An identifying number or code unique to the individual, that is issued to the individual by a common carrier and enables the individual to be charged by means of a phone bill for charges incurred independent of where the call originates.

This figure shows how a calling card can be used initiate calls through a telephone network. In this example, the customer uses the telephone number on the pre-paid calling card to initiate a call to a switching gateway. The gateway gathers the calling card account information by either prompting the user to enter information or by gathering information from the incoming call (e.g. prepaid wireless telephone number). The gateway sends the account information (dialed digits and account number) to the real time rating system. The real-time rating system identifies the correct rate table (e.g. peak time or off peak time) and inquires the account determine the balance of the account associated with the calling card. Using the rate information and balance available, the real time rating system determines the maximum available time for the call duration. This information is sent back to the gateway and the gateway completes (connects) the call. During the call progress, the gateway maintains a timer so the caller cannot exceed the maximum amount of time. After the call is complete (either caller hangs up), the gateway sends a message to the real time rating system that contains the actual amount of time that is used. The real time rating system uses the time and rate information to calculate the actual charge for the call. The system then updates the account balance (decreases by the charge for the call)

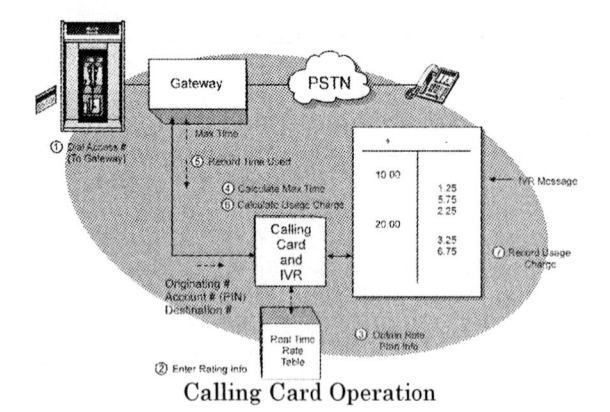

Calling Channel (CC)

Calling Card Operation

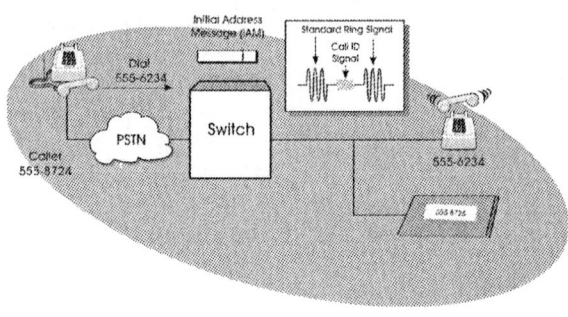

Calling Line Identification (CLID) Operation

Calling Channel (CC)-A radio channel a wireless telephone system that is used to setup calls to mobile phones. In some wireless systems (such as the NMT system), during peak usage periods, a calling channel may be used as a voice channel and no control channel will be available.

Calling Line Identification (CLI)-A service which displays the calling number prior to answering the call that allows telephone customer to determine if they want to answer the call. The calling number may be used by the telephone device to look-up a name in memory (e.g. mom) and display the name along with the phone number.

This figure shows the calling number identification operation. Calling number identification operation starts with the reception of a call. When the call is received, the initial address message (IAM) contains the calling party number of the incoming call. The IAM may contain additional information such as the text name of the calling party. This example shows that the local switching system extracts this information and combines this information with the ring signal (using different frequencies and amplitudes) and sends it to the customer during the alerting (ringing) process. If the customer has the appropriate display equipment, the calling number information is display as the telephone rings.

Calling Line Identification Display (CLID)-Calling line identification display (CLID) identifies for display the originating or calling phone number.

Calling Line Identification Presentation (CLIP)-The transmission of a calling number to the receiver of a call. The calling number may originate from the caller's equipment or it may be created by the network.

Calling Line Identification Restriction (CLIR)-The ability of a caller to block the delivery of their telephone number from being displayed on a caller identification device. The CLIR may be on a per call basis or per line (continuous) block. CLIR provides the ability for the caller to remain anonymous to the called person.

Calling Name Delivery-A feature that enables a customer to view the name of the calling party as well as the date and time of the call on a display device.

Calling Number Display Blocking-A feature that gives calling customers the option to change the caller ID status of their number on a per-call basis. When the calling directory number is private, the caller ID display will not be available to the called party.

Calling Number Identification (CNI)-Calling number (and/or name) identification is a service that provides the number and/or name of the incoming to the telephone user prior to the user's answering of the call. This allows the customer to view the telephone number of the person who is calling before deciding to accept the call.

Calling Party Pays (CPP)-A communication service that bills the calling party for the delivery of their call through a network (such as a mobile communication network or freephone service) or for the providing of information (such as a news service.)

Calls Per Second (CPS)-A measurement of how many calls can be processed in a communication system per second of time.
CALNET-California Network
CAM-Content Addressable Memory
CAM-Call Applications Manager
CAM-Computer-Aided Manufacture
CAM-Controlled Attachment Module
CAMA-Centralized Automatic Message Accounting
CAMA-ONI-Central Automatic Message Accounting-Operator Number Identification
CAMA/LAMA-Centralized Automatic Message Accounting/Local Automatic Message Accounting
CAMEL-Customized Applications For Mobile Enhanced Logic
Campaign Management-The process usually utilized for direct marketing (i.e. direct mail or phone sales), to manage the conceptualization, prioritization, planning, execution, and post campaign measurement of activities.
Camping-This process where a mobile phone tunes into and locks onto a radio channel and waits for incoming calls. This is also known as the idle state.
Campus Switch-A switch used within a campus backbone. Campus switches are generally high-performance devices that aggregate traffic streams from multiple buildings and departments within a site.
CAN-Cancel
Cancel Message-A message that is used to inform a person or device that a previous message or service that has been requested has been cancelled. The cancel message is defined in session initiation protocol (SIP).
Candelas (CD)-A unit of measure for luminous intensity (photons per second).
CAP-Carrierless AM/PM System
CAP-Carrierless Amplitude And Phase
CAP-Computer Aided Professional Publishing
Cap Code-An electronically embedded code in a paging unit used for addressing different functions by a transmitter. It is similar to an ESN in cellular terminal equipment.
Capacity Planning-Capacity planning applies to proactively assessing systems, whether network or computer based systems, and forecasting the needed future growth of the system. Capacity planning in network management can be a function of the Configuration or Performance Management level of the FCAPS model.

CAPC-Competitive Access Provider Capacity
CAPCS-Cellular Auxiliary Personal Communication Services
CapEx-Capital Expenses
CAPI-Common ISDN API
CAPI-Cryptography Application Program Interface
Capital Cost Model-A financial cost model that recognizes the differences in regulatory accounting versus tax accounting for telecommunications networks and other system.
Capital Expenses (CapEx)-A term CapEx used to define the expenses such as the purchase of land, buildings and, most importantly, the build out of network capacity in a telecommunications system.
CAPs-Competitive Access Providers
Capstan-(1-recording system) A motor-driven shaft or spindle used to control (and establish) a constant speed of the magnetic media as it moves through the recording path. A servo circuit controls the speed of the capstan relative to a reference signal. Any error between the capstan velocity (as determined by sensors) and the reference produces a correction signal that is fed back by the servo to increase or decrease capstan rotation. (2-cable reel) A rotating drum or cylinder used for pulling cables by exerting traction upon a nope or pull line passing around the drum.
CAPTAIN-Character And Pattern Telephone Access Information Network System
Caption-Text or titles that are inserted in a video program.
Capture Effect-An effect associated with the reception of frequency-modulated signals in which, if two signals are received on the same frequency, only the stronger of the two will appear in the output. The complete suppression of the weaker carrier occurs at the receiver limiter, where it is treated as noise and rejected.
Carbon Fiber Reinforced Plastic (CFRP)-A strong and light type of material, that is sometimes used in electronic products such as laptop computers and wireless telephones.
Cardioid Pattern-A general heart-shaped pattern associated with directional microphones and directional transmitting antennas.
CARE-Customer Account Record Exchange
CARE/ISI-Customer Account Record Exchange/Industry Standard Interface
Caribbean Broadcasting Union (CBU)-A broadcasting association located in the Caribbean.

CARL-Computerized Administrative Route Layouts

CAROT-Centralized Automatic Reporting On Trunks

CARP-Cache Array Routing Protocol

Carriage-The established procedures of a cable TV system regarding the carrying of signals of local TV stations on its channels.

Carried Load-The average volume in a traffic system. Carried load equals offered load when all calls are served. Carried load is less than offered load when some calls are denied service.

Carried Traffic-That part of the traffic offered to a group of servers that successfully seizes a server. Carried traffic equals offered traffic less the overflow traffic.

Carrier-(1-signal) A sine wave that can be modulated in amplitude, frequency or phase for the purpose of carrying information. (2-company) A business organization providing telecommunications service, such as a radio common carrier or a telephone service provider. (3-frequency) The frequency of the radio carrier signal. (4-level) The radio energy (power) of a carrier signal, typically expressed in decibels in relation to some nominal (reference) level.

Carrier Access Billing System (CABS)-A system that is used by network access providers to bill carriers for their customer's access to the network facilities.

Carrier Controlled Squelch-Carrier controlled squelch is a process where the audio output of a radio receiver is muted until the reception of an incoming RF signal that is above a predetermined carrier signal strength level. Carrier controlled squelch allows a radio user to avoid listening to noise or weak interference signals of distant radio transmissions that occur on the same frequency.

This diagram shows how a carrier controlled squelch system mutes the audio until the reception of an incoming RF signal is above a predetermined carrier signal strength level. In this example, the base station is receiving two signals from one desired user and one distant interfering user. This diagram shows that the squelch level has been set so the audio will be muted unless the incoming radio signal strength is above the background noise and above the level of RF signals from undesired distant users.

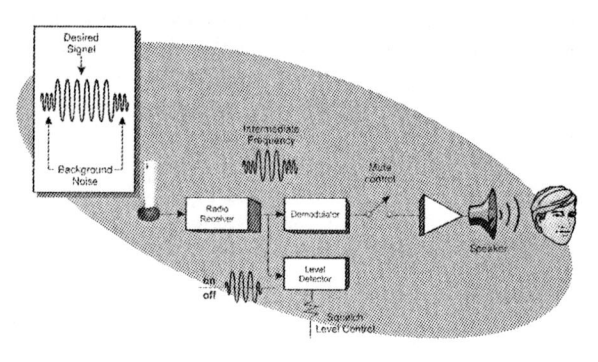

Carrier Squelch System

Carrier Detect (CD)-A control signal on a communication line (such as an RS-232 serial bus) that indicates the device has detected a carrier (modem) signal.

Carrier Identification Code (CIC)-A 3-digit code that uniquely identifies a telecommunications carrier within the North American Numbering Plan (NANP). The CIC is indicated by an XXX in a Carrier Access Code where X can be any digit, 0 through 9. After an XXX code has been assigned to a carrier, the code is retained for use with either Feature Groups B (950A-0XXX, 950-1XXX) or D (10XXX) throughout the area served by the NANP. (See also: Carrier Access Code, pre-subscription, primary interexchange carrier.

Carrier Interconnection Plan-A long-term plan for inter-connecting interexchange and international carriers to exchange carrier intraLATA networks. This plan also can be called the Exchange Access Plan. The term equal access commonly is used to refer to the features provided by this plan.

Carrier Noise Level-The noise level resulting from undesired variations of a carrier in the absence of any intended modulation.

Carrier Sense-In Ethernet, the act of determining whether the shared communications channel is currently in use by another station.

Carrier Sense Multiple Access (CSMA)-A system access process that allows multiple radios or stations to gain the attention (service) from a system. During the CSMA process, each station "listens" to a channel to sense whether it is busy prior to attempting access to the system. If the channel is not busy, the station will randomly attempt access. An access request collision detection process is used to allow the requesting station to recognize if the system response is from its access request as a result of a request from another station. If the station determines that the response is not resulting from its service request, that station will wait for a random period before it attempts to transmit a service request again.

Carrier Sense Multiple Access/Collision Avoidance (CSMA/CA)-A network access method in which each device signals its intent to transmit data before it actually does so. This keeps other devices from signaling information, thus preventing collisions between the signals of two or more devices. CSMA/CA is used at Home Radio Frequency (HomeRF) networks, which operates in the same unlicensed 2.4 GHz ISM band as the Bluetooth specification.

Carrier Sense Multiple Access/Collision Detection (CSMA/CD)-A carrier sense multiple access transmission scheme in which transmission resulting in collisions are followed by the transmitting stations backing off the network a random amount of time before attempting to retransmit. CSMA/CD is used as the basis of Ethernet networks. When multiple collisions occur for the same packet, Ethernet stations typically back off in exponentially increasing large random time amounts to further reduce the probability of collision. Collision avoidance algorithms are also common in Ethernet, whereby a station will listen for silence on the media before transmitting.

Carrier Sensitive Routing (CSR)-Selecting the call routing of a communication session based on carrier preferences. The preferences used may include cost, quality, and congestion.

Carrier Shift-(1 - modulation) A method of keying a radio carrier for transmitting binary data that consists of shifting the carrier frequency in one direction for a mark signal and in the opposite direction for a space signal. (2 - envelope) A condition resulting from imperfect amplitude modulation whereby the positive and negative excursions of the envelope pattern are unequal, thus effecting a change in the power associated with the carrier. Carrier shift may be positive or negative.

Carrier System-A system that carries many individual signals over a shared facility by multiplexing, that is, combining those signals for transmission. The most common transmission technique is time-division multiplexing, in which each information channel uses the transmission medium for assigned time intervals. Frequency-division multiplexing, in which each information channel occupies an assigned portion of the frequency spectrum, also has been used. Carrier systems range in size from two to many thousands of individual channels.

Carrier to Adjacent Channel (C/A) Interference Ratio-The amount of interference level from an adjacent signal in comparison to the desired carrier signal. The C/A ratio is commonly expressed in dB. Different types of systems can tolerate different levels of adjacent channel interference dependent on the modulation type and error protection systems.

Carrier To Interference Signal Ratio (C/I)-The amount of interference level from all unwanted interfering signals in comparison to the desired carrier signal. The C/I ratio is commonly expressed in dB. Different types of systems can tolerate different levels of interference dependent on the modulation type and error protection systems.

Carrier to Noise Ratio (C/N)-The amount of interference level from noise signal energy in comparison to the desired carrier signal. The C/N ratio is commonly expressed in dB. The C/N measurement is commonly used at the edges of mobile communication systems as noise is the only interference with the reception of a radio signal. This is called a "noise limited system."

Carrier-To-Interference Ratio (C/I)-The ratio of the desired radio carrier signal with respect to the combined interference due to other radio channel interfering signals.

Carrierless Amplitude And Phase (CAP)-Carrierless amplitude and phase (CAP) modulation is very similar to QAM modulation. The difference is the continuous shifting of phase (or signal mix) of the carrier signal level. CAP modulation was designed to help to reduce the effects of crosstalk and to simplify the signal processing of modulated signal. CAP transmits data signals on a single high bandwidth modulated carrier.

CARS-Cable Television Relay Service

CARS-Community Antenna Relay Service

Carterfone Decision-A Federal Communications Commission (FCC) decision in 1968 that allowed customers to directly connect their own customer-provided equipment (CPE) to telephone company networks. Prior to this decision, some telephone companies prohibited the attachment of such equipment to their networks. Equipment that is to be connected to telephone networks must conform to industry standards and FCC regulations to ensure the protection of the telephone network from potential damage that may result from defective CPE.

CAS-Channel Associated Signaling or Call-Path Associated Signaling
CAS-Centralized Attendant Service
CAS-Communicating Application Specification
CAS-Channel Associated Signaling
Cascade-(1-circuit) To connect the output of a device or system to the input of another device. (2-Windows) The arranging of program windows so they overlap with each other.
Cascaded-An arrangement of two or more circuits in which the output of one circuit is connected to the input of the next circuit.
Cascaded Codecs-The connection of one coder/decoder through another coder/decoder. The cascading of coders (such as speech coders used in mobile communication systems) usually results in significant degradation of voice quality.
CASE-Computer Aided Software Engineering
CASE Tools-Computer Aided Software Engineering Tools
Cassegrain Antenna-One of several types of reflective antenna systems used for satellite communications. The Cassegrain type uses a paraboloidal primary reflector with a secondary hyperboloidal reflector. Efficiency of the antenna system is improved because less microwave energy is permitted to spill past the edge of the primary reflector.
Cassette-A self-contained package of reel-o-reel blank or recorded magnetic tape that is continuous and may be rewound on demand.
CAT-Category
CAT5-Category 5
CAT7-Category 7
Category 5 (CAT5)-Category 5 unshielded twisted-pair wiring commonly used for10baseT and 100BaseT Ethernet networks and rated by the EIA/TIA
Catenet-A collection of networks (typically LANs) interconnected at the Data Link layer using bridges. Also known as a bridged LAN.
CATI-Computer Assisted Telephone Interviewing
CATLAS-Centralized Automatic Trouble Locating and Analysis System
CATNIP-Common Architecture for Next Generation Internet Protocol
CATS-Consortium for Audiographic Teleconferencing Standards
CATS-Calling Card & Third Number Settlement
CATV-Community Access Television
CATV-Cable Television

CAU-Controlled Access Unit
CAV-Constant Angular Velocity
CAV-Component Analog Video
CAVE-Cellular Authentication and Voice Encryption
CAVE-Cave Automatic Virtual Environment
Cavity-A resonant device, usually drum-shaped or cylindrical, that acts as a filter in a radio frequency system.
Cavity Coupling-Method of extracting or introducing microwave energy to or from a resonant cavity.
CB-Citizens Band
CBAC-Context Based Access Control
CBB-Customer Convenience Block
CBC-Cipher Block Chaining
CBCH-Cell Broadcast Channel
CBCP-Certified Business Continuity Professional
CBD-Central Business District
CBD-Central Billing District
CBDS-Connectionless Broadband Data Service
CBE-Certified Banyan Engineer
CBEMA-Computer Business Equipment Manufacturers Association
CBEMA-Computer And Business Equipment Manufacturers Association
CBF-Computer-Based Fax
CBH-Component Busy Hour
CBL-Common Business Line
CBQ-Class Based Queuing
CBR-Case-Based Reasoning
CBR-Constant Bit Rate
CBS-Columbia Broadcasting System
CBS-Certified Banyan Specialist
CBT-Computer Based Training
CBT-Core Based Trees
CBU-Caribbean Broadcasting Union
CBUD-Call Before U Dig
CC-Call Control
CC-Conference Calling
CC-Calling Channel
CC-Company Code
CC-Carbon Copy
CCB-Common Carrier Bureau
CCBM-Came Clear By Magic
CCBS-Customer Care And Billing System
CCC-Clear Channel Capability
CCC-Communications Competition Coalition
CCC-Clear Coded Channel
CCC-Call Content Connection
CCCH-Common Control Channel

CCD-Census County Division
CCD-Charge Coupled Device
CCD-Charge-Coupled Device
CCD-Charged Coupled Device
CCDA-Cisco Certified Design Associate
CCDN-Corporate Consolidated Data Network
CCDP-Cisco Certified Design Professional
CCF-Consumer Consultative Forum
CCFL-Cold-Cathode Fluorescent Lamp
CCH-Connections per Circuit per Hour
CCIA-Computer and Communications Industry Association
CCIE-Cisco Certified Internetworking Expert
CCIR-International Radio Consultative Committee
CCIRN-Coordinating Committee for Intercontinental Research Networks
CCIS-Common Channel Interoffice Signaling
CCITT-Consultative Committee for International Telephony And Telegraphy
CCL-Carrier Common Line
CCMA-Call Centre Management Association
CCNA-Cisco Certified Network Associate
CCNA-Customer's Carrier Name Abbreviation
CCNP-Cisco Certified Network Professional
CCP-Certified Computing Professional
CCR-Customer Controlled Reconfiguration
CCR-Current Cell Rate
CCRS-Centrex Customer Rearrangement System
CCS-Common Channel Signaling
CCS-Common-Channel Signaling
CCS-Centum Call Seconds
CCSA-Common Control Switching Arrangement
CCSA ACCESS-Common Control Switching Arrangement ACCESS
CCSD-Command Communications Service Designator
CCT-Continuity Check Tone
ccTLD-Country Code Top Level Domain
CCTV-Closed-Circuit Television
CCU-Communication Control Unit
CD-Call Delivery
CD-Carrier Detect
CD-Campus Distributor
CD-Cell Delineation
CD-Candelas
CD-Commitment Date
CD-Compact Disc
CD-E-Compact Disc Erasable
CD-ROM-Compact Disc-Read Only Memory
CD-ROM-Compact Disc—Read-Only Memory
CD-V-Compact-Disc Video

CD-WO-Compact Disc Write Once
CDA-Communications Decency Act of 1996
CDA-Certified Database Administrator
CDC-Call Data Connection
CDDD-Customer Desired Due Date
CDDI-Copper Distribution Data Interconnection
CDDI-Copper Distributed Data Interface
CDE-Common Desktop Environment
CDE-Solaris Common Desktop Environment
CDEV-Control Panel Device
CDF-Cutoff Decrease Factor
CDF-Channel Definition Format
CDF-Combined Distribution Frame
CDFS-Compact Disc File System
CDG-CDMA Development Group
CDI-Compact Disc-Interactive
CDIA-Certified Document Imaging Architect
CDL-Coded Digital Locator
CDLC-Cellular Data Link Control
CDLRD-Confirming Design Layout Report Date
CDM-Code Division Multiplexing
CDMA-Code Division Multiple Access
CDMA-IS-95
CDMA2000-Code Division Multiple Access 2000
CDMAone-A name for the IS-95 CDMA standard, used by the CDMA Development Group.
CDO-Codial Office
CDO-Community Dial Office
CDP-Customer Demarcation Point
CDP-Cisco Discovery Protocol
CDPA-Capture Division Packet Access
CDPC Cell Boundary-Cellular Digital Packet Data
CDPD-Cellular Digital Packet Data
CDPD Forum-Cellular Digital Packet Data
CDR-Clock/Data Recovery
CDR-Call Detail Record
CDR Exclude Table-Call Detail Record
CDR Processing-Call Detail Record
CDR Server-Call Detail Record Server
CDSA-Common Data Security Architecture
CDSL-Consumer Digital Subscriber Line
CDV-Compressed Digital Video
CDV-Cell Delay Variation
CDVCC-Coded Digital Verification Color Code
CDVCC [Field, 12 Bits]-Coded Digital Voice Color Code
CDVT-Cell Delay Variation Tolerance
CE-Connection Endpoint
CE Mark-European Conformity Mark
CEBus-Consumer Electronics Bus

CED-Called Equipment Identification Tone
CED-Capacitance Electronic Disc
CED-Called Station Identification
CEEFAX-A service developed in England by the BBC engineers launched a teletext service in 1974 called CEEFAX. These text broadcasts appeared and continue today (though there are many services available from other broadcasters) as pages of information on the TV one can access by punching in a three-number code on one's remote control and use of the Fast Keys (as we mentioned earlier). The "home page" (for want of a better term) of CEEFAX is 100.
CEF-Candidate Eligibility Filtering
CEF-Cable Entrance Facility
CEI-Comparable Efficient Interface
CEI-Connection Endpoint Identifier
Cell-(1 - cellular system) A radio coverage area associated with a fixed-location cellular or PCS radio tower that is interconnected with other cells to provide radio coverage to a larger geographic area. The term cell is often visualized as a hexagon as a relative building block depicting the ability of a cellular system to continually split so the system capacity can continually increase as new customers are added to the system. (2 - battery) A primary (disposable) or secondary (rechargeable) unit that stores energy for the supply to electrical or electronic equipment. (3- fuel) An electrochemical cell that produces electricity from the chemical energy of a fuel and an oxidant.
Cell Broadcast-The broadcasting of messages to one or a group of cell sites so that mobile radio receivers that are located anywhere within the coverage area of those cell sites can receive the broadcast messages.
Cell Broadcast Channel (CBCH)-A logical channel used to broadcast short messages.
Cell Coverage Area-The radio coverage area provided by a cell site that has a signal strength that is above a minimum acceptable area.
Cell Extender-A cellular system repeater that retransmits base station channels. A cell extender could be considered nothing more than a linear amplifier.
Cell Loss Priority (CLP)-A field contained in the header of an ATM packet (cell) that identifies the cell's discard priority in the event that network congestion occurs. A value of 1 indicates the cell can be discarded and a value of 1 indicates the cell has the highest priority.

Cell Rate-The cell rate is the number of cells per second (53 byte ATM packets) that are required to be passed through a switch for a particular call or communication session.
Cell Rate Margin (CRM)-The difference between a assigned cell rate (bandwidth allocation) and the sustained cell rate in an ATM system.
Cell Site-A transmitter-receiver tower, operated by a wireless carrier (typically cellular or PCS), through which radio links are established between a wireless system and mobile and portable units.
The photo image is a cell site self supporting radio tower near Peoria, Illinois for a wireless PCS system. This cell site raises the antenna high to 195 feet to allow cellular telephones to communicate with the system at greater distances.

Cell Site
Source: Rohn Industries, Inc

Cell Splitting-The process of dividing one cell site of a wireless telephone system into two or more new cell sites to provide additional capacity within the region of the original cell site.
This diagram shows the process of cell splitting that is used to expand the capacity (number of channels) of a mobile communication system. In this example, the radio coverage area of large cells sites are split by adjusting the power level and/or using reduced antenna height to cover a reduced area. Reducing the radio coverage area of a cell site by changing the

RF boundaries of a cell site has the same effect as placing cells farther apart, and allows new cell sites to be added.

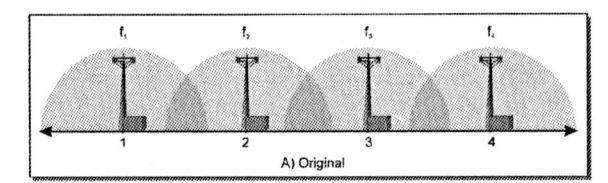

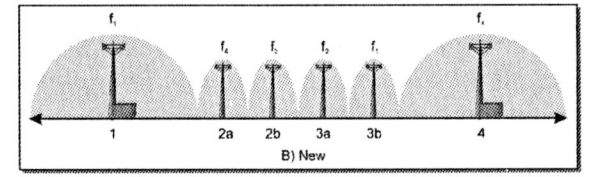

Cell Splitting Operation

Cell Switching-A term that describes how a mobile communication system transfers a communication path from one radio transmitter site (cell) to another cell as a mobile communication device (mobile telephone) moves throughout a cellular system.

Cell Type-A parameter identifying how the unit handles a neighbor cell. The cell type is set as either normal, preferred, or non-preferred. It is used by the phone to determine which reselection algorithm will be used in the control-channel reselection process. (See A/B Switching)

Cellular-A wireless local telephone system that divides geographic areas area into small radio areas (cells) that are interconnected with each other. Each cell coverage area has one or several transmitters and receivers that communicate with mobile telephones within its area.

Cellular Auxiliary Personal Communication Services (CAPCS)-Wireless PBX and residential services for mobile telephones. See IS-94.
Also an expanded service for the AMPS cellular system that allows business and residential services.

Cellular Carrier-A radio common carrier that provide cellular telephone service.

Cellular Digital Packet Data (CDPD)-Cellular Digital Packet Data (CDPD) is a wide area data transmission technology developed for use on cellular phone frequencies. CDPD uses unused cellular channels to transmit data in packets. This technology offers data transfer rates of up to 19.2 Kbps, better error correction and quicker call set up than using modems on an analog cellular channel.

This picture shows the Sierra Wireless AirCard 300, a cellular digital packet data (CDPD) wide-area wireless network (WAN) interface card. The AirCard 300 connects a laptop, handheld computer or PDA device directly to the Internet, intranet, and corporate applications without the need for a wireless phone or landline. It provides mobile users instant wireless access in most major cities in the United States and cities in other countries that have CDPD wireless data networks.

CDPD Wireless Data Card
Source: Sierra Wireless

Cellular Geographic Service Area (CGSA)-A geographic area in cellular telephony of frequency assignments in which virtually any number of users can be accommodated with a fixed number of allocated frequencies.

Cellular Intercarrier Billing Exchange Roamer (CIBER)-A billing standard designed to promote inter-carrier roaming between cellular telephone systems. The CIBER format is developed and maintained by CiberNet. Cibernet is owned by the Cellular Telecommunications Industry Association (CTIA).

This figure shows some of the information (fields) contained in a type 22 CIBER record. This example shows that the type 22 Ciber record field structure has been updated from the previous type 20 record structure to include additional fields that allow for

Cellular Line Interface (CLI)

telephone number portability (enabling telephone number transfer between carriers). This list shows that fields in the Ciber record primarily include identification of airtime charges, taxes, and interconnection (toll) charges.

Sample Charge Record

Type 22 Record - Sample of Fields

- Home Carrier SID/BID
- MIN/IMSI
- MSISDN/MDN
- ESN/IMEI
- Serving Carrier SID/BID
- Total Charges and Taxes
- Total State/Province Tax
- Total Local Tax
- Call Date
- Call Direction
- Call Completion Indicator
- Call Termination Indicator
- Caller ID
- Called Number
- LRN
- TLDN
- Time Zone Indicator
- Air Connect Time
- Air Chargeable Time
- Air Rate Period
- Toll Connect Time
- Toll Chargeable Time
- Toll Carrier ID
- Toll Rate Class

Cellular Intercarrier Billing Exchange Roamer (CIBER) Structure

Cellular Line Interface (CLI)-A device or assembly that converts signals from a cellular telephone to telephone line signals. A CLI allows standard telephone devices to be connected to and operate a mobile telephone. The CLI decodes touch tone signals and creates the dialing digits and automatically initiates a call (initiates the SEND button) after the dialing is completed. The CLI also creates the ring signal when an incoming call is received.

This pictures shows the CellSocket produced by WHP Wireless that can converts cellular or PCS mobile telephone signals into telephone lines signals that can be used with standard telephones.

Cell Socket
Source: WHP Wireless

Cellular Mobile Carriers (CMC)-The service providers that are licensed by the FCC to provide service in the CGSA. There are two CMCs authorized for each CGSA, the A and B side.

Cellular Mobile Radiotelephone Service (CMRS)-The Federal Communications Commission's (FCC's) description of a high-capacity mobile and portable radio service that divides the available frequency spectrum into discrete channels that have frequencies that are reused in smaller cells covering a geographic service area. In the Telecommunications Act of 1996, the definition of CMRS included enhanced specialized mobile services (ESMR) that provide interconnection to the public switched telephone network (PSTN).

Cellular Network-Cellular systems are comprised of a set of radio towers that are strategically distributed over a geographical area in order to provide a continuous service coverage area. A mobile switching center (MSC) provides the switching and control functions necessary to connect calls from the public switched telephone network (PSTN) to the individual mobile telephones. The MSC also manages the radio resources within the entire network.

Cellular Operator-The owner and/or operator of a cellular network

Cellular Paging-Cellular (mobile) paging is a process of delivering a call alert (command for a ring signal) via a cellular communication system. When the mobile radio receives a page message, it begins to alert the user that an incoming call is waiting to be received. If the user responds (usually by pressing SEND or GO,) the mobile radio will respond to the system with an answer message and a communication channel can be assigned.

Cellular Radio-A mobile radio system having two distinguishing properties compared to other mobile radio systems: 1) Re-use of the same radio frequency/frequencies in different cells of the system. 2) Handoff (or handover) of a call in progress, typically between two adjacent cells. Each cell has a distinct radio coverage area (aside from some intentional overlap of radio coverage at the boundaries between adjacent areas) and base antennas. Property 1 distinguishes cellular radio from earlier systems like Improved Mobile Telephone Service (IMTS), that use the same frequency throughout an entire city or large service area. Property 2 distinguishes cellular radio from trunked radio, which has reuse (property 1) but not handover. Some people use the term Personal Communication System

(PCS) for certain types of cellular radio systems because they operate on different radio frequency bands (1.9 GHz vs. 800 MHz, for example), or because they are digital vs. analog, or because of the particular service provider who owns and operates the base infrastructure equipment.

Cellular Radio Switching Office-The electronic switching office which switches calls between cellular (mobile) phones and wireline (I.e. normal wired, landline) phones. The switch controls the "handoff" between cells and monitors usage. Different manufacturers call their equipment different things (MTSO, MTX, etc.), as usual. The most widely used generic term is mobile switching center (MSC).

Cellular Subscriber Station (CSS)-Another name for mobile phone. This is the preferred term to network providers when Mobile Stations may be fixed in location.

Cellular Switching Office-The hub controller that controls a cellular communication network. A common name for a cellular switching office is the mobile switching center (MSC)

Cellular System- A fully automatic, wide-area, high-capacity RF network made up of a group of coverage sites called cells. As a subscriber passes from cell to cell, a series of handoffs maintains smooth call continuity.

This figure shows the basic parts of a cellular mobile communication system. The mobile telephone has the ability to tune in to many different radio channel frequencies or codes. The base station commands the mobile telephone on which frequency to use in order to communicate with another base station that may be from two to fifteen miles away. The base station routes the radio signal to the MSC either by wire (e.g. a leased telephone line), microwave radio link, or fiberoptic line. The MSC connects the call to the public switched telephone network (PSTN) and the PSTN then connects the call to its designation (e.g. office telephone.)

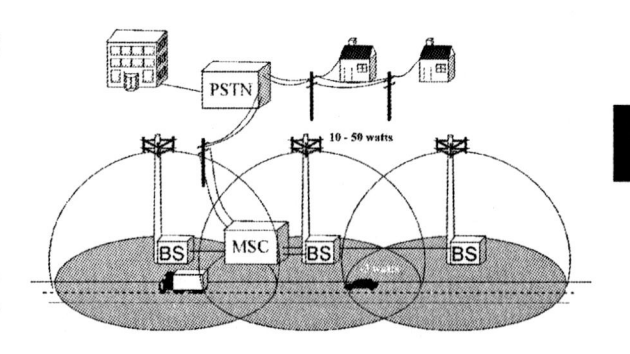

Cellular System

Cellular Telecommunications Industry Association (CTIA)-The Cellular Telecommunications Industry Association, a trade organization of cellular telephone operators.

Cellular Test Set-A test Instrument that provides a simulated cellular base station signal to a cellular telephone of a specification type (e.g. TDMA, CDMA or GSM) and measures the performance (such as transmission power, frequency and audio quality) of the cellular telephone.

This photo image shows a portable cellular test set that can be used to test the functionality and performance of several types of mobile telephones.

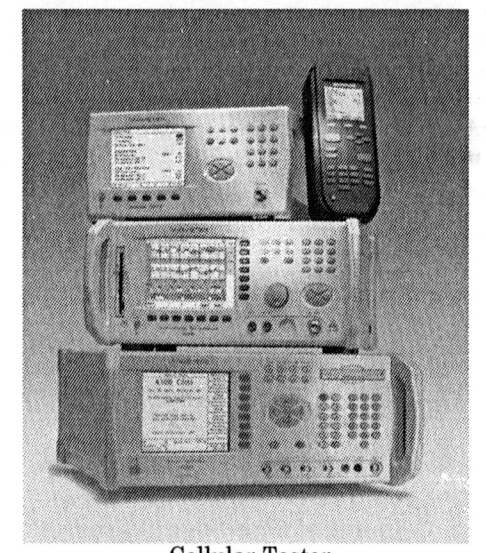

Cellular Tester
Source: Willtek Communications

CELP-Code Excited Linear Predictive
CELP-Code Excited Linear Prediction
CEMH-Controlled Environment Man Hole
Center Tap-A connection made at the electrical center of a coil.
Central Battery System-A centrally located battery system that is used to supply power to communication lines that are part of a telephone network (such as a local exchange network).
Central Billing District (CBD)-The billing district that is at the center of a city or network.
Central Office (CO)-The name commonly used in North America to identify the switch in a telephone network that connects local customers. The international name is "public exchange" or "telephone exchange." CO actual refers to the building that contains the switching and interconnection equipment.
Central Office Area-A geographic area served by an end office.
Central Processing Unit (CPU)-The electronic components of a computer that interpret instructions, perform calculations, move data in main computer storage, and control input/output operations. Often this expression refers to a mainframe or midrange machine.
Centralized Attendant Service (CAS)-A centralized group of customer service operators (attendants) who answer incoming calls for multiple telephone systems.
Centralized Automatic Message Accounting (CAMA)-A system for the recording of detailed billing information at a central location other than an end office, usually at a tandem.
Centralized Automatic Reporting On Trunks (CAROT)-A system that automatically and routinely reports on the performance of trunk lines to ensure high availability and to avoid potential failures. The reporting system may check for return loss, line noise, timing precision, transmission and call processing performance criteria.
Centralized Control-A system that relies on a central intelligence to coordinate the overall operation of the system.
Centralized Message Data System (CMDS)-A system that allows interexchange of billing and usage data between telephone exchange networks. CMDS has four parts: (1) Centralized Message Data System (CMDSI) which provides for collect services (collect, third party number, and calling card); (2) Carrier Access Billing System (CABS) that bills interexchange carriers (IXCs) for local exchange access services; (3) 800 Service Usage; and (4) Meet Point Billing which involves the billing of access services (via CABS) provided to two or more interexchange carriers.
Centralized Routing-Centralized routing uses a coordinated network control to determine the routing of packets of data through a communication network. Centralized routing switches are not very intelligent (e.g. "dumb") and the system switching can be more vulnerable to central system failures.
Centralized Trunk Test Unit (CTTU)-A test system that test trunks through a data link on a switch from a centralized location.
Centrex-Centrex is a service offered by a local telephone service provider that allows the customer to have features that are typically associated with a private branch exchange (PBX). These features include 3 or 4 digit dialing, intercom features, distinctive line ringing for inside and outside lines, voice mail waiting indication and others. Centrex services are provided by the central office switching facilities in the telephone network.
This figure shows a typical Centrex system. This diagram shows that the EO switch is equipped with Centrex software. Individual ports from the switch are connected to individual telephones at a company. The public telephone company programs the Centrex features for specific companies into the switch software. The Centrex software monitors the ports so advanced features such as abbreviated dialing can be performed. This example shows that a telephone that is used in a Centrex system can dial a 4 digit dialing number to reach a telephone connected within the companies Centrex telephone network.

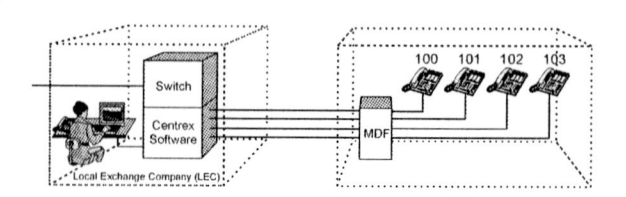

Centrex System

Centronics Printer Connector-A parallel wire connector traditionally used for computer printers, named for the first manufacturer to use this particular connector for a printer.

Centum Call Seconds (CCS)-A measurement of communication trunk usage that equals 100 seconds of continuous usage. Because the standard time interval for communication network engineering is based on activity over an hour, the system load is expressed in hundred call seconds (CCS) for one hour. A single hour has thirty-six hundred call seconds (equal to one erlang).

CEO-CABS End Office
CEO File-CABS End Office File
CEPT-European Committee of Post and Telegraphs
CEPT-Conference Of European Postal And Telecommunications Administrations
CEPT Format-European Conference of Postal and Telecommunications Administations
CER-Cell Error Ratio
CERB-Centralized Emergency Reporting Bureau
CERT-Computer Emergency Response Team
CES-Circuit Emulation Switching
CESID-Caller Emergency Service Identification
CEV-Controlled Environmental Vault
CF-Crest Factor
CF-Collection Function
CFA-Carrier Facility Assignment
CFA-Connecting Facilites Arrangement
CFA-Connecting Facility Assignment
CFA-Carrier Failure Alarm
CFB-Call Forwarding on mobile subscriber Busy
CFB-Call Forwarding-Busy
CFCA-Communications Fraud Control Association
CFD-Call Forwarding-Default
CFM-Cubic Feet per Minute
CFNA-Call Forwarding-No Answer
CFNRc-Call Forwarding on mobile subscriber Not Reachable
CFNRy-Call Forwarding on No Reply
CFR-Code Of Federal Regulations
CFR-Confirmation To Receive Frame
CFRP-Carbon Fiber Reinforced Plastic
CFU-Call Forwarding-Unconditional
CFU-Call Forwarding Unconditional
CFUC-Call Forwarding UnConditional
CFV-Call For Votes
CG-Character Generator
CGA-Color Graphics Adapter
CGA-Carrier Group Alarm
CGI-Common Gateway Interface
CGI-Bin-Common Gateway Interface-Binary
CGM-Computer Graphics Metafile
CGSA-Cellular Geographic Service Area
CGSA Restriction-Cellular Geographic Service Area

Challenge Handshake Authentication Protocol (CHAP)-A security protocol that performs an authentication procedure to identify of a user before they are given access to service. The CHAP protocol provides random numbers that are used with previously known information to calculate an authenticated response. The calculated authenticated response information is transferred rather than the passwords or other secure data.

Change Point-The point at which a cable run changes from one type of cable to another.

Changeback-In the Signaling System 7 protocol, the procedure of transferring signaling traffic from one or more alternative signaling links to an available signaling link.

Changeback Code-In the Signaling System 7 protocol a field in the signaling network management messages that facilitates the changeback procedure. The code helps distinguish among messages relating to different changeback procedures performed at the same time relative to the same signaling link. Changeover in the Signaling System 7 protocol, the procedure for transferring signaling traffic from one signaling link to one or more different signaling links when the link in use fails or must be cleared of traffic.

Channel-(1 - general) In general, a stream of information transmitted as part of a distinct communication, or conversation, or for a particular purpose or end use. One channel may be distinguished from other channels by the time of occurrence of the transmission, by the format or organization of its content, by the frequency of a carrier signal used to transmit it, or by some secondary property such as the type of error detecting code used for it, or by other properties. Due to the advance of technology, a single channel may at some time historically be modified so that it carries multiple channels. For example, when the only distinguishing feature of two radio signals was their carrier frequency and each one carried only one conversation, each one was described as a channel. Different carrier frequencies were designated by distinct channel numbers. At a later date, time division multiplexing was used with radio signals to distinguish 3 or 8 distinct conversations on one modulated radio carrier frequency. Confusion arose because the 3 or 8 distinct conversations, each using a designated time slot in time division multiplexing, were described as separate channels. At the same time, the entire signal (comprising 3 or 8 channels) was also described by some as a channel. In some cases the reader must read carefully, with awareness of the changing historical meanings of some terms. Regarding this particular example, the entire conglomeration of 3 or 8 time slots is described best as a "modulated carrier" and not as a channel.. (2 - broadcasting) A portion of the radio frequency spectrum assigned to a particular broadcasting station. (3- signal path) A transmission path between two or more termination points. The term channel can refer to a 1-way or 2-way path. (4 - video effects) A digital effects processing path for video. (5) A single unidirectional or bi-directional path for transmitting or receiving, or both, of electrical or electromagnetic signals. Line is another name.

This diagram shows the difference between physical and logical channels. In this example, a physical channel transports information between two points using electrical signals. The physical channel is divided into frames that contain various fields (groups of information within the frame). This diagram shows that the frames on the physical channel are divided into 4 logical channels; 3 logical channels for data and one logical channel for control. The exact relationship between the frame structure and the logical channels is called mapping. This example shows two different mapping examples. In the first example, the bits in the information portion of each frame are equally divided. In the second mapping example, more bits are proportioned to channel 2. This results in a lower data transfer rate for channels 1 and 3 while channel 2 has a higher data transfer rate.

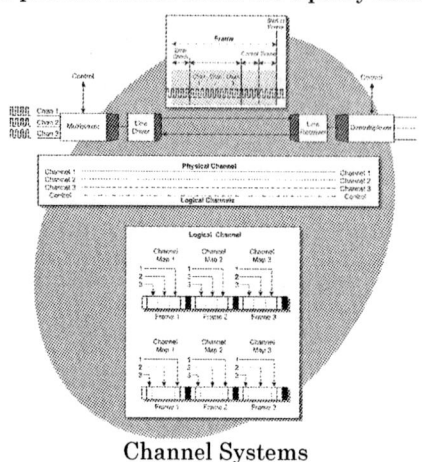

Channel Associated Bit Signaling-See Associated Bit Signaling

Channel Associated Signaling-A signaling method in which the signaling information necessary for the traffic carried by a single channel is transmitted in the channel itself or in a signaling channel permanently associated with it.

Channel Associated Signaling (CAS)-Channel associated signaling (CAS) is the transmission of signaling information within the voice channel. With CAS, circuit state is indicated by one or more bits of signaling status sent repetitively and associated with that specific circuit. E1 CAS uses the ABCD signaling bits of channel (timeslot) 16. T1 CAS robs a bit from every 6th frame of a multiframe. In both cases, these signaling bits set conditions such as On Hook, Off Hook, Coin Control, and Dial Pulsing.

Channel Busy Tone-An audible signal indicating that a call cannot be completed because all switching paths or toll trunks are busy, or that equipment is blocked. The tone is applied 120 times per minute. The channel busy tone also is called fast busy, all trunks busy, or reorder tone.

Channel Capacity-The amount of data or channel transmission capability of a communication channel.

Channel Coding-Channel coding is a process where one or more control and user data signals are combined with error protected or error correction information. After a sequence of digital data bits has been produced by a digital speech code or by other digital signal sources, these digital bits are processed to create a sequence of new bit patterns that are ready for transmission. This processing typically includes the addition of error detection and error protection bits along with rearranging of bit order for transmission.

Channel Collision-The result of two cellular phones trying to use the same channel at the same time. Cellular networks provide safeguards to inhibit this.

Channel Combiner-A device or filter assembly that allows several modulated carrier signals (physical channels) to be grouped on to the same transmission channel or antenna system.

Channel Systems

Channel Encoder-A device that converts a signal into a form suitable for transmission over the communications channel.

Channel Establishment-(1-general) The process of connecting a communication channel (physical and/or logical) (2-Bluetooth) Procedure for establishing a Bluetooth channel (a logical link) between two Bluetooth devices using the Bluetooth File Transfer Profile Specification. Channel establishment starts after link establishment is completed and the initiator sends a channel establishment request.

Channel Extension-A system that enables peripheral equipment, such as high-speed tape spoolers, printers, and terminals, to be connected to mainframe computers many miles away.

Channel Interleaving-Channel interleaving is the process of offsetting radio channel frequency spacing at nearby radio towers in a wireless communication system. This allows radio service provides to place more radio channels in each tower, reducing the total number of towers needed to provide a number of communication channels in a geographic area. Channel interleaving reduces the interference between nearby channels because the radio energy for both the nearby channel and interleaved channel is primarily concentrated at the center of the each radio channel frequency band. There is a reduced occurrence (minimized interference) of both modulated carriers fully occupying their frequency bands at the same time. In essence, it is acceptable to overlap in to each other's channel frequency band, just not at the same time.

This diagram shows how radio channels can be interleaved (offset in frequency) at nearby radio towers to allow mobile radio service provides to place more radio channels in each tower (reducing the total number of towers needed). This example shows that the radio signal energy is primarily concentrated at the center of the radio channel and that the signal energy decreases at the signal is propagated from the tower. The use of frequency offsetting (interleaving) radio channels at nearby radio towers minimizes the amount of interference between channels compared to channels located at the same frequency.

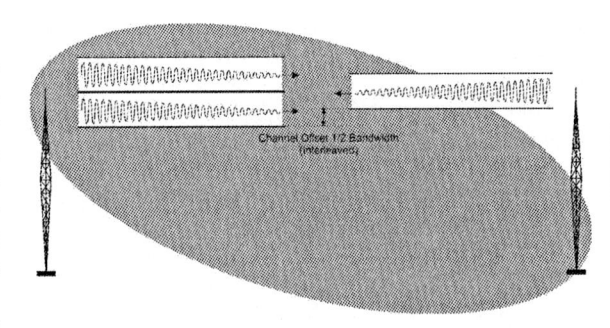

Channel Interleaving System

Channel Management-Formal process utilized to manage the creation, staffing, tasking, and measurement of sales and customer support channels.

Channel Mapping-A relationship that allows multiple communication channels to be assigned to other transmission channel formats. An example is mapping multiple DS3 channels to the frames within a synchronous optical network (SONET) transmission channel.

This diagram shows how multiple communication channels can be mapped (related to) specific portions of another communication channel (payload). Mapping allows several communication channels to share the same physical or logical data transmission medium.

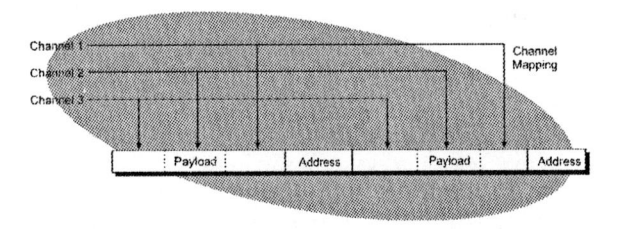

Channel Mapping Operation

Channel Modulator

Channel Modulator-A channel modulator is used to convert video signals into television broadcast channels. Channel modulators are used in cable-TV networks to convert a video program signal (such as CNN or MTV) and converts it with an RF carrier frequency for a television channel that is distributed through the CATV network. The modulator converts both video and audio signals. The frequency of this channel modulator carrier determines the television channel number (I.e., 2 to 99) that the program will be received on by subscribers.

This photograph shows UHF, television channel modulator, model number 7644. The features of this modulator include phase locked loop (PLL) frequency synthesized signal for stable channel frequency that can be set with a DIP switch. This modulator can be used with direct broadcast satellite (DBS) receivers and security cameras. It has an output frequency that may be set by 1 MHz increments, allowing CATV as well as off-air channel plans.

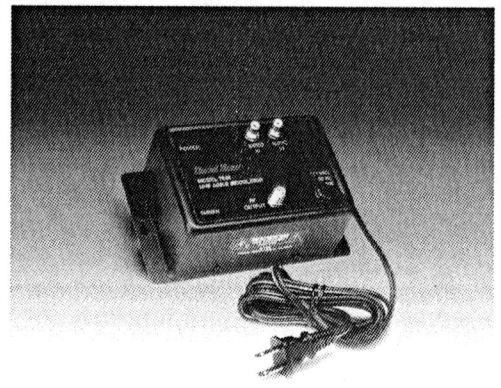

UHF Channel Modulator
Source: Channel Master

Channel Multiplexing-Channel multiplexing is a process that divides a single transmission path into several parts that can transfer multiple communication (voice and/or data) channels. Multiplexing may be frequency division (dividing into frequency bands), time division (dividing into time slots), code division (dividing into coded data that randomly overlap), or statistical multiplexing (dynamically assigning portions of channels when activity exists).

Channel Occupancy-The time that a radio channel is used to carry traffic.

Channel Quality Message (CQM)-Messages sent on a communication system (such as a mobile communication system) that provide the remote connection (e.g. base station) with channel quality information. Channel quality information may include carrier level received signal strength indication (RSSI) and bit error rate (BER).

Channel Rate-The data rate at which information is transmitted through the channel or communications media, typically stated in bits per second (BPS).

channel Re-use-The re-use of the same radio channel frequency at a distant radio tower in a wireless network. The minimum separation distance for the towers is determined by several factors including the power level and ability of the modulation type to reject interference. See also: Co-channel Interference.

Channel Reliability-The percentage of time a channel is available for use in a specific direction during a specified period.

Channel Service Unit (CSU)-A CSU is the hardware that is used to assign communication channels from one or more data terminal equipment (DTE) devices to logical channels on a multi-channel communication circuit. The CSU can be customer premises equipment (CPE) or provided by the telecommunications service provider. The CSU is often combined with a data service unit (DSU).

Channel Spacing-The spacing in a radio frequency band (in Hertz) between adjacent radio carrier signals. It is measured from the center of one channel to the center of the next adjacent channel.

Channel Splitter-A device or assembly that can split the signal energy into multiple outputs (taps) or separate communication channels (de-multiplex).

Channel Switching-The switching or rearrangement of dedicated channels that may use electronic equipment, such as digital cross-connect systems or circuit switches.

Channel Time Slot-A repetitive period of time or number of bits related to a frame or group of information on a physical channel that is used for a specific channel.

Channel Translator-A device or system that is used to change the channel frequency of one channel to another channel. Channel translators are commonly used to convert cable system upstream channels (from the customer to the system) into downstream channels (from the system to the customer).

Channel Unit-Equipment that provides transmission and signaling functions for one voice or data channel out of the channels that are available. A channel unit plugs into a carrier channel bank. (See also: CSU/DSU.)

Channelized Carrier-A communication line (carrier) that is divided into multiple logical communication channels. Users may be allowed to access some or all of the logical channels.

CHAP-Challenge Handshake Authentication Protocol

Character Generator (CG)-A device used to generate graphics, letters, or numbers in a video format. The characters are subsequently keyed over program video or background video.

Character Interval-The total number of unit intervals, including synchronizing, information, error checking, and control bits, required to transmit any given character in a communications system.

Character Reader-A device capable of reading typescript and producing a machine-readable output without operator assistance.

Character Set-A set of different characters such as letters, numbers, and symbols that are agreed upon to represent characters in a computing system.

Charge-(1-electric) The product of electric current and time. The SI unit of electric charge is the coulomb, named for a 19th century French physicist. One coulomb is the product of one ampere of current and one second of time. There are approximately 6,250,000,000,000,000,000 electrons in one coulomb of electric charge. To state that in another way, the electric charge of one electron is negative and its magnitude is 1.6 E-19 coulomb.(2-verb) The process of replenishing or replacing the electric charge in a capacitor, secondary cell or storage battery. (3-cost) Charge is the name for a price or fee or cost of some goods or services. The verb form of this word is the action of computing or applying a cost.

Charger-A device used to recharge a battery. Types of charging include: (1) constant-voltage charge, (2) equalizing charge, and (3) trickle charge.

Chassis Ground-A connection to the metal frame of an electronic system that holds the components in place. The chassis ground connection serves as the ground return or electrical common for the system.

Chat Room-Chat rooms are real-time communication services that allow several participants (typically 10 to 20) to interact act through the use of text messaging. Chat rooms may be public (allow anyone to participate) or private (restricted to those with invitations or access codes.)

Cheapernet-A slang term for thin-wire coaxial Ethernet (10BASE2).

Checksum-A number that is calculated from a block of data or sequence of data bits that is sent along with the data and used by the receiver to detect if transmission errors occurred. The receiver will use the same calculation process on the received data to calculate an compare the appended check sum with the value it has calculated. In some cases, the checksum can be used to correct some of the bit errors.

CHI-Concentration Highway Interface

CHILL-CCITT High Level Language, a software programming language (often used as a pseudo-language for purely descriptive purposes) standardized by the ITU to describe telephone switch software. Related to SDL.

China Wireless Telecommunication Standard Group (CWTS)-An association in China that oversees the creation of wireless telecommunications standards.

Chip On Board (COB)-A group of electronics components in very small plastic packages that are directly mounted on a printed circuit (PC) board. Unlike integrated circuits (ICs) that have a sturdy protective shell, COB circuits use the structure of the PC board for durable mounting and connection. Because COB circuits eliminate a mounting case and COB assemblies can be placed on both sides of a PC board, the density of components can high and the thickness of the PC board assembly can remain small.

Chirp-A signal of time-varying frequency.

Choke-An inductor that passes direct current but attenuates or blocks alternating current. A choke often is installed on a direct-current power lead to prevent undesirable alternating currents from reaching and interfering with the powered equipment.

Chroma Key-A video key effect in which one video signal is inserted as a replacement for a particular color in another video signal. For example, a weatherman stands in front of a blue wall with a camera focused on him. The camera signal feeds a chroma keyer, which detects the blue in the blue wall and replaces it with video from another camera, such as video of a weather map. The finished key makes the weatherman appear to be standing in front of the weather map.

Chromatic Dispersion

Chromatic Dispersion-A physical effect that limits data rates in optical fibers. The speed of light in a fiber is dependent on the wavelength of the light. Even apparently single wavelength sources, such as high performance lasers, emit pulses made up of a small range of wavelengths. As the pulse travels down the fiber, some wavelengths travel more slowly than others. As a result, the pulse broadens. If it broadens too much or the interval between pulses becomes too small, the receiver will not be able to distinguish separate pulses. The effect is called Chromatic Dispersion because wavelength determines the color of light (especially visible light).

Chromaticity-The attribute of light combining hue and saturation, independent of intensity. The color perceived is determined by the relative proportions of the three primary colors. Chromaticity is the color quality of light defined by wavelength and purity.

CHRP-Common Hardware Reference Platform

CHS-Cylinder-Head Sector

Churn-The process of customers disconnecting from one telecommunications service provider. Churn can be a natural process of customer geographic relocation or to may be the result of customers selecting a new service provider in their local area.

Churn Posture -The churn posture defines the default strategic position that upper management desires to communicate to others (within the company and to outsiders such as investors) regarding their values related to customer churn. Examples of churn posture include low or high value of churn management (customer retention).

CI-Clear Indication/Clear Request

CI-Circuit Identity

CI-Cell Identity

CI-Congestion Indicator

CIBER-Cellular Intercarrier Billing Exchange Roamer

CIC-Circuit Identification Code

CIC-Carrier Identification Code

CICS-Customer Information Control System

CID-Connection Identifier

CIDB-Calling Line Identification Delivery Blocking

CIDR-Classless Internet Domain Routing

CIDR-Classless Inter-Domain Routing

CIF-Common Intermediate Format

CIF-Cells in Frames

CIF-Cost, Insurance and Freight

CIF-Cells in Flight

CIF-AD-Cells In Frames-Attachment Device

CIFS-Common Internet File System

CIG-Calling Subscriber Identification

CIID-Card Issuer Identifier Code

CIM-Customer Information Manager

CIM-Common Information Model

CIO-Chief Information Officer

CIP-Carrier Identification Parameter

Cipher-An algorithmic transformation of data typically used to prevent unauthorized viewing or alteration of the original data. There are many different types of ciphers. One of the most common is based upon the Digital Encryption Standard, DES, algorithm.

Cipher Key (Kc)-(1-general) A secret key that is used in the authentication process. (2-GSM) A code key created by the GSM encryption algorithm that results from the key code Ki and a random number that is sent by the system. Kc is used by the GSM network as part of the authentication process.

CIR-Committed Information Rate

Circuit-(1 - communication) Any communication path through which any information can be transferred. (2 - electronics) A combination of electrical processing components that perform a process (such as signal amplification) or function (clock display processor).

Circuit Access Code (CAC)-A computer-generated code identifying a trunk group.

Circuit Assignment-A process that identifies, reserves, or designates a partially or wholly inventoried equipment item to a circuit.

Circuit Bonding-The combining of multiple circuits to one or more communication channels to increase reliability through circuit redundancy or to share data services.

Circuit Data-The continuous flow (or continuous connection) between two data devices.

Circuit Identification Code (CIC)-An information code that identifies a circuit between a pair of SS7 exchanges, for which signaling is being performed (14 bits in the ANSI version and 12 bits in the ITU version ISDN User Part).

Circuit Layout Record Card-The card containing a circuit layout record and related information.

Circuit Miles-The route miles of circuits in service, deter-mined by measuring the length in miles of an actual path followed by a transmission medium.

Circuit Noise-The noise that is measured or heard across an audio pair between the tip and ring conductors in a telephone system.

Circuit Switching

Circuit Provision-The process used by carriers to determine the need for trunks and special-service circuits.

Circuit Provision Center-A center that assigns facilities and equipment for message trunk circuits and designs special-service circuits and carrier systems. A circuit provision center also generates and maintains circuit records and inventory and assignment records for all interoffice facilities and equipment.

Circuit Provisioning-The process used by a service provider (network operator) that provides a circuit connection between two (or more) points.

Circuit Reliability-The percentage of time a circuit is avail-able to the user during a specified period of scheduled availability.

Circuit Switch-A device or assembly that receives signals on input ports and provides a continuous connection (may be a physical or logical connection) to output ports.

Circuit Switched Data-Circuit switched data is a data communication method that maintains a dedicated communications path between two communication devices regardless of the amount of data that is sent between the devices. This gives to communications equipment the exclusive use of the circuit that connects them, even when the circuit is momentarily idle.

To establish a circuit-switched data connection, the address is sent first and a connection (possibly a virtual non-physical connection) path is established. After this path is setup, data is continually transferred using this path until the path is disconnected by request from the sender or receiver of data.

This figure shows a circuit switched data system. In this figure, a computer is sending a data file through a circuit switched data communications network to a home office computer. To start the data file transfer, the computer sends the destination address (address 202.196.22.45) to the data network. The destination address is used to program the switches between the points on which ports they will receive the data and which ports they will send the data. As soon as all the switching connections are made, the computer can start sending data to the office. Throughout the connection, this path will be maintained through the initial path (the same switch ports) without any changes.

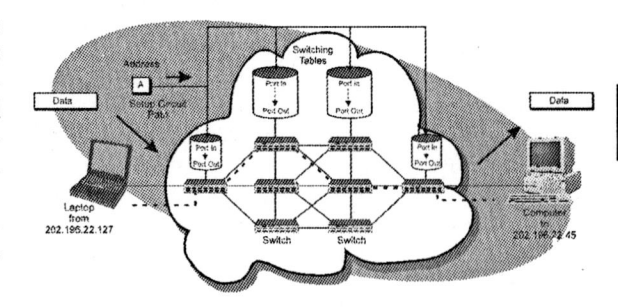

Circuit Switched Data Operation

Circuit Switching-A process of connecting two points in a communications network where the path (switching points) through the network remains fixed during the operation of a communications circuit. While a circuit switched connection is in operation, the capacity of the circuit remains constant regardless of the amount of content (e.g. voice or data signal) that is transferred during the circuit connection.

This figure shows how circuit switching is used for voice communication. In this example, a telephone is dialing a telephone that is connected to a distant switch. When the user dials the telephone, the dialed digits are captured and used to program the circuit switches between the two telephones. Each switch then has assigned input ports and output ports and each switch only adds a small amount of transfer time between ports. After the all the

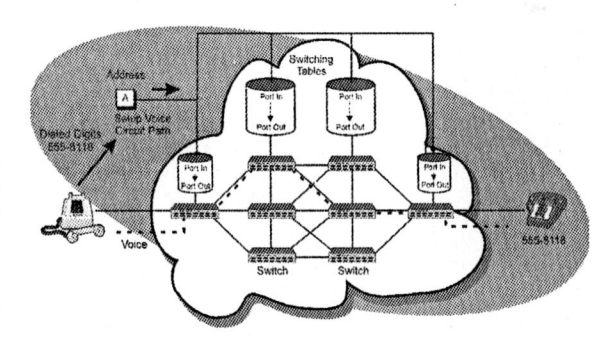

Circuit Switching Operation

Circuit Validation Test (CVT)

switching connections are made, an audio path can be connected between. Throughout the connection, this path will be maintained through the initial path (the same switch ports) without any changes.

Circuit Validation Test (CVT)-A SS7 signaling procedure used to ensure that the communication path between two exchanges have sufficient and consistent translation data for placing a call on a specific circuit.

Circuit Work Location-Any work location associated with a given circuit, such as the service address for a customer termination, for the central office or intermediate locations and for remaining terminations.

Circular MIL-The measurement unit of the cross-sectional area of a circular conductor. A circular mil is the area of a circle whose diameter is one mil, or 0.001-inch.

Circular Orbit-The path of a satellite in which the distance between the centers of mass of the satellite and the primary body is constant.

Circularly Polarized-The characteristic of a radio wave or lightwave whose plane of polarization rotates as the wave propagates in a forward direction.

CIS-Contact Image Sensor
CISC-Complex Instruction Set Computing
Cisco Discovery Protocol (CDP)-The Cisco Discovery Protocol or CDP is a proprietary discovery protocol developed by Cisco Systems, Inc. CDP is implemented on most of Cisco's network devices by default. CDP is a layer 2 discovery protocol. Cisco devices send out CDP packets every 30 seconds and are Layer 2 multicast packets (MAC address: 01-00-0c-cc-cc-cc).
CISSP-Certified Information Systems Security Professional
CIT-Computer Integrated Telephone
CIV-Cell Interarrival Variation
CIVDL-Collaboration for Interactive Visual Distance Learning
CIX-Commercial Internet Exchange
CIX-Commercial Internet eXchange Association
CJD-Certified Java Developer
CJP-Certified Java Programmer
CKR-Circuit Reference
Cl-II-Second Computer Inqukry
Cladding-Material that surrounds the core of an optical fiber. The cladding redirects (reflects or refracts) light so that it can travel through the fiber instead of escape from the fiber core. Cladding also protects the fiber core against scattering due to surface-contaminant impurities. In plastic-clad silica fibers, the plastic cladding also can serve as the coating.

Claims-Claims are the specially written sentences at the end of patent documents which set forth the metes and bounds of the patent monopoly. The claims describe the scope of protection of the patent. Claim drafting is subject to a myriad of specific requirements for form and style which vary among countries and regions. Claims of identical language may mean different things in different countries.

Clamp (Clamping)-(1-circuit) The device or process that restores the dc component of a signal. A video clamp circuit, usually triggered by horizontal synchronizing pulses, reestablishes a fixed dc reference level for the video signal. Some damp circuits clamp sync tip to a fixed level, and others clamp back porch (blanking) to a fixed level A major benefit of a clamp is the removal of low-frequency interference, especially power line hum. (2-mechanical) A device used to couple parts together mechanically.

Clamp-On Ammeter-An ammeter (current meter) that can clamp onto a wire to measure its current flow.

Clamping-Clamp
CLAS-Centrex Line-Assignment Service
CLASS-Custom Local Area Signaling Services
CLASS-Custom Local Access Signaling Services
CLASS (TM) Services-A group of telecommunication services that provide selective-call screening, alerting, and calling-identification delivery functions. Bellcore owns the trademark for CLASS (TM) services.

Class 4-A voice communications switching system used to interconnect local telephone switching systems. The class 4 switching system was one level above the class 5 end office switching system. Class 4 switches are also known as "Tandem Switches."

Class 5 Electronic Switching System-A classification of a switching system that is used by local telephone service providers. A class 5 switch is the last point in the network prior to the customer. Class 5 switches usually can handle from 10,000 to 100,000 customers.

Telica's Plexus ™ 9000 is a carrier-grade switch that includes all the modules of a softswitch-based solution, media gateway, media gateway controller and signaling gateway, to provide Class 4/5 switching for the PSTN as well as ATM and IP networks.

These various softswitch components may be deployed in either a distributed, centralized or integrated configuration to meet a wide range of network requirements.

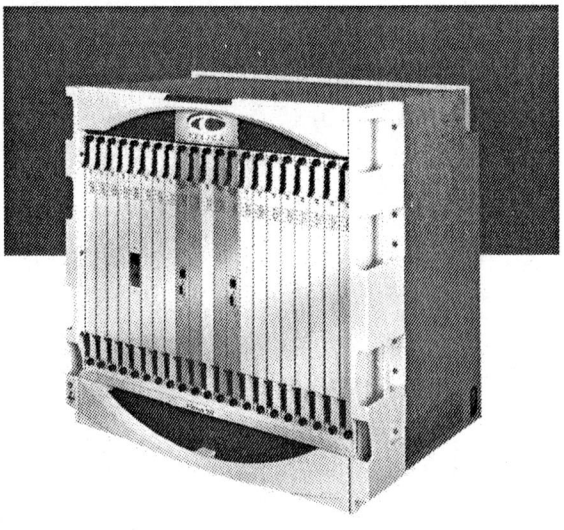

Class 5 Next Generation Switching System
Source: Telica Inc.

Class A Amplifier-An amplifier that is operating in class A (fully linear).
Class AB Amplifier-An amplifier that is operating in class AB. Class AB amplifiers can amplify between 180 degrees to 360 degrees of the input signal (somewhat non-linear).
Class B Amplifier-An amplifier that is operating in class B. Class B amplifiers can amplify a signal for 180 degrees of its input cycle (non-linear amplification for particular portions of the input signal).
Class Based Queuing (CBQ)-A priority scheduling method that gives preferential treatment traffic communication sessions that have different priority levels for each class of communication session (such as real-time audio traffic and near-real time video). CBQ allocates variable bandwidth for each class of traffic so all applications continue to operate under heavy traffic conditions.
Class C Amplifier-An amplifier that is operating in class C. Class C amplifiers can amplify a signal for less than 180 degrees of its input cycle (non-linear amplification for a majority of portions of the input signal).
Class Of Service-(1-multimedia) The type of service associated with a particular application or communication session. The class of service usually requires a specific quality of service (QoS) level. (2-telecommunications) Categories of services that are provided by tariff for charging customers for the particular service they select. Examples of class services include flat rate, coin, toll free (800) service, and PBX. (3-SS7) Services provided by the Signaling Connection Control Pant (SCCP) to its users.
Class of Service (CoS)-Class of Service is a way of managing traffic in a network by grouping similar types of traffic. 802.1p, Type of Service (ToS), and DiffServ are three main CoS technologies.
Classless Inter-Domain Routing (CIDR)-An routing solution that does not require the use of the upper parts of a routing address to be used to determine the specific ("class") of network. The CIDR address includes the number of bits used in the address mask. CIDR is pronounced "cider."
Classless Internet Domain Routing (CIDR)-In response to the limitations of class A, B, and C IP addresses, the Internet community implemented CIDR (pronounced "cider"). CIDR allows IP addresses to be broken down into subnets on arbitrary bit boundaries. CIDR networks are described with a "slash". For example, 192.168.128.0/17 represents a subnet specified by the high-order 17 bits of the given IP address. When a router adds this route to its forwarding table, any packet with a destination IP address that matches the high-order 17 bits of that address will be treated as members of the same subnet and forwarded accordingly.
CLC-Carrier Liaison Committee
CLCI-Common Language Circuit Identification
CLD-Competitive Long Distance carrier
CLE-Customer Location Equipment
Clear Channel-(1-radio, especially radio broadcasting) A carrier frequency that is licensed to only one transmitter in the entire nation, thus ensuring no radio interference. (2-digital telephone systems) A digital channel, typically 64 kb/s, for the subscriber that can carry any binary bit stream without restriction, including a string of binary zeros.
Clear Channel Capability (CCC)-A characteristic of DS1 transmission in which the 192 information bits in a frame can carry any combination of zeros and ones.
Clear Defective Pair-A facility modification that requires the repair of a defective wire pair for a spe-

cific service order or related line and station transfer.

Clear Text Authentication-An authentication process that relies on a user identification name and password information without encrypting the information (relatively low security).

Clear To Send (CTS)-A control signal on a communication line (such as an RS-232 serial bus) that indicates that the transmission of communication information can proceed.

Clearing House-A clearing house is a company or association that performs data and financial settlements functions between service providers. The clearing house receives data records from its member providers; after a validation process, the data is forwarded to the provider who will bill the customer. An additional role of the clearing house is to handle financial settlements between Service Providers. Clearing houses are particularly important for international roaming, particularly due to the need for conversion between different data record formats used by the various service providers, and the handling of currency exchange rates.

Clearinghouse-A clearinghouse is a company or association that transfers billing records and/or performs financial clearing functions between carriers that allow their customers to use each other's networks. The clearinghouse receives, validates and accounts for telephone bills for several telephone service providers. Clearinghouses are particularly important for international billing because they convert different data record formats that may be used by some service providers and convert for the currency exchange rate.

Clearinghouses provide a variety of services including processing proprietary records (e.g. switch records) into formats understandable by the member carriers' billing systems, validate charges from carriers with intersystem agreements, and extract unauthorized or un-billable billing records. Clearinghouses transfer messages in a standard format such as exchange message record (EMR), cellular inter-carrier billing exchange roamer (CIBER), or transferred account process (TAP) format. The EMR format is often used for billing records in traditional wired telecom networks and the CIBER and TAP formats are used for wireless networks. The records may be exchanged by magnetic tape or by other medium such as electronic transfer or CD ROM.

Cleaving-The controlled breaking of an optical fiber so that the end surface is smooth.

CLEC-Competitive Local Exchange Carrier

CLI-Calling Line Identification

CLI-Command Line Interface

CLI-Command Line Interpreter

CLI-Cumulative Leakage Index

CLI-Cellular Line Interface

Click to Dial or Click to Call-Click to Dial is a SIP service that allows a user who is viewing a web page to click a link on that web page to initiate a voice over Internet call. The link contains an embedded address (URL or IP address) that connects to a call server along with the necessary software (such as SIP) that allows for the setup and connection of the call. Click to dial service is similar in concept to the 'mailto:' link that can launch a user's email software when selected.

CLID-Calling Line Identification Display

Client-A computer, hardware device or software program that is configured to request services from a network. Client also may refer to the codec or terminating device located at one end of a network node.

Client/Server Architecture-A form of distributed computing in which each application is viewed as a series of inter-dependent tasks bring accomplished by several different computers linked on a network For example, a personal computer can be a client while it is linked to a remote processor acting as a server. The various tasks performed by the client is considered to be front end operations and those performed by the server are backend operations

This diagram shows the configuration of a client/server network. A client/server network is a form of distributed computing in which each application is viewed as a series of inter-dependent tasks bring accomplished by several different computers linked on a network For example, a personal computer can be a client while it is linked to a remote processor acting as a server. The various tasks performed by the client is considered to be front end operations and those performed by the server are backend operations.

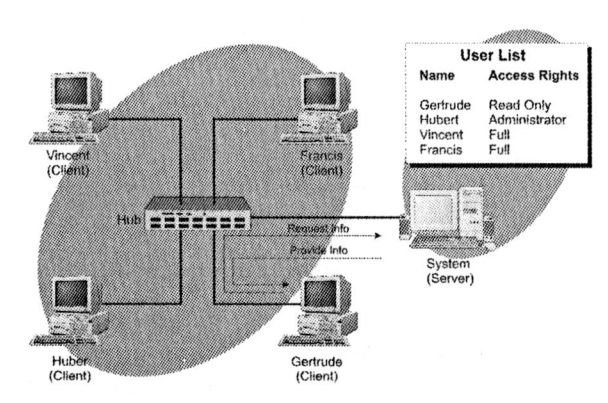

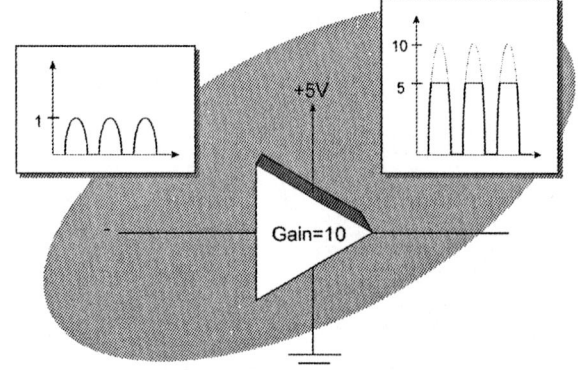

Client/Server Network

Clipping Operation

Clip-(1-insert) The trigger point or range of a key source signal at which the key or insert takes place. (2-control) Clip also may refer to the control that sets this action. To produce a key signal from a video signal, a clip control on the keyer control panel is used to set the clip code. (3-test) A flexible connector that can self-attach to an electronic component or assembly test point.
CLIP-Calling Line Identification Presentation
Clipper-A limiting circuit which ensures that a specified output level is not exceeded. It does this by restricting the output waveform to a maximum peak amplitude.
Clipping-Signal distortion that results from the non-linear (non-equal) processing of a signal through a device or circuit (such as an amplifier) where some of the signal is lost due to the maximum or minimum processing ability of the device or circuit.
This diagram shows how a signal may experience clipping if its level exceeds the maximum capabilities of an amplifier. In this example, an input signal varies from 0 to 1 Volt is applied to an amplifier that has a gain of 10. Because this amplifier has a maximum output voltage of 5 volts, when the input signal exceeds 0.5 Volts, the output signal reaches a maximum (clipped) to 5 Volts.

Clipping Level-An electronic limit beyond which unwanted distortion will occur in the audio or video signal
CLIR-Calling Line Identification Restriction
CLLI-Common Language Location Identification
CLLI Code-Code Common Language Location Identifier Code
CLNP-Connectionless Network Protocol
CLNS-Connectionless Network Service
CLOB-Central Limit Order Book
Clock Frequency-The master frequency of periodic pulses that are used to synchronize the operation of equipment.
Clock Jitter-Undesirable random changes in clock timing signal (phase deviation.)
Cloning-Duplicating certain information necessary for the connection of wireless devices in a fraudulent manner. Cloning the ESN's of existing customers is an example of cloning.
Closed Loop Power Control-A process of controlling the transmission power level for the mobile radio using the power level control commands from another transmitter that is receiving its signal (e.g. from a radio base station). Closed loop power control is used in systems where power control is required (such as in cellular systems).
This diagram illustrates how closed loop power control can accurately fine tune to received signal power level at an antenna. In this example, as the received signal power as sensed by the base station increases or decreases, the base station sends a power control command signals the mobile tele-

Closed Numbering Plan

phone to increase or reduce it's transmit power level. The combined open and closed loop adjustments precisely control the received signal power at the base station.

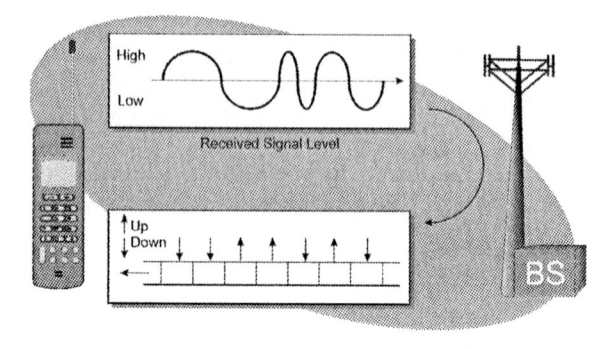

Closed Loop Power Control Operation

Closed Numbering Plan-A numbering plan in which all local numbers comprise the same number of digits, all area or zone codes comprise a fixed number of digits, etc. The North American Numbering Plan (NANP) is nominally a closed numbering plan. All local numbers are 7 digits in length. All area codes are 3 digits in length. There are some few exceptions to this rule, since certain short codes are used, such as 0 for the operator/attendant, 911 for emergency service, etc. See also the antonym Open Numbering Plan.

Closed User Group (CUG)-(1- access restriction) A group of directory numbers sharing an access restriction such that any directory number can reach others in the group but cannot access outside numbers. (2- cellular system) Advanced features such as 4-digit dialing authorized for a closed group of users of the service. (3 - X25 protocol) In the X.25 packet-switching protocol, a facility indicating a virtual grouping of terminals that can communicate only with other members of that group. The feature can be extended to a closed user group with outgoing access, or a closed user group with incoming access.

Closed-Circuit Television (CCTV)-A private (closed circuit) network of security that display images on one or more television (video) monitors.

CLP-Certified Lotus Professional
CLP-Cell Loss Priority Field
CLP-Cell Loss Priority
CLR-Cell Loss Ratio
CLR-Circuit Layout Record
CLS-Certified Lotus Specialist
CLTP-Connectionless Transport Protocol
Cluster-(1-data communications) A group of storage devices, processors, or computers that share a common bus or network and function as a single system or sub-system within a larger network. These devices are often coordinated by a cluster controller. (2-memory device) The amount of memory storage capability within a sector on a diskette or hard disk. (3-SS7) In the Signaling System 7 protocol, a set of signaling points that are identifiable as a group within the signaling-point code-address spare.
CLUT-Color Look-Up Table
CLV-Constant Linear Velocity
CM-Configuration Management
CM-Cable Modem
CMA-Communications Managers Association
CMAC-Control Mobile Attenuation Code
CMC-Common Messaging Calls
CMC-Cellular Mobile Carriers
CMCI-Cable Modem to CPE Interface
CMDS-Centralized Message Data System
CMIP-Common Management Information Protocol
CMIP/CMIS-Common Management Infomation Protocol/Common Management Information Services
CMIS-Common Management Information Services
CMISE-Common Management Information Service Element
CMOL-CMIP Over Logical Link Control
CMOS-Complementary Metal Oxide Semiconductor
CMOS RAM-Complementary Metal Oxide Semiconductor Random Access Memory
CMOS Setup-Complementary Metal Oxide Semiconductor Setup
CMOT-Common Management Information Protocol over TCP/IP
CMOU-Conversation Minutes Of Use
CMR-Common Mode Rejection
CMR-Call Management Record
CMR-Cell Misinsertion Rate
CMRFI-Cable Modem to Radio Frequency Interface
CMRR-Common Mode Rejection Ratio
CMRS-Commercial Mobile Radio Service
CMRS-Cellular Mobile Radiotelephone Service

CMS-Call Management System
CMS-Content Management System
CMS-Call Management Server
CMS-Cryptographic Message Syntax
CMS-Call Management Services
CMTRI-Cable Modem Telco Return Interface
CMTS-Cable Modem Termination System
CMTS-Cellular Mobile Telephone System
CMTS-DRFI-Cable Modem Termination System-Downstream RF Interface
CMTS-NSI-Cable Modem Termination System-Network Side Interface
CMTS-URFI-Cable Modem Termination System-Upstream RF Interface
CMYK-Cyan, Magenta, Yellow and BlacK
CNA-Centralized Network Administration
CNA-Certified Novell Administrator
CNE-Certified Network Engineer
CNE-Certified Novell Engineer
CNEPA-Certified Novell Engineer Professional Association
CNET-The first generation analog cellular radio system used in Germany and other parts of the world.
CNET-Centre National d'Etudes de Telecommunication
CNI-Calling Number Identification
CNI-Certified Novell Instructor
CNIP-Calling Number Identification Presentation
CNIP-Calling Name Identification Presentation
CNIR-Calling Number Identification Restriction
CNIR-Calling Name Identification Restriction
CNM-Customer Network Management
CNO-Corporate Networking Officer
CNP-Certified Network Professional
CNR-Carrier To Noise Ratio
CNR-Combat Net Radio
CNR-Customer Not Ready
CNR-Complex Node Representation
CNRI-The Corporation for National Research Initiatives
CNS-Complimentary Network Services
CNS-Complementary Network Service
CNX-Certified Network Expert
CO-Central Office
CO-Centrex Central Office
CO Lines-Central Office Lines
CO Simulator-Central Office Simulator
Co-axial Lightning Suppressor-A device or assembly that protects a coaxial cable by shorting surge voltages on the inner conductor to ground (shield).

Co-Channel-A communication channel that operating on the same carrier frequency at a nearby transmission site. The ability to use channels on the same frequency at other transmission sites is made possible due to the attenuation of the signal because of distance. This is referred to as frequency reuse.

Co-Channel Interference-Radio signal interference that results when two radio channels that are operating in the same geographic area on the same frequency interfere are not separated by enough distance.
This diagram shows that co-channel interference occurs in mobile communication systems that have transmitters that operate in nearby geographic regions to reuse frequencies. This example shows that mobile communication systems can allow a certain amount of co-channel interference. This example shows that the amount of co-channel interference that can be tolerated in a narrow band analog system is approximately 2% (17 dB below the carrier level).

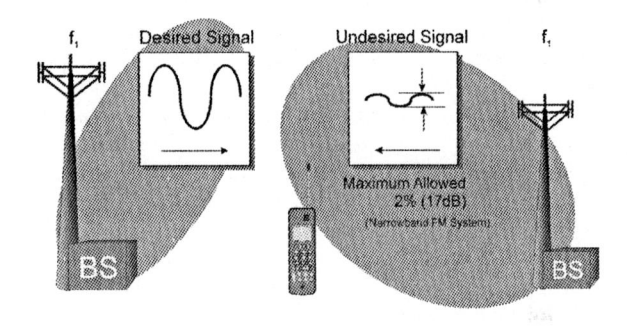

Co-Channel Interference Operation

Co-location-The location of equipment from multiple carriers at the same facility. Co-location commonly refers to the location of a competing telephone service provider's equipment at a local telephone company's switching facilities. This enables providers of interstate or competing telecommunications service providers to connect their facilities directly to those of a local exchange carrier (LEC).

Co-marketing

Co-marketing-The joint marketing of products or services. Co-marketing of products or services commonly involves the manufacturer or service provider giving marketing allowances or incentives to retailers or other sales focused companies for specific marketing promotions in their market areas.

Co-Op-Co-Operative Advertising

Co-Operative Advertising (Co-Op)-An amount of funds, percentage of sales, or marketing allowance that is provided to a distributor or retailer for their advertising of specific products or services. To receive the co-operative advertising funds, the distributor or retailer may be required to provide proof that the advertising was performed and paid for.

COAM-Customer Owned And Maintained

Coating-A protective material (usually plastic) applied to an optical fiber immediately after drawing to preserve its mechanical strength and to cushion it from external forces that can induce microbending losses.

Coax-Coaxial cable

Coaxial cable (Coax)-A multi conductor cable comprising a central wire conductor surrounded by a hollow cylindrical insulating space of air, or solid insulation, or mostly air with spaced insulating disks, finally surrounded by a hollow cylindrical outer conductor. Invented by Lloyd Espenschied in the 1930s for transmission of wideband television and radar signals, it is also used for other purposes. The hyphenated form of the name is preferable to avoid confusion with the English word "coax." Versions using more than one internal conductor are called "tri-ax," "quad-ax" for three total conductors or four total conductors respectively. Co-ax exhibits lower radiated electromagnetic power losses because the electromagnetic field is confined inside the outer conductor. The form with the hyphen is preferable to avoid confusion with the English word coax.

This figure shows a cross sectional view of a coaxial cable. This diagram shows a center conductor that is surrounded by an insulator (dielectric). The insulator is surrounded by the shield. This diagram shows that during transmission, electric fields extend perpendicular from the center conductor to the shield and magnetic fields form a circular pattern around the center conductor.

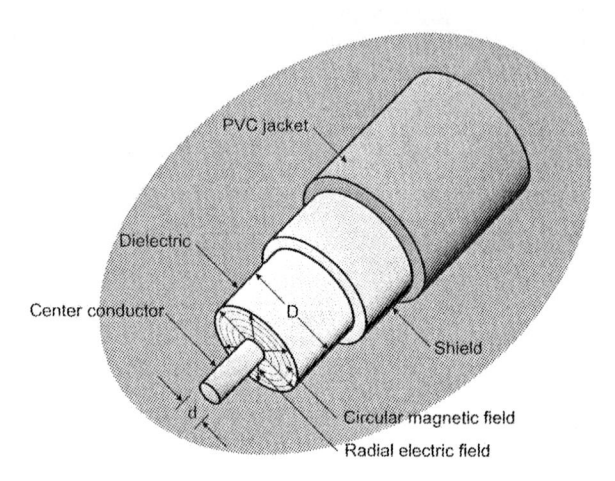

Coaxial Cable Diagram

COB-Chip On Board
COBOL-Common Business Oriented Language
COBRA-Common Object Request Broker Architecture
COC-Central Office Connection
COCUS-Central Office Code Utilization Survey
COD-Connection Oriented Data
Code Character-The representation of a discrete value or symbol in accordance with a specified code.
Code Common Language Location Identifier Code (CLLI Code)-CLLI codes are used to provide unique identification of facilities (equipment and cables) between any two interconnected CLLI coded locations. The CLLI code is mnemonic code that can be a maximum of 38 characters. An example of a CLLI code is 115T3NYCNY20DALTX. This example shows a T-3 carrier is connected between New York City, New York and Dallas, Texas.
Code Conversion-In address signaling, the alteration of leading received digits in order to direct a call through subsequent switching offices.
Code Division Multiple Access (CDMA)-A system that allows multiple users to share one or more radio channels for service by adding a unique codes to each data signal that is being sent to and from each of the radio transceivers. These codes are used to spread the data signal to a bandwidth much wider than is necessary to transmit the data signal without the code.
Code Division Multiple Access 2000 (CDMA2000)-CDMA2000 is an evolved version of the IS-95 code division multiple access (CDMA) sys-

118   www.ITMag.Com

Code Division Multiplexing (CDM)

tem that uses wider bandwidth radio channels and enhanced packet transmission protocols to provide for advanced high-speed data services. The CDMA2000 system uses multiples of the standard IS-95 radio channels. These multiples are 3,6,9 or 12 times the standard 1.25 MHz wide bandwidth. These wider bandwidths allow for user data transmission rates of up to 2 Mbps. CDMA2000 is also known as 1XRTT.

This figure shows how the CDMA2000 system uses multiples of the standard IS-95 radio channels to provide for 3rd generation broadband data transmission services. These multiples are 3,6,9 or 12 times the standard 1.25 MHz wide bandwidth. This diagram shows that to upgrade 2nd generation IS-95 systems to 3rd generation capability, IS-95 radio channels in the base station can be enhanced with new protocols and new 3rd generation mobile telephones are required.

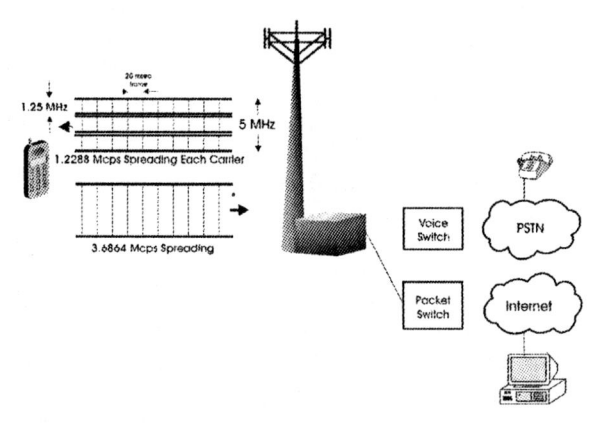

CDMA2000 IP Radio Node
Source: Airvana Inc.

Code Division Multiplexing (CDM)-The process of multiplexing several signals by allowing separate signals to be applied to a common transmission physical channel or line by using specific code sequences to uniquely identify each signal.

This figure shows how a single direct sequence spread spectrum communication channel can have several channels. In this example, there are 3 different code patterns that are used for communication channels. When a receiver uses the reference code, a direct sequence spread spectrum system can build a mask as shown in this figure for each conversation allowing only that information which falls within the mask to be transmitted or received.

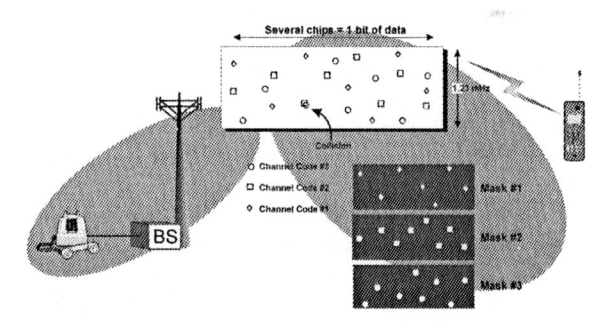

Code Division Multiple Access 2000 (CDMA2000) System

This photo image shows an Airvana IP Radio Node (IP-RN8000). It is a compact IP-based CDMA2000 1xEV-DO radio node. It can operate in the 800, 1800, and 1900 MHz frequency bands. The IP-RN 8000 reduces network operating costs by allowing the operator to choose the lowest cost data backhaul available. In addition, it is easy to install, maintain, and manage, making it the most cost-effective option for deploying 1xEV-DO service a short time frame.

Code Division Multiplexing (CDM) Operation

Code Domain-The process of examining the power level, phase etc of each of the Walsh Codes in a CDMA system.

Code Excited Linear Prediction (CELP)-A lossy compression technique for speech, sometimes pronounced like "kelp" or "selp". In general, a CELP decoder produces a segment of audio samples by filtering an excitation signal through a "linear predictor" filter, also sometimes called the synthesis filter. The excitation is one of several possibilities as indicated by a code in the bitstream, hence the term "code excited". The linear predictor filter typically has an order of about 10 coefficients, the values of which are sent in the bitstream in quantized form. At the CELP encoder, the linear predictor coefficients may be derived using the Levinson-Durbin algorithm, after which the encoder does a search over some possible excitation codes for a best match. CELP codecs are a type of vocoder. CELP-based standard speech codecs include G.728 (16 kbps), the GSM CELP codec (13 kbps), and the Federal Standard 1016 codec (4.8 kbps).

Code Set-The complete set of representations defined by a code or by a coded character set.

Codec-Coder/Decoder

Codec Negotiation-Codec negotiation is the process where two (or more) communication devices negotiate for which coder/decoder (CoDec) devices they will use during a communication session. Communication devices use codec negotiation when they have access to multiple audio, data, and video codecs (data compression devices). The codec negotiation process may include the use of customer preferences (e.g. use of low-bit rate codecs due to bandwidth costs), preferred codecs due to quality of service (QoS) requirements (e.g. good audio or video quality), or availability of compatible codecs.

This diagram shows how two voice gateways negotiate for codec selection in a voice over data network using user assigned preferences and options determined by equipment availability. In this example, the calling gateway sends a connection request message to the remote gateway. This connection request indicates that the calling gateway prefers to use a G.711 64 kbps speech codec because it has enough bandwidth and the caller prefers a high quality audio signal. Unfortunately, the receiving gateway cannot accept the request for the G.711 codec because its access bandwidth is low speed (28 kbps). The receiving gateway sends back a request to use the G.722 codec. When the calling gateway receives this request, it declines the request because it does not have a G.722 codec available. It then requests the G.729 codec (industry standard minimum) and the receiving gateway confirms the request and both devices use the G.729 codec.

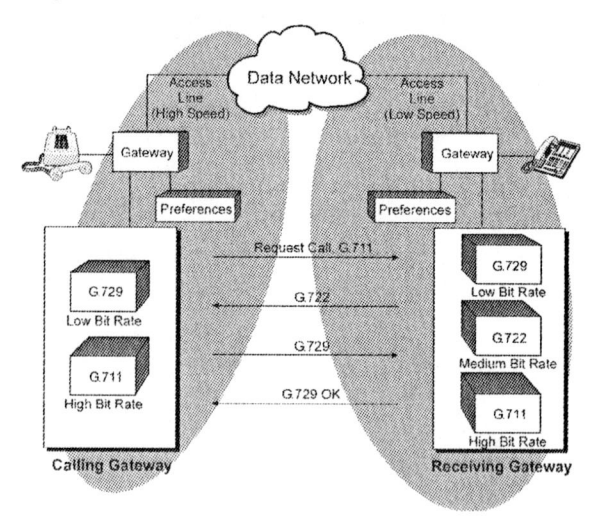

Codec Negotiation Voice Operation

Coded Digital Locator (CDL)-A DTC burst field used to notify the phone of the location of the DCCH.

Coded Digital Verification Color Code (CDVCC)-A unique code word that is continually transmitted to and from mobile radios with the voice information on a radio channel. The CDVCC is used to identify particular mobile radios and base stations that may be operating on the same frequencies in different locations. Although the radio transmission plans of cellular and PCS systems are not supposed to have interference from radios operating on the same frequency due to separation by distance, due to geographic variances (such as mountains) and other factors, it is sometimes possible that mobile radios in distant locations that are operating on the same frequency can interfere with each other. Because the CDVCC code for each of these radios is different so the cellular or PCS can determine if the received radio signal is from an interfering mobile radio and if it is interfering, mute the audio and take corrective action (such as assign the correct mobile radio to a new radio frequency).

Coded Speech-A speech signal that has been changed into a standard digital code form.
Coded Trunks-Trunks that are coded for identification so they can be identified as they hunt (search) for connections.
Codepoint-In the Signaling System 7 protocol, the single coding of a value within an information element.
Coder/Decoder (Codec)-A technique for compressing information to a fewer number of bits for more efficient transmission and storage (coding), and subsequently recovering the original data (decoding). Normally the term codec applies only to compression of human-perceived signals such as speech, audio, images, or video; and it usually refers to lossy compression.
Coding-(1 - digital) A process of changing digital bits to include error protection bits and/or signaling bits prior to the sending or storing of the information. (2 - software) The process of writing instructions or commands for software programs.
COER-Central Office Equipment Reports
COG-Centralized Ordering Group
Coherence-The correlation between the phases of two or more waves.
Coherence Time-The time over which a propagating light beam may be considered coherent.
Coherent-The condition characterized by a fixed phase relationship among points on an electromagnetic wave.
Coherent - Waveform or Light-Sinusoidal electromagnetic, acoustic or light waves is coherent when all parameters (frequency and phase) are predictable and in a fixed relationship at all points in time or space, particularly over the area in a plane perpendicular to the direction of propagation, or over time at a particular point in space. Two sine waves that have a fixed phase relationship with each other are described as mutually coherent. Coherent sine waves can produce constructive or destructive interference and diffraction. See: Interference.
An incoherent or non-coherent light source, such as an incandescent light bulb, has an optical waveform that is continually changing in frequency (not a single frequency sine wave) and even during short intervals when the frequency is approximately constant, the phase (time of each sine wave cycle peak compared to a uniform clock signal) is not constant. In non-technical discourse, the word "incoherent" describes speech that is incomprehensible because of inability of the speaker to speak properly, or an intentional use of nonsense syllables (so-called "double talk" or "double Dutch").
Coherent Network-A network in which inputs, outputs, signal levels, bit rates, digital bit stream structures, and signaling information all are inherently compatible throughout the network.
Coherent Pulse-The condition in which a fixed phase relationship is maintained between consecutive pulses during pulse transmission.
Coin Collect Message Type-A message charged to the coin station at which it terminates.
Coin Paid - Message Type-A message sent and paid from a coin station.
Coin Supervisory Trunk Group-A communication trunk that permits call processing features so an operator can check the operation of a pay telephone (e.g. stuck coins) and to allow the recording (coin entry) of overtime charges.
Coin Telephone-A telephone that allows the collection of coins. William Gray invented the coin telephone.
Coin-Zone Trunk Group-A trunk group that enables cord switchboard operators to supervise charge arrangements for coin dialing to destinations beyond the basic coin-rate zone.
Coinless Public Telephone Service-A service for use of a coinless public telephone for originating calls only. The phone is served by a single-party, loop-start line with no extensions and no custom calling features.
COLD-Computer Output to Laser Disk
Cold Boot-The startup process that begins when a equipment (such as a computer) or device is first powered on. A cold boot is used to start an operating system from a known good point (before other programs are activated).
Cold Joint-An inadequately heated soldered connection, leaving the wires held in place by rosin flux, not solder. A cold joint sometimes is referred to as a dry joint.
Collaboration-The process of two or more people working together on a shared medium (such as a word processor file or sketch whiteboard). Collaboration usually enables a group of people to work together in real-time to share screen displays, documents, and video images.
Collapsed Backbone-A method of interconnecting networks by using a switch or router as a central relay device.

Collection-The methods and procedures used by the service provider to receive payments due from a customer (also known in Europe as "chasing up".) Sometimes also refers to the methods used to minimize the risk that a customer will end up in arrears.

Collections-Collections are activities that a service provider performs to receive money from their customers. Ideally, all customers will receive their bills and pay promptly. Unfortunately, not all customers pay their bills and service providers must have a progressive collection process in the event a customer does not pay their bill.

When customers are first added to a system, they are rated on the probability that they will pay their bills. This is accomplished by using information on their application and reviewing the credit history as provided by an independent credit reporting agency.

The collection process for delinquent customers usually starts by sending a reminder messages to the customer be mail or recorded audio message. If initial attempts to collect are unsuccessful, more aggressive collection activities will progress that include restricted calling, service disconnection and sending or selling the uncollected invoice to a collection service.

Collision-Data collisions occur when two or more data stream attempt to occupy a telecommunications media at the same time causing each stream to be corrupted. Typically, each transmitting host that detects the collision will wait for some period of time and try again. This situation is most often associated with Ethernet and its access method carrier sense multiple access with collision detection (CSMA/CD).

Collision Domain-The set of stations among which a collision can occur. Stations on the same shared LAN are in the same collision domain; stations in separate collision domains do not contend for use of a common communications channel.

Collision Fragment-The portion of an Ethernet frame that results from a collision. On a properly configured and operating LAN, collision fragments are always shorter than the minimum length of a valid frame.

Color Code-(1-general) The word "color" is used in both the literal and figurative sense in telecommunications. The literal meaning occurs in component identification methods using visible colors. The figurative meaning is used for situations in which different component frequencies are not descriptive of actual visible light. One figurative meaning is used for different wavelengths of infrared light (different "colors") used in WDM and DWDM fiber optics. Electromagnetic radiation in the infrared frequency range is not actually visible to the human eye. A noise or an artificial random signal waveform that has uniform power density across the audio frequency or radio frequency spectrum or bandwidth of interest is called "white" noise . A similar signal that has higher power density at the low frequency (longer wavelength) part of its spectrum is called "pink noise" by analogy to pink visible light. (2-cable colors) the twisted-pair cable color code (PIC) using various body and stripe colors on wire insulation to distinguish various different wire pairs in a cable. (3-fiber optic) Fiber-optic color code using distinct jacket colors to distinguish different optical fibers in a multi-fiber cable. (4-Resistors and other component values) The resistor/capacitor color code using a distinct color stripe to represent each decimal digit of a component's resistance or capacitance value. (5-channel code) A special case is the entirely figurative meaning in terms such as Digital Verification Color Code, used in IS-136 cellular systems to distinguish different radio signals that use the same carrier frequency but are intended for use in different cells. The term "color" is used there in an entirely figurative sense meaning "identification."

Color Depth-In uncompressed digital images, the number of bits per pixel. For example, a color image that uses 8 bits for each color uses 24 bits per pixel. For a low color depth like 8 bits per pixel, a color palette may be used to map each value to one of a small number of more precisely represented colors, for example to one of 256 24-bit colors.

Color Field-In the NTSC system, the color subcarrier is phase-locked to the line sync so that on each consecutive line, subcarrier phase is changed 180 degrees with respect to the sync pulses. In the PAL system, color subcarrier phase moves 90 degrees every frame. In NJSC, this creates four different field types; in PAL there are eight To make clean edits, alignment of color field sequences from different sources is crucial.

Color Frame-In NTSC, a 4-field sequence that exists between the occurrence of exact coincidences of the color subcarrier signal. This span results from the uneven number of cycles of subcarrier per line (227.5) and per 525-line frame (119437.5). In PAL video, a similar sequence of eight fields exists because of the more complex phase structure of the PAL signal.

Color Gamut-The entire range of values a component signal or a combination of component signals may take on that are reproducible at the display device. Some component formats are interdependent (Y, R-Y, and B-Y), and the valid color gamut cannot be evaluated by looking at a single component alone.

Color Palette-The hard-coded colors that a video card can display on a computer.

Color Space-A set of components of a color, such as RGB or YUV. Any visible color can be represented as a set of values corresponding to the components of a color space.

Color Space Conversion-The process of converting a color from the representation in one color space to a representation in a different color space. A color conversion can be computed as a constant linear relationship (a matrix multiplication) followed by clipping to ensure the resulting color values lie within the valid ranges. Color space conversion is required for encoding or decoding images or video in many cases. For example, a JPEG image or a frame of MPEG video is decoded into a pixel-map of YUV color values, which need to be color-converted to RGB color values if the image is to be displayed on an RGB display device. In some operating systems, a graphics display API allows specifying display colors using an application-specified color space so that the color space conversion can be done by the video display hardware when supported.

Color Subcarrier-The frequency that carries the color information in the baseband composite video signal. in NTSC, the color subcarrier is 3,579,545 Hz, usually rounded off in text to 3.58 MHz.

Color Timing-The synchronization of the burst phase of two or more video signals. Proper color timing ensures that no color shifts occur in the picture when the signals are mixed in a switcher or other video device.

Colour-British English spelling of color.

COLT-Cell Site On Light Truck

COM-Continuation of Message

COM-Common Object Model

COM-Component Object Model

Com Port-Communication Port

Comb Filter-An electric filter circuit that passes a series of frequencies and rejects the frequencies in between, producing a frequency response similar to the teeth of a comb. A comb filter may be used on encoded video to select the chrominance signal and reject the luminance signal, thereby reducing cross chrominance artifacts, or conversely, to select the luminance signal and reject the chrominance signal, thereby reducing cross luminance artifacts. Comb filtering successfully reduces artifacts, but may cause a certain amount of resolution loss in the picture.

Combination Trunk-A trunk line that can operate as a direct inward dial (DID) or direct outward dial (DOD) trunk. This allows them to receive (ring-in) and send calls.

Combined Paging And Access (CPA)-The combination of sending paging messages and access channel assignment messages to a mobile radio on a single control channel. On the AMPS control channel, there is a data field that indicates if the paging and access channels are combined.

Combiner-(1-radio frequency) A device that is used to combine several channels on to a common transmission line or antenna system. (2-video manipulator) In a video or image processing system, a device or assembly that controls the interaction between video channels or images. The combiner prioritizes the display of video sources or images and controls the type of graphic transitions that can occur between them.

COMET-PreConditions Met

Command Line Interface (CLI)-Any interface between the user and the command processor that allows you to enter commands from the keyboard for execution by the operating system.

Command Sequence (Cseq)-An identification header (group of data fields) that contains information about the sequence number of a command that was issued and the identification code of the communication session. The use of a Cseq header ensures duplicate command messages do not get processed for a specific communication session. Command sequence is defined in the session initiation protocol (SIP) specification.

Committed Information Rate (CIR)-Committed information rate (CIR) is a guaranteed minimum data transmission rate of service that will be available to the user through a network. Applications that use CIR services include voice and real time data applications. CIR can be measured in bits per second, burst size, and burst interval.

Some service providers allow users to transmit data above the CIR level. However, when data is transmitted above the CIR level, some of the data may be selectively discarded if the network becomes congested.

Committee For European Electrotechnical Standardization-Committee For European Standardization

Committee For European Standardization (Committee For European Electrotechnical Standardization)-A European standards organization also known collectively as the Joint European Standards Institute.

Common-A point that acts as a reference for circuits, often equal in potential to the local ground.

Common Air Interface (CAI)-The specifications for a radio communication system that standardize the communication between radio transmitter(s) and radio receiver(s) within that system. If the radio equipment is designed to meet the common air interface (CAI) specifications, theoretically equipment or software developed by one company should be capable of operating with other equipment or software designed by a different company. It's defined in terms of Access Method, Modulation Scheme, Vocoding Method, Channel Data Rate and Channel Data Format.

Common Carrier-(1-general) A company that carries goods, services, or people from one point to another for the public. (2-telecommunications) A company that provides communications services and typically is subject to a regulatory agency.

Common Channel Signaling (CCS)-A signaling technique in which signaling information relating to a multiplicity of circuits (trunks), is conveyed over a separate single channel by addressed messages. Common channel signaling system #7 ("SS7") is the primary system used for interconnection of telephone systems. SS7 sends packets of control information between switching systems.

The SS7 network is composed of its own data packet switches, and these switching facilities are called signal transfer points (STPs). In some cases, when advanced intelligent network services are provided, STPs may communicate with signal control points (SCPs) to process advanced telephone services. STPs are the telephone network switching point that route control messages to other switching points. SCPs are databases that allow messages to be processed as they pass through the network (such as calling card information or call forwarding information).

This diagram shows the basic structure of the SS7 control signaling system. The SS7 network is composed of its own data packet switches, and these switching facilities are called signal transfer points (STPs). In some cases, when advanced intelligent network services are provided, STPs may communicate with signal control points (SCPs) to process advanced telephone services. STPs are the telephone network switching point that route control messages to other switching points. SCPs are databases that allow messages to be processed as they pass through the network (such as calling card information or call forwarding information). Messages originate and terminate at a service switching point (SSP.) A SSP is a part of the end office (EO) switching system. End offices are sometimes called central offices (COs).

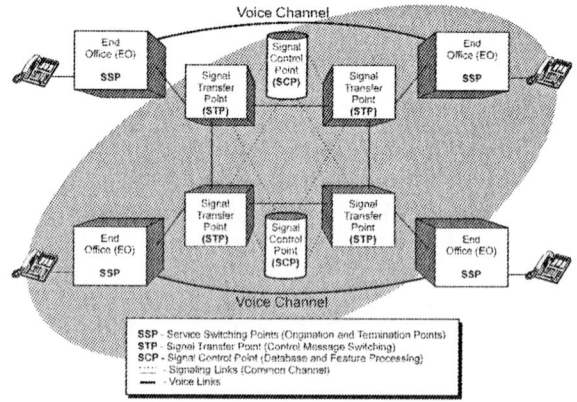

Common Channel Signaling System

Common Control-A system in which items of control equipment are commonly shared (such as network control signaling).

Common Control Channel (CCCH)-A group of logical control channels that support the establishment and maintenance of communication links between mobile phones and radio towers (base stations).

Common Equipment-Any apparatus used by more than one unit of equipment, or any equipment used by more than one channel; that is, equipment common to two or more channels.

Common Intermediate Format (CIF)-A video format, adopted by the Specialists Group of the International Telegraph and Telephone Consultative Committee (CCFIT), for transmission of video signals on Integrated Services Digital Networks (ISDNs). The standard CIF has 288 lines of 360 pixels each for luminance and half as many lines and pixels for each of two chrominance, or color, components. A second version, 1/4 CIF, has one-fourth that resolution.

Common Language Location Identification (CLLI)-A standard code used by telecommunications systems to identify specific locations of a switching office or network element. A CLLI code is composed of 11 alphanumeric characters. The first four characters of the CLLI code are an abbreviated place name. Characters 5 and 6 are state abbreviations. Positions 7 and 8 identify a specific building, and 9,10, and 11 represent a particular piece of equipment.

Common Mode-Signals identical with respect to amplitude, frequency, and phase that are applied to both terminals of a cable and/or both the input and reference of an amplifier.

Common Mode Hum-Power line interference (usually 60 Hz) that appears on both terminals of a cable with the same phase, amplitude, and frequency.

Common Mode Noise-Unwanted signals in the form of voltages appearing between the local ground reference and each of the power conductors, including neutral and the equipment ground.

Common Mode Range-The amplitude of the common mode signal that can be applied to the two differential inputs of an amplifier and maintain its performance. (See also: differential amplifier.)

Common Mode Rejection (CMR)-A measure of how well a differential amplifier rejects a signal that appears simultaneously and in phase at both input terminals. As a specification, CMR usually is stated as a decibel ratio at a given frequency.

Common Open Policy Service (COPS)-The COPS protocol allows a system to implement policy decisions by allowing a client to obtain system configuration and parameter information from a policy server. COPS is defined in RFC 2748.

Common Return-A return path that is common to two or more circuits, and that returns currents to their source or to ground.

Common System-A system that shares power, interconnections, or environmental support for network elements associated with transmission and switching.

Common Trunk-Trunks within a telephone systems that are accessible to all groups of service (trunk) grades.

Common-Channel Signaling Network-The separate control signaling network that sets up, maintains and disconnects communication paths that is used to transfer customer voice and data information. The control signals are routed via specialized network connection points (nodes) that do not interact directly with the customer voice or data information.

Commonwealth Broadcasting Association-An association of public broadcasting authorities of Commonwealth countries, numbering about 50.

Communicating Application Specification (CAS)-An application programming interface (API) that was developed by Intel and DCA to allow software developers to integrate communication functions (such as fax and telephone services) into their software programs.

Communication Center-A facility responsible for the reception, transmission, and delivery of information..

Communication Services-Communication services are the processes that transfer information between two or more points. Communication services may involve the transfer of one type of signal or a mix of voice, data, or video signals. When communication services only involve the transport of information, they are called bearer services. When communication services involve additional processing of information during transfer (such as store and forward), they are known as teleservices.

Communications-The conveyance of information, including voice, images, and/or data, through a transmission channel without alteration of the original message.

Communications Assistance For Law Enforcement (CALEA)-A statute that was enacted by the U.S. congress in 1994 to define requirements of telephone service providers to provide wiretap capabilities to law enforcement agencies. To attach a listening device to a communication line, the law enforcement agency must have a surveillance order from a court of competent jurisdiction.

Communications Channel-The medium and Physical layer devices that convey signals among communicating stations.

Communications Medium-The physical medium used to propagate signals across a communications channel (e.g., optical fiber, coaxial cable, twisted pair cable).

Communications Processor-A processor that manages and monitors the flow of voice calls and coverts (interfaces) to different communications protocols. A call processor may be part of an advanced intelligent network for call delivery.

Communications Satellite-An orbiting object that relays signals between ground-based communications stations.

Communications Satellite Corporation (COMSAT)-A corporation that provides the US portion of international satellite services between the United States and other countries. COMSAT is a part owner of the International Telecommunications Satellite Organization (INTELSAT) and the International Maritime Satellite Organization (INMARSAT) satellite systems. COMSAT was established by Congress in 1962.

Communications System-A collection of individual communications networks, transmission systems, relay stations, tributary stations, and terminal equipment capable of interconnection and interoperation to form an integral whole. The individual components must serve a common purpose, be technically compatible, employ common procedures, respond to some form of control, and, in general, operate in unison.

Communications Workers Of America (CWA)-A labor union established for the telecommunications and printing fields.

Community Access Television (CATV)-A system or process of delivering quality television reception by taking signals from a well-situated central antenna and delivering them to people's homes by means of a coaxial cable network.

Community Antenna Relay Service (CARS)-The 12.75 to 12.95 GHz microwave frequency band that the FCC has assigned to the CATV industry for use in transporting TV signals.

Community String-A password used with the SNMP protocol, SNMP community strings are used for both read only and read/write privileges. A community string is case sensitive, and may include some punctuation characters. The community string is sent along with SNMP operations like "get" and "set-request" packets. Read/write community strings are needed when issuing "set-requests".

Commutator-A circular assembly of contacts, insulated one from another, each leading to a different portion of the circuit or machine.

Companding-Companding is a system that reduces the amount of amplification (gain) of an audio signal for larger input signals (e.g., louder talker). The use of companding allows the level of audio signal that enters the modulator to have a smaller overall range (higher minimum and lower maximum) regardless if some people talk softly or boldly. As a result of companding, high-level signals and low-level signals input to a modulator may have a different conversion level (ratio of modulation compared to input signal level). This can create distortion so companding allows the modulator to convert the information signal (audio signal) with less distortion. Of course, the process of companding must be reversed at the receiving end, called expanding, to recreate the original audio signal.

This figure shows the basic signal companding and expanding process. This diagram shows that the amount of amplifier gain is reduced as the level of input signal is increased. This keeps the input level to the modulator to a relatively small dynamic range. At the receiving end of the system, an expanding system is used to provide additional amplification to the upper end of the output signal. This recreates the shape of the original input audio signal.

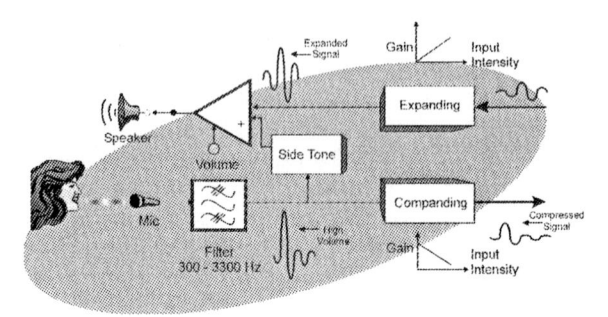

Analog Signal Companding and Expanding Operation

Compatibility-The ability of different systems to exchange information in usable form.

Compensation-(1-signal) The process of passing a signal through an element or circuit with characteristics the reverse of those in the transmission line, so that the net effect is a received signal with an acceptable level/frequency characteristic. (2-financial) The providing of financial or other rewards for performance of services or products.

Competitive Access Providers (CAPs)-A telecommunications service provider that offers competing services to an established (incumbent) telephone service provider. CAPs typically compete with a local exchange carrier (LEC). CAPs can provide service by reselling local service from the LEC.

Competitive Local Exchange Carrier (CLEC)-A telephone service company that provides local telephone service that competes with the incumbent local exchange carrier (ILEC).

Compile-To process a program from one language (usually a high-level language) into a language (machine language) that can be used by a computer.

Compiler-An executable program that converts a specific grammar of a computer programming language into the necessary machine level instructions needed to carry out the operations specified in the statements on a specific type of microprocessor. These grammars are typically called high-level programming languages and the compiler will insure that the language constructs obey the rules specified by the particular grammar. Compilers are also capable of linking together multiple files containing machine level instructions into a single executable program.

Most compilers do much more than this, however; they translate the entire program into machine language, while at the same time, they check your source code syntax for errors and then post error messages or warnings as appropriate.

Completing Field-A term designating those end offices at which a tandem can terminate, that is, complete calls.

Complex Tone-An aural stimulus whose free-air pressure is not sinusoidal over time. The waveform may or may not repeat periodically.

Complex Wave-A waveform consisting of two or more sine wave components. At any instant of time, a complex wave is the algebraic sum of all its sine wave components. Examples include voice and music signals.

Component-(1 - general) An assembly, or part thereof, that is essential to the operation of some larger circuit or system. A component is an immediate subdivision of the assembly to which it belongs. (2-SS7) in the Signaling System 7 protocol, the portion of the Transaction Capabilities Application Part that identifies a component type, provides correlation between components, specifies operations to be performed, and contains the parameters relevant to that operation.

Component Analog Video (CAV)-Signals, which represent the luminance and color information of a video picture. Each signal contains an analog voltage that varies with picture content. CAV also is referred to as analog component.

Component Video Signals-A set of signals, each representing a portion of the information needed to generate a full color image, for example RGB; Y, I, 0: or Y, R-Y, B-Y

Composite Analog-A complex signal (such as video) that is a composite of several analog signals.

Composite Blanking-The complete TV blanking signal, composed of both line rate and field rate blanking signals. (See also: line blanking, field blanking.)

Composite Color Signal-A signal consisting of combined luminance and chrominance information formed through frequency domain multiplexing. NTSC and PAL video signals are of this type.

Composite Sync (CS)-A video synchronizing signal that contains horizontal and vertical synchronizing information. Composite sync often is referred to simply as sync.

Composite Triple Beat-An intermodulation distortion often present in multichannel AM systems such as CATV and MMDS; characterized by horizontal streaking in a television picture.

Compound Error-Two or more errors on a data element or record.

Compress-(1-general) A process of converting information into a smaller form through the elimination of redundancy or the use of mathematical modeling of the information. (2-digital image) digital picture manipulator effect by which the picture is squeezed or made proportionally smaller.

Compressed Real-time Transport Protocol (CRTP)-Defined in RFC2508: Compressing IP/UDP/RTP Headers for Low-Speed Serial Links. Compressed RTP indicates that the RTP header is compressed along with the IP and UDP headers. The size of this header may still be just two bytes, or more if differences must be communicated. This packet type is used when the second-order difference (at least in the usually constant fields) is zero. It includes delta encodings for those fields that have changed by other than the expected amount to establish the first-order differences after an uncompressed RTP header is sent and whenever they change.

Compression-(1-digital) The processing of digital information to a form that reduces the space required for storage. (2-distortion) The altering of signal quality caused by the non-linearly conversion (compression) process in audio and video compression systems.

Compression Algorithm-The program or process that converts information into a smaller form (such as a smaller data file size).

Compromise Network-A network consisting of resistance and capacitance that is adjusted for an average impedance value for all circuits terminated on a switching system. It is used as a balancing network in conjunction with a hybrid, a test termination, or an idle circuit termination.

COMPSURF-Comprehensive Surface Analysis

CompTel-Competitive Telecommunications Association

CompTIA-Computing Technology Industry Association

Computer-A device, circuit, or system that processes information according to a set of stored instructions. Typically, an electronic computer consists of data input and output circuits, a central processing unit, memory storage, and facilities for interaction with a user, such as a keyboard and a visual monitor or readout.

Computer Aided Dispatch (CAD)-A computerized communication system that can coordinate and/or track mobile vehicles. CAD systems can be automated messaging devices to complex computer systems that display maps and vehicle positions on a computer monitor.

Computer Aided Software Engineering (CASE)-A process that uses software tools to assist the development of software programs. The use of CASE tools allows for the efficient use of functional requirements to develop software programs. CASE tools may simply manage the development process, validate software operation, or even automatically produce the code from functional specifications.

Computer Based Training (CBT)-The use of computers to assist the providing of training. Early CBT systems used text based screens and CD-ROM technology to provide for self paced learning experiences. CBT systems commonly provided training in small sections followed by a test section to ensure student learning prior to allowing additional sections to be presented. CBT is sometimes called computer-assisted instruction (CAI).

Computer Inquiry I,II,III-Studies by the Federal Communications Commission (FCC) of the impact new technologies could be expected to have on the telecommunications industry; also the orders that resulted from these studies.

Computer Language-The words and rules of construction for phrases and sentences that direct the operation of a computer.

Computer Peripheral-An auxiliary input, output, or storage device under the control of a computer.

Computer System For Mainframe Operations (COSMOS)-Software that assigns and inventories central-office equipment to provide effective short-jumper frame management, assignment of facilities, and load balance in a switching system.

Computer Telephony (CT)-Computer telephony (CT) systems are communication networks that merge computer intelligence with telecommunications devices and technologies.

This figure shows a sample CTI system computer that contains a voice card. This voice card is connected to a multiple channel T1 line. The voice card connects digital PBX stations through the voice card to individual DS0 channels on the T1 line when calls are in progress. Several software programs are installed on this system that provide for call processing, IVR, ACD, voice mail, fax, and email broadcasting. The monitor shows a directory of extensions. The advanced call processing feature shows text names along with the individual extensions to allow callers to automatically search through a company's directory without the need to use an operator.

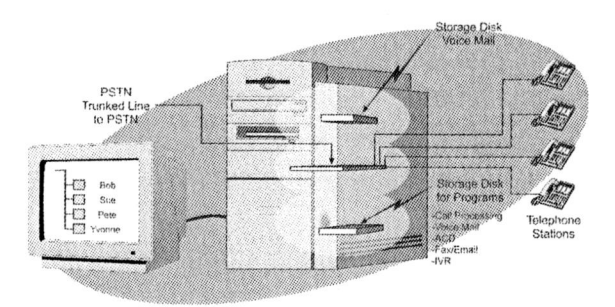

Computer Telephony (CT) System

Computer Telephony Integration (CTI)-CTI is the integration of computer processing systems with telephone technology. Computer telephony provides PBX functions along with advanced call processing and information access services. These services include, pre-paid telephony access control, interactive voice response (IVR), call center management, and private branch exchanges (PBX).

This diagram shows how a telephone system can be integrated with a computer system to provide for advanced call processing and information services. This diagram shows the core of the system is a voice card or PBX switch that is controlled by call processing software. The call processing software is customized with information about the system equipment it is operating with. This diagram shows that the computer telephone integration (CTI) system also interfaces to a company database to allow the call center to receive and update information based on both telephony commands (automatic number identification and user selected options) and customer service representative (CSR) screen commands.

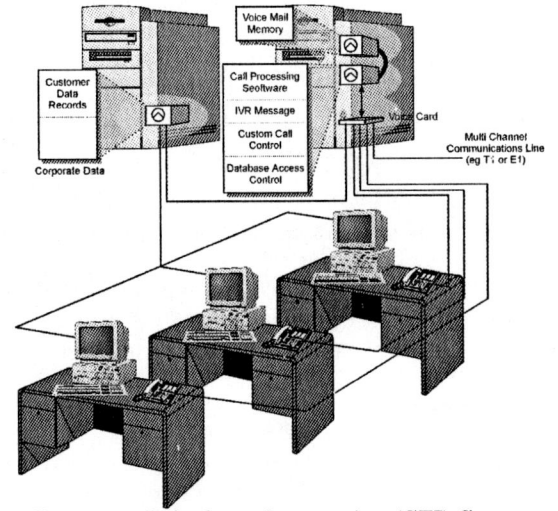

Computer Telephony Integration (CTI) System

This photograph shows a software system that is used in computer telephone systems. This software creates a screen (user interface) on a computer that allows a user to obtain telephone service using their computer (integrating computer and telephony services).

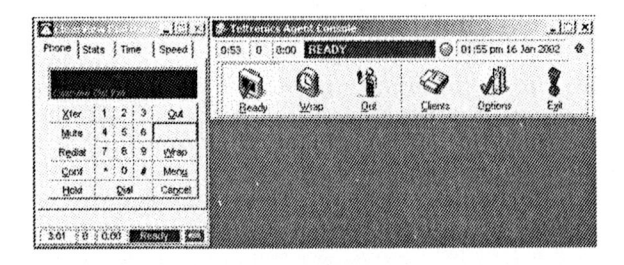

Computer Telephone Integration (CTI)
Source: Teltronics

Computer Terminal-Computer terminals are input, output, and processing devices that interface human operators to data communication systems. Computer terminals frequently consist of a keyboard, video display monitor, central processor unit (CPU), and communication circuitry that can connect the computer terminal with a data communication network.

The term "computer terminal" is often used to describe multiple types of devices including dedicated data "dumb" terminals, scientific workstations, and other types of computers that can communicate with other computers or a host computer.

Computer to PBX Interface (CPI)-An interconnection between a PBX's switching system and a host computer that allows the host computer to have control and access to the PBX switching system.

Computer-Aided Design/Computer-Aided Manufacture-A design system that uses computers to assist in the creation of diagrams and images.

Computer-Aided Dialing-The use of a computer system to assist in dialing telephone numbers. Computer aided dialing systems are often used for telemarketing services.

COMS-Circuit Order Management System

COMSAT-Communications Satellite Corporation

COMSEC-Communications Security

Concentration-The process of combining traffic from many low-usage communication lines to a lesser number of high-usage communication lines (trunks).

Concentrator-In a communication system, a concentrator combines multiple communication sources on to a higher capacity communication path. A concentrator usually provides more efficient communication capability between multiple low-speed channels to one or more high-speed channels.

Cone Of Protection-In reference to lightning, the space enclosed by a cone formed with its apex at the highest point of a lightning rod or protecting tower, the diameter of the base of the cone having a definite relationship to the height of the rod or tower. When overhead ground wires are used, the space protected is referred to as a protected zone.

Conference Bridge-(1-telephone) A telecommunications facility or service which permits callers from several diverse locations to be connected together for a conference call. The conference bridge contains electronics for amplifying and balancing the loudness of each speaker in a conference call so

Conference Call

everyone can hear each other and speak to each other. Background noises are suppressed and typically only the current two or three loudest speakers' voices are retransmitted to other participants by the bridge, while a speaker's own voice audio is not sent back to that speaker to avoid audio feedback, echo or "squealing" self-oscillation. (2- text) A facility to receive the character codes from the keyboards of multiple participants and retransmit this text to the display or printer of all participants in the conference. First used with electromechanical teletypewriters for private networks of automobile parts suppliers. Modern implementations on the Internet are typically called a "chat room."

Conference Call-The connection of three or more telephones to a telephone conversation.

This diagram shows how a conference call can uses a conference bridge to allow several users to effectively communicate in a conference call (3 or more users). This example shows that this conference bridge uses audio level detectors to determine the level of the microphone audio level for each conference call participant that is talking. As a person begins to talk, the conference bridge increases the gain on the microphone and decreases the gain on the speaker line. This process effectively dynamically reduces the background noise from non-participating members while providing good sound quality to participants that are talking.

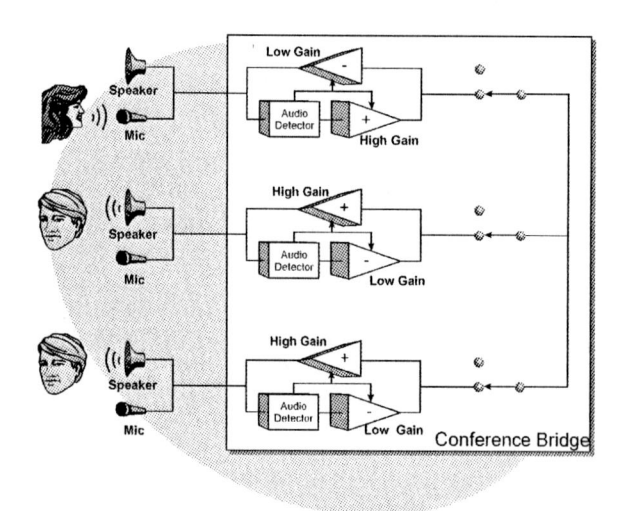

Conference Call Operation

Conference Calling (CC)-The process of connecting callers to a communication session that is shared between three or more callers (conferees).

Conference Of European Postal And Telecommunications Administrations (CEPT)-Original French language name "Conference Européenne des Administrations des postes et des télécommunications" abbreviated CEPT. A European standards body composed of national telecommunications administrators and official carriers. A standards organization for European telecommunications administrations, CEPT formerly issued technological standards. Today all relevant technological standards formerly issued by CEPT have been taken over by ETSI, and CEPT is devoted to tariff, legal and other non-technical topics.

Conferencing-The process of adding additional callers to join into a phone conversation or data conference.

Configuration-(1-system) A relative arrangement of interconnected equipment or software in a system. (2-file) The information used to adapt an equipment or software program to its environment (configuration).

Configuration Management (CM)-
Configuration management is one of the five functions defined in the FCAPS model for network management. With configuration management network operation is monitored and controlled. Hardware and programming changes, including the addition of new equipment and programs, modification of existing systems, and removal of obsolete systems and programs, are coordinated. An inventory of equipment and programs is kept and updated regularly.

Configuration Message-In the Spanning Tree Protocol, a BPDU that carries the information needed to compute and maintain the spanning tree.

Confirm Initial Traffic Channel-When the mobile station is assigned to a voice/traffic channel, it must confirm the message that it is being assigned within the cellular channel. If the assigned channel is not in the cellular band, it must re-initialize.

Conformance-(1-general) The ability or certification of a device or system to perform or act as defined to an agreed specification or process. (2- Bluetooth) When conformance to a Bluetooth profile is claimed by a vendor all mandatory capabilities for that profile must be supported in the specified manner (process-mandatory). This also applies for all optional and conditional capabilities for which support is indicated. All mandatory, optional, and conditional capabilities for which support is indicated are subject to verification as part of the Bluetooth certification program. See also Bluetooth Qualification Review Board.

Congestion-A condition that exists when the demands for service on a communications network exceed its capacity to deliver that service.

Congestion Notification-Congestion notification is a control flag signaling system that is used in a data transmission system to indicate the status of network congestion. Congestion notification allows data communication devices that are connected to the data network to send or delay packet transmission.

Forward congestion notification indicates to upstream data switching devices in a communication network that data that is being transmitted through congested switches and it is likely that some of the remaining data or packets may be discarded. The upstream switch can then change the discard priority accordingly.

Congestion Time-The time or probability that a system is congested over any time period.

CONN-Connect

CONN ACK-Connect Acknowledgement

Connect (CONN)-In the Integrated Services Digital Network (ISDN) Q.931 protocol, the layer 3 CONNect message that indicates call acceptance by a called party.

Connect Acknowledgement (CONN ACK)-In the Integrated Services Digital Network (ISDN) Q.931 protocol, the layer 3 CONNect ACKnowledge message.

Connect Day-The day of the week on which a connection is made.

Connect Time- (1 - cellular) That period of time a cellular phone is in radio contact with a cell site. The connect time is not the same as the duration of the conversation. (2 - telco) The local time at a calling party's location when a connection was made.(3 - general) The amount of time that a given circuit is in use.

Connecting-A phase in the communication between devices when a connection between them is being establishment phase is completed.

Connecting Link-A channel and/or facility used to connect a communication device (terminal) to a transceiver (transmitter/receiver) or communication device.

Connection Establishment-(1-general) The procedures used to establish a connection between communication devices. This may involve the establishment of physical connections, assignment of logical channels, and the selection and allocation of communication protocols. (2-Bluetooth) A procedure for establishing a connection between the applications on two Bluetooth devices. Connection establishment starts after channel establishment is completed. At that point, the initiating device sends a connection establishment request. The specific request used depends upon the application.

Connection Identifier (CID)-A unique name or number that is used to identify a specific logical connection path in a communication system.

This diagram shows how a connection identifier (CID) is used in a data communications network to subdivide frames that are transmitted through a communication network so the portions of data can reach their specific destination channels. This example shows that the CID is used to de-multiplex the frames (divide channels) that pass through an access device. In this example, 4 telephone channels are multiplexed onto frames that are sent through access device 1. These frames are sent on a permanent virtual circuit (PVC) to access device 2. Access device 2 uses the data link connection identifier 7 to route the frames to channel demultiplexer. The channel demultiplexer uses the channel identifier (CID) code to route the data to each specific digital telephone device.

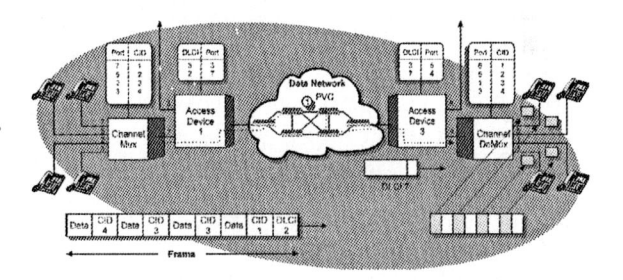

Connection Identifier (CID) Operation

Connection Oriented Data (COD)-A communication connection that allows the sequential delivery of its component parts (packets or frames). COD assures that a supported application such as voice or video will receive data with a minimum amount of delay.

Connection Section-In an SS7 system, the part of a Signaling Connection Control Part (SCCP) connection that is used between an end point and an intermediate point where a coupling must be provided between two adjacent sections.

Connection-Oriented-A communications model in which stations establish a connection before proceeding with data exchange and in which the data constitutes a flow that persists over time.

Connectioniess Network Service-A network service that transfers information between end users without establishing a logical connection or virtual circuit.

Connectionless-A communication connection that allows the delivery of its component parts (packets or frames) by a variety of network connection paths. Connectionless service helps assure that packets of data will arrive even if a portion of the network is disabled or disconnected. This interconnection allows data information to start transferring without first establishing a connection and without immediate acknowledgment of receipt. Sometimes it is (imprecisely) called datagram. Internet protocol is an example of connectionless service.

Connectionless Data Service-Service at a given layer of the OSI Reference Model in which there is no connection setup phase.

Connectionless Protocol-A packet-switching format that requires that each packet in a series carry complete ad-dressing information so that it can find its own way through a network without using logical channels.

Connectionless Switching-A communications model in which stations can exchange data without first establishing a connection. In connectionless communications, each frame or packet is handled independently of all others.

Connectionless Transport Protocol (CLTP)-A protocol that manages end-to-end transport data addressing and error correction to allow routing of packets through a network. CLTP is only a routing protocol and does not guarantee the delivery of data. It is the open systems interconnect (OSI) protocol equivalent to user datagram protocol (UDP).

Connector Block-A block of insulating material supporting terminals that can be used to connect electric conductors.

CONP-Connection-Oriented Network Protocol

CONP-Connection Oriented Network Protocol

Console-The computer access device that allows an operator to communicate with or control a network. A console commonly consists of a display monitor and keyboard.

Consolidated Carrier-A communications service provider that integrates several communication services. Also known as an integrated carrier.

Constant Bit Rate (CBR)-A class of telecommunications service that provides an end user with constant bit data transfer rate. CBR service is often used when real time data transfer rate is required such as for voice service.

Constant Voltage Source-A source with low, ideally zero, internal impedance, so that voltage will remain constant, independent of current supplied.

Constant-Current Source-A source with infinitely high output impedance so that output current is independent of voltage, for a specified range of output voltages.

Constant-Voltage Charge-A method of charging a secondary cell or storage battery during which the terminal voltage is kept at a constant value.

Constellation-A method of displaying multiple information parts of an signal. A constellation may display phase and amplitude on an x-y axis. Using the final parameter of time where the signal has a cycle of one rotation around the x-y axis, points on the constellation may display information elements (logic 1's and 0's).

Consultant-A person who borrows your watch, then charges you for the time of day. see also Guru.

Consultation Hold-A PBX feature that allows a telephone extension to place a call on hold while communicating on another line to receive a consultation before returning to process the call that is on hold. This consultation is often to determine the routing of the call (if the consulted person desires to receive the call).

Consultative Committee for International Telephony And Telegraphy (CCITT)-Original French language name "Comité Consultatif International Télégraphique et Téléphonique" abbreviated CCITT. A part of the United Nations Economic Scientific and Cultural Organization (UNESCO), based in Geneva, that develops worldwide telecommunications standards. Standards (called recommendations) were published every four years in complete sets. Each section of the standards appeared in a book having a letter code such as G for coding/decoding, Q for signaling, I for ISDN, V for modems and multiplexers, etc. Individual standards within each book are designated by a number following the letter, as in Q.931 or I.451 for ISDN call processing signals (appears in two different books for historical reasons), V.32 for certain modems and error correcting codes, X.25 packet networks, T.30 for facsimile, etc. In 1993, after a reorganization, the organization's name was changed to International Telecommunication Union-Telecommunications Sector (ITU-T or just ITU), and even though ITU now creates recommendations and standards, you will still hear the CCITT standards mentioned. ITU standards are still identified by letter-number codes of the same form, but they are no longer published in a full book, nor are they issued on a rigid four year schedule. Instead, each standard appears when a revision is deemed necessary. ITU standards can be purchased as a downloaded document via the ITU Internet wet site www.itu.int

Consumer Demand Curve -Economic model which defines the different level of service that an individual
customer will demand at a different price/service level.

Consumer Electronics Bus (CEBus)-A communications transmission system that transfers data on local (typically residential) power lines. The standard for CEBus is EIA IS-60. CEBus works similar to an Ethernet data network system.

Contact Noise-A noise resulting from current flow through an electrical contact that has a rapidly varying resistance, as when the contacts are corroded or dirty.

Contact Resistance-The resistance at the surface when two conductors make contact.

Contaminated Trunk Group-A trunk group that carries both local exchange and interexchange carrier traffic.

Contention-A condition which exists when two or more devices attempt to transmit at the same time using a shared channel.

Contention Control-A system for allocating channels within a communication system (such as a Ethernet data network or cellular telephone system) where requests for service from two or more communication devices may occur simultaneously.

The control portion usually involves the sensing of collisions and assignment of random delays for repeated access attempts.

Contingency Planning-Planning of a course of action such that the plan will be invoked only if the contingency materializes.

Continuation Application-An application for patent claiming priority to a pending application for patent which is identical to the co-pending application, but claiming a different invention.

Continuation-in-Part Application-An application for patent claiming priority to a pending application for patent to which new matter (such as an improvement to the invention) has been included.

Continuity-(1 - circuit) A continuous path for the flow of current in an electric circuit.

Continuity Check-A physical electrical check made to a circuit or circuits requested by a network (such as the SS7 network) to verify that an acceptable path (for data, speech, etc.) exists for transmission.

Continuity Test-A quick, general test that is used to measure copper cable plant using either an ohm meter, circuit tester or tone and probe kit. With the circuit power off, attach leads of meter to both conductors and confirm a low or near zero ohms reading. In addition both individual conductors should be tested in relation to a ground source. Simply attach one meter lead to electrical ground and the other to either the tip or ring side. A high or infinite ohms(resistance) reading should be achieved.

Continuous DTMF-Continuous Dual Tone Multi-Frequency

Continuous Dual Tone Multi-Frequency (Continuous DTMF)-A process for some telephones (especially mobile telephones and PBX telephones) that allows a dual tone multifrequency (DTMF) signal to be sent as long as the key is depressed. Some telephones send DTMF for a predefined time limit (such as 100 msec). Short DTMF tones may not allow access to services such as voice mail and answering machines. However, long DTMF tones that are sent over a poor communication line (such as an analog radio channel) may be recognized as multiple digits if temporary lapses or distortion occurs during DTMF transmission.

Continuous Transmission-A mode of operation where the mobile does not cycle its power level down when the modulating signal amplitude is low or off.

Continuous Wave (CW)-An electromagnetic signal in which successive oscillations of the waves are identical and have a constant or unvarying frequency and amplitude. Continuous wave usually refers to the output of a device (e.g., an optical fiber laser) which is turned on, but which is not modulated with a signal. Historically, in the first half of the 20th century, this term was used in radio to describe a radiotelegraph signal typically produced by an electronic (vacuum tube) oscillator, in distinction to the irregular, non-constant amplitude and non-constant frequency signal produced by an electric spark or arc oscillator.

Continuous Wave Laser-A laser in which the coherent light beam is generated continuously, as is normally needed for optical fiber communications systems.

Continuous Wave Radio Telegraphy-A radio system in which an unmodulated carrier is transmitted for a mark signal (and nothing transmitted for a source signal). At the receiving end, a beat frequency oscillator is used to make the continuous wave signal audible.

Continuously Variable Slope Delta (CVSD)-A type of delta modulation in which the size of the steps of the approximated signal is progressively increased or decreased as required to make the approximated signal closely match the input analog signal.

Continuously Variable Slope Delta Modulation (CVSD)-A type of delta modulation in which the size of the steps of the approximated signal is progressively increased or decreased as required to make the approximated signal closely match the input analog wave.

Contrast-The range of light-to-dark values of an image that are proportional to the voltage difference between black and white voltage levels of the video signal. The contrast control adjusts video gain (white bar, white reference).

Contrast Ratio-In video applications, the ratio of the highlight output level divided by the low light output level. In theory the contrast ratio of a TV system should measure as much as 300:1. In reality, there are several limitations, including the CRT itself. Light from adjacent elements contaminate the area of each element, and room ambient light further contaminates the image emitted from the CRT. Well-controlled viewing conditions should yield a practical contrast ratio of 30:1 to 50:1.

Control Bits-Bits contained within a sequence of data bits ("bit stream") that are used to control the transmission of information. Control bits are non-data transmission overhead bits.

Control Bus-In routing switchers, the interconnecting communications path between control panels or devices and the routing matrices.

Control Busy Hour-In a telecommunications network, the hour with the highest usage. (See also: busy hour.)

Control Center-A centralized work location from which managers and support staff administer the bulk of central office daily work requests. Examples include network terminal equipment centers, switching control centers, and frame control centers.

Control Channel-A radio channel in system that is dedicated to the sending and/or receiving of controlling messages between the base station and mobile radios. The control messages may include paging (alerting), access control (channel assignment) and system broadcast information (access parameters and system identification).

Control Channels-An allocation of frequencies in a communication system (such as a cellular telephone network) that are used to convey control information rather than voice traffic.

Control Character-A character in a computer program whose occurrence in a particular context initiates, modifies, or stops an action that affects the recording, processing, or interpretation of data.

Control Keys-Keys on a computer keyboard that, when pressed simultaneously, invoke control functions, for example, the use of "Control-D" to terminate a computer work session. One of the keys generally is marked Control or CTRL, and the other is an alphabetic character.

Control Mobile Attenuation Code (CMAC)-Used in a message sent to the mobile from the base which assigns the mobile to an absolute (specified) power level. This is important in small diameter cells where the mobile must access the system at a low power level to prevent co-channel interfering with other control channels.

Control Office-An exchange carrier center or office responsible for the installation and maintenance of a given access service furnished to an access customer.

Control Packet-A packet of data that is sent in a communication system that is used to control the transmission (link layer) or higher- level communication layer.

Control Plane-The portion of a network system that is involved in the setup, management, and termination of communication sessions and services.

Control Unit-A part of the microprocessor whose function is to read and decode instructions from the memory.

Controlled Environmental Vault (CEV)-A controlled environmental vault (CEV) is an underground room that is used to contain electronic and/or optical equipment under controlled thermal and humidity conditions.

This photograph shows the installation of a controlled environment vault (CEV) that is about the size of a small bathroom and provides access through a manhole. This CEV is designed and constructed of materials that help ensure a good environment for electrical assemblies and cables.

Controlled Environment Vault
Source: Hartford Concrete Products, Inc

Controlled Load Service-Controlled-load service provides a variable bandwidth for each communication session that varies based on factors including the amount of network activity (e.g. heavy traffic) and quality of service requirements (e.g. real-time compared to non-real time communication application).

Controlled Rerouting-In the Signaling System 7 protocol, the controlled transfer of signaling traffic from an alternative signaling route to the normal signaling route, when it becomes available.

Controller/Sequencer-An electronic assembly that receives and executes instructions from memory to allow the coordination of other electronic assemblies or processing communications data flow.

Controlling RNC (CRNC)-The controlling RNC is responsible for the load and congestion control of its own cell, and also executes the admission control and code allocation for new radio links to be established in those cells.

CONUS-Contiguous United States

Convention-A generally acceptable symbol, sign, or practice in a given industry.

Convergence Billing-An all inclusive bill that combines charges for: local, long distance, data, Internet, and possibly utilities (water, gas, electric) and cable TV. Convergent Billing generally implies one all-encompassing view of the Customer for all subscribed services, and a unified view of the product portfolio to enable cross-product packages.

Convergence Time-The amount of time it takes an echo canceller to train (learn) how to cancel echoes from the incoming audio signal.

Convergent-A process of integrating services such as local telephone service, mobile communications, cable television, and long distance together onto a single system. These services and/or systems may be truly integrated or loosely tied together (sometimes called "stapled") so allow a customer to have a single billing and customer care access point.

Convergent Market Cross Mapping-Technique used to compare the populations of customers "owned" by different product line managers, in order to determine the potential value that a sharing of those customers can have for each product manager.

Convergent Network-A network incorporating wireless, optical and copper transmission media and multiple protocols. In the past, networks have often been built for a specific purpose and/or based on a specific technology. In the future, more and more interconnections among networks are anticipated to maximize communications options.

Conversation-As used in Link Aggregation, a set of traffic among which ordering must be maintained.

Conversation Minute Miles-The product of the total number of message minutes carried on a trunk or circuit group and the average route miles of the trunk or circuit group.

Conversation Minutes-The actual time during which a customer uses a connection for voice transmission.

Conversation State-A state (mode) of a mobile telephone phone when it tuned to a voice or traffic channel and is providing a voice path for a call. The mobile telephone may provide additional information (such as MAHO0 during the conversation state. When the conversation ends, the phone will enter the control channel scanning and locking state.

Conversion Factor-A factor applied to known data to adjust it and make it appear as other data. For example, reported minutes can be converted to conversation minutes through the use of a conversion factor.

Conversion Rate-(1-Internet web) A measure of the people who log on to a web page and select a process or purchase via that web site. A high conversion rate percentage usually indicates more valuable a web site is to its visitors.

Conversion Timing-An agreed-upon scheduled date and time to begin a cutover in a coordinated conversion.

Convolutional Coding-A error correction process, which uses the input data to create a continuous flow of error protected bits. As these bits are input to the convolutional coder, an increased number of bits is produced. k is the length of the code. Convolutional coding is often used in transmission systems that often experience burst errors such as wireless systems.

This figure shows a convolutional coder that continuously receives the data bits in sequence to create a new digital signal that combines both the original information and new error protection bits. This example shows a 1/2 rate convolutional coder that generates two bits for every one that enters.

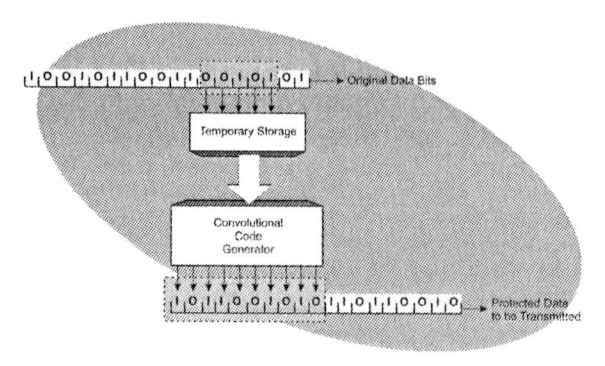

Convolutional Coding

Coordinated Restoral-An equipment or service restoral process that is coordinated. Coordinated restoral systems can sequentially or simultaneously restore equipment and or services.

COP-Cable Organizer Panel

COP-Committee Of Principals

Copper Cross Connect-A copper cross connect system allows access lines (copper lines) to be connected to several different DSL modems. There are two key reasons to use a copper cross connect system. The first reason is to allow a copper wire access line to be connected to different digital subscriber line modems. This could be because the customer may upgrade to a new type of modem (e.g. ADSL to VDSL) or if a DSL modem fails, a spare DSL modem could be connected to the customer's line. The second reason is to allow the access line to be connected to a DSL modem only when a connection is required. This would allow a DSL service provider to install lesser number of modems in a system than they have customers for.

Copper Distribution Data Interconnection (CDDI)-CDDI is an adaptation of the FDDI protocol for use on copper cable.

COPS-Common Open Policy Service

COPW-Customer Owned Premises Wire

Copy Port-Synonymous with Mirror Port.

Copyright-A copyright is a monopoly which may be claimed for a limited period of time by the author of an original work of literature, art, music, drama, or any other form of expression - published or unpublished.

Copyrights are Intellectual Property Rights which give the owner, or assignee, the right to prevent others from reproducing the work or derivates, including reproducing, copying, performing, or otherwise distributing the work.

Copyrights can, in many countries be claimed without registration. Most countries, or regions, however, have a copyright office where copyrights can be officially registered.

COR-Circuit Order Record

CORBA-Common Object Request Broker Architecture

CORD-Cibernet On-Line Roaming Database

Cordless-Cordless systems are short range wireless telephone systems that are primarily used in residential applications. Cordless telephones typically use radio transmitters that have a maximum power level below 10 milliwatts (0.01 Watts). This limits their usable range of a 100 meters or less.

Cordless Telephone System-A system consisting of two transceivers, one a base station that connects to the public switched telephone network and the other a mobile handset unit that communicates directly with the base station. Transmissions from the mobile unit are received by the base station and then placed on the public switched telephone network. Information received from the switched telephone network is transmitted by the base station to the mobile unit.

Because cordless telephone systems do not as a rule have a dedicated control channel to provide information, the cordless handset and base station continuously scan all of the available channels (typically 10 to 25 channels). This figure shows the basic cordless telephone coordination process. This diagram shows that when the cordless phone or base station desires to transmit, the unit will choose an unused radio channel and begin to transmit a pilot tone or digital code with a unique identification code to indicate a request for service. The other cordless device (base station or cordless phone) will detect this request for service when it is scanning and its receiver will stop scanning and transmit an acknowledgement to the request for service. After both devices have communicated, conversation can begin. When another nearby base station detects the request for service, it will determine that the message is not intended for it and will not process the call and scanning will continue.

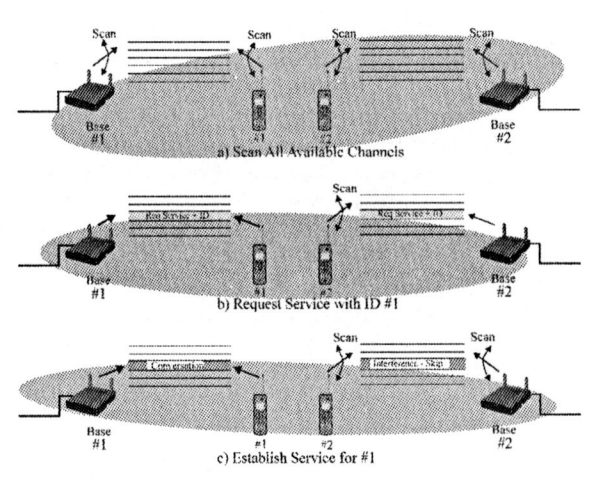

Cordless Telephone Operation

Cordless Telephony Profile-A "3-in-1 phone" solution for providing an extra mode of operation to cellular phones, using Bluetooth wireless technology as a short-range bearer for accessing fixed-network telephony service via abase station. The 3-in-1 phone solution can also be applied generally for wireless telephony in a residential or small-office environment, such as cordless-only telephony or cordless telephony services in a multimedia PC. See also Intercom Profile.

Cordless Telephony, Second Generation (CT2)-The second generation of cordless telephone technology that allows cordless telephones to access small coverage area public radio systems in places such as airports and railway stations. The CT2 system does not normally allow the receiving of calls in a public area.

Cordless Telephony, Second Generation, Enhanced (CT2+)-An enhanced version of the CT2 system that allows for incoming and outgoing calls when roaming.

Cordless Telephony, Third Generation (CT3)-The third generation of cordless telephone technology that allows cordless telephones to access small coverage area public radio systems in places such as airports and railway stations. The CT3 system does allow the receiving of calls in a public area.

CORE-Council of Registrars

Core Area-The urban center of a municipality such as the downtown region. Core area also can refer to the portion of an urban wire-center area characterized by its large buildings.

Core Size-A primary specification for an optical fiber, stated in microns. The core size does not include cladding. The core size determines the end surface area that accepts and transmits light.

Core Switch-A backbone switch that interconnects to other core switches and edge switches. Core switches maintain information about virtual paths that are connected through the network.

Corona-A bluish luminous discharge resulting from ionization of the air near a conductor carrying a voltage gradient above a certain critical level.

Corporation For Open Systems International-A user and vendor forum of computer and communications companies whose goal is to ensure a selection of interoperable, multi-vendor products and services operating under Open Systems Interconnection (OSI), Integrated Services Digital Network (ISDN), and related international standards.

COS-Corporation for Open Systems International
CoS-Class of Service
COSE-Common Open Software Environment
COSINE-Cooperation for Open Systems Interconnection Networking in Europe
Cosmic Noise-The random noise originating outside the earth's atmosphere.
Cosmic Rays-Charged particles (ions) emitted by all radiating bodies in space.
COSMOS-Computer System For Mainframe Operations
CoSN-Consortium for School Networking
Cost Of Clearing Blockage-In multiple outside plant, the average cost of rearranging a network to fulfill an inward service order each time a blockage occurs.
Cost Of Service-The total cost of providing utility service and includes operating expenses, depreciation, taxes, and a rate of return adequate to service investment capital.
Cost Per Plant Unit-An average cost for performing work on a plant unit within a plant classification, such as placing a telephone pole.
Cost Per Thousand (CPM)-The cost for each thousand units associated with media such as TV time, radio spots, print ads, etc. For example, a television ad might sell for $200 CPM. An ad for television show with 2,000,000 viewers will therefore cost $200 x 2,000 or $400,000 dollars.
Cost Per Work Unit-An average cost for performing a work unit within a plant classification, such as placing a telephone pole.
Cost-Based Pricing-A method of setting rates for telephone services based on the costs for providing those services to the exchange carrier.
Costa Model -The Costa Model is a representation used to help define the way that a telephone companies (telcos) use different departments and functions to optimally attract and keep customers. Key layers of the Costa Model include awareness, familiarity, attraction, preference and selection (sales).
COT-Central Office Terminal
COT-Continuity Check Message
COT-Customer Originated Trace
coulomb (C)-Unit of electric charge equal to one ampere second. Named for Charles Coulomb, 19th century French physicist. The standard unit of electric quantity or charge. One coulomb is equal to the quantity of electricity transported in one second by a current of one ampere.

Coulomb's Law-The principle that the forces of attraction and repulsion of electric charges are directed along a line between the charges. The forces are inversely proportional to the square of the distance between the charges, and proportional to the product of their magnitudes. (Named for the French physicist Charles-Augustine de Coulomb, 1736-1806.)
Counter-Poising-A bonding and grounding method used for isolated aerial cable facilities that requires outside plant telephone company (telco) personnel to connect a # 6 gauge (AWG)ground wire to the cable strand at one pole and bury 18" deep into earth ground. The #6 ground wire is then lashed to the strand along the section to the next pole and clamped to the strand at that pole or connected to a splice closure when available.
Country Code-A 1-, 2-, or 3-digit number that identifies a country or numbering plan to which international calls are routed. The first digit is always a world zone number. additional digits define a specific geographic area, usually a specific country.
Couple-The process of linking two circuits by inductance, so that energy is transferred from one circuit to another.
Coupler-A device that is used to transfer energy from one transmission medium to another transmission medium. A coupler is typically used to provide a sample of a transmission signal to another device. A coupler may also be used to insert a signal from one transmission line to another transmission line.
Coupling Factor-A figure representing the combined effect of the various couplings between a power circuit and a communications circuit in producing noise in the communications circuit.
Court of Appeals of the Federal Circuit (CAFC)-The United States Court of Appeals of the Federal Circuit. Established in 1982 to promote uniformity in patent law. Appeals of the decision of district courts are considered by one of 12 appeals courts of the CAFC.
COV-Control Over Voice
Coverage-(1 - general) The geographical area over which the signal strength of a given radio frequency is available for service. (2- broadcasting) The "footprint" of a broadcast station's signal, usually measured in terms of percentage of homes covered in a designated geographic area or in terms of the signal strength. (See also: penetration.)

Coverage Area-A geographical area that has a sufficient level of radio signal strength from a transmitting tower to provide an acceptable level of signal reception. An acceptable level of signal reception may be determined by signal to noise ratio (for analog systems) or bit error rate (for digital systems).

Covert-Adjective used to describe undercover operations by government agents. "Covert" communications are generally encrypted.

COW-Cell Site On Wheels

COW Interface-Character Oriented Windows Interface

CP-Consolidation Point

CP-Control Point

CP/M-Control Program for Microcomputers

CPA-Combined Paging And Access

CPA-Cost Per Action

CPC-Calling Party Control

CPCH-Uplink Common Paging Channel

CPCN-Certificate of Public Convenience and Necessity

CPCS-Common Part Convergence Sublayer

CPCS-SDU-Common Part Convergence Sublayer-Service Data Unit

CPDA-Certified PowerBuilder Developer Associate

CPDP-Certified PowerBuilder Developer Professional

CPE-Customer Premises Equipment

CPI-Cost Per Inquiry

CPI-Computer to PBX Interface

CPI-C-Common Programming Interface For Communications

CPL-Call Processing Language

CPLD-Complex Programmable Logic Device

CPM-Continuous Phase Modulation

CPM-Cost Per Thousand

CPN-Calling Party Number

CPNI-Customer Proprietary Network Information

CPODA-Compression Priority Demand Assignment

CPP-Calling Party Pays

CPR-Continuing Property Record

CPS-Calls Per Second

CPS-Characters Per Second

CPTS-Certified Performance And Tuning Specialist

CPU-Central Processing Unit

CPU Space-A protected memory space addressed only by the CPU itself. The CPU space is used for processor internal functions or vectored exception processing.

CPUG-Call Pickup Group

CQM-Channel Quality Message

CR-Call Rate

CR-Call Register

CR-Critical

CR-Carriage Return

CR-LDP-Constraint-based Routed Label Distribution Protocol

Cradle Switch-Synonym for Switch Hook.

Craft Access Terminal-A handheld personal computer (PC) terminal that contains an internal modem that is used by telephone company (telco) personnel for daily work order processing, accessing subscriber line records and remote loop testing.

Cramming-Cramming is the fraudulent addition of charges for services that were not agreed to by the end customer.

Crapplet-A Java applet that does not operate as desired.

Crash-A complete or partial failure of a hardware device or a software program or operation.

CRC-Cyclic Redundancy Check

CRC-Cyclical Redundancy Check

CRC Character-Cyclic Redundancy Check Character

CRCC-Cyclic Redundancy Check Code

CREDFACS-Conduit, Risers, Equipment space, Ducts and Facilities

Credit Score-A measure of a customer's credit worthiness that is based on information obtained from an independent credit bureau. Sometimes also termed "External Score".

CREN-Corporation for Research and Educational Networking

CRF-Connection Related Function

CRF-Cell Relay Function

Crimp Termination-A terminal connected to a conductor by high-pressure crimping of a lug onto a wire.

CRIS-Customer Record Information System

Critical Angle-In a fiber optics transmission system, the critical angle is the maximum angle at energy from an external light source can enter into the transmission medium.

Critical Frequency-In radio propagation, the limiting frequency below which a wave component is reflected by, and above which it penetrates through, an ionosphere layer.

Critical Report Date-(1-testing) The date on which point-to-point testing is completed. (2-service order) The date on which implementation groups must report that all documents and material have been received to carry out a service order.

Critical Section-The feeder section with the shortest lifetime in a route.

CRM-Cell Rate Margin

CRM-Customer Relationship Management

CRNC-Controlling RNC

CRO-Complete with Related Order

Cross Compiler-A compiler that runs on a host computer and outputs machine level instructions and generates executable programs for a target microprocessor / computer that is different than that running on the host. This is typically done to support new architectures or smaller micro-controller based systems that have insufficient resources to host their own compilation software.

Cross Connect Switch-A (typically telephone channel) switch that connects pre-designated channels in different links on a semi-permanent basis. Cross connect switches do NOT set up a different switched channel path for each individual call dialed by the originating subscriber.

Cross Link (C-Link)-A "C" (cross) link connects signal transfer points (STPs) that perform identical functions into a mated pair within an SS7 network. A "C" link is used only when an STP has no other route available to a destination signaling point due to link failure(s). Note that SCPs may also be deployed in pairs to improve reliability; unlike STPs, however, mated SCPs are not interconnected by signaling links.

Cross Modulation-The interference experienced when a carrier signal becomes modulated by an unwanted signal, as well as being modulated by its desired signal.

Cross Sell-Marketing activity that is designed to encourage customers to buy different products from the same company. In telecommunications, cross sell could include selling wireless or ISP service to wireline customers.

Cross Talk-A problem where the audio from one communications channel is imposed on another channel.

Cross-Office Transit Delay-The time it takes for data or a message to pass through a telephone exchange office.

Crossbar-A switching system that uses a centrally-controlled matrix switching network of electromagnetic switches which work with magnets and which connect horizontal and vertical paths to establish a path through the network. Crossbar switches are circuit switches, typically in the form of voice PBXs and Central Offices (COs). Crossbar was known for its reliability, at least in comparison to earlier Step-by-Step (SxS) electromechanical switches, but is now largely obsolete because it takes up a lot of space and isn't programmable. The first crossbar switch was a central office installed in Brooklyn, NY in 1937.

Crossbar Switches (XBar)-Crossbar switches used mechanical arms to physically connect to wires (or busses) together. These mechanical arms ("Crossbars") connect horizontal and vertical bars together to connect input and output lines together. Magnets are used to open and close the crossbar switch contacts.

This figure shows how a crossbar switching system can use a matrix of lines (busses) to allow any input line to be connected to any output line. When a connection needs to be made, a mechanical switch connects one of the busses with the other busses. The disadvantage of this system is that the number of mechanical switches for connecting each input port to an output port exponentially increases with the number of ports that require connection. For example, a switch with 10 inputs and 10 output lines requires 100 switches. A switch that has 20 inputs and 20 outputs requires 400 switches.

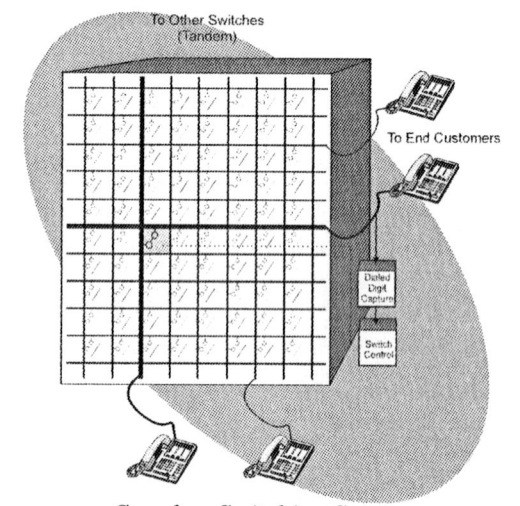

Crossbar Switching System

Crossover Cable-Another name for a null modem cable. A cross over cable reverses the transmit and receive communication lines to allow the direct connection of computers without the need for hubs, switches, or routers.

Crosstalk-The undesired leakage of a signal from one communications channel to another.

This diagram shows that crosstalk occurs when signal energy from one transmission line is transferred (coupled) to the other. This diagram shows that crosstalk can be divided into two categories: near end crosstalk (NEXT) and far end crosstalk (FEXT). NEXT results when some of the energy that is transmitted in the desired direction seeps into one (or more) adjacent communication lines from the originating source. FEXT occurs when some of the digital signal energy leaks from one twisted pair and is coupled back to a communications line that is transferring a signal in the opposite direction. Generally, NEXT is more serious than FEXT as the signal interference levels from NEXT are higher.

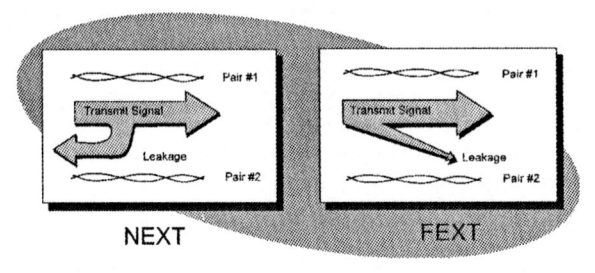

Crosstalk Operation

Crowbar-A short-circuit or low-resistance path placed across the input to a circuit.
CRP-Command Repeat
CRR-Candidate Reselection Rules
CRS-Call Relay Service
CRT-Cathode Ray Tube
CRTC-The Canadian Radio-Television And Telecommunications Commission
CRTP-Compressed Real-time Transport Protocol
Cryptology-The science of hidden, disguised, or encrypted communications. Cryptology embraces communications security and communications intelligence.
Crystal-(1-general) A solidified form of a substance that has atoms and molecules arranged in a symmetrical pattern. (2-piezoelectric) A crystal, such as quartz, that will generate a voltage when excited. (3-quartz) A piezoelectric crystal cut from natural quartz. (4 - X-cut) A crystal with its major flat surfaces cut so that they are perpendicular to the electrical (X) axis of the original quartz crystal.(5-XY-cut) A cut crystal that has characteristics similar to those of the X-cut and the Y-cut crystals. (6-Y-cut) A crystal with its major flat surfaces cut so that they are perpendicular to the mechanical (y) axis of the original quartz crystal.
Crystal Controlled Oscillator-An oscillator that couples a crystal's piezoelectric-effect to a tuned oscillator circuit so the crystal frequency adjusts (pulls) the oscillator frequency to its the crystals natural resonant frequency.
Crystal Filter-A filter that uses piezoelectric crystals to create resonant or anti-resonant circuits that allow desired frequencies (signals) to pass through.
CS-Capability Sets
CS-Composite Sync
CS-Convergence Sublayer
CS-1-Capability Set 1
CS-ACELP-Conjugate Structure-Algebraic Code Excited Linear Prediction
CS-CDPD-Circuit Switched-Cellular Digital Packet Data
CSA-Cailpath Services Architecture
CSA-Carrier Serving Area
CSA-Canadian Standards Association
CSC-Customer Service Center
CSCD-Circuit Switched Cellular Data
CSDC-Circuit Switched Digital Capability
CSE-Certified Solutions Expert
Cseq-Command Sequence
CSFI-Communications Subsystems For Interconnection
CSI or CSID-Called Subscriber Identification
CSID-Calling Station Identification
CSLIP-Compressed Serial Link Internet Protocol
CSMA-Carrier Sense Multiple Access
CSMA/CA-Carrier Sense Multiple Access/Collision Avoidance
CSMA/CD-Carrier Sense Multiple Access/Collision Detection

This figure shows the operational function of carrier sensing multiple access with collision detection (CSMA/CD). CSMA/CD is a carrier sense multiple access transmission scheme in which transmission collisions are followed the transmitting stations backing off the network a random amount of time before attempting to retransmit. CSMA/CD is used as the basis of Ethernet networks.

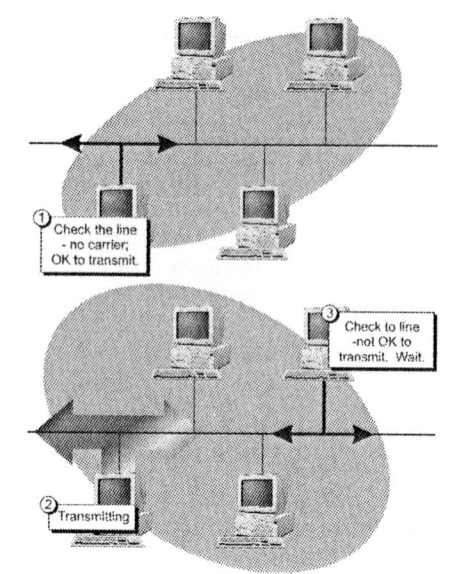

CSMA/CD Operation

CSO-Central Service Organization
CSP-Competitive Service Provider
CSP-Commerce Service Provider
CSP-Carriage Service Provider
CSP-Certified Service Provider
CSQP-Customer/Supplier Quality Process
CSR-Customer Service Representative
CSR-Customer Service Record
CSR-Cell Switch Router
CSR-Carrier Sensitive Routing
CSRG-Computer Systems Research Group
CSS-Capability Sets
CSS-Cascading Style Sheets
CSS-Cellular Subscriber Station
CST-Computer Supported Telephony
CSTA-Computer Supported Telecommunications Application
CSTA-Computer Supported Telephony Application
CSTD-Cable System Terminal Device
CSU-Channel Sharing Unit
CSU-Channel Service Unit
CT-Computer Telephony
CT-Call Transfer
CT-2-Cordless Telephony Generation 2
CT-2+-Cordless Telephony Generation 2-Plus
CT1-Cordless Telephony Generation 1
CT2-Cordless Telephony, Second Generation
CT2+-Cordless Telephony, Second Generation, Enhanced
CT3-Cordless Telephony, Third Generation
CTAS-Carrier Test Access Switch
CTCA-Canadian Telecommunications Consultants Association
CTCSS-Continuous Tone-Controlled Squelch System
CTD-Cell Transfer Delay
CTE-Connected Telecommunications Equipment
CTE Directive-Connected Telecommunications Equipment Directive
CTI-Computer Telephony Integration
CTIA-Cellular Telecommunications Industry Association
CTIP-Computer Telephony Interface Products
CTN-Consumers' Telecommunications Network
CTN-Call Tracking Number
Ctrl-Control
CTS-Clear To Send
CTS-Communication Transport System
CTTS-Coax to the Curb
CTTU-Centralized Trunk Test Unit
CTY-Console Teletype
CU-Centrex Customer Premises
Cu Interface-The electrical interface between the universal mobile telephone system subscriber identity module (USIM) and the mobile equipment (ME).
CUA-Common User Access
Cue-(1-magnetic tape) The operation through which a magnetic tape is stopped at a predetermined position ahead of an edit in-point. (2-performer) A program feed to a performer, which permits monitoring of other performers and program status.
CUE-Certified Unicenter Engineer
CUG-Closed User Group
Curie Point-The temperature at which piezoelectric properties cease.
Curing Threshold-The amount that a customer is required to pay in order for the collections process to be suspended. This amount is not necessarily the full amount due to-date and is dependent on: the type of customer, collectible amount, and behavior score.
Current Amplifier-A low output impedance amplifier capable of providing high current output.
Current Carrying Capacity-A measure of the maximum current that can be carried continuously without damage to components or devices in a circuit.

Customer Provisioning

Current Probe-A sensor, damped around an electric conductor, in which an induced current is developed from the magnetic field surrounding the conductor. For measurements, the current probe is connected to a suitable test instrument.

Current Transformer-A transformer-type of instrument in which the primary carries the current to be measured, and the secondary is in series with a low-current ammeter. A current transformer is used to measure high values of alternating current.

Custom Calling Services-Custom local area signaling services (CLASS) are telephone service features available in a local access and transport area (LATA) that are primarily based on information that can be processed inside the telephone network. CLASS features include call forwarding, caller identification, and three-way calling.

Custom Local Access Signaling Services (CLASS)-A set of telephone services and enhanced features available in a local access customers that may include calling number delivery or calling name delivery (CND), message waiting, and other features.

Customer-An entity that has the ultimate responsibility for: signing contracts, making payments, allowing use of the services. Sometimes, this term also refers to prospects, or potential prospects.

Customer Account Record Exchange (CARE)-Procedure used for the exchange of customer records between the local exchange carrier (LEC) and the long distance (LD) carrier primary interexchange carrier selected (PIC'ed) by the customer.

Customer Care And Billing System (CCBS)-A system that provides customer account tracking, service feature selection, billing rates, invoicing and details.

Customer Control And Management-The capability of a customer to remotely reconfigure special-service circuits using control circuits, software, and remotely configurable network elements.

Customer Data Rate-The data transmission rate that the customer receives.

Customer Desired Due Date (CDDD)-The date the customer desires delivery or operation of a product or service.

Customer Dialed - Operator Serviced-Calls that are partially handled by an operator. Services performed include recording of credit card numbers, collect calls, coin collection, and person-to-person calls. This function is also called "0+" dialing.

Customer Group-A common term for a group of customers with similar characteristics (e.g. business customers.)

Customer Interconnection Record-A record that is created at a control point for inter-connection.

Customer Loop-A dedicated communications channel, usually a pair of wires and/or a digital loop carrier channel, between a customer's telephone and a serving central office.

Customer Network Management-All activities associated with the planning, operation, administration, and maintenance of the communications network of a corporate customer. This term also refers to an arrangement that enables customers to manage their own networks, for example, to move Integrated Services Digital Network (ISDN) access channels from circuit-switched to packet-switched service.

Customer Originated Trace (COT)-A Custom Local Area Signaling Services (CLASS) feature that allows a subscriber to initiate a call trace request message. This call trace feature temporarily stores the dialed digits and alerts the telephone service operator to "tag" the originator's number to allow authorities to investigate the originator of the unwanted or unauthorized call. Some of the call trace activation codes include *57 on a touchtone phone or the dialing of 1157 on a rotary (pulse) phone. If the call trace of the last call was completed successfully, an announcement should be heard. The service operator will usually release the call trace information to law enforcement agencies and a signed authorization from the subscriber may be required.

Customer Premises Equipment (CPE)-All telecommunications terminal equipment located on the customer's premises, including telephone sets, private branch exchanges (PBXs), data terminals, and customer-owned coin-operated telephones.

Customer Problem Handling -Customer Problem Handling is the organizational unit responsible for taking trouble reports from customers and making sure that the problems are resolved.

Customer Provisioning-Customer provisioning is the process of delivering products and/or services to the customer. It is the operational aspects of delivering the service, from setting up accounts, user authentication, billing/return policies, channel deployment, packaging and customer support.

Customer Rating

Customer Rating-The process of rating the credit worthiness of a customer to determine what services they are authorized to use and the amount of deposit the customer may be required to pay.

Customer Record Information System-An electronic data processing system for keeping customer records and billing information, generally using magnetic tape.

Customer Relationship Management (CRM)-A process or system that coordinates information that is sent and received between companies and customers. CRM systems are used to schedule activities, allocate resources, and help control the sales activities within a company.

Customer Retrial-A customer's subsequent attempt, within a measurement period, to complete a call.

Customer Self Care-Customer self care is the process of a allowing the customer to review and/or activate and disable services without the direct assistance of a customer service representative (CSR). Customer self care can be as simple as providing account billing information to the customer by telephone through the use of an interactive voice response (IVR) system to providing interactive service activation menus on an Internet web site.

This figure shows that a customer self-care system can be used to review product or service options, check billing records, and change customer feature options (called "provisioning"). This diagram shows that the customer can contact the billing and customer care system via a gateway. The gateway may contain Internet web access (for graphic displays) or interactive voice response (IVR) systems to allow the customer to select their account, receive billing and customer care information, and possibly change feature options.

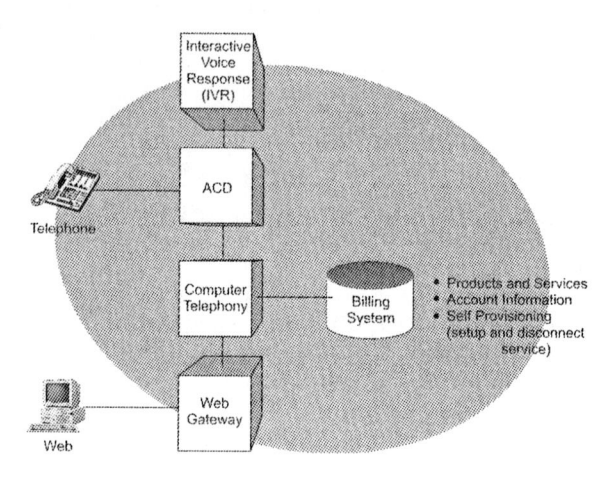

Customer Self Care Operation

Customer Service Class-The customer category of telecommunications services, including business, residence and public.

Customer Service Representative (CSR)-A company representative who manages customer communication for account inquiries, complaints, follow-up support, or other service related issues.

Customer Trouble Report Analysis Plan-An administrative procedure for recording trouble reports, furnishing a database from which analyses can be performed, and summarizing closed trouble data for exchange carrier administrative reports.

Customer Value Assessment -The analysis of how much value a customer provides to a company.

Customized Applications For Mobile Enhanced Logic (CAMEL)-Applications that operates on a "services creation node" in a GSM network. CAMEL allows a network operator to develop specialized services using advanced intelligent network (AIN) systems.

Cut-(1-video) A transition between two pictures that is instantaneous, without any gradual change. (2-crystal) The orientation of a crystal with respect to its electrical and mechanical axes.

Cut-Off Call-A call that has been disconnected by any means other than user intended.

Cut-Off Date-A date after which major redesign or reprogramming of a project cannot be achieved without serious effects on costs or completion date.

Cut-Off Frequency-(1-general) The frequency above or below which the output current in a circuit is reduced to a specified level. (2-radio frequency) The frequency below which a radio wave fails to penetrate a layer of the ionosphere at the angle of incidence required for transmission between two specified points by reflection from the layer.

Cut-Off Wavelength-(1-general) The shortest wavelength at which only the fundamental mode of an optical signal is capable of propagation. (2-theoretical) The shortest wavelength at which a single mode can propagate in a single-mode fiber. At wavelengths below cutoff, several modes propagate, and the fiber is no longer single-mode but multi-mode.

Cutback Gamma-The coefficient used to scale bandwidth when a long fiber is cut to a shorter length.

Cutback Technique-A method of measuring the attenuation in an optical fiber by cutting it about one meter from the transmitter, measuring the power at the cut-point without changing the launching conditions, rejoining the fiber, then measuring the received power at the output end.

Cutoff Date-The last date that transactions will be included in the current billing cycle.

CV-Code Violation
CV-Combined Voice
CVD-Chemical Vapor Deposition
CVF-Compressed Volume File
CVP-Co-operative Voice Processing
CVS-Computer Vision Syndrome
CVSD-Continuously Variable Slope Delta Modulation
CVSD-Continuously Variable Slope Delta
CVT-Circuit Validation Test
CW-Call Waiting
CW-Continuous Wave
CWA-Communications Workers Of America
CWA-Communication Workers of America
CWCG-Copper Wire Counterpoise Ground
CWID-Call Waiting Identification
CWML-Compact Wireless Markup Language
CWTS-China Wireless Telecommunication Standard Group
cXML-Commerce Extensible Markup Language

Cyan-Also called "process blue," one of the three subtractive primary colors.

Cyan, Magenta, Yellow and BlacK (CMYK)-The three primary subtractive colors used for printing ink or photographic color transparencies and prints. Black is used with these three subtractive primary colors because available inks or pigments do not inherently produce a solid black when all three are combined. Instead they produce a muddy brown, and the black ink is used to supplement the desired black areas. For the additive primary colors, see RGB.

Cyberspace-A term that is commonly used to describe an interconnected pubic data network (such as the Web) that has tools available that allow users to find and retrieve data from the network.

Cyclic Redundancy Check (CRC)-An error-checking technique in which bytes at the end of a packet are used by the receiving node to detect transmission problems. The bytes represent the result of a calculation performed on the data portion of the packet before transmission. If the results for the same calculation on the received packet are not equal to the transmitted results, the receiving node can request that the packet be resent. (See also: cyclic redundancy check code.)

An error detection and/or correction method that is used to determine if a series of data bits were received correctly during transmission. To setup a CRC error checking process, the original bits of data are supplied to a CRC generator. The CRC generator uses a specific mathematical formula to create a new group of data bits called the CRC check sum (bits). The CRC check bits are typically appended to the data bits that are being sent. The receiver of the message compares the result of the CRC generator on the receiver end to determine if the bits were received without error. In some cases, CRC check bits can be used to help correct some bits that were received in error during transmission, this is called Forward Error Correction (FEC) or Error Correction Code (ECC).

Cyclical Redundancy Check (CRC)-A process that uses a mathematical formula used to process and checks the accuracy of a digital transmission over a communications link. The process involves the processing of data information by the sending computer using one or several formulas to calculate a value (the CRC) from the information data. This value is added (appended) to the data message that is sent. The receiving device computes a value using the information received. If the two CRCs match, this indicates the transmission was not in error (or

Cypher Key

reduced probability of error). If the CRC does not match, receiving computer may ask the sending computer to retransmit the data. The CRC process is sometimes called a redundancy check because each transmission includes additional bits that represent the same information as the information data.

Cypher Key-A key used for the encryption process of data information.

D

D Bit-Delivery Confirmation

D Channel-This is the out-of-band control channel associated with ISDN bearer channels, (see B Channel), and is used to establish, monitor, terminate and provide enhanced telephone services. The D channel provides a 16kbps or 64kbps packet-mode connection between a serving switch and a customer's premises. The channel carries signaling and control information for B channel activity and also can carry user data in the form of packets.

D Flip-Flop-A type of bistable circuit with a single trigger input terminal, the P input

D Link-Diagonal Link

D&F-Demand And Facility Chart

D-1-The generic name used to describe a digital component video recording system. The D-1 system records digital component video to the CCIR-601 standard on 19 mm magnetic tape.

D-3-A digital composite video recording format for NTSC or PAL. A 1/2-inch medium is used. The concept is similar to the D-2 format.

D-AMPS-Digital Advance Mobile Phone Service

D-Inside Wire-Direct-Inside Wire

D-Layer-The lowest and most weakly ionized atmospheric layer affecting the propagation of radio waves. The D-layer occurs during daylight hours and is approximately 50 to 90 lcm above the surface of the earth.

D-Link-Diagonal Link

D-t-D-Digital-To-Digital Transfer

D/A-Digital To Analog Conversion

D/A Converter-Digital to Analog Converter

D/R-Distance to Reuse Ratio

DA-Distribution Amplifier

DA-Directory Assistance

DA-Destination Address

DA-Desk Accessory

DA-Discontinued Availability

DA-Pulse Distribution Amplifier

DAA-Data Access Arrangement

DAB-Digital Audio Broadcast

DAC-Digital-To-Analog Converter

DAC-Dual Attached Concentrators

DACD-Digital Automatic Call Distributor

DACL-Discretionary Access Control List

DACS-Digital Access Cross-Connect System

DAEMON-Disk and Execution Monitor

DAK-Deny Any Knowledge

DAL-Dedicated Access Line

DAMA-Demand Assigned Multiple Access

DAML-Digital Added Main Line

Damped Oscillation-An oscillation exhibiting a progressive diminution of amplitude with time.

DAP-Directory Access Protocol

DAP-Data Access Point

daraf-Informal but widely used name for the unit of reciprocal electrical capacitance, the reciprocal of the farad. ("farad" spelled backwards)

DARPA-Defense Advanced Research Projects Agency

DARS-Digital Audio Radio System

Dart Leader Stroke-The initial discharge that largely determines the path taken by a lightning flash. A dart leader develops continuously, and a stepped leader develops in short steps. Both are followed by a high-current discharge flash, usually in the reverse direction to the leader.

DAS-Dual Attached Stations

DAS Tape-Data Acquisition System

DASD-Direct-Access Storage Device

Dashboard-A dashboard is a screen or an online location where you can get all the information you need an account or service. A dashboard usually offers the user the ability to manage account information (such as changing features or preferences for an account).

DASS-Direct Access Secondary Storage

DASS1-Digital Access Signaling

DASS2-Digital Access Signaling System No. 2

DAT-Digital Audio Tape

Data-Frame

Data Acquisition System (DAS) Tape-The magnetic tape that is used to record information, communication, time change, tape management, and billing records for a cellular carrier system.

Data Base Administration System-An operations support system that maintains a database of billed-number screening data. The system also handles auditing and fraud reporting as well as other duties, including calling card, collect, and bill-to-third number information.

Data Bus-An electrical or optical path through which data is exchanged between different systems or parts of systems. In fiber optics, an optical waveguide is used as a common trunk to which several terminals can be interconnected through optical couplers.

Data Communications-Data communication is the transmission and reception of binary data and other discrete level signals that can be represented by a carrier signal that can represent the discrete (usually on-off levels) for signal transmission. There are two basic types of data communications: circuit-switched data and packet-switched data. Circuit-switched data provides for continuous data signals while packet-switched data allows for rapid delivery of very short data messages.

Data Communications Equipment (DCE)-The equipment that establishes, maintains, and terminates a data connection, as well as performs the signal conversion and coding required for communication between data terminal equipment (DTE) and a data circuit.

Data Communications Exchange-The hardware that enables data to be transmitted, received, switched, in real time to several different destinations over several different routes and channels, simultaneously.

Data Compression-A technique for encoding information so that fewer data bits of information are required to represent a given amount of data. Compression allows the transmission of more data over a given amount of time and circuit capacity. It also reduces the amount of memory required for data storage.

Data Coupler-An interface (interconnection) device that connects a data device (such as a computer) via a telephone device (such as a telephone or mobile radio) with a telecommunications network. A coupler typically includes circuitry that protects the telecommunications network from damage that may result from failure of the coupler or data device that is connected to the data coupler.

Data Encryption Standard (DES)-A data encryption algorithm in the public domain. It encrypts information in 16 stages of substitutions, transpositions, and nonlinear mathematical operations.

Data Line Interface-An interface, assembly, or connection point where a data line is connected to a telephone network.

Data Line Monitor-A device that connects to a data line and measures the performance of the data. This includes the data signal levels, the addressing accuracy, and protocol performance. The data line monitor only measures the data, it does not modify the information that is being transmitted.

Data Link (DL)-A transmission path between communications devices. The data link includes all equipment and signals in the connection.

Data Link Connection Identifier (DLCI)-A temporary channel identifier used in a communication system to identify a specific circuit along with its required communication parameters (such as peak data rates). The DLCI in a frame relay system is 10 bits. It is pronounced ("dill-see").

This diagram shows how a data link connection identifier (DLCI) is used in a data communications network to route information packets through a network and local data system to reach their destination. This example shows that the DLCI is used access devices to determine which virtual circuit is being used. This diagram shows how a computer that is connected to access device 3 is sending a print file to the printer connected to access device 1 using permanent virtual circuit (PVC) 1. Access device 2 is coordinating a digital voice connection to access device 1 using switched virtual circuit (SVC) 2. This example shows that the DLCI is only used by the access device. The data network uses its own routing tables to provide a virtual path connection through the network.

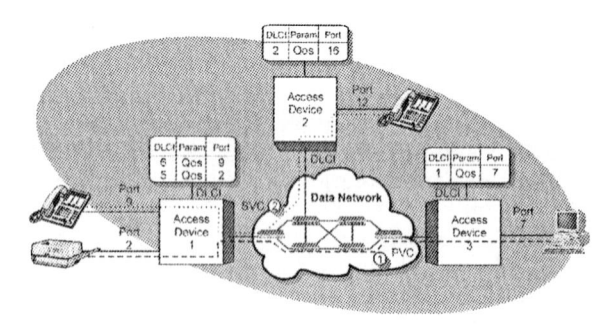

Data Link Connection Identifier (DLCI) Operation

Data Link Layer-Layer 2 of the seven-layer Open Systems Interconnection (OSI) protocol model that facilitates the detection of and recovery from transmission errors. A data link connection is built on one or more physical connections.

Data Local Exchange Carrier (DLEC)-A service provider that specializes in transferring data. A DLEC is a competitive local exchange carrier (CLEC) that competes with other carriers in the local area such as the local exchange carrier (LEC). DLECs commonly use DSL for data transmission.

Data Mart-A specialized form of data warehouse. Data marts are typically smaller and more focused in purpose than the larger, more generalized data warehouses.

Data Mining Model-A particular type of statistical model constructed with the use of a statistical analysis product that uses data mining technology.

Data Model-An abstract construct defining the classes, types, and relationships between the data to be stored in a data base. Data Models are typically used to define how a database will be constructed in order to guarantee builders the best possible organization of data and indexes.

Data Network-Data networks is a system that transfers data between network access points (nodes) through data switching, system control and interconnection transmission lines. Data networks are primarily designed to transfer data from one point to one or more points (multipoint). Data networks may be composed of a variety of communication systems including: circuit switches, leased lines and packet switching networks. There are predominately two types of data networks, broadcast and point-to-point.

This figure shows the basic types of data networks. This diagram shows several types of local area networks (LANs) including Ethernet, Token Ring and FDDI. It also shows that small networks can be interconnected to form wide area networks. Data networks can be private networks or public networks. It is also possible to encrypt (protect) data information that is transmitted on public networks to form virtual networks.

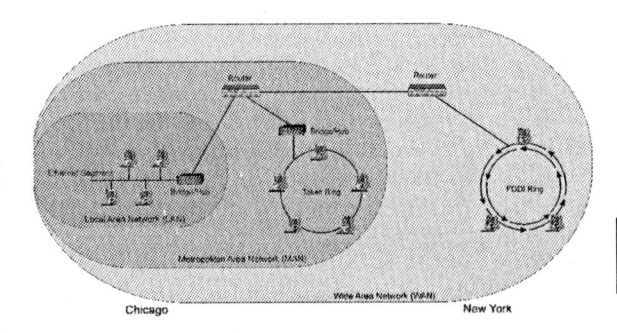

Data Networks

Data Over Cable Service Interface Specifications (DOCSIS)-A standard used by cable television systems for providing Internet data services to users. The DOCSIS standard was developed primarily by equipment manufacturers and CATV operators. It details most aspects of data over cable networks including physical layer (modulation types and data rates), medium access control (MAC), services, and security.

Data Packet-In Internet Protocol (IP) networks, the smallest amount of data that can packaged and transmitted from source to destination. In more general communications terms, any group of data which has been organized and labeled for transmission on a network or serial line. Each networking technology has minimum and maximum packet sizes which it supports.

Data Port-A connection point that allows a computer or device to connect to other devices.

Data Rate-The transfer rate of a communications line or channel that send or receives data. Data rate is usually measured in bits per second (bps).

Data Service Unit (DSU)-DSU is equipment that acts as an interface between a customer's data terminal equipment and a data communication network. DSU are the digital equivalent of the analog modem and are translation codecs (COde and DECode) coupled with a network termination interface (NTI). DSU/CSU's operate only in a digital environment. DSU are often used with channel service units (CSUs) to reformat and assign communication circuits to specific channels on a multiplexed (multi-channel) transmission line. The DSU formats the data for transmission on the data network

Data Service Unit/Channel Service Unit (DSU/CSU)

and ensures that the data network operator's required data formats are provided.

Data Service Unit/Channel Service Unit (DSU/CSU)-Devices that combine the functionality of data service units (DSU) and channel service units (CSU) to adapt data from user communication systems to communication lines with multiple channels. The DSU portion as an interface between a customer's data terminal equipment and a data communication network. DSU are the digital equivalent of the analog modem and are translation codecs (COde and DECode) coupled with a network termination interface (NTI). The CSU portion is used to coordinate communication from one or more data terminal equipment (DTE) devices to logical channels on a multi-channel communication circuit. This picture shows a D-SERV DSU/CSU from Kentrox which is an example of a combined data service unit (DSU) and channel service unit (CSU). This DSU/CSU allows for interfacing to T1 or FT1 (fractional, or partial) communication lines. It is ideal for high-speed Internet or remote office connectivity and supports Frame Relay and leased-line services. The D-SERV is available in stand-alone or shelf plug-in versions.

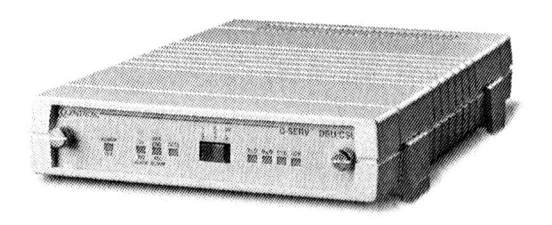

Data Service Unit (DSU) and Channel Service Unit (CSU)
Source: Kentrox

Data Set Ready (DSR)-A control signal on a communication line (such as an RS-232 serial bus) that indicates a device is ready to communicate.

Data Sink-The equipment that stores data after broadcast.

Data Speed-The data transfer rate of a signal. Data speed is also called baud rate or bit rate.

Data Stream-A continuous flow of digital information (data).

Data Switching Exchange (DSE)-The switching system that switches data traffic. In a cellular or PCS system, the DSE is typically located at the mobile switching center (MSC) facility.

Data Telephone-A data telephones is a telecommunication device that integrates analog telephone functions with a data communication interface. Because many data telephones use voice over Internet (VoIP) protocols, they are often referred to as an Internet telephones.

Data Terminal-Data terminals are data input and output devices that are used to communicate with a remotely located computer or other data communication device. Data terminals frequently consist of a keyboard, video display monitor, and communication circuitry that can connect the data terminal with the remotely located computer.

Data Terminal Equipment (DTE)-In a data communications network, the data source, such as a computer, and the data sink, such as an optical storage device. (See also: data sink, data source, network channel terminating equipment, channel service unit, data service unit.)

This diagram shows data terminals that are connected through data communication equipment (DCE) to allow a user to receive and send communication through a network to a remote computer. In this diagram, the data terminal allows the user to view information on a monitor and enter information through the keyboard. This data terminal is connected through data communication equipment (a modem) that converts the data terminal's digital signals into an analog form (audio signal) that can be sent through the telephone line to a remote computer.

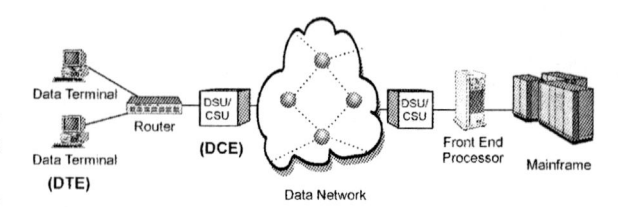

Data Terminal Equipment (DTE)

Data Terminal Ready (DTR)-A control signal on a communication line (such as an RS-232 serial bus) that indicates a device is ready to receive communication information.

Data User Part (DUP)-The User Part in an SS7 message specified for data services.

Data Warehouse-(1-information management) A information management service that stores, analyzes, and processes information that is derived from transaction systems. (2-system) A specialized database system, dedicated to the storage and retrieval of information for purposes of analysis and business intelligence investigation.

Data Word-In data communications, a character string, binary element string, or bit string that is considered as an entity.

Database-Databases are collections of data that is interrelated and stored in memory (disk, computer, or other data storage medium.) Database systems are typically accessed and controlled by computer terminals that are connected to the same data network as the database system.

Database Administrator (DBA)-A person who is responsible for the organization, design, and implementation of a company's databases.

Database Dip-A process of an information search within a database. An example of a database dip used in a telecommunication system is the searching in a signaling control point (SCP) database to find the actual telephone associated with a toll free (800) or freephone (0800) telephone number.

Database Grooming-The modification of an existing database to sort information and remove redundant or unwanted information.

Database Management System (DBMS)-A system that is controls access to, organization of, security, and application interfaces to information data.

Datagram-In packet switching, a self-contained data packet representing a portion of a message. A datagram is sent independently through a network to its destination. It is reassembled with other packets into the original message.

Datagram Service-Service at the network layer in which successive packets may be routed independently from end to end. There is no call setup phase. Datagrams may arrive out of order. In Internet Protocol parlance, datagram service generally implies the use of UDP or non-assured delivery.

Dataport-(1 - mobile radio) A connection port on a mobile radio that allows data devices such as fax machines to be connected to the mobile radio. (2 - channel unit) One channel unit from a digital channel bank that connects data signals from customers to a digital data network without the use of a digital multiplexer.

Date of Conception (Invention)-Date of Conception is relevant in the United States where patents are awarded on a first-to-invent basis. Date of Conception has no meaning where patents are awarded on a first to file basis.
Contrary to common belief, the date of conception is not the moment when an invention was first thought of, rather the date of conception occurs when a complete, working concept of the invention is completed. The date of conception is that date when a person skilled in the art would be able to reduce the invention to practice without undue experimentation.

Date of Invention-Date of Invention is only relevant where patents are awarded on a first-to-invent basis. Date of Invention has no meaning where patents are awarded on a first to file basis.
The date of invention is the date when the invention is actually, or constructively, reduced to practice. The filing of an application for patent is considered a "constructive" reduction to practice.
In certain circumstances, the date of invention can be the same as the date of conception, if it can be shown that the inventors worked diligently and without undue delay to reduce the invention to practice.

DATU-Direct Access Test Unit
DAVIC-Digital Audio Video Council
Day Rate Period-The period of time for which day rates apply. It may vary for different services, including interstate, intrastate, and overseas services. (See also peak time).
dB-Decibel (See also: Bel, decibel)
dB-decibel
DB-Dummy Burst
DB Connector-Data Bus Connector
DBA-Database Administrator
DBA-Dynamic Bandwidth Allocation
DBCS-Double-Byte Character Set
dBd-Decibels dipole
dBi-Decibels isotropic
DBI-Don't Believe It
DBIP-Dedicated Borderline Interconnection Percentage
DBM-Decibel ratio of Watts to one milliwatt
dBm-Decibels Relative to One milliwatt
DBMS-Database Management System

DBS-Direct Broadcast Satellite
DBU-Dial Back-Up
DC-Direct Current
DC-Direct Control
DC Block-Direct Current Block
DC Power-Direct Current Power
DC/MA-Dynamic Channel Multiple Access
DCA-Defense Communication Agency
DCA-Document Content Architecture
DCA-Dynamic Channel Allocation
DCC-Data Country Code
DCC-Data Communications Channel
DCC-Digital Compact Cassette
DCC-Digital Color Code
DCCH-Digital Control Channel
DCCH-Dedicated Control Channel
DCD-Data Carrier Detect
DCD-Dynamically Configurable Device
DCE-Data Communications Equipment
DCE-Data Circuit Terminating Equipment
DCE-Distributed Computing Environment
DCE-X.25 Data Communications Equipment
DCF Mode-802.11 Distributed Coordination Function
DCG-Dispersion Compensation Grating
DCH-Dedicated Channel
DCM-Digital Circuit Multiplication
DCME-Digital Circuit Multiplication Equipment
DCN-Disconnect Frame
DCOM-Distributed Component Object Model
DCPSK-Differentially Coherent Phase Shift Keying
DCS-Digital Cross Connect System
DCS-Distributed Call Signaling
DCS-Digital Command Signal
DCS-1800-Digital Cellular System 1800 MHz
DCS-1900-Digital Cellular System 1900 MHz
DCT-Digital Cordless Telephone
DCT-Discrete Cosine Transform
DCTE-Data Circuit Terminating Equipment
DCTI-Desktop Computer Telephone Integration
DCV-Digital Compressed Video
DDA-Domain Defined Attribute
DDC-Direct Department Calling
DDCMP-Digital Data Communications Message Protocol
DDD-Direct Distance Dialing
DDD-Automatic Overflow to Direct Distance Dialed
DDE-Dynamic Data Exchange
DDI-Direct Dialing Inward

DDM-Distributed-Data Management
DDN-Defense Data Network
DDN NIC-Defense Data Network Network Information Center
DDP-Distributed Data Processing
DDP-Datagram Delivery Protocol
DDS-Digital Dataphone Service
DDS-Digital Data System
DDS-Digital Data Storage
DDS-Dataphone Digital Service
DDSD-Delay Dial Start Dial
DE-Discard Eligibility
DEA-Digital Exchange Access

Dead Spot-An area within a wireless system (typically cellular or PCS) where, for one reason or another, signals do not have a sufficient level to an acceptable level of communications. See also: Signal Fading.

Debug-The process of removing problems or errors within a software program or system.

DEC-Digital Equipment Corporation

Decapsulation-The process of removing protocol headers and trailers to extract higher-layer protocol information carried in the data payload. See also Encapsulation.

decibel (dB)-A measurement that expresses the ratio of two amounts of power by use of the logarithm. The decibel is 1/10th the amount of a Bel, which was a measurement named after Alexander Graham Bell. The formula for expressing a number related to the power ratio, in decibels, is "10 log (P2/P1)" where "log" represents the common or base 10 logarithm. P2 is a second power level, and P1 is a first power level. P1 and P2 are both expressed in the same power unit, typically in watts.

The decibel is a non-linear measurement. Examples using power: 3 decibels (.3 Bels) indicates that P2 is approximately 2 times P1; 10 decibels (1 Bel) indicates that P2 is exactly 10 times as large as P1; and 20 decibels (2 Bels) indicates that P2 is exactly 100 times as large as P1.

Since power can also be expressed as voltage squared divided by resistance, the logarithmic decibel measure of the ratio of two voltages V2 and V1, measured at two points in a system where the local ratio of voltage to current are R2 and R1 respectively, then the formula for expressing a power ratio in decibels is "20 log (v2/v1) - 10 log (R2/R1)". An equivalent formula is "20 log (V2/V1) + 10 log (R1/R2)". For situations in which R2 and R1 are equal, these last two formulas reduce to "20 log (V2/V1)".

The decibel unit was introduced into the telephone industry about the year 1910. The power ratio of the input power to the output power for a 1 kHz test tone, for 1 mile (1.6 km) of number 19 (American wire gauge) copper wire, is almost exactly 1 dB. At that time, 19 gauge wire was widely used in the telephone industry (but was later almost completely replaced with 22, 24 or 26 gauge wire), and it was customary for many engineers and technicians to describe the power ratio of one mile of 19 gauge wire by the inappropriate name "1 mile of signal loss." Use of the decibel gave a formal basis and a mathematically consistent formula for this practice. Use of a logarithmic method of expressing the power ratio allows the engineer or technician to add the logarithmic dB units for multiple "miles" of wire, rather than multiplying the actual numeric power ratios. The dB unit has also been applied extensively to audio power ratios and light power ratios. Remember that a dB is an expression of a ratio, and not an absolute power unit. When the denominator power term is taken to be a pre-agreed unit of power, then dB can be used to express the logarithm of an absolute power level. Examples of such absolute expressions are dBm, dBW, and acoustic decibels.

Decibels dipole (dBd)-A measurement of signal gain used in radio antenna design. Specifically, dBd refers to the ratio of radiated signal power at the vertical angle of highest power density compared to the radiated power density of a horizontally isotropic dipole antenna. A horizontally isotropic antenna radiates equal amounts of power density for all azimuthal or horizontal directions. The radiated signal power varies only with a change in vertical angle.

Decibels isotropic (dBi)-A description of antenna gain relative to a hypothetical spherically isotropic antenna. No real antenna is totally spherically isotropic in absolutely all directions (both vertical and horizontal). This figure dBi describes the ratio of radiated power density in the direction of maximum signal power density, compared to the radiated power density (in any direction) of an isotropic radiating antenna.

Decibels Relative to One milliwatt (dBm)-The power of a signal referenced to 1 mW (milliwatt).

Decipherment-The decoding of data that is encrypted with a cipher key.

Decoding-In telecommunications, the process of converting encoded words into the original signal, possibly via an intermediate step of pulse amplitude modulation (PAM). In pulse code modulation (PCM), this is the conversion of 7-bit (D1 type) or 8-bit (including D2, D3, D4, and D5 type) pulse code modulation (PCM) words to analog signals. Decoding is the inverse of encoding. (See also: coder/decoder quantizing, sampling, encoding.)

DECT-Digital Enhanced Cordless Telephone

Dedicated Access Line (DAL)-A communication line that has its data transmission capability reserved (dedicated) for a specific user.

Dedicated Bandwidth-A configuration in which the communications channel attached to a network interface is dedicated for use by a single transmitter or receiver and does not have to be shared.

Dedicated Control Channel-A channel in a radio network used solely for system control. For the AMPS system, there are 21 dedicated control channels for each carrier. A mobile radio will search for control channels first prior to attempt access to a system.

Dedicated Control Channel (DCCH)-A channel (radio or logical) that is used coordinate and control access to a wireless system.

Dedicated Control Channels-(1 - general) A designated radio control channel where mobile radios first look to coordinate radio channel access. (2 - cellular) A set of 42 cellular radio channels (21 for each cellular carrier) that have been designated as control channels that cannot be used as voice channels. For the non-wireline service (A side), the control channels are 312 to 333. For the cellular wireline service carrier (B side), these channels are 334 to 354.

Dedicated Media-A configuration in which the physical communications medium used to connect a station to either a hub or another station constitutes a point-to-point link between the devices.

Dedicated Plant Assignment Card-An outside plant record that gives the particulars of a distribution plant served by a dedicated outside plant control point, or serving area interface. The card lists the address of every housing unit served from the control point or interface, as well as the telephone number, central office equipment, type of service, feeder and distribution pair numbers, and the address of the distribution terminal. The dedicated plant assignment card generally has been replaced by mechanical loop assignment systems, such as the Facilities Assignment and Control System (FACS).

Dedicated Token Ring-A data communication system specified by IEEE 802.5r that allows for full-duplex communication between two token ring devices, typically an end station and a switch. Using this protocol, standard Token Ring devices negotiate their ability to disable the token passing protocol. This can be done only when exactly two stations are present on a LAN segment. By enabling full-duplex communications, neither station must wait for a token, and is therefore able to transmit at 16 Mbps continuously, resulting in an aggregate of 32 Mbps by both stations. DTR is typically used by switch-to-switch and switch-to-server connections.

Dedicated Trunk-A communication trunk that is connected directly through to a particular phone or hunt group. Dedicated trunks bypass attendant consoles.

Default Port-A physical or logical port that is pre-configured or defined to be the communication port for traffic when a port assignment is not provided or when a process fails. Applications and protocols in communication systems such as the Internet use default ports so they can start specific applications or protocol communication when a packet is first received. For example, when a IP packet is received with the port number 21, it specifies file transfer protocol (FTP) should be used.

Default Setting-Default

Defects Per Million (DPM)-Defects per million (DPM) is the number of calls lost per million calls processed. DPM is particularly useful for measuring the network availability of switched virtual circuits (SVC) services in a multiservice switch, where connections are constantly made, sustained, and torn down.

Defense Advanced Research Projects Agency (DARPA)-The central research and development agency of the U.S. Department of Defense.

Defense Communications System-A communication system that is used in a defense systems network

Defense Data Network-A communication data system that is used in a defense data systems network

Defense Switched Network-A communication information system that is used in a defense systems network that allows the changing of communication path through a switching system.

Deflection-The control placed on electron direction and motion in CRTs and camera tubes by varying the strengths of electrostatic (electric) or electromagnetic fields.

Defocus Effect-A digital picture manipulation term meaning a controlled blurring of the image.

Degeneration-(1-recording) The loss of quality on a videotape, typically resulting from multiple generations of copying the material. (2-amplifier circuit) The process of reducing the gain of an amplifier stage by applying negative feedback (feedback that is 90 degrees out of phase) to the input.

Deglitcher-A circuit used to limit the duration of switching transients in a digital system.

DEK-Data Encryption Key

DEL-Delete Idle

Delay-The amount of time it takes for a signal to transfer or for the time that is required to establish a communication path or circuit.

Delay Circuit-A circuit designed to add time delay to a signal passing through it by a specified amount.

Delay Equalizer-A network that adjusts the velocity of propagation of the frequency components of a complex signal to counteract the delay distortion characteristics of a transmission channel.

Delay Line-A device that delays a signal by a very accurate amount of time. They are used in synchronizing electronic signals for computing, telecommunications and radar applications. SAW devices are often used for delay purposes.

Delay Spread-A product of multipath propagation where symbols become distorted and eventually will overlap due to the same signal being received at a different time. It becomes a significant problem in mountainous areas where signals are reflected at great distances.

Delay Time-The sum of waiting time and service time in a queue.

Delay Variation-The difference, in microseconds, between the maximum and minimum possible delay that a packet will experience as it goes out over a channel. This value is used by applications to determine the amount of buffer space needed at the receiving side in order to restore the original data transmission pattern.

Deliberate Churn -Deliberate churn is a type of voluntary churn. Deliberate churn occurs with the customer decides to terminate service because they have found a more attractive service elsewhere.

Delinquency Threshold-The amount owed by the customer that will trigger the collection process to be initiated.

DELNI-Digital Ethernet Local Network Interconnect

Delta-In a fiber communication system, delta is the normalized refractive index difference of a fiber. Delta is approximately equal to the difference in the indices of refraction of the core and the cladding divided by the index of the core.

Delta Frame-A type of data frame used in CD-ROM applications. The delta frame contains only differences between the current frame it represents and the frame that precedes it. A delta frame tends to be considerably smaller than an intra-frame. A stream of temporally compressed data must be interspersed with key frames (also called intra-frames, or stand-alone frames) in addition to delta frames, so that audio/video synchronization can occur.

Delta Modulation (DM)-A form of digital modulation where voltage changes of an analog signal are changed into a fixed difference of value (or delta), and a +/- sign. For adaptive delta modulation, differences in the voltages are not fixed, and vary depending on past history.

Demand Analytics-Collection of data mining models that can be utilized to help define how changes in operational expense (OPEX) spending (advertising, marketing, customer service etc.) will impact the customers demand for telecommunications services.

Demarc-Demarcation Point

Demarcation Point (Demarc)-The physical and electrical boundary between an end user's telecommunication equipment and the telecommunications network. The demarcation point establishes point of ownership and accountability.

DeMilitarized Zone (DMZ)-A demilitarized zone (DMZ) is trusted part of a communications network, typically behind a firewall, that allows unrestricted access and transfer of information between devices. Information that passes in or out of a DMZ may be delayed and filtered. This may cause challenges with real-time communication applications such as IP telephony and media streaming.

Demodulation-The process of recovering an information signal that has been previously modulated on a radio, optical or other electromagnetic carrier signal.

This diagram shows how a demodulator converts a modulated carrier signal into an information signal. This diagram shows that the demodulator compares the modulated carrier (carrier with the changes) to an unmodulated carrier (pure carrier signal) to produce the information signal (representing only the changes of the carrier signal).

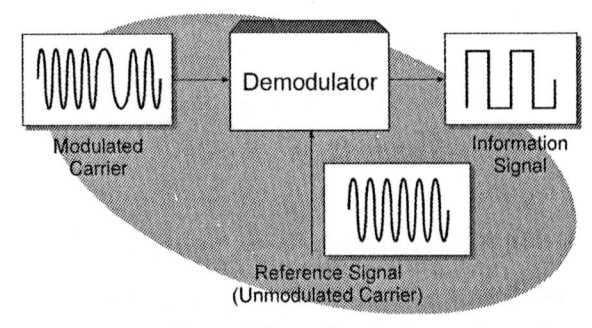

Demodulator Operation

Demultiplexer (DeMux)-(1-general) A device used to separate two or more signals that were previously combined by a compatible multiplexer and are transmitted over a single channel. (2-digital) A circuit or device for separating two or more signals received sequentially from a single communications channel into individual bit streams, and sending the bit streams to their respective destinations. in analog systems, a demultiplexer separates two or more signals received in a specified frequency spectrum into individual channels.

This diagram shows how demultiplexing can extract two or more low speed channels from a higher speed communication channel. In this diagram, there is a single 64 kbps communication signal that is supplied to a demultiplexer circuit. The demultiplexer gathers and routes 8 bits data to a specific port during each 125 usec time slot. This example shows that the data on each port is sent out (clocked out) at 8 kbps.

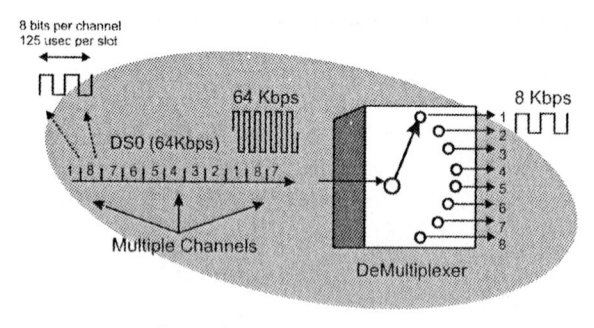

Demultiplexing Operation

DeMux-Demultiplexer
DEN-Directory Enabled Network
Denial Of Service-A process that inhibits or reduces the ability of authorized users from gaining access to communications systems through the continual transmission of service requests or messages that disable communication sessions.
Dense Wave Division Multiplexing (DWDM)-A version of fiberoptic communication that combines many optical channels on a single fiber, typically used to increase the data transmission capacity of previously installed fiber. Dense wave division multiplexing provides a significant increase in capacity compared to wavelength division multiplexing (WDM) that combined up to four different optical wavelengths on a single fiber. As of year 2001, DWDM systems support 8 to 80 different wavelengths on one fiber, providing the capability of transferring over 1 trillion bits of data per second. Each optical channel (wavelength) is capable of providing approximately 10 Gb/s (OC-192 9.3 Gb/s data rate).
Deny Any Knowledge (DAK)-A claim by a customer that a call was not made (as in "I deny all knowledge of this call").
Department Of Communications (DOC)-A government agency of countries throughout the world that sets policies and rules regarding telecommunications within their country. In the United States, the FCC is the equivalent of the DOC in some other countries.
Depletion Region-The area within a semiconductor that is depleted of mobile carriers (holes and electrons) by the application of a reverse bias voltage. N6 current flows in an external circuit unless carriers are moving in the depletion region. Also called depletion layer.

DER-Distinguished Encoding Rules
Derating Factor-An operating safety margin provided for a component or system to ensure reliable performance. Typically, a derating allowance also is provided for operation under extreme environmental conditions, or under stringent reliability requirements.
DES-Digital Encryption Standard
DES-Destination End Station
DES-Data Encryption Standard
DESI Strip-Designation Strip
Design Around (Patent)-Designing around is a legal exercise in avoiding existing patent claims. Designing around includes a legal analysis of the scope of the existing claims to ensure that a newly developed product does not infringe.
Design Patent-A design patent is a patent issued on non-functional ornamental features of a product. Design patents typically have a shorter duration than a Utility Patent.
Desktop Management Interface (DMI)-The process of managing desktop workstation hardware and software components automatically, often from a central location. The DMI is an application programming interface (API) that enables a software program to collect information about a computer environment.
Destination Address (DA)-The address for whom a message is intended.
Destination Point Code (DPC)-A part in the label of an SS7 signaling message which uniquely identifies, in a signaling network, the destination point of the message. In the form network-region-signaling point.
Destination Service Access Point (DSAP)-The logical address assigned to the service entity where data must be received when it arrives at its destination. This information may be included data portion of the 802.3 data transmission frame.
Detailed Billing-A type of billing in which the details of each message, such as the date of call, number called, and charge, are listed as separate items on a toll statement that is included with a customer's bill.
Detailed Regulatory Monthly Allocation System-A software system that uses data from the Plug-In Inventory Control System (PICS) to develop central-office investments in the categories required by the FCC's Jurisdictional Separations Procedures. Each month, the investments in PICS are matched to an accounting cost record to provide

cumulative investment in the required separations categories. These investments are put into the various separations allocations used by operating telephone companies.

Detariffed-A billing and regulatory term designated by the FCC and state public utility commissions that often refers to deregulated rate structures on subscriber-owned inside premise wiring and related CPE.

Detector-A device that senses the presence or level of a signal. A detector senses the presence of electromagnetic or optical energy and create an electrical signal that represents the level of the sampled signal.

Developed Portion Of A Wire Center-In long-range outside plant planning, land that is either completely built up, built up with scattered vacant lots, vacant but surrounded by developed land, or a cluster of more than 200 living units in an otherwise undeveloped part of a wire center. Such areas are also called developed areas.

Deviation-(1-general) A departure from a standard or specified value. (2-modulation) The peak difference between the instantaneous frequency of an AM signal and the carrier frequency. (3-phase) The peak difference between the instantaneous angle of a phase-modulated wave and the angle of the carrier.

Deviation Ratio-The ratio of the maximum frequency deviation of a FM system to the highest modulating frequency.

Device Descriptor-A device descriptor contains general information about a data communication device. It usually includes information that applies to the device configuration and communication settings. Each device in a USB system has only one device descriptor.

DF-Delivery Function
DFA-Doped Fiber Amplifier
DFB-Distributed Feedback Laser
DFE-Discrete Feedback Equalizer
DFI-Digital Facility Interface
DFS-Distributed File System
DFT-Discrete Fourier Transform
DFT-Direct Facility Termination
DGPS-Differential Global Positioning Service
DHCP-Dynamic Host Configuration Protocol
DIA-Document Interchange Architecture
DIA/DCA-Document Interchange Architecture/Document Content Architecture

Diagnostic Records-Records relating to a call or communication session that was processed by a call manager or a call server.

Diagnostics-A program, often built into a system, that tests the functionality of the system and reports the results. Diagnostic systems that are separate and simply monitor the operation of the subject system are considered to be non-intrusive.

Diagonal Link (D-Link)-A "D" (diagonal) link connects a secondary (e.g., local or regional) signal transfer point (STP) pair to a primary (e.g., inter-network gateway) STP pair in a quad-link configuration in an SS7 network. Secondary STPs within the same network are connected via a quad of "D" links.

Dial Around-The process of dialing a phone number via a pay phone via a toll free for freephone access number. Dial around has the effect of reducing the toll charges that may be collected by a pay telephone.

Dial By Name-The ability to dial a person by spelling their name out on the telephone or communication device keypad.

Dial In Tie Trunk-A trunk line that may be accessed by dialing in and using an access code. This trunk line is then seized for a dedicated transmission path. This allows the caller to use advanced features of communication equipment without concern of the trunk being released.

Dial Line Connection-Dial lines, plain old telephone service (POTS), are 2-wire, basic line-side connections from an end office with limited signaling capability. Because dial lines are line-side connections, call setup time may be longer than those connections that employ trunk-side supervision.

Dial Long-Line Circuit-A circuit that is used to extend the range of a communication line.

Dial Plan-Dialing Plan

Dial Pulse (DP)-Dial pulse (DP) signaling senses and counts the changes in current flow, such as from a rotary dial telephone, to allow the user to send address information (dialed digits) to the telephone system.

Dial Tone-A signal tone provided from a local telephone service provider that indicates that the telephone network is ready to send a call (to receive dialed digits). The dial tone signal is usually a combination of 350 Hz and 440 Hz signals.

Dial Up Line-(1-general) A communication line that is established by sending the dialed digits to the network that allow the connection to be made.

Dial-Up Network (DUN)

The term dial up is commonly used with the process of manually connecting a computer to the Internet through a standard telephone line. (2-humor) An ancient method of connecting to the Internet using MODEMS that signal over a telecom system known as the plain old telephone service or POTS. This method of communication was outlawed in the mid 2000s as part of the Computers With Disabilities Act.

Dial-Up Network (DUN)-A software portion of Microsoft Windows 95,98,NT,and 2000 that allows the user to connect the computer to a data network (such as the Internet). Because DUN is actually a process of establishing and maintaining a communications session, DUN is sometimes used for establishing connections on "always-on" circuits (such as DSL).

This figure shows how a dial connection is used to connect to the Internet. In this example, a user has a computer with a modem card in it. The dial-up connection software installed in the computer instructs the modem to dial a telephone number for the Internet service provider (usually a local telephone number). The computer's modem simulates operation as if it were a standard telephone and sends dialed digits (usually by touch-tone) to the telephone company. The telephone company recieves the dialed digits and connects the call to the number dialed. The dialed number if for the Internet service provider which usually has several modems (a pool of modems) available that can answer the call. When the call is answered by an available modem, a communication link (usually a PPP connection) is established between the computer's modem and the ISP modem. After the link is setup, the digital data from the computer is converted by the modem to an analog form. This analog signal is transferred through the telephone company to the ISP modem where it is converted back into a digital form. This digital form is further processed by the ISP into a form that can be sent through the Internet.

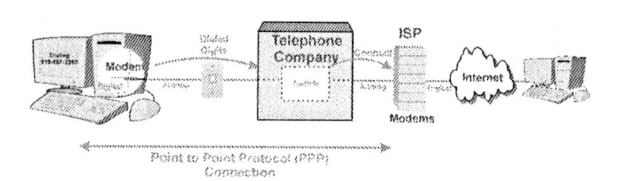

Dial-Up Connection

Dialed Number Identification Service (DNIS)-A call identification service typically provided by a toll free (800 number) network. The DNIS information can be used by the PBX or automatic call delivery (ACD) system to select the menu choices, call routing, and customer service representative information display based on the incoming telephone number.

Dialing Plan (Dial Plan)-A dial plan (also called a dialing scheme) is a systematic use of certain prefix digits to dial a destination via user selected routing. An example is the use of the dialed prefix "9" from within a PBX to first select an outside local telephone line so that the originator can then dial a (typically 7 digit) local city telephone number. Similarly, a PBX may use the dialed prefix "8" to select a tie line to another PBX. A dialing plan differs from a numbering plan by being used inside a particular private telephone system, and also the specifics of different dialing plans are different in different PBXs or different private networks in the same country, while a numbering plan is uniform throughout an entire country.

DIB-Directory Information Base

Dictionary Attack-A process of attempting to gain access to a communications resource by sequencially cycling through a list of words (such as a dictionary) that are likely to be used for passwords or account information.

DID-Direct Inward Dialing

DID director-Direct Inward Dialing director

DIEL-Disabled and Elderly People

Differential Encoding-Differential encoding is a process of transmitting information that is produced by a logical combination of the current data bit and the previous data bit.

Differential Pulse Code Modulation (DPCM)- (1-general) A modulation method that uses pulse coding which represents the difference in successive voltages samples of an analog signal. Each individual sample is approximated by a discrete set of values. (2-video) A modulation technology used for compressing video signals so they can be transmitted in the available capacity of an Integrated Services Digital Network (ISDN) channel.

Differential Quadrature Phase Shift Keying (DQPSK)-A type of modulation that alternates between 4 different phase shifts to represent the digital information symbol signal. During one symbol, these shifts are +/- 45 and +/- 135 degrees and during the alternate symbol, these shifts are +/- 90 and +/- 180 degrees.

Differentiated Services (DiffServ)-A protocol that identifies (tags) different types of data with data transmission requirement flags (e.g. priority) so that the routing (switching) network has the capability to treat the transmission of different types of data (such as real-time voice data) differently.

Diffraction Grating-A device that separates light into its component wavelengths. Diffraction is the change in direction of light (or other group of waves) when it encounters an edge. A grating is a series of parallel lines, grooves or edges. A diffraction grating uses parallel lines to reinforce the changes in direction of the light. Diffraction of light from one edge is difficult to observe, but from thousands of lines at the right separation, it can be observed easily. Diffraction gratings used in optics typically have thousands of parallel lines per inch. When light encounters such a grating, each of its constituent wavelengths changes direction by a slightly different amount. Sunlight reflected off a compact disk (which has a surface covered with a series of closely spaced, parallel lines) can thereby be broken down into the colors of the rainbow.

DiffServ-Differentiated Services

Diffuse Reflection-The scattering effect that occurs when lightwaves, radio waves or sound waves strike a rough surface.

Diffused Base Transistor-A transistor with a non-uniform base region produced by diffusion.

Diffused Junction Transistor-A transistor in which both the emitter and collector electrodes have been formed by diffusion of an impurity into the semiconductor wafer.

Diffusion-(1-waves) The spreading or scattering of a wave, such as an optical wave or sound wave. (2-semiconductor) The process through which the crystalline structure of semiconductor material can be doped so that either positive or negative carriers effectively transport electric current through the material.

Digest Authentication-An authentication process that processes a user identification code and password identification without sending the raw data through the communication network to validate the true identity of the user. The digest authentication process begins by the challenger (usually the service provider) sending a random number (nonce value) that is used by the user (usually the client) along with other previously known information (secret key) to calculate a response using an encryption algorithm. The end result is sent back to the challenger who also has access to the previously known information and random number that is used to calculate the same result. If the result matches, the authentication passes.

Digital-Digital electronic devices use electric currents or voltages that are intentionally restricted to take on a limited set of values for their intended use, rather than allowing continuous variation of the current or voltage. Typically, only two voltage values are used, for example, having values of approximately zero or five volts. Because undesired signals (noise, interference) are much smaller (typically less than 0.1 volt, for example) the digital signals often can be transmitted or recorded without errors because the presence of small deviations of the signal do not confuse the device from correctly interpreting the voltage as definitely representing one of the two intentional voltage levels. When a digital coded representation of an analog signal is used, and the digital part of the system does not introduce any errors, the only degradation of the signal is due to the inherent inaccuracy of the initial encoding device (codec) that converts the signal from analog to digital representation. This inaccuracy can be controlled by the design of the codec. This figure shows a digital signal that is in the form of a series of bits and these bits are combined into groups of 8 bits to form Bytes (B). In this example, the bits 01011010 are transferred in 1 second. This results in a bit (transmission) rate of 8 bps.

Digital Access Cross-Connect System (DACS)

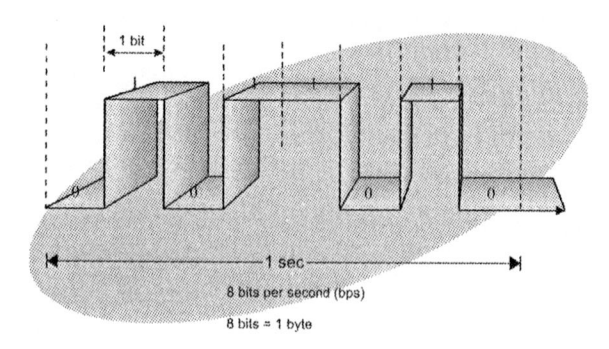

Digital Signal

Digital Access Cross-Connect System (DACS)- A digital switching system that interconnects specific communication channels (time slots) between digital multiplexed lines (usually t-carrier lines).

Digital Added Main Line (DAML)-A local loop access line system that uses ISDN digital transmission to provide two communication circuits on a standard copper wire pair. DAML differs from the standard ISDN basic rate interface (BRI) as the D signaling channel is not included. The D channel is simply not used. The line control (e.g. off-hook) is sensed by DAML modem. The modem creates the line control signaling messages. DAML allows a telephone service provider to add more lines on existing copper access lines without the need to add ISDN software upgrades to their switching offices.

Digital Advance Mobile Phone Service (D-AMPS)-A name that is sometimes used to describe IS-136 time division multiple access technology that evolved from 30 kHz radio channel AMPS technology.

Digital Audio Broadcast (DAB)-An audio broadcasting system that transmits voice and other information using digital transmission. The DAB signal is typically shared with additional digital information on a single digital radio channel.

Digital Audio Radio System (DARS)-A radio system that provides audio programming via satellite transmission. It is similar to Direct Broadcast System (DBS) TV systems. The initial application of DARS was by CD Radio in 1990. It is also called Digital Audio Broadcasting (DAB) outside the U.S.

Digital Audio Tape (DAT)-A digital audio tape storage format that uses a cartridge that has a 4 mm wide metal coated tape. A common helical scan method can be used to provide storage capacities of over 3 GB.

Digital Broadcasting-The process of transmitting the same digital data signal to all users that are connected to the digital broadcast network. Digital broadcast signals may be encoded in a way that only some of the users may be capable of decoding digitally broadcast messages (e.g. a specific pay-per-view movie channel).

Digital Cable Television System-A system that distributes television (and other information services) via a cable television distribution system in digital modulated form.

Digital Cellular-An industry term given to the new cellular technology that transmits voice information in digital form. This differs from Analog cellular in that the method of transmission for voice/data information is by means of digital signals.

This figure shows a basic digital cellular system. This diagram shows that there typically is only one type of digital radio channel called a digital traffic channel (DTC). The digital radio channel is typically sub-divided into control channels and digital voice channels. Both the control channels and voice channels use the same type of digital modulation to send control and data between the mobile phone and the base station. When used for voice, the digital signal is usually a compressed digital signal that is from a speech coder. When conversation is in progress, some of the digital bits are usually dedicated for control information (such as handoff). Similar to analog systems, digital base stations have two antennas to increases the ability to receive weak radio signals from mobile telephones. Base stations are connected to a mobile switching center (MSC) typically by a high-speed telephone line or microwave radio system. This interconnection may allow compressed digital information (directly from the speech coder) to increase the number of voice channels that can be shared on a single connection line. The MSC is connected to the telephone network to allow mobile telephones to be connected to standard landline telephones.

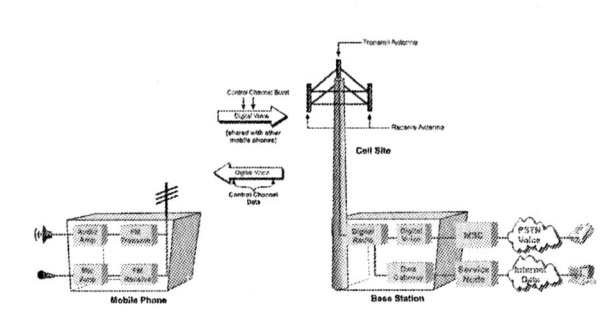

Digital Cellular System (2nd Generation)

Digital Cellular System 1800 MHz (DCS-1800)- A version of the 900 MHz GSM cellular system that has been modified to allow operation at 1800 MHz. The DCS-1800 system is primarily used in countries other than the United States and Canada.

Digital Cellular System 1900 MHz (DCS-1900)- A version of the 900 MHz GSM cellular system that has been modified to allow operation at 1900 MHz. The DCS-1900 system is primarily used as the PCS system in the United States and Canada.

Digital Compression- Digital compression is a process that uses a computing device (such as a digital signal processor) to analyzes a digital signal and create a new data signal that represents of the original signal using a lesser number of digital bits. Digital compression allows more information to be transmitted on a communication channel.

Digital compression devices use mathematical formulas and code book tables to compress the data. Mathematical formulae transform the original signal into it characteristic parts such as frequency and amplitude. Code book tables contain blocks of high occurrence information (such as particular tones used in fax machines). When transmitting digital information that has been compressed, only the parameters (such as the frequency, amplitude and code book word) are sent on the transmission channel. When the digital information is received, the compression process is reversed by a decoder to produce the same (or similar) initial signal.

Digital Control Channel (DCCH)- A control-channel that used in the IS-136 TDMA specification and other digital communication system that is used to control mobile radio channel assignment and to provide for enhanced call processing features.

Digital Enhanced Cordless Telephone (DECT)

Digital Controlled Squelch Systems- Digital squelch systems mute the audio of a radio receiver unless the incoming radio signal contains a specific digital code. Digital controlled squelch allows a radio user to avoid listening to noise or interference signals of all other radio transmissions that occur on the same frequency that do not have the correct code mixed in with the audio signal.

This figure shows how a digital squelch system operates. In this diagram, each mobile radio is assigned a unique digital squelch code. This squelch code is part of the digital message that is received. If the digital squelch code matches the prestored digital squelch code, the mobile radio will un-mute (connect) the audio path of the mobile radio.

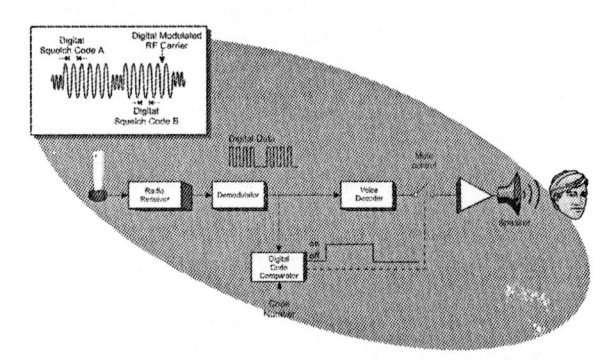

Digital Squelch System

Digital Cordless Telephone (DCT)- Digital transmissions in cordless technology helps eliminate RF noise and external interference as well as increases range. the newest of these is 900 MHz Digital Spread Spectrum technologies.

Digital Desktop- A desktop workplace for the employee that consists primarily of digital devices. The digital desktop digital devices are usually a computer, printer, and digital telephone.

Digital Enhanced Cordless Telephone (DECT)- Digital enhanced cordless telephones (DECT) is A low power wireless telephone system that is primarily used in business office (PBX) and residential (cordless) environments. It is often used as an extension to cellular radio and public telephone systems. DECT is also called digital European cordless telephone.

The DECT system is a time division duplex (TDD) system. The standard DECT system has a channel bandwidth of 1.728 MHz and a channel data rate of 1152 kbps. Each channel is divided into 24 time

Digital Land Mobile Radio (DLMR)

slots. Two time slots are dedicated for each user to allow full duplex operation. This provides each user with 32 kbps for voice or data information. By combining time slots, higher data rates can be achieved for other applications such as wireless local area network (WLAN). The DECT system uses very low power radio of 250 milliwatts peak.

The DECT system has different parameters when used in North America. These modifications include a more narrow bandwidth and reduced data transmission rate.

Digital Land Mobile Radio (DLMR)-Digital land mobile radio (LMR) systems are traditionally private or public safety systems that allow communication between a base and several mobile radios using digital modulation technology. DLMR systems can share a single frequency or use dual frequencies. DLMR services commonly include various types of voice and data services for police, taxi, fire and other types of dispatch services.

This figure shows a typical digital two-way radio system. In this example, a digital mobile is connected to a data display in a mobile vehicle. The radio transceiver has a transmitter and receiver bundled together. Some digital systems combine the control channel with a traffic channel while others use dedicated control channels. Digital systems can either use a single digital radio channel for each user or can divide a single radio channel into time slots for use. The use of time division allows for several users to simultaneously share each radio channel.

Digital Loop Carrier (DLC)-A high efficiently digital transmission system that uses existing distribution cabling systems to transfer digital information between the telephone system (central office) and a users telephone and/or computer equipment. A DLC system usually includes a high-speed digital line (e.g. T1) from a central office and a remote digital terminal (RDT). The RDT converts the high-speed digital line to low speed lines (analog or digital) for routing to the end customers.

Digital Microwave-A microwave transmission system that transfer digital information through the modulation of a microwave carrier signal. The type of modulation used may be amplitude, frequency or phase shift, but the digital signal is used as the source of modulation information.

Digital Modulation-A modulation process where the amplitude, frequency or phase of a carrier signal is varied by the discrete states (On and Off) of a digital signal.

This figure shows different forms of digital modulation. This diagram shows ASK modulation that turns the carrier signal on and off with the digital signal. FSK modulation shifts the frequency of the carrier signal according to the on and off levels of the digital information signal. The phase shift modulator changes the phase of the carrier signal in accordance with the digital information signal. This diagram also shows that advanced forms of modulation such as QAM can combine amplitude and phase of digital signals.

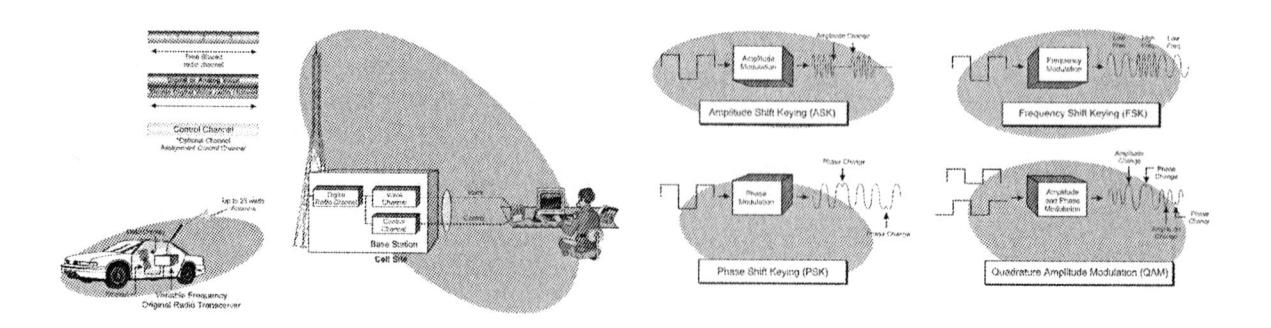

Digital Land Mobile Radio System

Digital Modulation Operation

Digital Multiplex System (DMS)-The trade name of a line of digital telephone office switches from Nortel Networks (formerly Northern Electric, then Northern Telecom). There are DMS-10 and DMS-100 end subscriber switches, DMS-200 and DMS-250 transit/tandem switches, DMS-MTX cellular/wireless switches and DSM-300 international gateway switches (used to connected between North America and other countries).

Digital PBX (DPBX)-A PBX system that uses digital transmission for control and audio signals. Most PBX systems available in the late 1980s were digital PBX systems.

Digital Power Line-Digital power line is a term that refers to the sending of digital information through electric power lines.

Digital Rights Management (DRM)-A system of copy protection or copy control for digital media. DRM relies on the media being encrypted, and the decryption key being controlled. For example a DRM-enabled video player application might only decrypt and play a video file for a user who had purchased a small "digital rights" file containing a key. Since a DRM-enabled player system is decrypting the media data at some point within its software and/or hardware in order to show it to the user, DRM ultimately relies on security through obscurity and therefore not ultimately secure.

Digital Service Unit (DSU)-A device that interconnects the customers digital telephone equipment to a telephone network.

Digital Service, Level 0 (DS-0)-A 64,000 b/s channel, the worldwide standard public telephone industry bit rate for digitizing one voice conversation. There are 24 DS-0 channels in a 1.544 Mb/s DS-1 digital multiplex bit stream, and 30 DS-0 traffic channels plus two additional DS-0 channels used for synchronization and signaling (a total of 32 DS-0 channels) in a 2.048 Mb/s E-1 digital multiplex bit stream.

Digital Signal-Digital signals consist of a series of ones and zeros, most often represented in telecommunications signals by two different voltages. For example a +5 Volt level could represent a logical 1 (one) and 0 Volt level could represent a logical 0 (zero). The ones and zeros are called bits. Several bits (usually eight) are grouped into a byte and each byte is defined to have a specific meaning, such as a specific letter on a keyboard. Digital signals are used to represent specific levels on an analog signal. While a digital signal cannot represent every point on an analog wave, they can come close enough to be almost indistinguishable. Digital signals are much easier to process by computer systems and they are able to resist the effects of noise better than analog signals.

Digital Signal 1 (DS-1)-The primary rate telephone industry digital multiplexing system used in North America and Japan. It combines 24 DS-0 (64 kb/s) channels and a single 8 kbit/s synchronizing bit stream for a total of 1.544 Mb/s. A different primary rate multiplexing system that combines 30 DS-0 channels with a 64 kb/s channel for signaling and another 64 kb/s channel for synchronization and control, for a total of 2.048 Mb/s, is used elsewhere. The 2.048 Mb/s system is sometimes named E-1, or MIC, or CEPT Primary Rate Multiplexing. Both the 1.544 Mb/s DS-1 system and the 2.048 Mb/s system are recognized ITU standards. The 2.048 Mb/s system was designed to have some improvements and a slightly larger channel capacity than the 1.544 Mb/s DS-1 system, and was intentionally incompatible, a result attributed by some industry observers to a motive of protecting the European market from imported product competition. (Later, manufacturers in different countries changed their objectives from the former strategy of setting intentionally incompatible standards in different countries to the present strategy of setting internationally compatible standards in all countries.) DS-1 is not a trade name of any manufacturer. T-1 is effectively a synonym of DS-1 in North America, and was originally a trade name of just one manufacturer, but today it is widely used for all compatible products regardless of manufacturer.

Digital Signal 3 (DS-3)-A standard digital transmission line that is divided into twenty eight DS1 (T1) channels. The gross transmission rate for a DS3 channel is 44.736 Mbps. A single DS3 provide for 672 standard (64 kbps) voice channels.

Digital Signal Level (DSx)-Digital signal (DS) transmission is a hierarchy of digital communication channels and lines that range from 64 kbps to 565 Mbps. Lower level DS structures are combined to produce higher-speed communication lines. There are different structures of DS levels used throughout the world with significant variations between North American and European systems. DSx has been used to represent the digital transmission standards where the "x" denotes which service is under discussion.

Digital Signature Processing

Digital Signal Processing-Digital signal processing is the manipulation of digital signals into other forms using computing circuits or systems. Digital signal processors use software programs to allow them to perform complex signal processing operations such as filtering, modulation, data compression, and shaping of the information (such as digital audio signal) that are represented by digital signals.

Digital Signal Processor (DSP)-An integrated circuit (chip) that is designed specifically for high-speed manipulation of digital information. DSP chips operate using software programs to allow them to perform complex signal processing operations such as filtering, modulation, data compression, and information processing. The use of DSPs in communication circuits allows manufacturers to quickly and reliably develop advanced communications systems through the use of software programs. The software programs (often called modules) perform advanced signal processing functions that previously complex dedicated electronics circuits. Although manufacturers may develop their own software modules, DSP software modules are often developed by other companies that specialize in specific types of communication technologies. For example, a manufacturer may purchase a software module for echo canceling from one DSP software module developer and a modulator software module from a different DSP software module developer.

This figure shows typical digital signal processor that is used in a digital communication system. This diagram shows that a DSP contains a signal input and output lines, a microprocessor assembly, interrupt lines from assemblies that may require processing, and software program instructions. This diagram shows that this DSP has 3 software programs, digital signal compression, channel coding, and modulation coding. The digital signal compression software analyzes the digital audio signal and compresses the information to a lower data transmission rate. The channel coding adds control signals and error protection bits. The modulation coding formats (shapes) the output signal so it can be directly applied to an RF modulator assembly. This diagram also shows that an optional interface is included to allow updating of the software programs that are stored in the DSP.

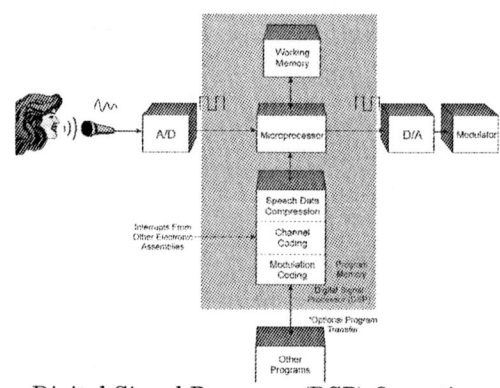

Digital Signal Processor (DSP) Operation

Digital Signal, level 1 Combined (DS-1C)-A digital multiplexing system in North America and Japan. The total bit rate is 3.152 Mb/s. It comprises two T-1s (two J-1s in Japan), at 1.544 Mbps each, which are interleaved to support 48 DS-0 channels. The additional 64 kb/s is overhead used to support additional signaling and control requirements. DS-1C is seldom used, outside of limited telco applications. It has mostly been supplanted by DS-2 multiplexing, which has 96 channels. There is no European equivalent and DS-1C is not included in ITU standards.

Digital Signaling Tone (DST)-A tone that is sent on the analog radio channel to indicate a change in status (e.g. end call).

Digital Signature-A number calculated from the contents of a file or message using a private key and appended or embedded within the file or message. The inclusion of a digital signature allow a recipient to check the validity of file or data by decoding the signature to verify the identity of the sender.

Digital Speech Interpolation (DSI)-In addition to multiplexing through channel division, statistical multiplexing can also be used by distributing transmission of a communications channel over idle portions of multiplexed channels. An example of statistical multiplexing is digital speech interpolation (DSI). DSI is a technique that dynamically allocates time slots for voice or data transmission to a user only when the have voice or data activity. This increases the system capacity as transmission for other users can occur when others are silent.

This figure shows the process of multiplexing using DSI. This diagram shows a communication circuit that has 96 independent communication channels (one communication link that has 96 time slots).

164

Digital Subscriber Line Access Multiplexer (DSLAM)

The DSI system monitors the activity of each voice conversation (a voice channel) using a voice activity detector (VAD). The VAD is an electronic circuit that senses the activity (or absence) of voice signals. This is used to inhibit a transmission signal during periods of voice inactivity.

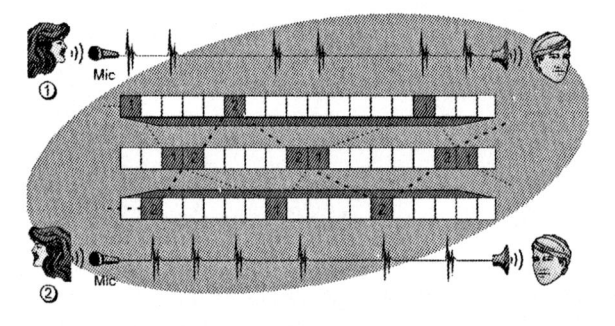

Digital Speech Interpolation (DSI) Operation

Digital Subscriber Line (DSL)-Digital subscriber line is the transmission of digital information, usually on a copper wire pair. Although the transmitted information is in digital form, the transmission medium is usually an analog carrier signal (or the combination of many analog carrier signals) that is modulated by the digital information signal.

This figure shows a basic DSL system. This diagram shows that the key to DSL technologies is a more efficient use of the 1 MHz of bandwidth available on a single pair of copper telephone lines. A DSL system consists of compatible modems on each end of the local loop. For some systems, the DSL system allows for multiple types of transmission on a single copper pair. This includes analog or ISDN telephone (e.g., POTS) and digital communications (ADSL or VDSL). This diagram shows that there are basic trade offs for DSL systems. Generally, the longer the distance of the copper line, the lower the data rate. Distances of less than 1,000 feet can achieve data rates of over 50 Mbps.

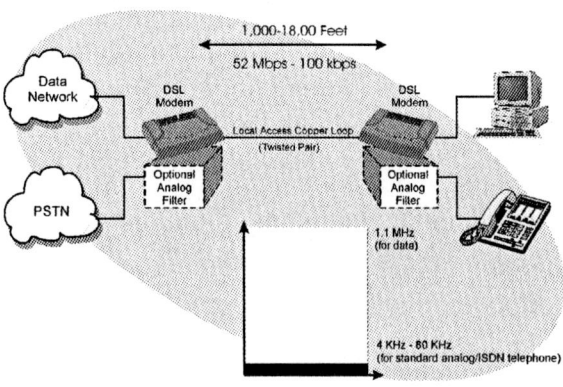

Digital Subscriber Line (DSL) System

This photograph shows a modem that connects a computer to a high speed digital subscriber line (DSL) at extended distances. This modem (ReachDSL model) is acceptable for businesses, home and private networks surroundings.

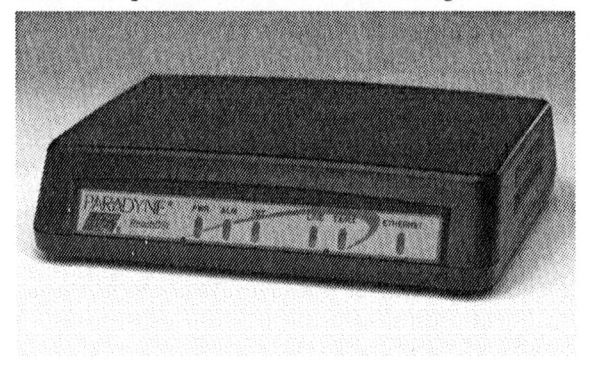

DSL Modem
Source: Paradyne

Digital Subscriber Line Access Multiplexer (DSLAM)-An electronic device that usually holds several digital subscriber line (DSL) modems that communicate between a telephone network and a end customers DSL modem via a copper wire access line. The DSLAM concentrates multiple digital access lines onto a backbone network for distribution to other data networks (e.g. Internet).

This picture shows a PL-2000 i-Slam Universal Services Access Multiplexter (USLAM) assembly. This assembly (equipment rack) is usually installed at a telephone company or at facilities (e.g. enclo-

Digital Subscriber Line Modem (DSL Modem)

sures) owned by the telephone company) and it contains the DSL modems that communicate with the modems used by the end customer.

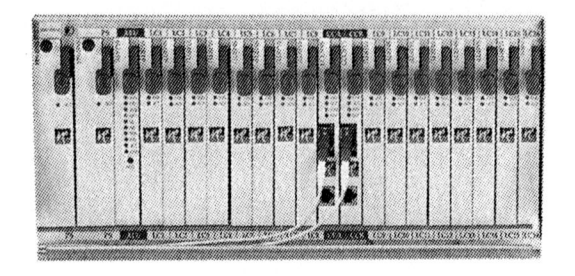

Digital Subscriber Line Access Module (DSLAM)
Source: Next Level Communications

Digital Subscriber Line Modem (DSL Modem)- An electronics assembly device that modulates and demodulates (MoDem) digital subscriber line (DSL) signals. DSL signals are usually transmitted on a twisted pair of copper wires. A DSL modem may be in the form of an internal computer card (e.g. PCI card) or an external device (Ethernet adapter). Most DSL modems have the ability to change their data transfer rates based on the settings that are programmed by the DSL service provider and as a result of the quality of the communication line (e.g. amount of distortion).

Digital Telephony-Digital telephony is a communication system that uses digital data to represent and transfer analog signals. These analog signals can be audio signals (acoustic sounds) or complex modem signals that represent other forms of information.

Digital Television (DTV)-A process or system that transmits video images through the use of digital transmission. The digital transmission is divided into channels for digital video and audio. These digital channels are usually compressed. Video compression commonly uses one of the motion picture experts group (MPEG) standards to reduce the data transmission rate by a factor of 200:1.

Digital Traffic Channel (DTC)-A radio channel that carries digital information that is divided into several logical channels. Logical channels on a digital traffic channel (DTC) are created by dividing the digital channel into time slots or groups that represent several different logical channels (such as voice and signaling channels).

Digital Verification Color Code (DVCC)-A unique code that is continuously sent on a communication channel to indicate a different between two or more channels that may be operating on the same frequency. DVCC is the digital equivalent of the supervisory auditory tone (SAT). The receiver of the channel continuously decodes and compared the received DVCC to the assigned DVCC to determine if it is receiving a co-channel interfering signal instead of the desired signal.

Digital Video-Digital video is a sequence of picture signals (frames) that are represented by binary data (bits) that describe a finite set of color and luminance levels. Sending a digital video picture involves the conversion of a scanned image to digital information that is transferred to a digital video receiver. The digital information contains characteristics of the video signal and the position of the image (bit location) that will be displayed.

This figure demonstrates the operation of the basic digital video compression system. Each video frame is digitized and then sent for digital compression. The digital compression process creates a sequence frames (images) that start with a key frame. The key frame is digitized and used as reference points for the compression process. Between the key frames, only the differences in images are transmitted. This dramatically reduces the data transmission rate to represent a digital video signal as an uncompressed digital video signal requires over 50 Mbps compared to less than 4 Mbps for a typical digital video disk (DVD) digital video signal.

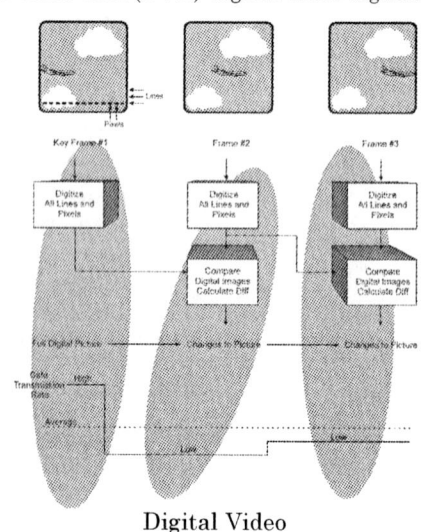

Digital Video

Digitization

Digital Video Broadcast (DVB)-A standard that uses phase alternate line (PAL) video display format to deliver digital television signals.

Digital Video Interactive (DVI)-A digital video transmission system that allows real time (or near real time) interaction to allow changing of the information content. DVI combines digital video and audio and allows the computer user to control the operation of the media display. DVI is a registered trademark of Intel.

Digital Video Recorder (DVR)-A device that stores video images in digital format.

This diagram shows the basic process that is used for digital voice compression process. In this diagram, a digital audio signal (64 kbps PCM signal) is continuously applied to a digital signal analysis device. The analysis portion of the speech coder extracts the amplitude, pitch, and other key parameters of the signal and then looks up related values in the code book for the portion of sound it has analyzed. Only key parameters and code book values are transmitted. This results in data compression ratios of 4:1 to over 16:1.

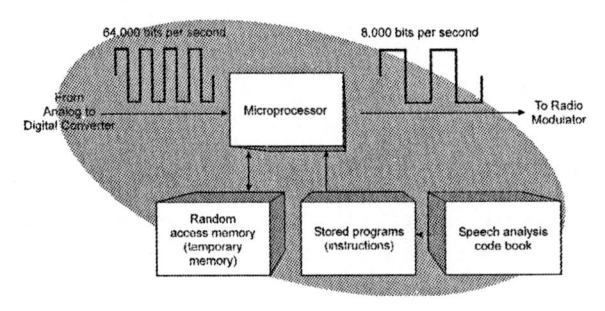

Digital Voice Compression Operation

Digital Voltmeter-A voltmeter that displays its readings in a digital format, either by LCDs or by a digital output signal supplied to another instrument.

Digital-To-Digital Transfer (D-t-D)-The process of transferring digital information (audio, video, or data) from one machine to another in the digital domain. The advantage in D-to-D transfers is that no signal degradation occurs because this process bypasses all of the analog circuitry which can degrade overall performance.

Digitization-Digitization is the conversion of analog into digital form. To convert analog signals to digital form, the analog signal is digitized by using an analog-to-digital (pronounced A to D) converter. The A/D converter periodically senses (samples) the level of the analog signal and creates a binary number or series of digital pulses that represent the level of the signal.

The common conversion process is Pulse Code Modulation (PCM). For most PCM systems, the typical analog sampling rate occurs at 8000 times a second. Each sample produces 8 bits digital that results in a digital data rate (bit stream) of 64 thousand bits per second (kbps).

Digital bytes of information are converted to specific voltage levels based on the value (weighting) of the binary bit position. In the binary system, the value of the next sequential bit is 2 times larger. For PCM systems that are used for telephone audio signals, the weighting of bits within a byte of information (8 bits) is different than the binary system. The companding process increases the dynamic range of a digital signal that represents an analog signal; smaller bits are given larger values that than their binary equivalent. This skewing of weighing value give better dynamic range. This companding process increases the dynamic range of a binary signal by assigning different weighted values to each bit of information than is defined by the binary system.

Two common encoding laws are Mu-Law and A-Law encoding. Mu-Law encoding is primarily used in the Americas and A-Law encoding is used in the rest of the world. When different types of encoding systems are used, a converter is used to translate the different coding levels.

This figure shows the basic audio digitization process. This diagram shows that a person creates sound pressure waves when they talk. These sound pressure waves are converted to electrical signals by a microphone. The bigger the sound pressure wave (louder the audio), the larger the analog signal. To convert the analog signal to digital form, the analog signal is periodically sampled and converted to a number of pulses. The higher the analog signal, the larger the number of pulses. The number of pulses can be counted and sent as digital numbers. This example also shows that when the digital information is sent, the effects of distortion can be eliminated by only looking for high or low levels. This conversion process is called regeneration or repeating. This regeneration progress allows digital signals to be sent at great distances without losing the quality of the audio sound.

Digitizing Tablet

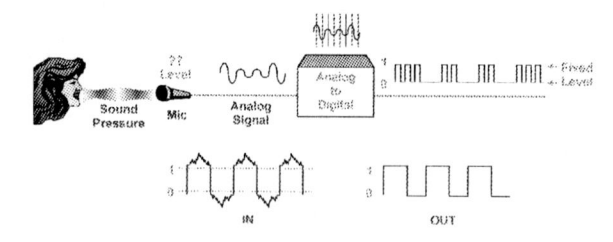

Digitization Process

Digitizing Tablet-Digitizing Pad
DIL-Direct In-Line
DILEP-Digital Line Engineering Program
DIM-Document Image Management
DIMM-Dual Inline Memory Module
DIN-Deutsche Institut fur Normung
DIN-Deutsche Industrie Norm
Diode Laser-A semiconductor device that creates coherent light using a special diode junction.
DIP-Document Image Processing
DIP-Dual Inline Pin
DIP-Dedicated Inside Plant
DIP Switch-Dual In-Line Package Switch
This diagram shows a dual inline package (DIP) switch that is mounted on an interface card. In this example, the DIP switch is used to select the operating mode of a device. While this DIP switch is a slide switch, DIP switches can also be rocker switches and DIP switches are often grouped for the operators convenience.

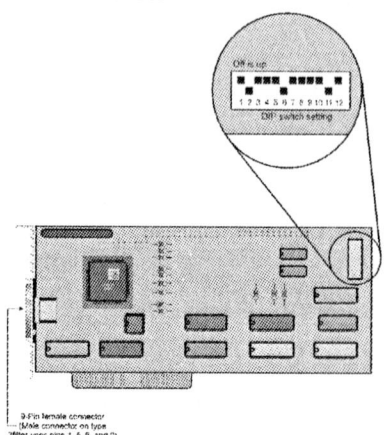

Dual Inline Package (DIP) Switch

Diplexer-(1-mobile communication) A device that enables the outputs of two wireless base stations or antenna to be combined onto a single feeder cable, which allows co-siting of mobile communication systems. (2-television) A device that combines the transmission of audio and video signals over a common channel.
This photograph shows a diplexer that allows the outputs of GSM 900 and GSM 1800 wireless base stations to be combined onto a single feeder cable, resulting in a 50% savings in cable costs.

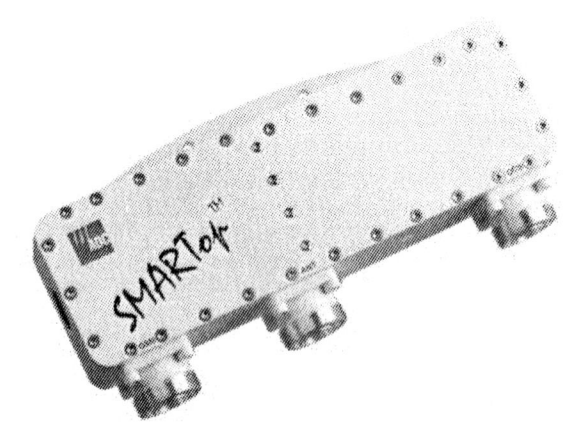

Diplexer
Source: ADC

Dipole Antenna-An antenna that has two poles for transmission.
DIRECT-Directory Project
Direct Broadcast Satellite (DBS)-A satellite with enough range and power to be received by small dish antennas suitable consumer home use. DBS can be sent to both direct individual homes as well as received by communities by means of retransmission over a small TV station or cable TV system. In the late 1990's, the DBS marketplace become a formidable competitor to the traditional cable industry. DBS systems provider digital-quality pictures and have the potential to offer high speed interactive services. By using digital compression technology, DBS systems can offer a greater number of channels than analog cable systems to both PCs and TVs. DBS systems can also be customized to provide unique services for limited video on demand (VOD), near video-on-demand (NVOD) and interactive pay-per-view channels.

Direct Current Power (DC Power)-Electrical power that is supplied in a form of direct current (DC). DC power is usually supplied from a regulated (fixed voltage) source such as a power supply or batteries and is used to supply to assemblies or signal processing circuits.

Direct Distance Dialing (DDD)-The automatic completion of customer dialed long distance or toll calls in response to signals from a customer's telephone where no operator assistance is required.

Direct Inward Dialing (DID)-Direct Inward Dialing (DID) connections are trunk-side (network side) end office connections. The network signaling on these 2-wire circuits is primarily limited to 1-way, incoming service. DID connections employ different supervision and address pulsing signals than dial lines. Typically, DID connections use a form of loop supervision called reverse battery, which is common for 1-way, trunk-side connections. Until recently, most DID trunks were equipped with either Dial Pulse (DP) or Dual Tone Multifrequency (DTMF) address pulsing. While many wireless carriers would have preferred to use Multifrequency (MF) address pulsing, a number of LEC's prohibited the use of MF on DID trunks.

Direct Inward Service Access (DISA)-The process of how incoming calls are handled to a telephone system.

Direct Modulation-Modulation of a carrier (such as a laser or LED light source) by the controlling (modulating) its output power or signal and therefore, controlling the amount of light it emits.

Direct Outward Dialing (DOD)-A feature that allows private telephone systems users (PBX or Centrex) to directly call public telephone numbers without the need to use an attendant or operator. The use of "dial 9" to get an outside line is a direct outward dialing feature.

Direct Sequence Spread Spectrum (DSSS)-A transmission technique in which multiplies an information signal with a sequence code (or information signal) to allow the modulation signal to occupy a frequency bandwidth that is much wider than is necessary to represent the information signal alone. The DSSS system then uses the multiplying code to decode the received signal.

Direct Station Select (DSS)-A process of using a button or softkey on a telephone station (such as a PBX extension) to select another specific extension or station to communicate with.

Direct Talk-A family of IBM voice processing products introduced in the summer of 1991. According to IBM, its IBM CallPath DirectTalk product line lets business automate routine operations and also provide callers with easy access to many kinds of information over the telephone — at any hour of the day and with greater accuracy. Businesses can raise the level of service they provide and do it with fewer people and with greater efficiency.

Directed Retry-Directed retry is a tool for increasing the traffic-handling capacity of a cellular system. It is a software program which routes traffic to alternate cell sites in a cellular telephone system when high traffic conditions make hand-off to the preferred cell site impossible.

Directional Antenna-A directional antenna focuses the transmitted or received signal in a specific direction. Directional antennas are used to provide signal gain in a specific direction while reducing the signal levels (that may cause interference) in other directions.

This diagram shows how the energy of an antenna can be focused (directed) to a particular area. This diagram shows the focusing of the beam into a main lobe and that the transmission patterns of directional antennas usually result in the creation of unwanted side lobes.

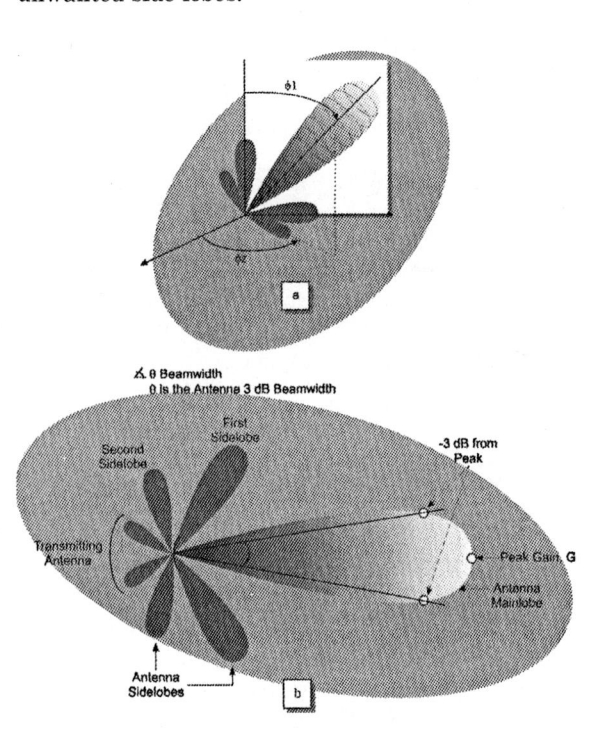

Directional Antenna Diagram

Directional Coupler

Below is a picture of a Yagi directional antenna that is used to focus or gather radio energy in a specific direction. The use of multiple antenna elements on this antenna provides a relatively wide frequency range (bandwidth) while providing signal strength gain in a specific direction.

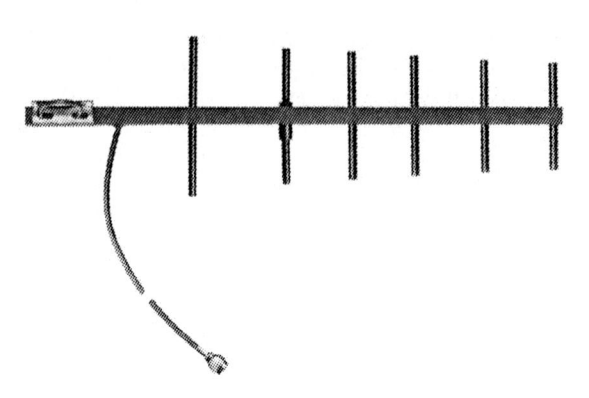

Yagi Directional Antenna
Source: Antenna Specialists, A Division of Allen telecom Inc.

Directional Coupler-A device that provides a sample of transmitted energy as the signal passes through in a particular (forward direction). Signals that pass through the directional coupler in the opposite direction (e.g. reflected signals) are isolated from the sample port.

This diagram shows how a directional coupler allows a signal to pass through while it directs (samples) a portion of an input signal to a coupled output port. This example shows how a 20 dB directional computer (20 dB = factor of 100 times) allows 99% of the RF signal to pass through the main output while directing 1% (10 mWatts) of the signal to the coupled output.

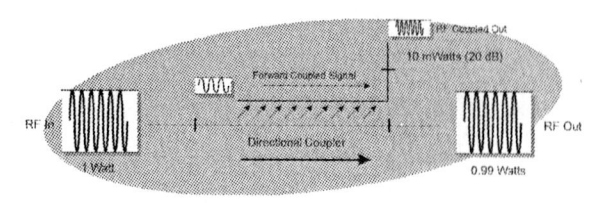

Directional Coupler

This photograph shows a directional coupler that is used in communication networks. This coupler allows a majority of the signal to flow through while directing (coupling) a portion to a separate connection port.

Directional Coupler
Source: PCI Technologies Inc.

Directional Filter-A device or assembly that combines high-pass and/or low-pass filters that is used to separate a frequency range in a portion of transmission bands in a specific direction of transmission. For example, a directional filter can be used in a bi-directional system to extract a particular signal in a specific bandpass frequency range in a single direction.

Directories URL-An address that is preprogrammed into or used by devices (such as an IP Telephone) where information is kept regarding directly listings.

Directory Gatekeeper-A gatekeeper that is used in a large voice over data communications network as a central information point for other gatekeepers. A directory gatekeeper allows an administrator to manage the configuration of the network without having to configure many local area systems.

DIS-Digital Identification Signal
DIS-Draft International Standard
DISA-Direct Inward Service Access
DISA-Data Interchange Standards Association, Inc.
DISA-Defense Systems Information Agency
DISA-Direct Inward System Access
Disaster Recovery-The processes that are used to restore services after a significant interruption (disaster) in communications systems. Disaster recov-

ery processes usually occur after events such as fires, floods, or earthquakes. However, disaster recover may also occur after critical equipment failures or information corruption that occurs from software viruses.

Discard Eligibility (DE)-A control flag system to indicate the essential nature of the packet's data that is transmitted through a packet data network. The DE flag(s) allow systems to selectively discard data packets or frames that are non-essential. This process allows some data transmission systems to send more data than is agreed to (dynamic bandwidth). If the network is not congested, it may allow the extra packets of data to reach their destination.

Disconnect for Non-Payment (DNP)-A transaction transmitted to the Network Operations system requesting that a customer's service be disconnected due to non-payment.

Discontinuous Reception (DRx)-A process of turning off a radio receiver when it does not expect to receive incoming messages. For DRx to operate, the system must coordinate with the mobile radio for the grouping of messages. The mobile radio (or pager) will wake up during scheduled periods to look for its messages. This reduces the power consumption which extends battery life. This is sometimes called: sleep mode.

This diagram shows how discontinuous reception (DRX) sleep mode can be used to extend the battery life in portable radio devices. In this example, the mobile telephone registers on a wireless network and informs the system of its sleep mode capability. The system responds by assigning the mobile telephone to a paging group. Paging groups are brief periods (typically 20-200 msec) that a mobile telephone must be awake to monitor for it's paging messages. There are usually ten or more paging groups that allow the mobile telephone to sleep for up to 90% of the time.

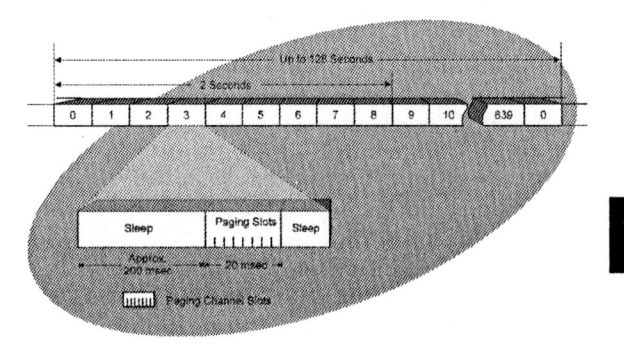

Discontinuous Reception (DRx) Sleep Mode Operation

Discontinuous Transmission (DTx)-The ability of a communications system to inhibit transmission when no or reduced activity is present on a communications channel. DTx is often used in mobile telephone systems to conserve battery life.

Discrete Cosine Transform (DCT)-A kind of Discrete Fourier Tranform (DFT) constrained to use real-valued input and output, as opposed to the general DFT which allows complex values. There exists a fast implementation analogous to the FFT. The DCT assumes that the input and output sequences have even symmetry. Because the DCT uses only real values it is more efficient for working with real-valued sequences such as audio samples or image pixel values. The DCT and its variations are building blocks of many audio and video transform codecs.

Discrete Fourier Transform (DFT)-A Fourier transform applied to a periodic sequence of complex values (known as samples) at discrete times. The result of this transform is a periodic sequence of complex values at discrete frequencies. Because the input and output of the DFT are periodic sequences, each can be represented by only one period of samples. Since the DFT is a mapping of one finite sequence to another finite sequence, it can be implemented by computer in a straightforward way, unlike a general Fourier transform of an arbitrary continuous function.

Discrete Multi-Tone (DMT)-A data communications process that transfers a high speed data communication channel by dividing it into several narrow sub-channels and sending them independently

Disk Mirroring

through frequency divided channels. When the sub-channels are received, the low speed parts are recombined to create the original high-speed data transmission signal.

The advantage of sending several sub-channels is the ability to independently adjust the transmission levels of each sub-channel signal. Because the frequency response of the line can vary and distortions can occur on specific frequencies (where only a few sub-channels may be affected). DMT is used in DSL systems as it adapts well to the hostile environment of copper wire transmission.

This figure shows a discrete multitone transmission system (DMT) system. In this diagram, a high-speed data signal is divided into several low speed data signals. Each low speed data signal modulates a sub-channel. The sub-channels are combined and supplied to the copper wire. At the receiving end, each sub channel is received and decoded. The sub-channel data signals are re-combined to recreate the original high-speed data signal.

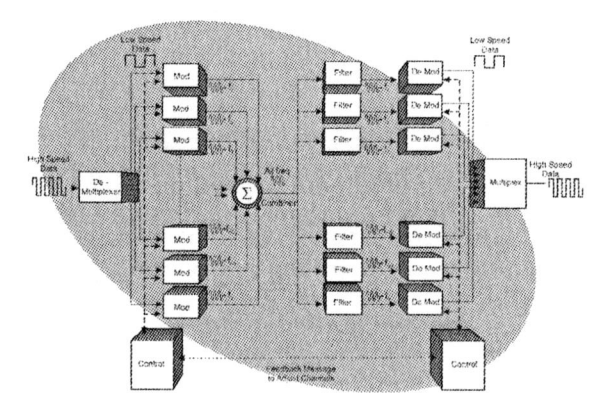

DMT Transmission System

Disk Mirroring-A data protection strategy that uses redundancy of information on two (or more) storage devices (disks) to allow for real time backup and recovery of information. The process of disk mirroring is the storage of the same information on two disks. One disk is used as the primary source of information and the other disk is used when a failure is detected on the primary disk.

This figure shows the process of mirroring information on two disk storage devices to increase the reliability of information storage. Disk mirroring is also known as redundant array of inexpensive disks (RAID). This diagram shows that disk Mirroring is performed by a RAID controller card that controls both primary and secondary disk storage devices. Information is stored to both hard disks simultaneously. If the primary hard drive fails, the controller will automatically begin to use the secondary hard disk and alert the user that one of the hard disks has failed.

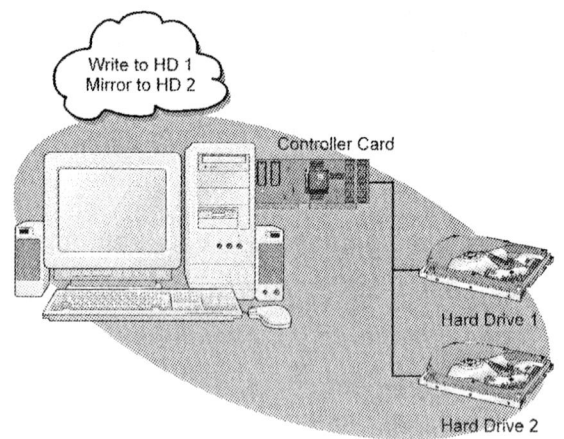

Disk Mirroring Operation

Dispatch-Dispatch radio service allows a central controller (dispatcher) to send dispatch assignments to one or more receivers (typically many mobile radios). Dispatch radio systems normally involve the coordination of a fleet of users via a dispatcher. All mobile units and the dispatcher can usually hear all the conversations between users in a dispatch group. Dispatch operation involves push-to-talk operation by a group of users on the dispatch system. Using trunking technology, there can be several different dispatch groups that operate on the same system. Dispatch systems usually charge a fixed monthly usage fee for each user (mobile radio) without any airtime usage charges.

This figure shows a typical two-way SMR dispatch radio system. In this diagram, a single radio channel is shared by several mobile radios that operate within the radio coverage limits of a high power radio base station. A dispatcher communicates to all the mobile radios by using a high-powered radio transmitter. Because the mobile radios mounted in vehicles transmit at a lower power than the dispatcher's transmitter, this SMR system has several receiving antennas. When a mobile radio is communicating with the dispatcher, the receiver selects which antenna has the best signal (called a voting receiver).

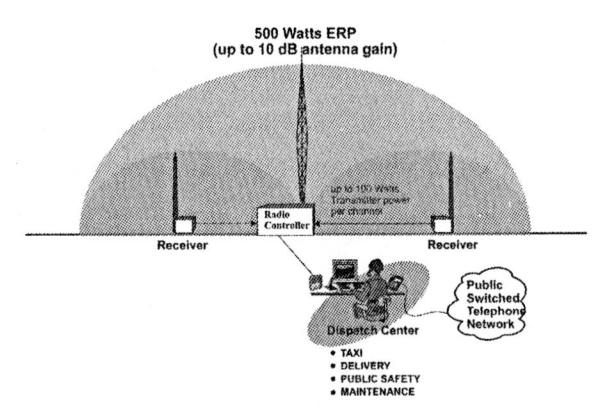

DMT Transmission System

Dispersed Control-A system that coordinates the distribution the call processing or switching intelligence to multiple control parts of a system or network.

Dispersion-(1-general) The dependence of an optical signal on a specific chromatic or wavelength parameter. (2-chromatic) Distortion of pulsed transmitted signals that occurs in an optical fiber system that is caused by multiple wavelengths of light that travel different paths (different distances) within the fiber due to differences in refractive indexes of the fiber to different optical wavelengths. (3-microwave) The scattering of microwave radio signal energy by small obstructions (such as raindrops).

Dispersion Compensation-A tactic for reducing the effects of chromatic dispersion in a long optical fiber run. Most optical fiber exhibits positive dispersion in which shorter wavelengths travel more slowly than longer ones. By passing the light through a length of fiber designed to have negative dispersion so that the longer wavelengths travel slowly compared to the short ones, the short waves can catch up with the long ones, allowing optical pulses to return to their initial shape and DWDM systems to have their signals traveling together. This design of a long stretch of standard fiber followed by a short stretch of negative dispersion fiber can be repeated several times, as long as there is adequate amplification available.

Dispersion Limited Operation-The condition where the limiting factor in a fiber optic transmission system is limited by the amount of pulse distortion (dispersion) rather than the attenuation (amplitude) of the received optical signal.

Dispersion Shifted Fiber (DSF)-A type of optical fiber fabricated to have uniform wave speed for optical signals over the infrared wavelength to be used. See: Dispersion

Dispersion-Shifted Fiber-A type of optical fiber that is engineered such that chromatic dispersion cancels out waveguide dispersion. This effect can be achieved only for a small range of wavelengths for any given fiber. In step index single mode fiber, they cancel each other out at about 1330 nm, one of the commonly used transmission windows for optical networks. By increasing wavelength dispersion, the zero-dispersion point was moved to 1550 nm, another important optical wavelength. Nonzero-dispersion fiber has its zero dispersion point moved out beyond the optical windows to about 1600 nm in order to avoid third order effects in the fiber, such as four wave mixing.

Disruptive Technology-A new technology that significantly reduces or eliminates the value of existing technology implementations by performing or providing the service faster and at lower cost.

Distance Learning-Distance learning is the process of providing educational training to students at locations other than official learning centers (schools). Distance learning has been available for many years and is now used in elementary education (grades K-12), higher education (college), professional (industry), government training and military training. In the early years, distance learning was provided through the use of books and other printed materials and was commonly referred to as "correspondence courses".

Distance learning has evolved through the use of broadcast media (e.g. televisions) and moved onto individual or small group training through the availability of video based training (VBT) or computer based training (CBT). These systems have developed to interactive distance learning (IDL) as the computer allowed changes in the training.

Distance to Reuse Ratio (D/R)-The minimum distance separation requirement in a wireless communication network that a frequency may be re-used without causing significant interference with a radio channel from another radio tower within the network.

Distance Vector Multicast Routing Protocol (DVMRP)

Distance Vector Multicast Routing Protocol (DVMRP)-A protocol that assists a router in the selection of routes based on the distance between connections. DVMRP keeps track of the distance of other connections between routers in a network. DVMRP uses Internet Group-Management Protocol (IGMP) to transfer routing information with neighboring routers.

Distance-Vector Routing Protocol-A routing protocol which mathematically computes routes using a measurement of distance. This measurement is known as the distance vector and can be based on a link's speed or other characteristic of the link. Each router utilizing a D-V protocol periodically transmits all or a subset of its routing table, in a routing-update message, at a regular interval to each of its adjacent routers. As this routing information proliferates through the network each router recomputes the distance to other routers by adding each link's D-V weight. The routing update messages identify new destinations as they are added to the network, convey link failures information, and calculate distances to all known destinations.

Distance vector routing protocols are often contrasted with Link-State routing protocols, such as Open Shortest Path First (OSPF) which requires each router to send only its local connection information, not too its neighboring routers, but instead to all routers in the internetwork.

In short, link-state algorithms require much more intelligence at each router but only require small update messages to be sent to each router. Distance Vector algorithms send large updates, but only to adjacent routers.

D-V protocols are easier to implement than Link-State protocols, however, D-V protocols are less resilient and take longer to converge. Routing Information Protocol (RIP) is the most common D-V routing protocol in use. More information about D-V, sometimes called Bellman-Ford algorithms can be found in RFC 1058, the RIP protocol specification.

Distinctive Ringing-A service feature that alerts a customer via a special ring (usually short, long or rapid ring) that an incoming call is received that has a different purpose or priority from others that are received on that same telephone line. Distinctive ringing is used for sharing multiple phone numbers on a single line or for priority ringing.

This diagram shows the operation of a telephone system that has distinctive ringing feature. This diagram shows a single telephone that is assigned two different telephone numbers even though the telephone operates on one telephone number (one switch port). In this example, when an incoming call is received for the registered number 555-6234, it is re-directed (forwarded) towards the actual destination number 555-1234 along with information that allows the system to uniquely identify the call with a dual ring (2 rings in the 2 second ring period). When calls are received to 555-1234, the ring is a single 2 second/4 second cadence. This allows the receiver of the call to determine which telephone number was dialed by the distinctive ring sound.

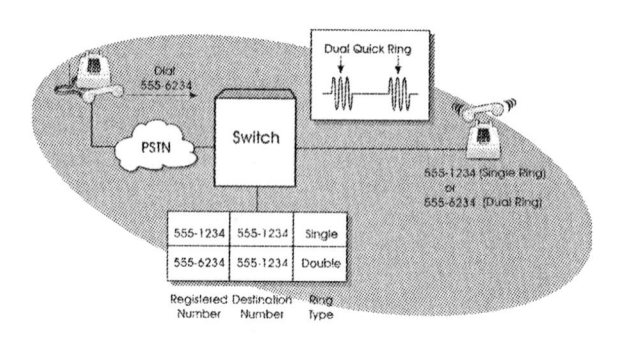

Distinctive Ringing Operation

Distortion-(1 - general) The inaccurate reproduction of a signal caused by changes in the signal waveform. The difference between the wave shape of an original signal and the signal after it has traversed the transmission circuit. (2 - delay) The distortion caused by the later arrival of higher-frequency components of a complex waveform as a result of the slower travel speed of higher-frequency components. (3-envelope delay) The distortion caused by a delay of the envelope or group of signals passing through a network. (4 - frequency) The changes in the relative amplitudes of different frequency components of a complex wave form. (5 - harmonic) The distortion caused by the creation of harmonics of a fundamental frequency. (6 - intermodulation) The distortion produced when two or more waves (or a complex waveform involving two or more frequencies) pass through a nonlinear device that produces sum-and-difference modulation product frequencies. (7 - linear) A distortion that is independent of the signal amplitude. (8 - nonlinear) A distortion that is dependent on signal amplitude. (9 - optical waveguide) The signal distortion caused by three primary mechanisms: waveguide dispersion, material dispersion, and profile dispersion. In addition, the signal may suffer degradation from intramodal distortion and multimode distortion. (10 - phase-frequency) The distortion resulting from the difference between phase delay at a given frequency and at a reference frequency.

Distributed Call Center-A call center that has several telephone workers (customer service representatives) that are located at different places. Distributed call centers usually allows the worker to have access to information systems even though they may be independent located at remote places (such as home workers).

Distributed Control-A system that distributes the call processing or switching intelligence to multiple parts of the system or network.

Distributed PBX-A PBX system that has multiple switching systems that are interconnected with each other.

Distributed Routing-Distributed routing uses intelligent routers or switches that forward packets toward their destination rather than using centrally coordinated rout9ing paths. This can provider for a more robust network as each switch or router makes its own routing decisions without the need for a centralized control center.

Distribution Cable-A cable or cabling system that is used to transfer signals from a central location (e.g. a central office or the head end of a CATV system). to end customers.

Distribution Channels-Organizational units (internal and external), dedicated to the process of distributing products and services directly to customers through retail or interpersonal facilities.

Distribution Network-(1-general) A system of cables and terminals to that interconnects communication devices. (2-cable television) The portion of a cable television system that links the head end to the end customers televisions. (3-telephone) Cables from a main telephone switching or distribution junction that usually contain from 25 to 200 pairs of wires.

Distribution Service (DS)-A service of distributing the same information to multiple receivers that are connected to a network.

DIT-Directory Information Tree

Diversity Combiner-A circuit or device that can combine two or more signals from the same information source (such as a delayed or reflected signal) so it can create an output signal that is better quality than any of the individual contributing signals.

Divisional Application-A divisional application is an application for patent filed in response an examiner's restriction requirement which claims priority to a pending application for patent which is identical to the co-pending application, but claiming a different invention. A divisional application is identical to a continuation application except that it is filed in response to an examiner's determination that the application contains more than one invention which would require a separate search.

DIW-Type D Inside Wire
DIX-Digital/Intel/Xerox
DIX-Ehternet II
DL-Data Link
DLC-Direct Line Console
DLC-Data Link Control
DLC-Digital Loop Carrier
DLCF-Delay Length Call Factor
DLCI-Data Link Connection Identifier
DLE-Data Link Escape
DLEC-Data Local Exchange Carrier
DLL-Data Link Layer driver
DLL-Dynamic Link Library
DLMR-Digital Land Mobile Radio
DLPBC-Dual Loop Port Bypass Circuitry
DLPI-Data Link Provider Interface
DLR-Design Layout Report
DLS-Data Link Switching
DLSw-DataLink Switching Workgroup
DLTU-Digital Link Trunk Unit
DM-Delta Modulation
DM-Disconnect Mode
DMA-Designated Market Area
DMA-Direct Memory Access
DMA Channel-Direct Memory Access Channel
DMD-Differential Mode Delay
DMH-Data Message Handler
DMI-Digital Multiplexed Interface
DMI-Desktop Management Interface
DMI-BOS-Digital Multiplexed Interface-Bit Oriented Signaling
DMPDU-Derived MAC PDU

DMS-Digital Multiplex System

DMS Switching System-A class of digital electronic switching systems manufactured by Northern Telecom. DMS is a registered trademark of Northern Telecom.

DMT-Discrete Multi-Tone

DMTF-Desktop Management Task Force

dmW-Digital Milliwatt

DMX-Digital Matrix Switch

DMZ-DeMilitarized Zone

DN-Directory Number

DNA-Dynamic Node Access

DNA-Distributed Network Administration

DNA-Digital Network Architecture

DND-Do Not Disturb

DNIC-Data Network Identification Code

DNIS-Dialed Number Identification Service

DNP-Disconnect for Non-Payment

DNPA-Data Numbering Plan Area

DNR-Dialed Number Recorder

DNR-Dynamic Noise Reduction

DNR-Dynamic Network Reconfiguration

DNS-Domain Name Server

Do Not Disturb-A call processing feature that indicates to the telephone system (such as a PBX system) that the telephone user does not desire to receive calls. The do not disturb feature usually disables the audio alert and may activate other call routing services such as voice mail or call blocking.

Do Not Disturb (DND)-A call processing feature that prevents calls from alerting (ringing) to the caller. The DND may also block call alerting features such as conditional call forwarding, call waiting, and message notification.

DOA-Dead On Arrival

DOC-Dynamic Overload Control

DOC-Department Of Communications

DOCSIS-Data Over Cable Service Interface Specifications

Document Type Definition (DTD)-A language that describes the content of a standard generalized markup language (SGML). The DTD defines the structural component of a document as distinct from the actual data or content of the document.

DOD-Direct Outward Dialing

DOD Master Clock-Department of Defense Master Clock

DOE-Direct Order Entry

Domain Name-The unique name that identifies a specific Internet host. A domain name is associated with one (or more) IP addresses through the use of Domain Name Service (DNS).

Domain Name Server (DNS)-The data processing device that translates text and numeric names for an Internet addresses. A DNS uses a distributed database containing addresses of other DNS servers that may contain the Internet address.

Domestic Satellite (DOMSAT)-A satellite system used for domestic communications generally within the continental United States, Alaska, Hawaii, Puerto Rico, and the Virgin Islands.

Domestic Satellite Carrier-A common carrier that provides communications services within the United States via that owned or leased satellite facilities.

DOMSAT-Domestic Satellite

Dongle-Dongle Key

DOP-Dedicated Outside Plant

Dopant-An impurity, usually in the form of individual atoms, added in small quantities to a material to change some of its properties. Boron and phosphorous atoms are common dopants in semiconductors. Germanium and fluorine atoms are common fiber dopants. Erbium atoms are used in the most common type of dopant used in fiber amplifiers.

Dope-To add small amounts of impurities to a material so that some of its properties are changed. Semiconductors are doped in order to slightly change their electrical properties so that they can be used in electronic devices. Fibers are doped to change their index of refraction and other properties.

Doped Fiber-Fiber that has had a small amount of an impurity added to it to change its properties. Most impurities degrade fiber performance but some can improve it or even add new capabilities. For instance, germanium (which is similar to silicon) can be used to increase the refractive index of the fiber core while fluorine can decrease the index of the cladding. Erbium and praseodymium (rare earth elements) are used as dopants to enable stimulated emission in fibers so that they can be used as amplifiers.

Doped Fiber Amplifier (DFA)-A device used to amplify signals across a broad range of wavelengths in an optical network. The ability to amplify more than one wavelength at a time made Wavelength Division Multiplexing (WDM) practical.

A DFA consists of a pump laser, a length of doped fiber and often an optical isolator. The pump laser is used to generate metastable electrons in the doped fiber to prepare it for stimulated emission. The isolator prevents the pump laser light from entering the optical fiber used for transmission,

where it would cause noise in the signals or damage other optical equipment. When light enters the DFA, its photons cause stimulated emission of more photons of the same wavelength from the metastable electrons, thereby amplifying the signal (as well as any noise). See also EDFA.
Doping-The process of adding impurity atoms (dopants) to a material to change its properties. For instance, doping of semiconductors is done to change their electronic properties.
Doppler-An offset frequency that is the result of a moving antenna relative to a transmitted signal.
Doppler Shift-Doppler Effect
DOSA-Distributed Open Systems Architecture
DOSA-Direct Outward Station Access
Dot Crawl, Chroma Crawl-Cross-Luminance
Double Play-Double play refers to providing of two main services such as voice and data, data and video, or video and data on one network. For cable MSOs, this usually means building out the next generation network to DOCSIS 2.0 specifications, for Carriers this often means building out fibre or VDSL (very fast DSL) networks. Usually it is the larger MSOs and Telecom Carriers that roll out triple play services, and the advantage is that they can sign customers to a bundle of two services, thereby increasing revenue and customer loyalty.
DOV-Data Over Voice
Downlink-(1- Satellite) The portion of a communication link used for transmission of signals from a satellite to a mobile or fixed receiver. (2- cellular system) The radio link from the base station to the mobile station.
Download-The transfer of a program or of data from one computer to another.
Downstream-The direction of transmission, usually from a network to an end customer.
Downtime-An amount of time that a communication network or computer system is not available to users. Downtime usually occurs from hardware failure, software crashes, or operator errors.
DP-Dial Pulse
DP-DCS Proxy
DP-Call Detection Points
DPA-Demand Protocol Architecture
DPA-Digital Port Adapter
DPBX-Digital PBX
DPC-Destination Point Code
DPCCH-Dedicated Physical Control Channel
DPCM-Differential Pulse Code Modulation
DPDCH-Dedicated Physical Data Channel

dpi-Dots Per Inch
DPI/DPU-Direct Pick-Up Interference
DPM-Digital Picture Manipulator
DPM-Defects Per Million
DPMI-DOS Protected Mode Interface
DPMS-Display Power Manager Signaling
DPNSS-Digital Private Network Signaling System
DPO-Dedicated Pair Out
DPO-Digital Pulse Origination
DPS-Digital Passage Service
DPSK-Differential Phase Shift Keying
DPT-Digital Pulse Termination
DPU-Director of Public Utilities
DPU-Dynamic Path Update
DQDB-Distributed Queue Dual Bus
DQoS-Dynamic Quality Of Service
DQPSK-Differential Quadrature Phase Shift Keying
DRAM-Dynamic Random Access Memory
DRB-Digital Radio Broadcasting
DRCS-Digital Radio Concentrator System
Drift Space-The area in a Klystron tube in which electrons drift at their entering velocities and form electron bunches.
Driver-(1-general) An electronic circuit that supplies an isolated output to drive the input of another circuit. (2-fiber optic) The electric circuit that drives the light-emitting source, modulating it in accordance with an intelligence bearing signal. (3-software) A software module that controls an input/output port or external device, such as a keyboard or a monitor.
DRM-Digital Rights Management
Drop And Insert-D-1
Drop Wire-A drop wire is the wire or pairs of wires that are connected between a customer's premises and a nearby network line. Although the first drop wires were connected from a telephone pole to a building, drop wires can be buried or aerial.
Drop/Add-A microwave link or other form of communication system whereby signals are dropped to customers along the transmission path. Conversely, a channel can accept and add new signals at intermediate points.
Dropped Calls-Cellular telephone calls that are inadvertently disconnected from the system because of interference, inadequate coverage or lack of capacity.
DRP-Distribution and Replication Protocol
DRS-Digital Reference Signal
DRS-Digital Reconfiguration Service

DRU-Digital Remote Unit
DRx-Discontinuous Reception
Dry Cell-A type of disposable battery (primary cell) that uses a paste electrolyte instead of a liquid.
Dry Circuit-A communication line that does not provide electrical power with the information signal. The use of a dry line requires the terminating equipment to supply its own power.
DS-802.11 Distribution Service
DS-Distribution Service
DS-Distributed Single Layer Test Method
DS-0-Digital Service, Level 0
DS-1-Digital Service, Level 1
DS-1-Digital Signal 1
DS-1C-Digital Signal, level 1 Combined
DS-3-Digital Signal 3
DS-4-Digital Signal 4
DS0-Digital Signal 0
DS3 PLCP-DS3 Physical Layer Convergence Protocol
DSA-Dial Service Assistance
DSA-Distributed Systems Architecture
DSA-Directory System Agent
DSAP-Destination Service Access Point
DSAT-Digital Supervisory Audio Tones
DSB-Digital Sound Broadcasting
DSBSC-Double Sideband Suppressed Carrier
DSBTC-Double Sideband Transmission Carrier
DSC-Digital Selective Calling
DSCH-Downlink Shared Channel
DSCP-DiffServ Code Point
DSDC-Direct Service Dialing Capability
DSE-Distributed Single-Layer Embedded Test Method
DSE-Data Switching Exchange
DSF-Dispersion Shifted Fiber
DSI-Digital Speech Interpolation
DSL-Digital Subscriber Line
DSL bonding-The combining of multiple DSL communication lines to provide for higher data rate. For example, if eight 1.5 Mbps DSL lines are combined, the data transmission rate is 12 Mbps. This is also called inverse multiplexing.
DSL Bridge-A device that translates the protocol between a DSL modem and a DSL network. A DSL only translates the protocol and does not assign a separate address to the end user.
DSL Concentrator-An interface that allows more local loop telephone lines to share a digital subscriber line access multiplexer (DSLAM) that is allowed by the number of DSL modems that are installed in the DSLAM. The DSL concentrator acts as a mini-switch connecting the local loop to the DSL modem when data service is requested.
DSL Forum-A forum that was started in 1994 to assist manufacturers and service providers with the marketing and development of DSL products and services. The DSL forum was previously called the ADSL forum.
DSL Modem-Digital Subscriber Line Modem
DSLAM-Digital Subscriber Line Access Multiplexer
DSMCC-Digital Storage Media Command and Control
DSn-Digital Service Hierarchy
DSN-Distributed Systems Network
DSN-Defense Switch Network
DSO-Digital Signal Level Zero
DSOM-Distributed System Object Model
DSP-Domain-Specific Part
DSP-Digital Signal Processor
DSP Modem-Digital Signaling Processor Modem
DSR-Data Set Ready
DSS-Direct Station Select
DSS-Direct Station Selection Console
DSS-Direct Satellite Service
DSS-Digital Switched Service
DSS-Direct Station Select Intercom
DSS2 Setup-Digital Subscriber Signaling #2
DSSS-Direct Sequence Spread Spectrum
DSSS Mode-802.11 Direct Sequence Spread Spectrum
DST-Digital Signaling Tone
DSTB-Digital Set-Top Box
DSTN-Double Super Twisted Nematic
DSU-Digital Service Unit
DSU-Data Service Unit
DSU/CSU-Data Service Unit/Channel Service Unit
DSVD-Digital Simultaneous Voice And Data
DSX-Digital Signal Cross-Connect
DSx-Digital Signal Level
DSX-1-Digital Cross-Connect 1
DSX-2-Digital Cross-Connect 2
DSX-3-Digital Cross-Connect 3
DT-Direct Termination
DTAP-Direct Transfer Application Part
DTC-Digital Trunk Controller/ Digital Transmit Command
DTC-Digital Traffic Channel
DTD-Document Type Definition
DTE-Data Terminal Equipment
DTE-DCE Rate-Data Terminal Equipment/Data Communications Equipment Rate

DTH-Direct To Home
DTL-Designated Transit List
DTLU-Digital Trunk and Line Unit
DTM-Dynamic Synchronous Transfer Mode
DTMF-Dual Tone Multi-Frequency
DTMF Cut Through-A process that allows the interruption of a service (such as an announcement message) when the system detects a DTMF tone that a caller has pressed. This feature allows for better control of interactive voice response (IVR) and auto attendant systems.
DTMF Decoder-A device or process that converts dual tone multi-frequency (DTMF) signals into another form (such as data digits).
DTO-Direct Termination Overflow
DTR-Data Terminal Ready
DTS-Digital Termination System
DTS-Dial Tone Speed
DTU-Digital Test Unit
DTV-Digital Television
DTx-Discontinuous Transmission
DUA-Directory User Agent
DUAL-Distributed Update Algorithm
Dual Attached Concentrators (DAC)-A data communication device that is used in a FDDI network that services multiple data communication devices and can automatically reconfigure the routing of data in the event of a break in the fiber line. The DAC normally routes both directions of the fiber ring (clockwise and counterclockwise) through its connections. The DAC inserts and extracts data for multiple data terminals (or other communication device) that use the FDDI network.
Dual Attached Stations (DAS)-A data communication device that is used in a FDDI network that can automatically reconfigure the routing of data in the event of a break in the fiber line. The DAS normally routes both directions of the fiber ring (clockwise and counterclockwise) through its connections. It inserts and extracts data from a data terminal (or other communication device) with the FDDI network.
Dual Band-A mobile radio that is capable of accessing radio channels on two bands of frequencies (such as cellular and PCS).
Dual Fiber Cable-A fiber cable that contains two separate single-fiber cables.
Dual Mode Cellular-Typically the combination of analog cellular and a digital cellular functionality into one mobile telephone. It automatically works in both analog and digital cellular systems, where digital is available.

Dual Seizure-The condition which occurs when two exchanges (switches) attempt to seize the same circuit at approximately the same time. (See also: glare.)
Dual Tone Multi-Frequency (DTMF)-DTMF signaling is a means of transferring information from a user to the telephone network through the use of in-band audio tones. Each digit of information is assigned a simultaneous combination of one of a lower group of frequencies and one of a higher group of frequencies to represent each digit or character. There are 8 tones that are capable of producing 16 combinations; 0-9, *, #, A-D. The letters A-D are normally used for non-traditional systems (such as the military telephone systems).
This diagram shows how dual tone multi-frequency (DTMF) tones can be used to send dialing information from a telephone to a telephone system. There are 8 different frequencies that can be combined to represent 16 keys. The keys A-D are not usually included on standard telephone sets. To represent each button, two tones are combined. In this example, the button 3 is pressed, followed by a pause, then button 2 is depressed. Button 3 is represented by the combined tones 1477 Hz and 697 Hz. Button 2 is represented by the combined tones 1336 Hz and 697 Hz. To determine if the user is finished dialing, a timer is used. When the user has stopped dialing, the digits can be sent to the call processing section of the telephone system to initiate the call.

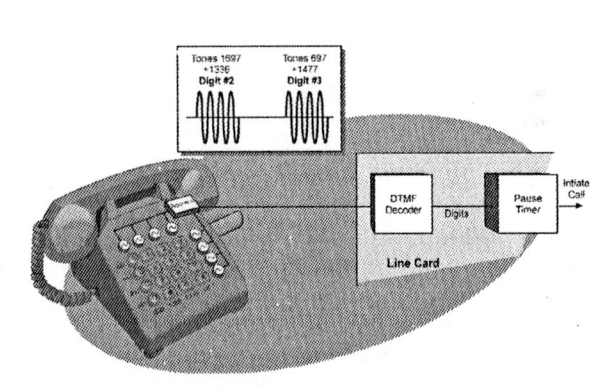

DTMF Dial Operation

DualCam

DualCam-A digital camera that can take pictures and be used as a WebCam to provide still or continuous images to the Internet.

This photograph shows the Clicksmart 510 DualCam that can capture and store images as a camera and also be connected to a personal computer to send images to the Internet and work as a WebCam for video conferencing. The DualCam contains removable memory cards that allows it to operate as a portable digital camera. It can capture video images and sound clips that can be downloaded to the computer for printing or transfer through the Internet.

DualCam
Source: Logitech

DUART-Dual Universal Asynchronous Receiver Transmitter

Dub-(1-copy) To copy information on one videotape or audiotape to another. (2-a copy) A copy of a videotape or audiotape.

Duct-(1-general) A pipe or conduit, installed underground or in a building, whose purpose is to protect the cables installed therein. (2-height) The height above the Earth of the lower boundary of an elevated propagation duct. (3-metal floor wiring) One of various proprietary schemes of cable ducting to provide flexibility in equipment installation in office areas. (4-nest) A number of cable ducts provided for and laid in one trench. (5-propagation) A layer of cold air under warm air, experienced in some areas, that causes microwave signals to propagate further than normally possible. (6-surface) A radio duct whose lower boundary is the surface of the earth. (7-thickness) The difference in height between the upper and lower boundaries of a tropospheric radio duct.

Duct Liner-A small diameter sub-duct into which fiber optic cable is placed. Duct liners usually are made from polyethylene or polyvinyl chloride and may have a smooth or corrugated wall.

DUE-Dumb User Error

Due Date - Billing-The date by which payment due must be received before the collections process is triggered. This date is typically dependent upon the invoice date, and is typically different for each cycle.

Dumb Switch-A term commonly applied to a time division multiplexed (TDM) telecommunications switch that only can connect lines to each other based on instructions a controller or other computer.

Dumb Terminal-A computer display terminal that serves as a slave to a host computer. A dumb terminal has a keyboard for data entry and a video display, but no computing power of its own.

Dummy Burst (DB)-A burst of information that contains no user data information. Dummy bursts are used to fill a time slot or frame with information to ensure a continuous flow of data is being sent to a channel or time slot.

DUN-Dial-Up Network

Dunning-Unique treatment of customers for the purpose of collection of service charges or account balances.

DUP-Data User Part

Duplex-The transmission of voice, data signals that allows simultaneous 2-way communication.

This figure shows the different types of duplex systems. Frequency division duplex is a communications channel that allows the transmission of information in both directions (not necessarily at the same time) via separate bands (frequency division). Time division duplex (TDD) is a process of allowing two-way communications between two devices by time-sharing. When using TDD, one device transmits (device 1), the other device listens (device 2) for a short period of time. After the transmission is complete, the devices reverse their role so device 1 becomes a receiver and device 2 becomes a transmitter. The process continually repeats itself so data appears to flow in both directions simultaneously. It is also possible to combine FDD and TDD.

Dynamic Host Configuration Protocol (DHCP)

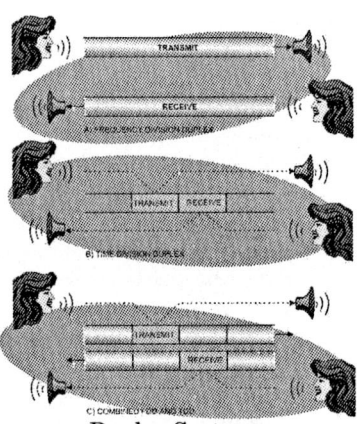

Duplex Systems

Duplexer-A combined filter device that permits a transmitter and receiver to share the same antenna assembly by using filters with different frequency bands. The use of a duplexer prevents transmitter power output to the antenna from transferring to sensitive receiver assembly.

This diagram shows how a duplexer allows one antenna to be connected to a transmitter and receiver at the same time. This example shows that the transmitter and receiver use different frequencies and the duplex filter (duplexer) contains two bandpass filter assemblies. This diagram shows that the bandpass filter allows the transmitter frequency to reach the antenna and that the receiver filter blocks the high-energy transmitter frequency from entering into the receiver. However, the duplexer's receiver bandpass filter does allow for receiver band frequencies (receiver channels) to enter into the receiver.

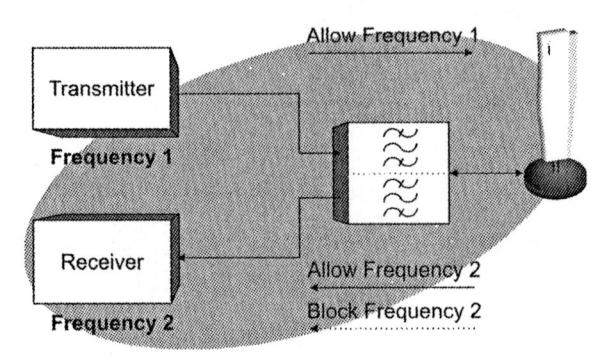

Duplexer Operation

DVB-Digital Video Broadcast
DVB-The Digital Video Broadcasting Group
DVB-MHP-Digital Video Broadcasting—Multimedia Home Products protocol
DVC-Digital Video Compression
DVCC-Digital Verification Color Code
DVD-Digital Video Disc
DVD+RW-Digital Video Disk ReWriteable
DVD-ROM-Digital Video Disc-ROM
DVD2-Digital Versatile Disk, 2nd Generation
DVI-Digital Video Interactive
DVMRP-Distance Vector Multicast Routing Protocol
DVR-Digital Video Recorder
DVRN-Dense Virtual Routed Networking
DWDM-Dense Wave Division Multiplexing
DWG-Drilled Well Ground
DWI-Data Warehousing Institute
DWMT-Discrete Wavelet Multi-Tone
DX-Duplex Signaling
DX Signaling-Direct Current Signaling
DXI-Data Exchange Interface
Dynamic Host Configuration Protocol (DHCP)-A process that dynamically assigns an Internet Protocol (IP) address from a server to clients on an as needed basis. The IP addresses are owned or controlled by the server and are stored in a pool of available addresses. When the DHCP server senses a client needs an IP address (e.g. when a computer boots up in a network), it assigned one of the IP addresses available in the pool.

This figure shows how a computer uses DHCP to obtain a temporary IP address when it requires an Internet communication session. In this example, the computer requests a connection with an Internet service provider (ISP) via a modem that is connected to a universal serial bus (USB) line. When the Internet service provider receives the request for connection, it assigns an IP address from is list of available IP addresses. The computer will then use this IP address for all of is communications with the Internet until it disconnects the connection to the ISP.

Dynamic IP Addressing

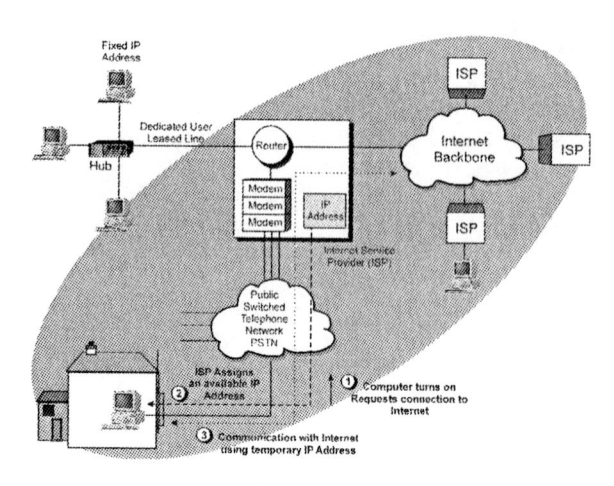

Dynamic Host Configuration Protocol (DHCP) Operation

Dynamic IP Addressing-A process of assigning and Internet Protocol address to a client (usually and end user's computer) on an as needed basis. Dynamic addressing is used to provide an enhanced level of security (no predefined address to use for hackers) and to conserve on the number of IP addresses required by a server. Also see DHCP.

Dynamic Link Library (DLL)-A feature of an operating system (e.g. Windows) that allow executable software code modules to be loaded on demand and linked when the applications begin to operate (run time). This enables the software code to access the latest parameters in its related software applications. DLL files are unloaded when they are no longer needed (when the application is closed).

Each DLL applications that is initiated is copied into the working memory of the computer. A key benefit of using dynamic linked libraries is that the executable program files are not as large because the frequently used routines can be put into DLL files.

Dynamic Load Balancing-A process that is used to evenly distribute incoming calls to customer service agents. Dynamic load balancing can be implemented in Automatic Call Distributor (ACD) systems.

Dynamic Power Control-The combination of self regulated power control and system power control. Self controlled power level is performed by sensing the received power level and increasing the transmitter power level as the received level decreases due to increased distanced from the base station transmitter. System controlled power control is a process of controlling the power level in a cellular system where the base station receiver monitors the received signal strength of a mobile telephone and control messages are transmitted from the base station to the mobile telephone commanding it to raise and lower its transmitter power level as necessary to maintain a good radio communications link.

Dynamic Random Access Memory (DRAM)-A type of memory that temporarily stores information and requires continual refreshing of the information. DRAM information is completely lost when the power and refreshing is removed.

Dynamic Range-(1 analog system) The range or extremes in amplitude, from the lowest to the highest pints that a system or radio is expected to operate. The dynamic range is typically expressed in decibels against a reference level. (2 - digital) The number of bits used to define the range of a given signal.

Dynamic Time Alignment-Dynamic time alignment is a technique that allows a radio system base station to receive transmitted signals from mobile radios in an exact time slot, even though not all mobile telephones are the same distance from the base station. Time alignment keeps different mobile radio's transmit bursts from colliding or overlapping. Dynamic time alignment is necessary because subscribers are moving, and their radio waves' arrival time at the base station depends on their changing distance from the base station. The greater the distance, the more delay in the signal's arrival time.

This diagram shows how the relative transmitter timing in a mobile radio (relative to the received signal) is dynamically adjusted to account for the combined receive and transmit delays as the mobile radio is located at different distances from the base station antenna. In this example, the mobile telephone uses a received burst to determine when its burst transmission should start. As the mobile radio moves away from the tower, the transmission time increase and this causes the transmitted bursts to slip outside its time slot when it is received at the base station (possibly causing overlap to transmissions from other radios.) When the base station receiver detects the change in slot period reception, it sends commands to the mobile telephone to advance its relative transmission time as it moves away from the base station and to be retarded as it moves closer.

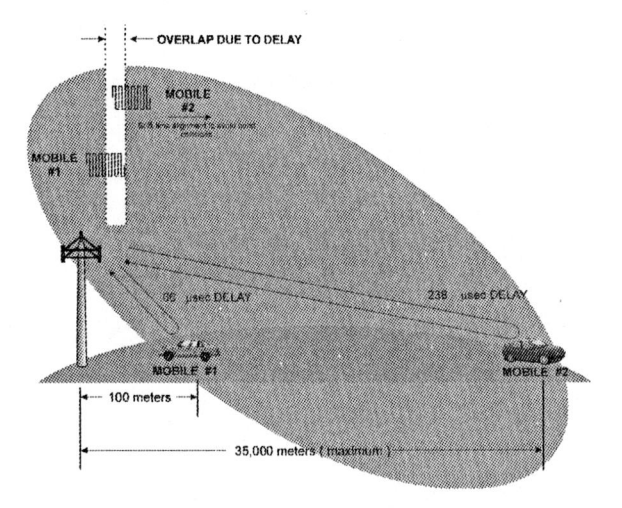

Dynamic Time Alignment

E

E-Exa
E & M Leads-The wire leads at a communication line interface that are used to control the connection setup, information transfer, and disconnection. These leads only transfer supervisory signals.

This figure shows how E & M signaling uses separate lines to send control messages between switching systems. In this example, two PBX systems are connected together using voice lines and E & M signaling lines. When a PBX system desires to use a communication line, it changes the voltage of the M line to alert the other PBX's E line. When the other PBX detects this change of state, additional control messages may follow to allow the switching units to communicate with each other and route calls to the correct extension or port.

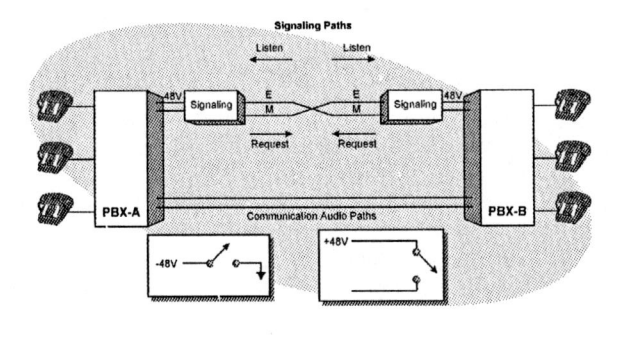

E & M Signaling

E Channel-Echo Channel
E Mail Gateway-Electronic Mail Gateway
E Purse-Electronic Purse
E&M-Ear and Mouth Signaling
E-911 Service-Enhanced 911 Service
E-BCCH-Extended Broadcast Channel
E-Carrier-A digital carrier system adopted by the international telecommunications union (ITU). Each digital signaling level supports several 64 kbps (DS0) channels. E-carrier was initially used in Europe and now is used throughout the rest of the world. The different digital signaling levels include;
- DS-1, 2.048 Mbps with 30 channels + 2 control channels
- DS-2, 8.448 Mbps with 120 channels
- DS-3, 34.368 Mbps with 480 channels
- DS-4, 139.264 Mbps with 1920 channels
- DS-5, 565.148 Mbps with 7680 channels

E-Check-Electronic Check
E-IDE-Enhanced IDE
E-Link-Extended Link
E-Mail Notification-E-mail notification is the process of sending a subscriber an email message that notifies them that information has changed (such as a new voicemail).
E-Nose-Electronic Nose
E-Rate-Electronic Rate
E-Stamp-Electronic Stamp
E.164 International Public Telecommunications Numbering Plan-The International Telecommunications Union (ITU), a division of the United Nations, has defined a world numbering plan recommendation, "E.164." The E.164 numbering plan defines the use of a country code (CC), national destination code (NDC), and subscriber number (SN) for telephone numbering. The CC consists of one, two or three digits. The first digit identifies the world zone. The number of digits used for telephone numbers throughout the world varies. However, no portion of a telephone number can exceed 15 digits. There are several "E" series of ITU numbering recommendations that assist in providing unique identifying numbers for telephone devices around the world.

This diagram shows the world (telephone) numbering plan recommendation, "E.164" developed by the International Telecommunications Union (ITU). This diagram shows the numbering plan divides a telephone number into a country code (CC), national destination code (NDC), and subscriber number (SN) for telephone numbering. The CC consists of one, two or three digits and the first digit identifies the world zone. This diagram shows that the local number can be divided into an exchange code (end office switch identifier) and a port (or extension) code.

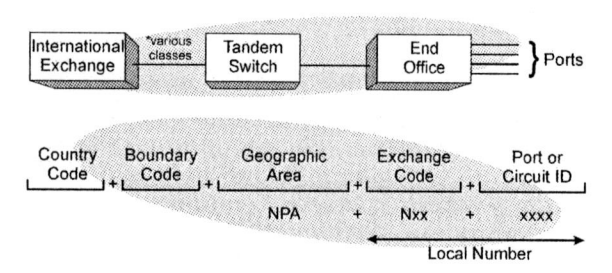

E.164 Telephone Numbering System

E/O-Electronic to Optical Conversion
E1-A communication line that was developed by European standards that multiplexes thirty voice channels and two control channels onto a single communication line. The E1 line uses 256 bit frames and transmitted at 2.048 Mbps.
E411-Enhanced 411
E911-Enhanced 911
EA-Extended Addressing
EAD-Engineering Administrative Data Acquisition System
EAEO-End Access End Office
EAN-European Article Numbering
Ear and Mouth Signaling (E&M)-A method of communication trunk signaling over a path separate line from the transmission path. The two leads of this separate path are called the ear (receive) and the mouth (transmit). As a memory aid, the leads are analogous, direction wise, to an ear and a mouth. E&M signaling also is used in special services applications. (See also: loop signaling.)
Although E&M signaling is commonly refer to as "ear & mouth" or "recEive and transMit", its origin actually comes from the term earth and magnet. Earth represents electrical ground and magnet represents the electromagnet used to generate tone. E&M signaling defines a trunk circuit side and a signaling unit side for each connection similar to the data circuit-terminating equipment (DCE) and data terminal equipment (DTE) reference type. Usually the PBX is the trunk circuit side and the telco, End Office (EO), or channel-bank is the signaling unit side.
EARN-European Academic Research Network

Earnings Before Interest, Taxes, Depreciation, and Amortization (EBITDA)-EBITDA is used by finance and investment analysts to gain a more accurate appraisal of the true net worth and/or cash flow of a company. EBITDA reflects the ability of a business to achieve profitability without the influence of financial positions that may vary between different companies.
EAROM-Electrically Alterable Read-Only Memory
EAS-Extended Area Service
EAS-Emergency Alert System
EATP/NPP-Equal Access Transition/Non-Premium Plan
EAX-Electronic Automatic Exchange
EB-Electronic Bonding
EB-Exabyte
Eb/No-The ratio of bit energy to a noise signal.
EBCDIC-Extended Binary Coded Decimal Interexchange Code
EBCDIC-Enhanced Binary Code For Digital Information Communication
EBD-Effective Bill Date
EBITDA-Earnings Before Interest, Taxes, Depreciation, and Amortization
EBONE-European Backbone
EBPP-Electronic Bill Presentation and Payment
EBS-End System Byte packet
EBS-Enhanced Business Service
EBS-Emergency Broadcast System
EBTS-Enhanced Base Transceiver System
EBU-European Broadcasting Union
EC-Exchange Carrier
EC-European Community
EC-Energy Communications
ECA-Exchange Carrier Association
eCash-Electronic Cash
ECB-Electronic Codebook
ECC-Error Correction Code
ECC-Error Correcting Code
ECC-Elliptic Curve Cryptography
ECC-Electronic Common Control Switches
ECC-Exchange Carrier Code
ECC RAM-Error-correcting Code Memory
ECCKT-Exchange Company Circuit
ECCS-Economic CCS
ECF-Embedded Code Formatting
ECH-Enhanced Call Handling
Echo-A type of transmission impairment in which a signal is reflected back to the originating source. In the transmission, the reflected signal often is attenuated and delayed, resulting in an echo.

Echo Canceling-Echo cancellation is a process of extracting an original transmitted signal from the received signal that contains one or more delayed signals (copies of the original signal). Echoes may be created in a baseband or broadband signal. When echoes occur on an audio baseband signal, it is usually through acoustic feedback where some of the audio signal transferring from a speaker into a microphone. When echoes occur on a broadband signal, it is usually the result of the same signal (such as a radio signal) that travels on different paths to reach its destination. In either case, echoed signals cause distortion and may be removed by performing via advanced signal analysis and filtering.

This diagram shows how an echo canceling system can remove unwanted echo signals in telecommunication circuits. In this diagram, the sources of echo signals include electrical mismatches in hybrid combining circuits and signal leakage in end user devices. The echo cancellation system uses a process that senses a complex audio signal and predicts the distortions that are produced by echo signals. This allows it to create an echo canceling signal (inverse signal of the echo signal) that is combined with the input signal so the distortion effects of the echo canceling can be removed

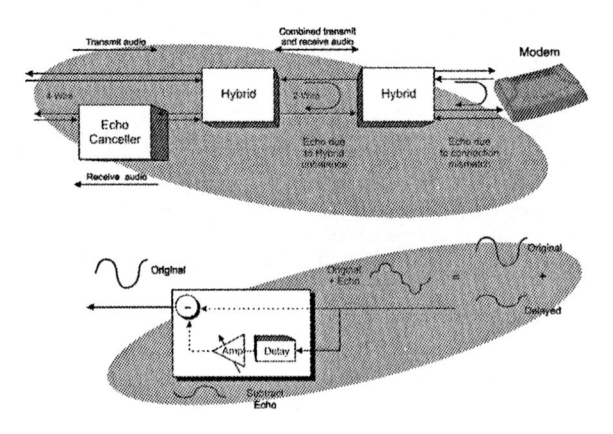

Echo Cancelling System

Echo Canceller-A signal processing device or circuit that reduces the effects of echo signals. This is performed by calculating an estimate of the expected echo signal and subtracting this estimate from the signal in which the echo appears. Echo cancellers are essential for communication systems that have long signal processing delays such as long distance voice, satellite and digital mobile telephony lines.

Echo Return Loss (ERL)-The amount of attenuation (loss) of a signal that is returned (echoed) back to its sender. This is how much lower the echo audio sounds compared to the original audio that was sent.

Echo Return Loss Enhancement (ERLE)-The additional amount of attenuation (loss) of a signal that is returned (echoed) back to its sender as a result of using an echo canceller.

ECL-Emitter Coupled Logic

ECM-Error Correction Mode

ECMA-European Computer Manufactures Association

ECMEA-Enhanced Cellular Messaging Encryption Algorithm

ECN-Electronic Communications Network

Ecommerce-Electronic Commerce

Economic Bypass-A form of bypass, the cost of which is lower than that of an equivalent telephone company service.

Economic Conditions Churn-Type of voluntary, incidental churn that occurs when customers loose their jobs or face other economic crises, and must terminate service.

Economic Geographic Spectral Efficiency-A measurement characterizing a particular modulation and coding method and its economics of implementation that describes how much information can be transferred in a given bandwidth within a cellular radio system having reuse of the same radio carrier frequency at the closest permitted reuse cells, per dollar (or other money unit) of system cost. This is often given as bits per second per Hertz per square km of service area per dollar of cost, or alternatively as bits per second per Hertz per cell per dollar of cost. The cost may either be equivalent total capital cost, or equivalent amortized cost. Regardless of how the cost figure is expressed, they should include both initial equipment and real estate and operating costs in order to be comprehensive. This is the overall figure of merit that can be used to made business decisions between different cellular base infrastructure systems. (See also: Geographic Spectral Efficiency or Spectral Efficiency)

ECP-Enhanced Call Processing

ECP-Extended Capabilities Port

ECPA-Electronic Communications Privacy Act
ECS-Electronic Commerce Services
ECS-European Communication Satellite
ECSA-Exchange Carriers Standards Association
ECTF-Enterprise Computer Telephony Forum
ED-End Delimiter
ED-CallED Party
EDA-Electronic Directory Assistance
EDACS-Enhanced Digital Access Communications System
EDDA-European Digital Dealers Association
EDE-Encrypt-Decrypt-Encrypt
EDFA-Erbium Doped Fiber Amplifier
EDGAR-Electronic Data Gathering Archiving and Retrieval
EDGE-Enhanced Data Rates For Global Evolution
EDGE-Enhanced Data Rates For GSM Evolution
EDGE Compact-Enhanced Data Rates For Global Evolution Compact
Edge Router-A device used to connect a private or enterprise network to a Service Provider's network. These devices are usually capable of more advanced routing protocols, such as BGP-4 and OSPF, and usually provide services such as tunneling, authentication, filtering, billing, traffic shaping and rate policing and network address translation. Depending on the service provider, the device may be owned and managed by the service provider or by the customer.
Edge Switch-A switch located at the boundary between two networks. The edge switch usually provides access an end users system to a carriers core or backbone network. Edge switches are sometimes called access nodes and service nodes.
EDI-Electronic Data Interchange
EDIA-Electronic Data Interchange Association
EDL-Edit Decision List
EDM-Electronic Document Management
EDO DRAM-Extended Data Output Dynamic Random Access Memory
EDO RAM-Extended Data Out RAM
EDP-Electronic Data Processing
EDRAM-Enhanced Dynamic Random Access Memory
EDSL-Extended Digital Subscriber Line
EDSS1-European Digital Subscriber Signaling no. 1
EDSS1-European Digital Signaling System no. 1
EDTV-Extended Definition Television
EE-End To End

EEHLLAPI-Entry Emulator High Level Language Applications Programming Interface
EEHO-Either End Hop Off
EEL-Enhanced Extended Link
EEMS-Enhanced Expanded Memory Specification
EEPROM-Electrically Erasable Programmable Read Only Memory
EES-Escrow Encryption Standard
EF-Entrance Facility
EF&I-Engineer Furnish And Install
eFAX-Electronic Facsimile
EFCI-Explicit Forward Congestion Indication
EFD-Event Forwarding Discriminator
EFF-Electronic Frontier Foundation
Effective Resistance-The increased resistance of a conductor to an alternating current resulting from skin effect, relative to the direct-current resistance of the conductor. Higher frequencies tend to travel only on the outer skin of the conductor, whereas dc flows uniformly through the entire area.
Effective Speed Of Transmission-The rate at which information is processed by a transmission facility, expressed as the average rate over some significant time interval. The quantity usually is stated as the average number of characters or bits per unit time.
Effective Transmission-A system of rating transmission performance based upon subjective tests of repetition rates.
Effective Two-Wire Channel-A channel that enables simultaneous transmission in both directions over the same two-wire path. Effective two-wire channels can be terminated with two or four-wire interfaces.
Effects-(1-video) The process of combining two or more video images to create a new composite image. (2-audio) The process of adding special audio elements, such as echo, to a program to enhance the overall aural impact.
Effects Memory-The capability of a video production switcher to store and recall effects created on the system through the use of computer control techniques.
Effects System-The portion of a video production switcher that performs mixes, wipes, and cuts between background and/or special effects key video signals.
EFM-Eight To Fourteen Modulation
EFM-Ethernet in the First Mile
EFP-Electronic Field Production
EFR-Enhanced Full Rate

EFRC-Extended Flat Rate Calling
EFS-Error Free Seconds
EFT-Electronic Funds Transfer
EGP-External Gateway Protocol
Egrep-Extended Global Regular Expression Pattern
Egress-(1-signal) A process where a strong signal inside a communication system leaves the transmission line or system. (2-network) The process of data or signals exiting from a communication network.
EHF-Extremely High Frequency
EHR-Enhanced Half Rate
EI-Emergency Interrupt
EIA-Electronic Industries Alliance
EIA-Electronic Industries Association
EIA-553 (AMPS)-The specification for the advanced mobile phone service (AMPS) cellular system.
EIDE-Enhanced Integrated Drive Electronics
EIDQ-European International Directory Enquiries Group
Eight To Fourteen Modulation (EFM)-A structured modulation code that expands blocks of 8 bits into blocks of 14 bits. This type of modulation allows for greater recording density. EFM is the code employed in compact disc (CD) systems.
EIGRP-Enhanced Interior Gateway Protocol
EIGRP-Interior Gateway Routing Protocol
EIP-Early Implementers Program
EIR-Excess Information Rate
EIR-Equipment Identity Register
EIRP-Equivalent Isotropically Radiated Power
EIRP-Effective Isotropic Radiated Power
EISA-Extended Industry Standard Architecture
EKTS-Electronic Key Telephone System
ELAN-Emulated Local Area Network
Elapsed Call Timer-A time display that indicates that amount of time that has elapsed since a call was initiated or received.
Elastic Buffer-A storage device that has capacity to store data and add variable amounts of delay before retransmitting the data. An elastic buffer smoothes out the variable transmission delay that digital audio signals may experience when transmitted through a packet switching network (such as the Internet).
ELEC-Enterprise Local Exchange Carrier
Electro-Absorption Semiconductor Modulator-A device used to modulate an optical signal. An electro-absorption modulator is essentially a laser that works in reverse. When this modulator is turned on, it absorbs photons and creates energized electrons within the modulator. When it is not on, it is transparent to light.
Electro-Optic Effect-A physical property of some materials. The index of refraction of such materials changes when an electric field is applied across it. Lithium niobate is one material that has this property. See also External Modulator and Lithium Niobate Modulator.
Electro-Optic Modulator-A device used to modulate an optical signal. An electro-optic modulator is made out of a material that changes its index of refraction when an electric field is applied across it. The device splits incoming light into two paths of identical length. If no field is applied to the device, the two lightwaves travel the two paths at the same speed and then undergo constructive interference when they meet at the output of the device. If a field is applied to one of those paths, the index of refraction changes and the light travels more slowly. The device is designed such that the lightwave passing through the path with the new index of refraction is 180 degrees out of phase with the unaffected lightwave and destructive interference takes place, canceling out the light.
Electro-Optic Waveguide Modulator-An optical waveguide device used to modulate an optical signal. The device passes or blocks light according to an applied field generated by the information signal that is to be transmitted by the light. An electro-optic modulator is made out of a material that changes its index of refraction when an electric field is applied across it. The device splits incoming light into two paths of identical length. If no field is applied to the device, the two lightwaves travel the two paths at the same speed and then undergo constructive interference when they meet at the output of the device. If a field is applied to one of those paths, the index of refraction changes and the light travels more slowly. Typically, opposite fields are applied to both paths, with the result that the two lightwaves undergo destructive interference at the output of the device. Polarization effects add complications. See also Lithium Niobate Modulator and Electro-Absorption Semiconductor Modulator.
Electromagnetic Coupling-The process of coupling two signals using electromagnetic energy.
Electromagnetic Energy-Energy with both electronic and magnetic components. In telecommunications, this energy is typically in the form of a

Electromagnetic Interference (EMI)

wave of varying electronic and magnetic fields. The electronic component is usually established by applying a varying voltage to a conductor or semi-conductor. The resulting motion of electrons generates a magnetic field.

Electromagnetic Interference (EMI)- Electromagnetic interference (EMI) is the radiation (transmission) of undesirable electro- magnetic waves from an electronic circuit or device. These EMI waves may interfere with the normal operation of other circuits or devices.

Electromagnetic Spectrum-The range of frequencies of electromagnetic waves. Radio and microwave are at the low frequency, long wavelength end of the spectrum, followed by the infrared light used in optical networks. Visible light represents a small portion of the spectrum at a higher frequency than infrared. Ultraviolet light is higher in frequency and shorter in wavelength than visible light. X-rays and gamma rays are still higher frequency portions of the electromagnetic spectrum.

Electromagnetic Wave-A periodic variation in electric and magnetic fields that travels through a medium, forming a wave. For instance, when a voltage is applied across a conductor, its electrons move. This electron acceleration results in an electric field, which induces a magnetic field. If the voltage is varied systematically, the direction and magnitude of both electric and magnetic fields will also vary systematically. The result is the apparent propagation of a wave through the conductor. An electromagnetic wave travels at the speed of light in a vacuum, or 186,300 miles per second (300,000,000 meters per second) and more slowly through other media.

Electronic Commerce (Ecommerce)-A shopping medium that uses a telecommunications network to present products and process orders. Also called "virtual mall."

Electronic Industries Association (EIA)-A trade association that develops standards for electronic components and systems and represents manufacturers of electronic systems and parts.

Electronic Key Telephone System (EKTS) -A small customer premises electronic telephone switching system with station sets designed to resemble station sets of the earlier electro-mechanical KTS systems, having, for example, a distinct lighted pushbutton to select each individual outside line or intercom link, etc. Another distinction of EKTS vs. KTS is that EKTS wiring uses only 4-wire or 6-wire cable connections between the station set and the central electronic key service unit. Because of this EKTS systems are sometimes called "skinny wire" key telephone systems, in contrast to the 50-wire cables used with older electro-mechanical KTS wiring.

Electronic Mail (Email)-A process of sending messages in electronic form. These messages are usually in text form. However, they can also include images and video clips.

Electronic Numerical Integrator and Computer (ENIAC)-An early computer system that was first built in 1944.

Electronic Programming Guide (EPG)-EPGs are an interface (portal) that allows a customer to preview and select from possible list of available content media. EPGs can vary from simple program selection to interactive filters that dynamically allow the user to filter through program guides by theme, time period, or other criteria.

Electronic Serial Number (ESN)-The electronic serial number of a cellular phone, which is embedded within the signal and transmitted every time a cellular call is placed.

Electronic Switching System (ESS)-A system that can connect incoming and outgoing digital lines together through the use of temporary memory locations. For an ESS system, a computer controls the assignment, storage, and retrieval of memory locations so that a portion of an incoming line (time slot) can be stored in temporary memory and retrieved for insertion to an outgoing line.

Electronic to Optical Conversion (E/O)-The process of changing an electronic signal into an optical one. Optical signals, or lightwaves, are used to transport information rapidly. However, many critical network processing steps can only be performed on electronic signals. After the electronic processing has been completed, the signal is changed back into an optical signal in order to be transported in the optical network. See also Optical to Electronic Conversion (O/E).

Electronic Voice Mail (EVM)-A system that stores messages in electronic form (usually digital audio) that can be saved, moved, and retrieved by the mailbox owner.

Electrostatic Coupling-The process of coupling two signals using electrostatic energy.

Element Management System (EMS)-EMS is becoming more synonymous with Network Management Station (NMS) as telephony and data

service providers come together to manage heterogeneous networks. EMS is used mainly in the telephony networks for provisioning versus the data networks.
Elementary Particle-Any of the particles of matter that cannot be further subdivided.
ELF-Extremely Low Frequency
Ellippso-A LEO satellite system that uses CDMA technology.
ELSU-Ethernet LAN Service Unit
ELT-Emergency Locator Transmitter
ELV-Expendable Launch Vehicle
EM-Element Manager
EM-End Of Medium
EMA-Electronic Messaging Association
Email-Electronic Mail
emall-Electronic Mall
EMBARC-Electronic Mail Broadcast to A Roaming Computer
Embed-To insert all of an information element or object directly into another information element or object.
Embedded Operations Channel (EOC)-A communications channel that is designed to be part of a communications circuit. The EOC allows for commands (usually system step, change and test commands) to be transferred without the need to interfere with an established communications link it is paired with.
EMC-Electromagnetic Compatibility
EME-Economic Modular Evaluation
Emergency Access to Attendant-A feature that allows calls to be directly routed to an attendant in an emergency situation. Emergency access usually requires an access code and it will interrupt the attendant from lower priority calls or services.
Emergency Ringback-A call return feature enables a public service answering position (PSAP) or emergency service attendant to call back to a caller who has terminated the call or left the phone off-hook. The emergency ringback service may produce a loud "howling" on the phone to get the attention of the caller.
EMF-Electromotive Force
EMF-Electromagnetic Force
EMI-Electromagnetic Interference
Emitter Coupled Logic (ECL)-A family of logic circuits in which transistor pairs are coupled together by their emitters. Such logic circuits can be made to operate at high frequencies.
EML-Element Management Layer

EMLPP-Enhanced Multi-Level Priority And Pre-Emption
EMM-Expanded Memory Manager
EMM-Entitlement Management Message Stream
EMP-Electromagnetic Pulse
Empirical-A conclusion that is based not on pure theory, but on practical and experimental work.
EMR-Exchange Message Record
EMR-Electromagnetic Radiation
EMRP-Equivalent Monopole Radiated Power
EMS-Enterprise Messaging Server
EMS-Expanded Memory Specification
EMS-Element Management System
EMT-Electrical Metal Tubing
ENA-Enterprise Network Accounting
Enablement (Patent)-Enablement means that an application for patent includes a description of the invention which sufficient for a person of ordinary skill in the art to construct and use the claimed invention without undue experimentation. Most countries have a requirement that an applicant for patent provide an enabling description.
Encapsulating Bridge-A bridge that encapsulates LAN frames for transmission across a backbone where the bridges networks are dissimilar.
Encapsulation Bridge-A bridge that takes the source data packet from one network and encapsulates the entire packet frame (address and data) into the data portion of the frame for the network the data is transitioning into.
Encipherment-The process of encrypting data with a cipher key.
Encoded Chroma Key-A chroma key that uses an encoded video signal instead of separate RGB or Y, Cr, Cb signals for deriving the key.
Encoding-(1-modulation) For digital modulation systems such as pulse code modulation (PCM), encoding is the process of converting the magnitude of a sample to a 7-bit code (DI-type), or an 8-bit code (D2- or later). Encoding is a transmit function. (See also: decoding, quantizing, sampling.) (2-data) In the context of data compression, as a gerund, the process of compressing the source data into compressed form. As a noun, the compressed data which is the result of the encoding process.
Encoding Tools-Software and/or hardware tools to create compressed digital media from source media. Encoding tools may also function as authoring tools.

Encryption

Encryption-Encryption is a process of a protecting voice or data information from being obtained by unauthorized users. Encryption involves the use of a data processing algorithm (formula program) that uses one or more secret keys that both the sender and receiver of the information use to encrypt and decrypt the information. Without the encryption algorithm and key(s), unauthorized listeners cannot decode the message. When the encryption and decryption keys are the same, the encryption process is known as symmetrical encryption. When different encryption and decryption keys are used (such as in a public encryption system), the process is known as asymmetrical encryption.

This diagram shows how encryption can convert non-secure information (clear text) into a format (cyphertext) that is difficult or impossible for a recipient to understand without the proper decoding keys. In this example, data is provided to an encryption processing assembly that modifies the data signal using an encryption key. This diagram also shows that additional (optional) information such as a frame count or random number may be used along with the encryption key to provide better information encryption protection.

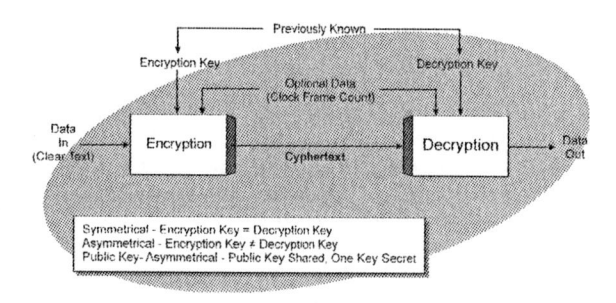

Encryption Operation

End Of Message (EOM)-A frame that indicates the end of a message that has been transmitting.

End of Procedure frame (EOP)-A frame that indicates the end of a procedure that has been running.

End Office (EO)-The end office interconnects calls between local customers and the telephone network. Each end office switch can usually supply service up to 10,000 customers. In larger areas (such as a city), established LECs may have several EO switches. The EO switches are interconnected using a higher level tandem switch. If is a significant amount of calls regularly processed between end offices, they may be directly connected via high-speed communication lines (trunks).

End Office Code-A 3 digit code that is assigned to a local end office. This code is used for billing information identification and routing within systems. N represents any number from 2 through 9, and X represents any number from 0 through 9. This code also is known as an Central Office Code.

End Station-A PC or host that allows communications between ATM end stations and LAN end stations.

End User-A customer or device that uses communications services.

End User Device-End user devices are communication adapters such as telephones, fax machines, or private telephone systems that adapt signals from a telecommunication system to a format (such as audio or visual) to a form that is suitable for an end user.

End User Licensing Agreement (EULA)-An agreement between an end-user of a product (usually a software product) and the owner of the product (e.g. software developer) that defines the terms of how the end user must abide by when they use a product. The end user is often prompted to enter into an EULA prior to installing and using a software application.

End-User Point Of Termination-The network interface that connects the end-user to the network.

Endoscope-A fiber-optic cable system that is used to return images from inside the human body.

Energized-The condition when a circuit is switched on or powered up.

Energy-Energy is the product of mass and half the square of velocity (for velocity values that are small relative to the speed of light). The SI unit of energy, consistent with electrical units, is the joule (also called the watt second), product of one kilogram of mass, the constant 1/2, and the square of a 1 meter per second velocity. Energy can also be described equivalently as the product of power and time, and the joule (or watt second) is the product of one watt of power with one second of time. The joule is named for James P. Joule, a 19th century Irish physicist (his name rhymes with foul).

In non-technical discourse, the four words force, energy, power and momentum are often unfortunately used as synonyms for each other. In science and technology, each of these words describes a distinct and separate physical quantity.

ENFIA-Exchange Network Facilities For Interstate Access

ENG-Electronic News Gathering

Enhanced 411 (E411)-Enhanced 411 (information) is a National Directory Assistance service that finds any listing in the U.S., Canada, or Puerto Rico and Entertainment Guides get you Movie Listings, Stock Quotes, Flight Times, Sports, Horoscopes, and More.

Enhanced 800 Services-A call processing service that allows for the routing of 800 (or other toll free or freephone numbers) to be routed to different locations based on other criteria such as day, time of day, or caller location.

This diagram shows an example how an Enhanced 800 number translation service operation allows a company to route calls to different service centers depending on the time of day and location of the caller. This diagram shows that a company has 3 offices; Boston, San Francisco, and Dallas and the company has a single toll free line 800-227-9681. Callers from east coast exchanges are routed to the Boston office from 9:00am - 5:00pm EST. Callers from west coast exchanges are routed to the San Francisco office from 9:00am - 5:00pm PST. All other calls are routed to the Dallas office. This allows the corporate office to handle calls after the regional offices are closed.

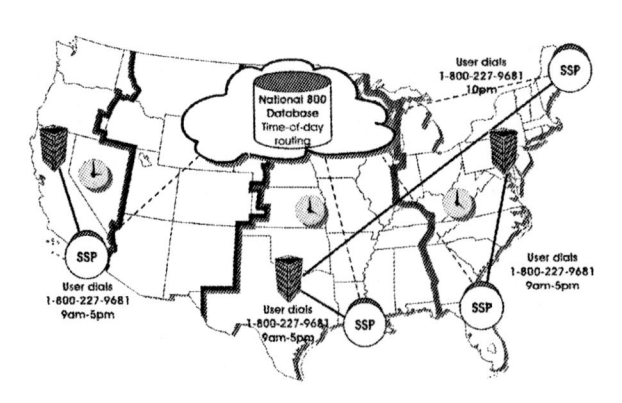

Enhanced 800 Service

Enhanced 911 (E911)-A emergency telephone calling system that provides an emergency dispatcher with the address and number of the telephone when a user initiates a call for help. The E911 system has the capability of indicating the contact information for the local police, fire, and ambulance agencies that are within a customers calling area.

Enhanced AMPS-IS-94

Enhanced Base Transceiver System (EBTS)-Base stations used in a SMR system that have been enhanced to provide services that are similar to cellular and PCS systems.

Enhanced Data Rates For Global Evolution (EDGE)-An evolved version of the global system for mobile (GSM) radio channel that uses new phase modulation and packet transmission to provide for advanced high-speed data services. The EDGE system uses 8 level phase shift keying (8PSK) to allow one symbol change to represent 3 bits of information. This is 3 times the amount of information that is transferred by a standard 2 level GMSK signal used by the first generation of GSM system. This results in a radio channel data transmission rate of 604.8 kbps and a net maximum delivered data transmission rate of approximately 474 kbps. The advanced packet transmission control system allows for constantly varying data transmission rates in either direction between mobile radios.

This diagram shows how a standard GSM radio channel is modified to use a new, more efficient modulation technology to create an a high-speed packet data EDGE system. The EDGE system users either 8 level quadrature phase shift keying (QPSK) modulation or the standard GMSK modulation (used by 2nd generation GSM systems.) This allows EDGE technology to be merged on to existing GSM systems as standard GSM mobile telephones will ignore the EDGE modulated time slots that they cannot demodulate and decode.

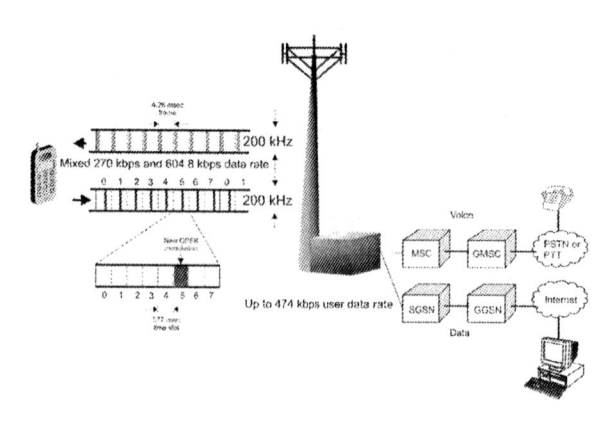

Enhanced Data Rates For Global Evolution Compact (EDGE Compact)

Enhanced Data rates for Global Evolution (EDGE) System

Enhanced Data Rates For Global Evolution Compact (EDGE Compact)-A version of EDGE that allows a increased frequency reuse through coordination of time slot transmission between adjacent cell sites to reduce interference to important control signals.

This figure shows an EDGE compact system. This diagram shows that the key concept of EDGE compact is the coordination of control messages between nearby base stations. Standard EDGE radio transceivers and EDGE capable mobile telephones are used in the system. The primary change coordination is the inhibiting of transmission on control time slots in nearby (alternate) cells.

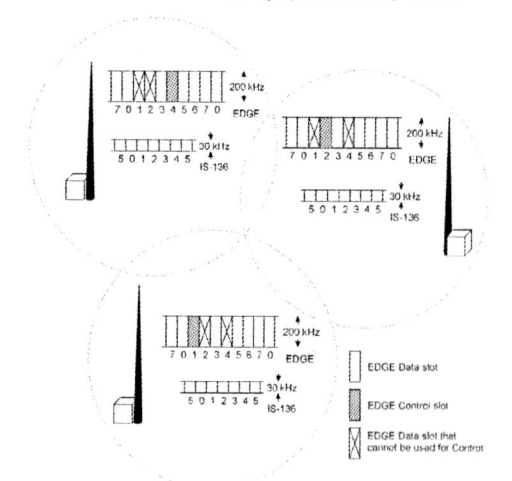

Enhanced Data Rates For Global Evolution (EDGE) Compact

Enhanced Data Rates For GSM Evolution (EDGE)-An extension to GSM networks to allow for increased capacity and packet data transmission using enhanced modulation (e.g. phase) and packet control technologies. EDGE is often referred to as cellular generation 2.5.

Enhanced Digital Access Communications System (EDACS)-A digital trunked radio system specification that is primarily used in public safety communication systems. EDACS systems add the ability to provide specific information about the location and status of emergencies.

Enhanced Full Rate (EFR)-A land mobile radio system that uses digital technology to provided for advanced voice and data services. EDACS provides information about the location of the emergency. For mobile telephones, enhanced emergency service involves the ability to location the mobile telephone.

Enhanced Half Rate (EHR)-An enhanced version of the half rate (HR) voice coder (vocoder) used in the GSM system. This enhanced version uses the same data rate as the standard half rate. However, the voice processing algorithms have been changed to improve the voice quality.

Enhanced Service Provider (ESP)-A provider of communication services that enhance the services provided by existing carriers. ESPs often provide information processing services such reformatting data or information management services. Because ESPs use existing communication services, ESPs do not typically file tariffs.

Enhanced Specialized Mobile Radio (ESMR)-A name given for the new technologies being implemented by Specialized Mobile Radio operators. These new technologies allow services which include dispatch, voice, paging, messaging as well as wireless data services.

This figure shows a sample ESMR system. In this system, ESMR mobile radios communicate with the system by first requesting access to the network on a control channel. The system responds to a valid request with a radio channel assignment. The ESMR system also has a switching system that can connect ESMR mobiles to each other or to other networks such as the public switched telephone network. The mobile switching center in an ESMR system must operate very quickly to allow two-way push-to-talk operation. When an ESMR customer presses the "push-to-talk" button, they must be connected to other radios in their group within about ¼ second or less. This is very different than a cellular phone system where the setup time of a telephone call can take 20 to 30 seconds. This diagram also shows that calls can be transferred between radio sites within the system (intra-system radio site handoff) or they can be transferred to radio sites in an adjacent ESMR system (inter-system radio site handoff).

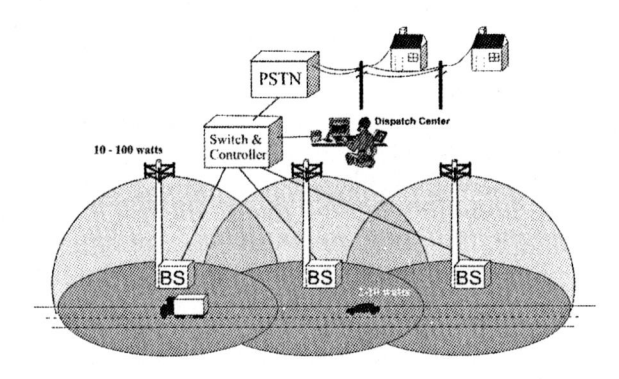

ESMR System Network Block Diagram

Enhanced Throughput Cellular (ETC)-A wireless data communications protocol that enhances the ability of modems to transfer data through wireless networks. ETC improves on the modem performance by adapting to the radio channel characteristics when transferring data and recovering from the random burst errors associated with radio transmission. To take advantage of ETC protocols, ETC capable modems must be installed at both ends of the wireless data connection.
Enhancing-The process of electronically adjusting the quality and sharpness of a signal or image. Enhancing also may refer to sweetening audio, for example, by adding laugh tracks and sound effects.
ENIAC-Electronic Numerical Integrator and Computer
ENOS-Enterprise Network Operating Systems
ENQ-Enquiry
ENTELEC-Energy Telecommunications and Electrical Association
Enterprise Network-The set of Local, Metropolitan, and/or Wide Area Networks and internetworking devices comprising the communications infrastructure for a geographically-distributed organization.
Enterprise Switch-A switch used within an enterprise backbone. Enterprise switches are generally high-performance devices operating at the Network layer that aggregate traffic streams from sites within an enterprise.
Entrance Facilities-The structure or conduit assembly that permits communication cabling to enter a building or facility.
Entropy-A measure of the amount of information in a message, which can be expressed in units of bits. If the probability of each possible variation of the message is known, the entropy can be calculated with a simple formula. The entropy of the message is always no greater than the number of bits used to represent the original message, so entropy coding techniques can generally be used to compress the information in the message into fewer bits.
Entropy Coding-A general term for lossless data compression techniques that take advantage of non-uniformity in the probability distribution of the data. The theoretically optimal compression performance of an entropy coding method would be, on average, to compress each symbol to a number of bits equal to the entropy of that symbol. Practical entropy coding techniques include Huffman coding and arithmetic coding, which can both achieve performance close to that theoretical optimum.
Enumeration-Enumeration is the process of discovering the communication parameters and configuration states of a data device that has been connected to a computer. When the computer (the host) discovers a new device, it determines what type of data transfers it anticipates transferring. Enumeration is a process used in the many computing systems including the Universal Serial Bus (USB) system.
EO-End Office
EO-Erasable Optical drive
EOA-End Of Address
EOB-End Of Block
EOC-Embedded Operations Channel
EOD-End Of Day
EOE-Electronic Order Exchange
EOF-End Of File
EOM-End Of Message
EOM-End Of Message Frame
EOP-End of Procedure frame
EOT-End-Of-Transmission
EOW-Engineering Orderwire

EPABX-Electronic Private Automatic Branch Exchange
EPD-Early Packet Discard
EPG-Electronic Programming Guide
EPIRB-Emergency Position Indicating Radio Beacon Station
EPLANS-Engineering, Planning and Analysis Systems
EPLD-Erasable Programmable Logic Device
EPN-Expansion Port Network
EPON-Ethernet Passive Optical Network
Epoxy-A liquid material that solidifies upon heat curing, ultraviolet light curing' or mixing with another material. Epoxy sometimes is used for fastening fibers together or for fastening fibers to joining hardware.
EPP-Enhanced Parallel Port
EPROM-Erasable Programmable Read Only Memory
EPS-Encapsulated PostScript
EPS-Encapsulated Postscript File
EPSCS-Enhanced Private Switched Communications Services
EQ-Equalization
Equal Access-A telephone service that provides a communication user with an equal choice of long distance carriers.
Equalization-A processes which modifies the receiver parameters to compensate for changing radio frequency conditions. Primarily used to compensate for multi-path propagation (see section RF channel).
Equalization (EQ)-(1 - general) Equalization is a process of adjusting received radio signal to counter the effects of distortion caused by multiple receptions of the same radio signal that have been delayed in time. (2 - audio) The process of improving the sound quality of an audio signal by increasing or decreasing the gain of the signal at various frequencies. (3-video) The process of altering the frequency response of a video amplifier to compensate for high-frequency losses in coaxial cable.
Equalize-The process of inserting in a line a network with transmission characteristics complementary to those of the line, so that when the loss or delay in the line and that in the equalizer are combined, the overall loss or delay is approximately equal at all frequencies.
Equalizer-1-general) A network that adjusts the frequency characteristics of a transmission circuit to allow it to transmit selected frequencies in a uniform manner. (2-device) A device installed in a signal transmission path to compensate for differences in time delay or in amplitude occurring in the transmission path for different frequency components of the signal. The purpose of the equalizer is to make the amplitude of all frequency components equal after the equalizer process (see: amplitude equalizer, graphic equalizer), or to make the time delay equal for all frequency components (see: Dispersion), or both. Some equalizers are self-adjusting adaptive equalizers. (3-absolute delay) A circuit or device within a network that is adds time delay to frequency components or circuits so the signals can be received from multiple sources at approximately the same time. (4-adaptive) An equalizer that constantly readjusts the amount of equalization (frequency or time delay) to adjust a distorted input signal to produce an equalized output signal. (5-delay) An equalizer that can delay the transmission of some frequencies so that the delay of all frequency components is approximately the same.
Equilibrium Length-The length of multimode optical waveguide, excited in a specified manner necessary to attain equilibrium mode distribution.
Equilibrium Mode Simulator-A device used to create an approximation of the equilibrium mode distribution. This distribution may be achieved by using selective mode excitation or mode filters, either with or without mode scramblers. (See also: equilibrium mode distribution.)
Equipment Cabinet-An enclosure assembly that allows the installation of electronic assemblies or components. Common equipment racks are 7 feet tall by 24 to 26 inches wide. The inside mounting area is often 19 or 22 inches wide (the rack). Some equipment cabinets come with power supplies and cooling fans.
Equipment Configuration-Equipment configuration is the process of sending information to a device that is used to adapt the equipment or software program to its environment (configuration).
Equipment Identity Register (EIR)-A database in a GSM telecommunications network that contains the identity of telecommunications devices (such as mobile phones) and the status of these devices in the network (such as authorized or not-authorized). The EIR is primarily used to identify mobile phones that may have been stolen or have questionable usage patterns that may indicate fraudulent use. The EIR has three types of lists;

white, black and gray. The white list holds known good IMEIs. The black lists holds invalid (barred) IMEIs. The gray list holds IMEIs that may be suspect for fraud or are being tested for validation.

Equipment Rack-A container that holds and interconnects electronic or electrical assemblies. A common size for an equipment rack is 19 inches (48.26 cm) wide. Electronic assemblies (called "modules") typically slide or are guided into the equipment rack where connectors located in the back of the equipment rack make contact with the modules. This diagram shows an equipment rack that is used for mounting and interconnecting network communication equipment. This diagram shows that several modules (line cards) can be mounted in each equipment rack. This diagram shows that the back (call the "backplane") of the equipment rack contains connectors that can interconnect the modules to each other and to cables from other equipment racks or devices.

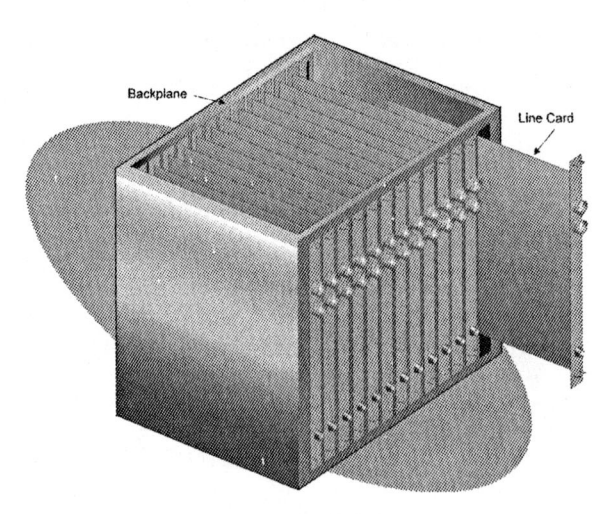

Equipment Rack Diagram

Equipment Room (ER)-A room or space that is dedicated for communication equipment and cable connection points. Equipment rooms may be centralized points that allow one or many companies to access telephone equipment, data communication equipment, and communication line connection points.

ER-Explicit Rate
ER-Equipment Room

Erase-The process of discarding, obliterating, or marking information as deleted from a storage medium.

Erbium Doped Fiber Amplifier (EDFA)-A device used to amplify signals across a broad range of wavelengths in an optical network. The ability to amplify more than one wavelength at a time made Wavelength Division Multiplexing (WDM) practical. EDFAs are used to boost signals at the transmitter, throughout the transmission line and at the receiver.

An EDFA consists of a pump laser, a length of doped fiber and often an optical isolator. The pump laser is used to generate metastable electrons in the doped fiber to prepare it for stimulated emission. The isolator prevents the pump laser light from entering the optical fiber used for transmission, where it would cause noise in the signals or damage other optical equipment. When light enters the EDFA, its photons cause stimulated emission of more photons of the same wavelength from the metastable electrons, thereby amplifying the signal (as well as any noise). Erbium doping is used because it causes the metastable electrons to have the right energy to produce photons with wavelengths from about 1520 nm to 1620 nm range, which is commonly used in long haul optical networks. The ability to amplify more than one wavelength at a time made Wavelength Division Multiplexing (WDM) practical.

ERL-Echo Return Loss

Erlang-A measurement unit of the average traffic usage of a telecommunications facility during a period of time (normally a busy hour) with reference to one hour of continuous use. The capacity of Erlangs is the ratio of time during which a facility is occupied to the time the facility is available for occupancy with reference to one hour. For example, a 12 minute call is 0.2 Erlangs. 1 Erlang equals 36 centum call seconds (CCS).

ERLE-Echo Return Loss Enhancement
ERMES-European Radio Messaging System
ERP-Effective Radiated Power
ERP-Enterprise Resource Planning

Error Concealment-When a decoder encounters erroneous or lost bits in the bitstream, the ability to generate an output signal with a minimal amount of perceptual distortion. One example of a technique to enable error concealment is to establish that codes within the bitstream that are close in Hamming distance should also cause perceptually

Error Correcting Code (ECC)

similar decoded output signals. Error concealment techniques are especially beneficial for bitstreams on high bit error rate links like mobile telephone networks.

Error Correcting Code (ECC)-A code in which each data signal conforms to specific rules of construction so that departures from this construction in the received signal can be detected automatically. This arrangement permits the automatic correction, at the receiving terminal, of some or all errors. Such codes require more signal elements than are necessary to convey the basic information.

This figure shows how block coding error detection and correction bits are added to the data to be transmitted (e.g. digital speech) by supplying blocks of data to a block code generator. To create an error protection code, the block cyclic redundancy check (CRC) parity generation divides a given block of data by a defined polynomial formula. The quotient and remainder are appended to the data stimulus to allow comparison when received. Using the same polynomial formula, the received data can be compared to the received error protection bits. In some cases, the formula can be used to fix some of the bits that were received in error.

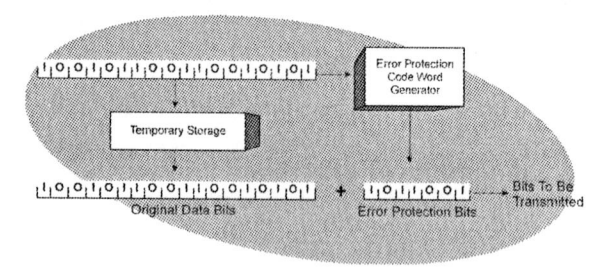

Block Error Coding

Error Correction-A process of using some data bits that are transmitted along with the data message to help correct bits that were received in error due to distorted radio transmission. Error correction is made possible by sending bits that have a relationship to the data that is contained in the desired data block or message.

Error Detection-The process of detecting for bits that are received in error during data transmission. Error detection is made possible by sending additional bits that have a relationship to the original data that can be verified. See also: error correction. This figure shows the basic error detection and correction process. This diagram shows that a sequence of digital bits is supplied to a computing device that produces a check bit sequence. The check bit sequence is sent in addition to the original digital bits. When the check bits are received, the same formula is used to check to see if any of the bits received were in error.

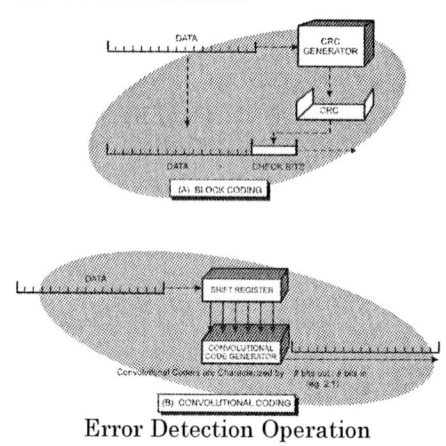

Error Detection Operation

Error Listing-A data set or printout of errors and adjustments. The listing can include totals, comparisons with other data, and a coded description of an error condition. An error listing is sometimes called an exception report.

Error Protection-The process of adding information to a data signal (typically by sending additional data bits) that permits a receiver of information to detect and/or correct for errors that may have occurred during data transmission.

Error Recovery-In the context of decoding compressed bitstreams, the ability of the decoder to recover from catastrophic bit errors or lost bits within the bitstream. For example, a decoder may need to terminate decoding the previous bitstream segment and completely resynchronize with a new synchronization point with the bitstream. The speed of error recovery may depend on both bitstream characteristics and decoder implementation.

Error Vector Magnitude (EVM)-Is a measurement of phase modulation accuracy of the transmitter as compared to predicted vectors.
ES-Errored Second
ES-End System
ES-IS-End System to Intermediate systems protocol
ESA-European Space Agency
ESC-Escape
Escalation-The process of taking a trouble call or ticket up through increasing levels of management until the problem gets resolved.
ESCON-Enterprise Systems Connectivity
ESD-Electronic Software Delivery
ESD-Electrostatic Discharge
ESD Protection-Electro-Static Discharge Protection
ESDI-Enhanced Small Device Interface
ESF-Extended Superframe
ESH-End System Hello paket
ESI-Enhanced Serial Interface
ESI-End System Identifier
ESI-Equivalent Step Index
ESI Profile-Equivalent Step Index Profile
ESMR-Enhanced Specialized Mobile Radio
ESMTP-Extended Simple Mail Transport Protocol
ESN-Emergency Service Number
ESN-Electronic Serial Number
ESN-Electronic Service Number
ESO-Equipment Superior to Operator
ESP-Enhanced Service Provider
ESP-Enhanced Serial Port
ESPRIT-European Strategic Program for Research in Information Technology
ESQ-End System Query packet
ESS-Electronic Switching System
ESS-802.11 Extended Service Set
Estimated Load-The usage activity of a system during a measurement period, as determined by available data and traffic theory.
ESZ-Emergency Service Zone
ET-Exchange Termination
ETACS-Extended TACS
ETB-End-Of-Transmission Block
ETC-Enhanced Throughput Cellular
ETDMA-Extended Time Division Multiple Access
Ethernet-Ethernet is a packet based transmission protocol that is primarily used in LANs. Ethernet is the common name for the IEEE 802.3 industry specification and it is often characterized by its data transmission rate and type of transmission medium (e.g., twisted pair is T and fiber is F).

Ethernet systems in 1972 operated at 1 Mbps. In 1992, Ethernet progressed to 10 Mbps data transfer speed (called 10 Base T). In 2001, Ethernet data transfer rates included 100 Mbps (100 BaseT) and 1 Gbps (1000 Base T). In the year 2000, 10 Gigabit fiber Ethernet prototypes had been demonstrated. Ethernet can be provided on twisted pair, coaxial cable, wireless, or fiber cable. In 2001, the common wired connections for Ethernet was 10 Mbps or 100 Mbps. 100 Mbps Ethernet (100 BaseT) systems are also called "Fast Ethernet." Ethernet systems that can transmit at 1 Gbps (1 Gbps = 1 thousand Mbps) or more, are called "Gigabit Ethernet (GE)." Wireless Ethernet have data transmission rates that are usually limited from 2 Mbps to 11 Mbps.
Originally created by an alliance between Digital Equipment Corporation, Intel and Xerox, Ethernet DIX, is slightly different than IEEE 802.3. In Ethernet the packet header includes a type field and the length of the packet is determined by detection. In IEEE 802.3, the packet header includes a length field and the packet type is encapsulated in an IEEE 802.2 header. Most modern day "Ethernet" devices are capable of using both protocol variation, however, older equipment was not able to do this.
This figure shows several types of Ethernet LAN systems and the approximate distances devices can be connected together in these networks. Thicknet Ethernet uses a low loss coaxial cable to provide up to 500 meters of interconnection without the need for repeaters. Thinnet systems use a relatively thin

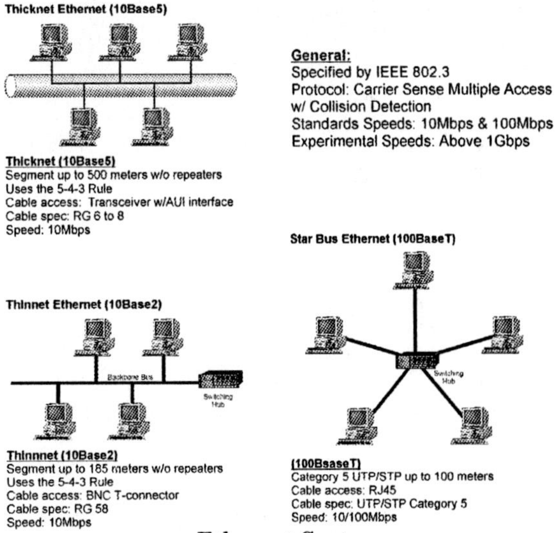

Thicknet Ethernet (10Base5)

General:
Specified by IEEE 802.3
Protocol: Carrier Sense Multiple Access w/ Collision Detection
Standards Speeds: 10Mbps & 100Mbps
Experimental Speeds: Above 1Gbps

Thicknet (10Base5)
Segment up to 500 meters w/o repeaters
Uses the 5-4-3 Rule
Cable access: Transceiver w/AUI interface
Cable spec: RG 6 to 8
Speed: 10Mbps

Star Bus Ethernet (100BaseT)

Thinnet Ethernet (10Base2)

Thinnnet (10Base2)
Segment up to 185 meters w/o repeaters
Uses the 5-4-3 Rule
Cable access: BNC T-connector
Cable spec: RG 58
Speed: 10Mbps

(100BaseT)
Category 5 UTP/STP up to 100 meters
Cable access: RJ45
Cable spec: UTP/STP Category 5
Speed: 10/100Mbps

Ethernet System

Ethernet Address

coaxial cable systems and the typical signal loss in this cable restricts the maximum distance to approximately 185 meters. 100 BaseT systems use category 5 UTP cable and the maximum distance is approximately 100 meters.

Ethernet Address-The unique non-changeable 48 bit address that identifies a specific hardware device in an Ethernet network. This includes routers, network interface cards (NIC's), printers, and bridges. A portion of the Ethernet address identifies the equipment manufacturer and a portion is a sequence number assigned by a manufacturer. The Ethernet address is also call the medium access control (MAC) address or hardware address.

Ethernet Frame Structure-The structure of data packet that is used in the Ethernet data network. The data packet frame length is variable and can go up to 1500 bytes. The fields within the data packet include a preamble for synchronization, 48 bit destination address, 48 bit source address, control fields, data packet, and error detection.

This diagram shows the structure of an Ethernet data frame. Ethernet frames begin with a preamble for synchronization, destination address, source address, control fields, data packet, and error detection. The data portion of the Ethernet frames has variable length up to 1500 bytes.

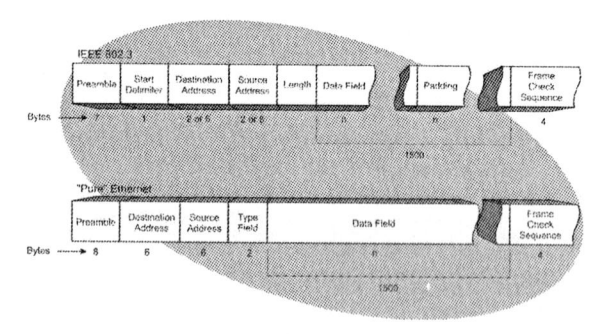

Ethernet Frame Structure

Ethernet Hub-In 10-Base-T and 100-Base-T Ethernet, a specific type of repeater which usually has between four and forty-eight ports. The hub allows Ethernet to be star-wired while still being a logical bus. This enables the IEEE 802.3 MAC protocol, based on collision domain (CSMA/CD), to be utilized while having the added advantage of centralized wiring.

This figure shows an Ethernet hub. This diagram shows that one of the computers has sent a data message to the hub on its transmit lines. The hub receives the data from the device and rebroadcasts the information on all of its transmit lines, including the line that the data was received on. The hub's receiver and transmit lines are reversed from the computers. This allows the computers that are connected to the hub to hear the information on their receive lines. The sending computer uses the echo of its own information as confirmation the hub has successfully received and retransmitted its information. This indicates that no collision has occurred with other computers that may have transmitted information at the same time.

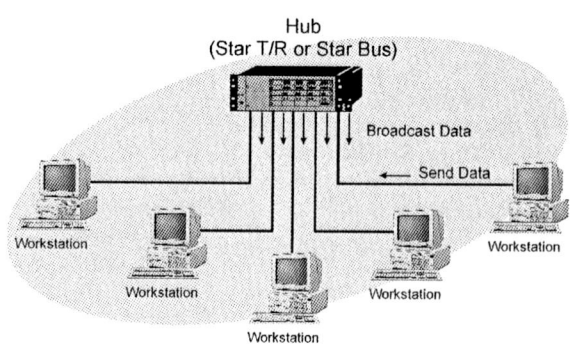

Ethernet Hub Operation

Ethernet in the First Mile (EFM)-Developed as standard IEEE 802.3ah, commonly known as Ethernet First Mile (EFM). This technology seeks to provide high-speed Ethernet-like communication over voice grade copper wire and fiber optic cables. "First mile" refers to the cables running from homes and business to the first device operated by a carrier or service provider outside of the customer premises. The technology leverages the ubiquity of Ethernet in the LAN with existing wire infrastructure already in place.

Ethernet Passive Optical Network (EPON)-A Passive Optical Network (PON) which utilizes Ethernet framing and protocols. EPONs allow service providers and customers to be required to own and understand a single networking technology: Ethernet.

EPONs are passive in that all devices in the physical cable plant from the service provider to the end-customer are passive. Passive splitters are used to direct light in fibers to multiple locations, much like coax splitters are used in cable TV networks. Likewise, precise timing and/or frequency control is utilized to allow multiple end-users to access the same upstream fiber. The key advantage of EPONs are their ability to provide high-speed access to many end-users with a single fiber, while not requiring any advanced electronics devices to be located in environmentally harsh locations such as atop telephone poles. EPONs are very similar in architecture, applicability and deployment challenges as cable modem networks based on DOCSIS.

Ethernet Switch-A device which relies on enhancements to the original Ethernet specification that transfer data directly between ports without the need to re-broadcast the information to all ports of the device. Switched Ethernet incorporates modified layer 2 (Data Link Layer) electronics in order to allow an individual 10 Mbps (for Ethernet) or 100 Mbps (for fast Ethernet) user to transfer data directly to an end segment.

This image shows a 48-port Ethernet switch. Unlike a hub, this devices switches data between port connections.

Ethernet Switched Hub
Source: 3Com

ETISALAT-Emirates Telecommunications Corporation
ETL-Extract, Transform, and Load
ETM-Electronic Ticketing Machine
ETM-Eight To Ten Modulation
ETN-Earning Telephone Network
ETN-Electronic Tandem Network
ETR-Estimated Time to Repair
ETS-Ethernet Terminal Server
ETS-ETSI Technical Specification
ETSI-European Telecommunications Standards Institute
ETSI Technical Specification (ETS)-A standard defined by the European Telecommunications Standards Institute (ETSI).
ETX-End Of Text Message
ETX-End Of Text
EUC-Extended Unix Code
EUCL-End User Common Line charge
EULA-End User Licensing Agreement
Eunet-European UNIX Network
Eureka-European Research Coordination Agency
EURO-ISDN-European Implementation of ISDN
European Digital Subscriber Signaling no. 1 (EDSS1)-The European variant of the SS7 (Signaling System 7) signaling and control protocol for use over the ISDN D (Delta) channel. EDSS1 was developed by ETSI (European Telecommunications Standards Institute). The ITU-T specifications Q.921 and Q.931 DSS1 (Digital Subscriber Signaling 1) are the baseline international specifications used throughout the world. The European specification from ETSI represents a modified form of Q.931, known as EDSS1 (European Digital Subscriber Signaling). Due to early acceptance within the industry, EDSS1 is the common variant of Q.931.

European Patent Application-An application for patent filed with the European Patent Office which is published 18 months after the date of filing. A1 Document - European patent application published with European search report. A2 Document - European patent application published without European search report (search report not available at the publication date). A3 Document - Separate publication of the European search report.

European Patent Office-The European Patent Office (EPO) grants European patents for the contracting states to the European Patent Convention (EPC), which entered into force in 1977. The EPO is the executive arm of the European Patent Organization, an intergovernmental body set up under the EPC, whose members are the EPC contracting states.

The EPO provides a single search, examination, and granting procedure for the member states. Granted European patents must, however, be registered as national patents in the countries where patent protection is desired.

Members of the EPO include: Austria, Belgium, Switzerland, Cyprus, Germany, Denmark, Spain, Finland, France, United Kingdom, Greece, Ireland, Italy, Liechtenstein, Luxembourg, Monaco,

Netherlands, Portugal, Sweden, Turkey, Bulgaria, Romania, Czech Republic, Estonia, Hungary, Slovenia, and Slovak Republic

European Patent Specification-A European Patent Specification, commonly called a "European Patent", is an application for patent which has been approved and granted by the European Patent Office. A European Patent can be used to obtain a national patent in any of the member countries of the European Patent Office. There is presently no single Europe-wide patent.

European Telecommunications Standards Institute (ETSI)-An organization that assists with the standards-making process in Europe. They work with other international standards bodies, including the International Standards Organization (ISO), in coordinating like activities.

EUROTELDEV-European Telecommunications Development

EUTELSAT-European Telecommunications Satellite organization

EUTP-Enhanced Unshielded Twisted Pair

EVA-Economic Value Added

Event-A process or data transfer that has initiated, changed, or completed in a communication system. Event records identify these changes and are commonly used for billing systems. Events that cause other processes to start are called trigger events.

Event - Billing-A change of equipment configuration or the use of a network resource that is recorded, usually for billing purposes. Event information is used by the billing system to determine the rate and amount of usage that will be billed to a customer. Events are sometimes referred to as Usage, CDR, xDR, Ticket, Message.

Event Log service-A Microsoft Windows service that logs important application, security, and system events into the event log. Event logs are also maintained on Network Management Stations (NMS) like syslog, SNMP traps, RMON events and alarms.

Event Management-The recording, organizing, and displaying events that occur in communication systems. These events may include the type of event such as a program activated by a software application or information that is generated by event triggers (some preset level). Event management is used to determine the status and changes (such as increased bandwidth or CPU usage) that are occurring in communication systems to help find potential problem areas before they cause system failures.

Events-(1-general) External information that is received that indicates a change in status within a communication system. (2-Bluetooth) Incoming messages to the L2CA layer along with any timeouts. Events are categorized as Indications and Confirms from lower layers, Request and Responses from higher layers, data from peers, signal Requests and Responses from peers, and events caused by timer expirations.

EVM-Error Vector Magnitude

EVM-Electronic Voice Mail

EVRC-Enhanced Variable Rate Vocoder

EW-Electronic Warfare

Examination-The process of qualifying an application for patent for grant. Also known as "substantive examination," this process includes an analysis of the claims with regard to Prior Art to determine whether or not the claims meet all statutory and regulatory requirements. Examination is performed by qualified patent examiners who have expertise in the technical areas where they work.

EXCA-Exchangeable Card Architecture

Exchange-(1-switch) The term used in some countries to refer to a network switch (See "Switch"). (2-company) A communication company or companies for the administration of communication service in a specified area, which usually embraces a city, town, or village and its environs, and consisting of one or more central offices, together with the associated plant, used in furnishing communication service in that area. The first exchange was installed in 1878.

Exchange Carrier (EC)-A company that provides local telecommunications services. Exchange carriers are generally regulated by a national regulatory agency for interstate or inter-regional services and are regulated by state or local agencies for internal local access and transport area (InterLATA) services.

Exchange Carriers Standards Association (ECSA)-Formed in 1980 to issue standards for the telephone industry, it was later supplanted by Alliance for Telecommunications Industry Solutions (ATIS).

Exchange Message Record (EMR)-Exchange message record (EMR) is a standard format for the exchange of messages between telecommunications systems. The EMR format is often used for billing records. The records may be exchanged by magnetic tape or by other medium such as electronic transfer or CD ROM.

Excitation-The current energizing field coils in a generator.

Exciter-(1-transmitter) The portion of a transmitter that generates a modulated RF signal, usually at the operating frequency (2-antenna) The driven element in an antenna that includes parasitic elements.

Executable Short Message-A message that is received by a Subscriber Identity Module card in a wireless system (such as a mobile phone system) that contains a program that instructs the SIM card to perform processing instructions.

This diagram shows how an executable short message can be sent by a system operator to add a new feature into a mobile phone. The executable short message is a program that is stored in the SIM card and interacts with the operation of the mobile phone to allow the new feature to operate. The system simply sends the file as an executable message directly to the mobile phone identification. When the complete executable short message has been received in the SIM card, it is stored in memory and this program can complete (run) instructs that allows the new feature to operate.

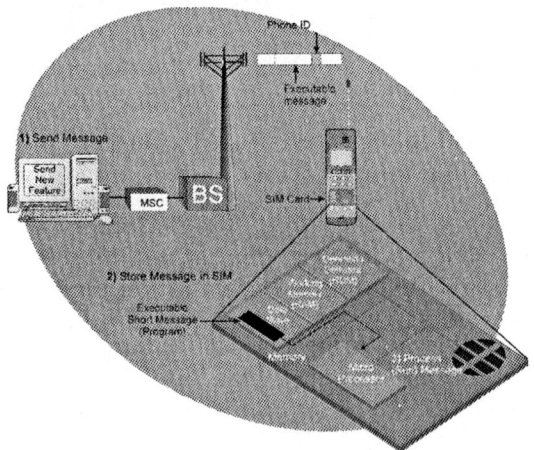

Executable Short Message Operation

Executable Short Message Platform-A system that can configure and deliver commands directly into the subscriber identity module (SIM) contained in a mobile telephone. These commands are processed directly by the SIM card.

EXM-Exit Message

EXOS-Extension Outside

Expandability-The ability of a system to supply processing or services without significant changes to its fundamental assemblies. Measures of expandability in communication include the maximum number of customers that can receive service, number of radio transmitter towers that can be connected to a switching system, and maximum data transfer rates on a communication line.

Expanding-Expanding is a system that increases the amount of amplification (gain) of an audio signal for smaller input signals (e.g., softer talker). The use of expanding allows the level of audio signal that leaves the modulator to have a larger overall range (lower minimum and higher maximum) regardless if some people talk softly or boldly. As a result of expanding, high-level signals and low-level signals output from a modulator may have a different conversion level (ratio of modulation compared to input signal level). This could create distortion so expanding allows the modulator to convert the information signal (audio signal) with less distortion. Of course, the process of expanding must be initiated at the transmitting end, called companding, to recreate the original audio signal.

Expansion Slots-A connection point and associated assembly mounting hardware that allows circuit cards or assemblies to be added to a computer or communication equipment. Examples of expansion slots include PCI connection in a desktop computer or a PCMCIA slot in a laptop computer.

Expert Witness-A person who is qualified to given an opinion or other information that can be used as part of legal proceedings.

Explicit Forward Congestion Indication (EFCI)-An explicit forward congestion indication (EFCI) is a field contained within an ATM cell header that is used to indicate other network elements (e.g. switches) that a switch is experiencing or will likely experience a congested state. The use of EFCI information allows other switches or systems to adapt or lower their cell rates.

Explorer Frame-In source routing, a frame used either to perform route discovery or to propagate multicast traffic. There are two types of explorer frames: Spanning Tree Explorers and All Routes Explorers.

Extended Broadcast Channel (E-BCCH)-A logical channel on a digital cellular system that is used to send non-critical system information (such as a candidate radio channel list of neighboring cell sites). See also: broadcast control channel (BCCH).

Extended Link (E-Link)

Extended Link (E-Link)-An "E" (extended) link connects an SSP to an alternate STP. "E" links provide an alternate signaling path if an SSP's "home" STP cannot be reached via an "A" link. "E" links are not usually provisioned unless the benefit of a marginally higher degree of reliability justifies the added expense.

Extended Simple Mail Transport Protocol (ESMTP)-An extended (enhanced) version of the original SMTP protocol to better support text languages, graphics, audio and video files. It is described in RFC 1651.

Extended Time Division Multiple Access (ETDMA)-An enhanced version of the IS-136 TDMA system. The ETDMA system adds a control channel that coordinates the dynamic time slot allocation to the standard 30 kHz channels used in IS-136 TDMA system. These channels are dynamically assigned based on voice activity detection.

Extender Board-An adapter board that extends a module outside of its frame to allow easier access to the module's components for troubleshooting and alignment.

Extensibility-The ability to upgrade existing systems or services without significant changes to the existing systems.

Extensible Markup Language (XML)-A software standard that is used to define exchangeable elements of a web (HTML) page. Extensible Markup Language was developed in 1996 by the World Wide Web Consortium (W3C). It is a widely supported open technology (i.e. non-proprietary technology) for data exchange. XML documents contain only data, and applications display that data is various ways. XML permits document authors to create their own markup for virtually any type of information. Therefore, authors can use XML to create entirely new markup languages to describe specific types of data, including mathematical formulas, chemical formulas, music and recipes.

Extension-(1-software) A set of commands or protocols that are used to extend the capabilities of another application or protocol. Software extensions are commonly used to rapidly extend the capabilities of an existing application or protocol without changing the underlying application or protocol. (2-telephone) An additional telephone connected to a line. Allows two or more locations to be served by the same telephone line or line group. May also refer to an intercom phone number in an office. (3-file name) The optional second part of a PC computer filename. Extensions begin with a period and contain from one to three characters. Most application programs supply extensions for files they create.

Extension Dialing-Extension dialing is a process that allows callers to enter the extension number of the person they are calling. The extension number may be entered by DTMF tones or by audio (voice) commands.

Extension Negotiation-The process of negotiating which software extensions will be used between devices or systems. Extension negotiation requires the communication devices to identify which software extensions they can use (and possibly their preferences for using such extensions) to define a common set of extensions that can be used between communication devices.

External Data Representation (XDR)-A standard information format that is used to define data that is transferred between networks or network parts. XDR was developed by Sun.

External F-ES-External Fixed End System

External Gateway Protocol (EGP)-EGP is an Internet protocol that is used to exchange routing information between switches and routers outside the network domain where it is located. An EGP can indicate if a given network is reachable. However does not make routing or priority decisions.

External Interface-A device that adapts internal networks to external networks. The use of an external interface allows devices and applications that are not directly connected to a network to request and receive some or all services from the network.

External Modulation-Modulation of a carrier signal source, such as a laser or LED, by varying its emitted signal. In direct modulation, the amount of light that is emitted is varied. In external modulation, the emission (such as light) is constant and some mechanism is used to alternately block and pass it. See also Electro-Optic Modulator and Electro-Absorption Semiconductor Modulator.

External Network-A network that is not part of a specific network domain (such as the Internet when compared to a company data network). External networks may allow direct or limited access to network information (routing and addressing) and control of network resources (bandwidth and QoS) depending on the interconnection type and network access (security) provisioning.

External RF Power Amplifier-A device that is capable of increasing power output when used in conjunction with, but not an integral part of, a transmitter.

External RF Power Amplifier Kit-An assembly or number of electronic parts, that can be attached to a transmitter as an external RF power amplifier, even if additional parts are required to complete assembly.

Extinction Ratio-The ratio of light transmitted when a modulation signal is on to that transmitted when it is off.

Extract-A process of removing information or data from another (usually larger) data source. Extracting often involves data pointers that indicate the start of the data block with the information data and processing steps for extracting and confirming the data integrity.

Extract, Transform, and Load (ETL)-The three processes involved in the preparation and initiation of a data warehouse. Extraction: pulling data from legacy systems. Transform: reformatting data for use in the data warehouse. Load: inserting the data into the warehouse.

Extrinsic Junction Loss-The junction losses caused by different geometric and optical parameter mismatch when two non-identical optical waveguides are joined.

F

F-farad

F-Framing Bit

F Framing Bit-The F or framing bit in the DS-1 (T-1) digital multiplexing system is usually described as the first bit in the frame pattern that comprises 193 bits. The following 192 bits in the frame comprise 24 channels or time slots, each time slot comprising 8 bits from one channel. In the very first historical 1961 version of DS-1 technology, called the D-1 system (now obsolete, the current version being D-4), the F bit was alternately set to 1 and then 0 and this simple pattern repeated in each two consecutive frames. The D-2 and all later versions of DS-1 required a multi-frame or super-frame sequence of 12 frames so that certain bits in the traffic channels could be used for robbed bit signaling in the 6th and 12th frames. In a 12-frame sequence this is the F bit pattern: 100011011100. Bear in mind that in the actual bit stream there are 192 bits of channel data (not shown here) between each two adjacent F bits shown in this list. In the 1980s, a new 24 frame sequence called Extended Super Frame (ESF) was introduced. It has several advantages and is now very widely used. ESF uses a sequence of 24 F bits as follows:

m e1 m 0 m e2 m 0 m e3 m 1 m e4 m 0 m e5 m 1 m e6 m 1

Note that only 6 of these are fixed values: 001011. The 12 m bits constitute a message channel at 4 kb/s. In many installations, this m channel is logically divided into two separate 2 kb/s channels, one for ZBTSI and one for general Operations, Administration and Management messages. The six e bits, represented by e1 through e6 in the list, comprise an error detecting code computed from all the bits in the previous 24 frames. These six e bits are computed as a cyclic redundancy check (CRC) code when the previous 24 frames are transmitted. At a receiving location, the same computation is repeated on the received data bits. If the value of the six bits computed at the receiving location does not match the six e bits transmitted over the link, there is at least one error and this can be indicated to repair staff craftspersons by means of alarms. In this way, ESF allows the system to constantly monitor for bit errors at the binary level, without being dependent on BPVs and without devoting a traffic (revenue producing) channel to the error detecting information.

F Port-Fabric Port

F-BCCH-Fast Broadcast Control Channel

F-ES-Fixed End System

F-Layer-(1-atmosphere) A layer of the atmosphere comprising two separate (2-data flow control) Flow-Control-Farameter Negotiation Facility in the Integrated Services Digital Network (ISDN) and the Public Packet-Switched Network (PPSN), the X.25 facility that permits the negotiation of packet sizes and window sizes in both directions of transmission.

F-Link-Fully Associated Link

F-PICH-Forward Pilot Channel

F-SYNC-Forward Sync Channel

F/R-CPHCH-Forward/Reverse Common Physical Channels

F/R-DPHCH-Forward/Reverse Dedicated Physical Channels

F2F-Face to Face

FA-Flexible Alerting

FAA-Federal Aviation Administration

FAB-Fulfillment, Assurance, and Billing

FAC-Feature Access Code

FAC-Forced Authorization Code

FACCH-Fast Associated Control Channel

FACCH/F-Full Rate Fast Associated Control Channel

FACCH/H-Half Rate Fast Associated Control Channel

FACN-Foreign Agent Control Node

FACS-Facilities Assignment And Control System

Facsimile (Fax)-A process of converting optical images into an electrical signals for transmission over communication systems. When the electrical signal is received, it is converted back to an optical format for display or printing of the original image. This process is commonly called FAXing.

Fax usually involves the transmission of still images (typically black and white with no intermediate shades of gray). The most widely used type of FAX service today is described by ITU standard T.30, Group 3. Group 3 facsimile uses a digital modem over a voice grade public telephone channel to transmit the FAX signal. Group 4 FAX also permits user-selectable pixel size and typically operates via a 64 kb/s ISDN channel. Group 1 was an early analog fax system, typified by the Xerox Telecopier. Group 2 was an early digital FAX system. Groups 1 and 2 are obsolescent, although some Group 3 FAX machines are backward compatible with Group 2 machines.

Fading-(1-radio signal) The variation in radio signal strength that results from changes in the characteristics of the radio transmission path. Fading can be caused by several different factors including signal reflection, ionosphere, and interference from other radio channels. (2-video) A progressive level of translucence (see-through) from one image or video screen to another.

FADS-Force Administration Data System

Fail Safe Operation-A type of system control that allows for the automatic operation of additional equipment or reconfiguration of existing equipment to prevents the improper functioning of communications or systems in the event of circuit loss or impairment.

Failure-A detected cessation of capability to perform a specified function or functions within previously established limits. A failure is beyond adjustment by the operator through means of controls normally accessible during routine operation of the system. (This requires that measurable limits be established to define "satisfactory performance.")

Failure Effect-The result of the malfunction or failure of a device or component

Failure In Time (FIT)-A unit value that indicates the reliability of a component or device over a period of time.

Failure Mode And Effects Analysis (FMEA)-An iterative documented process performed to identify basic faults at the component level and determine their effects at higher levels of assembly.

Fall Time-The amount of time that it takes a signal to decrease from a high level (e.g. peak) to its low level value. Fall time is usually measured when the signal decreases between 90% to 10% of maximum output.

False Colors Effect-A digital picture manipulator effect that permits user adjustment of colors in the picture.

Fan In-The greatest number of separate inputs acceptable to a single specified logic circuit without adversely affecting performance.

FAP-Fuse Alarm Panel

FAP-Formats And Protocols
FAQS-Frequently Asked Questions
Far End-The distant end to that being considered; not the end where testing is being carried out.
Far End Crosstalk (FEXT)-The leakage of signal that is coupled to a nearby cable or electronics circuit (called crosstalk) where the unwanted signal is received on the far end (remote end) of the cable. This diagram shows how far end crosstalk (FEXT) can cause interference at the distant end of a transmission line. FEXT occurs when some of the transmitted signal energy leaks from one twisted pair and is coupled back to a communications line that is transferring a signal in the opposite direction. Generally, FEXT has a lower level of signal interference as the crosstalk levels at the receiver end are lower than crosstalk at the sending (near) end.

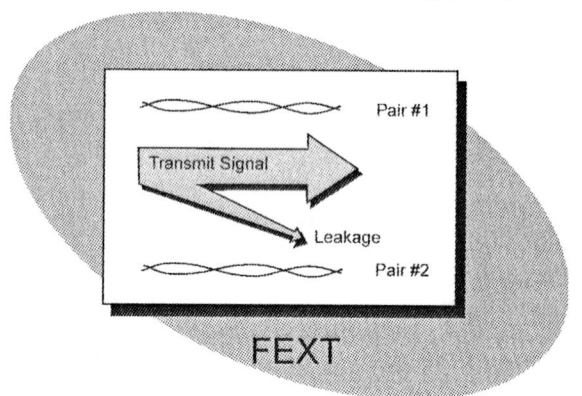

Far End Crosstalk (FEXT)

Far Field Region-(1-optical) The region, far from a source, where the diffraction pattern is equal to that observed at infinity. (2-radio frequency) The region of the field of an antenna where the angular field distribution is essentially independent of the distance from the antenna.

farad (F)-Unit of electrical capacitance. One ampere second per volt (or one coulomb per volt). Named for Michael Faraday, 19th century British scientist (1791-1867) .

Fast Associated Control Channel (FACCH)-In digital cellular radio, the FACCH is a logical channel on a digital traffic channel that is typically used to send urgent signaling control messages (such as a handoff or power control message). The FACCH channel sends messages by replacing speech data with signaling data for short periods of time. In GSM two special reserved bits are used to inform the receiving device if the data in the current time slot is digitally coded subscriber traffic or alternatively a FACCH message. In IS-136 a FACCH message is distinguished from digitally coded subscriber traffic because two different types of error protection coding are used for the two types of information.

Fast ATA-Fast AT Attachment

Fast Broadcast Control Channel (F-BCCH)-The information sent on this channel relates to system identification and parameters needed by a cellular phone to gain access to a system quickly. A broadcast logical channel used for mandatory, time-critical system information requiring a fixed repetition cycle. Also see broadcast channel (BCCH).

Fast Busy-A alert tone that indicates communication resources (such as switch trunks) are not available. A fast busy signal operates at twice the normal busy signal rate (120 tones/minute).

Fast Ethernet-The standard Ethernet protocol that transfers data at 100 Mbps capacity. Fast Ethernet is commonly called 100BaseT.

Fast Fourier Transform (FFT)-An alternative implementation of the Discrete Fourier Transform that eliminates significant computational redundancy.

Fast Frequency Shift Keying (FFSK)-A digital modulation process where each digit is represented a different (unique) frequency.

Fast IR-Fast Infrared

FAT-File Allocation Table

Fatigue-The reduction in strength of a metal caused by the formation of crystals resulting from repeated flexing of the part in question.

Fatigue Resistance Factor-A measure of the capability of a fiber optic cable to maintain its strength under stress. Higher fatigue resistance factor values often extend the lifetime of the fiber when it is subjected to stress.

FAU-Fixed Access Unit

Fault-A condition that causes a device, a component, or an element to fall to perform in a required manner. Examples include a short circuit, a broken wine, or an intermittent connection.

Fault Finder-A test set or other device that enables faults to be identified and localized.

Fault Location-A procedure of electrical tests made from a terminal station to determine the location (and sometimes the cause) of system malfunction.

Fault Management-Fault management is one of the five functions defined in the FCAPS model for network management. Fault management identifies the network problems, failures, events and corrects them. Fault management is the reactive form of network management. SNMP traps, syslog, and RMON typically is used in fault management.

Fault To Ground-A fault caused by the failure of insulation and the consequent establishment of a direct path to ground from a part of the circuit that, normally, should not be grounded.

Fault Tolerance-The ability of a network or subsystem to continue to operate in the event of a hardware or software failure. Fault tolerant systems are typically able to identify the fault and replace the failed component or sub-system with another equipment.

Fault Tolerant Network-A network designed or engineered to remain in operation in the event of system or component failures. For example, a fault-tolerant network might use alternate routing.

Fault Tree Analysis (FTA)-An iterative documented process of a systematic nature performed to identity basic faults, determine their causes and effects, and establish their probabilities of occurrence.

Fault-management, Configuration, Accounting, Performance, and Security (FCAPS)-Fault-management, Configuration Accounting, Performance, and Security is a categorical model of the working objectives of network management and is a standard adopted by the International Telecommunications Union (ITU). There are five levels, called the fault-management level (F), the configuration level (C), the accounting level (A), the performance level (P), and the security level (S).

FAX-Facsimile Communications

Fax-Facsimile

Fax Back-A service that allows callers to request information to be to them by fax. Fax back service usually involves the use of an interactive voice response (IVR) system to provide information to the caller of the fax back service. The user selects appropriate documents they desire to have sent, usually making selections with the keypad. The system then requests the caller to enter the destination fax number so the documents can be sent to the caller.

Fax Main (FxM)

Fax Mail (FxM)-A communications service that delivers documents in fax format by converting a document (such as a word processor document) to a fax image format delivering it to a fax machine.

Fax Mailbox-A portion of memory, usually located on a computer hard disk, that receives and sends fax images. The fax images are often in compressed digital image format.

Fax Modem-An adapter that is capable of both modem and facsimile transmission. A fax modem may be an internal card (such as a PCI card) or an eternal assembly. The fax modem can receive and convert digital fax images from their analog transmission form. The fax modem is capable communicating with both modem and fax protocols. There are several versions of modem and fax protocols and a fax modem may be capable of communicating with some (e.g. slow speed) or all of them.

This photo image is a fax modem, model 5686 that combines modem, fax transmission, and other advanced features into a external modem assembly. This modem also uses standard V.92 and V.90 communication technology.

operating manuals in fax mailboxes at the company. Each fax document that is stored is assigned to a fax mailbox. The verbal (audio) name that is associated with this fax mailbox is programmed into the interactive voice response (IVR) system. This allows the user to hear their options and to make a fax mailbox selection. This example shows how a caller dials a telephone number that this company uses for automatic fax back service (555-2345 in this example). When the phone system receives a call on this line, it automatically routes the call to the IVR system at extension 1001. The IVR system prompts the caller to select the fax mailbox and to enter the destination fax number (555-8111). The fax mailbox (usually a storage area on the computer hard disk) retrieves the data and sends it to a computer fax generator. The destination digits (destination fax number) are also sent to the fax generator. This allows the computer to create the fax from the fax mailbox data and send the fax to the destination fax machine.

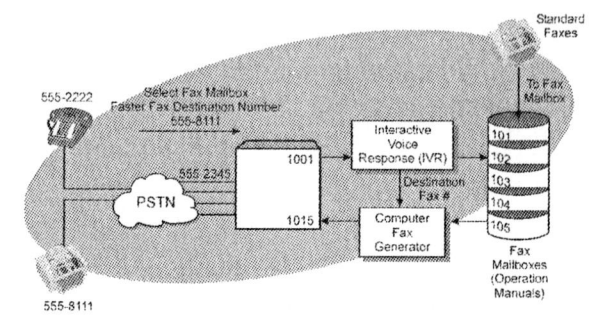

FaxModem
Source: U.S. Robotics

Fax on Demand (FOD) Operation

Fax On Demand (FOD)-A telecommunications transmission service that sends previously stored faxes to a user that has requested a specific fax message. For FOD service, the caller listens to the options available and requests that additional information is delivered by fax. FOD service is often used to deliver previously stored instructions or marketing materials.

This diagram shows how a fax on demand (FOD) system can store and automatically delivery faxes. The manager of this system has previously stored

Fax over Internet Protocol (FoIP)-A process of sending fax communication signals over the Internet. If the fax signal is in analog form (voice or fax, the signal is first converted to a digital form. Packet routing information is then added to the digital fax signal so it can be routed through the Internet.

Fax Profile-(1-general) Protocols or procedures that adapt an information format into a format suitable for facsimile communication. (2-Bluetooth)

Defines the protocols and procedures used by Bluetooth devices implementing the fax part of the usage model called "Data Access Points, Wide Area Networks." A Bluetooth cellular phone or modem may be used by a computer as a wireless fax-modem to send or receive fax messages.

FaxBIOS-An application programming interface (API) that allows faxes to be sent by software applications.

FBF-FeedBack Filter

FBG-Fiber Bragg Grating

FBU-Functional Business Unit

FC-Feedback Control

FC-Frame Control

FC and PC-Face Contact and Point Contact

FC Switch-Fibre Channel Switch

FC-AL-Fibre Channel Arbitrated Loop

FC-EL-Fibre Channel —- Enhanced Loop

FC-PH-Fibre Channel Physical standard

FCAPS-Fault-management, Configuration, Accounting, Performance, and Security

FCB-File Control Block

FCB-Frequency Correcting Burst

FCC-Federal Communications Commission

FCC Type Approval-A approval from the FCC that identifies the radio equipment manufactured has passed tests certifying it meets the minimum FCC requirements for that type of radio equipment. Most radio devices must meet several FCC specification requirements to receive FCC type approval. Companies typically use an independent testing lab to certify that equipment meets FCC requirements.

FCCH-Frequency Correction Channel

FCD-Frame Continuity Date

FCN-Function

FCOS-Feature Class of Service

FCOT-Fiber Control Office Terminal

FCS-Frame Check Sequence

FCS Error-Frame Check Sequence Error

FCSI-Fibre Channel Systems Initiative

FD-Floor Distributor

FDCCH-Forward Digital Control Channel

FDD-Frequency Division Duplex

FDDI-Fiber Distributed Data Interface

FDDI Frame Structure-Fiber Distributed Data Interface

FDDI-II-Fiber Distributed Data Interface II

FDI-Feeder Distribution Interface

FDL-Facilities Data Link

FDM-Frequency Division Multiplexing

FDMA-Frequency Division Multiple Access

FDS-Frequency Division Switching

FDTC-Forward Digital Traffic Channel

FDX-Full Duplex

FE-Extended Framing

FE-Functional Entity

Fear Factor-One or more reasons why potential customers avoid deciding to purchase a produce or service that may satisfy their business or personal needs.

Fear, Uncertainty, Doubt (FUD)-A marketing tactic that is used to discourage customers from buying their competitors.

Feature Access Code (FAC)-The codes assigned (usually numeric codes) that users may use to access the features of communication system.

Feature Group-A switched access service offered by a local exchange carrier to an inter-exchange carrier, in North America.

Feature Group A (FGA)-A line-side (end user side) access connection to a local exchange carrier end office with an associated local telephone number. The objective is line access to and from any interexchange carrier. This type of access is not used today but was the only type of access first available to competitive interexchange carriers. It requires the originator to dial an access number before dialing the destination number (and often an identification number as well), and does not have positive supervision signals for call timing and billing. The FGA line is the first point of switching in a telephone network.

Feature Group B (FGB)-A trunk-side connection for interexchange carriers access when the originator dials 950-XXXX, where XXXX represents a Carrier-specific Access Code. It also offers terminating access with multifrequency signaling. Although still used to some extent, this has largely been supplanted by dialing the 1010xx...x access prefix.

Feature Group D (FGD)-A trunk-side access to exchange carrier end-office switching systems and tandems. It provides the equal-access service that the former Bell operating companies must offer, as required by the Modification of Final Judgment. For an interexchange carrier, Feature Group D offers positive call connect and disconnect supervision for accurate call time billing, a uniform access code option (by dialing 1010XX...X), optional calling party identification, recording of access charge billing details, and pre-subscription to a customer-specified interexchange carrier.

Feature Service Access Code

Feature Service Access Code-A code, in the form -X or -XXX, used by customers for controlling access to custom calling services.
FEBE-Far End Block Error
FEC-Forwarding Equivalence Class
FEC-Forward Error Correction
FECN-Forward Explicit Congestion Notification
FED-Field Emission Display
Federal Open Systems User Council-A forum for federal organizations that helps to define government-wide requirements for a common application architecture and to develop strategies for addressing policy, management, and technical issues related to open systems.
Feed-(1-radio frequency) The wires, cable, or waveguide used to connect an antenna with its radio transmitter or receiver. (2-broadcast) A video (television) or audio signal source.
Feeder Cable-Feeder cables are one of several large cables that provides a physical connection between a central office and distribution cables that connect to end customers. Feeder cable usually is placed in underground conduit. A main feeder originates at a central office and usually follows a single route through a large, defined area. Branch feeders, or subfeeders, extend from a main feeder to provide facilities in a well defined segment of a main feeder area. Feeder cable often is referred to as feeder plant. When combined with distribution cables, it constitutes a customer loop.
Feeder Relief-The process of making more wire pairs available at a given demand point. This can be done either by adding new cables or by rearranging a network to make unused pairs available where needed.
Feeder Route Area-A geographic area served by feeder cables, secondary feeder cables, and distribution cables from a feeder route. The area is contained within the feeder route boundaries.
Feeder Route Boundaries-The boundaries outlining a geographic area served by feeder cables, secondary feeder cables, and distribution cables fed from a feeder route.
Feeder Route Schematic-An schematic drawing of the feeder lines in a communication network.
Feeder Sections-Linear segments of feeder routes defined so that the number of pairs in a route can be effectively and economically matched to present and future demand. Ideally, the feeder section should be associated uniquely with each allocation area.

FEFO-First Ended, First Out
Femto-A prefix meaning one quadrillionth.
FEN-Front End Network
FEP-Front End Processor
FEP-Fluorinated Ethylene Propylene
FEPS-Facility And Equipment Planning System
FER-Frame Error Rate
FERF-Far-End Receive Failure
FERF-Far-End Remote Failure
Fermal Principle-The principle that a ray of light traveling from one point to another follows the path that requires the least amount of time.
Ferroelectric Material-A crystalline material that exhibits, over a particular range of temperature, a remnant magnetic polarization that can be reoriented by application of an external electric field.
Ferrofluid-A fluid in which fine particles of iron, cobalt, or magnetite are suspended. Such fluids are super-paramagnetic (they exhibit no hysteresis effects) and can be used to provide liquid seals, held in position by a magnetic field. Such substances can be used, for example, on the drive shaft of a magnetic disk drive unit, keeping it dust-free.
Ferromagnetic Material-(1-hard) A material with low relative permeability and high coercive force so that it is difficult to magnetize and demagnetize. Hard ferromagnetic materials retain magnetism well, and are commonly used in permanent magnets. (2-soft) A metallic material with high relative permeability and low coercive force. Such a material is easy to magnetize and demagnetize. Soft ferromagnetic materials are suitable for use where rapid changes of flux are encountered, such as in transformers.
FEs-Form Effectors
FET-Field Effect Transistor
FEXT-Far End Crosstalk
FF-Form Feed
FFDI-Fast Fiber Data Interface
FFOL-Fiber Follow On LAN
FFSK-Fast Frequency Shift Keying
FFT-Fast File Transfer
FFT-Fast Fourier Transform
FG-Functional Group
FGA-Feature Group A
FGB-Feature Group B
FGC-Feature Group C
FGD-Feature Group D
FGD-Frame Ground
FHMA-Frequency Hopping Multiple Access

FHSS-Frequency Hopping Spread Spectrum
FHSS Mode-802.11 Frequency Hopping Spread Spectrum
FIB-Forward Indicator Bit
Fiber-Fiber optic cable is a strand of glass or plastic that is used to transfer optical energy between points. The size of most fibers is from 10 to 200 microns (1/100th to 1/5th of a mm). Optical fibers are typically used in a unidirectional mode (e.g., data moves in only one direction). Because of this, every transmission system requires at least two fibers (one for transmission and one for reception).
Fiber Amplifier-A device made from optical fiber and a pump laser that is designed to produce stimulated emission of photons at the signal wavelength, leading to an amplified signal. There are two major types of fiber amplifiers: Raman amplifiers and doped fiber amplifiers like EDFAs.
Fiber Bragg Grating-A type of optical filter that is formed in a fiber by periodically changing the index of refraction in a thin cross section of the fiber. These abrupt changes in index are spaced such that light of a specific wavelength is reflected by the resulting grating while other light passes through without interacting with the grating at all. As with other types of filters, fiber Bragg gratings can be used to select, divert, or focus light from within the fiber.
Fiber Cable-A cable that is constructed of an inner core of optical fibers (glass or plastic) that are covered by high-density polyethylene. Fiber cable may be wrapped with various types of strengthening materials including steel wire and high-density polyethylene.
This figure shows single mode and multimode fiber lines. This diagram shows that multimode fibers have a relatively wide transmission channel that allows signals with different wavelengths to bend back into the center of the fiber strand as they propagate down the fiber. The diagram also shows that single mode fiber has a much small transmission channel that only allows a specific wavelength to transfer down the fiber strand.

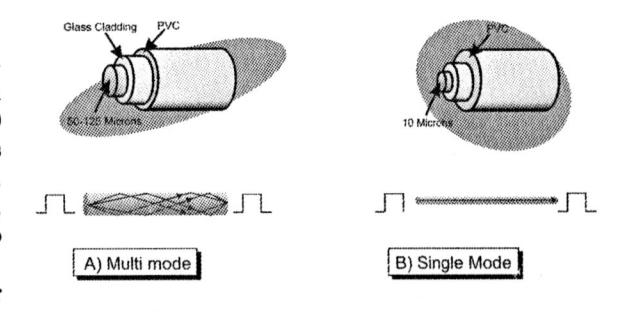

Fiber Optic Cable Diagram

Fiber Cut Off Wavelength-The wavelength at which the presence of the fiber second order mode introduces a measurable attenuation increase when compared with a fiber whose differential mode attenuation is not changing at that wavelength. This general definition applies for a short, uncabled singlemode fiber with a specified large radius of curvature. Because the fiber cutoff wavelength is influenced by the length of the link, the number and degree of bends, and other cabling considerations, the cut-off wavelength in practical applications must be viewed from a systems point of view. In general, the cable cut-off wavelength is less than the fiber cutoff wavelength.
Fiber Distributed Data Interface (FDDI)-Fiber distributed data interface (FDDI) is a computer network protocol that utilizes fiber optic or copper cable as the transmission medium to provide a token-passing, logical ring topology network operating at 100 Mbps. FDDI also provides for a mode of operation whereby two counter rotating rings are used to provide immediate fail-over and ring recovery should a fiber cut occur. FDDI was commonly used as a backbone network to interconnect lower speed Ethernet and Token Ring networks within an enterprise. The American National Standards Institute standard X3T12 defines the protocol.
This figure shows FDDI system that uses dual rings that transmit data in opposite directions. This diagram shows one dual attached station (DAS) and a dual attached concentrator (DAC). The DAS receives and forwards the token to the mainframe

Fiber Gain

computer. The DAC receives and token and coordinates its distribution to multiple data devices that are connected to it.

shows two optical network units (ONUs) that connect data networks together using fiber cable. This diagram shows that two fiber strands are needed: one for transmitting and one for receiving.

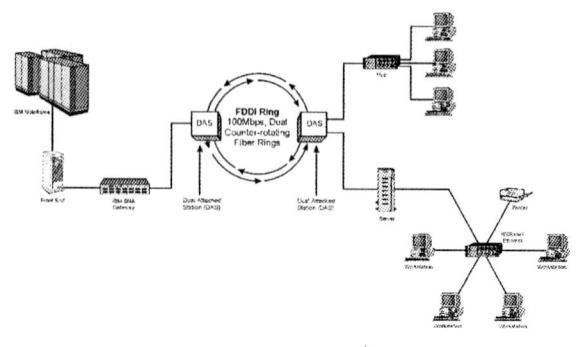

Fiber Distributed Data Interface (FDDI) System

Fiber Optic Transmission System

Fiber Gain-The increase in transport efficiency caused by multiplexing (or combining) optical signals so that only a few fiber links are needed to carry traffic for many customers. Fiber gain is the optical equivalent of pair gain in copper circuits.

Fiber Harness-In equipment interface applications, an assembly of a number of multiple fiber bundles or cables fabricated to facilitate installation into a system.

Fiber Hub-A termination point for optical fiber cables. A hub contains multiplexers and equipment that convert optical signals to electrical signals that are then carried to customer locations over physical pairs.

Fiber Optic Receiver-Equipment that detects a signal carried by an optical fiber and converts it to an electronic signal for processing. A receiver includes at least a photodetector, signal processing circuitry and a decision circuit (for digital signals) or a demodulator (for analog signals). The photodetector is usually a p-i-n diode or an avalanche photodiode. The processing circuitry typically includes amplification and filtering functions. The decision circuit or demodulator is used to recover the original, electronic signal from the transmitted signal.

This figure shows a fiberoptic communication system that is composed of two end-nodes and a fiber optic cable transmission medium. This diagram

Fiber Optic Transmission System (FOTS)-A term sometimes used by carriers to descript the generic application of SONET and SDH optical networks.

Fiber Optic Transmitter-A device that takes in an information signal (usually electronic) and converts it into an optical signal that it then launches into a network, typically via an optical fiber. A transmitter consists at least of a light source and a modulator (unless modulation is done directly).

Fiber Optic Waveguide - Optical Fiber-A thin (typically 125 to 250 microns diameter) strand of glass or optically transparent plastic of indeterminate length. The index of refraction of the core of the fiber and its surrounding cladding are engineered carefully to maximize internal reflection so that light entering the fiber will continue traveling down its center. The glass or plastic is designed to have minimal impurities and imperfections to minimize loss in the optical signal. See also multimode fiber, single mode fiber, and dispersion-shifted single mode fiber.

Fiber To The Curb (FTTC)-A distribution system that uses fiber optic cable to connect telephone networks to nodes that are located near homes or any business environment (near the curb). The fiber optic transmission is used to provide broadband services beyond the central office, all the way to the last 50-100 feet from the subscriber. The service pedestal is said to be "at the subscriber's curb."

Fiber To The Home (FTTH)-A distribution system that uses fiber optic cable to connect telephone networks to nodes that are located in the homes of customers. The fiber optic transmission is used to provide broadband services beyond the central office, all the way through the drop wire to the optical node that is located in the customers home.

Fiber Transmitter-A device or assembly that converts electrical signals to optical signals for transmission on a optical fiber line. Fiber transmitters can use carrier modulation (on-off) or sometimes use analog modulation (amplitude, frequency, or phase) to transfer the information signal (such as an RF signal) to a fiber communication line.

Fiber-Optic Splice Closure-A fiber-optic splice closure is used to cover (house) cables that have been spliced. These closures are designed specifically for fiber optic cable and they can be buried or placed in a manhole/vault environment.

This photo shows the Opti-Guard fiber optic splice enclosure unit that is used to cover optical cables that have been spliced. This device also has a connector to hold the fiber optic cable while it is spliced.

Fiber-Optic Splice Enclosure
Source: Mack Molding Company for AFL Telecommunications

Fiberoptic Cable-See Fiber Optic Cable

Fibre Channel-A networking standard developed by the American National Standards Institute (ANSI) X3T9 committee. Fibre Channel is commonly used to interconnect peripherals, mass storage systems, archiving and imaging systems, and sometimes high-speed engineering and design workstations.

Fibre Channel currently provides speeds of 1.06 Gbps, 2.12 Gbps and 4.24 Gbps, and was designed specifically with low latency applications, such as storage networking, in mind. Most commonly, Fibre Channel is used to transport SCSI protocol data units, but other networking technologies may be used.

Fibre Channel is supported on optical fiber as well as copper (twin-axial cable). Fibre Channel provides guaranteed and lossless delivery of data, and a credit based flow control methodology. Fibre Channel devices may be connected in point-to-point, looped, or switched (star) topologies.

FID-Field Identifier

Fidelity-The degree to which a system, or a portion of a system, accurately reproduces at its output the essential characteristics of the signal impressed upon its input.

Field Blanking-The blanking signals occurring at the end of each video field, used to make the vertical retrace invisible. Field blanking also is referred to as vertical blanking.

Field Effect Transistor (FET)-A transistor that controls a current path between two terminals by varying an electric field across the path (called a gate) that depletes the current carrying electrons in the channel (between the source and drain). The gate is perpendicular to the current path.

Field Emission-The emission of electrons from a surface, caused by a large external field rather than a heated filament effect.

Field Label-The label (name) of a field within a database record.

Field Programmable Logic Family (FPLF)-A family of TTL compatible programmable chips that provide the necessary links between large-scale integration (LSI) circuit blocks. The FPLF programming capability means that the logic can be altered to meet specific requirements. Devices of varying complexity are available: field programmable gate array (FPGA), field programmable logic array (FPLA), and field programmable logic sequence (FPLS).

Field Repeater

Field Repeater-The preset pattern of each successive frame of composite 525 and 625 television. Composite video (PAL, NISC, and SECAM) carry the color information on a subcarrier signal whose cyclic pattern repeats.

Field Upgrade-A product upgrade that takes place at the customer's site.

FIF-Fractal Image Format

FIFO-First-In-First-Out

FIGS-Figure Shift

File Server-A file server is a computer that stores data centrally for network users and manages access to that data. File servers may be dedicated so that no processes other than network management can be executed while the network is available, or non-dedicated so that standard user applications can run while the network is available.

File Transfer Access, And Management (FTAM)-The protocol of open systems interconnection (OSI) that enable users to transfer files between computers, regardless of manufacturer. The FTAM standards were developed by the International Standards Organization (ISO).

File Transfer Protocol (FTP)-A protocol that is used to manage the transfer of data files between computers and networks. Because FTP is a standard protocol, it permits the transfer of any type of data file between different types of computers or networks.

Filing Date-The date upon which a document must be filed after all computations of time authorized by this section have been made. In the United States, a filing date may be obtained by using a so-called express mail practice. The filing date of the first application for patent is considered as the "priority date" for subsequent applications.

Fill-(1-telecommunications) The ratio of the number of working or assigned pairs to the total number of pairs available in a cable or allocation area. A high fill number means that a cable or group of cables is almost out of spares and needs relief. The term also applies to channels in carrier systems. (2-video) In video keying, the fill is the video signal that is inserted into the hole cut in the background video by a key signal. (See also: key.)

Fill In Signal Unit (FISU)-An SS7 signal unit message that only contains only error control and delimitation information. This is transmitted when there are no message signal units (MSUs) or link status signal units to be transmitted. The continuous reception of this packet for periods that no messages are sent indicates the link is operation and allows the quality level (error rate) to be monitored. This figure shows the packet structure of the FISU in the SS7 system. This diagram shows the FISU packet has a single format structure. This sending of this simple packet structure allows the SS7 network to quickly detect if the signaling link experiences a failure when no other messages are being sent.

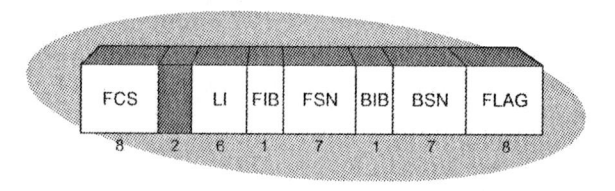

Audio Signal Filtering Operation

Filler-A material used to fill in gaps or voids, such as spaces between wires in a multipair cable.

Film Resistor-A type of resistor made by depositing a thin layer of resistive material on an insulating core.

Filter-(1-general) A network that passes desired frequencies but greatly attenuates other frequencies. (2-active) An RC filter that uses solid-state amplifiers to produce a desired frequency shaping characteristic. (3-band elimination) A band-stop or band-rejection filter that passes, with negligible loss, all signals except those in a specified band. (4-band-stop) A band elimination filter. (5-bandpass) A filter that greatly attenuates signals of all frequencies above and below those in a specified band. (6-capacitor-input) A common type of power-supply smoothing filter. The output from a rectifier is shunted by a large capacitor as the first element in the smoothing circuit. (7-cavity) A filter with precise characteristics for separating microwave frequencies using cavity resonance. (8-choke input) A low-pass power-supply smoothing filter with an inductance as its first element. (9-comb) A filter with several sharp band-stop sections for different frequencies. (10-composite) An m-derived filter made up of several filter sections, calculated to give the required impedance and sharp frequency changeover characteristics. (11-constant-k) A filter in which the product of the impedance of shunt components and of series components is a constant, independent of frequency. (12-crystal) A filter with sharp cutoff or changeover characteristics, obtained through the use of quartz crystal components in resonant circuits. (13-high-pass) A filter that attenuates signals below a specified frequency but passes with minimal attenuation all signals above that frequency. (14-LC) A filter with inductance (L) and capacitance (C) circuit elements. (15-longitudinal suppression) A filter designed to suppress unwanted noise signals flowing in the same direction on the two wires of a pair. (16-low-pass) A filter that greatly attenuates signals higher than a specified frequency, but passes with minimal attenuation all signals lower in frequency. (17-notch) A bandpass filter in which the upper cutoff frequency is twice the lower cutoff frequency. (18-power interference) A filter in series with the utility power input to a rectifier that passes the fundamental frequency of the power supply but greatly attenuates higher interfering frequencies. (19-software) A software routine that separates computer data according to specified criteria.

This figure shows typical audio signal processing for a communications transmitter. In this example, the audio signal is processed through a filter to remove very high and very low frequency parts (audio band-pass filter). These unwanted frequency parts are possibly noise and other out of audio frequency signals that could distort the desired signal. The high frequencies can be seen as rapid changes in the audio signal. After an audio signal is processed by the audio band-pass filter, the sharp edges of the audio signal (high frequency components) are removed.

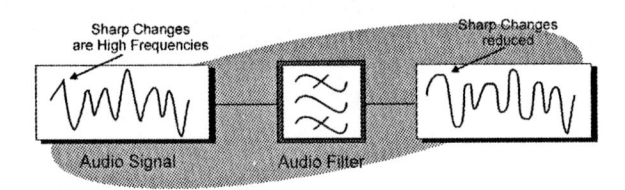

Audio Signal Filtering Operation

Filter Artifacts-Defects in the video picture caused by filtering. The most common artifacts appear as tinting and loss of resolution.

Filter Codec-A device or circuit that contains both filter and codec processing capabilities.

Filtered Noise-Noise signals that have been passed through a filter. This effects the power spectral density of noise signal that will have the same shape as the transfer function of the filter.

Filtering-Analog signal filtering is a process that changes the shape of the analog signal by restricting portions of the frequency bandwidth that the signal occupies. Because analog signals (electrical or optical) are actually constructed of many signals that have different frequencies and levels, the filtering of specific frequencies can alter the shape of the analog signal.

Filters may remove (band-reject) or allow (band-pass) portions of analog (possibly audio signals) that contain a range of high and low frequencies that are not necessary to transmit.

In some cases, additional signals (at different frequencies) may be combined with audio or other carrier signals prior to their transmission. These signals may be multiple channels (frequency multiplexing) or may be signals that are used for control purposes. If the signal that is added is used for control purposes (e.g. a supervisory tone that is used to confirm a connection exists), the control signal is usually removed from the receiver by a filter.

Filtering Database-A data structure within a bridge that provides the mapping from Destination Address to bridge port (in a D-compliant bridge), or from the combination of Destination Address and VLAN to bridge port (in a Q-compliant bridge).

Financial Posture -The strategic positioning established by upper management that defines the way that it wants to be perceived by investors, regulators and the public in financial terms. The indication of a strong financial posture would communicate to outsiders that the company was highly successful and cash rich and a weak posture would communicate the opposite. Financial posture objectives can be defined and controlled by the company through major marketing and public relations activities and they may be different than the actual financial performance of the company.

Find Me Service-A process of forwarding calls to a sequence of numbers so that the call will find the recipient. Follow me service could be password privileged to screen unwanted callers from being forward to private numbers such as home telephones or mobile phones.

FIPS-Federal Information Processing Standards

FIR-Finite Impulse Response Filter

Firefighting-A process of rapidly addressing many problems that are occurring in a system. Solving a specific problem is referred to as putting out a fire.

Firefighting Plan-A plan that addresses how to rapidly resolve critical problems within a network when normal procedures cannot be followed. A firefighting plan usually prioritizes which systems receive priority for corrective action (such as voice communication) and who will be responsible for the non-standard procedures to correct the problems.

Firewall-A firewall is a data filtering device that is installed between a computer server or data communication device and a public network (e.g. the Internet). A firewall continuously looks for data patterns that indicate unauthorized use or unwanted communications to the server. Firewalls vary in the amount of buffering and filtering they are capable of providing. An ideal (perfect) firewall is called a "brick wall firewall."

This figure shows how a firewall works. This diagram shows that a user with address 201 is communicating through a firewall with address 301 to an external computer that is connected to the Internet with address 401. When user 201 sends a packet to the Internet requesting a communications session with computer 401, the packet first passes through the firewall and the firewall notes that computer 201 has requested a communication session, what the port number is, and sequence number of the packet. When packets are received back from computer 401, they are actually addressed to the firewall 301. Firewall 301 analyzes the address and other information in the data packet and determines that it is an expected response to the session computer 201 has initiated. Other packets that are received by the firewall that do not contain the correct session and sequence number will be rejected.

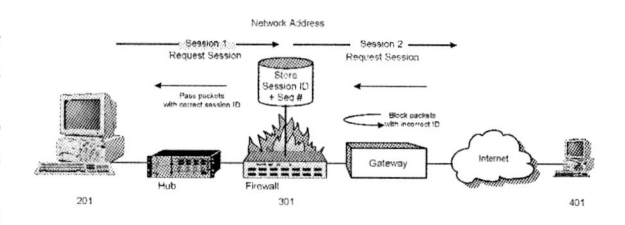

Firewall Operation

This photo shows a hardware (separate) firewall device that prevents unauthorized computers from accessing the computers attached to a network. This device is connected between the data network and the communication modem.

Firewall
Source: SonicWall

Firmware-Firmware is software program instructions that are stored in a hardware device that performs data manipulation (e.g. device operation) and signal processing (e.g. signal modulation and filtering) functions. Firmware is stored in memory chips that may or may not be changeable after the product is manufactured. In some cases, firmware may be upgraded after the product is produced to allow performance improvements or to fix operational difficulties.

First Application Region-For new services, the first regional company to use a new' network-based service. A network based service is one that requires the interaction of two or more types of network elements, possibly from different suppliers.

First Attachment-A hardware item comprised of porcelain, plastic, or steel fixtures that is affixed to a building to attach, buried or aerial drop wire.

First Attempt Load-An offered traffic load that excludes any load resulting from retrials.

First Choice Route-The first choice trunk group, or series of groups, between two switching systems.

First Generation (1G)-A term commonly used to describe the first technology used in a new application. In cellular telecommunications, the first generation used analog (usually FM) radio technology. For first generation cordless telephones, the first generation of products used single channel (using AM) radios.

First Mile-The signal path between a program origination site and its entry point to the communication network or a private satellite uplink. The first mile is usually a terrestrial RF link or a local telco loop.

First to File-In most regions of the world, patents are awarded to the person who first files an application for patent with a national or regional patent authority. First to file usually requires Absolute Novelty.

First to Invent-In the United States, patents are awarded to the person, or person(s), who first conceives an idea and reduces it to practice.

FISH-First In Still Here

FISK-Fax a Disk

FISU-Fill In Signal Unit

FIT-Failure In Time

FITL-Fiber In The Loop

FIVE-Format Independent Visual Exchange

FIX-Federal Internet Exchange

Fixed Charges-Recurring and non-recurring charges.

Fixed Count-The permanent connection of a group of cable pairs to the binding posts of a connecting block or terminal block at a terminal location.

Fixed Count Terminal-An outside plant cabling distribution terminal where only a portion of the entire cable count is readily accessible. The remainder of the cable count is behind the terminal posts and not readily visible or available for accessing facilities. Fixed count terminals are generally protected with carbon or other fusible station protection.

Fixed Reference Modulation-A type of modulation in which the choice of the significant condition for any signal element is based on a fixed reference.

Fixed Satellite Service (FSS)-A category of service devoted to point-to-point satellite communications.

Fixed Storage-A storage medium whose contents during normal operation are unalterable by a particular user.

Fixed Stuff R-Bits/Bytes-In the Synchronous Optical Net-work (SONET), bits and bytes that compensate for the differences between the bandwidth available in the Synchronous Transport Signal level 1 (STS1) and Virtual Tributary (VT) Synchronous Payload Envelopes, and the bandwidth required for actual payload mappings, such as DS1, DS1C, Ds2, and I1SSS. The data is defined to facilitate interworking with existing transmission terms or to enable interworking between fixed and floating Virtual Tributaries.

Fixed Terminals-A terminal interface device that is fixed in location.

Fixed Wireless-A communications service between two devices that are fixed in location.

FLAG-Fiberoptic Link Around The Globe

Flags-(1-computer) An indicator used to signal a condition or to mark information for further attention. The flag may be "raised" or lowered" through the results of computation, or specifically controlled through software. (2-ISDN, PPSN) In the Integrated Services Digital Network (ISDN) and the Public Packet-Switched Network (PPSN), a unique digital pattern that is used by a link layer to delimit frames. Each link layer frame starts with an opening flag and ends with a closing flag. (See also: Call Reference FIag). (3- SS7) In the Signaling System 7 protocol, a unique pattern on the signaling data link used to delimit a signal unit. The binary flag sequence used in ISDN, PPSN, SS7 and also X.25 and Frame Relay packet systems is the HDLC binary flag 01111110. To prevent false premature end of packet indication, the contents of every packet are processed before transmission to insert a binary 0 following any natural occurrence of 5 consecutive binary 1 bits. At the receiving end, a binary zero inside a packet is deleted if it follows 5 consecutive binary 1 bits, thus accurately restoring the original packet data.

Flapping-The continual changing of a network connection path that results from a intermittent circuit connection (or a similar condition) that indicates to the current connection path (the routers) into thinking there is a loss in connection or that a better connection path exists. This causes the continuous re-advertising of new available or unavailable packet routes within the network.

Flash-A system special service request feature that is used to indicate that a subscriber has a desire to recall a service function or to activate a custom calling feature (such as a call transfer request).

A flash feature service request can be created when the user initiates a short on-hook interval or through the sending of a special service request message. The short on-hook interval is created by a momentary operation of the telephone switch hook, in the midst of a prolonged off-hook period. This momentarily turns off the loop current in the subscriber loop. The duration of the current-off state must be typically more than 1 second but less than 2 seconds in most systems. The special service request message can be sent by a button on a telephone (such as a PBX telephone) sometimes labeled with such words as FLASH or BREAK or SERVICE, or by pressing the SEND or TALK key on a mobile telephone. In a mobile telephone or an ISDN telephone, a digital message is sent when the subscriber presses the button, instead of momentarily turning off the loop current.

This diagram shows how the flash signal (special service request) can be sent to the telephone system to activate additional call processing features. This example shows a flash feature is sent on an analog line by momentarily opening the current loop connection. When the loop current sensing circuit senses a brief open (no current flow) period, it creates a flash message that is sent to the call processing section of the telephone system. For digital telephones, the flash message is sent via a signaling message on the digital channel. This diagram shows that on an ISDN line, the flash message is sent on the D (signaling) channel.

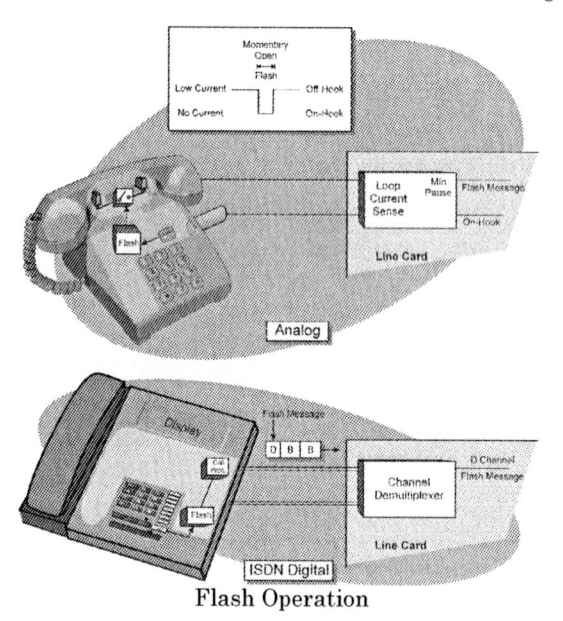

Flash [of Light]-An indicator light that flashes on and off with a cadence of approximately 2 flashes per second or slower is typically used to indicate a line that is on hold. The flashing cadence is slower than winking cadence.

Flash Request-A request to initiate a special processing function. Dual Mode cellular allows flash requests in both directions while analog systems only allow flash requests from the mobile phone to the cell site.

Flash ROM-Flash Read Only Memory

Flash Timing-The differentiation among hit, flash, and disconnect signals based on their time duration.

Flash With Info-A flash message (special call processing message) that is sent over a communication channel that provide an indication that the originator needs special processing along with some additional information (parameters) that are associated with the call processing request.

Flashover-An arc or spark between two conductors.

Flashover Voltage-The voltage between conductors at which flashover just occurs.

Flat Face Tube-The design of CRT tube with almost a flat face, providing improved legibility of text and reduced reflection of ambient light.

Flat Noise-A noise whose power per unit of frequency is essentially independent of frequency over a specified frequency range.

Flat Response-A performance parameter of a system in which the output signal amplitude of the system is a faithful reproduction of the input amplitude over some range of specified input frequencies.

Flat Room-A room that has been acoustically designed and/or treated to reproduce equally all sound in the audible spectrum. For a room that is not acoustically flat, the audio monitoring system may be conditioned through equalization to make the room seem flat.

Flash Operation

Flat Weighting-An amplitude and frequency characteristic that is flat (no variation over a specified frequency band). Flat weighting is used to measure noise on broadcast circuits. Flat noise power is expressed in dBrn or dBm. ionized layers, F1 and F2. Both layers exist during daylight hours but combine at night to form one layer from 200 to 400 km above the surface of the earth. These are the highest of the layers that refract radio waves.
FLEC-Forward Looking Economic Cost
FLEX-The FLEXTM paging technology is the newer two-way protocol developed by Motorola to increase the capacity of paging systems. The core FLEX technology is a one-way paging system that is capable of operating at variable data rates of 1,600, 3,200, 6,400 and up to 25,600 bits per second (bps).
Flex ANI-Flexible Automatic Number Identification
Flex Antenna-An antenna often used with portable telephones that consists of a length of stiff wire, usually with a fiberglass core, covered with an insulating material.
Flexible Key Assignment-A telephone system feature for customizing a station to the needs of a user by assigning different features to different keys on each set.
Flexible Line Ringing Assignment-A telephone system feature that enables calls to ring directly into specified departments.
Float-To operate a power load such as a telephone central office on a mains driven rectifier in parallel with a low-impedance storage battery. The battery is kept fully charged by the rectifier and is itself called upon only to provide power during temporary and short duration peaks for which the output is insufficient.
Floating-A circuit or device that is not connected to any source of potential or to ground.
Flocking-Assembling ad hoc teams of expert resources in response to high priority issues.
Flooding-The action of forwarding a frame onto all ports of a switch except the port on which it arrived. Normally used for frames with multicast or unknown unicast Destination Addresses.
FLOP-Floating Point Operation
Flow Chart-A graphic portrayal that shows the sequence in which functions are performed, from the beginning of a job to the end.
Flow Control-A hardware or software mechanism or protocol, that manages data transmissions when the receiving device cannot accept data at the same rate the sender is transmitting. Flow control is used when one of the devices communication cannot receive the information at the same rate as it is being sent; this usually occurs when extensive processing is required by the receiver and the receive buffers are running low. Examples are flow control algorithms are IEEE 802.3x used in Ethernet networks, Forward Explicit Congestion Notification and Backward Explicit Congestion Notification used in ATM networks and XON/XOFF used is RS-232 serial communication.
Fluorescence-The characteristic of a material to produce light when excited by an external energy source. No heat, or only a minimal amount, results from the process.
Flux Rise Time-In an LED or laser, the time required for the output radiation level to rise from 10 percent to 90 percent of its maximum level under the control of a specified driving pulse. The flux rise usually is specified in nano seconds or picoseconds.
Flyback (Retrace)-The election beam movement of the camera or TV monitor back to the starting point for the next line or field.
FM-Fault Management
FM-Frequency Modulation
FM Improvement Factor-The signal-to-noise ratio at the output of a receiver divided by the carrier-to-noise ratio at the input of the receiver. The FM improvement factor is obtained at the price of increased bandwidth in the receiver and the transmission path.
FMA-FRAMES Multiple Access
FMC-Fixed Mobile Convergence
FMEA-Failure Mode And Effects Analysis
FMIC-Flexible MVIP Interface Circuit
FMV-Fair Market Value
FN-Frame Number
FNC-Federal Networking Council
FNPA-Foreign Numbering Plan Area
FNPRM-Further Notice of Proposed Rulemaking
FO-Fiber Optics
FO cable-Fiber Optic Cable
FOA-First Office Application
FOB-Free On Board
FOC-Firm Order Confirmation
FOCC-Forward Analog Control Channel
Focused Overload-A condition of unusually high calling rates from many points to one point, for example, after a natural disaster or when a radio or TV station encourages mass calling.

Focusing Elements

Focusing Elements-Components used to control and confine an electron beam within an electronic device. In a device incorporating an electron gun structure, several elements are used to control the diameter of the beam. A focusing anode (plate) and several focusing coils may be used. In some devices, permanent magnets also are used to assist in maintaining correct beam diameter. In devices used to amplify microwave signals, such as klystrons, the focus system is critical because even a momentary change in the system can produce disastrous results and near instantaneous device failure.

FOD-Fax On Demand
FODU-Fiber Optic Distribution Unit
FOI Act-Freedom Of Information Act
FoIP-Fax over Internet Protocol
FOIRL-Fiber Optic Inter Repeater Link
Follow-Me Phone Service-A service that allows calls to be routed to a customer's choice of forwarding phone numbers. Follow-me service may be automatic (e.g. when a cellular telephone automatically registers with a visited systems) or manually set (e.g. when a customer calls in with a hotel phone number where they will be temporarily located).

This diagram show how follow-me service can allow calls to be automatically routed to one of three telephone numbers that a customer has provided to the follow-me system. In this example, the follow-me service is automatic. When the incoming call is received by the system, the system first tries the customer at the office number. Because there is no answer at the office, the system will then call the home number. When the user answers the call at home, the system will automatically move the home telephone number to the top of the follow-me calling list.

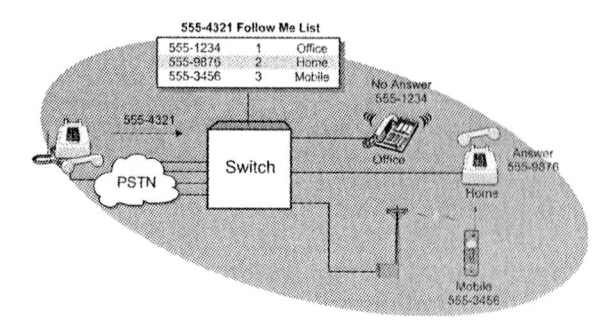

Follow-Me Phone Service Operation

Follow-Me Roaming-Follow-me calls are automatically forwarded to the area outside their home area where the mobile subscriber has registered. A cellular call generated to a user when the mobile subscriber has informed the local system that the user is roaming in another area.

FOMC-Federal Open Market Committee
FOMS/FUSA-Frame Operations Management System/Frame User Switch Access System
Footprint-(1 - satellite) The minimum radio coverage signal strength boundaries from a satellite over in a geographic area. A single satellite may have several antennas that have different footprints. (2 - equipment) The floor area occupied by a given piece of equipment.

Force - Mechanical-Force is the time rate of change of momentum. The SI unit of force, consistent with electrical units, is the Newton. One Newton is the product of one kilogram of mass and one meter per second per second of acceleration.

In non-technical discourse, the four words force, energy, power and momentum are often unfortunately used as synonyms for each other. In science and technology, each of these words describes a distinct and separate physical quantity.

Forced Rerouting-In the Signaling System 7 protocol, a procedure for transferring signaling traffic from one signaling route to another when the route in use fails or must be cleared of traffic.

Forecasted Load-The predicted load for a specified future period.

Foreign Agent Control Node (FACN)-A signaling control node used in 3rd Generation networks that manages several packet data serving nodes (PDSN).

Foreign Exchange Office (FXO)-Foreign Exchange Office (FXO) interface or channel unit that allows an analog connection (foreign exchange circuit) to be directed at the PSTN's central office or to a station interface on a PBX. The FXO sits on the switch end of the connection. It plugs directly into the line side of the switch so the switch thinks the FXO interface is a telephone. (See also: foreign exchange station.)

Foreign Exchange Station (FXS)-A type of channel unit used at the subscriber station end of a foreign exchange circuit. A foreign exchange station (FXS) interface connects directly to a standard telephone, fax machine, or similar device and supplies ring, voltage, and dial tone. (See also: foreign exchange office.)

Forking Proxy Server-A proxy server that forwards a communication session request to more than one device on behalf of the communication connection request.

Forklift Upgrades-An upgrade that requires the removal of old equipment (possibly by a forklift) and installation of new equipment.

Form Factor-(1-product) The physical shape and size of a product. (2-mathematic) The ratio of the root-mean-squire value of a periodic function to the average absolute value, averaged over a full period of the function.

Formula Translator (FORTRAN)-A computer language that was developed by IBM to allow programmers to code instructions in a manner similar to ordinary mathematical equations.

FORTRAN-Formula Translator

Forward Acting Code-A logically constructed code with enough redundancy to correct many errors without requiring retransmission.

Forward Analog Control Channel (FOCC)-The Analog control channel which goes from the cell site to the mobile unit.

Forward Analog Voice Channel (FVC)-The Analog voice/traffic channel which goes from the cell site to the mobile unit.

Forward Congestion Notification-Forward congestion notification indicates to upstream switching devices in a data communication network that data that is being transmitted through congested switches and it is likely that some of the remaining data or packets may be discarded. The upstream switch can then change the discard priority accordingly.

Forward Digital Traffic Channel (FDTC)-There are two separate control channels associated with the FDTC: the fast associated control channel (FACCH) and the slow associated control channel (SACCH).The FDTC is a digital channel from a base station to a cellular phone station used to transport signaling and user information. (See also fast associated control channel, slow associated control channel.)

Forward Error Correction (FEC)- FEC is a mathematical algorithm that is used to produce extra bits of data that are sent each packet. The sending of FEC bits allows for the correction of information bits that were lost or changed due to noise or other physical effects. The extra bits increase the data rate required for a given transmission by 5 to 25% depending on the level of correction required, so they are only used where error rates are high and retransmission is uneconomical.

Forward Explicit Congestion Notification (FECN)-A control bit carried within the overhead of a data packet in a frame-relay network that indicates network congestion exists in the forward (destination) direction of its flow. This control bit allows higher-level protocols within the data communications equipment (DCE) and data terminal equipment (DTE) to take appropriate bandwidth allocation action if necessary.

Forward Indicator Bit (FIB)-A bit inside an SS7 signal unit that indicates the start of a retransmission cycle.

Forward Pilot Channel (F-PICH)-Channel is continuously broadcast throughout the cell to provide timing and phase information.

Forward Scatter-The deflection, by reflection or refraction of an electromagnetic wave or signal, in a manner such that a component of the wave is deflected in the direction of propagation of the incident wave or signal.

Forward Sync Channel (F-SYNC)-A communications channel that provides information that allows receivers to acquire or determine the initial time synchronization with the system or channel.

Forward/Reverse Common Physical Channels (F/R-CPHCH)-The collection of all physical channels that carry information in a shared access, point-to-multipoint manner between the base station and multiple mobile stations.

Forwarding-The process of taking a frame received on an input port of a switch and transmitting it on one or more output ports.

Forwarding Delay-1. The amount of time that is used between the receiving of a packet or block of data to the time of is transmission through a data network (such as the Internet). 2. A measurement parameter used by the Spanning Tree Protocol that defines the delay that occurs between the transitions from a listening state to the learning state and from the learning state to the forwarding state.

Forwarding State-A stable state in the Spanning Tree Protocol state machine in which a bridge port will transmit frames received from other ports as determined by the bridge forwarding algorithm.

FOT-Optimum Working Frequency

FOTP-Fiber Optic Test Procedure

FOTS-Fiber Optic Transmission System

Four Wave Mixing (FWM)-A non-linear interaction between very high amplitude optical waves in optical fiber that produces light at other wavelengths/frequencies than the original light. The process of "mixing" is similar to the process of modulation, in which an electromagnetic wave is produced that has an instantaneous amplitude that is the product (rather than the sum) of the instantaneous amplitude of two source waves. The result of FWM is an undesired wavelength signal that degrades the performance of the fiber data transmission overall. See: Intermodulation.

Four-Wave Mixing-A nonlinear effect in optical networks resulting in cross talk, noise and/or loss. In an optical fiber carrying three or more wavelengths of light, if the power levels are high enough, three of the wavelengths will interact with each other and generate a fourth wavelength. If this new wavelength falls between the others, it adds noise to the signal. If it falls into one of the existing channels, it causes crosstalk.

Fourier Analysis-A mathematical process for transforming values between the frequency domain and the time domain. This term also refers to the decomposition of a time domain signal into its frequency components.

Fourth Generation (4G)-Fourth Generation wireless networks with bandwidth reaching 100 Mbps that allow for voice and data applications that will run 50 times faster than 3G. This capacity will enable three dimensional (3D) renderings and other virtual experiences on the mobile device.

FP-Feature Package

FPDL-Foreign Processor Data Link

FPGA-Field Programmable Gate Array

FPI-Formal Public Identifier

FPLF-Field Programmable Logic Family

FPLMTS-Future Public Land Mobile Telephone System

FPODA-Fixed Priority-Oriented-Demand Assignment

fps-Frames Per Second

FPT-Forced Perfect Terminator

FPU-Floating Point Unit

FQDN-Fully Qualified Domain Name

FR-Full Rate

Fractional T-1 (FT-1)-A digital transmission service that provides a customer with multiple 64 kbps channels but less than the full 24 channels offered by a T-1 channel.

FRAD-Frame Relay Access Device

Fragmentation-Fragmentation is a technique that divides a data packet into smaller data packets so that they can be sent through a network that can only transfer small data packets. Fragmentation occurs during network transmission. When these packets are received at their destination, they are reassembled to their original data packet size.

Frame-(1-general) A basic repeated bit pattern in time division multiplexing systems, and/or the time duration of this pattern. (2-frame relay) A packet of data in Frame Relay systems. (3-video) In video processing, a frame is a single still image within the sequence of images that comprise the video. Note that, in an interlaced scanning video system, a frame comprises two fields. Each field contains half of the video scan lines that make up the picture, the first field typically containing the odd numbered scan lines and the second field typically containing the even numbered scan lines. (4-equipment) A electronic rack that is used to interconnect and hold electronics assemblies. (4-audio) In audio processing, a frame is a group of audio samples that are processed together (to determine an instantaneous frequency spectrum for a codec like MP3, for example), though two adjacent frames might contain common samples in an overlap region.

Frame Error Rate (FER)-FER is calculated by dividing the number of frames received in error by the total number of frames transmitted. It is generally used to denote the quality of a digital transmission channel.

Frame Loss-The ratio of the number of frames that have been lost in transmission compared to the total number of frames that have been transmitted.

Frame Multiplexing-In a digital link the information bits used for different channels or different processes are grouped in a consecutive sequence of bits called a frame. In time division multiplexing systems, a frame has fixed time duration and contains a fixed number of bits. For example, in a DS-1 (T-1) digital multiplexing system a frame has 125 microseconds time duration and contains 193 bits. In some systems such as Frame Relay, the word frame refers to a packet of data for a single user or process, having non-fixed length and separated from preceding and following frames by special short sequences of bits used as separators, such as the HDLC binary flag sequence 01111110.

Frame Rate-The number of frames transferred over a period of time.

Frame Relay-Frame relate is a packet-switching technology provides dynamic bandwidth assignment. Frame relay systems are a simple bearer (transport only) technology and do not offer advanced error protection or retransmission. Frame relay were developed in the 1980s as a result of improved digital network transmission quality that reduced the need for error protection. Frame relay systems offer dynamic data transmission rates through the use of varying frame sizes.

Frame Relay-Frame Switching
This figure shows a frame relay system. This diagram shows a local area network (LAN) in San Francisco is connected to a LAN in New York. A virtual path is created through the frame relay network so data can rapidly pass through each frame relay switch as its path is previously established. When data is to be transferred through the LAN (e.g., a large image file), the data file passes through a FRAD that is the gateway to the frame relay network. The FRAD divides the data file from the LAN into variable length data frames. The FRAD sends and receives control commands to the frame relay network that allows the FRAD to know when and if additional data frames can be sent.

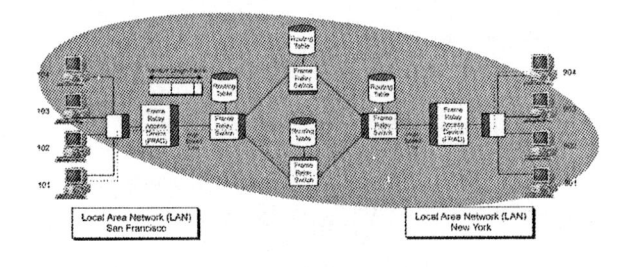

Frame Relay System

Frame Relay Access Device (FRAD)-A communications access device that converts data from a user's network into the format that is required by a frame relay network.

Frame Relay Network Device (FRND)-The FRND is a packet switch that also operates as a gateway to the frame relay network. The FRND passes frames it receives from the frame relay access device (FRAD) to other frame relay switches that forward packets toward their destination network. Frame relay switches have buffer memory that allows them to hold, prioritize packets before they are retransmitted. Packet switches can selectively discard packets if network congestion occurs. The FRAD and FRND provide information about the priority of the frames (e.g. non-essential discard eligibility) and status of the system (e.g. network congestion notification).

Frame Switch-A devices that forwards data frames based on layer 2 address contained within the received data frame. Frame switches may directly switch (cut-through) or operate as store and forward switches.

Frame UNI-Frame-based User Network Interface

Fraud Based Churn-A common form of involuntary churn. Fraudsters are customers who steal services without payment and who are terminated by the telephone company (Telco).

FRD-Fire Retardant

Free Space-An unbound medium for transmission of acoustic, electromagnetic, or photonic signal energy.

Free Space Path Loss-The amount of energy a signal is reduced as it is radiated (spread) into a space that is void of obstructions or interfering materials. Free space path loss from an omnidirectional antenna is approximately 20 dB per decade. That means for each 10 times increase in distance, the amount of energy decreases by 99% (100 times less).

Free Space Transmission-The theoretical transmission of a electromagnetic (radio) or optical communication signal through a vacuum with no absorption, no reflection, and no diffraction.

This figure shows two types of free space transmission systems: radio and optical. The microwave transmission system shows that some of the electromagnetic energy is absorbed by the water particles in the air. The optical transmission system uses a laser and photo-detector. The optical transmission system shows that some of the optical energy is scattered in other directions as it passes through smog and water particles.

Free Space Wave

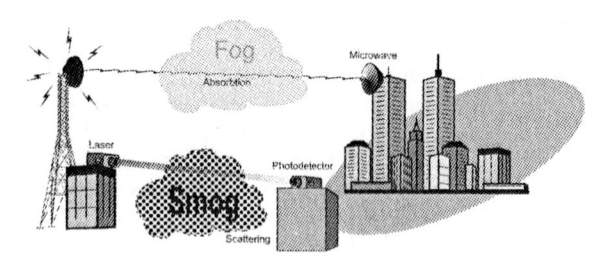

Free Space Transmission System

Free Space Wave-The portion of a radio wave signal that travels directly between antennas with no reflections or refractions.

Freephone-A service that allows callers to dial and telephone number without being charged for the call. The toll free call is billed to the receiver of the call. In Europe and other parts of the world, toll free calls are called Freephone and typically begin with 0-800. In the United States, freephone calls are called toll free calls and they are proceeded by an 1-800, 1-888 or 1-877 exchange.

Frequency-The frequency of an electrical or optical wave is the number of complete cycles or wavelengths that the wave has in a given unit of time (second). The standard measurement for this is number of cycles per second, also known as Hertz (the scientist, not the car company), abbreviated Hz.

This diagram displays how frequency is measured. In this example, there three cycles of a wave that are transmitted over a 1 second period. This equals a frequency of 3 Hertz.

Frequency Agile-Frequency agile systems allow devices to search and establish a communication link on a variety of different frequencies or channels. Frequency agile devices first search for a beacon signal on a pre-determined list of communication channels (set of frequencies) to determine which frequencies are available for communication. The frequency agile device may then attempt to attach itself to the network by registering with the system. This allows the system to know which channel (frequency) the device is listening to (camping on) in the event the system desires to send a message (page) to the communication device.

Frequency Agile Modem-A modem that can search and establish communications on a variety of different frequencies.

Frequency Correcting Burst (FCB)-A burst of information that is sent on a control channel that allows a receiver to adjust its reference frequency.

Frequency Correction Channel (FCCH)-A logical channel that provides a mobile radio with a frequency reference for the telephone system. For the GSM system, the FCCH only contain a frequency correction burst.

Frequency Division Duplex (FDD)-A communications channel that allows the transmission of information in both directions (not necessarily at the same time) via separate bands (frequency division).

This figure shows a frequency division duplex (FDD) system. In this system, the transceivers contain a transmitter and receiver that are operating on two different frequencies. One frequency is used to send signals in one direction and the other frequency is used to send signals in the opposite direction. FDD allows for the simultaneous audio communication between users.

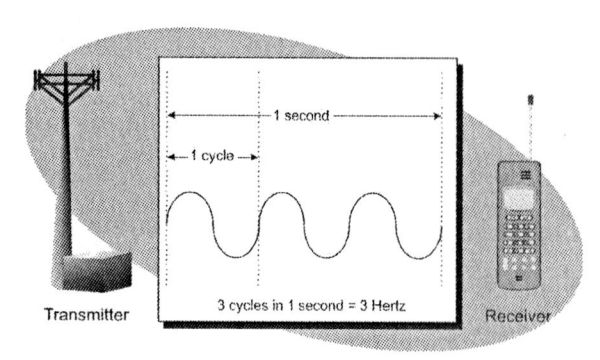

Frequency

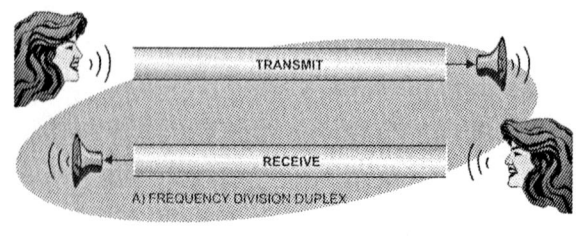

FDD Duplex System Operation

Frequency Hopping

Frequency Division Multiple Access (FDMA)- Frequency division multiple access is a process of allowing mobile radios to share radio frequency allocation by dividing up that allocation into separate radio channels where each radio device can communicate on a single radio channel during communication.

Frequency Division Multiplexing-The multiplexing of two or more signals into one output by assigning each signal its own bandwidth within a broad range of frequencies. Frequency division multiplexing is used to divide a frequency bandwidth into several smaller bandwidth frequency channels. Each of these smaller channels is used for one communications channel.

FM radio, broadcast television and some cellular telephone systems use frequency division multiplexing.

Frequency Division Multiplexing (FDM)- Frequency division multiplexing is used to divide a frequency bandwidth into several smaller bandwidth frequency channels. Each of these smaller channels is used for one communications channel.

This figure shows how a frequency band can be divided into several communication channels using frequency division multiplexing (FDM). When a device is communicating on a FDM system using a frequency carrier signal, it's carrier channel is completely occupied by the transmission of the device. For some FDM systems, after it has stopped transmitting, other transceivers may be assigned to that carrier channel frequency. When this process of assigning channels is organized, it is called frequency division multiple access (FDMA). Transceivers in an FDM system typically have the ability to tune to several different carrier channel frequencies.

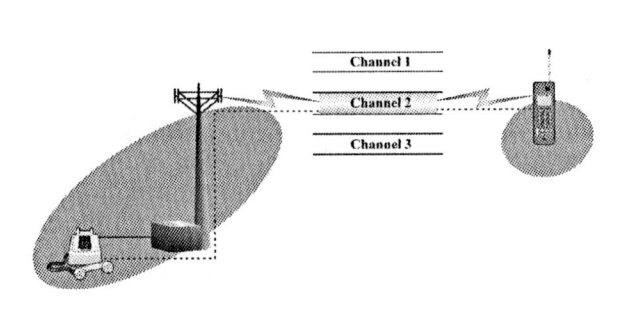

Frequency Guard Band-A bandwidth of frequency that is unused between two channels or bands of frequencies to provide a margin of protection against signal interference between energy that is transmitted outside the assigned frequency bands of channels operating near the guard frequency.

Frequency Hopping-A radio transmission process where a message or voice communications is sent on a radio channel that regularly changes frequency (hops) according to a predetermined code. The receiver of the message or voice information must also receive on the same frequencies using the same frequency hopping sequence. Frequency hopping was first used for military electronic countermeasures. Because radio communication occurs only for brief periods on a radio channel and the frequency hop locations are only known to authorized receivers of the information, frequency hopping signals are difficult to detect or monitor.

This diagram shows a simplified diagram of how a frequency hopping system transfers information (data) from a transmitter to a receiver using many communication channels. This diagram shows a transmitter that has a preprogrammed frequency tuning sequence and this frequency sequence occurs by hopping from channel frequency to channel frequency. To receive information from the transmitter, the receiver uses the exact same hopping sequence is used. When the transmitter and receiver frequency hopping sequences occur exactly at the same time, information can transfer from the transmitter to the receiver. This diagram shows that after the transmitter hops to a new frequency, it transmits a burst of information (packet of data). Because the receiver hops to the same frequency, it can receive the packet of data each time.

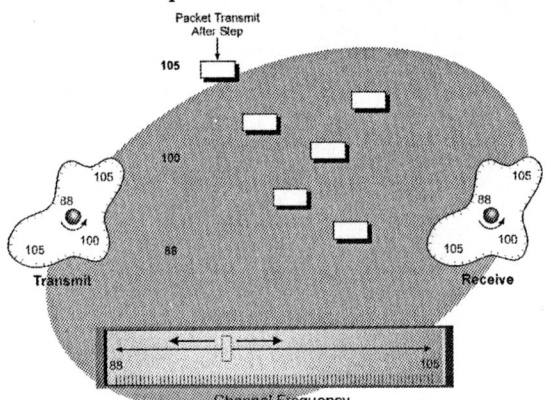

FDD Duplex System Operation

Frequency Hopping Operation

Frequency Hopping Multiple Access (FHMA)

Frequency Hopping Multiple Access (FHMA)- An access technology where mobile radios may share radio channels by transmitting for brief periods of time on a single radio channel and then hopping to other radio channels to continue transmission. Each mobile radio is assigned a particular hopping pattern and collisions that occur are random and only cause a loss of small amounts of data that may be fixed through error detection and correction methods.

Frequency Hopping Spread Spectrum (FHSS)- A communication process that uses multiple frequency channels that constantly change to transfer information. FHSS systems convert information (analog or digital data) into small portions of data (e.g. data packets) and each portion of information is transmitted over assigned and constantly changing frequency channels.

Frequency Modulation (FM)- Frequency modulation is the process of transferring an information signal onto a radio carrier wave by varying the instantaneous frequency of the radio carrier signal. In 1936, the inventor Armstrong demonstrated an FM transmission system that was much less susceptible to noise signals than AM modulation systems.

This figure shows how frequency modulation (FM) uses a modulation signal (audio wave) to change the frequency of the radio carrier signal as the voltage of the audio signal increases and decreases.

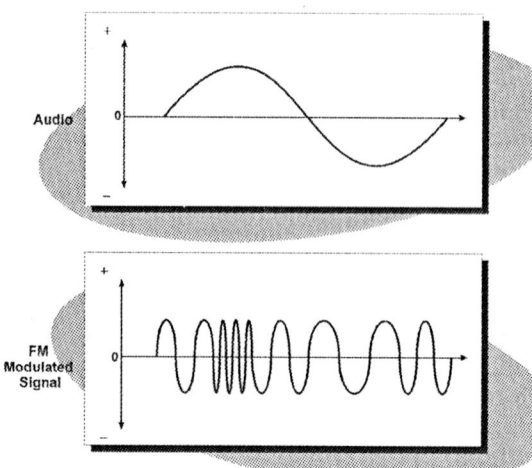

Frequency Modulation (FM) Operation

Frequency Shift Keying (FSK)- A form of frequency modulation in which the modulating signal shifts the output frequency between predetermined values to represent a digital signal. One frequency shift is used to represent a digital one (sometimes called a mark) and the other frequency shift represents a digital zero (sometimes called a space).

This figure shows a sample of frequency shift keying (FSK). In this diagram, each pulse from the digital signal (on top) creates a change in carrier signal frequency (on bottom). As the digital signal voltage is increased, the frequency of the radio signal changes above the center (unmodulated) carrier frequency. When the voltage of the digital signal is decreased, the frequency changes again so the frequency of the transmitted signal is below the center (unmodulated) carrier frequency.

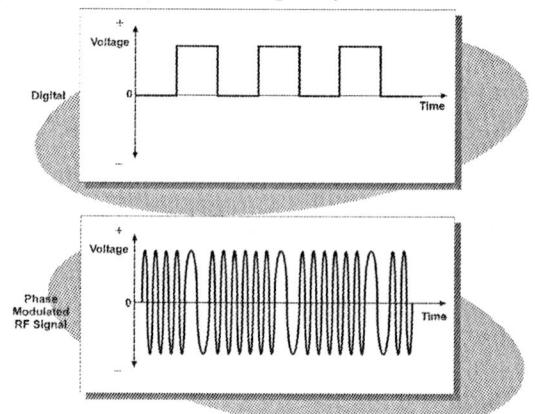

Frequency Shift Keying (FSK) Modulation

Frequency Spectrum- (1-general) The distribution of electromagnetic or acoustic energy as a function of frequency. (2-Fourier analysis) The amplitude and phase of each of the sine waves that compose a signal.

Frequency Synthesizer- A frequency synthesizer is a device or electronic circuit that is capable of producing a range of frequencies based on the settings (programming) of the synthesizer. Frequency synthesizers are usually capable of producing very accurate frequencies by comparing the programmed frequency to a precise reference frequency (usually controlled by a low frequency crystal). The synthesizer uses the programming (frequency setting) to control a dividing circuit (a counter) to samples the output frequency to the frequency of a reference crystal (the precise frequency of a crystal is controlled by its physical size). If the frequency changes above or below the reference frequency, it creates an adjustment signal (voltage) that is used to correct the output frequency.

Frequency Translating Repeater (FTR)-A radio repeater that receives on one frequency and re-transmits on another frequency.

Frequency-Agile-The capability of a transmission/reception system to operate across a broad range of frequencies.

Frequently Asked Questions (FAQS)-A list of questions and answers that are frequently asked about a particular topic or product. A list of FAQS is commonly provided on web sites to assist the visitor in finding information quickly without the need to call a customer service representative.

Fresnel Lens-An optical lens whose surface is broken up into many concentric annular rings.

Fresnel Reflection-A loss mechanism for optical signals traveling through a fiber optic network. When a lightwave reaches the end of a fiber or is connected to another component, it typically encounters a new index of refraction at the inter-face between the two media. This change in index causes reflections. The reflections can be reduced by adding antireflection coatings or index-matching gels between the two media.

Fresnel Reflection Range-The maximum ratio in decibels, as measured by an optical time domain reflectometer, of the optical power for the detector saturation level divided by the received one-way optical power from back scattered energy reflected off a perpendicular planar air/glass inter-face (with respect to the longitudinal axis of the fiber).

Fringe Area-The area outside the designated boundary of the communications system, such as a zone for radio, television, or mobile communication service may be provided.

FRL-Facility Restriction Level
FRMR-Frame Reject
FRND-Frame Relay Network Device
FRS-Family Radio Service
FRS-Frame Relay Service
FRTT-Fixed Round-Trip Time
FS-Form Separator
fs-Femtosecond
FSAN Forum-Full Services Access Network
FSF-Free Software Foundation
FSK-Frequency Shift Keying
FSL-Flexible Service Logic
FSN-Full Service Network
FSO-Foreign Switching Office
FSP-File Service Protocol
FSS-Fixed Satellite Service
FT-1-Fractional T-1
FT-3-Fractional T-3
FTA-Fault Tree Analysis
FTAM-File Transfer Access, And Management
FTAM-File Transfer And Access Management
FTB-Fade To Black
FTE-Full Time Equivalent
FTP-File Transfer Protocol
FTR-Frequency Translating Repeater
FTS-Federal Telecommunications System
FTS2000-Federal Telecommunications System 2000
FTSC-Federal Telecommunications Standards Commission
FTT Signal-Failure-To-Train
FTTC-Fiber To The Curb
FTTCab-Fiber To The Cabinet
FTTH-Fiber To The Home
FTTN-Fiber To The Neighborhood
FTTP-Fiber To The Premise
FUD-Fear, Uncertainty, Doubt
Fuel Cell-A chemical cell that produces electric energy from a chemical reaction.

Fulfillment-The process of preparing the network for customer access, taking orders for service, and initiating that service.

Fulfillment, Assurance, and Billing (FAB)-The three major functions performed by telecommunications companies. Fulfillment is the process of preparing the network for customer access, taking orders for service, and initiating that service. Assurance is the process of making sure that customers receive continuous and high quality service. Billing is the process receiving call detail and service usage information, grouping this information for specific accounts, producing invoices, and recording payments made for those invoices.

Full Access-An arrangement in which all traffic offered to a group of servers (trunks) has access to all the servers in the group.

Full Duplex-Full Duplex communication is the process of transferring of voice or data signals in both directions at the same time. Full duplex operation normally assigns the transmitter and receiver to different communication channels. When the communications system uses two different frequencies for simultaneous communication, it is called frequency division duplex (FDD). One frequency is used to communicate in one direction and the other frequency is required to communicate in the opposite direction.

The definition of full duplex becomes confusing when it is applied to the end result of simultaneous voice and data communication. This is because it is possible to provide information at the input and output of a communication system while not actually sending the information simultaneously in a communication system. When a communication system provides for simultaneous two way communication by time sharing, it is called time division duplex (TDD).

Full Period Service-In reference to the tariff of services, a service, circuit, facility, or piece of equipment that is continuously available for use by a customer.

Full Rate (FR)-(1- telephone) A customer data speed of 56 kbps. For the Integrated Services Digital Network (ISDN), 64 kbps is offered. (2- radio channel) The normal allocation of time slots or data rates for a mobile radio.

Full Rate Fast Associated Control Channel (FACCH/F)-The fast associated control channel used with a full rate traffic channel.

Full Rate TCH (TCH/F)-A traffic channel that uses the maximum number of slots per frame. For the GSM system, this is one slot per frame and for the IS-136 system, this is two slots per frame.)

Full Services Access Network (FSAN) Forum-A forum that was established in 1995 to help identify technologies and network architectures that can cost effectively provide narrowband and broadband telecommunications services.

Full Trunk Group-A normally high-usage trunk group on which no overflow to an alternate route is permitted. Enough trunks must be provided for acceptable service.

Full Wave Rectifier-A circuit or device in which positive and negative half cycles of the incoming ac signal are rectified to produce a unidirectional current (dc) through a load.

Fully Associated Link (F-Link)-An "F" (fully associated) link connects two signaling end points (i.e., SSPs and SCPs) on an associated link within an SS7 network. "F" links are not usually used in networks with STPs. In networks without STPs, "F" links directly connect signaling points.

Function-A process where that accepts one or more information inputs, or arguments, and produces a single output, or value, determined by the combination of the inputs and the formal specification of the function.

Functional Block-In defined process or service that contains one or more processes that interact with other defined functional blocks.

Functional Component-An established subroutine that constructs service logic programs in an intelligent network and can pass queries, responses, and instructions between network elements that are executing service logic programs.

Functional Group Addressing-A technique used for multicasting in Token Ring LANs in which specific bits within a 48 bit MAC Destination Address are associated with predefined functional entities. A frame so addressed will be received by all devices that have implemented the functions corresponding to the bits set within the address.

Functional Unit-An entity of hardware and/or software capable of accomplishing a given purpose.

Fundamental Frequency-The lowest frequency of a composite: signal; higher components often are harmonics or multiples of the fundamental.

Fundamental Mode-The lowest order mode of an optical waveguide, the only mode capable of propagation in a singlemode (monomode) fiber.

FUNI-Frame-based User-Network Interface

Fuse Wire-A thin-gauge wire made of an alloy that overheats and melts at the relatively low temperatures produced when the wire carries overload currents. When used in a fuse, the wire is called a fuse link.

Fused Fiber Splice-A splice accomplished by the application of localized heat sufficient to fuse or melt the ends of two lengths of optical fiber, forming, in effect, a continuous single fiber.

Fused Splice-A fiber splice that melts (fuses) the cable ends to each other. Also called a fusion splice.

Fusion Splice-The splicing together of optical fibers by the application of localized heat sufficient to fuse or melt the two ends, forming a solid continuous fiber. In this way, no reflection or refraction (no signal loss) can occur at the splice interface.

This photo shows the miniMass series 5000 fiber cable splice assembly that can fuse (splice) together optical fibers by the application of localized heat. This device can also be used as an insertion loss tester to verify the splice quality is acceptable.

Fusion Splice Machine
Source: Corning Cable Systems, LLC

Fusion Splicer-A device that is used to joining optical fibers together by fusing (melting) their cores together using a brief electric arc.

Future Public Land Mobile Telephone System (FPLMTS)-The third generation mobile telephone requirements as specified by the international telecommunications union (ITU). This system combines may of the features used in wired telephone systems with wireless networks. FPLMTS was renamed to IMTS-2000.

Fuzzy Logic-A branch of mathematics, used in some artificial intelligence computers, in which decisions are based on ideas and approximations rather than on mathematically rigid calculations.

FVC-Floating Virtual Connection
FVC-Forward Analog Voice Channel
FWA-Fixed Wireless Access
FWM-Four Wave Mixing
FX-Foreign Exchange Service
FX or FEX-Foreign Exchange
FXC-Fiber Switch Cross-Connect
FxM-Fax Mail
FXO-Foreign Exchange Office
FXS-Foreign Exchange Station
FZA-Fernmeldetechnisches Zentralamt

G

G-Giga

G.711-A standard analog to digital coding system (coded) that converts analog audio signals into pulse code modulated (PCM) 64 kbps digital signals. The G.711 is an International Telecommunications Union (ITU) standard for audio codecs. The G.711 standard allows for different weighting processes of digital bits using mu-law and A-law coding. The G.711 standard was approved in 1965.

G.721-A standard analog to digital coding system (coded) that converts analog audio signals into ADPCM 32 kbps digital signals. The G.711 is an International Telecommunications Union (ITU) standard. The G.721 standard has been superseded by G.726.

G.723-An International Telecommunication Union (ITU) standard for audio codecs that provides for compressed digital audio over standard analog telephone lines.

G.723.1-A low bit rate standard for audio codecs developed by the ITU in 1995. The compressed bit rates for G.723.1 include 5.3 kbps and 6.4 kbps. The frame length is 30 msec and the look ahead time is 7.5 msec.

G.726-An audio codec standard developed in 1990 by the ITU to encode the industry standard G.711 64 kbps signals into compressed bit rates. This compression standard uses ADPCM compression and includes data compression rates of 40, 32, 24, and 16 kbps.

G.728-A low-bit rate audio compression standard that uses code-excited linear prediction coding to produce high quality audio with a data transmission rate of 16 kbps.

G.729-A low bit rate speech coder that was developed in 1995. It has low delay due to a small frame size of 10 msec and look ahead of 5 msec. It has a relatively high voice quality level for the low 8 kbps data transmission rate. There are two versions of G.729: G.729 and G.729 A.

G.dmt-G.dmt is an ITU standards for asynchronous digital subscriber line (ADSL). G.dmt permits data transmission rates of up to 8 Mbps downstream and 1.54 Mbps upstream.

G.Lite-The limited version of asynchronous digital subscriber line (ADSL) technology that eliminates or reduces the need the installation of a splitter at the end customers location. The standard allows up to 1.5 Mbps downstream and 384 Kbps upstream.

G/T-Gain-Over Noise And Temperature

G3-Third Generation Mobile System

GA-Generally Available

GA-Go Ahead

ga-Gauge

GaALAs-Gallium Alluminum Arsenide

GAAP-Generally Accepted Accounting Principles

GAB-Group Access Bridging

GADS-Generic Advisory Diagnostic System

GAFF Chip-General Adaptive FIR Filter

Gaffs-Gaffs are hooks or climbing gear devices that are strapped and worn on the legs and underfoot and used in conjunction with belts, lanyards, and harnesses for accessing aerial telephone company (telco) plant.

Gain Frequency Characteristic-The gain-vs-frequency characteristic of a channel over the bandwidth provided. Gain frequency is also referred to as frequency response.

Gain-Over Noise And Temperature (G/T)-A relationship, in decibels, between the gain of a satellite receiving antenna system and the surrounding ambient noise. An increase in G/T corresponds with an increase in usable signal strength. G/T can be raised through the use of a larger receive antenna or an low noise amplifier (LNA) that has a lower noise temperature rating.

Galactic Radio Noise-A radio noise reaching the earth from outer space, in particular from stars in our own galaxy.

Galvanic-Pertaining to a device that produces direct current by chemical action.

Gamma-A relationship expressing the nonlinear characteristic between the input voltage and brightness on a video monitor. Gamma correction is a pre-distortion accomplished at the video source (the camera).

Gamma Rays-A form of high-frequency electromagnetic radiation, emitted by some radioactive elements.

Gamut-A range of voltages or signal levels allowed for a system to operate correctly. Signal voltages outside the range (that is, exceeding the gamut) may lead to clipping, crosstalk, or other distortions.

GAN-Global Area Network

Gang-To mechanically connect two or more circuit devices so that they all can be adjusted simultaneously. An example is the ganged capacitor in a superheterodyne radio receiver, which adjusts the input and oscillator stages at the same time.

Gang Tuning-A process of simultaneously tuning several different circuits by tuning a single shaft on which ganged capacitors are mounted.

Ganged-A device that is mechanically coupled, normally through the use of a shared shaft.

Ganged Capacitor-A variable capacitor with more than one set of moving plates linked together.

GAP-Generic Access Profile

GAP-Ground to Air Paging

GARP-Growth At a Reasonable Price

GAs-Gallium Arsenide

Gas Filled-A tube or glass envelope that contains apparatus and is filled with gas, either to improve the functioning of the apparatus or as an essential feature of the circuit itself.

Gassing-The production of gas at electrodes during electrolysis. For example, gassing occurs when a lead-acid storage battery is given an equalizing charge at a near fully charged state.

Gatekeeper (GK)-A gatekeeper is a server that coordinates access to other servers. The gatekeeper receives requests from clients, determines the destination server that it needs to communicate with, and coordinates access with that server. For packet voice systems, the gatekeeper translates user names or telephone numbers into physical address for H.323 conferencing.

This diagram shows the basic functions of the gatekeeper that controls a voice gateway. This diagram shows that the gatekeeper performs access control, address translation, services coordination, control signaling coordination, and billing recording. The gatekeeper receives requests for services from the voice gateway. The gatekeeper will check it's databases to ensure the caller is an authorized customer and what services the customer is allowed to use. If it is a request for an outgoing call, the voice gateway will then provide the gatekeeper with the destination IP address or dialed telephone number. The gatekeeper will search for the IP address of a remote gatekeeper in its database that is located near the called number. If there is no registered (pre-authorized) gatekeeper, the gatekeeper may send a broadcast request message to other gatekeepers in an attempt to find a gatekeeper that is willing to service the call. The Gatekeeper will then send a request message to the remote gatekeeper to request access to the destination telephone or communication device. If the destination gatekeeper authorizes access, the gatekeeper will then negotiate for the parameters for communication between the voice gateways (e.g. speech coder). The gateway will initiate a billing record of the call and its associated usage records.

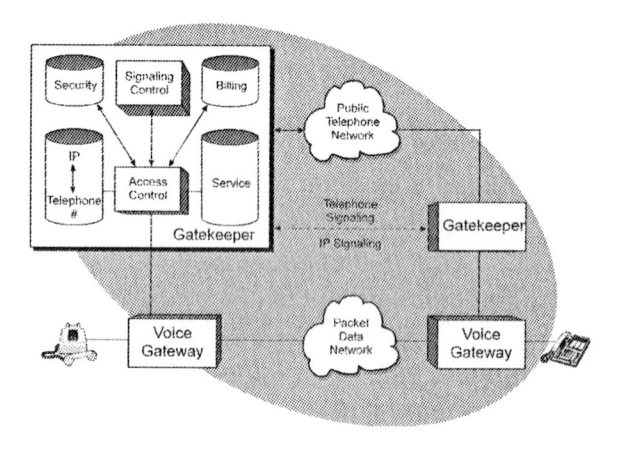

Voice Gatekeeper Operation

Gatekeeper Cluster-A group of gatekeepers that are linked together (possibly using GUP) to increase the reliability of a system.

Gatekeeper Update Protocol (GUP)-A proprietary protocol developed by Cisco to provide gatekeeper redundancy and load sharing. GUP can provide information about a gatekeeper's memory, CPU usage, number of endpoints that are registered, and available bandwidth. GUP is based on TCP.

Gateway-A gateway is a communications device or assembly that transforms data that is received from one network into a format that can be used by a different network. A gateway usually has more intelligence (processing function) than a bridge as it can adjust the protocols and timing between two dissimilar computer systems or data networks. A gateway can also be a router when its key function is to switch data between network points.

his figure shows how a gateway can convert large packets from a FDDI into very small packets in an ATM network. Not only does the gateway have to divide the packets, it must also convert the addresses and control messages into formats that can be understood on both networks.

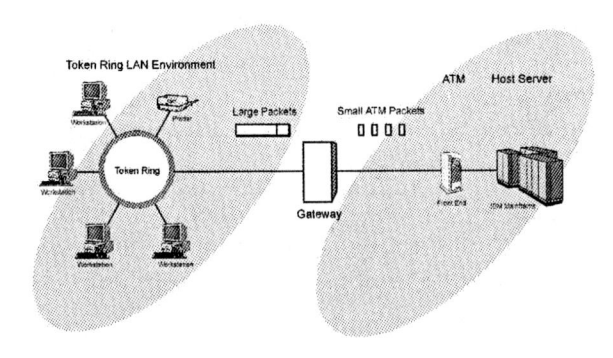

Gateway Operation

Gateway D-Gateway Daemon

Gateway GPRS Support Node (GGSN)-A packet switching system that is used to connect a GSM mobile communication network (GPRS Support Nodes) to other packet networks such as the Internet.

Gateway Location Protocol (GLP)-A protocol that was initially developed by a working group within the IETF to allow the gateway selection process in telephone and multimedia networks. This protocol is now called telephony routing over Internet protocol (TRIP).

Gateway Mobile Switching Center (GMSC)-A switching system that is used in a mobile communications network that also connects to other networks such as the public switched telephone network (PSTN).

Gating Pulse-A pulse that operates a logic gate.

Gauge (ga)-(1-wire) A measure of wire diameter. Under the American Wire Gauge (AWG) system, higher numbers indicate thinner wire. The AWG number indicates the number of times the wire has been "drawn" or pulled through successively thinner forming dies to make it thinner. The term Browne & Sharpe (B&S) gauge is a synonym for AWG. Browne & Sharpe is a manufacturer of measuring instruments. Sheet metal gauge, a similar number system indicating the number of times that the sheet metal has been rolled between rollers to make it thinner, has a different dimension for the same gauge number and for different materials. That is, 12 ga sheet copper is not the same thickness as 12 ga copper wire, nor is it the same thickness as 12 ga sheet iron. Note that outside the United States, copper wire and most other wires and sheet metals are expressed by their actual thickness in millimeters (mm) and the term gauge is not used. (2-instrument) A device that measures a value, such as pressure or temperature.

Gaussian Beam-A beam of light whose radial intensity distribution is Gaussian.

Gaussian Minimum Shift Keying (GMSK)-A form of frequency modulation in which the modulating signal shifts the output frequency between predetermined values. A form of MSK that uses gaussian low pass filtering of the binary data to reduce sideband energy.

GAZPACHO-Generation, Alignment, Zero [suppression], Polar, Alarm, Clock, Hunt, Office

GB-Gigabyte

GBH-Group Busy Hour

Gbps-Gigabits Per Second

GC-Gate Controller

GCAC-Generic Connection Admission Control

GCIDs-Global Call Identifiers

GCR-Group Call Register

GCRA-Generic Cell Rate Algorithm

GD-Graceful Discard

GDDM-Graphical Data Display Manager

GDF-Group Distribuiton Frame

GDI-Graphics Device Interface

GDOI-Group Domain of Interpretation

GDOP-Geometric Dilution of Precision

GDP-PI-GDP Price Index

GDS-Global Directory Service

GED-Global Engineering Documents

Geiger Counter-A measurement device that is used to detect ionizing radiation and to count particles from radiation.

General Packet Radio Service (GPRS)-A portion of the GSM specification that allows packet radio service on the GSM system. The GPRS system adds (defines) new packet control channels and gateways to the GSM system.

General Parameters

This figure shows some of the key GPRS network elements that include a gateway GPRS support node (GGSN), a serving GPRS support node (SGSN) and a GPRS backbone network (the Internet in this example.) This example shows that the GPRS system adds dynamic time slot control to the standard GSM radio system. To provide packet data service, the GPRS system, the SGSN provides the processes of switching and access control that is similar to a mobile switching center (MSC) and a visitor location register (VLR). However, the SGSN provides for switching and access control (authorization and tracking) based on packets of data rather than continuous connections. The SGSN registers and maintains a list of active packet data radios in its network and coordinates the packet transfer between the mobile radios. The GGSN is a packet switching system that is used to connect a GSM mobile communication network (GPRS Support Nodes) to other packet networks such as the Internet.

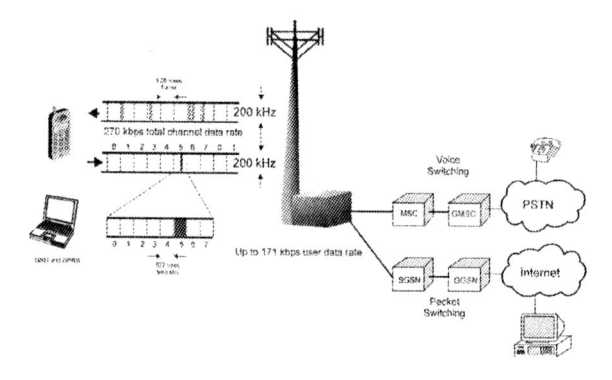

General Packet Radio Service (GPRS) System

General Parameters-In specification description language, the basic operating parameters of a unit or system.

General Purpose Interface (GPI)-(1-parallel) A parallel interconnection scheme that allows remote control of certain functions of a device. One wire is dedicated to each function. (2-computer) An interface for data processing equipment. The term usually refers to a serial connection (RS232 or RS422 format) between computer devices.

General Register-An internal addressable register in a central processing unit that can be used for temporary storage, as an accumulator, or for any other general-purpose function.

General Switched Telephone Network (GSTN)-A name used by the International Telecommunications Union (ITU) to describe telephone networks (such as public telephone networks) that may connect to packet telephony systems.

General Telemetry Processor (GTP)-A device that is used to receive and process telecommunications equipment alarming protocols.

Generalized Mark-up Language (GML)-Generalized mark-up language is a precursor to SGML.

Generation Loss-Loss caused by the copying of scenes from one videotape to another.

Generation, Alignment, Zero [suppression], Polar, Alarm, Clock, Hunt, Office (GAZPACHO)-The process used to terminate digital signals. It includes generation of frame codes, alignment of frames, zero string suppression, polar conversion, alarm processing, clock recovery, hunt during reframe, and office signaling.

Generic Flow Control Field (GFC)-A 4 bit field that is part of an ATM packet header. The GFC controls the flow between the user equipment and ATM edge switch. A GFC field informs the end-station of network information including congestion control.

Generic Object Exchange Profile (GOEP)-Defines the protocols and procedures used by the applications providing the usage models which need object exchange capabilities. Examples of such usage models include Synchronization, File Transfer, and Object Push. The most common devices using these usage models include notebook PCs, PDAs, smart phones, and mobile phones.

Generic Requirement (GR)-A specification document controlled by Telcordia (formerly Bellcore) that defines requirements of systems that connect within a telephone network.

Genlock Module-A module that can phase lock to another source of video or sync.

GEO-Geosynchronous Earth Orbit

Geographic Number Portability (GNP)-Geographic number portability involves the transfer of telephone numbers for telephone devices or services that are used outside the normal geographic boundaries of the service provider's original system or area. Geographic number portability allows a customer to keep their same area code when they move to new cities or other distant geographic regions.

Geographic Spectral Efficiency-A measurement characterizing a particular modulation and coding method that describes how much information can be transferred in a given bandwidth within a cellular radio system having reuse of the same radio carrier frequency at the closest permitted reuse cells. This is often given as bits per second per Hertz per square km of service area, or alternatively as bits per second per Hertz per cell. Modulation and coding methods that have high spectral efficiency often typically have very low geographic spectral efficiency and are not suitable for a cellular radio system. (See also Economic Geographic Spectral Efficiency or Spectral Efficiency)

Geometric Optics-The science that treats the propagation of light as rays. Rays are bent at the interface between two dissimilar media, or they may be curved in a medium whose refractive index is a function of position.

Geostationary Satellite-A satellite that orbits the Earth at the same relative speed that the earth rotate, resulting in the satellites fixed relative position to the surface of the Earth. Geostationary satellites have an altitude of approximately 22,300 miles (35,680 km) above the Earth when they are in Geostationary position.

This diagram shows that a geostationary satellite is launched from Earth to achieve an orbits of approximately 22,300 miles above the Earth. At this position, it appears to have the same relative position above the surface of the Earth. This diagram shows that the combined effects of speed and gravity actually keep the satellite in the same relative position above the Earth.

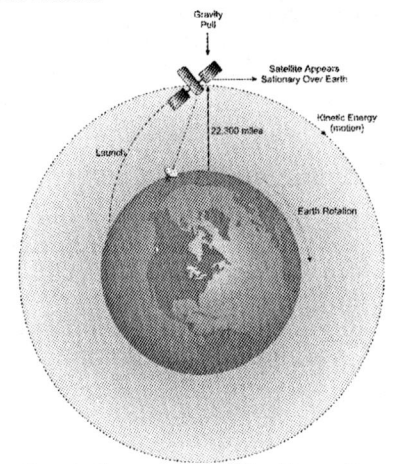

Geostationary Satellite Operation

Geosynchronous Earth Orbit (GEO)-A satellite system where the satellites are located approximately 23,500 miles above the Earth. GEO systems are unique because at the specific height of 23,500 miles, the rotation of the Earth matches the rotation of the satellite resulting in the appearance of a fixed location of the satellite relative to the surface of the Earth.

This figure shows a GEO satellite system. This diagram shows that the GEO satellite is connecting a telephone caller on land with a ship. The telephone call is routed through the gateway to the satellite. The satellite transponder converts the frequency and retransmits the signal back to earth. A satellite telephone on the ship receives the radio signal and converts it back to the original audio signal. The diagram also shows that a separate ground control facility is used to monitor and control the position of the satellite.

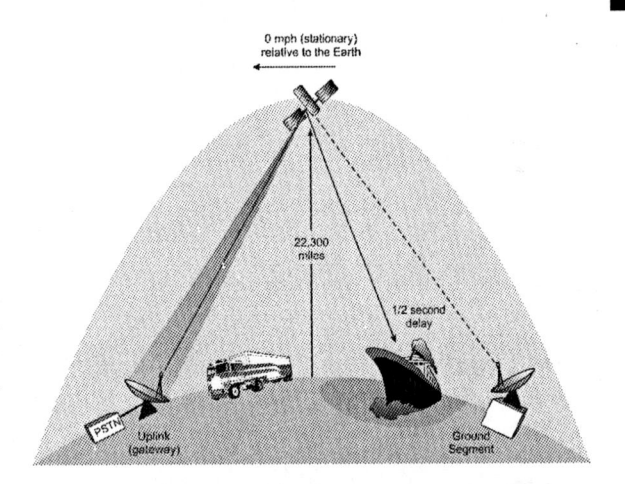

Geosynchronous Earth Orbit (GEO) Operation

GERBER File-A standard file format that is used to control plotters that produce film for printed circuit board production. The industry standard specification for GERBER files is RS-274-D. An expanded version of the standard has been created with the designation RS-274-X.

The GERBER format was first introduced in 1980. In 1991, the standard was enhanced to include embedded aperture codes. This allows each file to be self defining, needing no other accompanying documents for processing. Named for Joseph Gerber, inventor of early electromechanical line plotting/drafting machines.

Germanium-A metallic semiconductor material used in making transistors.

GET-Get is the simplest of the SNMP operations. The get operation is a request for an agent to return one or more objects or specific pieces of information. For each object requested (MIB object), the agent will return that object's value.

Get-Bulk-The get-bulk SNMP operation was added by SNMPv2 and improves performance of retrieving large amounts of SNMP data. This operation allows the agent to pack as many values in each get-response packet as will fit without exceeding the maximum SNMP packet size (The packet size is pre-defined or user defined on the agent).

Get-Next-The get-next SNMP operation allows a network management system (NMS) to request the "next" object or objects in the MIB supported by the agent.

Get-Response-The get-response SNMP operation is the packet type that SNMP agents send in response to receiving a get, get-next, get-bulk, or set-request packet.

GETS-Government Emergency Telecommunications Service

Getter-A metal used in vaporized form to remove residual gas from inside an electron tube during manufacture.

GFC-Generic Flow Control Field

GFCI-Ground Fault Circuit Interrupt

GFI-General Format Identifier

GFP-Ground Fault Protector

GGP-Gateway to Gateway Protocol

GGSN-Gateway GPRS Support Node

GH Effect-Gordon-Haus Effect

Ghost-A form of distortion, in a TV picture, in which a duplicate image is displayed, offset from the primary picture image. In VHF and UHF reception, ghosts result when the receiver antenna detects both the direct line of sight signal from the transmitter and one or more weaker signals reflected by nearby buildings, mountains, or other obstructions.

Ghost Image-The displaced image of a received TV picture, resulting from a secondary signal that is received slightly after the primary signal. Secondary signals can be caused by multipath propagation, such as a reflection from a nearby building. (See also: multipath propagation.)

GHz-Gigahertz

GIF-Graphics Interchange Format

Giga-A prefix that represents one billion units.

Gigabit-One billion bits of data.

Gigabit Ethernet Alliance-An alliance between a several companies that was formed in mid 1990s that is assisting in the creation of a Gigabit Ethernet system.

GIGO-Garbage In, Garbage Out

GII-Global Information Infrastructure

GIM-Group Identification Mark

GIP-Global Internet Project

GIS-Geographic Information Systems

GIX-Global Internet Exchange

GK-Gatekeeper

Glare Backout-The process of releasing a communication trunk that is seizure by one end office switch so that a call from another end office switch can be completed. This is also called glare release.

Glare Control-Glare is the conflict that occurs in the assignment of a communication trunk line (interconnection line) in a communication system when the line is accessed at both ends simultaneously. Glare is overcome through the use of glare hold and glare release processes.

Glare Release-A process of releasing a communication trunk that is simultaneously seized by two switches at the same time. Glare release results when one switch is given a lower priority and release the line. This is also called glare blackout.

Glass-A general term applied to a wide variety of chemical formulations. Glasses are amorphous solids.

Global-Applies to all users or devices connected to a network or system.

Global Call Identifiers (GCIDs)-A unique identifier (a tag) that is assigned to a telephone call to allow all events associated with that call to be easily identified and grouped. GCIDs were initially developed for computer telephony interface (CTI) systems.

Global Control-Global Channel

Global Engineering Documents (GED)-The company that manages and sells TIA and EIA industry standard specifications.

Global Navigation Satellite System (Glonass)-A Russian global positioning navigation system, similar to the US Global Positioning System (GPS).

Global Positioning System (GPS)-The Global Positioning System (GPS) is a network of 24 Navstar satellites that orbiting the Earth at 11,000 feet above the surface that provide signals that allow the calculation of position information. The transmit frequency for the GPS system is 1575.42 MHz. A GPS receiver compares the signals from multiple GPS satellites (4 satellite signals are usually used) to calculate the geographic position. In March of 1996, the military requirements for limited accuracy transmission were lifted. This allows the GPS system to provide very precise vehicle or device locations.

This figure shows a global positioning satellite (GPS) system. This diagram shows how a GPS receive receives and compares the signals from orbiting GPS satellites to determine its geographic position. Using the precise timing signal based on a very accurate clock, the GPS receiver compares these signals from 3 or 4 satellites. Each satellite transmits its exact location along with a timed reference signal. The GPS receiver can use these signals to determine its distance from each of the satellites. Once the position and distance of each satellite is known, the GPS receiver can calculate the position where all these distances cross at the same point. This is the location. This information can be displayed in latitude and longitude form or a computer device can use this information to display the position on a map on a computer display.

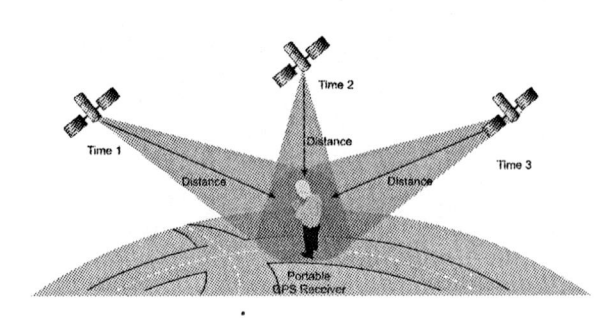

Global Positioning System (GPS) Operation

This photo image is a handheld GPS receiver that is capable of showing terrain maps using the position information gathered by the GPR receiver section.

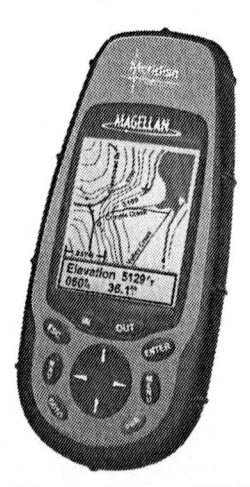

Global Positioning System (GPS) Receiver
Source: Magellan Meridian Platinum by Thales Navigation

Global System For Mobile Communications (GSM)-A digital cellular telephone system which originated in Europe and is now available in over most of the world. The GSM system uses 200 kHz wide channels that are divided into frames that hold 8 time slots. GSM was originally named Groupe Special Mobile.

Global Title (GT)-A routing name (such as customer-dialed digits) that does not contain explicit information to enable routing in a communication network. The global title is usually converted to an address that allows the network to setup or route information to its ultimate destination point.

Global Title Translation (GTT)-A process used in a common-channel signaling system (such as SS7) that uses a routing table to convert an address (usually a telephone number) into the actual destination address (forwarding telephone number) or into the address of a service control point (database) that contains the customer data needed to process a call.

Globalstar-A low Earth orbit (LEO) satellite system that uses code division multiple access (CDMA) technology.

Globus Toolkit-A toolkit for providing security, resource management, data management, fault detection, information infrastructure and communication to computational grid networks.

Glonass-Global Navigation Satellite System
GLP-Gateway Location Protocol
GMD-Gesellschaft fur Mathematik und Datenvarabeitung
GMDSS-Global Maritime Distress And Safety System
GML-Generalized Mark-up Language
GMPCS-Global Mobile Personal Communications Services
GMRS-General Mobile Radio Service
GMSC-Gateway Mobile Switching Center
GMSK-Gaussian Minimum Shift Keying
GMT-Greenwich Mean Time
GNE-Gateway Network Element
GNN-Global Network Navigator
GNP-Geographic Number Portability
GNP PI-GNP Price Index
GNSS-Global Navigation Satellite Systems
GO-MVIP-Global Organization for MVIP
GOEP-Generic Object Exchange Profile
Golay Sequential Coding (GSC)-In the 1970's, Motorola introduced a paging system called Golay Sequential Coding. The Golay format that is still used in many systems today allows numeric and text characters to be transmitted to paging devices. The Golay system uses 2 different signaling rates for addressing and data information. The addressing information is sent at 300 bps and the data information (numeric or text message) is sent at 600 bps.
GOLD-Global Online Directory
GOP-Group of Pictures
GOS-Grade Of Service
GOSIP-Government Open System Interconnection Profile
GOSIP-Government Open Systems Interconnection Profile
GP-Guard Period
GPA-General Purpose Adapter
GPB-Grand Pooh-Bah
GPC-Gateway Protocol Converter
GPF-General Protection Fault
GPI-GammaFax Programmers Interface
GPI-General Purpose Interface
GPI-Programmable General-Purpose Interface
GPRS-General Packet Radio Service
GPS-Global Positioning System
GPS Readings-Ground position satellite (GPS) locations involving latitudes and longitudes that are required when site surveying wireless prospects (e.g. tower or rooftop proposed sites).

GR-Generic Requirement
GR-303-A set of technical specifications that help define the next generation of digital loop carrier (DLC) interconnection.
Grade Of Service-(1 - general) The customer satisfaction level (estimated) associated with a particular aspect of service (such as voice quality). (2 - non-completed calls) A measure of the proportion of calls that cannot be completed because of limitations in the network capability to handle many simultaneous calls. (3 - subscriber features) A feature level granted to customers (typically in a cellular or PCS system) that allow access to basic or advanced services.
Graded Index Fiber (GRIN)-An optical fiber with an index of refraction that varies gradually across the diameter of the core of the fiber. As in step index fiber, the cladding has a lower index than that of the core in order to promote total internal reflection. The grading of the index in the core causes the light to shift direction gradually as it approaches the interface with the cladding. Most light never reaches the interface as a result, reducing losses to the cladding.
Graded Index Optical Waveguide-A waveguide that has a graded index profile. Because the refractive index is highest at the center of the core, light pulses that travel straight down the center do not move as fast as the ones that take long, zigzag paths. The result is reduced modal dispersion and increased maximum operating frequency.
Graded Index Profile-An index profile for a fiber optical cable that varies smoothly with the radius of the fiber.
Granularity-(1-network operation) The steps of network resource allocation. This is typically bandwidth allocation that is adjustable per communication session. (2-digital conversion) The increments (step size) of conversion from an analog signal to a digital signal.
Graphic User Interface (GUI)-The use of graphics (typically on a computer monitor display) to interface the output or requested input of a software application with a user. The use of buttons, icons and dynamically changing windows are typical examples of a GUI. Sometimes pronounced "gooey interface."
Graphics Interchange Format (GIF)-A graphics file data compression format that produces in relatively small graphics file. A GIF image may contain up to 256 colors and uses a lossless data compres-

sion method. A version of the GIF format GIF89a permits the addition of animation features, image interleaving, and the use of transparent backgrounds.

Graticule-A fixed pattern of reference marking used with oscilloscope CRTs to simplify measurements. The graticule may be etched on a transparent plate covering the front of the CRT; for greater accuracy in readings, it may be electrically generated within the CRT itself.

Grazing Path-A radio path that does not have a clean line of sight between the transmitting and receiving antennas; a path that grazes ground level.

GRE-Generic Routing Encapsulation

Great Understanding Relatively Useless (Guru)-A consultant or expert with a reputation for being helpful to other, less knowledgeable users.

Green Application-A service application or software program that is designed to efficiently use system resources.

GREP-Generalized Regular Expression Parser

GRIC-Global Roaming [For Reach] Internet

GRIC-Global Reach Internet Connection

Grid-(1-computing) A distributed computing infrastructure aimed at advanced scientific research. The Grid provides access control, computational resources, and resource discovery over large sets of data. This allows for the formation of a virtual organizations centered around a specific area of research where sharing data, computing resources and applications are important for collaboration. (2-tube) A mesh electrode within an electron tube that controls the flow of electrons between the cathode and plate of the tube. (The potential applied to a grid in an electron tube to control its center operating point. (3-control) The grid in an electron tube to which the input signal usually is applied. (4-screen) The grid in an electron tube, typically held at a steady potential, that screens the control grid from changes in anode potential. (5 - suppressor) The grid in an electron tube near the anode (plate) that suppresses the emission of secondary electrons from the plate.

GRID-Global Resource Information Database

Grid Security Infrastructure (GSI)-Security scheme used within the Globus toolkit that utilizes public key encryption, X.509 certificates and secure sockets layer to provide a single sign-on and authentication to distributed resources on a shared network.

GRIN-Graded Index Fiber

Gross Add-Refers to the total addition of a new subscribers to a communication system.

Ground Absorption-The loss of energy during the transmission of radio waves through absorption in the ground.

Ground Constants-The electrical parameters of the earth, such as conductivity and dielectric constant. These values vary with the chemical composition of the earth, moisture content, and frequency of interest.

Ground Field/Earth Interface-A configuration of metallic conductors, reds, and pipes used to establish a low-resistance contact with the earth.

Ground Loop-An undesirable circulating ground current in a circuit grounded via multiple connections or at multiple points.

Ground Plane-(1-circuit) A conducting material at ground potential, physically close to other equipment, so that connections may be made readily to ground the equip equipment at the required points. (2-radio) A conducting surface used to provide uniform reflection of an impinging electromagnetic wave. An example is the arrangement of buried wires at the base of an antenna tower.

Ground Plane Antenna-A unipole antenna with radials at the base forming a ground plane. For example, the radial elements in a mobile vehicle communication system are formed by the roof of the vehicle.

Ground Reflected Wave-A radio communication wave that is reflected from the ground or water in its path between the transmitting and receiving antennas.

Ground Return-A conductor used as a path for one or more circuits back to the ground plane or central facility ground point.

Ground Rod-A metal red driven into the earth and connected into a mesh of interconnected rods to provide a low-resistance link to ground.

Ground Rod Method-A method of grounding isolated cable, wireless towers, access points, and other telephone company (telco) facilities by burying five 6 foot long copper ground rods and bonding them together with # 6 AWG copper ground wire.

Ground Segment-The communications gateway that connects a satellites to the telephone network (or other type of network such as the Internet).

Ground Wave-A radio wave that is propagated over the earth and ordinarily is affected by the presence of the ground. Ground waves include all components of waves over the earth except ionospheric and tropospheric waves, and are affected somewhat by changes in the dielectric constant of the lower atmosphere. A ground wave also is known as a surface wave.

Ground Wire-A conductor used to extend a low-resistance earth ground to protective devices in a facility.

Grounded-The connection of a piece of equipment to earth via a low-resistance path.

Group Address-An address used to identify a set stations or devices as the destination for transmitted data. See Multicasting.

Group Call-A call from one mobile radio, telephone or dispatcher to a predefined group of receivers in a group that are capable of receiving and decoding the messages.

Group Call Register (GCR)-A network database that holds the attributes for the set-up and processing of voice group and broadcast calls. These include group call membership lists, priority authorization, locations of group callers.

Group Delay-(1-general) A condition in which the various frequency elements of a given signal suffer differing propagation delays through a circuit or a system. The delay at a lower frequency is different than the delay at a higher frequency, resulting in a time related distortion of the signal at the receiving point. (2-optical) The transit time required for optical power, traveling at the group velocity of a given mode, to traverse a specified distance. The measured group delay of a signal through an optical fiber has a dependence on specific wavelengths because of the each wavelength reacts differently to dispersion mechanisms in the fiber.

This figure shows how group delay can cause pulsed signals, such as in digital transmission system, can cause signal distortion. This diagram shows that a digital pulse signal is actually composed of many low, medium, and high frequency components. As the pulse is transmitted through the transmission line, some of the frequency components are delayed more than others. This results in a distorted pulse at the receiving end of the transmission line.

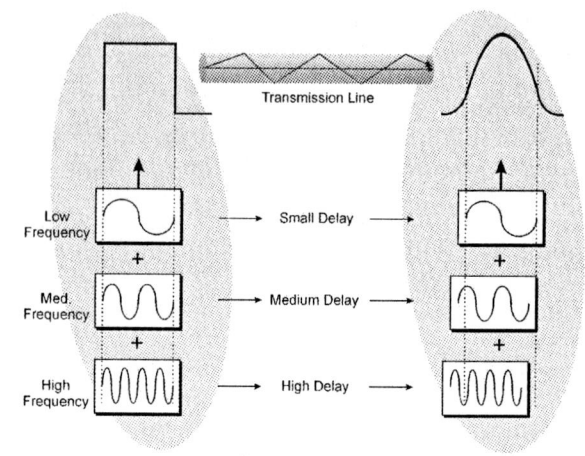

Group Delay

Group Delay Time-The rate of change of the total phase shift of a waveform with angular frequency through a device or transmission facility.

Group Domain of Interpretation (GDOI)-A protocol that allows users or devices to share and update cryptographic keys for group members.

Group Identification-A subset of the system identification (SID) that identifies a group of cellular systems.

Group Identity (GroupID)-A groupID or telephone number that indicates the group for which a group call or broadcast message is intended.

Group Index-The ratio of the speed of light in a vacuum to the group velocity of pulsed light in a transmission medium.

Group Index Of Refraction-The velocity of light in a vacuum divided by the group velocity of light in a fiber.

Group of Pictures (GOP)-In video compression, a group of pictures is an encoding of a minimal sequence of frames that contain all the information to be completely decoded. For all frames within a GOP that reference other frames (such as B-frames and P-frames), the frames so referenced (I-frames and P-frames) are also included within that same GOP.

Group Paging-A telephone call-handling feature that provides a quick way to contact a specific group (list) of users.

GroupID-Group Identity
GS-Group Separator
GS Trunk-Ground Start Trunk
GSC-Golay Sequential Coding
GSF-Generic Services Framework

GSI-Grid Security Infrastructure
GSM-Global System For Mobile Communications
GSM-Special Mobile Group
GSM PLMN-GSM Public Land Mobile Network
GSM PLMN Operator-A company that operates the GSM system.
GSM Public Land Mobile Network (GSM PLMN)-A generic name for all GSM mobile wireless networks that use earth base stations rather than satellites — the mobile equivalent of the PSTN.
GSO-Geostationary Satellite Orbit
GSST-General Subscriber Service Tariff
GSTN-General Switched Telephone Network
GT-Global Title
GTA-Government Telecommunciations Association
GTE-General Telephone and Electronics
GTF-General Trunk Forecast
GTG-Ticket Granting Ticket
gTLD-Generic Top Level Domain
GTN-Global Transaction Network
GTP-General Telemetry Processor
GTP-U-GPRS Tunnelling Protocol User Plane
GTT-Global Title Translation
Guaranteed Service-Guaranteed service provides a specific bandwidth with a set maximum end-to-end transmission delay time.
Guard Band-A portion of a resource (frequency or time) that is dedicated to the protection of a communication channel from interference due to radio signal energy or time overlap of signals. While guard bands protect a desired communication channel from interference, the guard band also uses part of the valuable resource (frequency bandwidth or time period) for this protection.
Guard Period (GP)-A time period that is a portion of a burst period where no radio transmission can occur. The guard period is used to protect adjacent burst from transmission overlap due to propagation time from the mobile radio to the base station.
Guard Time-A time allocated within a single time slot period in a communication system to help ensure variable amounts of transit times (e.g. from close and distant transmitters) do not cause overlap (collisions) between adjacent time slots. Transmission of information does not occur within the guard period.
GUI-Graphic User Interface
GUID-Globally Unique Identifier
Guide Cutoff-The lowest frequency that a particular waveguide size will propagate.

Guiding Billing Records-Guiding is a process of matching call detail records to a specific customer account. Guiding uses the call detail record identification information such as the calling telephone number to match to a specific customer account.
GUP-Gatekeeper Update Protocol
Guru-Great Understanding Relatively Useless
Guy Anchor-An anchor point (mount) where a antenna guy wire is attached.
Gyrator-A directional phase changer used in some types of transmission line.

H

H-Horizontal

H-henry

H channel-High-Speed Channel

H Drive-Horizontal Drive

H Phase-Horizontal Phase

H Rate-Horizontal Rate

H&V Lock Time-Horizontal and Vertical

H.225-H.225 call signaling is used to set up connections between H.323 endpoints (terminals and gateways), over which the real-time data can be transported. Call signaling involves the exchange of H.225 protocol messages over a reliable call-signaling channel. For example, H.225 protocol messages are carried over TCP in an IP based H.323 network.

H.235-The H.235 standard defines security procedures used for packet based communication systems that includes authentication, data integrity, privacy, and non-repudiation processes. Authentication is a mechanism to make sure that the endpoints participating in the conference are really who they say they are. Integrity provides a means to validate that the data within a packet is indeed an unchanged representation of the data. Privacy/Confidentiality is provided by encryption and decryption mechanisms that hide the data from eavesdroppers so that if it is intercepted, it cannot be viewed. Non-Repudiation is a means of protection against someone denying that they participated in a conference when you know they were there. These security "hooks" are introduced in H.323 Version 2.

H.245-A signaling control protocol that contains a library of transmission control messages for use in packet based multimedia communication systems. H.245 control signaling consists of the exchange of end-to-end H.245 messages between communicating H.323 endpoints. The H.245 control messages are carried over H.245 control channels. The H.245 control channel is the logical channel 0 and is permanently open, unlike the media channels. The messages carried include messages to exchange capabilities of terminals and to open and close logical channels.

H.248-H.248 is a control protocol specified by the ITU that uses text or binary format messages to setup, manage, and terminate multimedia communication sessions in a centralized communications system. This differs from other multimedia control protocol systems (such as H.323 or SIP) that allow the end points in the network to control the communication session. H.248 is also known as media gateway control protocol (MGCP) and it is the basis of the PacketCable NCS protocol.

H.261 Quarter Common Intermediate Format (QCIF)-An ITU standard for video encoding that was created in (1990).

H.263-An ITU standard for video encoding that improves on the H.261 version that was created in (1990).

H.320-A videoconferencing standard developed by the ITU-T for transmitting video and audio over circuit-switched digital networks. The H.320 system supports the use of video compression protocols H.261 and H.263.

H.323-H.323 is an umbrella recommendation from the International Telecommunications Union (ITU) that sets standards for multimedia communications over Local Area Networks (LANs) that may not provide a guaranteed Quality of Service (QoS). H.323 specifies techniques for compressing and transmitting real-time voice, video, and data between a pair of videoconferencing workstations. It also describes signaling protocols for managing audio and video streams, as well as procedures for breaking data into packets and synchronizing transmissions across communications channels.

H.323 Gateway-The H.323 gateway is a communications device or assembly that transforms audio that is received from a telephone device or telecommunications system (e.g. PBX) into a format that can be used by a data network. A voice gateway usually has more intelligence (processing function) than a bridge as it can select the voice compression coder and adjust the protocols and timing between two dissimilar computer systems or voice over data networks.

H.323 Packet-Based Media Communications Systems

This figure shows the basic structure of an H.323 system. This diagram shows that the H.323 system can interconnect standard telephones and data communication devices (multimedia computers) through the use of gateways and gatekeepers. Gateways convert the audio and multimedia information into formats that can be transmitted through the packet data network. Gatekeepers coordinate, authorize, and bill (if billing is required) access through the gateways. This diagram shows that when calls are initiated through the H.323 network, the gateway requests access from the gatekeeper. The gatekeeper reviews its database to determine if the request is authorized and may perform translation of dialed digits to a data (IP) address. The destination gatekeeper is then contacted and if it authorizes service, its associated gateway will be setup to translate the call to the communication device (e.g. telephone) that is receiving the call.

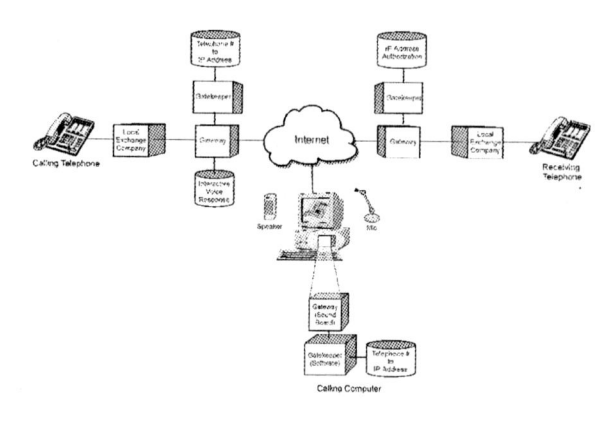

This table shows that the H.323 protocol suite is composed of several different types of protocols. Some of these protocols are dedicated for audio processing, data compression, and video transmission. Other protocols are used for control and special feature processing.

Audio	Data	Video	Transport
			H.225
			H.235
G.711	T.122		H.245
G.722	T.124		H.450.1
G.723.1	T.125	H.261	H.450.2
G.728	T.126	H.263	H.450.3
G.729	T.127		RTP
			X.224.0

H.323 System

H.323 System

H.323 Packet-Based Media Communications Systems-An ITU industry standard for multimedia communications that combines and coordinates multiple data compression and communication standards to allow audio, picture, and video transmission between users on packet switched networks. The H.323 system has four key components: terminals, gateways, gatekeepers, and multipoint control units (MCUs). The original name for H.323 was Visual Telephone Systems and Equipment for Local Area Networks.

H.323v1-The first version of H.323 Visual Telephone Systems and Equipment for Local Area Networks created by study group 16 in the ITU in 1995.

H.323v2-The second version of H.323 that changed the video conferencing focus of H.323 to multimedia communications. H.323v2 was released in 1998.

H.450 Supplementary Services-Supplementary Services for H.323, namely Call Transfer and Call Diversion, have been defined by the H.450 series. H.450.1 defines the signaling protocol between H.323 endpoints for the control of supplementary services. H.450.2 defines Call Transfer and H.450.3 Call Diversion. Call Transfer allows a call established between endpoint A and endpoint B to be transformed into a new call between endpoint B and a third endpoint, endpoint C. Call Diversion provides the supplementary services Call Forwarding Unconditional, Call Forwarding Busy, Call Forwarding No Reply and Call Deflection. The "hooks" for Supplementary Services are specified in H.323 Version 2.

HA-Home Agent

HA-Horn Alert

HAAT-Antenna Height Above Average Terrain

HAAT-Height Above Average Terrain

HACN-Home Agent Control Node

HAL-Hardware Abstraction Layer

Half Duplex-Half duplex communication provides the ability to transfer voice or data information in either direction between communications devices but not at the same time. The information may be transmitted on the same frequency or divided into different channels. When divided into different channels, one channel of frequency is used for transmitting and the other channel or frequency is used for receiving.

The use of different frequencies is common in half duplex radio transmission because the transmitter and receiver are commonly connected to the same antenna. If the same transmitter and receiver frequency were used, the high transmitter power would probably destroy the receiver circuitry.

Half Power Point-The point on a frequency spectrum, or on the radiation pattern of an antenna, at which the power of a signal is equal to one half of its maximum power. It often is called the 3dB point because the signal is approximately 3 dB less than maximum.

Half Rate (HR)-The process where only half the normal data rate (full rate) is assigned to a user operating on a communications channel (typically a cellular or PCS channel). By reducing the data rate, the number of users that can share the radio communications channel can be increased. For time division multiple access (TDMA) systems, half the number of times slots are assigned during each frame of transmission. This allows other radios to be assigned to the unused time slots. For coded systems, the data rate is reduced which decreases interference to other users, thus increasing the capacity of the system.

Half Rate Data TCH (TCH/H)-A traffic channel that uses half the maximum number of slots per frame. For the GSM system, this is a half slot per frame (one slot every other frame) and for the IS-136 system, this is one slot per frame.)

Half Rate Fast Associated Control Channel (FACCH/H)-The fast associated control channel used with a half rate traffic channel.

Half Wave Rectifier-A circuit or device that changes only positive or negative half cycle inputs of alternating current into direct current.

Half-Power Point - 3 dB Point-The point(s) on a frequency spectrum, or on the radiation pattern of an antenna, at which the power of a signal is equal to one half of its maximum power. It often is called the 3dB point because the signal there is approximately 3 dB less than maximum. The bandwidth of a filter (in kHz) or the beam width of an antenna radiation pattern (in degrees) is often measured between two 3 dB (half power) points, which is relatively easy to measure, but not always a meaningful bandwidth or beam width.

Hall Effect-The phenomenon by which a voltage develops between the edges of a current carrying metal strip whose faces are perpendicular to an external magnetic field.

Hamming Distance-The number of bits that are different between a pair of binary values.

HAN-Home Area Network

Hand Portable-A cellular telephone self contained in a lightweight, hand carried unit.

Hand Portable Unit (HPU)-A mobile radio that is able to be carried and operated by a user.

Handheld-A wireless telephone self-contained in a lightweight, handheld unit.

Handheld Device Markup Language (HDML)-HDML is an internet web browsing language that evolved from hypertext markup language (HTML). HDML is optimized for low bandwidth operation using limited screen size and limited user input keys. HDML allows soft keys to be dynamically defined to allow a phone and return the key presses to the network application.

Handhole-A plastic, steel, or tile enclosure used in buried cable distribution systems that includes a cover that is used as a splice or pull box. Handholes are smaller versions of manholes and do not enable outside plant personnel to enter for splice work. The splice closure or cable sheath is pulled out of the handhole, spliced or repaired above ground and then laid back in the enclosure. Handholes are common in residential and commercial subdivisions served by 200 pair or smaller cables between buildings.

HANDO-Handover

Handoff-A process where a mobile radio operating on a particular channel is reassigned to a new channel. The process is often used to allow subscribers to travel throughout the large radio system coverage area by switching the calls (handoff) from cell-to-cell (and different channels) with better coverage for that particular area when poor quality conversation is detected. Handoff (also called handover) is necessary for two reasons. First, where the mobile unit moves out of range of one cell site and is within range of another cell site. Second, where the

Handover (HANDO)

mobile has requested a cellular channel with different capabilities. This might mean assignment from a digital channel to an analog channel or assignment from an analog channel to a digital channel.

Handover (HANDO)-(1- GSM) A process where a mobile telephone operating on a particular frequency and time slot is reassigned to a radio frequency and/or time slot. See also: Handoff.

Handover Access Burst-The transmission of an access burst during the handover process. This allows the cell site to determine and adjust the burst transmission timing of the mobile station.

Hands-Free-An audio interface to a communication device (such as a mobile telephone) that permits the user to listen and talk without holding the handset to their ear. Hands-free operation is required in some countries and states for operation as a safety feature.

Handset Churn -Type of voluntary churn unique to, and common in the wireless industry. Handset churn occurs when a customer changes carriers in order to get a newer and better (and often free) handset.

Handshake-The process of exchanging information between communication devices to ensure a connection can be established between both devices.

Handshake Protocol-A series of messages exchanged between communication devices operating on a system to establish communications links for further transmissions of voice or data.

Hang Up-In ground-start supervision, an information signal indicating that a carrier is finished with a connection but is not ready to accommodate a new request for service.

Hanover Bars-An undesirable artifact of interlaced video scanning that shows scrolling bars.

Hard Clad Silica (HCS)-An optical fiber that has a silica core and a hard polymeric plastic cladding intimately bonded to the core.

Hard Handoff-This is a communication channel handoff (or handover) occurs when a connection with the initial server (e.g. cellular base station) is disconnected before the connection with the next communications access point is completed. Antonym, Soft Handoff.

Harmonic Analysis-The process of breaking down a complex wave into the sum of a fundamental and various harmonic frequency components.

Harmonic Analyzer-A test set capable of identifying the frequencies of the individual signals that makeup a complex wave.

Harmonic Emission-A spurious emission at frequencies that are whole multiples of those contained in the frequency band occupied by an emission.

HASP-Houston Automatic Spooling Program
HAVi-Home Audio Video
Hazardous Gas or Explosives Meter-A handheld indicator or metering device used by telco outside plant personnel/technicians for testing the presence of natural gas, gasoline vapors, and other combustible gases in central offices, vaults, CEV's, and manholes.
HBA-Host Bus Adapter
HBFG-Host Behavior Functional Group
HBS-Home Base Station
HCI-Host Controller Interface
HCI-Host Command Interface
HCL-Hardware Compatibility List
HCO-Hearing Carry Over
HCS-Hierarchical Cell Structure
HCS-Hard Clad Silica
HCS-Hundred Call Seconds
HCT-Hardware Compatibility Test
HDB3-High Density Binary or Bipolar 3
HDD-Hard Disk Drive
HDLC-High-Level Data Link Control
HDML-Handheld Device Markup Language
HDSL-High Bit Rate Digital Subscriber Line
HDSL2-High Bit Rate Digital Subscriber Line 2
HDT-Host Digital Terminal
HDTP-Handheld Device Transport Protocol
HDTV-High Definition Television
HDWDM-High Density Wave Division Multiplexing
HDWDM-Hyper Dense Wavelength Division Multiplexing
Head-A device that erases, records, or reads information from a hard disk or other media that is passed under it.

Headend-The part of a cable television system that selects and processes video signals for distribution into a cable television distribution network. A variety of equipment is used at the headend, including antennas and satellite dishes to receive signals, preamplifiers, frequency converters, demodulators and modulators, processors, and scrambling and descrambling equipment.

With the advent of the cable modem, the headend also houses sophisticated digital data communications equipment that is able to transmit computer communications over a standard television signal's spectrum (6 MHz in the U.S. or 8 MHz in other parts of the world).

This photo image is a headend equipment rack. This 7 foot tall equipment rack holds the satellite receivers, modulators, transcoders, along with the cable assemblies that interconnect them.

Headend
Source: Blonder Tongue

Headend Driver Amplifier-An amplifier that is used in a cable television network to increase the level of the RF signal for supply to the distribution network.

Header-(1-communications) The part of a data packet that contains address, routing, and origination information. (2- record) A record that contains administrative, physical, and electrical data describing a cable count, carrier facility, or type of equipment.

Header Compression-A process that removes redundant header (usually repetitive addressing and control data) information that is transmitted in the header of data packets.

Headroom-The difference, in decibels, between the typical operating signal level and a peak overload level.

Headset-A headset is a combination of a microphone and earpiece that is used to enable better audio conversations. Headsets may be connected to various types of devices by wires including telephone sets, wireless devices (cellular telephone or cordless telephone), or they may integrate wireless technology (such as Bluetooth) to allow them to directly connect to a wireless receiver.

A picture of a telephone headset, model Encore H91. This headset plugs into an amplifier that interfaces with the telephone system to allow the telephone user to more effectively communicate with customers.

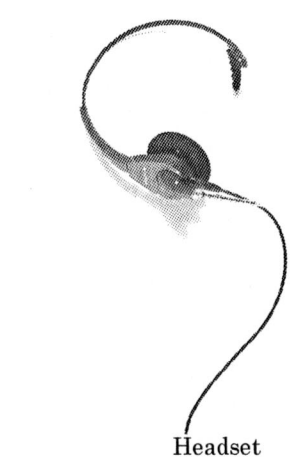

Headset
Source: Plantronics

Hearing Loss-When expressed as decibels, the difference between the measured threshold of audibility in an individual ear, and the normal or standard threshold at the same frequency. When expressed as a percentage, the value is equal to 100 times the hearing loss in decibels divided by the decibel difference between the normal or standard threshold of audibility and the normal or standard threshold of vibratory sensation.

Heat Coil- type of over-voltage protection. A temperature dependent resistor device used with cable protectors to guard central-office equipment against excessive current in a subscriber's line. The coil increases its electrical resistance when a fault condition causes excessive current to flow. See BORSCHT and PTC resistor.

Heat Coils-A protection device, generally located on the main distribution frame (MDF) in a central office to protect against higher than normal current flow by grounding out when a fault occurs.

Heat Loss-The loss of useful electric energy resulting from conversion into unwanted heat.

Heat Sink-A device that conducts heat away from a heat producing component so that it stays within a safe working temperature range.

Heater-In an electron tube, the filament that heats the cathode to enable it to emit electrons. In some tube types, the heater and the cathode are the same structure.

HEC-Header Error Control

Hecto-A prefix meaning 100.

HEHO-Head End Hop Off

Height Gain-For a given propagation mode of an electro-magnetic wave, the ratio of the field strength at a specified height to that at the surface of the earth.

henry (H)-Unit of electrical inductance, equal to one volt second per ampere. In a 1 henry inductor, a voltage of 1 volt occurs when the electric current changes at the rate of 1 ampere per second. Named after the 19th century American scientist Joseph Henry (1797-1878).

HEPA-High Efficiency Particulate Arrester

HERF-High Energy Radio Frequency Gun

Hertz (Hz)-A measurement unit for frequency that is equal to number of cycles per second. This measurement unit was named after the German physicist Herrich R. Hertz (1857-1894).

Heterodyne Frequency-The sum of, or the difference between, two frequencies, produced by combining the two signals in a modulator or similar device.

Heterodyne Wavemeter-A test set that uses the heterodyne principle to measure the frequencies of incoming signals.

HF-High Frequency

HFAI-Hands Free Answer on Intercom

HFC-Hybrid Fiber Coax

HFS-Hierarchical File System

HGC-Hercules Graphic Card

Hi-Cap-High Capacity

HI-FI-High Fidelity

Hierarchial Numbering-Multiple level numbering. An example is the telephone number made up of levels such as "Country Code," "Area Code," "Exchange Number" and "Line Number."

Hierarchical Cell Structure (HCS)-A structure in a cellular system that allows the separation of cells from a master control center (the MSC). Hierarchical cell structure distributes the switching and network control throughout the network to small cells that may be directly connected to the public switched telephone network (PSTN).

Hierarchical Computer Network-A computer network in which processing and control functions are performed at several levels by computers specially suited for the functions performed.

Hierarchical File System (HFS)-A file system that is structured like trees where files are associated with higher level files or directories (roots).

HIF-Host Interface Node

High Bit Rate Digital Subscriber Line (HDSL)-An all digital transmission technology that is used on 2 or 3 pairs of copper wires that can deliver T1 or E1 data transmission speeds. HDSL is a symmetrical service.

This diagram shows that the first application for HDSL used two pairs (and sometimes 3 pairs) of copper wire. Each circuit has an HDSL Termination Unit (HTU) on each end; an HTU-C (central office) and HTU-R (remote). This example shows that each pair of HDSL wires carries 784 kbps full duplex (simultaneous send and receive) data transmission. To carry the equivalent of a T1 line, two pairs of lines are used. It is also possible to carry an E1 line by using 3 pairs of copper wire. Although the framed transport for HDSL is different than for a T1 or E1 line, the HTU-C and HTU-R convert the protocols to standard T1 lines.

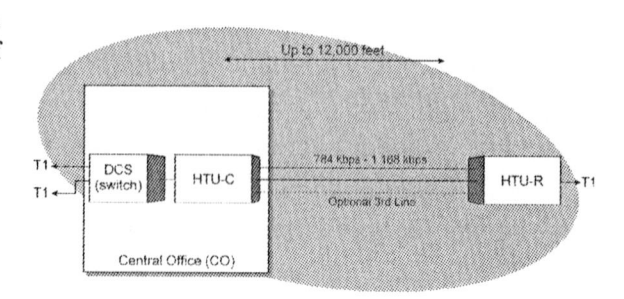

High Bit Rate Digital Subscribe Line (HDSL) System

High Bit Rate Digital Subscriber Line 2 (HDSL2)-A second generation of HDSL that offers several enhancements to HDSL data transmission. These improvements include the ability to transfer T1 or E1 data transmission rates over a single twisted-pair local loop instead of the two or three pairs of copper wire required for standard HDSL.

High Definition Television (HDTV)-High definition television (HDTV) is a TV broadcast system that proves higher picture resolution (detail and fidelity) than is provided by conventional NTSC and PAL television signals. HDTV signals can be in analog or digital form.

High Density Binary or Bipolar 3 (HDB3)-A line coding method used by 2.048 Mb/s digital multiplexers and related higher level multiplexers. This line coding prevents timing loss when excessive zeros are transmitted by substituting certain pre-determined BPVs instead of a string of four consecutive zero pulse values.

High Density Wave Division Multiplexing (HDWDM)-High density wave division multiplexing (HDWDM) is a form of wave division multiplexing that increases the number of optical network units (ONUs) per passive optical networks. In 2001, HDWDM increased PONs from 32 to 64 ONUs.

High Frequency (HF)-The band of frequencies within the limits of 3 MHz and 30MHz. Wavelengths in the HF band are between 100 m and 10 m (shortwave).

High Level Control-In data transmission, the conceptual level of control or processing logic existing in the hierarchical structure of a primary or secondary station that is above the link level, and upon which the performance of link level functions are dependent or are controlled.

High Pass Filter (HPF)-A device or circuit that passes signals of higher than a specified frequency but attenuates signals of all lower frequencies.

This diagram shows a highpass filter can be used to block low and mid frequency component parts of an input signal and allow high frequency components (signals) to pass through. In this example, both the low and mid frequency noise signals are highly attenuated by the bandpass filter while the desired high frequency is allowed to pass through the filter with minimal attenuation.

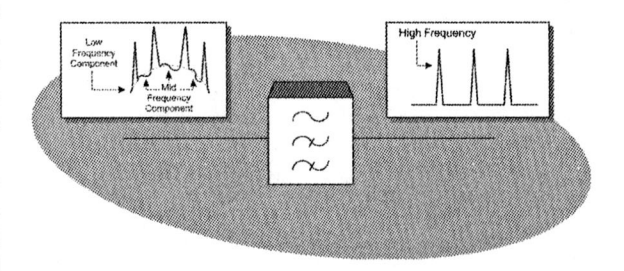

Highpass Filter Operation

High Speed Circuit Switched Data (HSCSD)-An enhancement to the GSM mobile communications system that combines up to four 14.4 Kbps channels (up to 4 slots per frame) to be combined to provide 57.6 Kbps data transfer.

This figure shows how GSM can provide HCSD service. This diagram shows that the HSCSD system combines multiple time slots to provide data services that are at higher data transfer rates than is possible using single slot operation. This example shows how data from the HSCSD system can be routed to a separate data network directly from a base station or a base station controller. The multiple the HSCSD system also permits asynchronous operation where there can be a higher data transfer rate in one direction (typically in the downlink when the user is downloading files or images from the Internet) than the other direction.

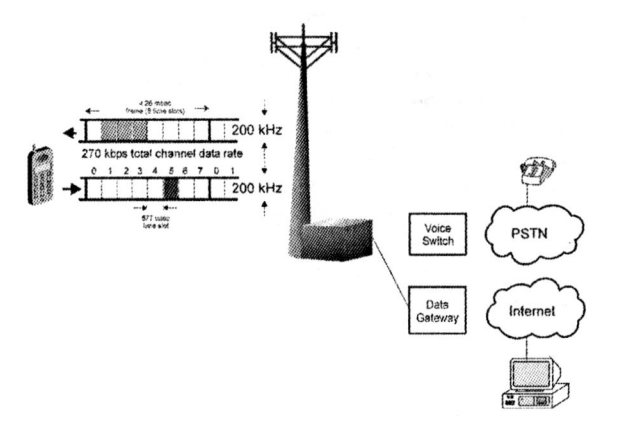

High Speed Circuit Switched Data (HSCSD) System

High Speed Serial Interface (HSSI)

High Speed Serial Interface (HSSI)-A standard serial data communications interface that can transfer data rates up to 52 Mbps. HSSI is sometimes used to connect DS3 (T3) lines to asynchronous transfer mode (ATM) connections.

High Tension-A high-voltage circuit or transmission system.

High-Level Data Link Control (HDLC)-An ISO communication protocol that is located in the data link layer in that delineates the beginning and end of a data frames. HDLC is used in X.25 packet switching communication networks to transfer bit-oriented, synchronous protocol that provides error correction at the data-link layer. HDLC systems allow messages to be transmitted in variable-length frames.

High-Speed Multimedia-High-speed multimedia usually refers to image based media such as pictures, animation or video clips. High-speed multimedia usually requires peak data transfer rates of 1 Mbps or more.

High-Tier-A wireless system that serves high-speed vehicular traffic and may have higher power levels.

Hijacking-A process of gaining security access by the capture of a communication link in mid-session after the session has already been unauthorized.

HIN-Handshake In

HIPPI-High Performance Parallel Interface

Historical Data File-A set of data that is altered infrequently and supplies basic data for processing operations, such as forecasting.

HIVR-Host Interactive Voice Response

HKSW-Hook Switch

HLLAPI-High Level Language Applications Programming Interface

HLR-Home Location Register

HMA-High Memory Area

HMAC-Hashed Message Authenticity Check

HMI Driver-Hub Management Interface Driver

HMM-Hidden Markov Method

HNPA-Home Numbering Plan Area

HO Tone-Handoff Tone

Hobbit-High-Order Bit

HOBIC-Hotel Billing Information Center

Hold-(1-general) A temporary mode of operation that is typically entered into by a device when there is no need to send voice or data information for a relatively long time. The hold mode allows the device audio to be muted or the transceiver to be turned off in order to save power. (2-Bluetooth) The Bluetooth hold mode is used to release devices from actively communicating with the master. This allows the devices to sleep for extended periods and allows the master control device to discover or be discovered by other Bluetooth devices that want to join other Piconets. With the hold mode, Piconet capacity can be freed up to do other things like scanning, paging, inquiring, or attending another Piconet sessions.

Holding Time-The amount of time that a line connection or other shared resource is in use for a call or call attempt.

Hole-A vacancy in the valence band of an atom where an electron normally would be. Such holes act, in effect, as positive charges (positrons) and are the majority carriers of current in p-type semiconductors.

Hole - Semiconductor-In the theory of semiconductor physics, a hole is a vacant energy level that occurs at relatively high temperatures, or as a result of shining light on the electrons in the valence band of a semiconductor solid, at the energy level where an electron normally would be. At extremely low temperatures, all electrons in the semiconductor fall back from the conduction band energy levels into the valence band, leaving no such holes. Such holes act, in effect, as positive charges (they drift in the same direction as the electric field, instead of opposite the electric field as electrons do) and are the majority carriers of current in p-type semiconductors.

The term "hole" was first proposed historically in 1927 by the late British physicist Paul A.M. Dirac as part of his theory of relativistic quantum mechanics. He proposed that space is filled with negative energy level electrons and the transfer of energy to one such electron causes it to change to a positive energy level electron, leaving behind in the negative energy levels a "hole" which has the characteristics of a positive electric charge. This process is called "electron-hole" pair production. The apparent positive charge particle with negative energy is called a positron. The term "hole" was later taken into semiconductor physics as well, but a Dirac positron can exist in empty space (vacuum), while a "hole" of the electrical type can exist only in a solid semiconductor.

Hole Current-The current in a semiconductor directly associated with the movement of holes.

Holographic Memory-The nonvolatile storage of information throughout the whole of a storage medium rather than in a specific location.

Home-The geographic location where communication service is based.

Home Agent (HA)-A router used to manage the current location and routing information for mobile users. A foreign agent exchanges information with the home agent to allow data packets to be forwarded to network where the user is currently operating.

Home Agent Control Node (HACN)-A signaling control node used that manages several home agent nodes.

Home Audio Video (HAVi)-A specification for home networks comprised of consumer electronics devices such as CD players, televisions, VCR's, digital cameras, and set-top boxes. The network configuration is automatically updated as devices are plugged in or removed. The IEEE 1394 protocol, also known as FireWall, is used to connect devices on the wired HAVi network at up to 400 Mbps.

Home Channels-The radio channels available within a given frequency band within the home mobile service area (MSA) of a cellular subscriber.

Home Location Register (HLR)-The part of a wireless network (typically cellular or PCS) that holds the subscription and other information about each subscriber authorized to use the wireless network.

Home Mobile Station-A mobile telephone that is operating in a wireless network where it has established home service.

Home Numbering Plan Area (HNPA)-The numbering plan area within which a calling line appears at a local switching office.

Home Phoneline Networking Alliance (HomePNA) Association-An association that assists in the development of phoneline networking standards that allow for data to be distributed to devices that are connected to telephone wiring in a home or business.

HomePNA-Home Phoneline Networking Alliance

HomePNA Association-Home Phoneline Networking Alliance

HomeRF-An industry working group that is assisting in the development of a local area RF communications that permits consumer devices such as computers, printers and fax machines to communicate with each other.

HomeRF Working Group (HRFWG)-An organization that was created to assist in the development and commercial introduction of interoperable wireless consumer devices. The HRFWG established an industry specification that allows for digital communication between computers, accessories, and other consumer electronic devices in small geographic areas (such as the home).

Homing Arrangement-The last choice trunk group between switching systems in a specific routing ladder.

Homochronous-Signals whose corresponding significant instants have a constant but uncontrolled phase relation-ship with each other.

Hook Flash (Hookflash)-A special feature service request signaling process that allows the user to momentary disconnect circuit disconnection (temporary off-hook state) as a means if indicating a special service request (e.g. call forwarding request).

Hookflash-Hook Flash

Hookswitch-The switch of a telephone set that closes and opens a customer local loop circuit. It is used to indicate control (supervisory) signals from the user. The hookswitch is usually operated by lifting the handset or returning it to its cradle (holder). A hookswitch is also called a switchhook.

Hoot and Holler-A multi-drop 2 or 4 wire circuit used with speakers where many people can listen to the speaker simultaneously. Generally provisioned at the 56 Kb/s level and used to link head office and remote city financial trading floors, railroad communications and wholesale parts auctioneers.

Hopoff-The process of transitioning from one network to another network. An example of hopoff is the transition from an H.323 network to a PSTN network.

Hopping Sequence Number (HSN)-The number that identifies the hopping sequence a mobile station should use when communicating with the system.

Horizon Angle-The angle, in a vertical plane, between a horizontal line extending from the center of an antenna and a line extending from the same point to the radio horizon.

Horizontal (Hum) Bars-A group of relatively broad horizontal bars, alternately black and white, that extend over the entire picture on a video display. The bars may be stationary or they may move up or down. Hum bars are caused by interference

Horizontal and Vertical (H&V) Lock Time

from the power line (utility) frequency or one of its harmonics.

Horizontal and Vertical (H&V) Lock Time-The amount of time required for a video receiver to lock to horizontal and vertical synchronization signals.

Horizontal Line-A single horizontal scan of a camera or beam. A given number of these video scans together form a frame of video. There are 525 interlaced lines per frame in NTSC and 625 lines per frame in PAL.

Horizontal Period-The length of time allocated for a complete horizontal line of video information.

Horizontal Phase (H Phase)-(1-general) The horizontal sync phase relationship of one piece of video equipment to another for studio timing purposes. (2- subcarrier) The phase of horizontal sync in relation to the subcarrier.

Horizontal Rate (H Rate)-The number of complete horizontal lines, including trace and retrace, scanned per second in a video system.

Horizontal Scan-A single horizontal scan line of a camera or electron beam on a picture tube. Multiple horizontal line scans are grouped together form a frame of video. There are 525 interlaced lines per frame in NTSC and 625 lines per frame in PAL.

Each horizontal scan signal contains a horizontal synchronizing pulse that identifies the start of each line of video. For color systems, the horizontal synchronization pulse is followed (during a short time interval called the "back porch") by a color burst (3.58 MHz in NTSC color systems, 4.43 MHz for PAL) that provides a phase reference signal for the phase sensitive decoder used to extract the three primary color signals from the composite phase modulated video waveform.

This diagram shows the electrical video signal that represents a single line (horizontal sweep) of video (there are 525 to 625 lines of video for most television systems). In this example, the horizontal video scan signal starts with a horizontal synchronizing pulse and this synchronization pulse also includes a color burst tone. Following the horizontal pulse, a composite video represents the intensity (brightness) of the image as the sweep moves across the picture tube. Because this is a color video signal, the composite signal represents the intensity of 3 colors. This diagram also shows that a horizontal blanking period occurs during the sync pulse to allow the picture tube beam to sweep back to the other side of the picture tube without being seen (blanked)

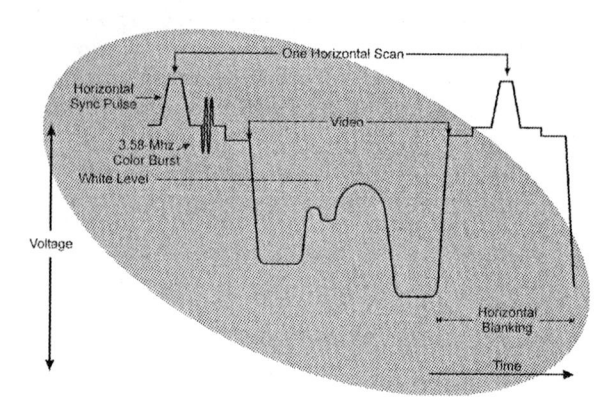

Horizontal Scan Operation

Horizontal Sync Pulse-The synchronizing pulse at the end of each video line that stimulates the start of horizontal retrace. This pulse signals the picture monitor to go back to the left side of the screen to trace another horizontal line of picture information.

Horn Antenna-A microwave antenna formed by flaring out the end of a waveguide.

Horn Feed-A horn antenna used in conjunction with a parabolic dish or reflector to obtain improved directional (antenna gain) performance.

Horn Gap-A lighting arrester utilizing a gap between two horns. When lightning causes a discharge between the horns, the heat produced lengthens the arc and breaks it.

Horn Reflector Antenna-An antenna that uses a focusing horn that reflects the radio signal energy into a focused beam.

Host-A computer processor or other type of data information processing device that is connected to a network that processes request and provides information services to remote users.

Host Controller Interface (HCI)-The host controller interface (HCI) is a standard communication protocol that is used to control a communication link to a data processing device. HCIs are used in many types of communication systems include universal serial bus (USB) and Bluetooth.

Host Name-The name that has been assigned to the computer responsible for one or several IP addresses.

Host Number-The part of an Internet address that identifies the node on the network or subnetwork that is being addressed.

Host Switching System-A switching system with centralized control over most of the functions of one or more remote switching units. The host usually provides trunk access to an exchange carrier's intraLATA network and access tandems.

Hosted Telephony-Hosted telephony is a managed IP Telephony communication service (also known as IP Centrex). Hosted telephony is a communications service for users who prefer to use a communication system that is managed by another company. The term "hosted" or "managed" is used instead of "Centrex" because an IP-based feature set is generally broader than what existing Centrex offerings provide. Furthermore, hosted or managed solutions do not just cater to the traditional Centrex market - typically campus-based, such as universities or government. The flexibility of IP allows for very compelling offerings to other markets, such as SOHO and greenfields.

Hosting-The providing of application services (such as virtual telephone service or the providing of web service) for the benefit of a client or customer.

Hot Patch-A method of patching a failed digital line, such as a T1 line, onto a spare facility. This technique does not use bridging repeaters. Service is interrupted briefly when the patches are removed. This type of patching is used only when bridging repeaters are not available for full-service patching.

Hot Spots-Regions in a cellular service area which, due to traffic jams and other heavy congestion receive excessive amounts of usage.

Hot Standby Routing Protocol (HSRP)-HSRP is a proprietary protocol developed by Cisco Systems. It is used to provide network redundancy for IP networks, especially for network edge devices or hosts. It allows user traffic to seamlessly recover from first hop (default gateway) router failures. Multiple routers on LAN segments are configured to communicate via HSRP status messages to provide this level of redundancy, thus a "virtual router" to end host devices. HSRP allows two or more routers to share the same IP address and MAC (Layer 2) address. The ACTIVE router in the HSRP group is the primary route out of the locally attached network. All other routers in the HSRP group act as STANDBY routers until the ACTIVE router goes away.

Hotline-A restricted calling class that forces a telephone (usually a wireless telephone) to be connected to an operator regardless of the digits actually dialed. Hotline is typically used when a telephone is first sold or activated to allow activation after the customer has provided the information to register for service or when the customer has not paid their bill.

Housing-An enclosure for carrier equipment and other electronic systems deployed in an outside plant. A housing may be above or below ground and may range in size from a small shelter to a large building containing many carrier systems.

HOUT-Handshake Out
HPA-High Power Amplifier
HPAD-Host Packet Assembler/Dissembler
HPC-Handheld Personal Computer
HPF-High Pass Filter
HPFS-High-Performance File System
HPO-High Performance Option
HPR-High Performance Routing
HPU-Hand Portable Unit
HR-Half Rate
HRFWG-HomeRF Working Group
HSCSD-High Speed Circuit Switched Data
HSD-High Speed Data
HSDL-High-speed Subscriber Data Line
HSDPA-High Speed Downlink Packet Access
HSLN-High Speed Local Network
HSN-Hopping Sequence Number
HSPD-High-Speed Packet Data
HSR-Harmonic Suppression Reactor
HSRP-Hot Standby Routing Protocol
HSSI-High Speed Serial Interface
HST-High Speed Technology
HSTR-High Speed Token Ring
HTCP-HyperText Caching Protocol
HTML-Hypertext Markup Language
HTTP-Hypertext Transfer Protocol

HTTP Pseudostreaming-A technique for streaming audio or video over the Internet using the HTTP protocol. When compared to streaming using other protocols, its advantages include being able to operate through any firewall that allows web traffic, and being based on a ubiquitous standard protocol. However in poor network conditions it is not suitable for real-time delivery since it is based on TCP. Many internet streaming systems can fall back to HTTP pseudostreaming as a last resort if connections using preferred protocols fail due to a firewall or other issues. Some simple Internet streaming systems support HTTP pseudostreaming exclusively. HTTP pseudostreaming cannot be used for true multicast streaming since Internet multicast is not TCP-based.

HTTPS-Hypertext Transfer Protocol Secure

Hub-A hub is a communication device that distributes communication to several devices in a network through the re-broadcasting of data that it has received from one (or more) of the devices connected to it. A hub generally is a simple device that re-distributes data messages to multiple receivers. However, hubs can include switching functional and multi-point routing connection and other advanced system control functions. Hubs can be passive or active. Passive hubs simply re-direct (re-broadcast) data it receives. Active hubs both receive and regenerate the data it receives.

This figure shows how a hub distributes communication to several devices in a network through the re-broadcasting of data that it has received from one (or more) of the devices connected to it. A hub generally is a simple device that re-distributes data messages to multiple receivers. However, hubs can include switching functional and multi-point routing connection and other advanced system control functions. Hubs can be passive or active. Passive hubs simply re-direct (re-broadcast) data it receives. Active hubs both receive and regenerate the data it receives.

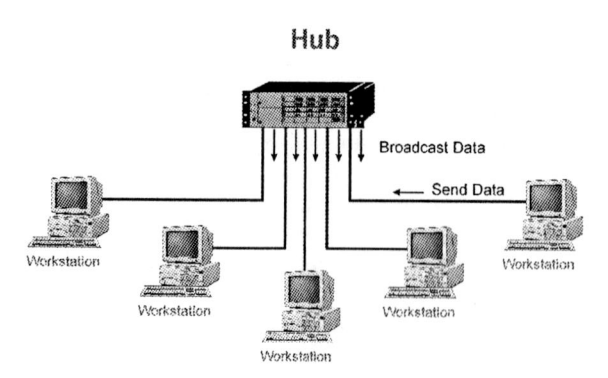

Hub_Operation

This is a picture of the SureStoreHub 10/100 Ethernet hub that is capable of both 10 Mbps and 100 Mbps operation.

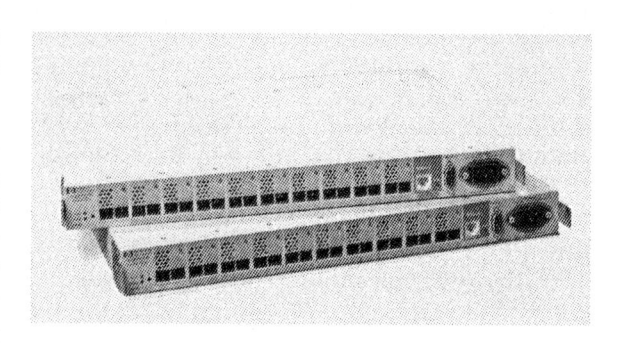

Hub
Source: Hewlett-Packard Company

Huffman Coding-A data compression technique that takes advantage of different probabilities of each data symbol occurring. More probable symbols use shorter codes while less probable symbols use longer codes, resulting in an overall average code length which is within one bit of the optimum for any compression technique. All codes are "prefix-free" that means that no valid codes can start with another valid code, which would be ambiguous when decoding.

Hum-Undesirable coupling of the 50 Hz or 60 Hz power sine wave into other electrical signals and/or circuits.

Hum Bars-Horizontal

Hum Rejection-The capability of a circuit to cancel interference in a video or audio signal, usually at the 50 Hz or 60 Hz power line frequency.

Human Factors Engineering-Engineering that applies the study of ergonomics to the design of equipment and software to create safe, easy-to-use systems.

Hunt Group-A hunt group is a list of telephone numbers that are candidates for use in the delivery of an incoming call. When any of the numbers of the hunt group are called, the telephone network sequentially searches through the hunt group list to find an inactive (idle) line. When the system finds an idle line, the line will be alerted (ringing) of the incoming call. Hunt lines are sometimes called rollover lines.

Hunting-A telephone call-handling feature that causes a transferred call to "hunt" through a predetermined group of telephones numbers until finds an available ("non busy") line.

This diagram shows the process of hunting (also called "roll-over") for an available telephone. This diagram shows that an incoming call enters into the telephone switching system and is attempted to be delivered to the main telephone line extension (or dialed telephone number port). The switching system is programmed with a hunt list that allows the switching system to determine where to redirect a call if it is unable to deliver a call. This example shows that the system first tries 1001 that is off-hook (unavailable). The hunt group table shows that the call should be routed to 1002. Because 1002 is also unavailable, the hunt group list shows instructs the switch to try extension 1003. This telephone (or port) is available and the call can be delivered (telephone rings).

the audio to an electrical signal. This signal travels down the line to hybrid #1. Hybrid #1 subtracts the energy from microphone #1 (the combination of both signals are actually on the line) and applies the different (audio from customer #2) to the speaker #1.

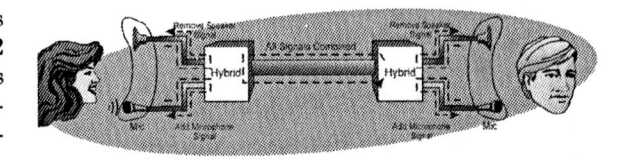

Hybrid Telephone Operation

Hybrid Backbone-A high-speed interconnection network that uses two (or more) types of transmission systems (such as SONET and Ethernet).

Hybrid Balance-The degree to which a hybrid termination prevents incoming energy on a four-wire circuit from being reflected in the opposite direction. Hybrid balance is measured in decibels of return loss, which is the difference between the levels of an input signal and its reflection.

Hybrid Cable-A cable television communication system that uses a combination of optical fiber and electric conductors.

This figure shows a typical cable distribution system that uses a combination of fiberoptic cable and coaxial cable for the local connection. This diagram shows that the multiple video signals from the head-end of the cable television system is converted into digital form to allow distribution through high-speed fiber cable. The fiber cable is connected in a loop around the cable television service area so that if a break in the cable occurs, the signal will automatically be available from the other part of the loop. The loop is connected (tapped) at regular points by a fiber node. The fiber node converts the fiber signals into RF television signals that are distributed on the local coaxial cable network. The coax network distributes the RF signals to homes in the cable television network.

Hunting Operation

HVQ-Hierarchical Vector Quantization
Hybrid-(1-Device) A device that combines transmit and audio signals from two-pairs of lines to one pair of lines. (2-Network) The combination of two different types of network technologies (such as fiber and coax) to form a combined (Hybrid) network.

This figure shows how a typical analog telephone transmission line operates. In this diagram, audio from customer #1 is converted to electrical energy by microphone #1. This signal is applied to the telephone line via the hybrid adapter #1. A portion of this signal is applied to the handset speaker to produce sidetone (so the customer can faintly hear what they are saying). This audio signal travels down the telephone line to hybrid #2. Hybrid #2 applies this signal to speaker #2 so customer #2 can hear the audio from customer #1. When customer #2 begins to speak, microphone number #2 converts

Hybrid CDPD

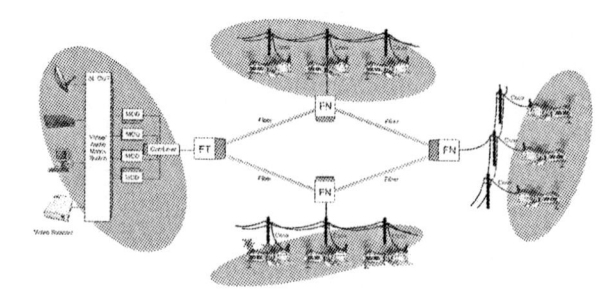

Hybrid Cable Operation

Hybrid CDPD-A combination of cellular packet switched data and circuit switched data systems.

Hybrid Coil-A coil that is used as a bridge to connect a 2-wire circuit (combined transmit and receive) to a 4-wire circuit (separate transmit and receive pairs).

Hybrid Connector-A connector that contains both electrical conductors and optical fiber connections.

Hybrid Fiber Coax (HFC)-The hybrid fiber coax (HFC) system is an advanced CATV transmission systems that uses fiber optic cable for the head end and feeder distribution system and coax for the customers end connection. HFC are the 2nd generation of CATV systems. They offer high-speed backbone data interconnection lines (the fiber portion) to interconnect end user video and data equipment. Many cable system operators anticipating deregulation and in preparation for competition began to upgrade their systems to Hybrid Fiber Coax (HFC) systems in the early 1990's.

Hybrid Integrated Circuit-A single integrated circuit that contains different types of components such as discrete (independent) components and integrated components on the same IC substrate (base).

Hybrid ITSP-An Internet telephone service provider that combines packet voice with switched voice services.

Hybrid Key System-A key telephone system (KTS) that combine some of the advanced features of PBX systems with traditional KTS features. An example of a hybrid key system is the ability to assign different telephone lines to each electronic telephones without the need to rewire the switching system.

Hybrid Local Network-A local communication network that consists of more than type of local network. For example, a hybrid local network may consist of Ethernet and Token Ring.

Hybrid Loss-The transmission signal loss that occurs when a signal passes through a communication hybrid (such as a 2 line to 4 line converter)

Hybrid System-A communication system that accommodates multiple types of signals or physical channels. Examples of hybrid systems include hybrid fiber coax (HFC) and combined analog and digital signal transmission.

Hybrid Tee-A waveguide junction that permits energy to be supplied in from multiple ports.

Hygrometer-An instrument that measures the relative humidity of the atmosphere.

Hyperband Channel-The CATV channels from 37 through 59. These channels are normally downconverted to frequencies above the VHF band to alleviate the excessive signal loss that would result if UHF channels were transmitted on the cable. In order for viewers to uses these channels, they must use a converter or have a cable ready television receiver.

Hyperframe-A frame structure made up of two superframes (primary and secondary). The hyperframe in an IS-136 TDMA system consists of 192 frames. In GSM, a hyperframe is 4096 superframes and requires over 3 hours to complete.

HyperLAN-A wireless local area network (WLAN) specification overseen by the European Telecommunications Standards Institute (ETSI). It operates in the 5.7 GHz industrial, scientific, and medical (ISM) frequency band. HyperLAN provides for data transmission rates of up to 54 Mbps. HyperLAN is uses the 802.3 Ethernet standard for it's fundamental structure.

Hypertext Markup Language (HTML)-A text based communications language that allows formatting and item selection features to be transferred independent of the type of computer system. HTML is primarily used for Internet communication.

Hypertext Transfer Protocol (HTTP)-A protocol that is used to transmit hypertext documents through the Internet. It controls and manages manage communications between a Web browser and a Web server.

Hypervideo-Hypervideo is a video program delivery system that allows the embedding of links (hotspots) inside a streaming video signal. This allows the customer (or receiving device) to dynamically alter the presentation of streaming information. Examples of hypervideo could be pre-selection of preferred advertising types or interactive game shows.

Hysteresis Loss-The loss in a magnetic core resulting from saturation of the core that causes non-linear state changes.

Hz-Hertz

I

I Frame-Information Frame
I&R-Installation and Repair
I-Carrier-A system operating at one of the standard levels in Japan's digital hierarchy. Each digital signaling level supports several 64 kbps (DS0) channels. J-carrier was initially used in Japan and now is used throughout several parts of the world. The different digital signaling levels include;
- DS-1, 1.544 Mbps with 24 channels
- DS-1C, 3.152 Mbps with 48 channels
- DS-2, 6.132 Mbps with 96 channels
- DS-3, 32.064 Mbps with 480 channels
- DS-4, 397.200 Mbps with 5760 channels

I-Commerce-Internet Commerce
I-frame-Intra-frame
I-MAC-Isochronous Media Access Control
i-Mode-Internet Mode
I-Node-Information Node
I-PNNI-Integrated-Private Network-To-Network Interface
I/O-Input/Output
I/O Address-Input-Output Address
I/O Device-Input/Output Device
I2C-Inter-Integrated Circuit
I2R Loss-The amount of power that is lost as a result of the effect of current passing through resistance. The energy lost is transformed into heat.
IA5-International Alphabet No.5
IAB-Internet Architecture Board
IACC-Interaural Cross Correlation
IAD-Integrated Access Device
IAD-Internet Addiction Disorder
IAHC-International Ad Hoc Committee
IAL-Intel Architecture Labs
IAM-Initial Address Message
IANA-Internet Assigned Numbering Authority
IAO-IntrAOffice SONET Signal
IAOS-Interim Automated Operator Services
IAP-Internet Access Provider
IAP-Initial Alignment Procedure
IAP-Intercept Access Point
IASG-Internetwork Address Sub-Group
IBA-Independent Broadcasting Authority
IBC-Initial Billing Company
IBDN-Integrated Building Distribution Network

IBG-Interblock Gap
IBM-International Business Machines
IBS-In-Door Base Station
IBS-Intelligent Battery System
IBX-Internet Business Exchange
IBX-Integrated Business Exchange
IC-Interchange Carrier
IC-Integrated Circuit
IC-interLATA Carrier
IC-Intermediate Cross-connect
IC/EC Factor-Interexchange Carrier/Exchange Carrier
ICA-Independent Computing Architecture
ICA-International Communications Association
iCal-iCalendar
ICANN-Internet Corporation for Assigned Names and Numbers
ICB-Individual Case Basis
ICD-International Code Designator
ICE-Information Content and Exchange
ICE Age-Information Communications And Entertainment Age
Ice Tray-A metal or plastic shield mounted over switch gear at the base of a wireless tower for it's protection from ice and other falling debris.
ICFA-International Computer Facsimile Association
ICI-Interexchange Carrier Interface
ICI-Incoming Call Identification
ICIT-International Center for Information Technologies
Icky PIC-Sticky Plastic Insulated Conductor
ICMP-Internet Control Message Protocol
iCOMP-Intel Comparative Microprocessor Index
ICP-Internet Cache Protocol
ICP-Integrated Communications Platform
ICP-Intelligent Call Processing
ICP-Integrated Communications Provider
ICP-Independent Communications Provider
ICQ-I Seek You
ICQ-I See You
ICR-Initial Cell Rate
ICR-Internet Call Routing node
ICS-Intercompany Settlements
ICS-Integrated Communications System
ICSA-International Computer Security Association

ICSC-Interexchange Customer Service Center
ICSC-Interexchange Carrier Service Center
ICTA-International Computer-Telephony Association
ICV-Integrity Check Value
IDC-Insulation Displacement Connection
IDCMA-Independent Data Communications Manufacturers Association
IDCS-Integrated Digital Communications System
IDDD-International Direct Distance Dialing
IDE-Integrated Development Environment
IDE-Integrated Drive Electronics
IDEA-Internet Development & Exchange Association
IDEA-International Data Encryptions Algorithm
iDEN-Integrated Dispatch Enhanced Network
Identification-Color Frame ID
IDF-Intermediate Distribution Frame
IDI-Initial Domain Identifier
IDL-Interactive Distance Learning
IDLC-Integrated Digital Loop Carrier
Idle Mode-A period of time that a mobile radio is not required to transmit or receive data. During the idle period, a mobile radio may measure the channel quality of other radio channels.
Idle Noise-Noise that occurs on an idle communication channel.
Idle URL-A URL address that is used by a device (such as an IP telephone) should the device come into an idle state. Using an idle URL allows for the display of customized status messages on device displays ("Lunch Special").
IDN-Integrated Digital Network
IDS-Intrusion Detection System
IDSL-ISDN Digital Subscriber Line
IDT-Interdigitated Transducer
IDT-Integrated Digital Terminal
IDTS-Integrated Data Test System
IDTV-Improved Definition Television
IDU-Interface Data Unit
IE-Information Elements
IEC-Interexchange Carrier
IEE-Institution Of Electrical Engineers
IEEE-Institute Of Electrical And Electronics Engineers
IEEE 802.1-An IEEE specification that defines the interconnection of IEEE 802 LANs at the Data-Link Layer (layer 2 of the OSI reference model). The standard specifies the operations of bridge devices, including MAC address learning and aging, packet forwarding services and the Spanning Tree Protocol. Spanning Tree Protocol is used to prevent loops in bridged networks, so that packets are not continually and repeatedly forwarded between sides of the bridge.
IEEE 802.11-The IEEE standard that specifies the MAC protocol for commonly available wireless LANs that use radio frequency for transmission.
IEEE 802.14-An IEEE standard that specifies a MAC protocol two-way data communication over cable TV (CATV) systems. The standard has generally been supplanted by another industry standard, CableLabs' Data Over Cable Service Interface Specification (DOCSIS).
IEEE 802.17-An IEEE standard that specifies a MAC protocol for a metropolitan area network which utilizes fiber rings and transports packets. Commonly known as a resilient packet ring, the protocol is intended to provide better metropolitan area networking for enterprises using other IEEE 802 protocols, such as Ethernet.
IEEE 802.2-This standard applies to most IEEE 802 LANs, regardless of their topology. Commonly known a Logical Link Control (LLC), it standardizes the commands and command formats for connectionless (type 1) or connection-oriented (type 2) service. These services define the sequence of communication between LAN nodes. LLC services were intended to provide assured delivery at layer 2 of the OSI reference model; however, in recent years and with the proliferation of TCP/IP, this service is often provided via TCP at layer 4.
IEEE 802.3-The IEEE standard that specifies the MAC protocol for a collision-based network commonly known as Ethernet . The 802.3 standard allows 10 Mbps or 100 Mbps data transmission rates. Because Ethernet can use CDMA/CD, users can share a network cable, however only one user can transmit data at a specific time.
IEEE 802.4-The IEEE standard that specifies the MAC protocol for a token passing network protocol which utilizes a bus topology. This protocol is popular in products used in manufacturing environments such as factories.
IEEE 802.5-The IEEE standard that specifies the MAC protocol for a token passing network protocol which utilizes a ring topology. This protocol was widely used by IBM products in the 1990's.
IEEE 802.6-The IEEE standard that specifies the MAC protocol for a metropolitan network based on a Distributed Queue Dual Bus (DQDB) protocol.

With line rates of 1.5 to 155 Mbps, the protocol has rarely been used; more commonly ring topologies, such as FDDI are used instead.

IEEE standards-The Institute of Electrical and Electronics Engineers (IEEE) provides industry guidance and consensus building enabling various vendors to work together to produce interoperating products. The results are specifications, or standards, which corporations agree to abide by to produce interoperating equipment. One of the most influential bodies in the networking industry, the IEEE 802 committee is responsible for LAN and MAN standards, such as 802.3 Ethernet and 802.11 wireless LANs.

IEN-Internet Experimental Note
IES-Inter-Enterprise Systems
IESG-Internet Engineering Steering Group
IETF-Internet Engineering Task Force
IF-Intermediate Frequency
IFFM-Incoming First Failure To Match
IFG-Interframe Gap
IFIP-International Federation for Information Processing
IFS-International Freephone Service
IFS-Installable File System
IG-Isolated Ground
IGFs-International Gateway Facilities
IGMP-Internet Group Management Protocol
IGP-Interior Gateway Protocol
IGRP-Interior Gateway Routing Protocol
IIH-IS-IS Hello, Intermediate System To Intermediate System "Hello"
IIOP-Internet Inter-ORB Protocol
IIS-Internet Information Server
IISP-Information Infrastructure Standards Panel
IISP-Interim Interface Signaling Protocol
IITF-Information Infrastructure Task Force
IIW-ISDN Implementors Workshop
IJ-Insulating Joint
IKE-Internet Key Exchange
IKP-Internet Keyed Payments Protocol
ILA-Intermediate Light Amplification
ILD-Injection Laser Diode
ILEC-Incumbent Local Exchange Carrier
Illumance-The amount of light that is received on a specific surface area.
Illuminated-An surface or geographic area to which a radio communications signal is directed and can be received.
ILMI-Interim Link Management Interface
ILS-Instrument Landing System
ILS-Input buffer Limiting Scheme
ILS-Internet Locator Service
ILSR-IPX Link State Router
IM-Intensity Modulation
IM-Instant Messaging
IM-Intermodulation
IMA-Inverse Multiplexing Over ATM
IMA-Interactive Multimedia Association
Image Frequency-A frequency on which a carrier signal, when heterodyned with the local oscillator in a superheterodyne receiver, will cause a sum or difference frequency that is the same as the intermediate frequency of the receiver. Thus a signal on an image frequency will be demodulated along with the desired signal and will interfere with it.
Imagery-Collectively, the representations of objects reproduced electronically or by optical means on film, electronic display devices, or other media.
Imaging-A process that may include converting and image into a different form (such as digital format), transferring, processing, storing, displaying, and reproducing (printing).
IMAP-Internet Messaging Access Protocol
IMAP-Internet Message Access Protocol
IMAP4-Internet Messaging Access Protocol 4
IMAS-Intelligent Maintenance Administration System
IMASS-Intelligent Multiple Access Spectrum Sharing
IMC-Internet Mail Consortium
IMD-Intermodulation Distortion
IMEI-International Mobile Equipment Identifier
IMEI-International Mobile station Equipment Identity
IMHO-In My Humble Opinion
IML-Incoming Matching Loss
Immunity-In telecommunications, a characteristic that permits equipment to operate normally when the equipment or any external lead or circuit is subjected to electromagnetic voltages, currents, or fields.
IMNSHO-In My Not So Humble Opinion
IMO-In My Opinion
IMP-Interface Message Processor
Impact Ionization-The ionization of an atom or molecule as a result of a high energy collision.
Impacted User Minutes (IUM)-Used as a network management measurement in conjunction with availability. Impacted User Minutes or IUMs is the total amount of unproductive user time due to

network or server downtime and performance issues.

Impairment-(1-circuit) The loss of quality of service provided by an individual circuit when its transmission units are exceeded or signaling functions (such as seizure, disconnect, and automatic number identification) are experiencing intermittent failures. (2-signal transmission) Any distortion that affects a transmitted waveform or is perceivable by a video or audio user.

IMPDU-Initial MAC Protocol Data Unit

Impedance Characteristic-A graph of the impedance of a circuit showing how it varies with frequency.

Impedance Irregularity-A discontinuity in an impedance characteristic, caused, for example, by the use of different coaxial cable types.

Impedance Matching-The adjustment of the impedance of adjoining circuit components to a common value so as to minimize reflected energy from the junction and to maximize energy transfer across it.

Improved Mobile Telephone Service (IMTS)- A mobile telephone service deployed in the United Sates in 1964 that provided automatic radio channel selection, direct customer dialing, automatic accounting, and full simultaneous duplex operation. IMTS was the precursor to cellular networks.

Impulse Current-A current that rises rapidly to a peak, then decays to zero.

Impulse Excitation-The production of an oscillatory current in a circuit by impressing a voltage for a relatively short period compared with the duration of the current produced.

Impulse Voltage-A unidirectional voltage that rises rapidly to a peak, then falls to zero.

Impurity-An atom of a given substance that has been intro introduced purposely into a material (typically a semiconductor), or that occurs naturally in the material. Impurities can have a considerable effect on the electrical properties of a material.

IMS/VS-Information Management System/Virtual System

IMSI-International Mobile Subscriber Identity

IMT-InterMachine Trunk

IMT-2000-International Mobile Telephony 2000

IMTA-International Mobile Telecommunications Association

IMTC-International Multimedia Teleconferencing Consortium

IMTS-Improved Mobile Telephone Service

IMUIMG-ISDN Memorandum of Understanding Implementation Management Group

IMUX-Inverse Multiplexer

IN-Intelligent Network

In Band Signaling-Signaling that occurs within the audio signal bandwidth. In band signaling while the conversation is in progress requires the users voice or data information to be momentarily interrupted or altered while signaling messages are being transferred. In band signaling is sometimes called blank and burst signaling.

Historically, the term "in-band" is related to the use of different frequency bands, but the term is also applied to signals in digital multiplexing that occur in the same time slot, better named "in-slot." Two examples of true in-band signaling are DTMF (touch tone) and MF. During the period of in-band signaling, the voice or data communication should be temporarily inhibited (muted) to allow the transfer of control messages without interference.

This figure shows how the basic process of in band signaling is used in analog cellular radio to deliver control messages sharing the same communication channel for voice and control signals. In this diagram, a radio base station desires to send a message to the mobile radio. The base initially sends a dotting sequence that indicates a synchronization word and message will follow. The mobile radio detects the dotting sequence. As a result, the mobile

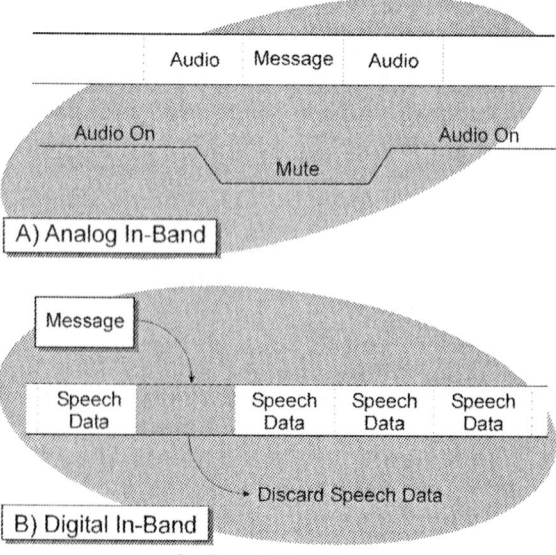

In Band Signaling

radio mutes the audio and begins to look for a synchronization word. The synchronization word is used to determine the exact start of the message. The mobile radio receives the message and on completion of the message, the mobile radio will then un-mute the audio and conversation continues. Because the sending of the message can be less than ¼ second, the user may not even notice a message have been received.

In The Broadcasting-Satellite Service-
Community Reception

In-Slot-See: in-band

INA-Information Networking Architecture

INA-Integrated Network Access

Inactive Signaling Link-A signaling link that has been de activated and, therefore, cannot carry signaling traffic.

INAP-Intelligent Network Application Part

Inbound Call Center-A call center (group of customer service agents) that receives telephone calls from customers. Inbound call centers (teleservice centers) are often used in response to advertisements and direct marketing campaigns. The call routing to inbound call centers can be fixed (established telephone lines) or dynamically controlled (based on activity or skills based routing.)

INC-International Carrier

incAlliance-Isochronous Network Communication Alliance

Incarnation Number-A unique name or number sent within a data unit to avoid duplicate data unit acceptance.

Incident Report-A report that lists the current unresolved problems (incidents) that have occurred within a communication system.

Incident Tracking-The process of recording and updating information about problems that occur in a communication system (incidents) and the steps taken to resolve the problem.

Incidental Churn -Incidental churn is a type of voluntary churn where the customer changes carriers because of changes in their situation or lifestyle, and where the change of carrier is only a secondary issue.

InCollects-Incollects are call detail records (CDRs) that are received by service provider A from service provider B for services provided by B to A's customers. An example of an Incollect is a roaming record.

Incoming Call Restriction-In telephone call-processing feature that disables a telephone from receiving incoming calls.

Incoming Peg Count-A measurement of the number of attempts, counted at the incoming end of a trunk group, that have seized a trunk in the group. Incoming peg count measurements generally are used where no peg count measurement is available at the originating end of a trunk group.

Incoming Selectors-In step-by-step switching, the terminating third, fourth, or fifth selectors for traffic coming into an office.

Incumbent Local Exchange Carrier (ILEC)-A telephone carrier (service provider) that was operating a local telephone system prior to the divestiture of the AT&T bell system.

InDate-Installation Date

Independent Broadcasting Authority (IBA)-A nonprofit public corporation in Britain that was established to set up and supervise commercial radio and TV operations. After the IBA builds and operates transmitter facilities, the IBS rents these services and facilities to those it authorizes.

Independent Software Vendor (ISV)-A vendor (company) that develops software that is independent of the computer hardware that the software operates on.

Index Of Refraction-The ratio of the speed of an electromagnetic wave in a vacuum to the speed of the same wave in another medium. The larger the index of a medium, the slower the speed of propagation within that medium. For light, the index of a medium is always equal to or greater than one.

Index of Refraction - Refractive Index-A physical characteristic of transparent materials defined by the ratio of the speed of light in a vacuum to the speed of light in the transparent material. High index material has a low speed of light and is said to be optically dense or have high optical density. Low index material has low optical density. Air has an index of refraction of about one. Optical fiber is usually designed to have high index material in its core and low index in its cladding.

Index-Matching Fluid-A fluid with a refractive index close to glass that reduces refractive index discontinuities in a fiber optic system.

Induce-To produce an electrical or magnetic effect in one conductor by changing the condition or position of another conductor.

Induced Charge-An electrostatic charge produced on one body when it is brought near another charged body.

Induced Current-The current that flows in a conductor cause a voltage has been induced across two points in, or connected to, the conductor.

Induced Voltage

Induced Voltage-(1-general) A voltage developed in a conductor when the conductor passes through magnetic lines of force. (2-interference) The undesired alternating-current voltage developed in telecommunications cables by an electric power system.

Induction Coil-(1-electrical components in general)A single coil of insulated wire wound around a core volume. The core volume is typically in the form of a rod or a toroid (donut, bagel). The core material is typically air or a ferromagnetic material such as iron or ferrite. Also called an inductor. (2- in a telephone set) A multi-winding transformer used as part of a directional coupler device in a telephone set as a directional coupler to separate the microphone and earphone electric waveforms that both flow together in opposite directions on the subscriber loop. The so-called "hybrid coils" used at the central office directional coupler for 2-wire to 4-wire conversion are similar, but traditionally described by that distinct name.

INE-Intelligent Network Element

Inert-An inactive unit, or a unit that has no power requirements.

Inert Cell-A type of primary power cell that contains all needed chemicals in dry form, and begins to operate only when water is added.

INET-Institutional Network

Infinite Line-A transmission line that appears to be of infinite length. There are no transmitted signal reflections back to the source from the far end because it is terminated in its characteristic impedance.

InFLEXion-InFLEXion is a high-speed paging systems that can transfer data at speeds of up to 112 kbps in the forward direction. These higher speeds support voice and computer applications. The voice messaging essentially downloads an electronic copy of a person's "voice mail" message to the pager. The user then "plays" back the message from the pager as if it was an answering machine. InFLEXion is also like a cellular system, in that it knows exactly which antenna is serving the pager and does not broadcast the "voice mail" to the entire system.

INFO-Information

Information (INFO)-In the Integrated Services Digital Network (ISDN), a Layer 3 information message. Customer premises equipment uses this message to provide additional address information. A stored-program control system uses the message to inform the equipment of certain actions.

Information Access Service-A service offering that gives many telephone callers simultaneous access to selected prerecorded messages or databases furnished by private entrepreneurs known as information providers. This service also is called Mass Announcement Network Service.

Information Facility-A facility whereby a data service user, by sending a predetermined address from the terminal installation, may gain access to general information regarding data communication services.

Information Highway-A term that refers to a common communication path (highway) for the transport of information. The world wide web (Web) is sometimes called the Information Highway.

Information Networking-The processing, delivery, and management of any type of information through public or private telecommunications networks.

Information Service-Information services involve the processing of information that is transferred through a communications system. Information services add value to information by generating, acquiring, storing, transforming, processing, retrieving, utilizing, or making available information via telecommunications. Examples of information services include fax store and forward, electronic publishing, text to voice conversion, and news services.

Information Sytsem-Information systems store, transfer, and process information for specific purposes. Information systems consist of hardware (usually computers) and software (data and applications) that add value to information by; generating, acquiring, storing, transforming, processing, retrieving, utilizing, or making available information via data and telecommunications connections. Examples of information systems include information storage, financial applications, order processing, web e-commerce, and engineering design.

Information URL-A URL address that is used by a device (such as an IP telephone) should the user select an information button or icon. Using an information URL allows for the display of appropriate data should an information button be pressed or selected.

Informs-Informs are synonymous with SNMP traps (from SNMPv1) or notifications (from SNMPv3). Events on network devices or SNMP agents send informs to the network management system when something of interest occurs on the agent. Informs were introduced in SNMPv2.

Infrared (IR)-Communication that uses non-visible infrared light (approximately 850-nanometers) for data transmission between communication devices. The data transmission rate can exceed 16 Mbps using IR transmission. Infrared signals must have an unblocked path between devices. Some wide area systems include reflection points that allow IR signals to reflect to reach their destination. The distance between many IR communication devices is commonly less than 3 feet.

Infrared Spectrum-Electromagnetic radiation in the wavelength range of 700 to 2000 nanometers, spectrally adjacent to the red portion of the visible spectrum at one end, and to the millimeter wavelength radio spectrum at the other. Sensed by a human being as heat, and not visible to the human eye. Used for optical fiber communication and by certain electronic equipment for short range cordless connections.

Infrared Transmission-Transmission of information through the use of a band of electromagnetic energy that is located between above microwave frequencies and below the low end of the visible part of the spectrum. Infrared communication may occur in free space or through an optical fiber. Data transmission rates for infrared transmission ranges from low speed data (e.g. for television remote controls) to high speed data (above 10 Mbps used for local area networks).

Infrasonic-Sound waves at frequencies too low to be heard by a human ear (below about 15 Hz).

Infrastructure-All parts of the communication systems, excluding the subscriber. This includes switches, radio carrier equipment, databases, and other network parts that enable telecommunications networks. In a traditional telephone network, the equipment includes the switches, multiplexers and other equipment used to manage the network and the copper cables that connect them together and to the users. In a cellular system, the equipment includes base stations and microwave links as well as wireline equipment. In an optical network, infrastructure includes optical analogues to the traditional switches and multiplexers, as well as the optical fiber among them and connections to the traditional and cellular networks.

ING-Calling Party

InGaAs-Indium Gallium Arsenide

Ingress-(1-signal) A process where a strong signal outside a communication system enters into a transmission line or system. (2-network) The process of data or signals entering into a communication network.

Ingress Filter-A qualification function implemented in an input port of a VLAN-aware switch. The Ingress Filter can be configured to discard any received frame associated with a VLAN for which that port is not in the member set.

Inhibit-A process or control signal that prevents a device, circuit, or system from operating.

INIM-ISDN Network Interface Module

Initial Address Message (IAM)-A message in the SS7 system that is used in the ISDN User Part to initiate trunk signaling between two SSP's.

Initial Alignment-A procedure that prepares a signaling link to carry signaling traffic either for the first time or after a failure.

Initial Alignment Procedure-An SS7 link procedure by which a signaling link becomes able to carry signaling traffic either for the first time or after a failure has occurred.

Initial Alignment Procedure (IAP)-A procedure that is used when a signaling link is activated for the first time or when a link connection is restored after a communication failure.

Initial Period-(1-billing) The unit of time for billing at the beginning of each message, as defined by a tariff. (2-default) The initial (default) time required before normal processing can occur.

Initial Signal Unit Alignment-The signal unit alignment process that is used when activating or restoring a signaling link.

Initial Time Delay Gap (ITOG)-The elapsed time between a direct sound arrival at a listener and the first reflection of significant loudness.

Initialization-A process of setting the initial values and settings of the software and electrical components within a circuit or equipment is when it is first activated.

Injection-The application of a signal to an electronic device or assembly.

Injection Laser Diode (ILD)-A semiconductor diode in which lasing (emission of coherent light) takes place within the PN junction.

Inline Power Patch Panel-An inline power patch panel is used to insert power on the non-data lines in a data communications system (such as an Ethernet data network). Inline power patch panels connect to a data network. It allows for data transmission to transfer directly through it unaltered. The inline power patch panel is used to supply power to IP telephones and other data communica-

INMARSAT

tion devices that can use these power pins. Only devices that are connected to the inline power patch panel receive power.

INMARSAT-International Maritime Satellite Organization

INMS-Integrated Network Management Services 2.0

INOS-Intelligent Network Operating System

INOW-Interoperability Now

INP-Interim Number Portability

INP Board-Intelligent Network Processor Board

INPA-Interchangeable Numbering Plan Area

Input-(1-signal) The waveform fed into a circuit, or the terminals that receive the input waveform. (2-data) Any data being sent to a computer from a user, another computer, or other equipment.

Input Impedance-The impedance presented at the input terminals of a circuit, device, or channel.

Input Looping-A circuit arrangement that permits the input of a device to be connected to another system downstream. The looping connector may or may not be buffered from the input signal.

Input Return Loss-The amount of signal that is returned from the circuit or device that is being supplied (return loss).

Input Transformer-A transformer at the input of a device or circuit to match the impedance of the device to that of the preceding stage, or to isolate the preceding stage from the device.

Inquiry-A process that requests specific information from a computer or communication device to determine its access code or availability for a communication session. Inquiry is a process in a Bluetooth system that is used to determine the address of other Bluetooth devices that are operating in the same area.

Inquiry Scan-A process that allows a communication device to listen (scan) for other devices that desire to discover its availability.

INREFS-Integrated Reference System

Insertion Loss-(1-general) The amount of signal loss that is caused by the insertion of a component or device, usually expressed in decibels (dB), as the ratio of input power to output power. (2-optical fiber coupler) The loss associated with that portion of the light that does not emerge from the nominally operational ports of an optical coupler device.

Inspection Lot-A collection of units of product from which a sample is drawn and inspected to determine conformance with acceptability criteria.

Installation Date (InDate)-The date on which a telecommunications link or equipment is installed, sometimes referred to as an indate.

Installed First Cost-The installed cost of an outside plant item when it is placed in service.

This figure shows how the instant activation process works. In this example, a customer selects an Internet telephone service provider (such as from the list www.ITSPdirectory.com) that offers the services they desire (such as voice mail and a specific area code). The customer directly enters the billing information along with payment information (a charge card). This information is added to the customer database and an account code is provided to the customer. When the customer receives the user identification code and password, they can immediately begin to use the service.

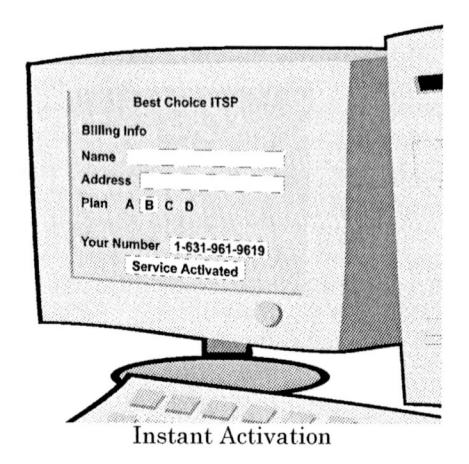

Instant Activation

Instant Messaging (IM)-A process that provides for direct connection between computers that are connected to a data communications network. Instant messaging (IM) service usually includes client software that is located on the communicating computers and an instant messaging server that tracks and maintains a list of alias names and their communication status. The IM server usually registers each client and links an address (usually an internet protocol address) so the clients can directly communicate with each other. The client software controls the presentation of information as it is sent directly between each computer.

This diagram shows the basic process used by instant messaging (IM) systems to allow IM users to directly communicate with each other. This diagram shows how IM server is primarily used as an address book for IM clients that want to directly communicate with each other. This diagram shows that IM clients sign on (register) and sign off (de-register) with the IM server each time they want to be able to send and receive messages to other IM members. The first step in this example shows that IM client "Buddy Steve" signs on with the IM server. The IM server captures the current Internet address from "Buddy Steve" and sends back a list of current IP addresses for other buddies identified by "Buddy Steve." Next, IM client "Ready to Play" signs on with the IM server and receives the IP address of "Buddy Steve." Because "Buddy Steve" and "Ready to Play" now have stored the current IP addresses of each other, they can directly communicate with each other without having to go through the IM server. This communication can involve the sending of messages or the continual sending of data (such as packet voice.)

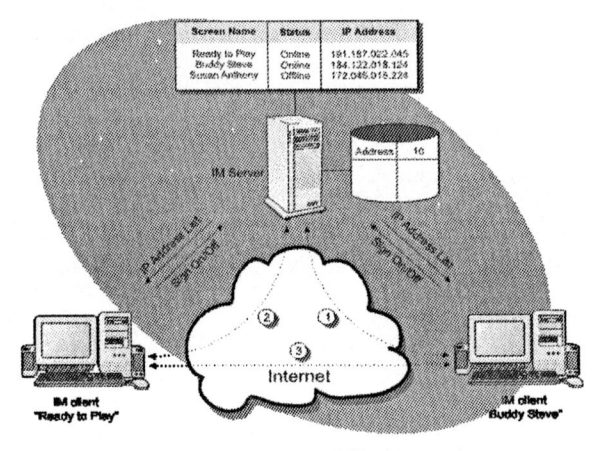

Instant Messaging (IM) Operation

Instantaneous Value-The value of a varying waveform at a given instant of time. The value maybe expressed in volts, amperes, or phase angle.

Institute Of Electrical And Electronics Engineers (IEEE)-An organization formed in 1963 that represents of electrical and electronics scientists and engineers. The IEEE resulted from the merger of the institute of Radio Engineers (IIRE) and the American Institute of Electrical Engineers (AIEE). Its has various societies that focus on key industry technical specialties (such as communications and robotics).

Instruction-A set of identifying characters designed to cause a processor to perform certain operations.

Instrument Landing System (ILS)-A radio navigation system that provides an aircraft with horizontal and vertical guidance information just before and during landing and, at certain fixed points. The ILS indicates the distance to the reference point of landing.

Instrument Landing System Glide Path-A system of vertical guidance embodied in the instrument landing system that indicates the vertical deviation of the aircraft from its optimum path of descent.

Instrument Multiplier-A measuring device that enables a high voltage to be measured using a meter with only a low-voltage range.

Instrument Rating-The range within which an instrument has been designed to operate without damage.

Instrument, Scientific and Medical Band (ISM)-An frequency band that is authorized for the use of instrument, scientific and medical radio devices. These bands of the electromagnetic spectrum include frequency ranges at 902-928 MHz and 2.4-2.484 GHz, which do not require an operator's license from the Federal Communications Commission (FCC) part 15.247 of FCC regulations or the regulatory agencies of other countries.

Insulate-The process of separating one conducting body from another conductor.

Insulating Joint (IJ)-Insulating joints (IJs) are openings in cable sheath that are grounded on each side and installed in central office vaults to minimize electrolysis and to added protection from outside power surges. IJ's are located as close as possible to the second upright from the conduit entrance. The intent is to maximize the distance from the insulating joint to the first vertical bend in the cable.

Insulator-(1-general) A material or device used to separate one conducting body from another. (2-guy) A double hole or loop shape. Insulator used in the guy-wire line of a tower or power pole. (3-stand off) A device, typically made of porcelain, used to support high-voltage conductors and components. (4-strain) A guy insulator used to provide separation between upper and lower sections of a guy-wire line.

INT-Induction Neutralizing Transformer
INT 14-Interrupt 14

Integrated Access Device (IAD)

Integrated Access Device (IAD)-A device that converts multiple types of input signals into a common communications format. IADs are commonly used in PBX systems to integrate different types of telephone devices (e.g. analog phone, digital phone and fax) onto a common digital medium (e.g. T1 or E1 line).

This figure shows an integrated access device (IAD) combines multiple types of media (voice, data, and video) onto one common data communications system. This diagram shows that three types of communication devices (telephone, television, and computer) can share one data line (e.g. DSL or Cable Modem) through an IAD. The IAD coordinates the logical channel assignment for device and provides the necessary conversion (interface) between the data signal and the device. In this example, the telephone interface provides a dialtone signal and converts the dialed digits into messages that can be sent on the data channel. The video interface buffers and converts digital video into the necessary video format for the television or set top box. The data interface converts the line data signal into Ethernet (or other format) that can be used to communicate with the computer. This diagram also shows that the IAD must coordinate the bandwidth allocation so real time signals (such as voice) are transmitted in a precise scheduled format (isochronous). The digital television signal uses a varying amount of bandwidth as rapidly changing images require additional bandwidth. The IAD also allocates data transmission to the computer as the data transmission bandwidth becomes available (what left after the voice and video applications use their bandwidth).

Integrated Digital Loop Carrier (IDLC)-IDLC systems are the integration of the integrated digital terminal (IDT) and remote digital terminal (RDT). The IDT is part of the local digital switch (LDS) and it acts like a concentrator to put more channels on a digital communications line. The IDLC system moves some of the switching services from the local switches into RDTs to increase the efficiency of communication lines between customers and the central office.

BellCore's (now Telcordia Technologies) GR-303 specification defines the interconnection of the LDS and RDT.

This diagram shows how an integrated digital loop carrier (IDLC) system can be installed in a local telephone distribution network to allow a 24 channel T1 line to provide service to up to 96 telephone lines. This diagram shows that a switching system has been upgraded to include an IDT and a remote digital terminal (RDT) has been located close to a residential neighborhood. The IDT dynamically connects access lines (actually digital time slots) in the switching system to time slots on the communications line between the IDT and RDT. The RDT is a local switch that can connect to up to 96 residential telephone lines. When a call is to be originated, the RDT connects (locally switches) the residential line to one of the available channels on the DS1 interconnection line. The IDT communicates with the RDT using the GR-303 standard.

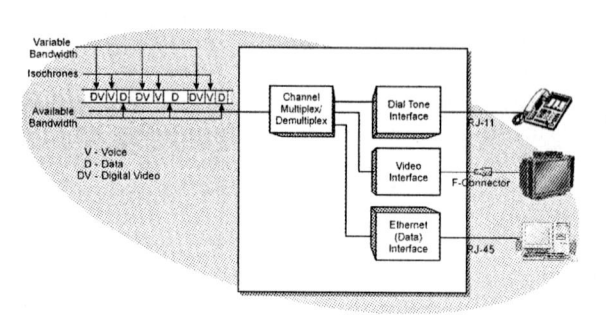

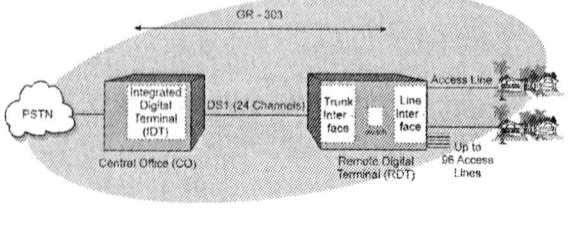

Integrated Access Device (IAD) Operation

Integrated Digital Loop Carrier (IDLC) System

Integrated Digital Network (IDN)-A network in which connections carrying digital signals are established by digital switching systems. (See also: Integrated Services Digital Network.)

Integrated Digital Terminal (IDT)-An electronic assembly that is part of a local digital switch (LDS) that coordinates communication with a remote digital terminal (RDT). The IDT concentrates some of the communication channels onto high-speed digital lines that are routed to RDTs. The RDTs demultiplex the digital line and assign channels to individual access lines.

Integrated Dispatch Enhanced Network (iDEN)-A digital radio system that provides for voice, dispatch and data services. iDEN was formerly called Motorola integrated radio system (MIRS). iDEN was deployed in 1996 for enhanced specialized mobile radio (E-SMR) service. The iDEN system radio channel bandwidth is 25 kHz and it is divided into frames that have 6 times slots per frame. The iDEN system allows 6 mobile radios to simultaneously share a single radio channel for dispatch voice quality and up to 3 mobile radios can simultaneously share a radio channel for cellular like voice quality.

The figure shows a typical integrated dispatch enhanced network (iDEN) system. This diagram shows that the iDEN system communicates with mobile stations in two modes: dispatch and cellular-like. The repeater transmitters are called enhanced base transceiver system (EBTS). The EBTS contains the radio transceivers that link the radio channel to the network communication lines. The control of the EBTS is provided by base station controllers (BSCs). BSCs can control one or more EBTS units. The BSC connects communication paths to either a mobile switching center (MSC) or to a metro packet switch (MPS). The MSC can connect (switch) voice calls to the public telephone network. The MPS routes dispatch calls (called direct connect) to a dispatch applications processor.

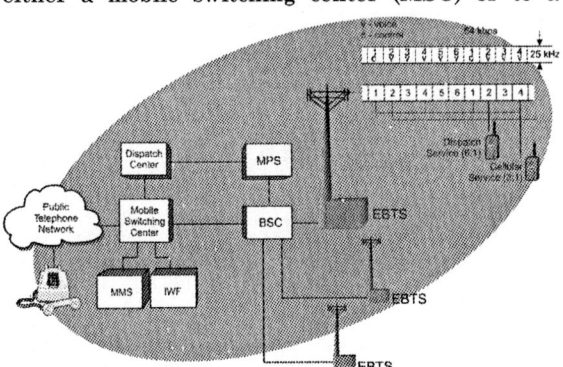

Integrated IS-IS-Integrated Intermediate System To Intermediate System

Integrated Market Planning -The process of combining market research, advertising, direct marketing, sales, and public relations activities into one, cohesive united operational plan.

Integrated Network Access (INA)-A network architecture for special services that extends digital carrier into a distribution loop to eliminate the need for an analog interface between distribution and interoffice facilities. The carrier terminates at either a remote terminal site or a customer's premises. Integrated Network Access eliminates the need for tandem digital-to-analog-to-digital conversions in a network.

Integrated Optical Circuit (IOC)-A monolithic optical circuit, composed of both active and passive miniaturized components, employing planar waveguides for coupling to optoelectronic devices and providing signal processing functions, such as modulation, multiplexing, and switching.

Integrated Optics-A class of optical devices that perform two or more functions integrated on a single substrate.

Integrated Service Control Point-An on-line, fault-tolerant database system that provides call processing information in real time in response to queries from service switching points (SSPs) and other network elements in the Advanced Intelligent Network (AIIN) architecture. In a typical configuration, the service control point software connects with network elements through a common-channel signaling (CCS) SS7 network.

Integrated Service Unit (ISU)-The combination of a digital service unit (DSU) and channel service unit (CSU) into one device. The ISU interfaces a customer's digital telephone equipment to the formats used by a telephone network by adapting digital protocols, electrical levels and physical connections.

Integrated Services (Intserv)-A network in which multiple types of traffic can flows with different quality of service (QoS) requirements. The two types of services defined in Intserv are guaranteed service and controlled local service. Guaranteed service provides a specific bandwidth with a set maximum end-to-end transmission delay

Integrated Dispatch Enhanced Network (iDEN) System

Integrated Services Control Point (ISCP)

time. Controlled-load service provides a variable bandwidth for each communication session that varies based on factors including the amount of network activity (e.g. heavy traffic) and quality of service requirements (e.g. real-time compared to non-real time communication application).

Integrated Services Control Point (ISCP)-A software system that integrates services control point (SCP) features that allow for the efficient deployment of customized intelligent network services.

Integrated Services Digital Network (ISDN)-A structured all digital telephone network system that was developed to replace (upgrade) existing analog telephone networks. The ISDN network supports for advanced telecommunications services and defined universal standard interfaces that are used in wireless and wired communications systems.

ISDN provides several communication channels to customers via local loop lines through a standardized digital transmission line. ISDN is provided in two interface formats: a basic rate (primarily for consumers) and high-speed rate (primarily for businesses). The basic rate interface (BRI) is 144 kbps and is divided into three digital channels called 2B + D. The primary rate interface (PRI) is 1.54 Mbps and is divided into 23B + D for North America and 2.048 Mbps and is divided into 30B + 2D for the rest of the world. The digital channels for the BRI are carried over a single, unshielded, twisted pair, copper wire and the PRI is normally carried on (2) twisted pairs of copper wire.

This diagram shows the different interfaces that are available in the integrated services digital network (ISDN). The two interfaces shown are BRI and PRI. These are all digital interfaces from the PSTN to the end customers network termination. Network termination 1 (NT1) equipment devices can directly connect to the NT1 connection. Devices that require other standards (such as POTS or data modems) require a terminal adapter (TA). This example shows that the NT2 interface works with the NT1 interface to allow the application layers (terminal intelligence) to communicate with the ISDN termination equipment.

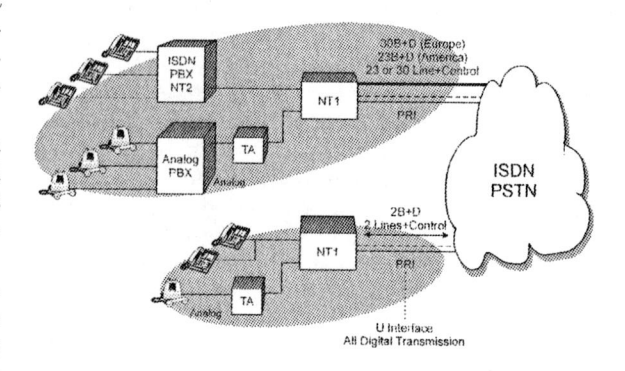

Integrated Services Digital Network (ISDN) System

This photo shows the MTA128NT model modem and other modems that are capable of interfacing ISDN lines to allow them to transmit and receive data. These modems convert ISDN signals into a format that can be understood by a computer bus or other type of standard data connection.

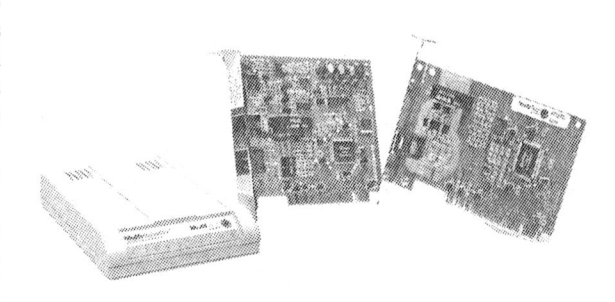

MultiModem ISDN
Source: Multi-Tech Systems, IncI

Integrated Signal Transfer Point (ISTP)-A signaling switch used in the SS7 common channel signaling network that integrates other functions (such as global title translation) with its signal transfer (message switching) functions. These transfer points are used to route signaling messages (packets) to other signaling transfer points or network parts.

Integration-(1-component) The production of complete and complex circuits on a single chip, usually of silicon. (2-services) The combining of different services (such as voice, data, and video) onto a common communication system.

Intellectual Property Rights-Often abbreviated as IPR, Intellectual Property Rights is a broad term encompassing patents, trademarks, service marks, copyrights, and trade secrets.

Intellectual Property Rights (IPR)-Legal rights that protect the innovations or creative content of proprietary to an individual, group or company. IPR is commonly associated with patent rights although it also applies to copyright and trademarks.

Intelligent Building-A building, often part of a commercial complex, in which local area networks, alarm circuits, and similar communications facilities are designed around a common communications strategy. Intelligent buildings are sometimes called smart buildings.

iIntelligent Call Processing (ICP)-The dynamic processing of calls based on information obtained from or selected by the caller. This information can be used by automatic call distribution (ACD) systems to route calls to the next available agent, agents with specific qualifications, or other call centers.

Intelligent Modem-A modem that performs certain functions under computer control. As an example, an intelligent modem automatically dials a selected telephone number and hangs up when it detects a busy signal and may re-dial the same number. It is also called a smart modem.

Intelligent Network (IN)-A telecommunications network architecture that has the ability to process call control and related functions via distributed network transfer points and control centers as opposed to a concentrated in switching system. (See. Advanced Intelligent Network.)

Intelligent Peripheral (IP)-A type of hardware that can be programmed to perform a new intelligent network capability in an SS7 network. IP's perform processing services such as interactive voice response (IVR), selected digit capture, and feature selection and account management for pre-paid services.

Intelligent Vehicle Highway System (IVHS)-Various systems that improve highway travel through the use of transportation management scheduling, travel demand forecasting, public transportation operation, electronic payment (toll booths), commercial vehicle tracking, emergency management and automatic vehicle control and safety monitoring.

INTELSAT-International Telecommunications Satellite Consortium

Intensity Modulation (IM)-A method of modulation used in fiber optic systems as a method of transmission in which an analog signal directly modulates the light source.

Inter Symbol Interference (ISI)-The appearance of part of a pulse symbol in a waveform appearing at an undesired time overlap with the following symbol(s). This is a result of undesired dispersion or time broadening of the pulses. ISI may be caused by multipath transmission where a single radio transmitted signal is reflected by objects (such as buildings) and part of the radio energy travels a path of different distance compared to another part of the signal (e.g. direct line of sight compared to reflected off a building or mountain). The interference results when the combined effect of multiple signals changes the decision points used to convert the radio signal back into its original digital form. ISI can be compensated by means of an equalizer. See: Dispersion, Equalizer.

Inter-Cell-Handoff

Inter-Cell Handover-A handover that transfers a call between different cells. See also intra-cell handover.

Inter-Piconet Capability-In a Bluetooth system, the capability of a master device to keep the synchronization of a Piconet while page scanning in free slots and allowing new members to join the Piconet. While a new unit is in the process of joining the Piconet, and until the master-slave switch is performed, operation may be temporarily degraded for the other members. A gateway that supports multiple terminals has the inter-Piconet capability. The terminals also may have the inter-Piconet capability. See also Piconet.

Inter-Tandem Trunk-A trunk that connects two tandem offices.

Interaction-The process by which a system accepts input, processes input requests, and (if necessary) returns appropriate response data to the originating terminal.

Interactive Television (ITV)-Interactive television has three basic types: "pay-per-view" involving programs that are independently billed, "near video-on-demand" (NVOD) with groupings of a single film starting at staggered times, and "video-on-demand" (VOD), enabling request for a particular film to start at the exact time of choice. Interactive television offers interactive advertising, home shopping, home banking, e-mail, Internet access, and games.

Interactive Voice Response (IVR)-IVR is a process of automatically interacting with a caller through providing audio prompts to request information and store responses from the caller. The responses can be in the form of touch-tone(tm) key presses or voice responses. Voice responses are converted to digital information by voice recognition signal processing. IVR systems are commonly used for automatic call distribution or service activation or changes.

This figure shows a sample IVR system that is used to route an incoming call. When this call is received by the PBX system, an initial voice prompt informs the user of the system along with initial menu options. The user selects and option. This results in the playing of another prompt indicating new menu options. The user enters the data for the option and the IVR system retrieves data and creates a new verbal response.

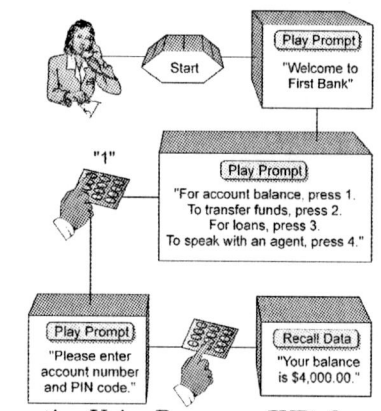

Interactive Voice Response (IVR) Operation

Interblock Gap (IBG)-A signal that is inserted in the helical track in the DAT format for the purposes of preventing adjacent track interference. This process also ensures that new data is written over the exact position of the previously recorded track during editing or re-recording.

Intercarrier Sound System-The method of transmitting the TV aural carrier at a fixed offset above the visual carrier, forming a composite signal. For NTSC, the separation of the two carriers is 4.5 MHz.

Intercept Access Point (IAP)-A place in a communication network where information and/or content is intercepted for the purpose of passing it to a law enforcement agency.

Intercept Trunk Group-A trunk group that provides information about called numbers that are unassigned, changed, disconnected, or placed on trouble intercept.

Intercom Call-In a Bluetooth system, intercom calls allows calls to directly communicate from one terminal toward another terminal. An intercom call between two terminals can be rapidly set up with gateway support if the two terminals are members of the same wireless user group (WUG).

Intercom Group-A group of telephone devices (the intercom group) that can be simultaneously accessed using an intercom feature. The use of the intercom group feature allow divisions or functional groups within a company to more effectively communicate with all the members of the group.

Intercom Profile-In a Bluetooth system, the Intercom profile defines the protocols and procedures that would be used by Bluetooth devices for implementing the intercom part of the usage model called "3-in-1 phone," also known as the "walkie-talkie" application of the Bluetooth specification.

Intercompany Settlements (ICS)-Financial settlements made between carriers for usage of each other networks (such as for collect calls or roaming charges).

Interconnect-The term for manufacturers who provide telephone equipment for and on consumer premises.

Interconnect Cable-A short-distance cable (generally less than 3 m) intended for use between pieces of equipment.

Interconnection-Interconnection commonly refers to the connection of telephone equipment or communications systems to another network such as the public switched telephone network. Government agencies such as the FCC or department of communications usually regulate the interconnection of systems to the public switched telephone network.

This figure illustrates some of the different types of private to public telephone system interconnection. This diagram shows some groups of phone lines (e.g., dial line, Type 1) that provide limited signaling information (line-side) that primarily interconnect the PSTN with private telephone systems. Another group of lines (Type 2 series) are used to interconnect switching systems or to connect to advanced services (such as operator services or public safety services). The interconnection lines (trunk-side) provide more signaling information. Also shown is the type S connection that is used exclusively for sending control signaling messages between switching system and the signaling system 7 (SS7) telephone control network.

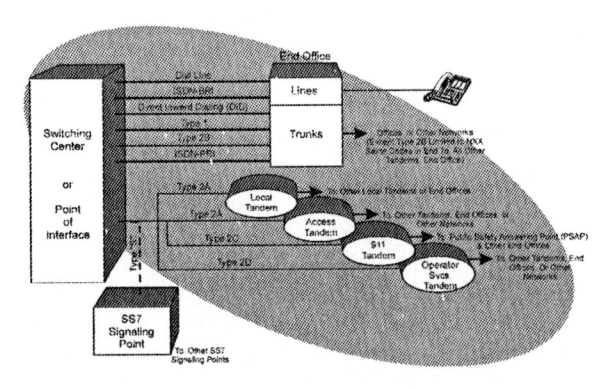

Private to Public Telephone System
Interconnection System

Interdigitated Transducer (IDT)-The transducer on a SAW device that converts electrical energy into mechanical vibrations and vice versa. The term "interdigitated" refers to their appearance, which is similar to interlaced fingers. The precise dimensions- length, width and thickness- of these fingers, along with their total number and the relative number on each half of the transducer determine the characteristics of the device. For instance, more fingers on a SAW bandpass filter result in a sharper filter but usually with more loss. Design of IDTs is complex and typically requires highly specialized software for high performance designs.

Interexchange Carrier (IEC)-A telephone service company that is provides long distance (interLATA, interstate, and/or international telecommunications service.

Interexchange Carrier (IXC)-Inter-exchange carriers (IXCs) interconnect local systems with each other. IXCs are also known as long distance carriers. In the US, from 1984 until 1997, IXC and LEC operating companies were legally required to refrain from engaging in directly competitive business operations with each other. Since 1997, one business entity can engage in both IXC and LEC business if it satisfies certain competitive legal rules. In Europe and throughout the rest of the world, the same PTT operators also usually provide inter-exchange service within their country. In any case, governments regulate how networks are allowed to interconnect to local and long distance networks.

For inter-exchange connection, networks as a rule connect to long distance networks through a separate toll center (tandem switch). In the United States, this toll center is called a point of presence (POP) connection.

This figure shows a diagram of an inter-exchange carrier network. This diagram shows that the IXC interconnects LECs and CLECs with teach other through POP switching points. Access lines connect the IXC POP switching centers with LEC and CLEC tandem switching systems. These interconnection lines are typically dedicated high-speed carrier transmission lines such as DS3 or OC3 lines.

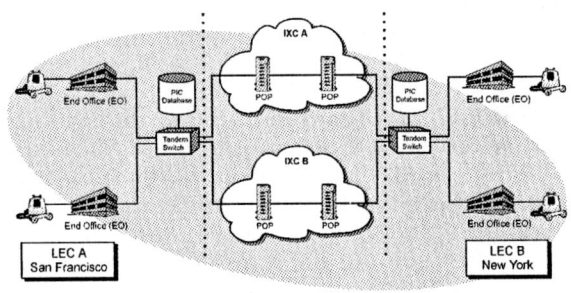

Inter-Exchange Carrier (IXC) Network Systems

Interexchange Circuit-A circuit or trunk between two exchanges that carries primarily message telecommunications service and WATS traffic, or a private line.

Interexchange Trunks-Multiple channel communication lines (trunks) that interconnect switches between different exchanges.

Interface Equipment-The conversion equipment that enables circuits designed to one set of characteristics to communicate efficiently with circuits designed to meet different protocols and specifications.

Interference-(1-undesired signal) An undesired signal that, in theory, can be turned off because it originates from a device that can, in principle, be turned off. An example is an undesired radio signal coming from another radio transmitter, or an undesired radio signal eliminating from a faulty neon sign or fluorescent light. Noise, in contrast, originates from fundamental physical mechanisms that cannot be turned off, such as the random thermal motion of electrons. Some authors lump the two types of undesired signals together under the name "noise," sometimes inappropriately. (2- combination of sine waves having the same frequency)This second meaning of the word "interference" typically occurs in the context of such phrases as "constructive interference," "destructive interference," or "interference and diffraction." These terms occur mostly, but not exclusively, in the context of radio waves and light. It describes the situation where (typically) two sinusoidal waveforms, having the same amplitude and frequency and a fixed phase relationship, are added together physically. Two sine waves that have a fixed phase relationship are described as mutually "coherent." If they exactly are in phase (zero degrees phase difference), the resulting signal has a higher power level than either sine wave alone (more precisely, double the amplitude or four times the power level of one sine wave alone). If they are not in phase, the resulting signal waveform has a lower power level, and in the special case where they are exactly out of phase (180 degree phase difference) the composite waveform has zero power. In a situation where the two sine waves have different phase difference at different places in space (because each sine wave has traveled a different distance from its own source location) the two sine waves will cancel in some locations and produce higher power (higher brightness for visible light) in other locations.

Interference - Constructive and Destructive- Interference is the interaction between two waves. The simplest, but very useful, case is when both waves have the same wavelength. If the waves are in phase with each other, the interference is constructive. That is, they are both positive at the same time and negative at the same time, so they add together. The resulting wave has the same phase but larger amplitude. If the waves are exactly out of phase with each other so that when one is positive the other is negative and vice versa, the interference is destructive. The two waves exactly cancel each other out and there is no resulting wave. This effect is very useful in optical components, such as thin film filters.

Interfering Source-An emission, radiation, or induction that is determined to be a cause of interference in a radio communications system.

Interframe Gap-The spacing between time-sequential frames.

Interim Link Management Interface (ILMI)-A interim specification that was developed by the ATM forum to allow network management functions between public networks, private networks, and end users. ILMI has some of the capabilities associated with simple network management protocol (SNMP.)

Interim Standard 124 Data Message Handler (IS-124)-A standard billing communication protocol that allows for the real time transmission of billing records between different systems. IS-124 messaging is independent of underlying technology and can be sent on X.25 or SS7 signaling links. The development of the standard is primarily led by CiberNet, a division of the cellular telecommunications industry association (CTIA).

Interim Standard 3 [Analog Cellular] IS-3-An interim standard issued by the EIA/TIA that specified the AMPS analog cellular system. This interim standard has now been officially accepted as the EIA-553 standard.

Interior Gateway Protocol (IGP)-IGP is an Internet protocol that is used to exchange routing information between switches and routers within the same domain network.

Interior Gateway Routing Protocol (EIGRP)- Enhanced Interior Gateway Routing Protocol (EIGRP) is an enhanced version of IGRP. The same distance vector technology found in IGRP is also used in EIGRP, and the underlying distance information remains unchanged. The convergence properties and the operating efficiency of this protocol have improved significantly. This allows for an improved architecture while retaining existing investment in IGRP

Interior Gateway Routing Protocol (IGRP)- IGRP is Cisco's Interior Gateway Routing Protocol used in TCP/IP and OSI internets. It is regarded as an interior gateway protocol (IGP) but has also been used extensively as an exterior gateway protocol for inter-domain routing. IGRP uses distance vector routing technology. The concept is that each router need not know all the router/link relationships for the entire network. Each router advertises destinations with a corresponding distance. Each router receiving the information adjusts the distance and

Intermediate Service Part

propagates it to neighboring routers. The distance data in IGRP is represented as a composite of available bandwidth, delay, load utilization, and link reliability. This allows tuning of link characteristics to achieve optimal paths.

Interlaced-An image display that uses alternating graphic lines (e.g. odd and even) such as the lines of a television picture display during each picture scan.

This diagram shows how the lines displayed on each frame are interlaced by alternating the selected lines between each image frame. In frame one, every odd line (e.g. 1,3,5, etc) is displayed. In frame two, every even line (e.g. 2,4,6, etc) is displayed. In frame 3, the odd lines are displayed. This process alternates very quickly so the viewer does not notice the interlacing operation.

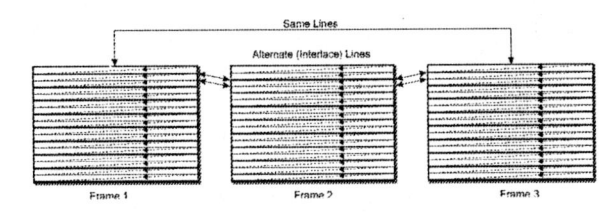

Interlaced Operation

interLATA-Telecommunication services that cross from a local access and transport area (LATA) into another LATA.

Interleaving-Interleaving is the reordering of data that is to be transmitted so that consecutive bytes of data are distributed over a larger sequence of data to reduce the effect of burst errors. The use of interleaving greatly increases the ability of error protection codes to correct for burst errors. Many of the error protection coding processes can correct for small numbers of errors, but cannot correct for errors that occur in groups.

This diagram shows that a block of data information may be distributed over multiple time slots or frames in a carrier line to distribute the effect of burst errors on the information signal. In this example, a block of digital audio is being transmitted through a radio channel. The digital audio is divided into blocks of 4 bits and the bits for each block is distributed (interleaved) over a communication channel. During the transmission, a lightning bolt creates a burst of electrical noise that disrupts 3 bits of data transmission. Because these bits are interleaved, the received data has burst errors that are distributed. This allows the audio to be continuously heard with a marginal amount of distortion instead of completely losing the audio during the burst errors.

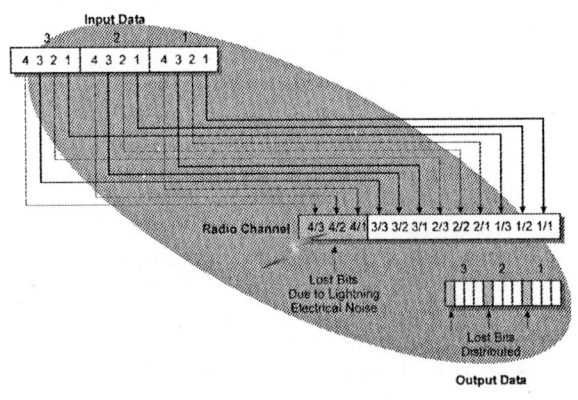

Interleaving Operation

Interlock-(1-circuit) A protection device or system designed to remove all dangerous voltages from a machine or piece of equipment when access doors or panels are opened or removed. (2-tape machines) The state of synchronous operation of two separate audio or video playback machines, achieved by a synchronizer that electronically compares the time code from the two machines and adjusts the speed of one or both until they are locked together.

InterMachine Trunk (IMT)-InterMachine Trunk (IMT) High capacity communication channels that directly connects switches to each other.

Intermediate Distribution Frame (IDF)-An intermediate cross connect that is used at telephone company (telco) or riser closets to interface horizontals, laterals, risers, and/or feeders to other distribution cabling. IDF's can be 110, Krone, 66, or RJ45 Patch Panels.

Intermediate Service Part-An element of Transaction Capabilities Applications Part (TCAP) in an SS7 system that supports connection-oriented messages. It represents OSI layers 4 to 6.

Intermediate Station-Synonymous with internetworking device.

Intermittent-A non-non-continuous recurring event, often used to denote a problem that is difficult to find because of its unpredictable nature.

Intermodulation (IM)-The generation of unwanted radio or optical signal frequencies in a receiver or in a device such as a diode, or an optical medium such as glass, as a result of non-linear "mixing" of input radio or optical signals. This is typically caused when one or more input signal power level(s) is/are high enough so that the non-linear effects cause a mathematical multiplication of the two signals. This can be described mathematically by considering a very simple non-linear input-output relationship. Assume that the output of some non-linear device is proportional to the square of the input, v = x squared. If the input x consists of two voltages (u+w), then when we square this we get three terms: u squared, 2 times the product of u and w, and w squared. That middle term shows the mathematical product result, which produces IM.

Intermodulation is the same technological process as modulation, but it produces undesired output frequency components instead of desired frequency components. One scientist has compared this difference in terminology in the use of "modulation" and "intermodulation" to the two names "soil" and "dirt." If you are a botanist, and you prepare soil to grow plants in pots, that is a favorable name. But if the soil spills on an expensive carpet, you call it "dirt," an unfavorable name!

Intermodulation Distortion (IMD)-The distortion that results when mixing two input signals in a nonlinear system. The resulting output contains new frequencies that represent the sum and difference of the input signals and the sums and differences of their harmonics. IMD also is called intermodulation noise.

Intermodulation Noise-In a transmission path or device, the noise signal that is contingent upon modulation and demodulation, resulting from nonlinear characteristics in the path or device.

Internal Resistance-The actual resistance of a source of electric power. The total electromotive force produced by a power source is not available for external use; some of the energy is used in driving current through the source itself.

International Callback-International callback is a call processing service that reverses the connection of calls. International callback service is popular in countries that have high tariffs (fees) for outgoing (originating) international calls and have low tariffs for incoming (received) international calls. This process is divided into the call setup (dial-in) and callback stages. The international caller dials a number that provides access to the international callback service. This number may be local in the visited country or be an international number. The international callback gateway receives the call and prompts the caller to say or enter (e.g. by touch tone) the international number they desire to be connected to and the number they want the callback service to connect to. The international callback center then originate calls to both numbers and connects the two individuals to each other.

International Gateway Facilities (IGFs)-Systems or equipment that provide access between telephone systems in different countries. International gateways may convert SS7 and other signaling formats between different signaling formats. These include ANSI standards, ITU standards, national variants of SS7 signaling standards, MF signaling, and R2 signaling. International gateways may also provide for transcoding services between mu-LAW PCM and A-LAW PCM speech coding.

This figure shows how two national SS7 systems interconnect using an international gateway between an ANSI based end office SSP in North America and an ITU based end office SSP in Asia. This example shows that an ANSI based SS7 system require address translation and circuit identifier code format changes as the messages are passed between the systems. The ANSI 24 bit destination point code (DPC) and origination point code (OPC)

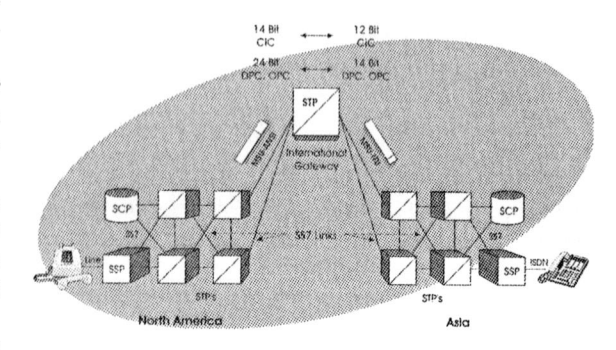

International Gateway Facilities

addressing must be translated to 14 bit DPC and OPC codes for the ITU system. It also shows that the 14 bit ANSI CIC code used in ISUP messages must be translated to 12 bit CIC codes used by the ITU system.

International Maritime Satellite Organization (INMARSAT)-An organization that was established by an international treaty to provide satellite services for wide area mobile communications. INMARSAT is based in London, England.

International Mobile Equipment Identifier (IMEI)-An electronic serial number that is contained in a GSM mobile radio. The IMEI is composed of 14 digits. 6 digits are used for the type approval code (TAC), 2 digits are used for the final assembly code (FAC), 6 digits are used for the serial number and 2 digits are used for the software version number.

International Mobile Subscriber Identity (IMSI)-The unique identification number for a GSM mobile telephone customer. The IMSI is stored in the Subscriber Identity Module (SIM).

International Mobile Telephony 2000 (IMT-2000)-The name given to third generation wireless systems by the international telecommunications union (ITU).

International Private Network Service (IPNS)-An international standardized private line service that typically ranges from 9.6 kbps up to 2 Mbps (E1).

International Radio Consultative Committee (CCIR)-Standards organization established by UNESCO originally having the French language name "Comité Consultatif International des Radiocomunications." As a result of a 1993 reorganization, it was supplanted by the International Telecommunication Union - Radio Sector (ITU-R or simply ITU). It is responsible for the study of technical and operating questions relating to radio communications. Its findings, if adopted as treaty agreements, have the effect of international law. (See also: CCITT, Consultative Committee for International Telephony And Telegraphy)

International Record Carrier (IRC)-A communication carrier that transfers data between international locations.

International Signaling Point-A signaling point which belongs to the international signaling network.

International Signaling Point Code-The part in the label of a CCITT#7 or SS7 message that uniquely identifies each signaling point belonging to the international signaling network. It consists of a sub-field for the signaling area/network code (11-bit) and a sub-field that identifies a signaling point in a specific area or network (3-bit).

International Standard Book Number (ISBN)-A unique identifying number used in the book trade for each title and edition. Included in this dictionary to clarify that it is different from Integrated Services Digital Network (ISDN).

International Telecommunication Union (ITU)-A specialized agency of the United Nations established to maintain and extend international cooperation for the maintenance, development, and efficient use of telecommunications. The union does this through standards and recommended regulations, and through technical and telecommunications studies. Based in Geneva, Switzerland, the ITLI is composed of two consultative committees: the International Radio Consultative Committee (CCIIR) and the Consultative Committee for International Telephony And Telegraphy (CCITT).

International Telecommunications Union - T (ITU-T)-A sector of the International Telecommunications Union (ITU) that focuses on telecommunications.

International Toll Free Service (ITFS)-International toll free is a service that allows callers to dial and telephone number from other countries without being charged for the call. The toll free call is billed to the receiver of the call.

Internet-A public data network that interconnects private and government computers together. The Internet transfers data from point to point by packets that use Internet protocol (IP). Each transmitted packet in the Internet finds its way through the network switching through nodes (computers). Each node in the Internet forwards received packets to another location (another node) that is closer to its destination. Each node contains routing tables that provide packet forwarding information. The Internet evolved from ARPANET and was designed to allow continuous data communication in the event some parts of the network were disabled.

This figure shows the basic structure of the Internet. This diagram shows that the Internet is composed of users (end points), Internet service providers (ISPs), network service providers (NSPs), and network access points (NAPs). Computers are connected to the Internet via an ISP. The ISP receives data from the computer, reformats it (if necessary), and forwards it to the destination computer in its network. If necessary, the data packets may be routed to an NSP, which will route the data packets to a destination ISP that is outside the NSP's own network. Eventually, packets reach their destination ISP that forwards the packets to the user.

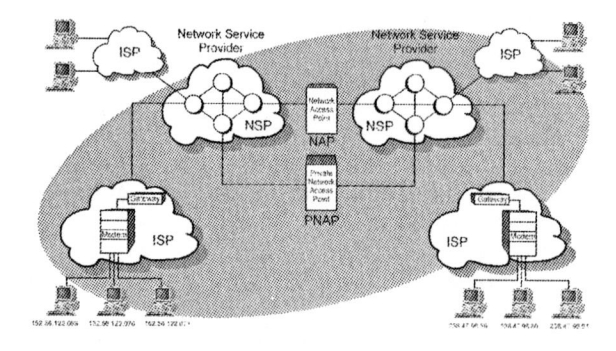

Internet System

Internet Access Provider (IAP)-A company that provides an end user with data communication service that allows them to connect to the Internet. Internet access providers are also called Internet service providers (IAPs.)

Internet Address-A unique binary digital number that identifies a specific connection point within the Internet. An internet address is 32 bit for version 4 and 128 bits for version 6.

Internet Architecture Board (IAB)-A technical advisory group that is part of the Internet Society (ISOC) that manages Request for Comments (RFCs) publication standards and documents. The IAB also serves as an board to hear appeals and provides other services to the ISOC.

Internet Assigned Numbering Authority (IANA)-The authority that is responsible for assignment and coordination of Internet addresses and key parameters such as protocol variables and domain names.

Internet Control Message Protocol (ICMP)-A protocol that is used to report errors that occur during transmission of IP datagram packets. Routers send ICMP to the sender of the datagram to indicate when packets are undeliverable.

Internet Engineering Task Force (IETF)-An organization that assists in the development and coordinates protocol standards that are used on the Internet.

Internet Fax-The process of sending faxes through the Internet. Sending faxes through the internet can be performed unaltered (Internet is a bearer service) or it can interact using Internet fax protocol (IFP) to increase the reliability of complete fax information transfer.

Internet Group Management Protocol (IGMP)-An Internet protocol that is used to report their information on group memberships to neighboring multicast routers. IGMP is defined in RFC 2236.

Internet Keyed Payments Protocol (IKP)-A protocol that is used to process secure payments over the Internet. iKP uses the RSA public-key encryption system to process transactions of financial nature. iKP allows a buyer and a seller interact with a third party (such as a merchant processing company) to securely process transactions

Internet Message Access Protocol (IMAP)-A protocol that defines the access and storage procedures for electronic mail (e-mail) messages. IMAP is used with simple mail transfer protocol (SMTP) to move e-mail messages between email servers and mailboxes. IMAP is defined in RFC 2060.

The IMAP protocol defines message headers to allow the recipient to better search, select, and download specific messages or parts of messages. IMAP also includes security authentication procedures.

Internet Mode (i-Mode)-A wireless information service that provides information to wireless users (primarily mobile telephone users). It was developed and initially offered by NTT DoCoMo in Japan on February 22, 1999.

Internet Number-The dotted-quad address used to specify a certain system within the Internet.

Internet Paradigm-The sending of information over a common communication system (Internet protocol) as compared to sending information over separate dedicated networks (such as telephone, data network, and cable television systems).

Internet Protocol (IP)-A low-level network protocol that is used for the addressing and routing of packets through data networks. IP is the common language of the Internet. The IP protocol only has routing information and no data confirmation rules. To ensure reliable data transfer using IP protocols, higher level protocols such as TCP are used IP protocol is specified in RFC-791.

This protocol defines the packet datagram that hold packet delivery addressing, type of service specification, dividing and re-assembly of long data files and data security. IP protocol structure is usually combined with high-level transmission control protocols such as transaction control protocol (TCP/IP) or user datagram protocol (UDP/IP).

Internet Protocol Address (IP ADDRESS)-The address portion of an Internet Protocol (IP) packet. For IP version 4, this is a 32-bit address and for IP version 6, this is a 128 bit address. To help simplify the presentation of IPv4 addresses, it is common to group each 8 bit part of the IP address is a decimal number separated from other parts by a dot(.), such as: 207.169.222.45. For IPv6 it is customary to represent the address as eight, four digit hexadecimal numbers separated by colons, such as 1234:5678:9000:0D0D:0000:5678:9ABC:8777.

This diagram shows how different types of data network addressing systems. This diagram shows a end-to-end data connection may transfer through many different networks and each network may use a different addressing system. This example shows that an end-user uses an Internet address to connect to a remote data device that has its own Internet address. The Internet address and its data is carried through the entire end-to-end communication in the data parts of each network packet. The first path is connects the user to a company Ethernet network. The computers network interface card (NIC) that has a 48 bit address unique to the Ethernet. Each packet that travels in the company's Ethernet network has it's own Ethernet address. Each packet of data from the end user includes the 32 bit Internet address. This packet (datagram) is encapsulated (stored) as part of the data message after the Ethernet address. The company's network is connected to an ISP by a high-speed frame relay connection. The frame relay access device (FRAD) has a unique identifier to the ISP. The ISP connects the data connection via asynchronous transfer mode (ATM) to the ASP.

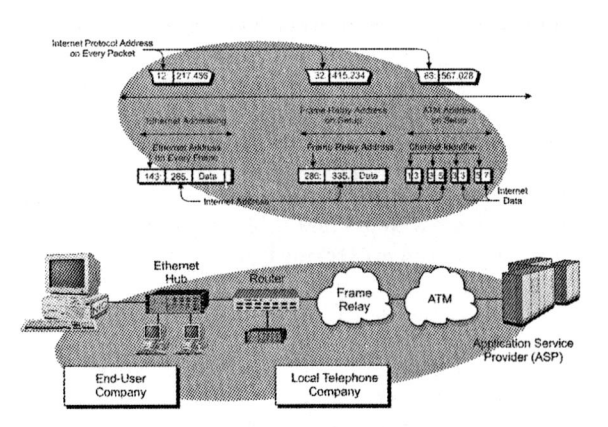

Internet and Network Numbering System

Internet Protocol Broadcast (IP Broadcast)-A data packet that uses a frame address mask of 255.255.255.255 to identify it is intended for broadcast distribution. This allows devices within the network to identify broadcast messages and inhibits routers from constantly circulating packets through the network. The use of the address mask inhibits the normal transmission of the data packet through routers because routing protocols use the zeros at the end of the subnet mask number to identify the subnet. Because of the subnet mask (11111111.11111111.11111111.11111111 equals 255.255.255.255), the end of the address does not contain any zeros.

Internet Protocol Centrex (IP Centrex)-IP Centrex is the providing of Centrex services to customers via Internet protocol (IP) connections. IP Centrex allows customer to have and use features that are typically associated with a private branch exchange (PBX) without the purchase of PBX switching systems. These features include 3 or 4 digit dialing, intercom features, distinctive line ringing for inside and outside lines, voice mail waiting indication and others.

Internet Protocol Configuration (IPConfig)-An Internet Protocol application is used to program Internet Protocol configuration information such as the host IP address, gateway IP address, and subnet mask.

Internet Protocol Datagram (IP Datagram)-The packet of data that contains addressing header (IP header), it's associated control header (such as a TCP header), and the data associated with the

Internet Protocol Detail Record (IPDR)

packet. The IP header contains addressing information (source and destination address) and basic control information (such as time to live). Additional headers (such as a TCP header) contain control or additional routing information (such as a port number) related to a specific application for which the packet is related to. The data portion of the datagram has a variable length.

Internet Protocol Detail Record (IPDR)-A data record containing information related to an IP-based communication session. This information usually contains identification information of the users of the service, types of services used, quantity measurement unit type (e.g. kilobytes or time), quantities of services used, Quality of Service parameters, and the date/time (usually relative to GMT) the services were used.

Internet Protocol Header (IP Header)-The addressing portion of an IP packet that contains the routing and control information for the packet. The IP header is used by routers to determine the route the packet must be forwarded to help it reach its destination. The IP header contains the version number of the protocol (e.g. IPV4 or IPV6), source and destination address information, and other control (e.g. type of service) information.

This diagram shows the IP header field structure for version 4 IP. The IP header is located at the beginning of each IP datagram packet. The IP packet header starts with a version number that indicates which IP version is being used. The version indicates the field structure that is used for this packet (e.g. 32 bit addressing compared to 128 bit addressing). The Hlen field identifies the length of the packet header. A service type field identifies that type of service that the IP packet is being used for. This type of service field allows for differentiated service (DiffServ) handling of packets that require real time (e.g. voice) or reliable (e.g. data) packet transfer. The total length field identifies the total length of the packet (with data). A time to live protocol field is used to allow the network to discard packets that travel through too many switching points preventing the possibility of infinite loops. Each packet contains the source IP address and the destination IP address.

IP Packet Header Version 4 Structure

Internet Protocol Next Generation (IPng)-A common term used for the next generation of Internet Protocol.

Internet Protocol Private Branch Exchange (IPBX) or (IP PBX)-A private local telephone system that uses Internet protocol (IP) to provide telephone service within a building or group of buildings in a small geographic area. IPBX systems are often local area network (LAN) systems that interconnect IP telephones. IPBX systems use a IP telephone server to provide for call processing functions and to control gateways access that allows the IPBX to communicate with the public switched telephone network and other IPBX's that are part of its network. IPBX systems can provide advanced call processing features such as speed dialing, call transfer, and voice mail along with integrating computer telephony applications. Some of the IPBX standards include H.323, MGCP, MEGACO, and SIP.

IP PBX represents the evolution of enterprise telephony from circuit to packet. Traditional PBX systems are voice-based, whereas their successor is designed for converged applications. IP PBX supports both voice and data, and potentially a richer feature set. Current IP PBX offerings vary in their range of features and network configurations, but offer clear advantages over TDM-based PBX, mainly in terms of reduce Opex (operating expenses).

Internet Protocol Security (IPSec)-A part of the Internet Protocol that helps to ensure the privacy of user data. IPSec is part of the next generation internet, IPv6. IPSec is defined in RFC 1827.

Internet Protocol Suite-A combination of network protocols that have designed to interoperate with each other to provide a common data communication language that is used on the Internet. The layers of protocol suite include physical layer, network (or routing) layer, transport (or session) layer, and application layer.

This protocol suite is overseen by the Internet Engineering Task Force (IETF). Key protocols included in the Internet Protocol Suite include Internet Protocol (IP), Transaction Capabilities Protocol (TCP), and User Datagram Protocol (UDP). There are many other protocols that are part of the Internet Protocol suite.

Internet Protocol Telephony (IP Telephony)- IP telephone systems provide voice or multimedia communication services through the use Internet protocol (IP) networks. These IP networks initiate, process, and receive voice or multimedia communications using IP protocol. These IP systems may be public IP systems (e.g. the Internet), private data systems (e.g. LAN based), or a hybrid of public and private systems.

Internet Protocol Telephony (IPTEL)- A process of sending voice telephone signals over a data network (such as the Internet) using Internet protocol.

Internet Protocol Version 4 (IPv4)- A revision of Internet protocol that uses 32 bit addressing.

Internet Protocol Version 6 (IPv6)- A network packet routing protocol that uses 128 bit address. IPV6 is an enhanced version of Internet protocol version (4 IPv4) that was developed primarily to correct shortcomings of IPv4 such as the 32 to bit address that limited the maximum number of devices that could be addresses and to extend the capabilities of IP to meet the demands of the future such as improved quality of service (QoS) capabilities. IPv6 addresses are denoted as 8 hexadecimal numbers, separated by colons. A typical address will look like this:

0800:5008:0000:0000:0000:1005:AABC:AD46

A short-hand notation that replaces one set of consecutive zeros with colons (::) may also be used. The above address can also be denoted by:

0800:5008::1005:AABC:AD46

IPv6 utilizes a hierarchical address, called an Aggregatable Global Unicast Address Format (AGUAF). In this format, each IP address is built by concatenating a Top-Level Aggregation (TLA) ID, a Next-Level Aggregation ID (NLA) and a Site-Level Aggregation ID (SLA) and an interface ID. This enables the IPv6 address space to be assigned in a logical manner by multiple address assignment authorities while still guaranteeing that all hosts have unique IP addresses and the addresses can be used to easily route packets without requiring switches to maintain enormous routing tables.

The IPv6 header has been simplified with the introduction of extension headers. The basic IPv6 header contains just seven fields such as hop count and destination IP address. Also included is a Next Header field, which points to the next header in the packet. This greatly simplifies the logic required to parse the packet size in hosts and routers. IPv6 natively supports functions to discover neighbors, assign IP addresses dynamically and to identify multicast participants.

The most useful application of IPv6 is in next generation wireless phone networks. Many companies are moving to IP based networks to replace existing cellular technology. Over the next few years the need for IPv4 and IPv6 to co-exist will be an important factor in the deployment of IPv6 networks.

Internet Relay Chat (IRC)- An Internet protocol that was developed to allow for instant messaging (IM) between members of a chat group.

Internet Reliability- The ability of the Internet to consistently provide data transmission between points that are connected to the Internet.

Internet SCSI (iSCSI)- A protocol used to build storage networks based on TCP/IP for transport. The protocol scales the SCSI command set used by direct attached storage so that it may be used on traditional networking technologies such as Ethernet.

iSCSI allows the creation of a storage network using familiar networking technology, while gaining the benefits of SANs.

Internet Service Provider (ISP)- A company that provides an end user with data communication service that allows them to connect to the Internet. An ISP purchases a high-speed link to the Internet and divides up the data transmission to allow many more users to connect to the Internet.

Internet Signaling Transport Protocol (ISTP)- A signaling protocol that used by PacketCable networks to provide SS7 type signaling capabilities.

Internet Society (ISOC)- An international organization assists the develop of Internet technologies and applications. The Internet society is composed of many companies that share an interest in the development of the Internet and services for the Internet. The ISOC also oversees and coordinates

Internet Telephone (IP Telephone)

the activities of various Internet working groups. This includes the Internet Research Task Force (IRTF), Internet Architecture Board (IAB), the Internet Engineering Task Force (IETF), and the Internet Assigned Numbers Authority (IANA).

Internet Telephone (IP Telephone)-A telephone device that is specifically designed to communicate through the Internet without the need for a voice gateway. Internet telephones contain embedded software that allows them to initiate and receive calls through the Internet using standard protocols such as H.323 or SIP.

This figure shows the Opti 400 model telephone an Internet telephone that can directly connect to a high-speed data line (such as the Ethernet data connection on a DSL or cable modem line) without the need to connect to a computer. This telephone converts the voice to IP data signals and provides advanced call features that are found on standard telephones.

IP Telephone
Source: Siemens Corporation

Internet Telephony-Telephone systems and services that use the Internet to initiate, process and receive voice communications.

Internet Telephony Service Provider (ITSP)- Internet Telephony Service Providers (ITSPs) are companies that provide telephone service using the Internet. ITSPs setup and manage calls between Internet telephones and other telephone type devices.

An ITSP coordinates Internet telephone devices so they can use the Internet as a connection path between other telephones. ITSPs are commonly used to connect Internet telephones or PC telephones to telephones that are connected to the public telephone network. This is accomplished by using gateways. Gateways convert packets of audio data from the Internet into standard telephone signals.

This figure shows how an ITSP sets up connections between Internet telephones and telephone gateways. The ITSP usually receives registration messages from an Internet telephone when it is first connected to the Internet. This registration message indicates the current Internet address (IP address) of the Internet telephone. When the Internet telephone desires to make a call, it sends a message to the ITSP that includes the destination telephone number it wants to talk to. The ITSP reviews the destination telephone number with a list of authorized gateways. This list identifies to the ITSP one or more gateways that are located near the destination number and that can deliver the call. The ITSP sends a setup message to the gateway that includes the destination telephone number, the parameters of the call (bandwidth and type of speech compression), along with the current Internet address of the calling Internet telephone. The ITSP then sends the address of the destination gateway to the calling Internet telephone. The Internet telephone then can send packets directly to the gateway and the gateway initiates a local call to the destination telephone. If the destination telephone answers, two audio paths between the gateway and the Internet telephone are created. One for each direction and the call operates as a telephone call.

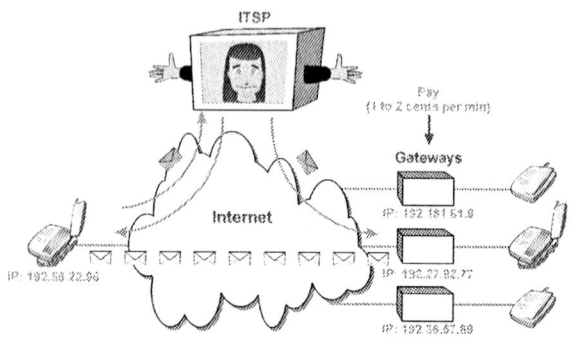

Internet Telephony Service Provider (ITSP) Operation

Internet2-A second generation of the Internet that uses a high-speed backbone communications network. The Internet system is a result of the Next Generation Internet (NGI) initiative that is sponsored by the United States government. Internet2 is seen as the way to deliver multimedia content (e.g. video on demand) through the Internet.

Internetwork Operating System (IOS)-An operating system that is part of the CiscoFusion architecture that can provide centralized integrated, automated installation and management of Internet and intranet networks.

Internetwork Packet Exchange (IPX)-A data communication protocol that is a trademark of Novell. It is part of Novell's NetWare's protocol stack that is used to transfer data between the server and workstations within the network. IPX data packets are encapsulated (both address and data are stored in the payload) and carried by the packets used in local area networks (such as Ethernet or Token Ring).

IPX packet header contains 30 bytes of information that includes the address information (network, node, and socket addresses) of the source and the destination. It is followed by the data area that can vary in length from 0 bytes (only the header) to 65,535 bytes. Many networks limit the maximum packet size to approximately 1500 bytes.

Internetwork Protocol-The protocol used to move frames from originating source stations to their ultimate target destinations (through routers, if necessary) across an internetwork. IP, IPX, and DDP are all examples of internetwork protocols.

Internetworking Device-A device used to relay frames or packets among a set of networks (e.g., a bridge or router).

interNIC-Internet Network Information Center

Interoffice Channel (IOC)-A communications link between two end office switching centers or between two points of presence (POPs) for interexchange carriers (IXCs).

Interoffice Facilities (IOF)-Communication lines or trunk (multi-channel lines) that interconnect switching systems. The facilities include the channel multiplexing and de-multiplexing equipment along with physical copper, wireless or fiberoptic lines.

Interoperability-The condition achieved among communications and electronics systems or equipment when information or services can be exchanged directly between them, between their users, or both.

InterPBX-The direct connection of calls between PBX switching systems through the use of dedicated connection lines (tie lines).

Interrupt-An event signal that is used to inform a processing system that suspension of the operation is required so another sequence of instructions can be processed. Interrupts are usually created by device under a processors control, such as an accessory device, that interrupts normal processing to gain rapid processing response. Interrupts usually cause software processing routines to branch from their current processing steps to an interrupt service routine (ISR).

Interrupts are sometimes classified to different types. These include internal hardware, external hardware, and software interrupts. Because there can be several types of interrupts in a system, interrupts can be prioritized. Interrupts used by Intel products allows up to 256 prioritized interrupts. Of these, the first 64 interrupts are reserved for use by the system or the operating system.

Intersystem Handover-A process where a mobile radio operating on cell site in one mobile system is reassigned to a new channel on a cell site in another mobile system. Intersystem handover requires signaling messages to be transmitted between the different systems.

Intersystem Signaling-Control signaling that occurs between systems.

Intersystem Signaling 41 (IS-41)-The application entity that dedicated to the communication aspects of intersystem signaling (such as in the SS7 system) that is used for the AMPS (analog), IS-136 TDMA, and IS-95 CDMA mobile communication systems.

Interworking-The process of adapting the communications between two different types of networks. This may include circuit switched, packet switched or messaging services.

Interworking Function (IWF)-A functional part of a telecommunications network that is used to process and adapt information between dissimilar types of network systems.

Intra-Cell-Handoff

Intra-frame (I-frame)-In video compression, a frame encoded with no reference to any other frames. In MPEG and H.261/H.263 video codecs I-frames are encoded in a way similar to JPEG for still images. Also called I-pictures.

Intranet-A private network that is used within a company to provide company information to employees. Intranets may be connected to vendors and customers through private data connections or via public Internet connections. When Intranets are connected to the Internet, they are commonly connected through firewalls to protect the company's internal data.

Intrastate Service-A broad category that includes all telecommunications services and/or activities that are covered by an intrastate or state tariff. Such services normally include state toll, local exchange, extended area service, multiple message unit, most optional calling service plans, and access services under state jurisdiction.

Intrinsic Junction Loss-In optoelectronics, the total loss resulting from the joining of two identical optical waveguides. Factors influencing this loss include spacing loss, alignment of the waveguides, Fresnel reflection loss, and end finish.

Intrinsic Semiconductor-An i-type semiconductor, with equal concentrations of holes and electrons under conditions of thermal equilibrium. An intrinsic semiconductor is either "pure" (with no intentional dopant materials added) or has equal numbers of both P and N type dopant materials.

Intserv-Integrated Services

Invention-Date of Conception

Invention Disclosure-Most companies require an inventor to submit an "invention disclosure" describing an invention and providing supplementary information such as the project to which the invention is connected.

An invention disclosure is not a formal requirement in order to obtain a patent, but it does form part of the chain of evidence which can be used in the United States to prove date of invention.

Inventive Step-According the Article 56 of the European Patent Convention, "an invention shall be considered as involving an inventive step if, having regard to the state of the art, it is not obvious to a person skilled in the art."

Inventive Step is analogous to the US requirement that inventions be Non-Obvious.

Inventor's Notebook-An inventor's notebook is a bound, numbered, book containing information as to the conception of ideas and their subsequent development up to and including reduction to practice. A properly kept inventor's notebook can be useful in proving the date of invention in the United States.

Inventory Available Date-The date on which equipment and facilities are disconnected and made available for reuse.

Inverse Multiplexer (IMUX)-A device that divides a single telephone or data communication channel into two or more channels to be transported over multiple communication links. Inverse multiplexing may be in the form of frequency division (e.g. multiple radio channels on a coax line), time division (e.g. slots on a T1 or E1 line), or code division (coded channels that share the same frequency band) or combinations of these.

Inverse Multiplexing-The combining of information signals received on multiple communications channels to form a higher speed communication channel than is possible on a single independent communication channel. Inverse multiplexing has been used on wireless communication systems to allow high-speed digital video signals to be sent over cellular radio channels that have a limited maximum data transmission rate.

Inverse Multiplexing Over ATM (IMA)-A standard process of dividing a high speed data channels into multiple lower speed data channels (such as inverse multiplexing high-speed ATM channel over two or more T1 circuits).

Inversion-(1-signal) The change in the polarity of a signal or pulse, such as from positive to negative. (2-scrambling) A form of speech scrambling used to ensure the privacy of a transmission. A voice signal is mixed with a higher-frequency audio signal, and only the difference frequency is transmitted.

Invite Message-A message that is used to invite a person or device to participate in a communication session. The invite message is defined in session initiation protocol (SIP).

Invoice Record-A telecommunications data record that contains control counts or totals that describe an accompanying data set. Excluded are indexes and header and trailer information.

This figure shows a sample invoice. This invoice statement provides the customer account information (name, address, and account number), invoice charge totals, along with detailed billing information. This example shows that the customer pays recurring charges (monthly fees) plus additional charges such as taxes and communication costs that are not part of their rate plan. The detailed charges identify the category of the charge (rate), the amount of usage (time), and any additional charges (surcharges) that may apply.)

IP Centrex System

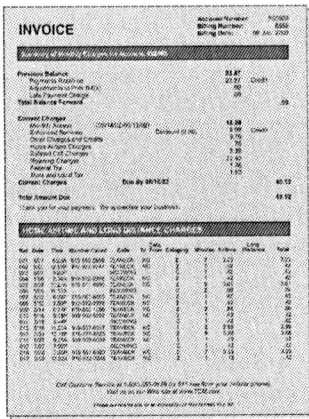

Invoice Statement

Invoke Component-A request message that is part of the Transaction Capabilities Application Part (TCAP) in the SS7 system that requests that an operation be performed at a receiving node.
Involuntary Churn-A disconnection of service that occurs when a carrier decides to disconnect service from an existing customer regardless if the customer wants to disconnect service or not.
INWATS-Inward Wide Area Telephone Services
INWATS-Inward Wide Area Telephone Service
IO-Information Outlet
IOC-Integrated Optical Circuit
IOC-Interoffice Channel
IOD-Identification Of Outward Dialing
IOEF-Instantaneous Override Energy Function
IOEngine-Input/Output Engine
IOF-Interoffice Facilities
ION-Integrated On-Demand Network
Ion Implantation-A technique used in the manufacture of integrated circuits in which the surface of a semiconductor is bombarded in order to implant ions into the lattice of the crystal.
Ionization-A procedure by which an atom is given a net charge by adding or subtracting an electron.
IONL-Internal Organization of the Network Layer
Ionosphere-The ionosphere is the region of the atmosphere that extends from 50 km to 400 km above the surface of the Earth. Because this region has radio characteristics that are different than the Earth's surface, radio waves between 2 MHz and 50 MHz are bent (refracted) back towards the Earth. The actual distance between the transmitter and where these radio waves return to the Earth can range from 500 to 3000 miles. Because the ionosphere characteristics can change during the day, this refraction distance can also change.
IOP-Interoperability
IOPS-Internet Operators Group
IOR-Index of Refraction
IOS-Internetwork Operating System
IOTP-Internet Open Trading Protocol
IP-Internet Protocol
IP-Intelligent Peripheral
IP-Information Provider
IP-Intellectual Property
IP ADDRESS-Internet Protocol Address
IP Broadcast-Internet Protocol Broadcast
IP Centrex-Internet Protocol Centrex
IP Centrex System-A system that provides Centrex services to customers using Internet protocol (IP) connections. IP Centrex allows customer to have and use features that are typically associated with a private branch exchange (PBX) without the purchase of PBX switching systems.
This figure shows a simplified IP Centrex system that provides telephone services to IP telephones that are connected to a company's local area data network. This diagram shows that the IP Centrex telephone system consists of IP telephones located at the customers location, a data connection that can connect the IP telephones to the IP Centrex system, a Centrex call processing system (call server), and gateways that allow the IP Centrex service provider to interconnect calls to the public telephone network. Each IP telephone has its own network data address and it registers with the IP Centrex call server when it is connected to the com-

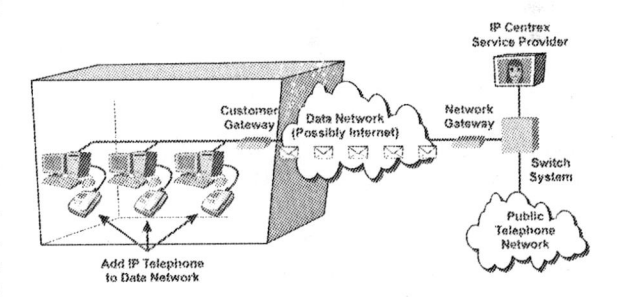

:IP Centrex Operation

www.ITExpo.Com

pany's data network. When calls are received from the PSTN at the IP Centrex system, the call server looks in it's databases to find the associated IP telephone address (data address) and this address is used to alert the IP telephone of an incoming call. When calls are originated from the IP telephone, the dialed telephone number is passed to the IP Centrex call processing system. This system determines if the call is routed within the data network or if the voice gateway must be used to connect the call to the PSTN.

IP Connectivity-The ability to setup communication sessions using Internet protocol (IP).

IP Datagram-Internet Protocol Datagram

IP Datgram-A packet of data that contains the ultimate network destination along with some control information and travels within (encapsulated) in the data portion of other network packets that are used to transport the IP datagram towards its ultimate destination.

IP Enabled PBX-A non-IP PBX (such as a tradional TDM PBX) which allows for the support of IP phones or IP interconnect through the use of a special line or trunk card which converts TDM signals to IP signals. This is a common approach in migrating users of TDM PBXes to IP telephony technologies. This term often implies that IP to IP telephony audio may pass through a TDM switch.

IP Header-Internet Protocol Header

IP Multicast-An Internet protocol that is used to broadcast the same message to multiple recipients. An IP multicast message is transferred to all the members within pre-defined group.

IP PBX-Internet Protocol Private Branch Exchange

IP Precedence-IP Precedence utilizes the 3 precedence bits in the Type-of-Service (TOS) field in the IP header to specify class of service assignment for each IP packet. IP Precedence provides considerable flexibility for precedence assignment including customer assignment (e.g. by application) and network assignment based on IP or MAC address, physical port, or application. IP Precedence enables the network to act either in passive mode (accepting precedence assigned by the customer) or in active mode utilizing defined policies to either set or override the precedence assignment. IP Precedence can be mapped into adjacent technologies (e.g. Frame Relay or ATM) to deliver end-to-end QOS policies in a heterogeneous network environment. Thus, IP Precedence enables service classes to be established with no changes to existing applications and with no complicated network signaling requirements.

IP Telephone-Internet Telephone

IP Telephony-Internet Protocol Telephony

IP Tunnel-A logical path (a tunnel) between connection points through a data network using Internet protocol. This logical path allows data to freely flow between the connection points (entry and exit point) regardless of the underlying type of data or physical types of networks that are used to complete the tunnel connection (e.g. Ethernet LAN, ATM).

iPBX System-Internet protocol private branch exchange (IPBX) systems use Internet protocols to provide voice communications for companies. IPBX systems can be separate from data network or they may share the data network systems. When the iPBX system shares the local area network (LAN), it may be called LAN Telephony or TeLANophy.

This figure shows an iPBX system that shares a company's local area data network. This diagram shows that a iPBX telephone system consists of IP telephones, a data network, a call processing system, and a voice gateway to the public telephone network. IP telephones convert audio into digitized packets that are transferred on the call server. Each IP telephone has its own network data address. The call server communicates with IP telephones over the same high-speed LAN data network that communicates with computers. When calls are received from the PSTN, the call processing system looks in the database to find the associated IP telephone address (data address) and this address is used to alert the IP telephone of an incoming call. When calls are originated from the IP telephone, the dialed telephone number is passed to the call processing system. This system determines if the call is routed within the data network or if the voice gateway must be used to connect the call to the PSTN.

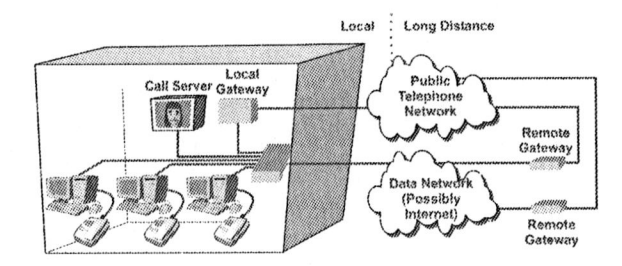

Internet Protocol Private Branch Exchange (iPBX) Operation

IPC-ISDN To POTS Converter
IPC-Interprocess Communications
IPCH-Initial Paging Channel
IPCI-Integrated Personal Computer Interface
IPConfig-Internet Protocol Configuration
IPCP-IP Control Protocol
IPDR-Internet Protocol Detail Record
IPDS-Intelligent Printer Data Stream
IPE-Intelligent Peripheral Equipment
IPG-Interpacket Gap
IPLC-International Private Line Circuit
IPM-Interpersonal Messaging
IPM-Interpersonal Message
IPND-Integrated Public Number Database
IPng-Internet Protocol Next Generation
IPNS-International Private Network Service
IPO-Initial Public Offering
iPOD-IP Phone over Data
IPP-Internet Print Protocol
IPR-Intellectual Property Rights
IPRS-Internet Protocol Routing Service
IPS-Inches Per Second
IPS7-Internet Protocol Signaling 7
IPSec-Internet Protocol Security
IPSO-Internet Protocol Security Option
IPT Gateway-IP Telephony Gateway
IPTC-Internet Telephony Solution for Carriers
IPTEL-Internet Protocol Telephony
IPU-Intelligent Processing Unit
IPv4-Internet Protocol Version 4
IPv5-Internet Protocol Version 5
IPv6-Internet Protocol Version 6

IPX-Internetwork Packet Exchange
IPX Address-A logical path (a tunnel) between connection points that uses Internet protocol to carry data between these points. This logical path allows data to freely flow between the connection points (entry and exit point) regardless of the underlying type of data or the different types of physical connections that are used by the tunnel.
IPX Header Structure-The header portion of an IPX packet that the routing that contains the protocol version number, address information, and routing control information.
This diagram shows that the IPX packet header follows the media access control (MAC) header and it contains a checksum for ensuring data integrity, packet length, transport control (number of routers a packet can cross before being discarded,) the type of packet type based on the service that created the packet, destination network address, destination node address, destination socket address, address of the source network, source node address, and the source socket address.

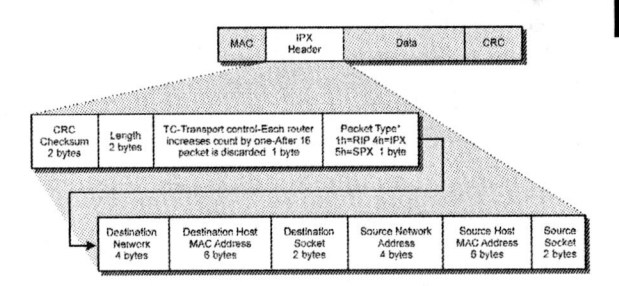

IPX Header Structure

IPX-SPX-Internetwork Packet Exchange
IPX/SPX-Internetwork Packet Exchange/Sequenced Packet Exchange
IPXCP-IPX Control Protocol
IPXODI-Internetwork Packet Exchange Open Data-Link Interface
IPXWAN-IPX Wide-Area Network
IR-Investor Relations
IR-Infrared Radiation
IR-Infrared
IR-Internet Numbers Registry

IR Loss-The conversion of electric power to heat caused by the flow of electric current through a resistance.
IRAC-Interagency Radio Advisory Committee
IRAC-International Radio communications Advisory Committee
IRAC-Interdepartment Radio Advisory Committee
IRAM-Intelligent RAM
IRC-Internet Relay Chat
IRC-International Record Carrier
IRD-Integrated Receiver/Descrambler
IrDA-Infrared Data Association
IRF-Inherited Rights Filter
Iridium-A low Earth orbit (LEO) satellite communications system that uses time division multiple access (TDMA) technology.
Iris-A restriction of the opening into a waveguide that affects its susceptance properties.
IRL-In Real Life
IRL-InterRepeater Link
IrLAP-Infrared Link Access Protocol
IRM-Inherited Rights Mask
IROB-In Range of Building
IRP-I/O Request Packet
IRQ-Interrupt Request
IRTF-Internet Research Task Force
IRU-Indefeasible Right of Use
IS-Interim Standard
IS-Information Separator
IS-124-Interim Standard 124 Data Message Handler
IS-135 Data Specification-An interim standard issued by the EIA/TIA that specifies the data transfer services for the IS-136 TDMA specification.
IS-136 (TDMA)-An interim standard issued by the EIA/TIA that specifies one of the TDMA systems developed in North America. Originally named IS-54, IS-136 combines the features of the IS-54B digital cellular mobile telecommunication systems with a digital control channel (DCCH). The IS-136 specification is divided into two primary documents; IS-136.1 that contains the digital control channel and IS-136.2 that contains the modified IS-54 B specification.
IS-136 (US TDMA)-An Interim standard issued by the EIA/TIA that specifies one of the TDMA systems developed in North America. Originally named IS-54, IS-136 combines the features of the IS-54B digital cellular mobile telecommunication systems with a digital control channel (DCCH). The IS-136 specification is divided into two primary documents; IS-136.1 that contains the digital control channel and IS-136.2 that contains the modified IS-54 B specification. See also IS-54 or IS-54B.
IS-41-Intersystem Signaling 41
IS-54 (TDMA)-A TDMA cellular system interim standard specification that was first developed and deployed in North America. This system was developed as an upgrade to the 30 kHz channel AMPS analog system. This industry specification has gone through various revisions and has now evolved to the IS-136 specification.
IS-54B-Revision B for the original North American TDMA cellular standard. See also: IS-136.
IS-54C (TDMA)-Revision C of the IS-54 TDMA digital system standard.
IS-641 (Speech Coding)-An interim standard issued by the EIA/TIA that specifies the speech coders used for US digital cellular.
IS-826-Interim Standard 826
IS-88 (NAMPS)-An interim specification for the 10 kHz narrowband AMPS (analog) mobile communication system.
IS-94 (Enhanced AMPS)-The specification for cellular auxiliary personal communication services (CAPCS) that enhances AMPS cellular systems to allow wireless PBX and home cordless services.
IS-95 (CDMA)-The specification for the CDMA cellular system developed in the United States.
IS-IS-Intermediate System To Intermediate System
ISA-Integrated Services Architecture
ISA-Industry Standard Architecture
ISAKMP-Internet Security Association and Key Management Protocol
ISAKMP/Oakley-Internet Security Association and Key Management Protocol/Oakley
ISAM-Indexed Sequential Access Method
ISAPI-Internet Server Application Program Interface
ISBN-International Standard Book Number
ISC-International Switching Carrier
ISCP-Integrated Services Control Point
iSCSI-Internet SCSI
ISD-International Subscriber Dialing
ISDL-Integrated Services Digital Line
ISDN-Integrated Services Digital Network
ISDN-Notify In The Integrated Services Digital Network

ISDN-NTI Integrated Services Digital Network
This diagram shows the basic structure of an ISDN S frame. This diagram shows that the 250 msec frame is divided into fields that include F bits for framing, two B channels, one D signaling channel, an E channel to allow bits to be echoed back for network management and test purposes, an S channel for local signaling, and other logical channels. The field structure also shows that some of these bits are interleaved (sub-divided) through various parts of the S frame structure.

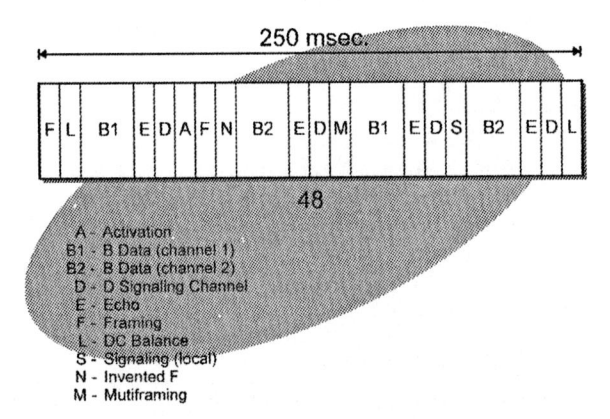

ISDN S Frame Structure

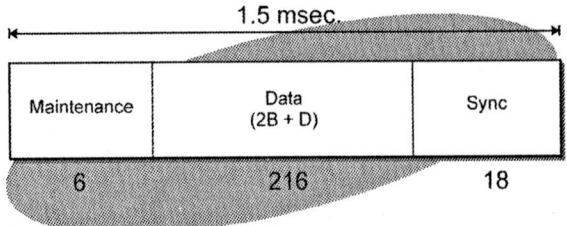

ISDN U Frame Structure

ISDN S Interface-A 4 wire digital interface between end user terminal equipment (e.g. telephones and fax machines) and the network termination point in the building. Two wires are used for a transmit signal and 2 wires are used for receiving signals. The S interface allows up to 7 devices to share the S interface communication line. The S interface is the multi-device version of the T (single device) interface.

ISDN To POTS Converter (IPC)-A device that converts an ISDN basic rate interface (BRI) into a POTS analog telephone interface.

ISDN U Interface-A 2 wire subscriber loop that transports a duplex ISDN 160 kbps digital signal between the ISDN central office and the termination point.
This diagram shows the basic structure of an ISDN U frame. This diagram shows that the 1.5 msec frame is divided into fields that include maintenance bits, user and control data (two B data channels + D signaling channel), and a synchronization field.

ISDN User Adaptation Layer (IUA)-ISDN User Adaptation layer is used to transport ISDN user signaling (Q.931) over IP between two signaling endpoints. The use of an IUA protocol eliminates the use of the MTP protocol portion in a signaling system.

ISDN User Part (ISUP)-The functional part of SS7 protocol that provides the call processing control signaling functions that are required to support basic communication services. Although is it based in the Integrated Services Digital Network (ISDN) signaling functions, ISUP is used for analog and digital call processing functions.

ISDN-UP-Integrated Services Digital Network User Part

ISFMU-Extended Superframe Monitoring Unit

ISG-Incoming Service Group

ISI-Inter Symbol Interference

ISIS7-Internal Switch Interface System

ISL-InterSatellite Link

ISM-Instrument, Scientific and Medical Band

ISN-Initial Sequence Number

ISO-International Standards Organization

ISO 9000-Quality management standards that are defined by the international standards organization (ISO) to help companies define and improve their production and service processes. These standards require companies to define their own product and service processes and demonstrate to independent auditors that their processes provide the ability to track the quality of the products or services they provide.

ISOC-Isochronous

ISOC-Internet Society
Isochronous (ISOC)-A communication process that sends data between communication devices in continuous form (equal transmission time for all data). Isochronous signals are used in systems that require continuous data to be sent at specific time intervals (such as digital audio communication systems).
Isochronous Ethernet (IsoEnet)-A system that allows Ethernet systems to co-exist with isochronous systems (such as T1 or E1 lines). The IsoEnet consortium was formed in 1992 and is specified in IEEE 802.9a.
Isochronous User Channel-The channel used for time-bounded information like compressed audio.
ISODE-ISO Development Environment
IsoEnet-Isochronous Ethernet
Isolated Cable-A cable that is energized or fed with "live" Central Office facilities via drop wire. The cable's grounding integrity from the CO is interrupted and isolated and requires a special bonding and grounding process that utilizes either conventional ground rod or counter-poising methods.
Isolated Ground Plane-A set of connected circuit frames that are grounded through a single connection to a ground reference point. That point and all parts of the frames are insulated from any other ground system in a building.
Isolator-An RF device that allows radio signals to pass through in one direction but it does not allow signals to pass through in the opposite direction. An isolator is used to protect a transmitter assembly from reflected signals.
This diagram shows how an isolator allows a signal to pass through in one direction and restricts the signal flow in the opposite direction. When a signal is reflected from the antenna back towards the transmitter, the reflected signal is absorbed by the isolator.

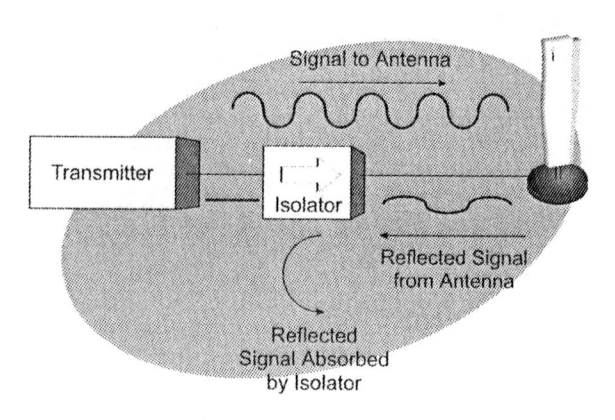

Isolator Operation

Isophasing Amplifier-A timing device that corrects for small timing errors.
Isotropic-A quantity exhibiting the same properties in all planes and directions.
ISP-Information Service Provider
ISP-Integrated Service Provider
ISP-Internet Service Provider
ISR-International Simple Resale
ISR-Intermediate Session Routing
ISSI-InterSwitching Interface
ISSI-Inter-Switching System Interface
IST-Independent Sideband Transmission
ISTP-Internet Signaling Transport Protocol
ISTP-Integrated Signal Transfer Point
ISU-Integrated Service Unit
ISUP-ISDN User Part
ISV-Independent Software Vendor
IT-Information Technology
ITA 1-International Telegraph Alphabet #1
ITA 2-International Telegraph Alphabet #2
ITA 3-International Telegraph Alphabet #3
ITA 4-International Telegraph Alphabet #4
ITA 5-International Telegraph Alphabet #5
ITAA-Information Technology Association of America
ITAR-International Traffic in Arms Regulations
ITB-Intermediate Block Character
ITC-Independent Television Commission
ITC-Independent Telephone Company
ITCA-International TeleConferencing Association
Iterative Loop-A repeated group of instructions in a software routine.
ITESF-Internet Traffic Engineering Solutions Forum
ITFS-International Toll Free Service

ITFS-Instructional Television Fixed Station
ITG-Integrated Telemarketing Gateway
ITI-Information Technology Industry Council
ITI-Interactive Terminal Interface
ITO-International Telecommunications Organization
ITOG-Initial Time Delay Gap
ITORP-IntraLATA Toll Originating Responsibility Plan
ITS-Intelligent Transportation System
ITS-Institute for Telecommunications Sciences
ITS-International Teleproduction Society
ITSP-Internet Telephony Service Provider
ITSP System-Internet Telephony Service Providers (ITSPs) are companies that provide telephone service using the Internet. ITSPs setup and manage calls between Internet telephones and other telephone type devices.
An ITSP coordinates Internet telephone devices so they can use the Internet as a connection path between other telephones. ITSPs are commonly used to connect Internet telephones or PC telephones to telephones that are connected to the public telephone network. This is accomplished by using gateways. Gateways convert packets of audio data from the Internet into standard telephone signals.
ITT-International Telephone and Telegraph
ITU-International Telecommunication Union
ITU-R-International Telecommunications Union-Radiocommunication Sector
ITU-T-International Telecommunications Union - T
ITV-Interactive Television
IUA-ISDN User Adaptation Layer
IUC-Interval Usage Code
IUM-Impacted User Minutes
IUT-Implementation Under Test
IV-Initialization Vector
IVDM-Integrated Voice and Data Multiplexer
IVDS-Interactive Video and Data Service
IVDT-Integrated Voice/Data Terminal
IVHS-Intelligent Vehicle Highway System
IVR-Interactive Voice Response
IW-Inside Wire

IW-Inside Wiring
IWF-Interworking Function
IWTA-International Wireless Telecommunications Association
IX-Internet Exchange
IXC-Interexchange Carrier
IXC Mileage-Inter-exchange Mileage
IXN-Interconnection Revenue

J

J-joule
J Box-Junction Box
J-Hook-(1-waveguide) In length of waveguide that has one end turned 180 degrees (J), commonly used with antenna assemblies. (2-mounting hardware) A pole hardware item in the shape of a letter J that is used to attach dropwire to a pole or building.
J-TACS-Japan Total Access Communication System
J2ME-Java Version 2 Mobile Edition
Jack Field-The location of jacks or jack strips on a piece of equipment, or on a panel or panels in a central office.
JAE-Java Application Environment
JANET-Joint Academic Network
JATE-Japan Approvals Institute for Telecommunications Equipment
Java-An object-oriented programming language that works with a wide variety of computers. Created by Sun Microsystems, Java adds animation and interactivity to Web pages, granted you have a Java-enabled browser. Java technology is a portable, object-oriented language that is well suited for web-based (Internet) and platform-independent applications. Java is a high level language and is architecturally neutral as it can operate on almost any underlying operating system.
Java Application Environment (JAE)-The development of applications using the Java programming language and its development kits.
Java Telephony Application Programming Interface (JTAPI)-An industry standard that defines the application interface between computers and telecommunications devices based on the Java programming language. The JTAPI standard allows computers to control private telephone systems (such as PBX systems).
Java Version 2 Mobile Edition (J2ME)-A compact version of Sun's Java technology targeted for embedded consumer electronics.
JavaScript (Jscript)-Javascript is a scripting language designed to allow the development of active online content on Web severs (web sites). Javascript was created by Netscape Communications and Sun Microsystems that allows developers to add a specific information processing capabilities Web pages.

JavaTel-Java Technology Toolkit
JavaTel API-Java Telephony Application Programming Interface
JB7-Jam Bit 7
JBIG-Joint Bitonal Image Group
JBOD-Just a Bunch Of Disks
JCL-Job Control Language
JDBC-Java Database Connectivity
JDC-Japanese Digital Cellular
JDK-Java Developer's Kit
JECF-Java Electronic Commerce Framework
JEDEC-Joint Electronic Devices Engineering Council
Jeopardy-A condition resulting from any schedule change that is likely to cause a service request to be completed later than a committed due date. Failure to update the status of an order on or before a critical report date can result in a jeopardy condition. The term applies to both installation and repair jobs.
JEPI-Joint Electronics Payments Initiative
JES-Job Entry Subsystem
JF-Junction Frequency
Jini-A connection technology developed by Sun Microsystems that provides simple mechanisms which enable devices to plug together to form an impromptu community. -The Jini system allows each device within a community system to provide services that other devices in the community may use without any required planning, installation, or human interaction. These devices provide their own interfaces, which ensures reliability and compatibility. Jini works at high (application and session layer) protocol layers.
JIT Compiler-Just-In-Time Compiler
Jitter-(1-general) Jitter is a small, rapid variation in pulse arrival time of a substantially periodic pulse waveform resulting typically from fluctuations in the wave speed (or delay time) in the transmission medium such as wire, cable or optical fiber. When the received pulse waveform is displayed on an oscilloscope screen, individual pulses appear to jitter or jump back and forth along the time axis. (2-packet) The short-term variation of transmission delay time for data packets that usually results from varying time delays in transmission due to different paths or routing processes used in a packet communication network.

Jitter Buffer-The jitter buffer receives and adds small amounts of delay to packets so that all the packets appear to have been received without varying delays. Jitter buffers allow for the smoothing out of digital audio signals that experience variable transmission delay across a network (such as the Internet).

This diagram shows how a jitter filter can remove the variable transmission delay for packets that experience variable transmission time through a packet switched network. This diagram shows that packets are delayed variable amounts (delay 1-3). The jitter filter receives the packets and stores the packets in memory until a specific start time. The jitter filter has a clock that provides specific start times for the transmission of the pulse. This fixes the amount of delay to an anticipated maximum amount.

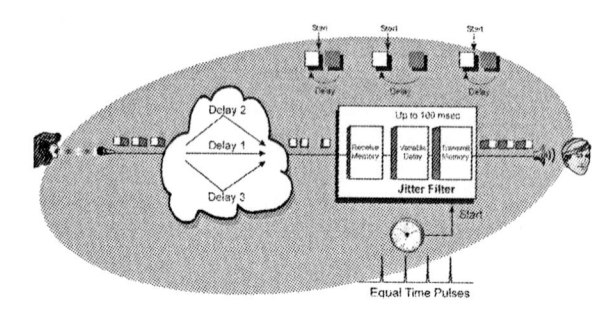

Jitter Filter Operation

Job Control Language (JCL)-A computer language that links an operating system and applications programs in order to define processing jobs and executable programs, and to provide for a job-to-job transition.

Jog-Jogging

Jogging (Jog)-The process of moving a videotape forward or backward one field or frame at a time.

Joint Access Costs-The costs associated with network access facilities that are used for services from two (or more) access systems (such as local and long-distance connections.) Included are costs for basic termination, installation labor, inside wiring, drop wire, a subscriber loop, and all non-traffic-sensitive equipment in a local central office. Joint access costs include those that are incurred regardless of whether a customer makes a call. Also included are those expenses that vary with the number of customers rather than the amount of use, such as commercial, directory monthly billing, and testing.

Joint European Standards Institute-The Committee for European Standardization/Committee for European Electrotechnical Standardization

Joint Technology Committee (JTC)-A working committee under the auspices of the International Standards Organization (ISO) with the goal of defining a standard for digital compression and decompression of still images for use in computer system.

Joint Use-A mutual agreement among two or more telephone, power, cable television, or other utility companies to use common poles, trenches, and similar facilities.

Josephson Effect-The tunneling of pairs of electrons, without breaking apart, between two superconducting films separated by a thin insulating barrier of a non-superconducting metal. A Josephson junction switches faster and uses less power than devices made of silicon or gallium ansenide.

joule (J)-The standard unit of work (unit of energy) that is equal to the work done by one Newton of force when the point at which the force is applied is displaced a distance of one meter in the direction of the force. One joule is also equal to the energy conveyed by a power level of one watt acting for one second, so a joule is equivalent to a "watt second." 3 600 000 joules is a kilowatt hour (kWh) of energy. The joule is named for the 19th century Irish physicist James Prescott Joule (1811-1889).

Journalization-The process of booking billing charges & credits to the appropriate financial accounts. A journal is the interface between the billing system and the general ledger.

Joystick-An electromechanical control level, similar to the control stick on an aircraft, used for hand positioning graphic images on a video or computer monitor.

JPEG-Joint Photographic Experts Group

JPEG-Baseline Sequential Joint Photographic Experts Group

Jscript-JavaScript

JTAG-Joint Test Action Group

JTAPI-Java Telephony Application Programming Interface

JTC-Joint Technology Committee

JUGHEAD-Jonzy's Universal Gopher Hierarchy Excavation And Display

Jumbo Frame-A frame longer than the maximum frame length allowed by a standard. Specifically used to describe the dubious practice of sending 9-Kbyte frames on Ethernet LANs.

Jump-A class of instruction that causes a software program to move forward or backward to a specific location.

This diagram shows how jumper connectors can be used instead of switches. This diagram shows a jumper connector that is a small plastic and metal connector that completes a circuit across two (or more) pins. The jumpers in this diagram are used to select circuit options from a set of several user-definable options.

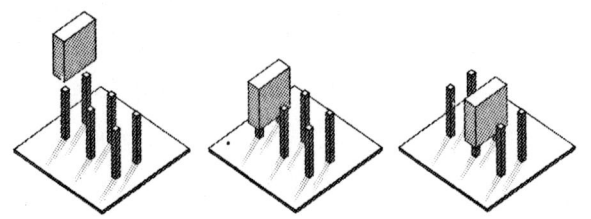

A jumper is used to connect various pairs of pins in an array of six pins.

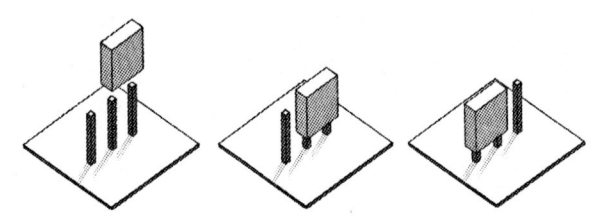

Jumper Diagram

Junction Frequency (JF)-The highest radio frequency that can be propagated in a particular mode between specified terminals by ionospheric refraction alone.

Junction Transistor-The most common type of transistor, in which current flow depends on both the majority and minority carriers with two junctions between n-type and p-type material. The result is a sandwich form, either n-p-n or p-n-p. The three external connections each go to a section of the sandwich: the rise to the center, the collector to one end, and the emitter to the other end.

K

k-kilo
K-Kelvin
K-1024
K-factor-(1-tropospheric radio propagation) The ratio between the effective and the actual earth radius. (2-ionospheric radio propagation) A correction factor applied in calculations involving a curved layer that is a function of the distance and real height of a reflection. (3-video) A specification rating method that assigns a higher factor to video disturbances that cause the most observable picture degradation.
kbps-kilobits Per Second
KBps-Kilobytes per second
kbyte-Thousand bytes.
Kbyte-Kilobyte
Kc-Cipher Key
KDC-Key Distribution Center
Keep It Simple Stupid (KISS)-A philosophy that states that simple things or processes are better than complex methods. KISS is commonly applied to business practices.
KEK-Key Encryption Key
Kennelly-Heaviside Layer-The ionospheric E-layer.
Kernel-A kernel is a software part of an operating system that is fundamental to its overall operation. The kernel software continually manages specific processes such as system memory, the file system, and disk operations. A kernel may also run processes such as the communication between applications, coordinating the input and output of information within an application. Once loaded, the kernel software usually remains stored and operates from computer memory and is typically not seen by the users of the system.
Kerr Cell-An electronic light intensity modulator or fast acting photographic shutter that exploits Faraday rotation of the plane of polarization of an electromagnetic wave (or light or infra-red) caused by an electric field applied in a transverse direction to an appropriate transparent material (solid or fluid) carrying the wave. The rotation of the plane of polarization causes the intensity of the wave to be changed when it passes through a fixed position polarizing filter.

Key Frame Effect-A video effect consisting of a series of effects "snapshots" called key frames. When the overall effect is replayed, the video processing system automatically and gradually dissolves from one key frame to the next. A process called interleaving defines what happens between key frames. The result is an animation effect.
Key Performance Indicator (KPI)-Metrics (measurements) established by upper management to determine the accomplishment of objectives set for organizational units (e.g. Marketing KPIs may include Net Customer Adds, Net Churn, Total Revenue, etc.).
Key Service Unit (KSU)-The central operating unit of a key telephone system (KTS) or non-PBX/ACD telephone system (small customer premises telephone switch).
Key Set-A telephone set that is part of a key telephone system (KTS). A key set usually has several buttons that allow the user to select call hold, line connection, intercom and other private telephone system features.
Key Tag-The parameter defining one of several encryption codes or methods.
Key Telephone System (KTS)-Key telephone systems are (usually small) multi-line private telephone network that allows each key telephone station to select one of several telephone lines, place a line on hold, and call via an intercom circuit between key telephones. Key systems contain a central key service unit (KSU) that coordinates status lights and lines to key telephones ("Key Sets"). Early KTS system technology was based on electro-mechanical relay hardware. They required all the outside telephone lines to be connected to all of the key telephone sets in the installation. In addition, two additional pairs of wire were used in conjunction with each telephone line, one pair for the A/A1 connection indicating if that line is off hook at that particular key telephone set, and another pair to operate a small light to indicate the status of that line. Consequently, each key telephone set was connected to the central KSU via a thick cable containing 50 wires (25 pairs). Newer KTS systems typically use only 4 wires to connect the electronic KSU to each electronic key telephone set, and are often called "skinny wire" key systems. Modern electron-

ic key systems are small microprocessor controlled switching systems and have some of the same advanced call processing features such as call hold, busy status, multi-line conference, abbreviated dialing, and station-to-station intercom that are available in a larger PBX.

This figure shows a typical key telephone system. This diagram shows telephones wired to a key service unit (KSU) that is connected to the PSTN. The KSU allows the telephones to have access to the outside lines to the PSTN. The KSU controls lights on the telephone sets, intercom access, and call hold.

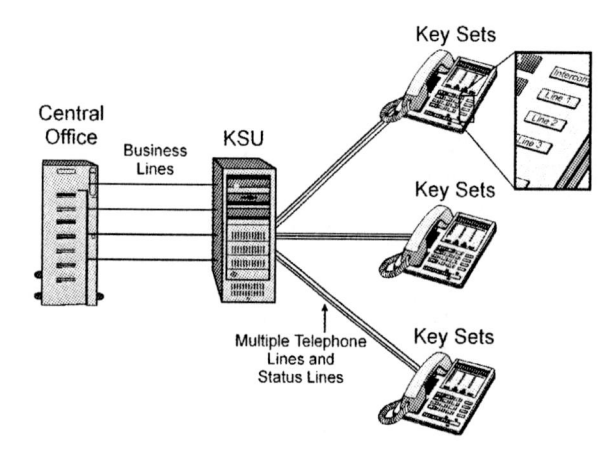

Key Telephone System (KTS) Service Unit
Source: Panasonic

Key Telephone System (KTS) Operation

This photo shows the KX-TA1232 Advanced Hybrid Telephone System main switching control unit that is used in a key telephone system (KTS). This switching unit controls the switching and access between telephone extensions and public telephone lines.

Key Video-The video key fill signal, key source, or both.

Keyboard-A physical device that allows a user to enter data to a computer or other electronic device. A keyboard usually consists of individual key switches that are assigned one (or several) alphanumeric characters and/or functions.

Keyer-An electronic circuit that creates a signal to control a video multiplier based on selective information contained in a video signal.

Keypad-A physical interface device (a group of keys) that allows for a user to input data to a computer or other electronic device.

kg-kilogram

kHz-kilohertz

Ki-(1-GSM) A code key used by the GSM encryption algorithm to help authenticate a user. The Ki key is unique to each user and is stored in the subscriber identity module (SIM) card in a protected memory area.

Killer App-Killer Application

Killer Application (Killer App)-a software application of such great importance to the end customer that it alone motivates the customer to buy the entire system. One example is a computer game having devoted users.

kilo (k)-Prefix indicating 1000 units in the metric system (Decimal 1000). Note the use of a lower case k.

kilobit-A quantity equal to one thousand bits.
kilobits Per Second (kbps)-A measure of data transmission equal to one thousand bits per second.
kilogram (kg)-The unit of mass. It is equal to the mass of the international prototype of the kilogram kept at the International Bureau of Weights and Measures laboratory at Sévres (a suburb of Paris) France. The prototype is a cylinder of iridium-platinum alloy 39 mm in diameter and 39 mm in height. Approximately equal to the mass of a liter of water, a cube of water having each edge of 100 mm length. See also mass.
kilohertz (kHz)-A unit of measure of frequency equal to one thousand hertz.
kilovar-A unit of measure equal to one thousand volt-amperes.
kilovolt (kV)-A unit of measure of electric voltage equal to one thousand volts.
kilowatt (kW)-A unit of measure equal to one thousand watts.
Kilowatt Hour (kWh)-kWh is a widely used unit of energy, equal to 1000 watts of power per hour . See also joule.
Kiosk Billing System-A uniform billing system used by France Telecom in its deployment of mass-market videotex services. The system takes its name from the kiosk, or newsstand, where information is purchased on an as needed basis.
Kirchoff's Laws-(1-current) In a circuit node there is as much current flowing into the node as there is flowing away from it. (2-voltage) In a closed electric circuit, the algebraic sum of the applied voltages and the voltage drops is equal to zero.
KISS-Keep It Simple Stupid
Klystrode-An amplifier device for UHF-TV signals that combines aspects of a tetrode (grid modulation) with a klystron (velocity modulation of an electron beam). The result is a more efficient, less expensive device for many applications. Klystrode is a trademark of EIMAC, a division of Varian Associates.
Knee-In a response curve, the region of maximum curvature.
Knee Voltage-A voltage transient on a battery during a charge cycle that indicates the battery if fully charged.
Knowledge Neighborhood -Term used to define a group of related departments and disciplines. Groups within the telco who share a common vocabulary, view of the world and information resources. For example, all groups involved in customer relationship management form a CRM Knowledge Neighborhood.

Kokai-An unexamined Japanese patent application.
Kokoku-An examined and allowed Japanese patent application.
KPI-Key Performance Indicator
KS Number-Kearney System Number
KSR-Keyboard Send Receive
KSU-Key Service Unit
KTS-Key Telephone System
KTU-Key Telephone Unit
Ku-band Radio-Frequencies in the range of 15.35 GHz to 17.25 GHz, typically used for satellite telecommunications.
kV-kilovolt
KVM Switch-Keyboard/Video/Mouse Switch
kW-kilowatt
kWh-Kilowatt Hour
KWH-KiolWatt Hour

L

L-Pad-A volume control that presents to the source approximately the same impedance at all settings. The device consists of an L network arranged so that both of its elements are adjusted simultaneously.

L1-Level 1 Cache

L2-Level 2 Cache

L2CAP-Logical Link Control And Adaptation Protocol

L2TP-Layer 2 Tunneling Protocol

LAC-Location Area Code

LAD-Life After Death

LADC Leads-Local Area Data Channel

Ladder Network-A type of filter with components alternately across the line and in the line.

LADT-Local Area Data Transport

LAES-Lawfully Authorized Electronic Surveillance

Lag-The difference in phase between a current and the voltage that produced it, expressed in electrical degrees.

Lagging Current-A sine wave current that lags behind the alternating electromotive force that produced it. That is, the positive peak of current occurs later in time than the positive peak of voltage. A circuit that produces a lagging current is one containing inductance alone, or whose effective impedance is inductive.

Lagging Load-A load whose combined inductive reactance exceeds its capacitive reactance. When an alternating voltage is applied, the current lags behind the voltage.

LAI-Location Area Identity

LAMA-Local Automatic Message Accounting

Lambda-The Greek symbol for "L" that is used to represent wavelength, for instance in optical and electrical signals. Wavelength is the distance one cycle of a wave travels. Wavelength can be calculated by dividing the speed of the wave by its frequency. For example, in free space (in a vacuum) the wavelength of an electromagnetic wave is equal to: (300,000,000 meters/sec)/ frequency (Hz). The wavelength of a radio or lightwave traveling in vacuum at 300 MHz is 1 meter.

LAN-Local Area Network

LAN Access Profile-In a Bluetooth system, the LAN access profile defines how the Point-to-Point Protocol (PPP) is supported with regard to LAN access for a single Bluetooth device, LAN access for multiple Bluetooth devices, and PC-to-PC communication using PPP networking over serial cable emulation. See also Point-to-Point Protocol.

LAN Emulation (LANE)-An application (adaptation) of the asynchronous transfer mode (ATM) high-speed packet data specification that allows the ATM network to transparently operate as a local area network (LAN) system. The use of LANE allows the connection of standard local area networks (LANs) such as Ethernet and token through an ATM network. LANE translates the services between the two types of networks and is transparent to higher-level applications that use LAN networks.

The major part of LANE include broadcast and unknown server (BUS) to manage broadcast and multicast addresses, LAN emulation client (LEC) to adapt (map) between the 48 bit Ethernet media access control (MAC) addresses and ATM addresses, LAN emulation configuration server (LECS) to manage the LAN emulation clients, and a LAN emulation server (LES) that controls and coordinates the entire LANE system.

LAN Emulation Client (LEC)-A process used for LAN Emulation (LANE) in an asynchronous transfer mode (ATM) system to handle the adaptation of Ethernet data frames into ATM packets. The LEC is used by the each node that connects the LAN (e.g. Ethernet) system to the ATM network.

LAN Emulation Server (LES)-The controlling server in LAN Emulation (LANE) system operating on an asynchronous transfer mode (ATM) network. The LES controls and coordinates the emulated LAN. It is responsible for mapping addressing between LAN 48-bit MAC addresses and ATM addresses. The LES works with a broadcast and unknown server (BUS) to distributes broadcast and multicast packets.

LAN Segmentation-The practice of dividing a single LAN into a set of multiple LANs interconnected by bridges.

LAN Switch-A switch that allows for segmentation of a LAN passing frames between segments only when necessary. Ethernet switches are the most common and are used to reduce congestion and collisions.

LAN Telephony (TeLANophy)

LAN Telephony (TeLANophy)-Local access network (LAN) telephony (sometimes called TeLANophy) use LAN systems to transport voice communications.

This diagram shows a LAN telephony system. This diagram shows that a LAN telephone system consists of LAN telephones, a data network, a LAN call processing system, and a voice gateway to the PSTN. LAN telephones convert audio into digitized packets that are transferred on the LAN to the call processing computer telephony integration (CTI) system. Each LAN telephone has its own network data address. The call processing system communicates with LAN telephones over the same high-speed LAN data network that communicates with computers. When calls are received from the PSTN, the call processing system looks in the database to find the associated LAN telephone address (data address) and this address is used to alert the LAN telephone of an incoming call. When calls are originated from the LAN telephone, the dialed telephone number is passed to the call processing system. This system determines if the call is routed within the data network or if the voice gateway must be used to connect the call to the PSTN.

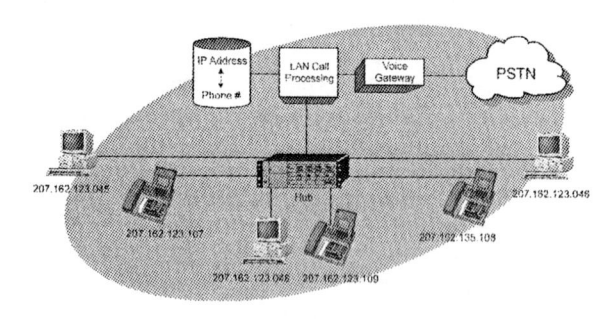

LAN Telephony

Land Mobile Radio (LMR)-Land mobile radio (LMR) systems are traditionally private systems that allow communication between a base and several mobile radios. LMR systems can share a single frequency or use dual frequencies. LMR in the United States is regulated by the FCC in part 90, Private Land Mobile Radio Services, includes various types of private radio services including police, taxi, fire and other types of two-way and dispatch services.

This figure shows a traditional two-way radio system. In this example, a high power base station (called a "base") is used to communicate with portable two-way radios. The two-way portable radios can communicate with the base or they can communicate directly with each other.

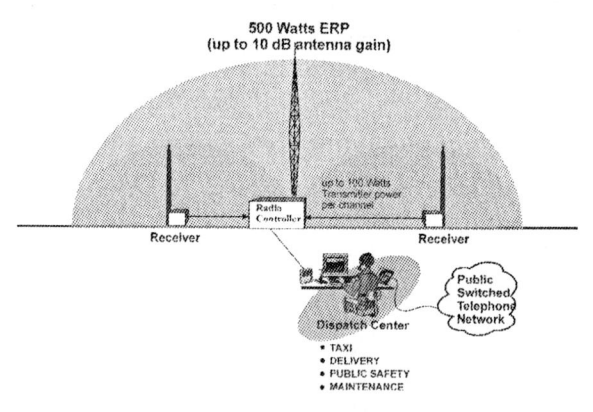

Land Mobile Radio (LMR) System

LANDA-Local Area Network Dealers Association
Landline-A conventional domestic or business telephone circuit. The term landline applies to telephone lines are either buried or carried just over the ground.
Landline Network-The communications infrastructure that generally is associated with the public switched telephone network. (See also: landline.)
LANE-LAN Emulation
Language-(1-general) A set of symbols, characters, conventions, and rules used for conveying information. (2-programming) A set of commands and associated parameters that can be decoded to represent specific sequences of computer processing instructions.
LAP-Link Access Procedure
LAPB-Link Access Protocol Balanced
LAPB-Balanced link access procedure
LAPB Extended-Link Access Procedure Balanced Mode Extended
LAPD-Link Access Protocol on D Channel
LAPM-Link Access Procedure For Modems
Laser-An electronic device that produces coherent (synchronized) light (visible or infrared.) The optical radiation is produced by light amplification by stimulated emission of radiation. Lasers produce light in various portions of the electromagnetic spectrum including visible, infrared, and ultraviolet regions.

To produce optical energy, some of the atoms in a laser's material are excited, either by electric discharge through a gas or by current through a solid-state diode. This transitions their energy resulting in the emission of a stream of photons. Some lasers can only create pulses of light and others are continuously operated. To transfer information signals onto a laser beam, the optical signal can be modulated by turning the laser on and off.

LASER-Light Amplification By Stimulated Emission Of Radiation

Laser Chirp-A shift of the center wavelength of a laser during single pulse operations as a result of laser instability.

Laser Diode (LD)-A junction diode that emits electromagnetic radiation or light when injected electrons under forward bias recombine with holes in the vicinity of the junction. The laser diode is used to transmit light signals over fiber optic cables. See also Injection Laser Diode (ILD).

Laser Exciter-A device that supplies a signal that modulates a laser driver in an optical communication system.

Laser Head-In fiber optics, a module containing active lasing material, a resonant cavity, and other components, all within a single enclosure.

Laser Resonator-An arrangement of reflecting mirrors inside a laser that produces a single beam of coherent light.

Lashing-The attachment of a cable to a support strand by wrapping steel wire or dielectric filament helically about the two lines.

Lasing Medium-Laser

LASS-Local Area Signal Service

LAST-Local Area Systems Technology

Last Call Return-A telephony service that allows a telephone user to automatically call back the phone number of the last received incoming call. Last call return is normally accomplished by the customer entering the service code (e.g. "*69").

Last Choice Route-The last choice trunk group, or series of such groups, between two switching systems. This term also can refer to a final trunk group or only-route trunk group.

Last Choice Trunk Group-A final trunk group or only-route trunk group in a hierarchical network.

Last Mile-The last portion of the telephone access line that is installed between a local telephone company switching facility and the customer's premises.

Last Number Redial-The ability for a telephone to remember and dial the last dialed telephone number.

Last Trunk Busy-The condition in which the last available trunk in a trunk group is busy.

LAT-Local Area Transport

LATA-Local Access And Transport Area

Latch-An electronic circuit that holds a digital signal after it has been selected. To latch a signal means to hold it.

Lateral Cables-Sections of cable between feeder and distribution networks. The feeder cable can come from underground conduit systems, buried cable systems, or aerial cable systems and extend to a distribution network through serving area interfaces, cross connect terminals, or (rarely now) control, access, or splice points.

Latitude-An angular measurement of a point on the earth above or below the equator. The equator represents 0, the north pole +90 and the south pole -90'.

Launch-(1-satellite) The process of lifting a satellite from earth into orbit. (2-electromagnetic wave) The process of transferring energy from a feeder or waveguide to an antenna.

Launch Angle-(1-general) The angle at which a light beam emerges from a given surface. (2-fiber optic) The angle between the input radiation vector and the axis of the fiber or fiber bundle.

Launch Fiber-A short length of optical fiber that couples the light from an optical source into the optical fiber of a communications system.

Launch Numerical Aperture (LNA)-The numerical aperture of an optical system used to couple (launch) power into an optical waveguide. LNA is one of the parameters that determine the initial distribution of power among the modes of an optical waveguide.

Launch Power-The amount of light actually coupled into an optical fiber from the light source. The launch power typically is expressed in decibels above one milliwatt, or microwatts.

Layer-(1- LAN) A collection of related network processing functions that constitutes one level of a hierarchy of functions. (2-video) A single video image that is processed so that it can be inserted into a final composite image. There may be other layers in the image, which can be prioritized as to location. (3-computer) An overlay which can be used to place information, so that CAD/CAM drawings can be logically subdivided, for viewing or hardcopy purposes. (4-boundaries) A group of one or more entities contained within an upper and lower logical boundary. Layer (N) has boundaries to the layer (N + 1) and to the layer (N - 1).

Layer 2 Switch-Synonymous with bridge.

Layer 2 Tunneling Protocol (L2TP)-An protocol that is used for to allow a secure communication path, a virtual private network link, between computers. It is an evolution of earlier point-to-point tunneling protocol (PTPP) as it offers more reliable operation and enhanced security. L2TP enables private communication lines through a public network. L2TP was developed via the Internet engineering task for (IETF).

Layer 3 Switch-Synonymous with router.

Layer 4 Switch-A router that can make routing policy decisions based on Transport layer Information (e.g., TCP port identifiers) encapsulated within packets.

Layer Interface-The boundary between two adjacent layers of the protocol model.

Layer Management Entity (LME)-An protocol layer interface used in the SS7 system that converts a protocol layer (e.g. MTP2) to a Layer Management Interface (LMI) format that can communicate with a Management Information Base (MIB) associated with a device or system.

Layer Management Interface (LMI)-The interface between a management information base (MIB) and the protocol layers of a communication system, such as the SS7 system.

Layer Service-A capability of the (N) layer and the layers beneath it, which is provided to (N + 1) entities, at the boundary between the (N) layer and the (N + 1) layer.

Layer Service Elements-An indivisible component of the layer service made visible to the service user via layer primitives.

Layer Service Primitives-A means for specifying in detail the adjacent layer interactions.

Layer-2-Data Link Layer

Layered Network Architecture-A network structure that divides the network communication functions into layers that perform specific functions. The precise definition of each layer allows products to be developed for specific functions by different companies. Different types of networks may have different layer types. Examples of layered architecture include the 7 layer open systems interconnect (OSI) model and the 4 layer Internet model.

Layered Protocols-Protocols that are designed to communicate with higher or lower level protocols in a communication network. Each layered protocol performs a specific function and each layered protocol has specific ways to pass information to protocols that in layers directly above or below it.

Layout-(1-location) A proposed or actual arrangement or allocation of equipment or physical position of components on a circuit board. (2-process) The design the physical layout of an electronic circuit board.

LBI-Load Balance Index

LBRV-Low Bit Rate Voice

LBT-Listen Before Talk

LC Circuit-An electric circuit with both inductance (L) and capacitance (C) that is resonant at a particular frequency.

LC Ratio-The ratio of inductance to capacitance in a given circuit.

LCA-Local Calling Area

LCD-Liquid Crystal Display

LCI-Logical Channel Identifier

LCM-Line Concentrating Module

LCN-Logical Channel Number

LCR-Line Concentration Ratio

LCR-Least Cost Routing

LCRIS-Loop Cable Record Inventory System

LCS-Live Call Screening

LCU-Line Control Unit

LD-Laser Diode

LD-Long Distance

LD-CELP-Low Delay-Code Excited Linear Prediction

LDAP-Lightweight Directory Access Protocol

LDF-Local Distribution Frame

LDM-Limited Distance Modem

LDN-Listed Directory Number

LDS-Local Digital Switch

LDS-Local Digital Services

LDS-Local Distribution Service Station

LE_ARP-LAN Emulation Address Resolution Protocol

LEA-Law Enforcement Agency

Lead-An electrical wire, usually insulated.

Leader Stroke-In lightning, the first stroke, which usually determines the path to be followed by the return stroke, where most of the energy is carried.

Leading Current-A sinewave current that leads the alternating electromotive force that produces it. That is, the positive peak of current occurs earlier in time than the positive peak of voltage. A circuit that produces a leading current is one containing capacitance alone, or whose effective impedance is capacitive.

Leading Load-A reactive load in which the reactance of capacitance is greater than that of inductance. Current through such a load leads the applied voltage causing the current.

LEAF-Large Effective Area Fiber

Leakage Resistance-The resistance of a path from a circuit to ground through which leakage current flows.

Leaky Cable-A cable that is designed to deliberately leak RF energy. A leaky cable often is used to provide radio coverage in a shielded area, such as a tunnel or basement

Leaky Ray-In an optical waveguide, a ray for which geometric optics would predict total internal reflection at the core boundary but suffers loss by virtue of the curved core boundary.

Leaky VLAN-A VLAN that may, under certain boundary conditions, carry frames that do not belong to that VLAN. Leaky VLANs typically result from devices that improperly implement the VLAN association and/or forwarding rules either to reduce cost or as a result of sloppy design practices.

Leap Second-A time step of one second, used to adjust Coordinated Universal Time to ensure approximate agreement with international universal time. An inserted second is called a positive leap second, and an omitted second is called a negative leap second.

Learning Process-The process whereby a bridge builds its filtering database by gleaning address-to-port mappings from received frames.

Leased Line-Leased lines are telecommunication lines or links that have part or all of their transmission capacity dedicated (reserved) for the exclusive use of a single customer or company. Leased lines often come with a guaranteed level of performance for connections between two points.

Leased Network-A data network using circuits or channels leased from a telephone company (telco) or other telecommunications carrier and dedicated to use solely by the lessee.

Leased Service-The exclusive use of any channel or combination of channels designated to a subscriber.

LEC-Light Energy Converter
LEC-Linear Echo Canceller
LEC-Local Exchange Carrier
LEC-LAN Emulation Client

LECID-LAN Emulation Client Identifier
LECS-Local Area Network Emulation Configuration Server
LED-Light Emitting Diode

Legacy System-A communication system or network that satisfies specific business needs using technology or equipment that has become obsolete or is incompatible with new industry standards. To extend the life of existing investments in legacy systems, new technologies or systems are often designed to communicate with legacy systems.

Lempel-Ziv-Welch (LZW)-An algebraic digital data compression algorithm originally published by Abraham Lempel and Jacob Ziv, and implemented in a practical software program by the late Terry A. Welch. Used in the .GIF file format (developed by CompuServe), and a similar scheme is used by the V.42bis modem data compression standard, in PKZip (developed by the late Philip Klein) and other file compression programs.

Length-The number of bits or bytes in a packet, data block, information field, record, or other variable length block of information.

Length Indicator-In common-channel signaling, a 6-bit field that differentiates between message, link status, and fill-in signal units. When the binary value of an indicator is less than 63, it indicates the length of a signal unit.

LEO-Low Earth Orbit
LERG-Local Exchange Routing Guide
LES-LAN Emulation Server

Level-(1-general) The strength or intensity of a given signal (2-crosstalk) The power of the crosstalk signal compared with a reference signal. (3-peak) The maximum applied sound or signal amplitude. (4-speech) The energy of speech measured in volume units (VU), and typically displayed on a VU meter. (5-speech power) The acoustic power in human speech. (6-transmission) The ratio of the power of a test signal at one point to the test signal power applied at another point in the system used as a reference. (7-video routing) An independently controllable spectrum of signals within a routing switcher. Typically, a routing switcher has a video level and one or more audio levels.

Level Alignment-The adjustment of transmission levels of single links and links in tandem to prevent overloading of transmission subsystems.

Level Setting-Adjustment of video or audio signal levels.

Levinson-Durbin Algorithm-A computational procedure for determining the linear prediction coefficients for use in a system such as the CELP family of speech codecs. The algorithm does a recursive computation starting with the samples of the original audio signal within the frame to determine the coefficients.

Lexicographical Ordering-Lexicographical ordering means that the next item (MIB object) must always have a greater OID (Object Identifier) than the last as defined in the SMI of the MIB for Network Management.

LF-Line Feed

LF-Low Frequency

LFFV-Low Power TV

LGN-Logical Group Node

License Fee-A fee that is charged for the use of a product, service, or asset. License fees can be a fixed fee, percentage of sales, or a combination of the two.

Licensed bands-Licensed frequency bands give the licensee (service provider or service user) the authority to user the radio spectrum within their licensed frequency band according to the requirements of the license. These requirements may include a type of service (such as paging or mobile telephone service), channel types (single or multiple channels), and power levels within a specific geographic area (amount of signal strength allowed).

Licensed Frequency Band-A frequency band that requires a license from a regulatory agency or owner of the frequency band in a geographic area for permission to transmit radio signals in that area.

Licensee-The holder of license that permits the user to operate a product or use a service. In telecommunications, a licensee is usually the company or person who has been given permission to provide or use a specific type of communications service within a geographic area.

LIDB-Line Information Database

Life Safety System-A system designed to protect life and property such as emergency lighting, fire alarms, smoke exhaust and ventilating fans, and site security.

Life Test-A test in which random samples of a product are checked to see how long they can continue to perform their functions satisfactorily. A form of stress testing is used, inducting temperature, current, voltage, and/or vibration effects, cycled at many times the rate that would apply in normal usage.

Life Time-(1-general) The estimated time over which a product, assembly, or a communication cable can be used before it must be replaced. (2-cable) the end of life for a cable reached when all spare pairs or a specified maximum percentage of the total number of pairs are in use. The lifetime of pair groups within a cable can be estimated by dividing the number of spare pairs by the forecasted growth rate.

Lifeline Service-A communication service that is considered a "Lifeline" in case of emergency. Communication service that assures a person can call for assistance or be contacted.

Lifetime Value-Lifetime value is a technique utilized in the retail and banking industry to define how

valuable a customer will be to the company, over their lifetime. For telephone companies (telcos), lifetime value calculations are usually not valid or useful because of the vagaries in pricing and product structures over the average lifetime of an individual. As an alternate, many telco's use customer value assessments.

LIFO-Last In First Out

Light-Electromagnetic radiation visible to the human eye. The visible wavelength of light ranges from approximately at 400 nm to 700nm. Commonly, the term is applied to electromagnetic radiation in most fiber optic communication systems. An electromagnetic radiation with wavelengths from 400 nm (violet) to 740 nm (red), propagated at a velocity of roughly 300,000 km/s (186,000 miles/s), and detected by the human eye as a visual signal in the optical communication field, the term also includes the much broader portion of the electromagnetic spectrum that can be handleddied by the basic optical techniques used for the visible spectrum. This extends the definition of light from the near-ultraviolet region of approximately 300 nm through the visible region, and into the mid-infrared region of 3.0 to 30 nm.

Light Amplification By Stimulated Emission Of Radiation (LASER)-A Laser is a device that emits coherent light of essentially one wavelength in a narrow beam. Lasers can be made using gaseous, liquid or solid state. In optical networks, solid state or semiconductor lasers are used as high performance light sources. Photons are generated in the semiconductor by application of a voltage. Photons with the right wavelength, phase, and direction of travel are selected by an optical cavity in the laser. See also Optical Cavity, LED. Laser light is monochromatic (it has only one frequency or "color," although infrared or ultraviolet lasers produce optical frequencies that are not visible to the human eye), in contrast to white light that has, at the other extreme, many frequency components, and the phase of the sine wave electromagnetic waveform maintains a fixed (coherent) timing relationship over a long interval of time. The peak amplitudes of each cycle of a coherent wave form occurs at absolutely uniform time intervals. Lasers create monochromatic coherent light by combining the light radiation due to the oscillation of individual electrons in individual atoms as these electrons change their electric charge configuration from a higher energy level to a lower energy level. First these electrons are "pumped" up to a higher energy level by a primary source of power. We call these the "excited" electrons. One way to excite electrons is to accelerate the atoms in a gas by applying a high voltage to electrodes at two ends of a container of the gas, so that some atoms collide with each other and transfer their kinetic (motion related) energy to some of the electrons, or alternatively by shining a non-coherent light source of higher frequency (shorter wavelength) than the desired output light frequency on a solid material. Some excited electrons in an atom subsequently "fall" to a lower energy level, and when doing this they emit light at a frequency proportional to the difference in energy between the high and low energy levels. The energy difference (E2-E1) is related to the frequency of the light, f, by the formula $E_2-E_1 = hf$, where h is Planck's constant. Excited electrons "fall" naturally, but at unpredictable times, from the higher to the lower energy level for no apparent reason. One way to make even more electrons fall from the high energy level to the low energy level is to shine a light on these atoms, that light having the same frequency f as the expected output light frequency. This latter process is called stimulated emission. Because the gas or solid is made up of atoms having the same energy level structure, all the atoms then emit light at the same frequency. By placing two parallel reflecting mirrors at two ends of the material, and precisely locating the distance between these mirrors so that their distance is an integral number of wavelengths of the light in question, a standing wave of multiply reflected light is set up in the gas or solid. This is what makes the light emission from all the different electrons coherent. By making one of the two mirrors either partially reflecting and partly transparent, or by having a small transparent spot on one otherwise fully reflective mirror, some of the light is able to escape in a straight, monochromatic, coherent light beam. This is the laser beam. It can be guided into the core of an optical fiber for communication purposes. The light output of the laser can be turned on and off electrically to produce light pulses to convey digital information.

Light Emitting Diode (LED)-All semiconductor electrical diode junctions emit light as a result of their forward electric current. Electrons decrease their energy as they pass from the N to the P side of the diode, and this energy difference is equal to the energy radiated. While some diodes are made with an opaque material surrounding the junction, LEDs are made specifically with transparent material around the junction to allow the light to be visible from the outside. The color (frequency or wavelength) of the light is dependent upon the difference in electron energy level, which in turn is dependent on the amount of added material dopants used in the two parts of the diode. A low energy change corresponds to infrared light (not visible to the human eye), a medium energy change corresponds to red or yellow or green color, and a high energy change corresponds to blue or ultraviolet "color." Ultraviolet light is not visible to the human eye.

Light Energy Converter (LEC)-A photo-votalic semiconductor device that converts light energy into electrical energy.

Light Ray-The path of a given point on a wavefront. The direction of a light ray is generally normal to the wavefront.

Light Receiver-A photodiode or other transducer that is used for receiving optical signals.

Light Source-A generic term that includes lasers and LEDs, even though these may operate outside the visible light band.

Lightning-A flow of current between a charged cloud and the ground resulting from an electric discharge due to large potential differences between cloud charge and ground (or the lightning strike point).

Lightning Flash-An electrostatic atmospheric discharge. The typical duration of a lightning flash is approximately 0.5 seconds. A single flash is made up of various discharge components, usually including three or four high-current pulses called strokes.

Lightning Protector

Lightning Protector-(1-general) A device that limits impulse voltages from lightning to prevent damage to people and electronic equipment. Basic spark-gap ("carbon block" or gas tube) protectors installed in buildings are inadequate to protect modern electronic equipment. Supplementary lightning protectors (surge protectors) are frequently installed at the equipment to protect against excess voltages on both signal and AC connections. (2-rod) A lightning rod system that routes lighting voltages to ground.

This photo image shows a MAX 8 Com/Data Lightning Protector. This is a "surge protector" that protects telephone equipment from lightning-produced voltages brought to the equipment via either signal or AC connections. Combined ("Multiport") AC and signal protectors are recommended by the IEEE and EPRI to protect equipment that has AC and signal connections. The device shown uses Suregate™ technology to protect against excess AC supply voltages as well.

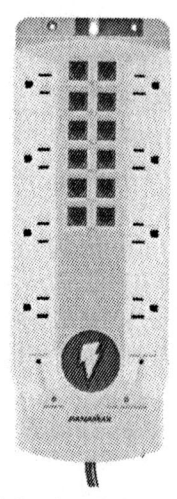

Lightning Protector
Source: Panamax

Lightweight Directory Access Protocol (LDAP)-A standard protocol that allows users to find other devices and services in a communication network. It provides directory services for LAN and the Internet. LDAP is a subset of the X.500 protocol that operates over TCP/IP.
LIJP-Leaf Initiated Joint Parameter
LILO-Linux Loader

LIM-EMS-Lotus Intel Microsoft-Expanded Memory Specification
Limited Access-An arrangement in which only some traffic offered to a group of servers has access to all the servers in the group.
Limiter Circuit-A circuit of nonlinear elements that restricts the electrical excursion of a variable in accordance with some specified criteria.
Line Alarm Indication Signal Code-A code generated by section terminating equipment (STE) upon the loss of an input signal, the loss of a frame, or an equipment failure.
Line Coding-The process of modulating and formatting data for transmission on a communications line.
Line Information Database (LIDB)-A DataStore application program that resides on a service control point (SCP) in the SS7 telephone signaling network and provides validation information for use in alternate billing services, such as telephone calling cards. LIDB data base contains up-to-date records of all working lines, including directory listing name, description of the type of dialing capability subscribed (rotary dial vs. touch-tone), calling card numbers, and other data required for validation services. The acronym LIDB often is pronounced "lid-bee."
Line Not Cutting (LNC)-A notation that itemizes the telephone numbers that will not be affected by a work order that involves changing plant facilities. This notation may be on an engineering work order that involves splicing "live" cable which has working subscribers.
Line Resistance-Copper cable has resistance (impedance) that is dependent on the size (diameter) of the cable. The resistance of the copper wire increases as the diameter decreases (gauge number increases). The higher the line resistance, the more of the signal energy is dissipated by the line and less energy is transferred to the receiving device.

This figure shows how line resistance attenuation and the wire size decreases. This diagram shows that cables with larger diameter copper wires are typically used to in the distribution system. As the distribution system nears its destination, the size of the wire often decreases.

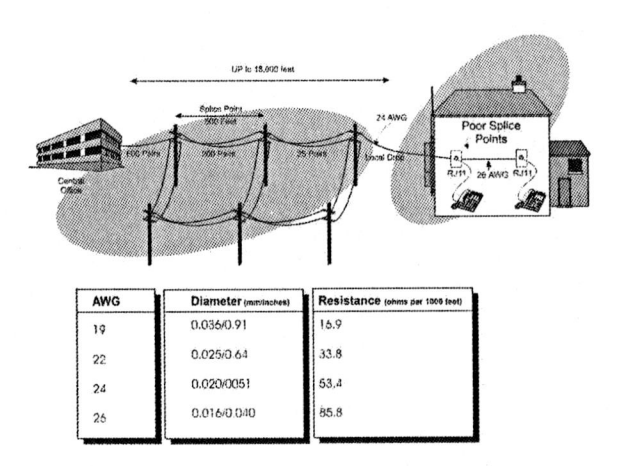

Line Resistance Attenuation

Line Side Connection-Line side connections are an interconnection line between the customer's equipment and the last switch (end office) in the telephone network. The line side connection isolates the customer's equipment from network signaling requirements. Line side connections and are usually low capacity (one channel) lines.

Line Signal-A signal sent over a line; included are call progress, supervisory, control, address, and alerting signals.

Line Source-(1-spectral) An optical source that emits one or more spectrally narrow lines, as opposed to a continuous spectrum. (2-geometric) An optical source whose active (emitting) area forms a spatially narrow line.

Line Station Transfer (LST)-A process of clearing communication line pairs for new subscribers in telecom distribution areas with marginal or depleted facilities. After a review by outside plant technicians and discussions with engineering, an existing customer may be switched to a new shorter length cable pair. This frees up the higher capacity pair as an available spare to serve the newly-signed customer at the far end.

Line Terminating Equipment (LTE)-A device or system that terminates a line in an optical system. LTE assemblies are used in a SONET network to originate and/or terminate Optical Carrier (OCn) signals. LTE equipment contains optical transmitter/receiver assemblies and an LTE assembly can decode, modify, creates the overhead control messages used in the optical network.

Line-Rate-(1-Transmission) The raw speed, in bits-per-second, of a transmission protocol on a particular media. (2-Device) A term used to specify a device, such as a switch, is able to operate with all ports at maximum speed and traffic levels. For example, an Gigabit Ethernet switch is said to be able to handle line rate if it can process packets on all ports simultaneously at a gigabit per second AND all the packets are 64-bytes (minimum packet size) in length.

Linear-(1-general) A circuit, device, or channel whose output is directly proportional to its input. (2-frequency) A circuit, device, or channel whose response is constant over a specified frequency range. (3-video effects) A straight-line motion path for objects being manipulated by a digital effects device.

Linear Combiner-A diversity combiner that adds two or more receiver outputs.

Linear Receiver-A radio receiver that operates in such a manner that the signal-to-noise ratio at the output is proportional to the signal level at the input, and/or to the degree of modulation.

Link Access Protocol Balanced (LAPB)-In the integrated Services Digital Network (ISDN) and the Public Packet-Switched Network (PPSN), the layer 2 data-link-layer procedures of the X.25 packet-switching protocol. LAPB is similar to the High-level Data Link Control (HDLC) asynchronous balanced mode. This allows a terminal to start a transmission without receiving permission from a system control unit.

Link Access Protocol on D Channel (LAPD)-In the Integrated Services Digital Network (ISDN), the primary data link-layer protocol on the D channel. LAPD is similar to LAPB except it uses a different framing sequence.

Link Establishment-(1-general) The process of establishing a communication link. This may involve the creation of a physical link and/or the creation of a logical channel on the physical link. (2-Bluetooth) A procedure is used to setup a physical link-specifically, an Asynchronous Connectionless (ACL) link- between two Bluetooth devices using procedures from the Bluetooth IrDA Interoperability Specification and Generic Object Exchange Profile.

Link Layer Discovery Protocol (LLDP)-A draft standard within the IEEE 802.1 working group which intends to provide a common way for devices connected to each other (say via ethernet) to identi-

Link Manager (LM)

fy a directly connected device. For example, an IP telephone using LLDP to a connected Layer 2 switch could share information about power requirements, VLAN configurations, and location information for E911 purposes.

Link Manager (LM)-The functional assembly that creates link setup, authentication, link configuration, quality of service (QOS) capabilities, and other management functions. The link manager in a Bluetooth system coordinates the different modes of operation (park, hold, sniff, and active).

Link Set-A set of signaling links in an SS7 network that connects a pair of adjacent nodes.

Link State Control (LSC)-Coordinates functions of the SS7 signaling link including signal unit delimitation, signal unit alignment, error detection, error correction, initial alignment, signaling link error monitoring and flow control.

Link Status Signal Unit (LSSU)-A signal unit (data packet) that contains information about the status of the link (e.g. failure or errors) in which it is transmitted in an SS7 network. This packet is specific to the link and is not transmitted through the network.

Link Supervision-(1-general) The process of monitoring the link status and managing physical and logical changes to the link. (2-Bluetooth) Each Bluetooth link has a timer used to link supervision. This timer is used to detect link loss caused by devices moving out of range, a device's power-down, or failure cases. The scheme for link supervision is described in Bluetooth's Baseband Specification.

Link-By-Link Signaling-A mode of network operation in which each office along a call route acts autonomously, forwarding all information requited to complete a call to the next office in the chain.

Link-State Routing Algorithm-Link-state algorithms (for example Open Shortest Path First) require each router to provide information about the state of its local links to all routers in the network. From this information, each router builds a map of the entire network.

In contrast, Distance-Vector algorithms require each router to send all or some portion of its routing table and rely on the propagation of these update messages for creation of the routing table. Link-state algorithms are less prone to routing loops, more scalable and more resilient than distance vector algorithms, however, they require more compute power and memory than distance vector algorithms.

Linked Operation-An SS7 operation invoked from one end of a Signaling Point that is linked to another operation previously invoked

LIP-Loop initialization Protocol

LIP-Large Internet Packet

Liquid Crystal-A material of low viscosity that matches the shape of the vessel in which it is contained (like a liquid) but has different refractive indices for light, depending on the path direction of the light through the material (like a solid crystal). Under the influence of an electric field, molecules align themselves in specific directions, changing the polarization plane and enabling characters to be made visible in a display panel.

LIS-Link Interface Shelf

LIS-Local Interconnection Service

Lissajous Pattern-The looping patterns generated by a CRT spot when the horizontal (x) and vertical (Y) deflection signals are sinusoids. The Lissajous pattern is useful for evaluating the delay or phase of two sinusoids of different frequencies. Named for the French physicist Jules Antoine Lissajous.

Listen Before Talk (LBT)-A process of listening to an access channel to determine if the channel is busy before attempting access to the channel or system.

Listing Services System-An interactive software system that manages customer listing information, including name, address, and telephone number. The data can be used to support customer and network services as well as to compose listings for use in white pages directories and other specialized products.

LIT-Line Insulation Test

Lithium Niobate - Lithium Niobate Modulator-A type of external modulator for optical networks. Lithium niobate is transparent at the 1300 nm to 1600 nm wavelengths commonly used in optical networks and also has a strong electro-optic effect. This unique combination has led to its extensive use as a material for external modulators.

Litz Wire-Litzendraht Wire

Litzendraht Wire (Litz Wire)-A braided wire, with individually insulated fine strands, that gives low resistance at high radio frequencies.

Live-(1-electric circuit) A device or system connected to a source of electrical potential. (2-acoustical) An area in which sound is not greatly absorbed by the walls and timings; the room, therefore, reverberates.

Live Streaming-In audio or video streaming, a stream for which the clients may not control the playback time of the media. That is, the clients may not control when the stream starts, pause the stream, skip to a different time within the presentation, and so on. Live streaming is often used for broadcast of an event happening in real time.
LL-Long Lines
LLB-Line LoopBack
LLB-Line Loop Back
LLC-Logical Link Control
LLC2-Logical Link Control 2
LLDP-Link Layer Discovery Protocol
LLDP-local loop demarcation point
LLF-Line Link Frame
LLWAS-Low Level Window Shear Alert
lm-Lumen
LM-Long Distance Marketer
LM-Link Manager
LMCS-Local Multipoint Communication System
LMDS-Local Multichannel Distribution Service
LME-Layer Management Entity
LMEI-Layer Management Entity Identifier
LMI-Local Management Interface
LMI-Layer Management Interface
LMOS-Loop Maintenance Operations System
LMR-Land Mobile Radio
LMS-Location And Monitoring Service
LMS-Local Message Switch
LMSI-Local Mobile Station Identity
LMSS-Land-Mobile Satellite Service
LMSS-Land Mobile Satellite Service
LNA-Launch Numerical Aperture
LNA-Low Noise Amplifier
LNB-low noise blockamplifier
LNB Converter-Low Noise Block Converter
LNC-Low Noise Converter
LNC-Line Not Cutting
LND-Last Number Dialed
LNNI-LANE Network-to-Network Interface
LNP-Local Number Portability
LNPA-Local Number Portability Administration
LNRU-Like New Repair And Update
LO-Local Operator
Load-(1-general) The work required of an electrical or mechanical system. (2-data) The process of inputting programs or data to a computer for storage or manipulation. (3-device output) A circuit or device that receives the output of an amplifier or transmission line. (4-generator) The amount of electric power taken from a generator. (5-magnetic tape) The process of placing a magnetic tape reel on a drive in preparation for recording or playback. (6-telecommunications) A volume of traffic that equals the sum of the holding times for a number of calls or call attempts. Such loads are expressed in either hundred call seconds or erlangs.
Load Box-A box or circuit that simulates a load (power sink) that is used to test the ability of a system to supply energy to a system that uses or absorbs energy.
Load Factor-The ratio of the average load over a designated period of time to the peak load occurring during the same period.
Load Generator-A device that can be connected to a line or trunk to originate or terminate simulated telephone calls for the purpose of verifying load-handling capability. A load generator also may be called a load box.
Load Line-A straight line drawn across a grouping of plate current/plate voltage characteristic curves showing the relationship between grid voltage and plate current for a particular plate load resistance of an electron tube.
Load Section-In a loaded cable, a uniform length of cable between two loading coils.
Load Set-In traffic engineering, the matrix of loads that results from the statement of a load from each specified pair of points in a network. A load set is further defined as a time consistent load set.
Load Transfer-In a common channel signaling system, the transfer of telecommunications signaling traffic from one signaling link to another.
Loaded Cable-A cable with uniformly spaced loading coils to improve transmission quality.
Loaded Loop-A cable pair with loading coils placed at intervals along its length.
Loading-(1-circuit) The addition of electrical inductance to a metallic transmission line to improve the frequency characteristics of the line. Loading a line increases the distance over which a quality signal can be sent. (2-antenna) The addition of an inductance to enable an antenna to be tuned to a frequency lower than its natural frequency. (3-multichannel communications) The insertion of white noise or equivalent dummy traffic at a specified level to simulate system traffic performance. (4-system) The total signal power of a multichannel system, expressed as the total of the average power on all channels, or as the per channel load that may be carried by all channels. (5-wind) The total ice

Loading Coil

and wind pressure allowed for in the design of a tower, pole, or line.

Loading Coil-A inductive device (temporary storage of energy in a magnetic field) that is installed in a telephone line to help enhance the frequency response of the line at specific audio frequencies. Unfortunately, loading coils significantly add distortion to high-speed data signals on those lines (such as DSL signals).

This diagram shows that there may be several installed audio loading coils on a single local loop line. Although these loading coils improve the audio frequency response, they must be removed to allow for high-frequency transmission for systems such as DSL.

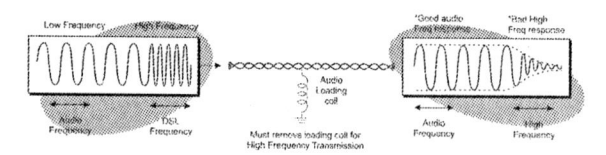

Loading Coils Operation

Lobe-(1-general) A representation of the transmission directional efficiency of a radio antenna; the larger the major lobe, compared with minor lobes, the more directive the system. (2-back) A lobe in an antenna radiation directivity pat-tern pointing directly away from (at 180' to) the intended direction. (3-front) The lobe in the required direction of an antenna radiation directivity pattern. The front lobe is the main or major lobe. (4-minor) Any of the lobes in the radiation directivity pattern of an antenna, except the major or front lobe. Also called side lobe.

LOC-loss of cell delineation

Local Access And Transport Area (LATA)-A geographic region in the United States where a local exchange carrier (LEC) is permitted to provide interconnected telephone service. LATAs were created as a result of the division of the company AT&T by the designated by the Modification of Final Judgment (MFJ). A LATA contains one or more local exchange areas, usually with common social, economic, or other interests.

Local Area Network (LAN)-Local area networks (LANs) are private data communication networks that use high-speed digital communications channels for the interconnection of computers and related equipment in a limited geographic area. LANs can use fiber optic, coaxial, twisted-pair cables, or radio transceivers to transmit and receive data signals. LAN's are networks of computers, normally personal computers, connected together in close proximity (office setting) to each other in order to share information and resources. The two predominant LAN architectures are token ring and Ethernet. Other LAN technologies are ArcNet, AppleTalk, and fiber distributed data interface (FDDI).

This figure shows several of the most popular LAN topologies and their configurations. Some data networks are setup as bus networks (all computers share the same bus), as start networks (computers connect to a central data distribution node), or as a ring (data circles around the ring). This diagram shows for popular types of LAN networks: Thinnet, Thicknet, token ring networks, and Ethernet star network.

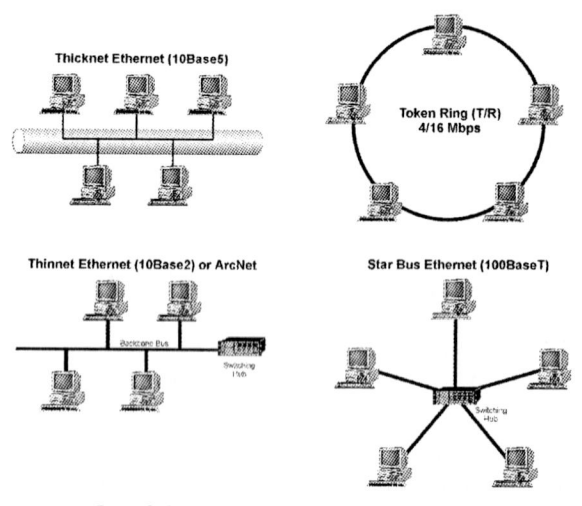

Local Area Network (LAN) Systems

Local Area Network [LAN] Telephone-A telephone that provides telephone services through the use of a local area network (LAN) system.

Local Area Network Emulation Configuration Server (LECS)-A process used for LAN Emulation (LANE) in an asynchronous transfer mode (ATM) system to manage LANE clients so they are assigned to specific servers.

Local Calling Area (LCA)-Applies only to originating minutes of use and foreign carrier (OHX) account and second dialtone (OHY) accounts. The file contains the subscriber line counts by interexchange carrier (IXC) for each end office in the local access and transport area (LATA). The line counts are used to calculate ratios (factors) that are then multiplied by the IXC's OHY actual or assumed originating minutes of use (MOU) in that LATA to assign MOU to end offices for reclassification. In billing, LCA usually refers to an area within which a customer (typically residential) is not charged for usage.

Local Control-A function of the mobile unit which has been designated to provide special features in addition to those specified by the cellular standard.

Local Digital Switch (LDS)-A digital switch that is the final switching point between the end customer and the public switched telephone network.

Local Exchange-Another term for a end office (EO) telephone switching system. The local telephone company is sometimes called the local exchange.

Local Exchange Carrier (LEC)-Local exchange carriers (LECs) or post and telegraph and telecommunications (PTT) companies provide telephone services directly to residential and business customers located within a localized geographic area. Typically, these telephone companies provide services via copper lines that extend from a local carrier's switching facilities to the end customer's premises equipment (CPE). This is referred to local loop.

Until the early 1990's, most countries had a single company that provided local telephone services. This company was either owned or highly regulated by the government. To increase competition and reduce telephone service prices to consumers, some governments have begun to allow other companies to provide basic (local) telephone service. These competitive local exchange company (CLEC) or competitive access providers (CAPs) provide alternative connections to the public switched telephone networks (PSTN). The established telephone companies are now called the incumbent local exchange carriers (ILECs),

Local Loop-The local loop is the connection (wireless or wired) between a customer's telephone or data equipment and a local exchange company (LEC) or other telephone service provider. Traditionally, the local loop (also called "outside plant") has been composed of copper wires that extend from the end office (EO) switch. The EO is the last switching office in the telephone network that connects customers to the telephone network. This diagram depicts a traditional local loop distribution system. This diagram shows a central office (CO) building that contains an EO switch. The EO switch is connected to the MDF splice box. The MDF connects the switch to bundles of cables in the "outside plant" distribution network. These bundles of cables periodically are connected to local distribution frames (LDFs). The LDFs allow connection of the final cable (called the "drop") that connects to the house or building. A NT block isolates the inside wiring from the telephone system. Twisted pair wiring is usually looped through the home or building to provide several telephone connection points, or jacks, so telephones can connect to the telephone system.

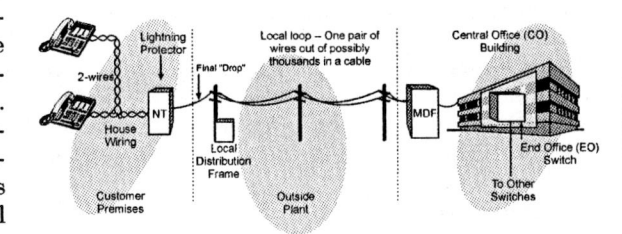

Telephone System Local Loop Operation

Local Measured Service-A method of charging customers based on actual usage. Factored into local measured service are the number of local messages, the duration of those messages, the time of day, and the distance within a local exchange area.

Local Multichannel Distribution Service (LMDS)

Local Multichannel Distribution Service (LMDS)-One of two wireless cable systems in use in the United States, LMDS stands for Local Multichannel Distribution Service. In 1997 new 28 GHz LMDS wireless cable systems entered into the FCC auction process. See also Multichannel Multipoint Distribution Service (MMDS).

LMDS is an asymmetric wireless service. It uses a small dish antenna that acts as both a receiver and sender in association with a hub station that is located up to 5 miles. Because LMDS systems use extremely high frequencies, this limits the range of the transmitted signal to 5 miles. The advantage to the limited distance is that LMDS signals from one antenna will not interfere with other antennas placed 10 or more miles apart. This allows the radio bandwidth to be reused (frequency reused) in a cellular like fashion.

LMDS uses approximately 1.3 GHz wide spectrum band at around 28 GHz. This provides a typical data rate for each LMDS channel of 1 Gbps.

This figure shows a LMDS system. This diagram shows that the major component of a wireless cable system is the head-end equipment. The head-end equipment is equivalent to a telephone central office. The head-end building has a satellite connection for cable channels and video players for video on demand. The head-end is linked to base stations (BS) which transmits radio frequency signals for reception. An antenna and receiver in the home converts the microwave radio signals into the standard television channels for use in the home. As in traditional cable systems, a set-top box decodes the signal for input to the television. Low frequency wireless cable systems such as MMDS wireless cable systems (approx 2.5 GHz) can reach up to approximately 70 miles. High frequency LMDS systems (approx 28 GHz) can only reach approximately 5 miles.

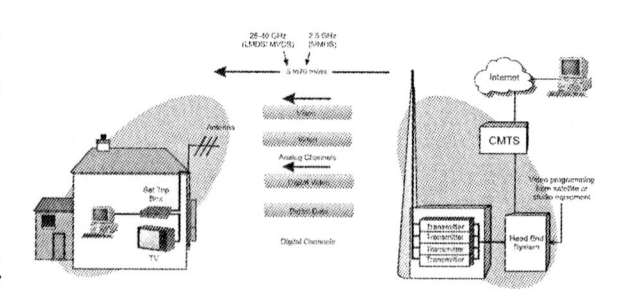

local multichannel Distribution Service (LMDS) System

Local Number Portability (LNP)-LNP is the process that allows a subscriber to keep their telephone number when they change service provider in their same geographic area. Local number portability requires that carriers release their control of one of their assigned telephone numbers so customers can transfer to a competitive provider without having to change their telephone number. LNP also involves providing access to databases of telephone numbers to competing companies that allow them to determine the destination of telephone calls delivered to a local service area.

This figure shows an example of the typical operation of local number portability (LNP). In this diagram, a caller in Los Angeles is calling someone in Chicago who has kept (ported) their old phone number when they connected their service to a competitive local exchange carrier (CLEC). This required the incumbent local exchange carrier (ILEC) to move (port) the telephone number to a LNP database. The line connected to the customer from the CLEC actually has a new telephone number (which the customer is not likely to be aware of). The LNP database associates the new number with the old number. This example shows how the call can be routed from a LEC in Los Angeles to the new telephone line in Chicago using the old telephone number. The call is routed from Los Angeles, through a long distance provider (IXC) who knows by the dialed area code that it needs to connect the call into a local telephone company in Chicago. Because there are several local telephone service providers in Chicago, the IXC must look first into a LNP database to see if the number has been ported to a different service provider. This LNP database (ported telephone number list) must be available to the next to last switch (called "N-1") before the call reaches the end office switch. This LNP database search instructs the last switch to the actual number used for the final connection. The call is then routed to the correct local switching office (new line) so the call can be completed.

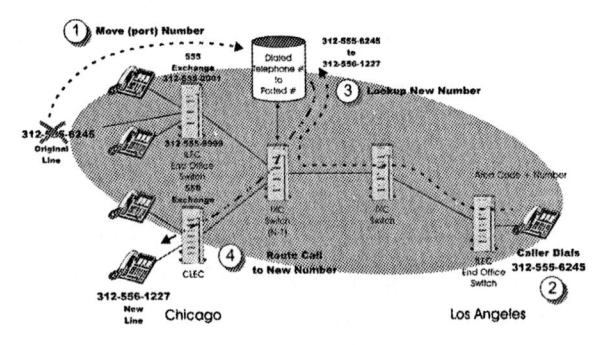

Local Number Portability (LNP) Operation

Local Service Request (LSR)-A form used by a Competitive Local Exchange Carrier (CLEC) to request local service form an Incumbent LEC (ILEC).

Local Switching System-A switching system that connects lines to lines, and lines to trunks in an end office. The system may be located entirely in a wire center or it may be geographically disposed, as in host remote configurations

Localization-(1-sound) The perception of sound as originating from a particular direction or distance. (2-troublshooting) The process of localizing equipment failures or below tolerance equipment.

Locating Receiver-In a cellular system, a locating receiver is a radio receiver that is located in a base station that can tune to any frequency in an allocated band to find a transmitting mobile radio. The locating receiver can determine the approximate energy level of the transmitting radio to determine if mobile radio requires a handoff to a new cell site that. A locating receiver is also called a scanning receiver.

Location Area Code (LAC)-(1- GSM) A code assigned by the system operator to identify specific areas of operation. These LAC identifiers can be used to indicate regions that have different billing codes or the types of authorized service features, and most importantly, to limit the number of cells used to page a mobile station when setting up a mobile destination call.

Location Routing Number (LRN)-A telephone number (e.g. 10 digit number) that is used to route calls to and end office switch that allows for the processing of portable (assignable) telephone numbers.

Location Server (LS)-Location servers provide information regarding the location of resources that are located within a network (such as the Internet or within a SIP system). Location servers are typically databases that maintain a binding (mapping) for each registered user. This binding maps the address of the user to one or more addresses at which the user can be currently reached. The Location Service supports user mobility within a communication system. In a SIP system, the Location Service database is updated as a result of SIP User Agents performing a registration.

LocDev-Local Device

Lock-To time synchronize two or more signals, lock to each other.

Lock Code-Wireless unit's built-in functionality which prevents unauthorized use by entering in the user-controlled lock code. It may lock out the keypad or prevent the unit from powering up altogether.

Locking Shift-In the Integrated Services Digital Network (ISDN), an information element indicating that the information elements that follow are in a different code set.

Log File-A file that contains a list of events that have occurred for a particular application or service. The log file is continually updated (added to) as new events occur. Log files are used to analyze problems that have or may occur with a particular application or service.

Log Normalization-Normalizing logs on Network Management Stations (NMS) is a method to organize log files and events, like syslog, snmp traps, RMON statistics, or polled MIBS. Normalizing the data saves on drive space and increases performance on the NMS.

Log Time-(1-general) The time at which a service or program was initiated or terminated. (2- video) new video source is placed on the program bus, usually recorded in the station log for FCC accounting and customer billing purposes.

Logarithmic Scale-(1-meter) A meter scale with displacement proportional to the logarithm of the quantity represented. (2-graph paper) A printed

Logging

graph paper with one or both of the grids on a logarithmic, rather than an arithmetic, scale.

Logging-The recording of data about events that occur in a time sequence. Logging can be used with syslog to monitor network events. Logging also applies to monitoring event, application, and system logs on Windows PC based systems. Logging is very useful in troubleshooting and correlating network and system environments and events.

Logic Analyzer-A test instrument that is used for monitoring computer system logical operating states and state sequences.

Logic Element-A device that performs a logical function, also known as a logical element or gate.

Logic Gate-The basic decision-making circuit used in digital equipment. A logic gate usually has two or more binary inputs and one binary output. Simple functions can be implemented by single gates, but several gates of different types, together with various forms of memory, often are combined to form complex decision-making networks.

Logic Trunked Radio (LTR)-Logic Trunked Radio (LTR) was developed by the E.F. Johnson Company. LTR uses sub-audible signaling and distributed control logic to provide for fast system access and voice channel assignments to any pre-assigned radio channel. LTR has become the de facto standard for analog trunked radio systems. The first trunked radio system was Clearchannel LTR®.

The Clearchannel LTR® system allows the automatic sharing of channels in a multiple repeater system. The LTR system uses distributed control, where the mobile radio shares the intelligence for radio channel access and assignment with the repeater (base station). Access on the LTR system can be made on any pre-assigned RF channel that is idle.

A transmitting repeater RF channel provides information to mobile radios by simultaneously transmitting information in a low bit rate data stream (sub-audio at 150 Hz, 300 bps) that coexists with voice information. This eliminates the need for a separate dedicated control channel. Each repeater maintains its own data stream and handles all accesses on its channel. In the LTR system, radio network access control is independently handled by the mobile radios. Repeaters that are transmitting constantly provide updated information about the status of the repeater (its ID), calls that are to be received by mobiles and the status of other repeaters at the site (e.g., if they are idle). Figure 3.1 shows a basic LTR system block diagram.

Every mobile radio used in an LTR system must be programmed for LTR signaling in that system. This information includes which RF frequencies to use and the home system identification number. Preprogramming the LTR radio allows the operation of LTR mobile radios to be much simpler than that of conventional radios. This is because many functions that would normally be performed by the user are automatically performed by the LTR system. The user only has to select which system to use (or group ID) and press the push-to-talk switch. If the system is not busy serving other mobile radios, speaking can begin. If the system is busy, an audio signal (beep tone) is usually provided.

Logical Channel-A portion of a physical communications channel that is used to for a particular (logical) communications purpose. The physical channel may be divided in time, frequency or digital coding to provide for these logical channels.

This figure shows how a single digital communication channel is divided into several different logical channels. In this example, there are four fields per frame. Field 1 contains system information, field 2 contains the channel status information (busy/idle), field 3 transfers channel assignment commands and field 4 is defined for other sub channels.

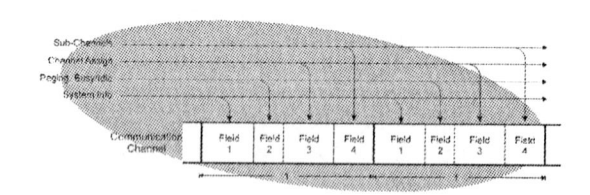

Logical Channels

Logical Link Control (LLC)-A lower layer protocol that is used to manage the link transmission. This layer provides access between the communications stack and the transmission medium. it can be used to provide data reformatting and repackaging functions to allow communication between different network types.

Logical Link Control And Adaptation Protocol (L2CAP)-A Bluetooth protocol that interfaces between the application layer and the physical layer in a Bluetooth communication system. This protocol performs logical channel multiplexing, packet segmentation and re-assembly (SAR), and group link management. The L2CAP allow applications to transmit and receive data packets of up to 64 Kbps in length.

Long Distance-Services charged at a toll rate, or services offered by interexchange companies for traffic that crosses LATAs (InterLATA). (See also: long-haul communications, toll.)

Long Distance (LD)-The connection of calls outside the local service calling area.

Long Haul System-(1-general) A communication system which includes a number of drop/add points, repeaters locations, over long distances that extend outside the local service area. (2-microwave) A microwave system that the longest radio circuit of tandem radio paths exceeds 402 km (250 miles). This diagram shows a terrestrial microwave system-connecting IXC switches in Philadelphia and New York City. The microwave signals are moved between the two switching offices through a series of relay microwave systems located approximately 30 miles apart. Microwave is a line-of-sight technology that must take the earth's curvature into consideration. Also note that microwave towers are not limited to only facing one or two directions. A single tower can be associated with several other towers by positioning and aiming additional transceiver antennas at other microwave antennas on other towers.

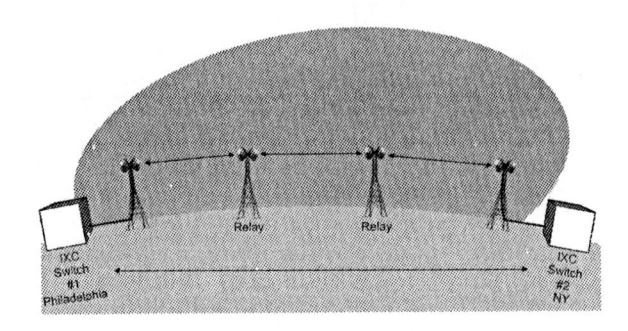

Long Haul Microwave

Long Persistence-A type of phosphor in a cathode ray tube that continues to glow after the original election beam has ceased to create light by producing the usual fluorescence effect.

Long Range Navigation (LORAN)-A radio positioning system that provides location information by using the time difference between reception of radio signals from two or more high power fixed transmitters.

Longitude-The angular measurement of a point on the surface of the earth in relation to the meridian of Greenwich (England). The earth is divided into 360, of longitude, beginning at the Greenwich mean. As one travels west around the globe, the longitude increases.

Longitudinal Time Code (LTC)-Time code information encoded as an audio like (FM) signal and recorded on audio channels of a videotape or audiotape recorder. LTC is readable at standard tape speed, and greater/slower than standard play speeds, but becomes unusable in still-frame mode.

Longwave-The band of radio signals with wavelengths longer than 600 meters (frequencies less than 500 Hz).

Look Ahead Preview-The output of a video switcher that permits the operator to observe an effect before it is aired.

Loop (Software)-The repetition of a group of instructions in a computer routine.

Loop Analyzer-A device that analyzes the performance characteristics of a local loop line.

Loop Assignment Center-An operations center that assigns customer loop facilities, telephone numbers, and central-office lines and equipment.

Loop Battery-A direct current voltage source applied between the conductors of a line and used for loop start, ground start, and loop reverse battery supervision. For loop start supervision, loop battery is used to detect a request for service. For ground start supervision, loop battery indicates that a request for service has been recognized. For loop reverse battery supervision, the loop battery polarity indicates the supervision state of the equipment (on-hook or off-hook) connected to one end of the loop.

Loop Gain-The total gain of all the active devices in a closed loop minus the losses of all the passive devices in the loop.

Loop Plant Improvement Evaluator (LPIE)-A system that analyzes the economics of proposed changes to facilities, such as serving area interface redesign and cable replacement.

Loop Pulsing-Signaling accomplished by the repeated opening and closing of a loop at the originating end of a circuit. Rotary telephone dials are loop pulsing devices.

Loop Signaling-Signaling protocols and processes used in a distribution network, or loop.

Loop Signaling System-Methods for sending signaling information over a communication loop. Loop signals can be transmitted by opening and closing the loop path, reversing the voltage polarity, or varying the line resistance.

Loop Start-A form of line supervision in which a service request is indicated to a network when a terminal enables loop current to flow.

Loop Wire-A wire that links several terminals or adjacent components.

Loop-Subscriber Line-Insulated wire pair connecting a subscriber's telephone set (station set) to a central office switch. In historical documents dating from circa 1890 and slightly later, the loop was an innovation by John Carty of AT&T, replacing the earlier practice (copied from telegraph systems of the 19th century) using a single wire between the central office and the subscriber, with earth/ground conductivity to complete the circuit. Historical single wire subscriber connections were less costly but had extremely variable total path resistance due to changes in soil moisture content, and also suffered from bad crosstalk due to use of the same shared earth conductor path for several simultaneous conversations.

Looped Clock-An option on digital terminals that enables a digroup transmit clock to be locked to a receive clock. The receive clock always is derived from an incoming DS1 bit stream.

Looping-(1-software) A programming technique by which a portion of a program is repeated until a certain result is obtained. (2-post production) The replacement of dialogue in post production. The term looping is derived from earlier film processing techniques that used loops of film and magnetic film stock to facilitate dialogue replacement.

Loose Construction-A type of fiber optic cable construction in which the fibers are permitted to float freely to relieve stresses and minimize bending losses.

LOP-Loss of Pointer

LORAN-Long Range Navigation

LORAN-Long Range Aid to Navigation

LOS-Line Of Sight

LOS-Loss of Signal

Loss Deviation-The change of actual loss in a circuit or system from a designed value.

Loss Variation-The change in actual measured loss over time.

Lossy Cable-A coaxial cable constructed to have high transmission-mission loss so that it can be used as an artificial load or as an attenuator.

Lot Size-A specific quantity of similar material or a collection of similar units from a common source; inspection work, the quantity offered for inspection and acceptance at any one time. The lot size may be a collection of raw material, parts, subassemblies inspected during production, or a consignment of finished products to be sent out for service.

Loudspeaker-A transducer (converter) that transforms audio electrical signal into sound waves (audible signals).

Loudspeaker Baffle-An assembly that is mounted on a loudspeaker to help focus the sound waves in a particular direction (such as to the front or side of the speaker).

Loudspeaker Paging-A feature on a communication system (such as a PBX) that permits a user to transmit their voice over one or several loudspeaker systems. Loudspeaker paging systems were commonly used to alert people in a geographic area (such as on a retail sales area) that they are receiving a call or they are needed at a specific location.

Louver-The slots or holes on the front of a loudspeaker that permit sound to pass, but provide mechanical projection to the device.

Low Earth Orbit (LEO)-A satellite system where the satellites are located approximately 500-1,000 miles above the Earth. LEO systems typically provide mobile satellite services (MSS) to handheld or mobile satellite telephones.

This Figure shows a LEO satellite system. In this diagram, a portable satellite telephone is communicating with a landline telephone. The satellite telephone communicates with the closest LEO satellite. Because LEO satellites fly very close to the surface of the earth, they go across the visible horizon in approximately 10 minutes in reference to a mobile satellite customer's location. When the first satellite moves out to the horizon, another LEO satellite becomes available to continue the call. However, robust network communications need to be in place to maintain calls (especially data transmission) within this period. Some systems will use satellite diversity to allow talking through more than one satellite at a time, avoiding call "dropouts" from signal blockage.

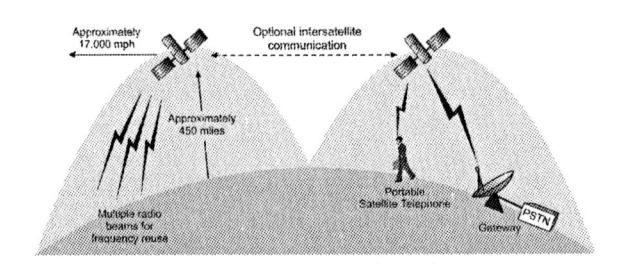

Low Earth Orbit (LEO) Operation

Low Level Language-A programming language that reflects the structure of a computer or that of a given class of computers. A low level language consists of instructions that are converted directly into machine code.

Low Noise Amplifier (LNA)-A sensitive preamplifier used at a focal point (the feedhorn) of a satellite antenna to strengthen the weak satellite signal. The most important parameter of the LNA is its noise temperature, as described in degrees Kelvin. In general, the lower the noise temperature, the better the signal quality. There is a generally a tradeoff between noise temperature of the LNA and the size of the satellite receive antenna. A higher noise temperature rating for an LNA requires a larger diameter antenna to maintain the same level of performance.

Low Noise Block Converter (LNB Converter)- A device that shifts a band of received frequencies to a different (usually lower) frequency band with a small amount of added (unwanted) signal noise. A common application of a LNB converter is the conversion of extremely high-frequency satellite receiver signals (such as the KU frequency band) to a lower frequency (e.g. C frequency band). The LNB converter is often located on or near the satellite receiver antenna to allow the transfer of lower frequency received signals (instead of extremely high frequency signals) for transfer from the satellite antenna (satellite dish) to a nearby head-end building using coax cable or other types of transmission line.

This photograph shows an LNB converter that mounts on a satellite antenna to convert the received band of frequencies to a lower frequency signal.

Low Noise Block (LNB) Converter
Source: TERK Technologies Corp.

Low Pass Filter (LPF)-A filter that passes frequencies below a frequency cutoff point. Lowpass filters are often used in telephone networks to pass audio frequencies below 4 kHz and block (attenuate) high frequencies.

This diagram shows a lowpass filter can be used to block mid and high frequency component parts of an input signal and allow low frequency components (signals) to pass through. In this example, both the mid and high frequency noise signals are highly attenuated by the bandpass filter while the desired low frequency is allowed to pass through the filter with minimal attenuation.

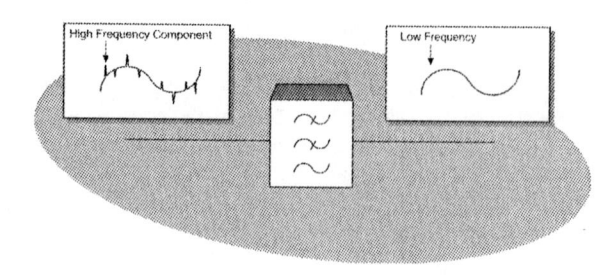

Lowpass Filter Operation

Low-Tier-A wireless system which uses low-power levels intended for pedestrians and other slow moving traffic.
Lower Sideband (LSB)-The sideband of an amplitude-modulated signal containing all frequencies below the carrier frequency.
LPC-Local Procedure Call
LPC-Linear Predictive Coding
LPF-Low Pass Filter
LPIE-Loop Plant Improvement Evaluator
LPTV-Low Power Television Service
LRC-Longitudinal Redundancy Check
LRN-Location Routing Number
LRQ-Sequential Location Request
LS-Location Server
LSA-Link State Advertisement
LSA-Leased Space Agreement
LSB-Lower Sideband
LSB-Least Significant Bit
LSC-Link State Control
LSL-Link Support Layer
LSO-Local Central Office
LSP-Link State Packet
LSR-Local Service Request
LSSU-Link Status Signal Unit
LST-Line Station Transfer
LT-Line Termination
LTA-Line Turn Around
LTC-Longitudinal Time Code
LTE-Line Terminating Equipment
LTL-Local Transport Location
LTR-Logic Trunked Radio
LTS-Long Term Support
LU-Logical Unit

Lumen (lm)-A unit of total visible light power output in all directions from a luminous object. Light used for this comparison has a visible spectrum distribution of power corresponding to the spectral radiation from a piece of "black" surface platinum at its standard (normal atmospheric pressure) melting/solidification temperature.
Luminance-The amount of visible optical power, measured in Lumens.
Luminance Border-A non-color, luminance-only fill video for key banners and drop shadows.
Luminance Key-A key effect in which the portions of a key source that are greater in luminance than the clip level cut a hole in the background video.
Luminance Nonlinearity-A video distortion in which the luminance gain of the TV system changes as a function of luminance amplitude. The resulting TV picture will display poor resolution between brightness levels in the nonlinear range.
Luminance Signal (Y)-The video signal that describes the amount of light in each pixel. Luminance is equivalent to the signal provided by a monochrome camera. It may be generated as a weighted sum of the RGB signals in accordance with the formula: $Y = 0.3R + 0.5G. + 0.11B$. Luminance is differentiated from brightness in that the latter is non-measurable and sensory. The color video picture in-formation contains two components: luminance (bright-ness and contrast) and chrominance (hue and saturation). Luminance is the photometric quantity of light radiation.
Luminous Flux-The amount of visible light intensity per square meter (or other area unit).
Lumped Constant-A resistance, inductance, or capacitance connected at a point, and not distributed uniformly throughout the length of a route or circuit.
Lumped Loading-The use of inductors, spaced at regular intervals along a transmission line, to improve the transmission characteristics of the line over a specific frequency band. (See also: loading coil.)

Luneburg Lens-A spherical antenna invented by R.K. Luneburg. It uses a solid sphere having a radially varying dielectric (index of refraction) property to focus electromagnetic power from a wide area of sources onto a small detector area when receiving, or vice versa when transmitting.
LVDS-Low Voltage Differentail Signaling
LWT-Listen While Talk
LZW-Lempel-Ziv-Welch

M

m-Milli

M-Mega

M Format-A component video format for use in videotape recorders. The signal set consists of separate Y, I, and Q signals. The terminology M refers to the way in which the tape is routed through the recording mechanism. M-format is a registered trademark of Panasonic.

M Regions-The areas of the surface of the sun that appear to be responsible for many of the electromagnetic disturbances experienced on earth.

M, m-(1- Metric prefix) Capital or upper case M represents "Mega" or one million (1,000,000). Small or lower case m represents "milli" or 1/1000 or 0.001. (2- Roman Numeral) The Roman Numeral M represents one thousand or 1000. (3- metric unit) Lower case m represents the length unit "meter."

M12 Multiplexer-A multiplexer that combines 4 DS-1 "tributary" bit streams into a DS-2 bit stream. Another name for a DS-2 or T-2 multiplexer. Less used today because of the economic advantage of an M13 mutiplexer.

M13 Multiplexer-A multiplexer that combines 28 DS-1 "tributary" bit streams into a DS-3 bit stream.

M23 Multiplexer-A multiplexer that combines 7 DS-2 "tributary" bit streams into a DS-3 bit stream. Another name for a DS-3 or T-3 multiplexer. Less used today because of the economic advantage of an M13 mutiplexer.

M2UA-MTP2 User Adaptation Layer

M3UA-MTP3 User Adaptation Layer

MAC-Moves, Adds, And Changes

MAC-Message Authentication check

MAC-Medium Access Control

MAC-Macintosh

MAC Address-The physical address of the device on the medium. An example of a MAC addresses is the 48 bit address of a device on the Ethernet.

MAC Address-Media Access Control

MAC Algorithm-A set of procedures used by communication devices to coordinate access to a shared communications medium or channel. Examples of MAC algorithms include CSMA/CD and Token Passing.

MACA-Mobile Assisted Channel Allocation

Machine Code-The instruction code designed into the hard-ware of a microprocessor. Machine code is the direct representation of the computer instruction in memory.

Machine Language-A low-level, native programming language to a specific type of computer or processor, whose instructions consist only of computer instructions. Machine language is a program of binary coded instructions stored in memory.

Macro-(1-computers) An abbreviation for macroinstruction, an instruction that generates a larger sequence of instructions for a computer. (2-video) A special function of some zoom lenses that permits an object to be in focus at closer than usual distances to the objective element. The function usually offers magnification of the object. (3-application) A set of stored keystroke sequences or processes that are grouped to allow the user to perform repetitive control or editing sequences of application commands.

Macro Virus-A macro command that attaches itself to application documents that are capable of running macro (multiple keystroke) commands. Macros are commonly used in Word processing or spreadsheet applications to execute repetitive commands and this macros to open, edit, and delete files on a computer.

Macrobend-A large fiber optic cable bend that can be seen with the unaided eye. Macrobends with a radius of 10 m or more often cause negligible signal loss.

Macrobend Loss-In an optical waveguide, that portion of the total loss attributable to macrobending.

Macroblock-In video compression, a macroblock is a region over which motion compensation from a reference frame may be applied. Typically a frame is divided into 16 by 16 pixel sized macroblocks, that is, groupings of four 8 by 8 pixel blocks.

Macrocell-A cell site providing coverage over a relatively large geographical area (radius 1-5 miles).

MAE-Metropolitan Area Ethernet

Mag Stripe-Magnetic Stripe

Magenta-A subtractive primary color, also known as "process red."

Magnet-A device that produces a magnetic field and can attract objects of iron, cobalt or nickel. The magnetic field developed around a magnet can attract or repel the fields of other magnets.

Magnetic Amplifier-An amplifier that uses magnetic fields to amplify signals.

Magnetic Card-A card with a magnetizable layer in which data can be stored.

Magnetic Disk-A memory device employing magnetic material coated on a circular base. Data is stored by changing the direction of magnetization of small localized areas or domains along concentric tracks on the surface of the disk. A read/write head moves radially to access any of the tracks.

Magnetic Drum-A large capacity memory storage device that operates on the same principle as a magnetic disk. The magnetic material is on the outer surface of a rigid cylinder, usually mounted vertically. The drum rotates at high speed while the fixed read/write heads access data as it passes beneath them.

Magnetic Field-An energy field that exists around magnetic materials and current-carrying conductors. Magnetic fields combine with electric fields in lightwaves and radio waves.

Magnetic Field Strength-The strength of a magnetic field at a point following the direction of the lines of force at that point.

Magnetic Flux-The field produced in the area surrounding a magnet or electric current The standard unit of flux is the Weber.

Magnetic Hysteresis-A property of magnets based on the fact that magnetization of a ferromagnetic material does not vary on a linear basis with the strength of the magnetic field applied.

Magnetic Ink Character Recognition (MICR)-A system developed by the American Banking Association to allow computers to read and sort documents such as checks. Standardized characters are printed with a magnetic ink. Special scanners read the ink, which also is visible to the eye (along the bottom edge of a check, for example). Often, MICR is used where accidental marks would contuse an optical character recognition (OCR) system.

Magnetic Leakage-The magnetic flux that does not follow a useful path.

Magnetic Pole-A point that appears from the outside to be the center of magnetic attraction or repulsion at or near one end of a magnet.

Magnetic Recording-A method of storing information in magnetic material, such as tape or magnetic disks. Metal particles are magnetically oriented in relation to the frequency and amplitude of the recorded signal. Analog recording stores information as varying frequency and amplitude changes that relate to the orientation of the particles. In digital recording, transitions with changing polarity at a fixed amplitude are recorded at saturation.

Magnetic Saturation-A level of magnetism at which all particles or domains of the magnetic material are aligned with the applied magnetic force. Any increase of the applied field will cause no further alignment of the particles.

Magnetic Storm-A violent local variation in the earth's magnetic field, usually the result of sunspot activity.

Magnetic Stripe (Mag Stripe)-A strip of magnetic material affixed to a badge, credit card, or other item on which data can be recorded and read.

Magnetic Stripe Card-A card that stores information on a magnetic strip.

Magnetic Tuning-The tuning of a microwave cavity-type oscillator by varying the magnetic flux density of a ferrite rod in the resonator.

Magnetism-A property of iron and some other materials, including conductors carrying an electric current, by which external magnetic fields are maintained, other magnets being thereby attracted or repelled.

Magnetization-The exposure of a magnetic material to a magnetizing current, field, or force.

Magnetomotive Force-The force that tends to produce lines of force in a magnetic circuit. The magnemotive force bears the same relationship to a magnetic circuit that voltage does to an electric circuit.

Magnetron-A high-power, ultra-high-frequency electron tube oscillator that employs the interaction of a strong electric field between an anode and cathode with the field of a strong permanent magnet to cause oscillatory electron flow through multiple internal cavity resonators. The magnetron may operate in a continuous or pulsed mode.

MAH-Mobile Access Hunting

MAHO-Mobile Assisted Handoff

MAHO-Mobile Assisted Handover

Mail Server-A host, with its associated network software, that offers electronic mail reception and (optionally) email forwarding service. Users may send messages to, and receive messages from, any other user in the system.

Mailbox (MBX)-A system for storage and transmission of electronic text messages. Mailboxes are often storage areas on computer hard disks that are managed by mail server computers that interconnect to data networks such as the Internet. Mailbox systems often provide notification of an incoming message and confirmation of delivery.

Main Distribution Frame (MDF)-The wire connection point (wire rack) that is located at or near the central switching that is the point where all local access loops are terminated. The MDF connects cable pairs to the line and trunk equipment terminals of a switching system. The frame also serves as a test point between individual telephone lines and central office equipment The vertical side carries the outside lines and protective devices. All connections to central office equipment are made on the horizontal side. The main distributing frame also is referred to as a mainframe.

Main Lobe-The main portion of a radiation pattern from an antenna.

Main Station-A telephone connected directly to a central office by either an individual or a shared line. The principal telephone of each party on a party line is a main station ion. Telephones that are connected manually or automatically to a central office through a private branch exchange (PBX) or extension telephone are not main stations, but usually are called equivalent main stations.

Mainframe-Computer systems that are used for handling large quantities of central data processing and information storage applications. Mainframe computers are used for applications including invoice creation, account reconciliation, and management information reporting.

Maintainability-The probability that a failure will be repaired within a specified time after it occurs.

Maintenance-Any activity intended to keep a functional unit in satisfactory working condition. The term includes the tests, measurements, replacements, adjustments, and repairs necessary to keep a device or system operating properly.

Maintenance Center-An operations center that administers all upkeep and repair work in an outside plant network.

Maintenance Fees-Periodic fees which must be paid over the life of a patent in order to keep the patent in force. Most countries require the payment of maintenance fees. Failure to pay maintenance fees can result in premature expiration of a patent.

Maintenance Measurements-Counts of events and their duration that provide information about the maintenance condition of a network element, especially a switching system. Maintenance measurements can include a subset of traffic measurements.

Maintenance Terminating Unit-The equipment located at a network interface that isolates a terminal from a network for testing purposes.

Major Lobe-The primary radiation pattern of an antenna.

Major Trading Area (MTA)-A geographic region within the United States where most of the area's distribution, banking, wholesaling is performed. The United States has been divided into 51 MTAs and personal communications services (PCS) licenses were granted based on MTA.

Make Busy-(1-general) The setting of a line, trunk, or switched equipment unit to make it unavailable for service. To anyone seeking a connection, the circuit appears to be busy. (2-automatic call delivery) The marking of a customer service representative line as busy ("busy out") so the system will not transfer calls to that phone.

Make Busy Leads-Terminal equipment leads at the network interface designated MB and MB1. The MB lead is connected by the terminal equipment to the MB1 lead when the corresponding telephone line is to be placed in an unavailable or artificially busy condition.

Make Interval-In dial pulse signaling, that portion of the pulse cycle during which the dial contacts are closed.

Malfunction-An equipment failure or a fault.

Malfunction Timer-A timer that runs separate from all other functions within a communication device. It continuously counts down and needs to be reset. If the mobile is operating correctly (without failure) this timer will be reset continuously and will not expire. A malfunction timer is used in mobile radios to turn off the transmitter in event of a failure in critical parts of the transceiver.

Malicious Call Trace-A process that allows the identification of the location of an undesired caller. Malicious call trace is activated after the recipient has informed the telephone company. Malicious call trace will work even if the unwanted caller's telephone number is blocked. For privacy purposes, the telephone company may only provide the unwanted caller's telephone number to the public safety authorities (such as the police) rather than directly to the recipient of the unwanted call.

MAN-Metropolitan Area Network

Man Machine Interface (MMI)-The interface between the user and a device or system.

Man Machine Language-A language designed to facilitate direct user control of a computer. A man-machine language contains inputs (commands), outputs, control actions, and procedures sufficient to ensure the performance of all functions relevant to the operation, maintenance, and installation testing of stored-program control systems.

Managed Object-An atomic element of an SNMP MIB with a precisely defined syntax and meaning, representing a characteristic of a managed device.

Management Information Base (MIB)- Management information bases (MIBs) are a collection of definitions, which define the properties of the managed object within the device to be managed. Every managed device keeps a database of values for each of the definitions written in the MIB. MIBs are used in conjunction with the simple network management protocol (SNMP) as well as RMON to manage networks. MIBs (referred to now as MIB-i) were originally defined in RFC1066.

Management Information Base Browser (MIB Browser)-A management information base (MIB) browser is a graphic user interface (GUI) that allows an administrator to review and change the stored configuration and operational parameters of equipments.

Management Information Base II (MIB-II)- MIB-II obsoletes the original MIB (MIB-I) definition defined in RFC1066. MIB-II is widely used in SNMP and RMON managed networks and was originally defined in RFC1156, but was made obsolete by the more well defined and utilized standard, RFC1213.

Management Information System (MIS)-A system that gathers, organizes, and processes information for a department or a company. MIS systems are developed and used by companies to manage its information needs.

Management Inhibiting-An SS7 procedure included in traffic management messages used to keep a signaling link unavailable to User Part generated signaling traffic, except for test and maintenance traffic.

Management Routing Verification Test (MRVT)-Testing that validates the routing tables in the SS7 network nodes. MRVT tests ensure that no routing loops or other routing anomalies are present.

Manchester Encoding-A digital encoding technique that divides each bit period into half periods. A negative to positive transition represents a binary 1, and a positive to negative transition represents a binary 0. The use of Manchester encoding allows for clock recovery as transitions occur on every bit transmission.

This diagram shows how Manchester encoding transfers digital information in the form of positive or negative transitions. This example shows that a logical 1 is indicated by a positive transition and a logical 0 is indicated by a negative transition. This example also shows that Manchester coding forces continual transitions during each bit period and these transitions can be used to synchronize the clock timing signal

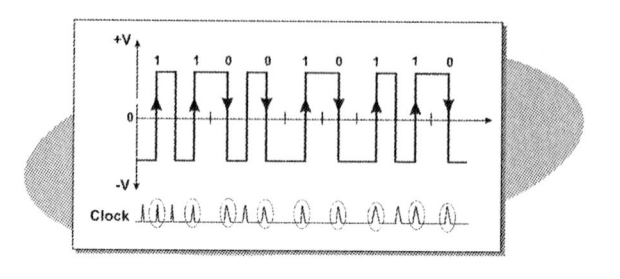

Manchester Encoding Operation

Mandatory Fixed Part-The part of a signaling message that contains those parameters that are mandatory and of fixed length.

Manhole-An access hole that allows entry of service personnel into a system or facility.

manifold-A hardware device that is used in cable pressurization systems to cross-connect 5/8" air pipe with 1/4" tubing to facilitate the distribution of dry air throughout a system.

Manipulation-(1-general) The modification or reformatting of information or data. (2-video image) In a video effects system, the various processes used to alter a video image, such as transformations and programmed effects.

Manometer-A metering device for measuring gas pressure.

Manufacturing Automation Protocol (MAP)- Standards for local area networks used in manufacturing applications on factory floors. The protocol follows the Open Systems Interconnection (OSI) communications model.

MAP- The relationship of a logical channel to a specific position in a transmission channel. The process of assigning logical channels to physical transmission channels is called mapping.

MAP- Mobile Application Part

MAP- Manufacturing Automation Protocol

MAPI- Messaging Application Programming Interface

Mapping- A process of assigning information to specific time, frame or code locations on communication channels or circuits. When the information is received, the mapping process can be used to extract the channels or information from the time, frame, or code positions as needed.

MAR- Memory Address Register

Margin- (1-performance) The difference between the value of an operating parameter and the value that would result in unsatisfactory operation. Typical parameters include signal level, signal to noise ratio (SNR), distortion, crosstalk coupling, and/or undesired emission level. (2-receiver) The signal power available to a receiver in excess of its design limit

Maritime Mobile Satellite Service- A radio service for ships at sea, using geostationary satellites in addition to shore stations.

Maritime Radiodetermination Service- A maritime radio communication service for determining the position, velocity, and/or other characteristics of an object, or the obtaining of information relating to these parameters, by the propagation properties of radio waves.

Maritime Radionavigation Satellite Service- A radio navigation-satellite service in which earth stations are located on board ships.

Mark- Mark represents a logical value of 1. Mark was defined from the closed circuit condition in a teletypewriter system that actuates a printer function. Mark is the opposite of a space.

Mark In- The point at which an edit on video tape begins, that is, the first frame that will be recorded.

Mark Out- The point at which an edit will end, that is, the first frame that will not be recorded.

Mark Signal- A sequence of marks (logical ones) that is sent before the start of a message or data block.

Mark Table- A video editing term meaning a list of edit in-points and out-points, and the sources to be used.

Marker Beacon- A transmitter in the aeronautical radionavigation service which radiates vertically a distinctive pattern for providing position information to aircraft.

Market Awareness -The first objective of a brand management or marketing campaign in the Costa Model that states customers must first be aware that a company or product exists and that it offers services they are interested in.

Market Convergence Management -A formal approach used by the managers of different telephone company (telco) product groups to combine their separate smaller customer populations into a much larger, shared pool, allowing the maximizing of revenues through cross selling, brand extension and churn proofing.

Market Demand Curve -Economic model which defines the level of service (number of ERLANGS) that the market will demand at a given price. The market demand curve is used to help set prices and define market strategies.

Market Development Funds (MDF)- The allocation of funds or sales credit allowances that are given by manufacturers as incentives to retailers to promote their products.

Market Familiarity- A second objective of a marketing activity in the Costa Model that makes the customer familiar with a company name or services and its availability.

Market Granularity -Market granularity defines the way that a company approaches its customers as a group. A telephone company with low granularity will typically divide customers into a small number of segments (Business, Small Business and Consumer for example), and then set prices, strategies and treatments for each. A company with a high granularity view will create many more segments, of varying sizes and values, and focusing more precisely on the needs of these smaller, better defined groups.

Market Maintenance Strategy -A defined market strategy that attempts to minimize the reduction of average revenue per user (ARPU) erosion and customer disconnection (Churn) but foregoes any attempts to attract new customers. Market maintenance strategies are usually used by incumbent local exchange carriers (ILECs).

Market Preference -The third phase of any marketing activity in the Costa Model that motivates the customer to prefer a product or service from one company as compared to a product or service a different company.

Marketing Campaign-Any of a broad range of marketing activities designed to send messages to customers about products, services, and options that the telco makes available. Marketing campaigns can be executed via advertising, direct marketing, public relations, place, or other media.

Marketing Channels-Departments, divisions, and external business partners that participate in the process of determining customer wants and needs and communicating how the company can provide these to the customer.

Marketing Program-A series of related marketing campaigns assembled to accomplish a single objective (i.e. a customer retention program could be a series of advertising, sales, and direct marketing campaigns with a related set of messages, concepts, and icons).

Markman Hearing-As a result of a 1996 decision by the United States Supreme Court (Markman v. Westview Instruments), claim construction is a question of "law" for the court to determine. A so-called Markman hearing precedes the trial of facts before a jury. The result of the Markman Hearing is a so-called "Markman Order" in which the judge decides, as a matter of law, what the claims mean.

Markov Model-A statistical model of the behavior of a complex system over time in which the probabilities of the occurrence of various future states depend only on the present state of the system, and not on the path by which the present state was achieved. This term was named for the Russian mathematician Andrei Andreevich Markov (1885-1922).

MARS-Multicast Address Resolution Server

MARTIAN-Mis-Addressed/Routed Transmission In A Network or Mis-Addressed/Routed Telepacket In A Network

MAS-Multiple Address System

Mask-(1-semiconductor) A device used in the production of thin-film circuits and other components as a means of restricting patterns or deposits. (2-video) A video key model that allows use of a wipe pattern, box shape, or external mask signal to prevent some undesirable portions of the key source from cutting a hole in the background. The key occurs only in the area covered by the mask pattern. Areas not covered by the mask pattern consist entirely of background video (no key). (3-binary) A code or binary sequence that is used to allow, bock, or modify specific bits that are being transferred through a system or assembly. Binary masks may use AND, OR, NOT, XOR logical operators to block, allow to pass, or modify the binary number.

Mask Invert-A video keyer mode similar to mask except that the sense of the mask is inverted so that the key appears in the area not covered by the mask pattern. The area covered by the mask pattern will consist entirely of back-ground video (no key).

Masker-A sound that reduces the subjective audibility of another sound. An example of a masker is the noise induced into open-plan offices to reduce worker distraction caused by speech intrusion from other work areas.

Masking-(1-semiconductor) The process of covering protected areas of a semiconductor prior to depositing materials on its surface or etching them away. (2-OTDR) A process by which the detector circuit In an optical time-domain reflectometer is shielded from high-power return pulses. (3-sound) The reduction in subjective audibility of one sound by another interfering sound.

Masking Level-The subjective raising of the audibility threshold, in decibels, for a given sound by another sound.

MASQ-IP Masquerade

Mass-The ratio of the force applied to an object in ratio to its resulting acceleration. Measured in kg. Proportional to but not equal to weight, although the two terms are loosely used as synonyms in everyday speech. See also weight.

Mass Announcement Network Service-A service that enables many telephone callers to access a selected, prerecorded message simultaneously. Also called Information Access Service.

Mast-A guyed structure meant to support one or more antennas.

Master-(1-media) An original recording data file, videotape, or audiotape. (2-system) A device within a system that is used to coordinate other devices. It is possible for devices within the network to change roles and become a master. An example of this is the Bluetooth system where the master coordinates the other devices within the Piconet.

Master Clock-An accurate timing device that generates a synchronous signal to control other clocks or equipment.

Master Oscillator-A stable oscillator that provides a standard frequency signal for other hardware and/or systems.

Master Slave Relationship-A relationship within a communication session that assigns the control coordination to the device that assumes the role of master. The master slave relationship can be permanently or dynamically assigned. The dynamic assignment of a master slave relationship is necessary in communication systems where each device can provide similar functions as the other.

This diagram shows two personal digital assistants (PDA) that establish a master slave relationship to allow communication. In this example, PDA 1 requests to send an electronic business card to PDA 2. In this example, PDA 1 attempts to establish a communication session with PDA 2 where PDA 1 is the master. PDA 2 is in the listening mode (typical when it is idle) and hears the request from PDA 1 to establish a communication session. PDA 2 accepts the slave role and follows the commands provided by PDA 1 to allow the transfer and acknowledgement of the business card data transfer. After the transfer is complete, PDA 2 requests to send an electronic business card to PDA 1. This time, PDA 2 attempts to establish a communication session with PDA 1 where PDA 2 is the master. PDA 1 is in the listening mode and hears the request from PDA 2 to establish a session. PDA 1 accepts the slave role and follows the commands provided by PDA 2 as the business card information is transferred.

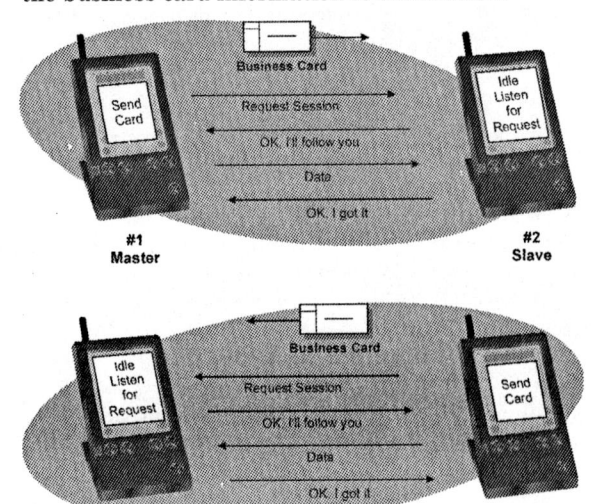

Master Slave Relationship

Master Station-A station in a multiple address radio system that controls, activates, or interrogates four or more remote stations. Master stations performing such functions may also receive transmission from remote stations.

Master/Slave-(1-video editing) A system in which one or more video tape recorders (VTR) slaves are controlled by another VTR master. (2-sync generator) A system in which several video sync generators (slaves) are controlled by one main sync generator (master). (3-Bluetooth) A relationship between devices where one device coordinates communication (the master) and the other device (the slave) follows the commands of the master.

MAT-Meridian Administration Tools

Matched Termination-A termination of the same impedance as the conducting medium that absorbs all incident power, thereby producing no reflected waves or match loss.

Matching-The connection of channels, circuits, or devices in a manner that results in minimal reflected energy.

Material Dispersion-A characteristic of fiber optic transmission in which the velocity of light through a glass fiber varies with the wavelength of the transmitted signal. Material dispersion can impair the bandwidth, information carrying capability and maximum transmission distance of the system by smearing or broadening digital pulses.

Material Scattering-In an optical waveguide, that part of the total scattering attributable to the properties of the materials used for waveguide fabrication.

Matrix-(1-general) A logical network configured in a rectangular array of intersections of input/output signals. (2-disk manufacture) Nickel electroplated onto a lacquer master, forming a negative image of it, and from which the metal ~' is produced. Matrix in this usage also is called a metal master. (3-electronics) A routing or switching array with multiple inputs and outputs. (4-mathematics) An arrangement of numbers representing the coefficients in simultaneous linear equations. (5-microphone technique) A circuit combining a unidirectional and a directional microphone into M-S stereo. (6-optical recording) A method of recording for playback channels onto two discreet optical tracks, also referred to as a 4-2A matrix. (7-TV receiver) A circuit that combines the luminance and color signals and transforms them into individual red, green, and blue signals. In a TV set, these signals then are applied to the picture tube grids.

Matrix Switch-A switching and control system that automatically shifts the flow of data from failed lines or vices into functioning equipment A matrix switch connects to both the front end processor of a computer network and the transmission lines that connect with remote sites.
Matte-A solid color video signal that can be adjusted for chroma, hue, and luminance to till of keys and borders.
MAU-Media Attachment Unit
MAU-Medium Attachment Unit
Maximum Busy Hour-The busiest hour of the busiest day of a normal week, excluding holidays, weekends, and special event days.
Maxwell's Equations-Four differential equations that relate electric and magnetic fields to electromagnetic waves. The equations are a basis of electrical and electronic engineering. Named for the Scottish physicist James Clerk Maxwell (18314879).
MB-Megabyte
MBCP-Master Business Continuity Professional
Mbit-Megabit
MBONE-Multicast Backbone
MBps-Mega Bytes Per Second
Mbps-Mega Bits Per Second
MBX-Mailbox
MC-Message Center
MC-Multicarrier Mode
MCA-Media Control Architecture
MCA-Microchannel Architecture
MCC-Mobile Country Code
McCaw Consent Decree-The proposed Consent decree filed on July 15, 1994, in the antitrust action styled U.S. v. AT&T Corp. and McCaw Cellular Communications, Inc. Civil Action N0. 94-01555, in the U.S. District Court for the District of Columbia. Such terms includes any stipulation that the parties will abide by the terms of such proposed consent decree until it is entered and any order entering such proposed consent decree.
MCF-Message Confirmation Frame
MCI-Media Control Interface
MCI-Malicious Call Identification
MCNS-Multimedia Cable Network System
Mcommerce-Mobile Commerce
MCP-Microsoft Certified Professional
MCSD-Microsoft Certified Solutions Developer
MCSE-Microsoft Certified Systems Engineer
MCT-Microsoft Certified Trainer
MCU-Multipoint Control Unit

MD5-Message Direct Number 5 algorithm
MDF-Main Distribution Frame
MDF-Market Development Funds
MDS-Multipoint Distribution System
MDS-Multipoint Distribution Service
MDS-Messaging Delivery Service
MDT-Mobile Data Terminal
ME-Mobile Equipment
Mean-In statistics, an arithmetic average in which values are added and the sum divided by the number of such values.
Mean Opinion Score (MOS)-Mean opinion score (MOS) is a measurement of the level of audio quality. The MOS is number that is determined by a panel of listeners who subjectively rate the quality of audio on various samples. The rating level varies from 1 (bad) to 5 (excellent). Good quality telephone service (called "toll quality") has a MOS level of 4.0.
Mean Output Power-The calorimetric power measured during the active part of transmission.
Mean Power-The power at the output terminals of a transmitter during normal operation, average over a time sufficiently long compared with the period of the lowest frequency encountered in the modulation.
Mean Time Between Failures (MTBF)-For a particular time period (typically rated in hours), the total functioning lifetime of an assembly or item divided by the total number of failures for that item within the measurement time interval.
Measured Load-The load that is indicated by the average number of busy servers in a group over a given time interval.
Measured Rate Service-A usage sensitive telephone service for which a customer pays a reduced monthly charge in exchange for a set amount of service. Usage beyond the set amount is billed at a specified rate.
Measurement-(1-general) A procedure for determining the amount of a quantity. (2-data) The output of a data collection system that indicates the load carried or service provided by a group of telecommunications servers.
MECABS-Multiple Exchange Carrier Access Billing
Mechanical Protection-The use of outside cable plant hardware to protect optical or copper cable sheath from manmade, animal or weather-related damage. Examples of this are tree guards, u-guards, and air vents.

Mechanical Splice-An optical waveguide splice accomplished by external fixtures or materials rather than by thermal fusion. Index-matching material may be applied between the two fiber ends.

Mechanized Loop Testing-An automated testing system that verifies the condition of a loop and identities any problems. The analysis of measured values is made available automatically for repair service personnel.

Media Access Control (MAC) Address-A physical layer address of a network element that is used to identify devices or assemblies within a network. For an Ethernet system, the MAC address is 48 bits (6 bytes) long. A MAC address is commonly called a hardware or physical address.

Media Binding-Creating a relationship of one or more media signals to a communication session. An example of media binding is synchronizing digital audio with a digital video signal.

Media Format-A method of containing audio, video, and/or other digital media within a file structure. Media formats are usually associated with specific standards like MPEG video format, or software vendors like Quicktime MOV format or Windows Media WMA format. In a few cases like MP3 files, the media "format" is little more than a single codec bitstream in a file. However in most cases the media format is not to be confused with the codecs used for any compressed bitstreams within specific files of that format.

Media Gateway (MG)-A network component which converts one media stream to another. In IP telephony this most commonly refers to a device which converts IP streams (such as audio) to the TDM or analog equivalent. A media gateway may interact with call controllers, proxies, and softswitches via proprietary or standard protocols such as MGCP, Megaco (H.248), and SIP.

There are two main types: Access gateways provide regular analog or primary rate (PRI) interfaces to a voice-over-packet (VoP) network. The inverse function is also available in VoB (voice over broadband) applications: calls are encoded digitally before entering the access network and are routed via conventional telephony once inside. Trunking gateways interface directly between the telephone network and a voice over packet (VoP) network in the core. Such gateways typically manage large numbers of digital virtual circuits.

This diagram shows the functional structure of a media gateway (MG) device. This diagram shows that this gateway interfaces between a public telephone network line side analog connection to a Internet packet (IP) data network connection. The overall operation of the voice gateway is controlled by a media gateway controller (MGC.) The MGC section receives and inserts signaling control messages from the input (telephone line) and output (data port). The MGC section may use separate communication channels (out-of-band) to coordinate call setup and disconnection.

Signals from the public telephone network pass through a line card to adapt the information for use within the media gateway. This line card separates (extracts) and combines (inserts) control signals from the input line from the audio signal. Because this audio signal is in analog form (another option could be an ISDN digital line side connection,) the media gateway converts the audio signal to digital form using an analog to digital converter. The digital audio signal is then passed through a data compression (speech coding) device so the data rate is reduced for more efficient communication. This diagram shows that there are several speech coder options to select from. The selection of the speech coder is negotiated on call setup based on preferences and communication capability of this media gateway and the media gateway it is communicating with. After the speech signal is compressed, the digital signal is formatted for the protocol that is used for data communication (IP packet.) This diagram shows that the call processing section of the media gateway is not part of the gateway. It is a separate controller that commands the gateway to insert messages in the media stream (in-band signaling) or it may communicate with the other gateway through another media gateway controller (MGC.)

Media Gateway Control (MEGACO)

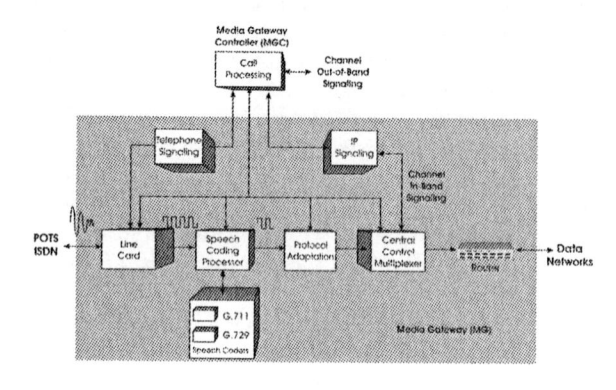

Media Gateway Operation

Media Gateway Control (MEGACO)-Media Gateway Control (MEGACO) is an IP telephony protocol that is a combination of the MGCP and IPDC protocols. MEGACO is specified in H.248.

Media Gateway Control Protocol (MGCP)- MGCP is a control protocol that uses text or binary format messages to setup, manage, and terminate multimedia communication sessions in a centralized communications system. This differs from other multimedia control protocol systems (such as H.323 or SIP) that allow the end points in the network to control the communication session. MGCP is specified in RFC 2705 and it was first drafted in 1998. MGCP forms the basis of the PacketCable NCS protocol.

Media Gateway Controller (MGC)-The media gateway controller is the portion of a PSTN gateway that acts as a surrogate call management system (CMS). The MGC controls the signaling gateway and the media gateway (MG). The protocols between the MGC and MG include media gateway control protocol (MGCP), MEGACO, and H.323. The MGC acts as a call agent coordinating sessions between devices. Signaling between MGCs (agents) may use SIP or H.323 protocol.

Media Hub-A communication device that distributes or adapts multiple types of communication media to one or several devices in a network through the re-broadcasting of data that it has received from one (or more) of the devices connected to it.

Media Player-A software application for playback of digital media such as audio and/or video. Media players may contain support for many media formats, many codecs, and local file playback as well as multiple network streaming protocols.

Media Processing-The processing of types of media such as playback of voice messages, recording of video, fax generation from computer screen, and speech recognition.

Media Relay-Media Relays are servers that provide distributed points of service for media that is rich with content (large quantities of data). They are placed in strategic locations, such as on the "edge" of networks, in the "middle mile" between the media source and destination (often consumer) devices. They are also used to address network address translation (NAT) and firewall traversal issues associated with VoIP traffic.

Media Server (MS)-The media server is an application server that provides common telephony features and/or specialized telephony capabilities. Examples of Media Servers include: Announcement Server: The announcement server provides network announcements to callers (e.g. "all circuits are busy"), Conference Server: The conference server provides support for n-way calling (e.g. 3-way calling), Voicemail/Unified Messaging Server: The VM/UM server provides support for combined voice and email communications, and CALEA server: As part of the U.S. Congress's Communications Assistance for Law Enforcement Act (CALEA) of 1994, the CALEA server enables service providers to meet legal requirements for lawful call intercept, providing call-identification information and/or call content to law enforcement based on court order.

The media server's central role is to process and manage media resources for the application server, and usually include both call processing and media processing. In addition to announcements, key functions include IVR, speech recognition, fax and video. Media servers come in two classes - hardware-based and software-based, with the former being dominant, especially among service providers.

Legacy-based media servers have provided announcements for decades, but IP media servers have recently evolved into a more integral nextgen network role. The media server's central role is to process and manage media resources for the application server, and usually include both call processing and media processing. In addition to announcements, key functions include IVR, speech recognition, fax and video. Media servers come in two classes - hardware-based and software-based, with the former being dominant, especially among service providers.

Mediation Device-A network device in a telecommunications network that receives, processes, reformats and sends information to other formats between network elements. Mediation devices are commonly used for billing and customer care systems as these devices can take non-standard proprietary information (such as proprietary digital call detail records) from switches and other network equipment and reformat them into messages billing systems can understand.

This figure shows a mediation system that takes call detail records from several different switches and reformats them into standard call detail records that are sent to the billing system. This diagram shows the mediation device is capable of receiving and decoding proprietary data formats from three different switch manufacturers. The mediation device converts these formats into a standard call detail record (CDR) format that can be used by the billing system.

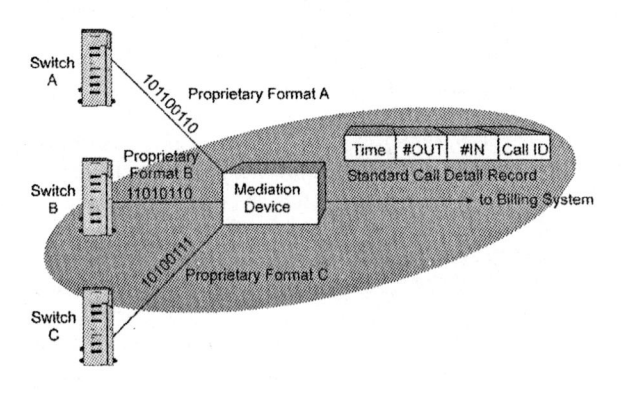

Mediation System Operation

Medium-(1-general) An electronic pathway or mechanism for passing information from one point to another. (2- data storage) A material or device on which data can be stored, such as magnetic tape. (3- transmission system) The structure or path along which a signal propagates, such as a wire pair, coaxial cable, waveguide, optical fiber, or radio path.

Medium Access Control (MAC)-A process used by communication devices to gain access to a shared communications medium or channel. Examples of MAC systems CSMA/CD and Token Passing. A MAC protocol is used to control access to a shared communications media (transmission medium) which attaches multiple devices. The MAC is part of the OSI Data-Link Layer. Each networking technology, for example Ethernet, Token Ring or FDDI, have drastically different protocols which are used by devices to gain access to the network, while still providing an interface that upper layer protocols, such as TCP/IP may use without regard for the details of the technology. In short, the MAC provides an abstract service layer that allows network layer protocols to be indifferent to the underlying details of how network transmission and reception operate.

The MAC protocol also defines the frame format, bit ordering and other characteristics to maintain a reliable network.

This diagram shows the key ways networks can control data transmission access: non-contention based and contention based. This diagram shows that non-contention based regularly poll or schedule data transmission access attempts before computers can begin to transmit data. This diagram shows that a token is passed between each computer in the network and computers can only transmit when they have the token. Because there is no potential for collisions, computers do not need to confirm the data was successfully transmitted through the network. This diagram also shows contention based access control systems allow data communication devices to randomly access the system through the sensing and coordination of busy status and detected collisions. These devices first listen to see if the system is not busy and then randomly transmit their data. Computers in the contention-based systems must confirm that data was successfully transmitted through the network, because there is the potential for collisions.

Medium Access Control [MAC] Address

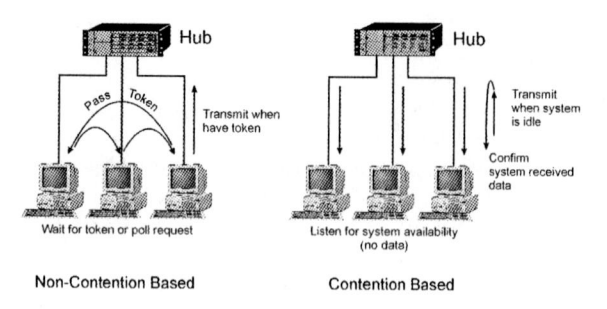

Medium Access Control (MAC) Operation

Medium Access Control [MAC] Address-The medium access control (MAC) address used to distinguish between units participating in a data network. MAC addresses are low-level address and are only associated with the MAC layer of the system that the data device is operating in.

Medium Earth Orbit (MEO)-A Mobile Satellite Services (MSS) system where the satellite(s) have orbit heights that range from about 1,000 to 6,500 miles above the Earth.

This Figure shows a MEO satellite system. In this diagram, several satellites circle the earth at several thousand miles per hour. In this example, a landline telephone is communicating with a portable satellite telephone. The telephone call is routed through the gateway to the satellite. The satellite transponder converts the frequency and retransmits the signal back to earth. The portable satellite telephone receives the radio signal and converts it back to the original audio signal

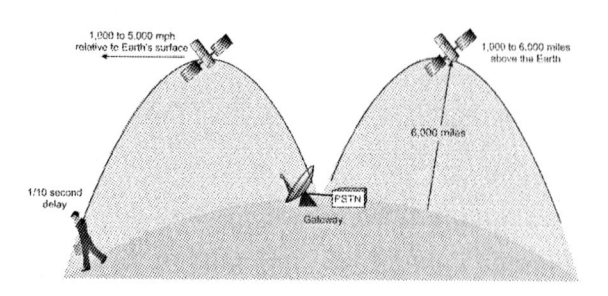

Medium Earth Orbit (MEO) Operation

Medium Frequency (MF)-The frequency band between 300 kHz and 3000 kHz (3 MHz). The wavelength of the MF band ranges from 10 m 100 m. The MF band also is known as medium wave.

Meet Me Conference-A telephone conference call arrangement, usually on a private PBX system that enables callers to use an access code to connect to a specific conference call.

Meet Point-A point, designated by two exchange carriers, at which one carrier's billing responsibility for service begins and the other's ends. There can be one or more meet points on a circuit.

Meet Point Billing (MPB)-Billing systems that must meet when unbundled network elements (UNE) or access services are provided by two or more providers, or by one provider in more than one state.

Mega (M)-A prefix meaning one million.

Mega Bits Per Second (Mbps)-A measurement of digital bandwidth where 1 Mbps =1 million bits per second (1,000,000 bits per second). The word "mega" is sometimes used to describe the nearest integral power of 2, namely 1,048,567.

Mega Bytes Per Second (MBps)-A measurement of the amount of information being transferred on a communications link in one second where 1 MBps =1 million bytes per second (1,000,000 8 bit bytes per second).

Mega Flops-An acronym for millions of floating point operations per second, a figure of merit for the processing speed of a computer system.

Megabyte (MB)-One million bytes, that is 1,000,000 bytes. Some authors use the word "mega" to describe the nearest integral power of 2, namely 1,048,567.

MEGACO-Media Gateway Control

Megahertz (MHz)-One million hertz, or cycles per second.

Mel-The subjective unit of pitch; 1000 mels is the apparent pitch of a 1 kHz tone 40 dB above the threshold of audibility.

Memorandum Of Understanding (MOU)-(1-general) A statement or agreement stating the objectives and scope of a project or plan. (2-GSM) A legal agreement between the GSM committee members to create the GSM network and its revisions. The first MOU was signed in 1987.

Memory Capacity-The total number of bits or bytes that can be stored in a device or assembly within a device.

Memory Effect-A condition in a battery where it memorizes its charge cycles. The memory effect is primarily caused when one or more of the charge cycles are not complete. Memory effect is present when the battery is provided with a full charge time interval and the battery will only supply energy for a lesser time due to the charge cycle. Memory effect was a significant problem for NiCd batteries until the early 1990's when the construction of NiCd batteries was changed that virtually eliminated the memory effect.

Memory IC-An integrated circuit incorporating from several hundred to several million logic cells for storing digital information.

Memory Management-The process used in a computing system that manages the assignment and use of memory storage. Memory management systems may involve the allocation of electronic memory (e.g. RAM and ROM), hard disk memory, and other memory storage systems (e.g. removable disk or tape).

Menu Screen-A computer monitor screen format that lists a set of options from which users make a choice.

MEO-Medium Earth Orbit

Meridional Ray-A ray that crosses the optical axis of an optical waveguide.

MES-Mobile Earth Station

MESFET-Metal Semiconductor Field Effect Transistor

Mesh Network-A data network where each communication device (typically computers) are interconnected to the other computers in the network.

Message-(1-general) Any idea expressed briefly in a plain or secret language and prepared in a form suitable for transmission by any means of communication. (2-data communications) A set of information, typically digital and in a specific code (such as binary or ASCII), carried from a source to a destination. (3- ISDN) In the Integrated Services Digital Network (ISDN), a set of layer 3 information that is passed between customer premises equipment and a stored-program control switching system for signaling. (4-telephone communications) A communication session or a successful call attempt.

Message-Frame

Message Body-The data information or message words that are contained in a communication message.

Message Center (MC)-A new node on the communications network which accommodates messages sent and received via short messaging service (SMS).

Message Center Time Stamp-In SMS, a feature that informs the recipient of the local time and date of when the message was accepted by the message center.

Message Discrimination-The process which decides for each SS7 incoming message, whether the signaling point is a destination point or if it should act as a signal transfer point (STP) for that message.

Message Distribution-The process of determining which User Part the signaling message is to be delivered to in an SS7 system.

Message Format-The rules for the placement of such portions of a message as its heading, address, text, and end.

Message Investigation-A generic term used to describe the processes and group(s) responsible for investigating call detail records (CDRs) that are rejected by the rating engine portion (rate selection and cost allocation) of a billing system.

Message Minute Miles-The total number of message minutes carried on a trunk or circuit group, multiplied by the average route miles.

Message Minutes-The connection time used by a customer for transmission.

Message Processing System (MPS)-The function within a billing system that processes the events recorded in the network. MPS is sometimes referred to as the Rating Engine. Typically the Rating Engine receives events from the network, reformats each event into an internal standard, identifies the customer to be billed (see "Guiding"), and assigns a rate (see "Rating") based on parameters such as: date, day-of-week, rate period, call type, jurisdiction, an others.

Message Recording-A message coding system used to distinguish customer dialed messages from operator completed messages.

Message Retrieval-The process of locating a message that has been entered in a telecommunications system.

Message Signal Unit (MSU)-A signal unit (data packet) that carries the signaling information (messages) that are transmitted through an SS7 network. This MSU packet contains control flags (fields) that indicates the protocol that is being transmitted (e.g. mobile application part or ISDN user part) along with a variable length information (message content) field.

Message Suffix

This diagram shows that the MSU in the SS7 system is a variable length SIF field that allows the MSU to carry many types of signaling packets. These include SCCP, ISDN-User Part, and OMAP messages.

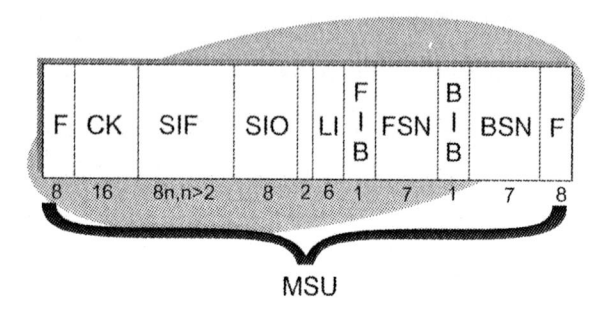

F	CK	SIF	SIO	LI	FIB	FSN	BIB	BSN	F
8	16	8n,n>2	8	6	1	7	1	7	8

MSU

Message Signal Unit (MSU) Structure

Message Suffix-In data transmission, the character indicating the end of a message.

Message Switching-A transmission method in which messages are sent to an intermediate point, stored temporarily, and transmitted later to a final destination. The destination of the message usually is indicated in an internal address field (or header) in the message itself.

Message Transfer Part (MTP)-The functional part of a common channel signaling system which transfers signaling messages as required by all the users. The message transfer part also contains, for example, error control and signaling security.

Message Type-A message that is defined according to how it is billed and the way in which it is paid. Message types include: sent paid noncom, third number, credit card, collect, special collect, coin sent paid, and coin collect. Message type also is called message class.

Message URL-An address that is preprogrammed into or used by devices (such as an IP Telephone) where information is kept regarding messages. The message URL may be associated with a message button or an icon on the display of an IP telephone.

Message Waiting Indicator (MWI)-A feature that informs a user that they have messages waiting in email, voice mail, or video mail. Optionally it may indicate how many mail messages are waiting without the user having to call their voice mailbox. MWI may use unique tone (rapidly changing dial time) or an indication on the telephone device (such as a light) as an indication of message waiting. MWI should not impact a subscriber's ability to originate calls or to receive calls. If the dial tone is altered to indicate a message is waiting, it will typically reset to a standard dialtone after a time period or after the user has re-established the connection. As a result, MWI may affect auto-dialers (such as modems) that sense for a dialtone signal before sending the dialed digits. Message waiting indication is also called message waiting notification (MWN)

Messaging-A telephone system feature that alerts station users, via a lighted lamp or other visual display, or by an interrupted dial tone, that messages are waiting.

Messaging Application Programming Interface (MAPI)-A standard program interface that was developed by Microsoft to allow the transfer of messages between software applications.

Messenger-A wire that runs in parallel to the conductors which acts as a supportive strand for the attached multi-pair cable in self-supporting cable or dropwires. It also can serve as a bonding or grounding facility on either end.

Metadata-Data that describes the attributes of other data. Commonly used in databases where the attributes of a particular column of data are defined. For example: the phone number column of the employee data table contains groups of numbers 3 or 4 characters in length separated by a -.

Metal Film Resistor-A resistor made of a metallic material deposited on film.

Metal Oxide Semiconductor (MOS)-A device whose operating characteristics are determined by conditions at the interface between a semiconductor layer, usually silicon, and an insulator layer, usually silicon dioxide.

Metal Oxide Varistor-A solid-state voltage damping device used for transient suppression applications.

Metal Rectifier-A rectifying device using metallic plates in contact with plates of other materials, typically copper on selenium or copper on copper oxide.

Metal Tape-A magnetic media formulation of fine, densely packed iron, cobalt, and nickel metal particles layered onto a thin base film. This construction improves performance, relative to conventional tape formulations. There are two types of metal tape: metal powder tape (a metal powder alloy consisting of cobalt and nickel mixed with binders and lubricants, and disbursed onto a base film); and metal evaporated tape (a metal powder alloy consisting of iron, nickel, and cobalt evaporated onto the base film).

Metastable-An energy level that is between stable and unstable. The metastable state is critical to the operation of a laser. In a laser, electrons in a stable electron state have their energy pumped up from an external source to an unstable level called the pump level. They rapidly lose some or all of that energy to heat. Those that lose all of it drop to the stable level. Those that lose part of it drop to the metastable level where they are trapped until a photon comes along and helps one to lose the rest of its energy in the form of a photon so it can drop to the stable level as well.

Meteor-A metallic or stone body that enters the Earth's atmosphere, burning up during entry. Meteors produce ionization in the atmosphere that can affect long distance radio communications.

Meteor Burst Link-A radio link capable of operating over long distances (2000 km) using the presence of meteor trails (about 100 km above the Earth's surface) to reflect radio transmissions back to earth, using frequencies in the 40 MHz to 50 MHz band. About 109 meteors of different sizes enter the atmosphere daily; transmission, which takes place in bursts when a suitable meteor trail has been located, is therefore fairly reliable.

Meter-(1-SI unit) The meter of length is now precisely defined as the length of the path traveled by light in vacuum during a the time interval 1/ 299 792 458 of a second. It can be equivalently defined as 1 650 763.73 wavelengths of the orange-red light from the isotope krypton-86, measured in a vacuum. Historically, the meter was 1/40 000 000 of a meridian of longitude passing through the poles, as determined in 1797 by a land survey of a partial meridian distance between Barcelona, Spain and Dunquerque, France. (2-ampere-hour) A device that integrates current and time to indicate the number of ampere-hours of power consumed by a load. (3-field strength) A combination radio receiver and meter (calibrated for use with a particular antenna) designed to give a direct reading of the strength of a radio signal at a given point. (4- instrument) A device for measuring the value of some quantity. (5- running time) A totaling clock that runs whenever a device is in operation. Such meters are used with various types of equipment so that maintenance work can be carried out at appropriate times.

Metric System-A decimal system of measurement based on the meter, the kilogram, and the second.

Metropolitan Area Network (MAN)-A MAN is a data communications network or interconnected groups of data networks that have geographic boundaries of a metropolitan area. The network is totally or partially segregated from other networks, and typically links local area networks (LANs) together.

This diagram shows a five node MAN connecting that connects several LAN systems via a FDDI system. This diagram shows that each LAN may be connected within the MAN using different technology such as T1/E1 copper access lines, coax, or fiber connections. In each case, a router provides a connection from each LAN to connect to the MAN.

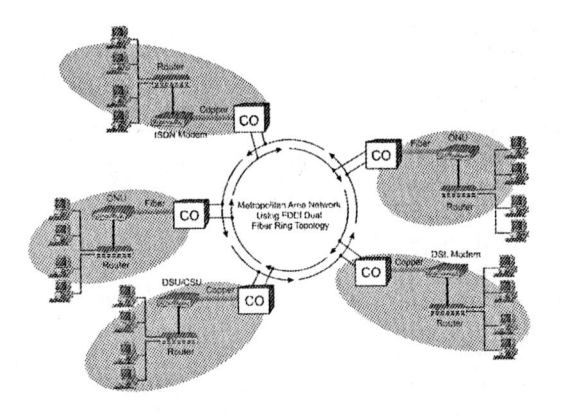

Metropolitan Area Network (MAN) System

Metropolitan Fiber Ring-A fiber optic network that provides high-speed local network capabilities for the connection of businesses and residences to long-distance carrier networks. The ring topology is used in the metropolitan area as it provides a protection switching capability that allows traffic to be quickly rerouted in the event of a fiber cut. SONET is the most widely deployed metropolitan ring architecture, but newer packet based technologies, such as IEEE 802.17 and other resilient packet rings are being deployed.

Metropolitan Service Area (MSA)

Metropolitan Service Area (MSA)-Metropolitan Service Area or Metropolitan Statistical Area. An area designated by the FCC for service to be provided for by cellular carriers. There are two service providers for each of the over 300 MSA's in the United States.

Metropolitan Statistical Area (MSA)-A geographic area in the United States that typically includes a city that has a population of at least 50,000 people that have similar economic and social characteristics. MSAs are defined by the federal government and used to gather and report statistics.

MF-Medium Frequency
MF-Multifrequency Signaling
MFC-Multifrequency Compelled
MFJ-Modification Of Final Judgment
MFS-Macintosh File System
MFSK-Multiple Frequency Shift Keying
MG-Media Gateway
MGA-Media Gateway Agent
MGC-Media Gateway Controller
MGCP-Media Gateway Control Protocol
MH-Modified Huffman Code
mho-Informal but widely used name for the unit of electrical conductance, the reciprocal of the ohm. ("ohm" spelled backwards) Equal to a siemens.
MHS-Message Handling Service
MHz-Megahertz
MIB-Management Information Base
MIB Browser-Management Information Base Browser
MIB Instance-A MIB instance is a suffix identifier associated with a particular MIB object. Usually the instance has a value of "0" when the MIB object is to return 1 value. The instance suffix identifier can increment as well like in gathering interface statistics, where there can be more than 1 interface and many values for 1 object. For example: In the case of the ifInOctets MIBobject it would look like: 1.3.6.1.2.1.2.2.1.10.<instance number> Where the instance number is the interface number for which you want the received octets for, so if you wanted the first interface then the MIB object definition would look like: 1.3.6.1.2.1.2.2.1.10.1
MIB-II-Management Information Base II
MIC-Message Integrity Check
MICR-Magnetic Ink Character Recognition
Micro Channel-A personal computer bus architecture introduced by IBM in some of its PS/2 series microcomputers. Micro Channel is incompatible with original PC/AT ISA architecture. Micro Channel is a registered trademark of IBM.

Microbend-A small bend in an optical fiber due to damage inflicted during installation or other handling of the fiber or to changes in its environment. If the bend is severe enough, light traveling through the core may impact the cladding at an angle beyond the critical angle, resulting in loss of light.

Microbend Loss-In an optical waveguide, that portion of the total loss attributable to microbending.

Microbending-In an optical waveguide, a sharp curving involving local axial displacement of a few micrometers, and spatial wavelength of a few millimeters. Such bends may result from waveguide coating, cabling, packaging, or installation.

Microbrowser-An Internet Browser designed for small display screens on smart phones and other handheld wireless devices.

Microcell-A radio coverage area that has a radius of between 200 feet and 1,000 feet.

Microchip-A common term for an integrated circuit component.

Microcode-A set of control functions that are performed by the instruction decoding and execution logic of a computer and define the instruction repertoire of that computer.

Microcomputer-A small-scale program or machine that processes information; it generally has a single chip as its central unit and includes storage and input/output facilities in the basic unit.

Microcracks-Submicroscopic flaws in the surface of glass fibers.

Microelectronics-A technology used to build integrated circuits and other small electronic devices and components, sometimes referred to as microminiaturization.

Microinstruction-An instruction of a micro-program.

Micrometer-One-millionth of a meter. Usually abbreviated as um.

Micron-A unit of length equal to one millionth part of a meter. Also called a micro-meter, written μm (not to be confused with a measuring device called a micrometer).

Microphone-A transducer that converts sound waves into electrical signals.

Microphonics-Undesirable noise introduced into an audio or video system by mechanical vibration of electric components.

Microportable-A lightweight handheld cellular or PCS mobile telephone.

Micropositioner-A device used to hold and align small parts, such as integrated circuits or optical fibers.

Microprocessor-A single package (normally a single chip) electronic logic unit capable of executing from external memory a series of general-purpose instructions contained in the external memory. The unit does not contain integral user memory although memory on the chip may be present for internal use by the device in performing its logic functions.

Microsecond (usec)-One millionth of a second.

Microvolts Per Meter (MPM)-A measure of the field intensity of a radio signal.

Microwatt (mW)-One millionth of a Watt.

Microwave-(1-radio) The portion of the electromagnetic spectrum between approximately 1 GHz and 100 GHz. (2-heating device) A radio oven that operates at approximately 2.4 GHz..

Microwave Dish-An antenna system that uses a parabolic-shaped reflector to reflect received signals to a specific focal point (signal feed element). Because of the high signal gain and the requirement for precise positioning, dish antennas are commonly used for transmission and reception from point-to-point microwave stations and fixed position GEO communications satellites.

This picture shows the Compact line model type of microwave dish. This microwave antenna is used in point-to-point microwave communication links.

Microwave Dish
Source: Radio Frequency Systems

Microwave Frequency-Any of the frequencies suitable for microwave communication. The most commonly used microwave frequencies in use range from approximately 1-10 GHz.

Microwave Landing System (MLS)-An instrument landing system operating in the microwave spectrum that provides lateral and vertical guidance to aircraft having compatible avionics equipment.

Microwave Link-A microwave link uses microwave frequencies (above 1 GHz) for line of sight radio communications (20 to 30 miles) between two directional antennas. Each microwave link transceiver usually offers a standard connection to communication networks such as a T1/E1 or DS3 connection line. This use of microwave links avoids the need to install cables between communication equipment. Microwave links may be licensed (filed and protected by government agencies) or may be unlicensed (through the use of low power within unlicensed regulatory limits).

The photo seen here is the Microstar microwave radio. This picture shows a small directional antenna that is connected to a 40 GHz transceiver for point-to-point microwave communication.

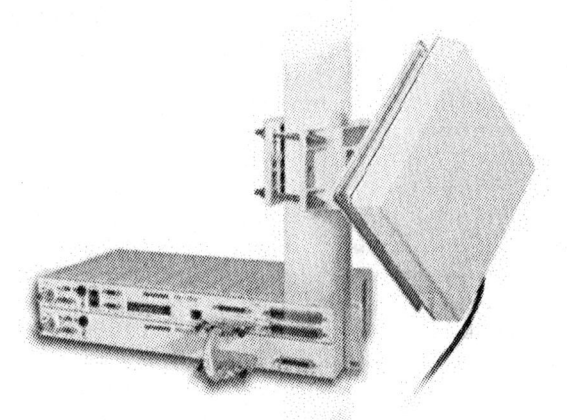

Microwave Radio
Source: Harris Microwave Division

Mid Air Meet-The point midway between separately owned radio transmission facilities at which responsibility for the radio system changes from one company to another.

Mid Section-In a lumped cable loading system, the middle of a load section between loading coils.

Mid Span Meet-Carrier spans of wire or optical fiber whose ownership changes at a demarcation point generally located at a terminal site, but sometimes located at a repeater or splice point. The spans are jointly maintained by its two owners.

Mid-Side Encoding-In stereo audio processing, an alternate representation of the stereo signal by taking the sum of right and left channels (the mid channel) and the difference of left and right channels (the side signal). Especially for audio compression, it is advantageous to encode the mid and side channels separately due to pychoacoustic effects (for example the complete loss of the side channel merely causes the signal to become monaural). This is analogous to the use of YUV color representation in image coding.

Middleware-Software that operates between the core application layer of a system and a lower layer of the network. An example of middleware is electronic programming guide (EPGs) that reside in cable converter boxes that allow a customer to select from a list of available video programs.

MIDI-Musical Instrument Digital Interface

Midrange-Frequencies in the range spanned by the human voice, from approximately 200 to 2000 cycles per second.

Midspan-In aerial telephone company (telco) outside cable plant, the midpoint between two telephone poles. Also in relation to digital loop carrier copper plant, any point on either transmit or receive pairs between repeater locations.

Mileage Band-A group of individual mileage steps, such as from 0 to 50 miles, or 50 to 100 miles, measured in airline miles and used to determine billing rates for telecommunications services. This billing rating process is also called "banding."

Milli (m)-A prefix used in a unit of measure meaning one thousandth.

Milliammeter-A measurement instrument that is used to quantify the amount of electric current flowing in a circuit.

Millihenry-A unit of measure equal to one thousandth of a Henry.

Million Instructions Per Second (MIPS)-A unit of measure for the millions of processing steps that can be accomplished by a computer processing device in one second.

Millisecond (msec)-One-thousandth of a second (0.001 S).

Millivolt (mV)-One-thousandth of a volt (0.001 V).

MIME-Multipurpose Internet Mail Extensions

MIN-Mobile Identification Number

Minimum Commitment Size-In outside plant administration, the smallest number of pairs that can be committed or recommitted to a permanently connected pair group.

Minimum Cost Routing-In automated facility planning, a circuit-routing scheme that determines a path through the network for each point-to-point demand for each year so that, when point-to-point demands are provided on these paths and the resulting capacity expansion problem is solved, the total cost of transmission facilities is minimized. Minimum cost routing is not related to least cost routing.

Minimum Point Of Entry (MPOE)-A minimum point of entry is a location where communication lines first enter the customer's building or facility. An MOPE can be a network interface termination (NT) or wiring closet inside the building.

Minimum Shift Keying (MSK)-A form of frequency modulation in which the modulating signal shifts the output frequency between predetermined values. Sometimes called fast frequency shift keying.

Minimum Usable Field Strength-The minimum value of the transmitted field strength necessary to permit a desired reception quality, under specified receiving conditions, in the presence of natural and man-made noise, but in the absence of interference from other transmitters.

Ministry of Posts and Telegraph 1327 (MPT 1327)-MPT1327 is an analog trunked private land mobile radio standard that is primarily used in Europe, Asia and developing countries in other parts of the world. The MPT1327 standard can be used to implement systems with only a few radio channels (even single-channel systems) to large interconnected networks. The MPT1327 system was developed to standardize land mobile radio equipment and services. The system is primarily used for public safety applications but it can be used for cellular like services.

The system has two types of radio channels; control channels and the traffic (voice or data) channels. Control channels can be dedicated or non-dedicated. Dedicated control channels are permanently available for sending and receiving control information. Non-dedicated control channels can be dynamically converted to a traffic channel. This as a rule occurs if all the other channels are in use. Radio channels are 25 kHz wide. They have a data signaling rate of

1200 bps and the modulation type is Fast Frequency Shift Keying (FFSK) subcarrier modulation. It is designed for use by two-frequency half-duplex radio units and a duplex base station. The system can have up to 1024 channel numbers and 32768 system identity codes. The MPT1327 system allows for voice, dispatch (group call), data, emergency and messaging services. Messaging services allow up to 184 bits of text data to be sent between units or to the control center. A unique MPT1327 service is the sending of up to 32 status messages that can be sent between units. Some of these messages are pre-defined while others can be specified by a user.

Minutes Of Use (MOU)-A measurement (usually billing related) of the number of minutes, actual or assumed, in traffic-sensitive (usage) equipment and facilities.

MIPS-Million Instructions Per Second

Mirror Site-A duplicate data or Internet Web site. Mirror sites are used to process communication traffic to local or regional areas as each mirror sites contain the same information as the each other mirror site. Mirror sites are also used for mission-critical applications that allows a company or user to continue processing information in the event of system failure or network loss due to natural disasters.

Mirroring-A fault prevention architecture in which a backup data storage device maintains data identical to that on the primary device, and can replace the primary if it fails.

MIRS-Motorola Integrated Radio System

MIS-Management Information System

Misframe-An error condition in which a line signal entering a digital terminal contains a data bit pattern that simulates an actual framing pattern. Such a simulated pattern can cause a terminal that has previously lost framing to lock falsely onto the undesired pattern.

Mix-(1-video) A transition between two video signals in which one signal is faded down as the other is faded up. (2-audio) The result of a audio mixing session, wherein various inputs, often from a multi-track recording are combined into fewer tracks. This may also refer to a particular set of level of the mixing equipment.

Mix Down-In audio recording, the combining of multiple sources or tracks into a lesser number.

Mixer-(1-general) A circuit used to combine two or more signals to produce a third signal that is a function of the input waveforms. (2-audio) An audio console used to switch and combine various audio sources to produce a finished output (3- broadcast) The studio control console or other unit used to combine or "mix" the various program elements into a final program that is sent to the transmitter. (4-receiver) The stage in a superheterodyne radio re receiver at which the incoming signal is modulated with the signal from the local oscillator to produce an intermediate frequency signal. (5 - video) A European term for production switcher. The complete term is vision mixer.

This diagram shows how a mixer combines two signals to produce a sum or difference frequency. This diagram shows this mixer contains a diode (non-linear device) that allows the two-incoming signals to interact with each other to produce the difference (subtractive) frequency and sum (additive) frequencies. The output of this mixer circuit contains a tuned circuit (resonant circuit) that only allows the difference frequency to transfer out of the mixer

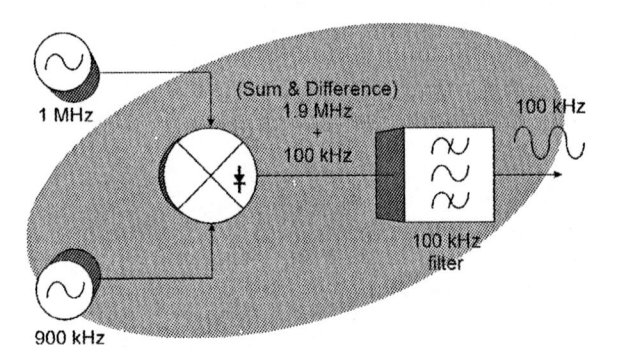

Mixer Operation

MJ-Modular Jack

MKS System of Units, see: Systém International (SI)-The Systém International or meter-kilogram-second version of the metric system.

MLID-Multiple Link Interface Driver

MLPP-Multi-Level Priority And Pre-Emption

MLS-Microwave Landing System

MM-Mobility Management

MMDS-Multichannel Multipoint Distribution Service

MMH-Multilinear Modular Hash

MMI-Man Machine Interface

MMJ-Modified Modular Jack

MMR-Modified Modified READ
MMS-Multimedia Messaging Services
MMU-Memory Management Unit
MNC-Mobile Network Code
Mnemonic-A memory aid in which an abbreviation or arrangement of symbols has an easily remembered relationship to the subject.
Mnemonic Address-A simple address code with some easily remembered relationship to the actual name of the destination, often using initials or other letters from the name to make up a pronounceable word.
MNOS-Metal, Nitride, Oxide Semiconductor
MNP-Microcom Networking Protocol
Mobile Access Hunting (MAH)-A telephone call-handling feature that causes a transferred call to "hunt" through a predetermined group of telephones numbers until finds an available ("non busy") line. The mobile systems searches a list of termination addresses sequentially for one that is idle and able to be alerted. If a particular termination address is busy, inactive, fails to respond to a paging request, or does not answer alerting before a time-out, then the next termination address in the list is tried. Only one termination address is alerted at a time. The mobile telephones in the group may be alerted using distinctive alerting. Additional calls may be delivered to the MAH Pilot Directory Number at any time.
Mobile Application Part (MAP)-A set of call processing messages, originally defined for use with GSM, for setup and control of wireless calls via the public switched telephone network. It is normally implemented in conjunction with SS7 call processing messages. The North American standard IS-41 is similar in principle but different in details.
This illustration is a functional block diagram of a wireless network and how it uses MAP protocols between the equipments. This diagram shows that the different versions of SS7 MAP are used between network elements in a wireless network. It also shows that MAP is not used in the radio link. Instead, the relative parts of MAP are transformed into commands that can be sent on the radio links.

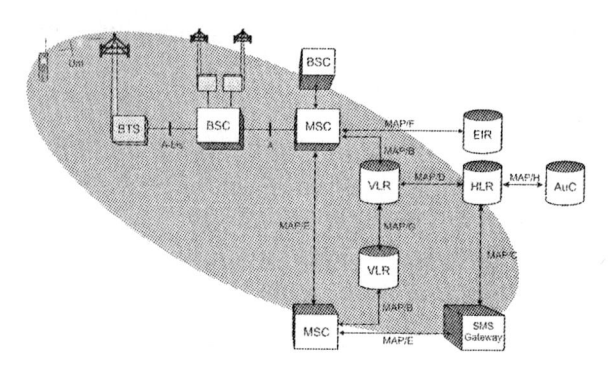

SS7 MAP Network

Mobile Assisted Channel Allocation (MACA)-In a mobile communication system (such as the IS-136 TDMA system), mobile assisted channel allocation is used to provide information such as signal-strength levels and channel quality levels at assigned frequencies during idle communication mode. During the system access procedure, the mobile phone includes a message containing these measurements. The receiving base station can use this information to assist in the channel assignment procedure.
Mobile Assisted Handoff (MAHO)-A process where a mobile phone assists in the base station in the decision to transfer the call (handoff/handover) to another base station. The mobile radio assists by providing RF signal quality information that typically includes received signal strength indication (RSSI) and bit error rate (BER) of its own and other candidate channels.
Mobile Assisted Handover (MAHO)-A process where a mobile phone assists in the base station in the decision to transfer the call (handoff/handover) to another base station. The mobile radio assists by providing RF signal quality information that typically includes received signal strength indication (RSSI) and bit error rate (BER) of its own and other candidate channels. MAHO is an official term of the GSM system.
Mobile Carriers-(1-service provider) Companies that provide mobile communication (e.g. cellular or PCS) services. (2-electrons) Free electrons and holes that move through a conductor or semiconductor.
Mobile Commerce (Mcommerce)-A shopping medium that allows wireless devices in a telecommunications network to present products to the customer and process orders.

Mobile Control Terminal-The fixed base radio station that controls a network of mobile radio stations. The control terminal sends out coded pulses to call the requested mobile, arranges for frequency/channel allocation, and provides supervisory services during a call.

Mobile Country Code (MCC)-An identity code that is assigned to identify the country of a GSM system.

Mobile Coverage Area-A geographical area that has a sufficient level of radio signal strength to allow two-way communication with mobile radios.

Mobile Earth Station (MES)-An earth station in the mobile-satellite service intended to be used for communication via satellite systems.

Mobile Equipment (ME)-(1)The equipment that the customer uses to originate and receive calls in a wireless network. (2) Radio terminal used for radio communication over the Um interface.

Mobile Equipment SIM Lock-A mobile phone feature that ensures a mobile phone will only work with one or a group of subscriber identity module (SIM) cards. The mobile phone is programmed with information that must match the information on the SIM card to enable operation. This information may include one or more of the following; IMSI, Group identifier (Service Provider identity) or PLMN identity (MCC+MNC). If a SIM card is inserted that does not match the stored information, the phone will be inoperative.

Mobile Identification Number (MIN)-The 10-digit number that represents a mobile telephones (mobile station) identity. It is divided into MIN1 and MIN2. MIN1 is the 7 digit portion of the number. MIN2 is the 3 digit area code portion of the number.

Mobile Message-A message placed to or from a mobile communications device (mobile station).

Mobile Network Code (MNC)-A code that is assigned to identify a GSM network operator in a specific country.

Mobile Originated Short Message Service (MOSMS)-The sending of a message from a mobile telephone to a system or to another user in a system.

Mobile Party Pays (MPP)-A billing process where mobile telephone user pays for the acceptance of delivery of a call to their telephone (typically a cellular or PCS network). See also Calling Party Pays.

Mobile Relay-A mobile transceiver (transmitter and receiver) that is used to repeat (retransmit) a signal received from a mobile radio. When setup as a mobile relay, the mobile radio receiver automatically activates the transmitter when a received signal of sufficient level or code is received.

Mobile Repeater Station-A mobile station authorized to retransmit automatically on a mobile service frequency, communications to or from hand-carried transmitters.

Mobile Reported Interference (MRI)-The reporting of interference signals from a mobile to its base station.

Mobile Satellite Service (MSS)-A form of wireless service that employs satellites as part of the wireless infrastructure and is capable of serving very large geographic areas. The use of MSS may be appropriate for areas that are economically not viable for land based radio towers or to provide wide area group call (dispatch type) services.

This is a picture of an Iridium model 9505 wireless phone that allows a user to communicate through satellites to standard telephones and other mobile satellite telephones. This telephone communicates with Iridium satellites that have the capability to automatically route calls between satellites to help reach their destination without the need to use International land based communication lines.

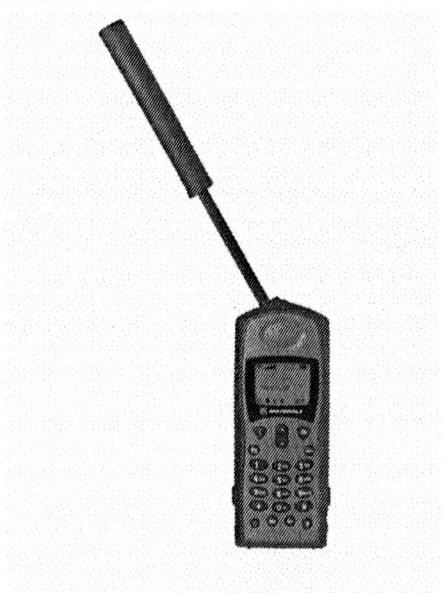

Satellite Telephone
Source: Iridium Satellite LLC

Mobile Service

Mobile Service-A service of radio communication between mobile and land stations, or between mobile stations.

Mobile Service Area (MSA)-A geographic area authorized for mobile radio service coverage that is usually divided into a number of smaller radio coverage areas (cells.)

Mobile Service Provider-A generic term used to describe companies, including cellular carriers, radio common carriers, and private carriers that provide mobile telecommunications service.

Mobile Station (MS)-A mobile radio telephone operating within a wireless system (typically cellular or PCS). This includes hand held units as well as transceivers installed in vehicles.

Mobile Station Class (MSC)-The classification of the power level capability of mobile radios. The use of a MSC code allows the processing of signals from mobile radios with different maximum power levels (e.g. high, mid-range, or a low power output) to be handled differently. The mobile radio identifies its power level capability by using the Station Class Mark (SCM) field.

Mobile Station Class Mark (MSCM)-A classification code for mobile stations that identifies their revision level, RF power capability, encryption algorithm, frequency capability and short message capability.

Mobile Station Identity (MSID)-A 34-bit number used in the IS-136 TDMA system that enables the current networking and authentication mechanisms to coexist with the digital control channel (DCCH.)

Mobile Subscriber ISDN (MSISDN)-The telephone number assigned to mobile telephones. This number is compatible with the North American Number Plan.

Mobile Switching Center (MSC)-Switching system that are used for mobile communication networks (cellular, PCS, and 3G.) The MSC was formerly called the mobile telephone switching office (MTSO).

The MSC consists of controllers, switching assembly, communications links, operator terminal, subscriber database, and backup energy sources. The controllers, each of which are powerful computers, are the brains of the entire cellular system, guiding the MSC through the creation and interpretation of commands to and from the base stations. In addition to the main controller, secondary controllers devoted specifically to control of the cell sites (base stations) and to handling of the signaling messages between the MSC and the PTSN are also provided. A switching assembly routes voice connections from the cell sites to each other or to the public telephone network. Communications links between cell sites and the MSC may be copper wire, microwave, or fiber optic. An operator terminal allows operations, administration and maintenance of the system. A subscriber database contains features the customer has requested along with billing records. Backup energy sources provide power when primary power is interrupted. As with the base station, the MSC has many standby duplicate circuits and backup power sources to allow system operation to be maintained when a failure occurs.

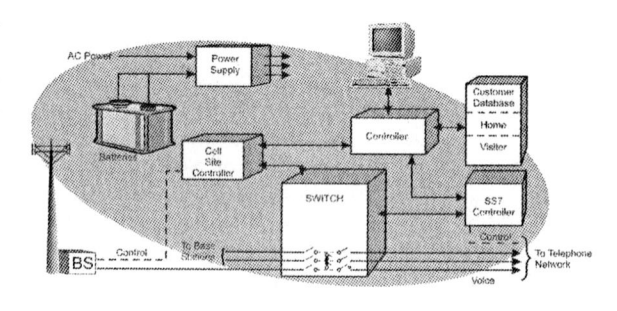

Cellular Mobile Switching Center (MSC)

Mobile System Identification Code (MSIC)-(1-GSM) The identification code of the GSM system network. The MSIC is continuously broadcast to allow mobile stations to identify the system in which they are operating.

Mobile Telephone Exchange-A switching center that controls calls in a cellular mobile radio network. A common name for mobile telephone exchange is the mobile switching center (MSC).

Mobile Telephone Service (MTS)-Mobile telephone service (MTS) is a type of service where mobile radio telephones connect people to the public switched telephone system (PSTN) or to other mobile telephones. Mobile telephone service includes cellular, PCS, specialized and enhanced mobile radio, air-to-ground, marine, and railroad telephone services. MTS was a name used for the first public mobile telephone system used in the United States.

This figure shows a mobile telephone system. The wireless network connects mobile radios to each other or the public switched telephone network (PSTN) by using radio towers (base stations) that are connected to a mobile switching center (MSC). The mobile switching center can transfer calls to the PSTN.

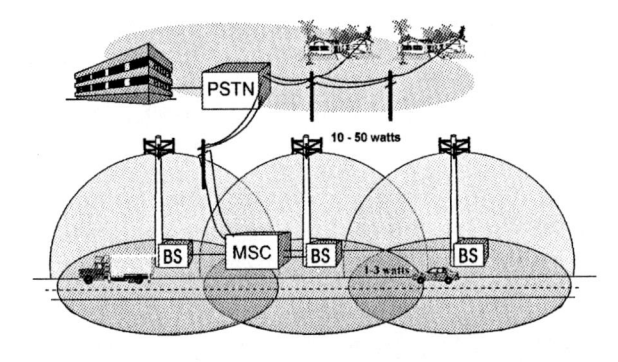

Mobile Telephone System

Mobile Telephone Switching Office (MTSO)-A cellular carrier switching system which includes switching equipment needed to interconnect mobile equipment with land telephone networks and associated data support equipment. See also mobile switching center (MSC).

Mobile Terminals-Transceivers (radios) that are mobile (as opposed to fixed position terminals.)

Mobile Terminated Short Message Service (MTSMS)-The sending of a message from a system or to another data device to a mobile telephone capable of receiving and displaying short messages.

Mobile Unit-The mobile radio transceiver that can be a handheld or car mounted transceiver. The mobile unit connects the user to the base station via RF (radio frequency). The mobile unit is also known as the "Subscriber" or "Mobile Station."

Mobile Virtual Network Operator (MVNO)-A Mobile Virtual Network Operator (MVNO) is a mobile communications service provider that resells the communication services of other wireless communication network operators. MVNO providers purchase airtime (minutes of use) in quantity and resell the airtime to customers they obtain and manage.

MVNO's may provide value added services such as information services, brand labeling, special sales support, and support of unique distribution channels. MVNO's attempt to position their services so customers do not recognize that the operator does not own a network. To provide advanced services, some MVNO operators may own network equipment that interfaces with wireless networks. This may allow MVNO's to have more control over customer databases and SIM cards.

Mobile-service Switching Center (MSC)-The central switch of a mobile or cellular radio infrastructure. Although this term is sometimes written "Mobile Switching Center" the MSC is not, itself, mobile or moveable. The older approximately synonymous terms Mobile Telephone Switching Office (MTSO) and Mobile Telephone Exchange (MTX) are occasionally used, but less so than MSC because they are also trade names of particular manufacturers.

Mobility Management (MM)-The processes of continually tracking the location of mobile telephones or devices that are connected to a communication system. Mobility management typically involves regularly registering telephones or communication access devices. Mobile telephones typically automatically register when the are first turned on (attach) and when they are turned off (detach).

MOD-Music On Demand

Modal Bandwidth-A bandwidth-limiting mechanism in multimode fiber optic cables that results from the different arrival times of the various modes. This mechanism also may be observed in singlemode fibers when operated at wavelengths below cut-off.

Modal Dispersion-A physical effect that limits data rates in optical fibers. In some types of optical fiber, multiple modes of the same wavelength of light can propagate down the fiber. Each mode has a characteristic speed in the fiber so each reaches its destination at a different time. A light pulse made up of multiple modes will therefore broaden as it travels down the fiber, causing it to overlap with others if the data rates get too high (and the pulses correspondingly too short). Single Mode Fiber is used to avoid this effect when long distances are required.

Modal Noise-Fluctuation of optical power in a fiber optic cable because of the interaction of power traveling in more than one mode.

Mode Coupling-In an optical waveguide, the exchange of power among modes. Mode coupling reaches equilibrium after propagation over a finite distance that is designated the equilibrium length.

Mode Filter-A device used to select, reject, or attenuate a certain mode or modes in a fiber optic cable.

Mode Mixing-The exchanging of energy between modes in a fiber optic system resulting from in homogenous fiber geometry.

Mode Of Failure-The physical description of the manner in which a failure occurs and the operating condition of the equipment or part at the time of the failure.

Mode Scrambler-A device used to couple power into all the modes of a fiber.

Mode Volume-The number of modes that a fiber is capable of guiding.

MoDem-Modulator/Demodulator

Modem Pool-A grouping of modems that are shared by users and other network devices to make a data call over the switched telephone network.

Modes - Wave Modes-Stable variations of a wave that can be propagated down a waveguide. Modes are subject to modal dispersion in optical fibers and so are undesirable in most applications. However, fibers that support only one mode tend to be expensive and fragile.

Modification Of Final Judgment (MFJ)-A 1982 settlement reached between AT&T and the Department of Justice (DOJ) that required AT&T to divest itself of exchange and exchange access telecommunications services, as well as Yellow Pages directory functions. As a result, the Bell operating companies were separated from AT&T Long Lines, Western Electric, and Bell Telephone Laboratories. The settlement was reached on Jan. 8, 1982, and was approved by a federal district court in Washington, D.C., on Aug.24, 1982. The MFJ replaced the 1956 Consent Decree and settled the 1974 antitrust case of the United States vs. AT&T.

Modified Huffman Code (MH)-A one-dimensional data compression technique that compresses data in an horizontal direction only and does not allow transmission of redundant data. Huffman encoding is a lossless data compression algorithm that replaces frequently occurring data strings with shorter codes. Often used in image compression.

Modular-a) Retain the existing definition but place before it "(1- general)" b) Append the following text to the end: "(2- telephone cord connector) Trade name for small molded plastic electrical connectors used on handset and mounting (wall) cords. The RJ11 connector is one example of a modular connector. Called Teledapt in Canada."

Modular Design-The process of designing systems or networks as groups of modules. The use of well defined modules (such as a printer module) allow for more simplified upgrading and troubleshooting.

Modular Engineering-An engineering process that provides communication lines in multiples of specific data rate multiples or quantities.

Modulate-The process of varying the characteristics of a carrier waveform to convey information supplied by a modulating signal.

Modulation-(1) The process of changing the amplitude, frequency, or phase of a radio frequency carrier signal (a carrier) to change with the information signal (such as voice or data). See also: AM modulation, FM modulation and phase modulation. (2-digital) A modulation process where a code represents the original analog signal. (3) A controlled variation of any property of a carrier wave for the purpose of transferring information.

Modulation Capability-The maximum percentage of modulation that can be successfully carried by a transmitter without introducing distortion which is deemed unacceptable.

Modulation Depth-The modulation factor expressed as a percentage.

Modulation Index-The ratio of the frequency deviation of the modulated signal to the frequency of the modulating signal.

Modulation Monitor-A test device that is used to measure characteristics of a modulated radio signal. This is typically the percentage of modulation.

Modulation Noise-Intermodulation distortion in an analog amplifier or modulator whose level varies as the input power changes or from other distortions.

Modulator-A device that transfers the intelligence in an information signal onto another signal that is used to better able to transport the information. A modulator modifies a carrier wave by amplitude, phase, and/or frequency as a function of a control signal that carries the intelligence.

Modulator/Demodulator (MoDem)-Modems are devices that convert signals between analog and digital formats for transfer to other lines. Data modems are used to transfer data signals over conventional analog telephone lines. The term modem also may refer to a device or circuit that converts analog signals from one frequency band to another. This figure shows how a data modem converts digital information into analog signals that can be transmitted on an analog communications network. In this example, the data signal comes from the a computer (called the data terminal equipment (DTE)), via an RS-232 serial data interface. The RS-232 data interface uses pre-defined signaling commands and data transmission rates to communicate with the data modem. The modem performs a digital-to-analog conversion and from the line to the DTE an analog-to-digital conversion.

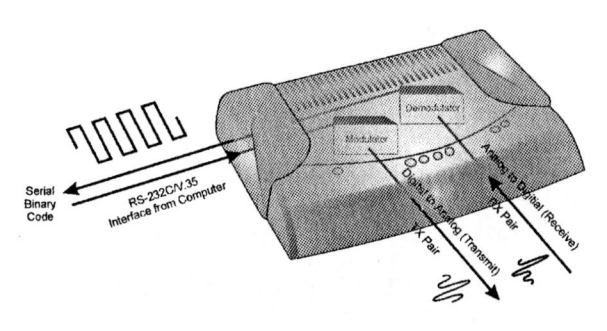

Data Modem Functional Operation

Module-(1-circuit) An assembly replaceable as an entity, often as an interchangeable plug-in item. A module is not normally capable of being disassembled. (2-software) A program unit that is discrete and identifiable with respect to compiling, combining with other modules, and loading.

Module Extender-An assembly that is used to extend the position of circuit board or assembly to permit servicing on that board. A module extender is usually a circuit board with connectors on each end.

MOE-Microsoft Office Expert

MOES-Microsoft Office Expert Specialist

MOH-Music On Hold

Moire-A video distortion in which a wavy pattern appears in the picture, caused by two similar high-frequency signals in the image that mix to create a visible low-frequency feat pattern.

Molecular Hydrogen Loss-The optical loss increase caused by the presence of hydrogen molecules in a fiber. This loss increase is reversible if the hydrogen is permitted to escape.

Momentum - Mechanical-Momentum is the product of mass and velocity. The SI unit of momentum, consistent with electrical units, is the product of one kilogram of mass and one meter per second of velocity. In non-technical discourse, the four words force, energy, power and momentum are often unfortunately used as synonyms for each other. In science and technology, each of these words describes a distinct and separate physical quantity.

Monitoring-The process of listening to or viewing a communication service for the purpose of determining its quality or whether it is free from trouble or interference.

Monochromatic-A single color or wavelength of light. Monochromatic is an idealized concept. In communication systems, optical radiation has a narrow band of wavelengths.

Monochrome-Monitors and other devices with low-pixel resolution presented in a single color, such as green or amber. Occasionally referred to as green screen.

Monochrome Signal-A single color video signal, usually a black-and-white signal. The term monochrome signal also may refer to the luminance portion of a composite or component color signal.

Monolithic Integrated Circuit-A microcircuit fabricated as a single component consisting of elements formed in or on a single semiconducting substrate by diffusion, implantation, or deposition.

Monomode Optical Waveguide-Another term for single mode optical fiber.

Monopole Antenna-An antenna mounting system that uses a single tower pole to raise the antenna to its' necessary height.

Monostable-A device that is stable in one state only. An input pulse causes the device to change state, but it reverts immediately to its stable state.

Montage Effect-In a digital picture manipulator, a recursive effect that develops over time; a composite picture made of several key frame pictures.

Monthly Recurring Cost (MRC)-A cost for service or equipment usage that is continuously charged on a monthly basis.

Moore's Law-Named after Gordon Moore who in 1965 predicted that the number of electronic devices that can be made upon a circuit chip would double every year.

MOPS-Microsoft Office Proficient Specialist

Morse Code-An information coding system that converts letters and numbers into sequences of short and long tones or pulses. Samuel F.B. Morse and Alexander Vail developed the Morse coding system in 1836.

This diagram shows the dots and dashes used to create letters and numbers for the modern version of the Morse code signaling. This modern version is called Continental Telegraph Code, and differs in several characters from Morse's original code The dots represent short bursts of audio or signals and the lines represent long bursts of audio or signals.

Morse Code

MOS-Mean Opinion Score
MOS-Metal Oxide Semiconductor
Mosaic Effect-In a digital picture manipulator, an effect in which the picture seems to be made up of a number of small squares or tiles.
MOSMS-Mobile Originated Short Message Service
Most Significant Bit (MSB)-In a digital word, the first bit in the word sequence defining the largest increment of resolution. In some cases, the MSB is a sign bit signifying the polarity of the word value.
Motherboard-A circuit board that accommodates plug-in cards or daughter boards and provides for interconnections between them. A motherboard also may provide input/output connections.

Motion Artifacts-Defects in a video picture that are evident during motion.

Motion Compensation-In video compression, a macroblock within a frame can be described as a difference from a macroblock-sized region in another reference frame. The spatial difference between the two macroblocks is called the motion vector, and presumably it implies probable motion of an object or camera perspective between the two frames. If indeed there is motion, encoding only the difference between the two macroblocks (hopefully requiring fewer bits) compensates for that motion.

Motion Decay-A digital picture manipulator effect in which objects in motion are blurred.

Motion Estimation-The process of searching a fixed region of a previous frame of video to find a matching block of pixels of the same size under consideration in the current frame. The process involves an exhaustive search for many blocks surrounding the current block from the previous frame. Motion estimation is a computer-intensive process used for CD-ROM applications to achieve high compression ratios and is found in some TV standards conversion systems.

Motion JPEG-A video codec that uses JPEG image compression on each frame of video. Since it does not take advantage of the high correlation between frames in a typical video sequence, it requires many times the number of bits for the same quality when compared to a codec with P-frames and B-frames.

Motion Path Animation-A method of choreographing a scene by specifying a "path" for an object and the number of frames over which the motion is to occur: Camera movement can be specified using this technique.

Motion Picture Experts Group (MPEG)-A working committee that defines and develops industry standards for digital video systems. These standards specify the data compression and decompression processes and how they are delivered on digital broadcast systems. MPEG is part of International Standards Organization (ISO).

Motion Picture Experts Group 1 and 2, Layer 3 (MP3)-A lossy audio codec standardized by the ISO/IEC Moving Picture Experts Group (MPEG) committee in 1992. MP3 is intended for high-quality audio (like music) and expert listeners have found some MP3-encoded audio to be indistinguishable from the original audio at bit rates around 192 kbps. The design of the Layer 3 (MP3) codec was constrained by backward compatibility with the Layer 1 and Layer 2 codecs of the same family. In 1997 the MPEG committee subsequently standardized Advanced Audio Codec (AAC), an improved but non-backward-compatible alternative to MP3.

Motion Picture Experts Group Level 3 (MP3)- A audio signal compression system that is an extension of the Moving Picture Experts Group standard for audio and video compression.

Motor-(1-electric) A machine that converts electric energy into mechanical energy. (2-series wound) An electric motor with its armature and field windings connected m series. (3-shunt-wound) An electric motor with its armature and field windings connected in parallel. (4-stepper) A type of rotary motor that converts pulses of direct current into rotary steps, one step per pulse. (5-synchronous) An ac motor that operates at a speed controlled by the frequency of the ac supply.

Motor Effect-The repulsion force exerted between adjacent conductors carrying currents in opposite directions.

Motorola Integrated Radio System (MIRS)-A land mobile radio system developed by Motorola that provides for voice, dispatch and data services. The MIRS system has evolved to become the integrated digital enhanced network (iDEN).

MOU-Memorandum Of Understanding

MOU-Minutes Of Use

Mounting-A support, such as a mounting plate, that holds and integrates a plug-in unit to other circuitry.

Mouse-A handheld device that translates movement or click-button instructions into corresponding movement of a cursor on a video display. Mouse movement, measured in mickeys, relates the on-screen distance a cursor moves to the distance the mouse moves across the desk.

Moves, Adds, And Changes (MAC)-These define a system administration activity responsible for reconfiguring telephone sets and computer workstations on an existing switch or host system. This also includes the reconfiguration of local area network (LAN) devices.

MP-Multifrequency Pulsing

MP3-Motion Picture Experts Group 1 and 2, Layer 3

MP3-Motion Picture Experts Group Level 3

MPB-Meet Point Billing

MPCD-Minimum Perceptible Color Difference

MPEG-Motion Picture Experts Group

MPEG Compression-The compression of video signals as the conform to the motion picture experts group (MPEG). There are various levels of MPEG compression; MPEG-1 and MPEG-2. MPEG-1 compresses by approximately 52 to 1. MPEG-2 compresses up to 200 to 1. MPEG-2 typically provides digital video quality that is similar to VHS tapes with a data rate of approximately 1 Mbps. MPEG-2 compression can be used for HDTV channels, however this requires higher data rates.

MPI-Media Platform Interface Libraries

MPLS-MultiProtocol Label Switching

MPM-Microvolts Per Meter

MPOA-Multiprotocol Over ATM

MPOE-Minimum Point Of Entry

MPP-Mobile Party Pays

MPS-Message Processing System

MPS-Multi-Page Signal

MPT 1327-A trunked radio system specification that is primarily used in public safety communication systems.

MPT 1327-Ministry of Posts and Telegraph 1327 This diagram shows that the radio channel for the MPT1327 system is 12.5 kHz and the basic parts of an MPT1327 system includes a repeater, TSC channel card, system control interface (SCI), regional control processor (RCP), switching matrix, system control terminal (SYSCON), and a message handling dispatcher (MHD). The TSC channel card is the logic intelligence for the trunked radio functions and independently controls each RF repeater. The SCI links the TSC cards to the outside facilities (e.g. switching facility). The SCI is used to transfer billing records, status reporting, system control, and mobile radio identity validation. An RCP is used for call setup management tasks between multiple sites. The RCP is also the gateway between the PSTN, PABXs and hard-wired dispatch consoles. The PCM switching matrix allows for the switching of speech or data information between radio sites, the PSTN, PABXs and voice mail systems. SYSCON integrates the subscriber database for customer billing, call traffic statistics, and the remote monitoring of equipment. The MHD contains a list of mobile radios that allow it to receive, store, process, and forward messages.

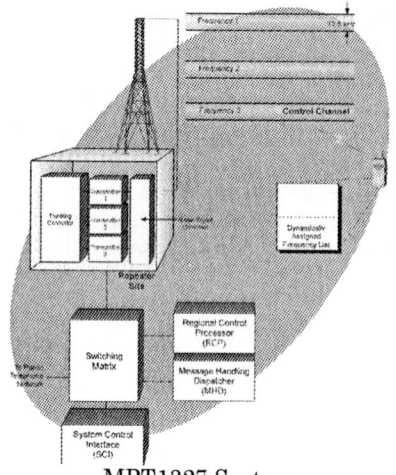

MPT1327 System

MR-Modified READ
MRC-Monthly Recurring Cost
MRI-Mobile Reported Interference
MRVT-Management Routing Verification Test
MS-Mobile Station
MS-Media Server
MS-DOS-Microsoft Disk Operating System
MSA-Mobile Service Area
MSA-Metropolitan Service Area
MSA-Metropolitan Statistical Area
MSAU-Multistation Access Unit
MSB-Most Significant Bit
MSC-Mobile-service Switching Center
MSC-Mobile Switching Center
MSC-Mobile Station Class
MSCM-Mobile Station Class Mark
msec-Millisecond
MSI-Modular Station Interface
MSI-Medium Scale Integration
MSIC-Mobile System Identification Code
MSID-Mobile Station Identity
MSISDN-Mobile Subscriber ISDN
MSK-Minimum Shift Keying
MSO-Multiple System Operator
MSS-Mobile Satellite Service
MSU-Message Signal Unit
MTA-Multimedia Terminal Adapter
MTA-Major Trading Area
MTBF-Mean Time Between Failures
MTF-Mean Time To Failure
MTP-Message Transfer Part
MTP level 3-SS7 Message Transfer Part

MTP Routing Verification Test-An SS7 procedure used to determine if the data of the MTP routing tables in the signaling network are consistent.
MTP3 User Adaptation Layer (M3UA)-M3UA MTP3-User Adaptation Layer is a protocol for supporting the transport of any SS7 MTP3-User signaling (e.g., ISUP and SCCP messages) over Internet Protocol (IP) using the services of the Stream Control Transmission Protocol. This protocol would be used between a Signaling Gateway (SG) and a Media Gateway Controller (MGC) or IP-resident Database.
MTS-Mobile Telephone Service
MTS-Multichannel Television Sound
MTS-Message Telecommunications Service
MTSMS-Mobile Terminated Short Message Service
MTSO-Mobile Telephone Switching Office
MTTR-Mean Time To Repair
Mu (μ) - Law-The type of non-linear digital voice coding (digital signal companding) that is commonly used in the Americas and other parts of the world. The U Law (pronounced Mu Law) coding process is used to compress the 13 bit sampling of a digitized audio signal into the equivalent of an 8 bit sample. It does this by assigning a non-binary (non-linear) value to each of the binary bits. Another non-linear voice coding system is the A Law coding system that is used in Europe and other parts of the world.
MUF-Maximum Usable Frequency
Multi User Software-An application designed for simultaneous access by two or more network nodes, typically employing file and/or record locking.
Multi Vendor Integration Protocol (MVIP)-A standard protocol used for telephony and data switching that allows the design of advanced telephony applications such as voice mail, PBX, faxback service and others that are interoperable with other hardware equipment and software programs.
Multi-Function Subscriber Identity Module (SIM) Cards-SIM cards that are capable of performing more functions than the providing the identity of a wireless telephone subscriber. These functions may include electronic cash, prepaid calling cards, security access card, medial record storage, electronic airline travel ticket and many other functions.
Multi-Party Call Conference-An enhanced telecommunications service that allows the (or more) users to be connected to the same call.

Multicast Streaming

Multibeam Satellite-A satellite that uses multiple directional antennas to allow the same frequency to be reused in different geographic locations.
-This figure shows a satellite system that uses spatial division multiple access (SDMA) technology. In this example, a single satellite contains several directional antennas. Some of these antennas use the same frequency. This allows a single satellite to simultaneously communicate to two different satellite receivers that operate on the same frequency. Usually beams that are separated by more than two or three half-power beamwidths can use the same frequencies, as shown in the figure.

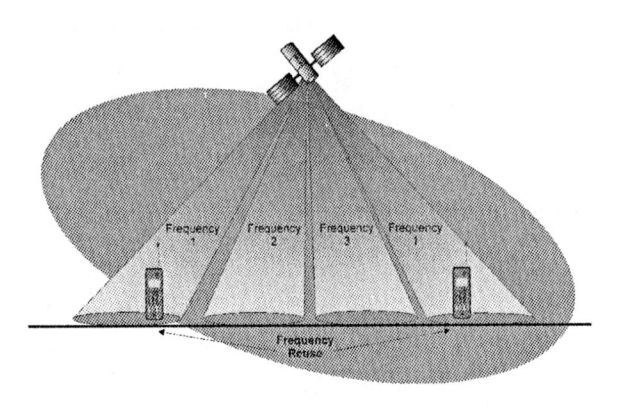

Multibeam Satellite Operation

Multicast-A communications service where a single message or information transmission contains an address (code) that is designated for several devices (nodes) in a network. Devices must contain the matching code to successfully receive or decode the message. An example of a multicast service is a pay per view television broadcast. While all the television broadcast receivers all receive the same radio signal, only the receivers with the correct code will be able to descramble the television signal.
This figure shows examples of how multicast services can be implemented. The first method uses encoded video broadcast transmission and encoded messaging to allow only a select group to view the received information. While all the television broadcast receivers all receive the same radio signal, only the receivers with the correct code will be able to descramble (decode) the television signal. The second method uses multicast routing in the Internet to store and forward data to an authorized group of recipients that are connected to its router. When a router in the Internet that is capable of multicast service receives a multicast message, it will store the message for forwarding. It then uses the multicast address to lookup a list of authorized recipients in its routing table. The stored message is then forwarded to the authorized receiving device or next router that is part of the multicast service

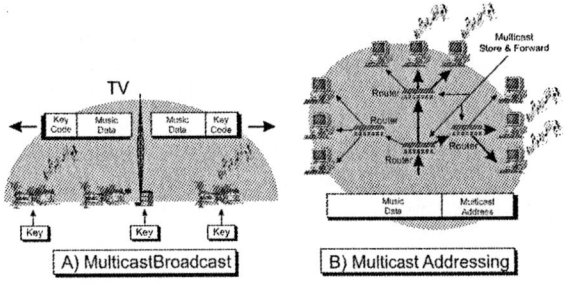

Multicast Operation

Multicast Address-A method of identifying a set stations as the destination for transmitted data. Also known as a group address. See Multicasting.
Multicast Backbone (MBONE)-A high-speed data communications system that interconnects the Internet that allows multicast services. The MBONE network is composed of interconnected multicast LANs.
Multicast Streaming-Streaming audio or video over the internet based on the IP multicast standard, analogous to a broadcast of the media. If implemented and deployed properly, multicast streaming can enable a single server to stream media to an unlimited number of clients. Since the number of clients receiving a multicast stream may be enormous, multicast streaming systems must be designed to avoid overwhelming the original source server with information from each client, such as requests to retransmit lost packets. Examples of techniques to overcome these problems include multicasting of redundant information, such as the use of forward error-correction codes.

Multicasting

Multicasting-Using multicast medium access control (MAC) or Internet protocol (IP) addresses, multiple end users may "tune" to the same stream of data as it is transmitted over the network. This is in contrast to a unicast transmission whereby multiple copies of the stream, each individually addressed to an end user, are transmitted over the network. Multicasting provides much more efficient use of the network resources, however, individual users are not able to use functions such as pause and fast-forward. Multicasting is very useful for non-interactive applications, such as monitoring many stock prices in real-time.

Multichannel Multipoint Distribution Service (MMDS)-One of two wireless cable systems in use in the United States, MMDS stands for Multichannel Multipoint Distribution Service. The FCC authorized access for wireless cable service to a series of channel groups, consisting of channels specifically allocated for wireless cable (the "commercial" channels) and the other channels were originally authorized for educational purposes. This service is called MMDS. See also Local Multichannel Distribution Service (LMDS).

Multichannel Television Sound (MTS)-Any system of aural transmission that utilizes aural baseband operation between 15 kHz and 120 kHz to convey information or that encodes digital information in the video portion of the television signal that is intended to be decoded as audio information.

Multichannel Video Programming Distributor (MVPD)-A person such as, but not limited to, a cable operator, a multichannel multi point distribution service, a direct broadcast satellite service, or a television receive-only satellite program distributor, who makes available for purchase, by subscribers or customers, multiple channels of video programming.

Multicoupler-A device that splits (couples) two or more channels from a signal source (such as an antenna) to allow the same signal to be shared by multiple devices.

This diagram shows a 16 port multicoupler device that allows an antenna signal to be connected to multiple receivers. This multicoupler is composed of one main 4 port (channel) coupler that supplies four additional 4 port (channel) couplers. This example shows that each time a signal passes through a coupler, the signal energy drops, usually in proportion to the number of ports (signal splits) for the multicoupler.

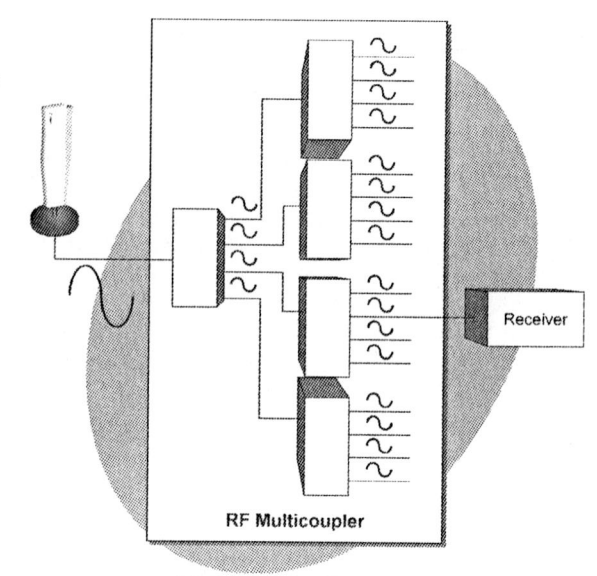

Multicoupler Operation

Multifiber Cable-An optical cable that contains two or more optical waveguides, each of which provides a separate information channel.

Multifiber Connector-An optical connector designed to mate two multifiber cables, providing simultaneous optical alignment of all individual waveguides. It is also called a multi-filter joint.

Multiframe-Multiple Timeframe

Multifrequency Compelled (MFC)-Multifrequency (MF) signaling is a type of in-band address signaling method that represents decimal digits and auxiliary signals by pairs of frequencies. MFC is a different version than MF signaling that is used in the United States that is used to indicate telephone address digits and control signals.

Multifrequency Pulsing (MP)-The transmission of address signals at voice frequencies. Ten decimal digits and five auxiliary signals are each represented by two of the following frequencies: 700, 900, 1100, 1300, 1500, and 1700 Hz. Multifrequency pulsing is used for interoffice signaling and requires a separate supervisory signaling system. Multifrequency pulsing also is referred to as 2-of-6 pulsing, multi-frequency signaling, and multi-frequency key pulsing.

Multifrequency Signaling (MF)-Multifrequency (MF) signaling is a type of in-band address signaling method that represents decimal digits and auxiliary signals by pairs of frequencies from the following group: 700, 900, 1100, 1300, 1500 and 1700

Hz. These audio frequencies are used to indicate telephone address digits, precedence, control signals, such as line-busy or trunk-busy signals, and other required signals.

This diagram shows how 5 different multi-frequency (MF) tones are used to send signaling (dialing digit) information on a communication line between end office (EO) switches. This diagram shows that the MF system uses 5 different frequencies that are combined to represent 10 keys; 700, 900, 1100, 1300, and 1500 Hz.

In this example, dialed digits are gathered from a telephone that is connected to the end office switch. A MF tone generator converts these digits to tones that are sent on a trunk line between end office switches. The receiving end office switch uses an MF receiver to convert the tones back to the digit information. This allows the receiving end office switch to determine which telephone (switch port) to connect to the communication trunk line.

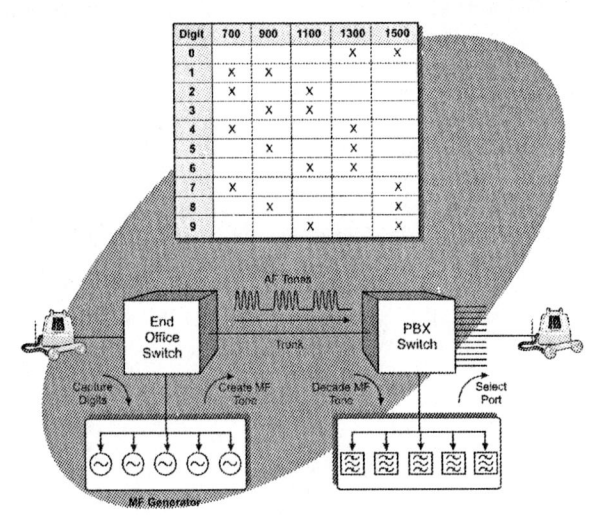

Multifrequency Signaling (MF)

Multihop-(1-passive) A long distance radio communication service that operates via more than one reflection from the ionosphere. (2-active) A microwave system with several repeaters between the two terminal stations.

Multilayer-A type of printed circuit board that has several layers of circuitry interconnected by electroplated holes from one plane to another.

Multilayer Dielectric Filter-An optical filter consisting of a sequence of thin layers of transparent material with controlled thickness and refractive indices.

Multilayering-The capability to layer several video sources together at one time. Multi-layering can reduce the number of recording passes required to create a complex effect.

Multimedia-Multimedia is a term that is used to describe the delivery of different types of information such as voice, data or video. Because Internet service is often used with broadband (high-speed) data services, it is possible to send multiple types of information at the same time.

This figure shows how multiple forms of media can be sent during an Internet telephone call. This example shows a single broadband connection can simultaneously allow telephone calls (Internet Telephone service), transfer data (such as browsing the web), and allow the display of video. In this scenario, a teacher is presenting a training session to students. Each student can see the professor on their television, they can see the course presentation on the computer monitor, and they can hear the professor by the audio on the computer speakers.

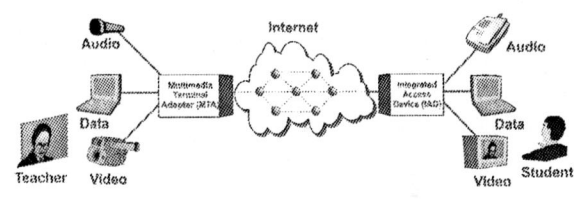

Multimedia Operation

Multimedia Computing-A term referring to the delivery of multimedia information via computer.

Multimedia Messaging Services (MMS)-A system that allows pager and short messaging service (SMS) messaging to include graphics, audio or video components.

Multimedia Terminal-An electronic instrument that is connected to a network that allows a user to

Multimedia Terminal Adapter (MTA)

communicate by voice, data or video communication. Multimedia terminals vary from personal computers with telephone software to dedicated Ethernet telephones.

Multimedia Terminal Adapter (MTA)-A customer premise device that connects the subscriber's telephone to a managed broadband IP network (HFC cable, ADSL, fiber, wireless) and call control elements in the network to deliver high-quality telephony services. Multimedia Terminal Adapters (MTAs) provide the codecs and all signaling and encapsulation functions required for media transport and call signaling.

These pictures show the InnoMedia MTA 3328-1 Multimedia Terminal Adapter (MTA) that is equipped with a single voice port that can be used with any analog telephone and deployed with the installed base of cable modems, ADSL modems, digital set-top boxes and digital home gateways. The device is simple to install and provides superior voice services by using advanced compression, echo cancellation, voice packet recovery algorithms, and packet prioritization technologies. The MTA 3328-1 supports remote provisioning, monitoring and testing and offers features like three-way calling and fax support.

Multimedia Terminal Adapter (MTA)
Source: InnoMedia Incorporated

Multimeter-A test instrument fitted with several ranges for measuring voltage, resistance, and current and equipped with an analog meter or digital display readout. The multimeter also is known as a volt-ohm-milliammeter (VOM.)

Multimode-(1-mobile telephones) The ability of a mobile telephone to operate on multiple types of radio system. For example, as a cellular telephone and a cordless telephone. (2-Fiber Optics) A term used to describe an optical waveguide that permits the propagation of more than one mode.

Multimode Distortion-In an optical waveguide, that distortion resulting from the superposition of modes having differential modal delays. The term multimode dispersion often is used as a synonym but is erroneous because the mechanism is not dispersive in nature.

Multimode Effect-A fiber optic transmission consideration relating to the difference in time required for various light signals to traverse the length of a multimode optical fiber.

Multimode Fiber-An optical fiber that can support many modes of a given wavelength of light. In general, multiple modes are undesirable because they are susceptible to dispersion effects. However, multimode fiber is less expensive to manufacture and less fragile than single mode fiber, so it is often used in less demanding applications.

This diagram shows how multimode fiber transmission uses a relatively wide (50-125 micron) fiber strand to allow several wavelengths of light to pass through the fiber. This example shows that relatively wide optical channel allows optical signals with different wavelengths to make it through the fiber. Some of these signals primarily travel through the center of the fiber while other signals travel from side to side (by refraction) through the fiber strand. Because multiple wavelengths can occupy the same fiber strand, this allows in a higher total bandwidth than single mode fibers.

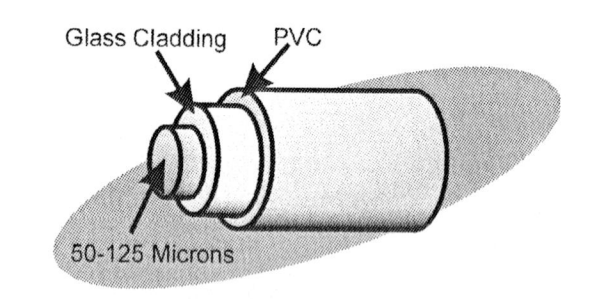

Multimode Fiber Operation

However, this also results in larger group delay as some signals must travel farther (zig-zag as opposed to direct) than others. As a result, this increases signal dispersion as compared to single mode fibers. Signal dispersion can be observed as pulse distortion (edge rounding) increases as the distanced of the fiber transmission increases.

Multimode Laser-A laser that produces simultaneous emission at two or more discrete wavelengths and/or in two or more transverse modes.

Multimode Optical Fiber-An optical fiber that has a large core diameter relative to the wavelength of the light it carries and thus transmits light pulses over many paths. Multimode fiber has greater loss than singlemode fiber and can carry less information, but it is easier to handle and splice.

Multipath-The result of propagation of a radio signals for which part of the signal energy is received before another part of the signal is received that is delayed in time. The delay is due to the extra travel time for the other part of the radio signal that may have been reflected from a building or mountain.

Multipath FadIng-The fading (dynamic reduction of signal level) of a radio communications signal at specific locations due to the combining of incoming signals that travel more alternate (multiple) paths. Multipath fading occurs because the path lengths differ and the incoming multipath signals cancel each other as specific points where the signal levels are inverted (opposite).

Multipath Propagation-A propagation effect resulting from the reception of signals that have taken two or more paths from a transmitter to a receiver. The effect can cause audio distortion in a radio receiver or ghost images in a TV set.

Multiple Access-The capability of a communications system to allow more than one user to access to one ore more channels in the system.

Multiple Address System (MAS)-A multiple address radio system is a point-to-multipoint communications system, either one-way or two-way, utilizing frequencies and serving a minimum of four remote stations. If a master station is part of the multiple address system, the remote stations must be scattered over the service area in such a way that two or more point-to-point systems would be needed to serve those remotes.

Multiple Frequency Shift Keying (MFSK)-A form of frequency shift keying in which multiple frequency codes are used in the transmission of digital signals.

Multiple System Operator (MSO)-A company that owns more than one telecommunications system that provides communications services. In the United States, MSO is the term that is commonly used to describe a company that owns and operates more than one cable television system.

Multiple Timeframe (Multiframe)-The combining of frames or portions of frames (such as a time slot) in a system that continuously sends frames of information to compose a one or more new channels of information.

Multiple Unit Steerable Antenna (MUSA)-A directional antenna made up of fixed elements. The direction of the main power lobe is varied by adjustment of the phase relationship of the different units.

Multiplex-(1-general) The use of a common channel to make two or more channels. This is accomplished either by splitting of the common-channel frequency band into narrower bands, each of which is used to constitute a distinct channel (frequency-division multiplex), by allotting this common channel to multiple users in turn, to constitute different intermittent channels (time-division multiplex), or by allowing the simultaneous transmission of channels using unique identification codes (code-division multiplex.) (2-frequency division) A multiplexing system in which different frequency bands are used by different channels, enabling many different channels to be carried by a single frequency bearer channel. (3-time division) A multiplexing system in which the original analog signals are converted into digital form. The digital signals (for each of many channels) are transmitted sequentially at different time instants. (3-code division) A multiplexing system in which the original signals are converted into digital form and multiplied by a unique identification code. The digital signals (for each of many channels) are transmitted in parallel using different code identifiers.

Multiplexer (MuX)-A device that conveys two or more telephone or data conversations or connections on a single channel or link. Multiplexing may be in the form of frequency division (e.g. multiple radio channels on a coax line), time division (e.g. slots on a T1 or E1 line), code division (coded channels that share the same frequency band) or combinations of these.

Multiplexing-A process that divides a single transmission path to parts that carry multiple communication (voice and/or data) channels.

Multiplier - Billing

Multiplexing may be time division (dividing into time slots), frequency division (dividing into frequency bands) or code division (dividing into coded data that randomly overlap).

This diagram shows how multiplexing can combine two or more low speed channels into one higher speed communication channel. In this diagram, there are eight 8 kbps communication channels that are supplied to a multiplexer. The multiplexer stores and sends 8 bits of each slow speed communication channel during each 125 usec time slot on the 64 kbps channel.

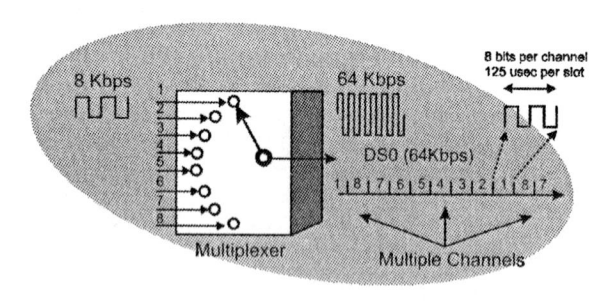

Multiplexing

Multiplier - Billing-The value used in determining a billed partial charge. For instance, if an average calculation is used, every month has an average of 30.417 days (365/12); therefore a customer who subscribes to a service on the 8th of the month will be charged (Y/30.417)x8 (where Y is the total charge for the month).

Multipoint Circuit-A specific-service circuit that has more than two terminations. User terminations may be in the same telephone exchange or widely separated.

Multipoint Control Unit (MCU)-A control system that allows the coordination and interaction between multiple communication devices. Usually, these devices are part of a conference call or multicast communication session.

Multipoint Distribution Service (MDS)-A one-way domestic public radio service rendered on microwave frequencies from a fixed station transmitting (usually in an omni-directional-directional pattern) to multiple receiving facilities located at fixed points.

Multipoint Link-A data communication link connecting two or more stations.

Multiprocessor-A processing method in which program tasks are logically and/or functionally divided among a number of independent central processing units, with the programming tasks being simultaneously executed.

Multiprogramming-Programming that enables a single central processing unit or computer to execute two or more interleaved programs concurrently.

MultiProtocol Label Switching (MPLS)-A network routing protocol that is based on switching through the use of tag labels. The MPLS standard is being developed by the IETF.

Multipurpose Internet Mail Extensions (MIME)-A data communication format this allows information blocks (such as binary images and multimedia data) to be sent with email messages that may be developed primarily for text (7-bit) characters. MIME is defined in RFC 1521.

Multiservice Switch-A switch that receives that provides multiple channel connections that can each have varying bandwidths and different levels of quality of service (QoS).

Multitasking-The process of switching from one task to another on a computer without losing track of either. Multitasking usually is accomplished by time slicing shared resources.

Multivibrator-(1-general) A relaxation oscillator with the outputs from each of two amplifying stages fed back, in phase, to the input stage of the other amplifier to produce oscillation. (2-free-running) A multivibrator that operates without external triggering or synchronization pulses. (3-one shot) A multivibrator that provides one output pulse for each input pulse received.

MUSA-Multiple Unit Steerable Antenna

Music On Hold (MOH)-A feature that connects a source of music to a telephone line that is on hold.

Musical Instrument Digital Interface (MIDI)-An industry standard connection format for computer control of musical instruments and devices.

Mute-A control function that turns the audio input or output of a device to silent (off) mode.

Muting-The process of inhibiting audio (squelching). Muting can be automatic (such as when interference is detected) or can be manually enabled (by the user).

Mutual Induction-The property of the magnetic flux around a conductor that induces a voltage in a nearby conductor. The voltage generated in the secondary conductor in turn induces a voltage in the primary conductor. The inductance of two conductors so coupled is referred to as mutual inductance.

MuX-Multiplexer

mV-Millivolt

MVI-Multi-Vendor Interactions

MVIP-Multi Vendor Integration Protocol

MVNO-Mobile Virtual Network Operator

MVPD-Multichannel Video Programming Distributor

MVS-Multiple Virtual Storage

mW-Microwatt

MWI-Message Waiting Indicator

N

n-Nano
N-Type Connector-A threaded connector with a bayonet center pin that is usually used with RG-8 coaxial cable.
NAB-National Association Of Broadcasters
NACD-Network Automatic Call Distribution
Nack-Negative Acknowledgement Message
NACN-North American Cellular Network
NAEC-Novell Authorized Education Center
Nailed Up Connection-The assignment of a long term (permanent) dedicated path created a network.
NAM-Negative Nonadditive Mix
NAM-Number Assignment Module
NAM Burner-A machine that stores ("burns") information into a number assignment module (NAM) chip. NAM chips were similar to subscriber identity modules (SIM) and were used in older analog AMPS phones.
Name Discovery-(1-general) A procedure that provides the initiating device with the device name of other connectable devices. (2-Bluetooth) The process of identifying other devices located nearby in a Bluetooth system. These Bluetooth devices within range that will usually respond to paging and requests for name identification information using service discovery protocol (SDP).
Name Resolution-A process of translating a text based name into a numeric address, such as Internet Protocol (IP) address. See domain name server (DNS).
Name Server-The data processing device that translates text names to address information (such as an Internet addresses).
NAMPS-IS-88
NAMPS-Narrowband Advanced Mobile Phone Service
NANC-North American Numbering Council
NANC-Number Assignment Number Changes
Nano (n)-A Metric preface representing 0.000 000 001.
Nanometer-A measurement of signal using wavelength of 0.000 000 001 meter.
Nanosecond (nsec)-A measurement of time using 0.000 000 001 of a second.
Nanotechnology-A marriage of engineering and chemistry where microscopic machines are built from atoms and molecules. Such microscopic machines could be programmed to attack cancer cells once injected into the blood stream or to "consume" hazardous materials and convert them into harmless waste.
Nanowband Emission-An emission having a spectrum exhibiting one or more sharp peaks that are narrow in width compared with the nominal bandwidth of the measuring instrument, and are far enough apart in frequency to be resolvable by the instrument.
NANP-North American Numbering Plan
NAP-Network Applications Platforms
NAP-Network Access Point
Narrowband-A communications channel of relatively small bandwidth compared to the information signal it is transmitting. Narrowband channels usually have a data transmission rate of 1 bit per Hz of available bandwidth.
Narrowband Advanced Mobile Phone Service (NAMPS)-A cellular system that allows the use of either 30 kHz or 10 kHz FM modulated (analog) channels. The NAMPS system was developed to increase the serving system capacity by allowing each base station to contain more channels (transceivers). NAMPS adds sub-band signaling operation to the signaling control channels used in the AMPS system.
Narrowband Channel-A transmission channel whose bandwidth can be wholly contained within the bandwidth of the information being transmitted (e.g. bandwidth a 4 kHz voice signal is carried in a radio channel with a frequency bandwidth of 4 kHz.
Narrowband PCS (NPCS)-Narrowband PCS is 3 MHz of bandwidth in the 900 MHz frequency band that is used for two-way messaging (paging) services in the United States. System operator licenses for the Narrowband PCS frequency channels were auctioned for sale by the FCC in 1995. Narrowband PCS is different from Broadband PCS that allows for two-way simultaneous voice as well as data communications.
Narrowcasting-The transmission of information to a audiences that have specific characteristics such as automobile owners of a specific type of car.
NAS-Network Access Server
NAS-Network Attached Storage

NASC-Number Administration And Service Center
NASTD-National Association Of State Telecommunications Directors
NAT-Network Address Translation
NATA-North American Telecommunications Association
National Association Of Broadcasters (NAB)-An association representing radio and Television stations as a lobbying group in interacting with the FCC.
National Electrical Code (NEC)-A document providing rules for the installation of electric wiring and equipment in public and private buildings, published by the National Fine Protection Association. The NEC has been adopted as law by many states and municipalities in the United States.
National Emergency Relocation Center-A secure location from which National Security Emergency Preparedness operations would be directed in the event of a national emergency.
National Institute Of Standards And Technology (NIST)-A non-regulatory agency of the Department of Commerce that serves as a national reference and measurement lab oratory for the physical and engineering sciences. Formerly called the National Bureau of Standards, the agency was renamed in 1988 and given the additional responsibility of aiding U.S. companies in adopting new technologies to increase their international competitiveness.
National Number-The telephone number identifying a calling subscriber station within an area designated by a country code.
National Security Agency (NSA)-The United States agency responsible for the development of cryptographic and other security measures.
National Signaling Point (NSP)-A signaling point (exchange or switch) which belongs to the national SS7 signaling network.
National SS7 Signaling Network-A common channel signaling network consisting of national signaling points and the connecting links. Included in the national network are gateway exchanges connected to the international signaling network.
National Telecommunications and Information Administration (NTIA)-A policy unit of the Department of Commerce which assigns frequencies in the spectrum used by the federal government. The NTAI also advises the President and Congress on telecommunications issues.
National Television System Committee (NTSC)-The industry group that established the standard for TV transmission currently in use in the United States, Canada, Japan, and other countries. The abbreviation NTSC is often used to describe the analog television standard that transmits 60 fields/seconds, 30 frames or pictures/second, and a picture composed of 525 horizontal scan lines, regardless of whether or not a color image is involved.
National Transcommunications Limited (NTL)-The authority that owns, operates, and maintains the terrestrial transmission services for all independent TV stations in the United Kingdom.
NAUN-Nearest Active Upstream Neighbor
NB-Normal Burst
NBS-National Bureau Of Standards
NC-Network Computer
NC-Normally Closed
NCC-National Coordinating Center
NCE Television Station-Qualified Loca Noncommercial Educational
NCE Television Station-Qualified Noncommercial Educational
NCH-Notification channel
NCIP-Novell Certified Internet Professional
NCOS-Network Class of Service
NCP-Network Control Program
NCP-Network Control Point
NCS-Network-Based Call Signaling
NCS-National Communications System
NCSA-National Center for Supercomputing Applications
NCSC-National Computer Security Center
NCTA-Non-Conversation Time Additive
NCTA-National Cable Television Association
NDIS-Network Driver Interface Specification
NDM-U-Network Data Management - Usage
NE-Network Element
NEAP-Novell Education Academic Partner
Near End Cross Talk (NEXT)-The leakage of signal that is coupled to a nearby cable or electronics circuit (called crosstalk) where the unwanted signal is received on the originating end (opposite direction) of the cable. NEXT is usually more troublesome than far end crosstalk, as the crosstalk signal levels of NEXT are higher.
This diagram shows how near end crosstalk (NEXT) can cause interference at the sending end of a transmission line. NEXT occurs when some of the transmitted signal energy leaks from one twist-

ed pair and is coupled back to a communications line that is transferring a signal in the opposite direction at the sending end. Generally, NEXT has a higher level of signal interference as the crosstalk levels at the transmitter end are higher than crosstalk that may occur at the receiving (far) end.

This diagram shows a near video on demand (NVOD) system. This NVOD system allows a customer to select from a limited number of broadcast video channels. These video channels are typically movie channels that have pre-designated schedule times. This system allows the user to unblock an encoded channel during pre-scheduled play times.

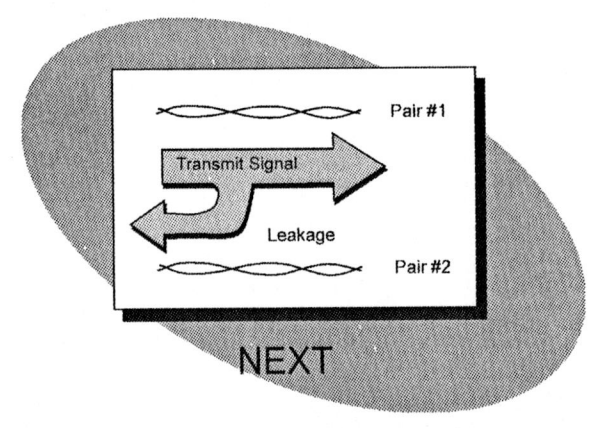

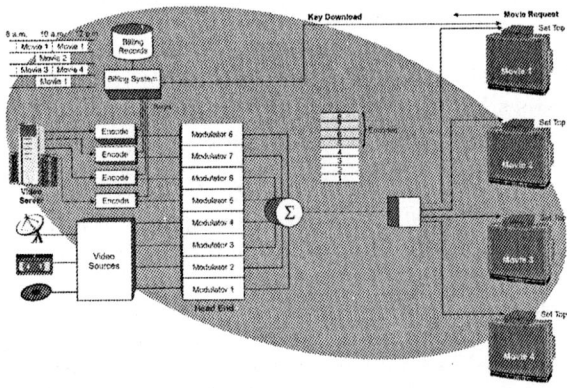

Near End Crosstalk (NEXT)

Near Video on Demand (NVOD) Operation

Near Field Region-(1-optical) The region close to a source, or aperture. The diffraction pattern in this region typically differs significantly from that observed at infinity, and varies with distance from the source. (2-antenna) The region of the field of an antenna between the close-in reactive field region and the far field region wherein the angular field distribution is dependent upon distance from the antenna.

Near Field Scanning-A technique for measuring the index profile of an optical fiber by illuminating the entrance face with an extended source and measuring the point-by-point radiance of the exit face.

Near Infrared-The part of the infrared near the visible spectrum, typically 700 to 1500 or 2000 nm; it is not rigidly defined.

Near Video On Demand (NVOD)-A video delivery service that allows a customer to select from a limited number of broadcast video channels when they are broadcast. NVOD channels have pre-designated schedule times and are used for pay-per-view services.

Nearest Active Upstream Neighbor (NAUN)-In Token Ring Networks, the medium access control (MAC) address of the network interface card immediately preceding a station in the ring. This address is used by stations during the beaconing and ring recovery procedures to isolate faults in the ring to specific locations, based on MAC address.

NEBS-Network Equipment Building System

NEC-National Electrical Code

Negative Acknowledgement Message (Nack)-A message that is responded by a communications device to confirm a message or portion of information has NOT been successfully received. If a communications device sends a NACK messages back to the originator, the system will typically re-send the message. See also acknowledgment message (ACK)

Negative Feedback-The return (feedback) of an output signal that subtracts (adds 180 degrees out of phase.) from the output signal. Negative feedback decreases the output signal amplitude and usually stabilizes the amplifier. This may result in reduced distortion and noise.

Negative Impedance-An impedance characterized by a decrease in voltage drop across a device as the current through the device is increased, or a decrease in current through the device as the voltage across it is increased.

Negotiation -The fourth phase of the marketing process in the Costa Model that motivates the customer to engage in negotiation with a company for the sale of products or services. Negotiation can be explicit (talking with the sales rep about options) or implicit (the customer chooses between options without direct interaction).

Nematic-A type of liquid crystal, used in display units, in which the molecules are aligned parallel to one another but with centers of gravity arranged at random.

NEP-Noise Equivalent Power

Neper-A unit for expressing the ratio of two quantities of electrical power. The number of nepers is the natural base logarithm of the ratio. The term is no longer widely used.

Nested Program-A software program that is included as a component of another larger program to fulfill a specific task.

Net Gain-The overall gain of a transmission circuit. Net gain is measured by applying a test signal of some convenient power at the local termination of a given communications circuit, measuring the power delivered at the other end of the circuit, and taking the ratio of the powers as expressed in decibels.

Net Loss-The overall loss of a transmission circuit. Net loss is measured by applying a test signal of some convenient power at the local termination of a given communications circuit, measuring the power delivered at the other end of the circuit, and taking the ratio of the powers as expressed in decibels.

Net Negative-Occurs when the number of gross adds for a given period is less than the number of existing customers who cancelled service for the same period.

NetBEUI-NetBIOS Extended User Interface

NetBIOS-Network Basic Input/Output System

NetBIOS Extended User Interface (NetBEUI)-A network device driver that is used as part of Microsoft's local area networks (LAN). The NetBEUI protocol communicates with the network interface card using the network driver interface specification (NDIS).

Netbits-Network Bits

Netcasting-A process of sending information directly to the desktop of a recipient list, usually through the Internet.

NetDDE-Network Dynamic Data Exchange

Netiquette-A set of unwritten rules (etiquette) that define the normal (socially acceptable) use of electronic mail (e-mail) and other network communication services.

NetWare-The Novell network operating system using internetwork packet exchange (IPX) and sequential packet exchange (SPX) protocols. Net Ware is a trademark of Novell.

Network-A series of points that are interconnected by communications channels, often on a switched basis. Networks are either common to all users or privately leased by a customer for some specific application.

Network Access-Electronic circuitry that determines which station may transmit next or when control a particular station may transmit. This circuitry may be centrally located or may be located in each of the network interface controllers.

Network Access Charge-A fee paid by an operator of a system for access to other network systems.

Network Access Point (NAP)-An key physical (layer 2) interconnection point that interconnects regional Internet systems and sub-systems. NAPs are public network exchange facility where Internet Service Providers (ISPs) can connect with one another in peering arrangements. The NAPs are a key component of the Internet backbone because the connections within them determine how traffic is routed. At the end of the 20th century, they were one of the points of Internet congestion. As a result of congestion of NAPs, several ISP have invested in private NAPs to interconnect to each other's network.

Network Access Revenue-Service revenue that results from charge for service access between carriers, such as when inter-exchange (IXC) carriers pay to connect through the local telephone access network infrastructure.

Network Access Server (NAS)-A server that coordinates access to network systems. NAS are used in MGCP systems.

Network Address-(1-general) A unique number associated with a network host that identifies it to other hosts and devices during network communication. (2-SS7) The Signaling System 7 signaling code that contains a network identification, a network duster, and network cluster member fields.

Network Address Translation (NAT)-A process that converts network addresses between two different networks. NAT is typically used to convert public network addresses (such as IP addresses) into private local network addresses that are not recognized on the internet. NAT provides added security as computers connected through public networks cannot access local computers with private network addresses.

This figure shows the basic operation of network address translation (NAT) system. In this diagram, the NAT receives a message with a desired pubic IP address (209.67.22.59) originating from a local computer with a private IP address of 10.01.01.01. The NAT translates this originating address to a public IP originating address. The he NAT then initiates as session with the Internet server (web site) using the network's public IP address 118.54.23.11 as the originating address. The Internet server receives the request for information and responds with data messages address to the NAT's public IP address. When the NAT receives these data messages for that particular communications session, they are translated to the local (private) IP address 10.01.01.01 and forwarded to the originating computer. If messages are received to the NAT's public IP address that are not part of a communications session that it knows about, the NAT will not route the messages to computers connected to the LAN.

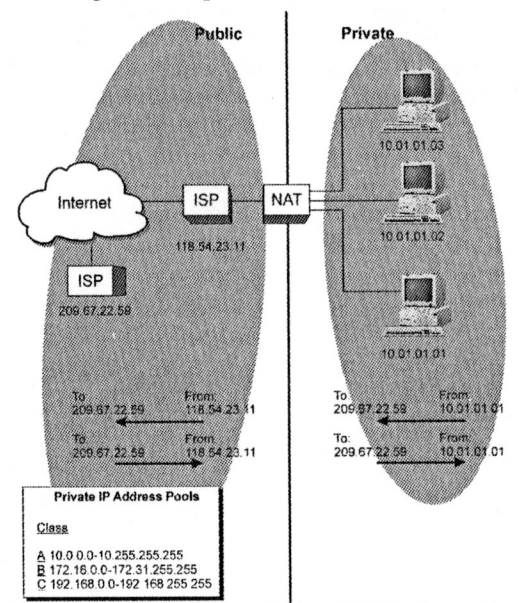

Network Address Translation (NAT) Operation

Network Administration Center-An operations center with administrative responsibility for local and tandem switching systems.

Network Administrator-(1-coordinator) A person responsible for managing the day-to-day operations of a network. (2-humor) The person blamed for all computing problems, whether they are related to the network or not.

Network Analyzer-A test instrument that receives, decodes, and analyzes data transmitted through a network. A network analyzer may be an integrated software program or separate hardware device.

Network Architecture-The design, physical structure, functional organization, data formats, operational procedures, components, and configuration of a network. Network architectures usually divide network functions into layers of software and hardware. Each network layer serves a specific purpose. There are often specific relationships between network layers that allows different manufacturers or equipment that operate at different layers to interoperate with each other (e.g. a router interfacing with a network hub.)

Network Attached Storage (NAS)-(1-product) A collection of mass-storage devices contained in a single chassis with a built-in operating system. Typically connected to a local area network, these devices usually support Network File System (NFS) and common Internet file system (CIFS) as a means to share data in a departmental or enterprise environment. NAS products are marketed to small and medium businesses as self-contained, plug-and-play, easy to operate storage expansions.

(2-architecture) Network Attached Storage is an architecture in which traditionally LAN-oriented technologies, such as Ethernet and TCP/IP are used to connect storage. NAS utilizes LAN technology in place of traditional storage protocols such as Fibre Channel and parallel-SCSI to produce large arrays of disk drives with a virtualized interface. One NAS disk array may by configured to appear as a single disk drive or as multiple volumes of varying sizes.

Network Automatic Call Distribution (NACD)-NACD is a is a call processing system that routes (distributes) incoming telephone calls to specific telephone sets or stations calls based on the characteristics of the call or network settings. These characteristics can include an routing on network congestion, time of day routing, and other criteria.

Network Basic Input/Output System (NetBIOS)

Network Basic Input/Output System (NetBIOS)-A transport layer protocol that connects stations using NetBIOS names. NetBIOS is commonly referred to as an un-routable protocol because, the address (NetBIOS name) is not divisible into a network portion and a host portion like the addressing used by Internet Protocol (IP). Recently however, vendors have implemented NetBIOS name caching, whereby the NetBIOS names of each packet are inspected and tables maintained much like layer 2 MAC addresses.

NetBIOS provides an Application Programming Interface (API) with a high-level, easy to use command set. NetBIOS was invented in 1984, long before wide-scale deployment of LAN technology. As a result, it is a limited protocol designed to prevent applications from having to understand the low-level details of network communication, at the expense of not being easily scalable to the size corporate networks commonly achieve.

Network Bits (Netbits)-The number of bits in an address that are used to identify a network or subnetwork.

Network Busy Hour-The hour in a given 24-hour period during which the total load carried by all trunks in a network is greater than the total load carried during any other hour.

Network Channel Code-An encoded description of a channel provided by a local exchange carrier from the point of termination at an interexchange customer's premises to a central office, or from that customer's premises to an end-user location or Centrex system.

Network Channel Interface Code-The code that identifies interface characteristics at an end user's network interface or an interexchange carrier's termination point. The code identifies the number of wires at the interface, the protocol, the nominal impedance, and a protocol option. The code applies to both switched and special access service and helps define the physical and electrical characteristics of a point of termination for an access service request, a service order, and circuit-provisioning processes.

Network Class of Service (NCOS)-The types of access or services that users are authorized to receive from a communication system.

Network Cluster-(1-trunk) A final trunk group, and all high-usage trunk groups with which it shares at least one common terminal and for which it is in the last choice route chain. (2-signaling point code) The field in the signaling point code structure that identifies groups of signaling points and individual transfer points of a signaling network.

Network Control-Network control is the transmission of signals or messages that perform call control, equipment configuration, or information management functions. Network control can be centralized or distributed. The control of public telecommunications networks is a centralized system as call processing is coordinated through a common channel signaling (CCS) network. The Internet uses distributed control as the switching information dynamically changes in packet switching centers (routers) throughout the Internet network.

Network Control Point (NCP)-A special applications processor that provides network access to a variety of centralized database services. The corresponding term in common channel signaling is Service Control Point.

Network Control Signaling-The transmission of signals that perform call control functions, such as supervision, address signaling, calling and called-number identification, and flow control.

Network Data Collection Center-An operations center that administers network data collection and supervises the operation and maintenance of the Engineering and Administrative Data Acquisition System.

Network Data Management - Usage (NDM-U)-The network data management - usage (NDM-U) is a standard messaging format that allows the recording of usage in a communication network, primarily in Internet networks. The NMD-U defines an Internet Protocol detail record (IPDR) as the standard measurement record.

The IPDR record structure is very flexible and new billing attributes (fields) are being added because Internet services are now offered in almost all communications systems. The NMD-U standard is managed by IPDR organization at www.IPDR.org.

Network Driver Interface Specification (NDIS)-An interface specification that was developed by Microsoft to provide a common set of rules for network adapters to interface with operating systems. NDIS is independent of hardware and types of network interface cards and it allows multiple protocol stacks to co-exist in the same computer.

Network Element (NE)-A facility or the equipment used in the provision of a telecommunications service. The term includes subscriber numbers, databases, signaling systems, and information sufficient for billing and collection or used in the transmission, routing, or other provision of a telecommunications service.

Network Equipment-The telecommunications equipment and facilities owned, installed, and maintained by a telephone company or service provider and that are part of a telecommunications network.

Network Equipment Building System (NEBS)-Generic construction requirements that help ensure the compatibility of network elements with central office systems. Included are requirements for equipment floor layout, fire and earthquake resistance, power and grounding compatibility, and heat dissipation.

Network Gateway (NGW)-A media and signaling adapter (gateway) used in a network to interface between different types of networks. A network gateway can convert both the media and signaling control messages between the systems.

Network Harms-Adverse effects on telephone company services, employees, or customers. The four basic harms are excessive signal power, hazardous voltage, improper network control signaling, and line imbalance.

Network Identity And Timezone (NTIZ)-A feature that allows a GSM network to transfer the serving system name and current time zone.

Network Indicator-In the Signaling System 7 protocol, the part of the sub-service field within a service information octet that can be used to differentiate between national and international signaling messages.

Network Integration-The joint provision of telecommunications services and the joint assumption of risk through a partnership arrangement among telephone companies. The expression often is used to describe both technical and economic integration.

Network Interface (NI)-(1-interface) The point of connection between customer premises equipment and a public switched telephone network. This is also called a standard network interface (SNI). (2-boundary) The physical and electrical boundary between two separately owned telecommunications systems.

Network Interface Card (NIC)-A network interface card (NIC) adapts data communication network protocol (such as Ethernet) to a data bus or data interface in a computer or data terminal. The NIC is installed between a computer network (such as the Ethernet) and a computer data bus (such as a PCI socket). The NIC is usually a PC expansion board connector and operating system. Software in the computer is installed and setup to recognize the NIC card.

Network Interface Device (NID)-A connection point between the end customers equipment and the telecommunications network. This is also called the demarcation point.

Network Interface Unit (NIU)-An electronic assembly that terminates a telecommunications line to an end user's facility. For optical networks, the NIU may terminate fiber and copper lines and convert the signals into analog (telephone) and digital (computer network or multimedia) signals.

Network Intrusion Detection Expert System-An expert system that provides the final connection between an end customers premises and the telecommunications network.

Network Layer-The Network layer performs the switching and routing of data through the network, controls the flow of data within the network, segments (divides) or reformats data packets if necessary between network types, and performs error control functions specific to the address decoding and routing functions. The network layer receives data for transmission from an upper layer (such as a transport or session layer) and converts it into network addressable data formats that can be transferred through a network or transmission line. An upper layer provides the network layer with the necessary addressing and network routing control requirements (e.g. priority codes) to allow the network layer to send data through the network. The location of the network layer within the protocol stack is usually above a physical layer and below a transmission or session layer. The network layer is layer 3 in the open system interconnection (OSI) protocol layer model.

Network Maintenance Center (NMC)-A facility that allows monitoring, testing and maintenance of a telecommunications network. The NMC is typically operational 24 hours a day, 7 days per week.

Network Management-Network management is the processes of configuring equipment in the network, the setup (provisioning) of services, system maintenance, and repair (diagnostic) processes. Network management systems are commonly composed of a network management server computer and network management software.

Network Management Center (NMC)

Network management systems usually include a set of procedures, equipment, and operations that keep a telecommunications network operating near maximum efficiency despite unusual loads or equipment failures.

This figure shows how network management is used to configure, setup services, and maintain networks. In this example, a network management server is transferring configuration parameters to a media gateway. The network manager is also setting up (provisioning) services for a call server to allow the call server to route calls through the gateway. After the gateway is setup, this example shows that the network management system will periodically receives messages from the gateway that contain system status information to ensure the network is operating within its desired specifications.

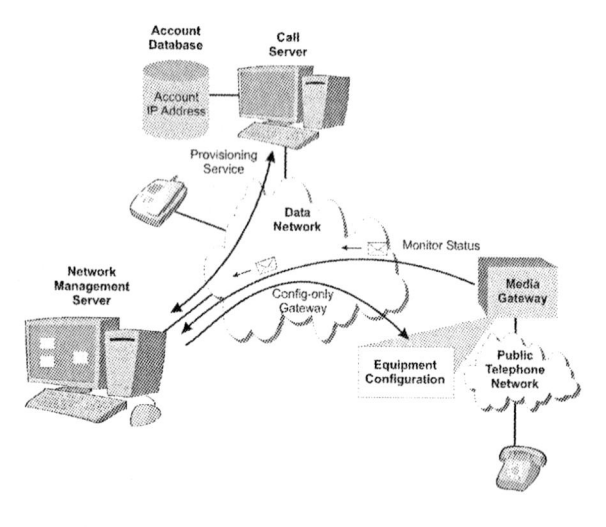

Network Management

Network Management Center (NMC)-An operations center that monitors and controls traffic flow to help ensure the most efficient and economical use of available network capacity. The center plans strategies and works to minimize the effects of disasters, abnormal traffic loads, and switching system or facility failures.

Network Management Protocol (NMP)-Communication protocols that were developed by AT&T to control network equipment and assemblies including modems data multiplexers.

Network Management Station (NMS)-A device that communicates with network management or SNMP agents throughout a network. Typically it comprises a workstation operated by a network administrator, equipped with network management software or other relevant applications that assist in the monitoring of the network. Some, if not all the components of the FCAPS model are usually designed into the NMS. NMS is also known as a network manager.

Network Monitor-A software program or graphic display that monitor and identifies network-related problems. Network Monitor tracks data as it moves through the network layer. It may insert, filter packets, or perform packet analysis.

Network Monitoring And Analysis System-A system that collects and analyzes information on network alarms and performance data and alerts a center if it detects trouble in loop, interoffice, or switching systems.

Network News Transport Protocol (NNTP)-The protocol that governs the transmission of network new, a threaded messaging system for posting messages to form newsgroup discussions.

Network Node Interface (NNI)-The defined (typically standardized) interface between functional elements (network nodes) in a system or network. NNI interfaces may add additional test functionality for connections between network elements and reduce access control functions as network physical connections are typically fixed for extended periods of time. See: user network interface (UNI)

Network Operating System (NOS)-A software program that manages communication with a network. The NOS oversees resource sharing and often provides security and administrative tools.

Network Operations Center (NOC)-(1-Surveillance) A center responsible for the surveillance and control of telecommunications traffic flow in a service area. (2-Service) A facility or organization responsible for maintaining, monitoring, and troubleshooting a network infrastructure. Responsible for applying the FCAPS model to the network.

Network Operator-A company that manages the network equipment parts of a communications system. A network operator does not have to be the service provider. Also see Service Provider and Reseller.

Network Order-Multi-octet values transmitted in the specific order (usually from most significant octet to least significant octet.)

Network Planning System (NPS)-An interactive computer program that assists strategic planners in the development of interoffice facilities and wire centers, and aids in the planning of traffic and distribution routes.

Network Port-A communication input/output access point to a network. The network port usually has specific network access protocols and security levels associated with it.

Network Processor-A programmable device (silicon chip) which is capable of receiving packets from one or more network interfaces and processing these packets. The processing of the packets includes address lookup, packet modification and transmission and is under software control. Network processors are special purpose microprocessors, specifically designed to meet the needs of systems that require each packet to be inspected in detail. By using network processors, system's designers and vendors are able to change the function of routers, switches and other devices in the field as new protocols are developed.

Network Program-(1-computer system) A program that operates on a server within a network. (2-distribution) Any program delivered simultaneously to more than one broadcast station regional or national, commercial or noncommercial.

Network Server Interface Specification (NSIS)-A specification for network interface cards (NICs) that is independent of hardware and protocol.

Network Servers-Hosts, and sometimes personal computers, may function as specialized types of nodes called network servers. These specialized nodes serve the other nodes by storing many of their files and running much of their common software.

Network Service Center (NSC)-An operations center that performs quality control functions related to a grade of network service. These functions include the completion of a call to a desired number, the capability of hearing and being heard, accurate billing of calls, and the capability of receiving incoming calls.

Network Service Part (NSP)-In Signaling System 7, the combination of the message transfer part and the signaling connection control part.

Network Service Part of SS7 Message-The combination of the Message Transfer Part (MTP) and the Signaling Connection Control Part (SCCP).

Network Service Provider (NSP)-Any company that provides network services to customers or devices.

Network Subsystem (NSS)-(1-general) The network parts of a communication system. network. (2-GSM) The system parts of a GSM network this includes the mobile switching center (MSC), home location register (HLR), visitor location register (VLR) and equipment identity register (EIR).

Network Termination (NT)-A final end point in a network that is usually owned by the network service provider. After the network termination, the equipment is commonly owned by the customer (called customer premises equipment - CPE). When the network termination (NT) is an active device, it typically has standard communications parameters such as protocols, timing and voltages to allow specific types of equipment to correctly communicate with the network.

Network Termination 1 (NT1)-A standard network termination in the ISDN network that adapts the physical characteristics of the ISDN network.

Network Termination 2 (NT2)-A standard intelligent network termination in the ISDN network that contains the intelligence (application layers) for ISDN termination equipment.

Network Termination Equipment (NTE)-Network termination equipment (NTE) are the devices used by end-user to access the network. In the traditional PSTN world, network termination equipment was generally confined to the telephone, headset or conference phone. The cellular industry expanded this to include cell phones and pagers. In IP Communications network termination equipment can include all these traditional devices as well as computers and PDAs.

Network Time Protocol (NTP)-NTP provides a method for synchronizing clocks on physically separated machines. NTP is a complex protocol that uses multiple methods to synchronize clocks on a computer to a more accurate time source. NTP is defined in RFC 1129.

Network Topology-The physical and logical relationships between nodes in a network, typically star, bus, tree, ring, or hybrid.

Network Tunneling-Network tunneling is the process that creates a secure communication path through the use of a virtual connection within a network link through the encryption of data that is transmitted between the virtual connection points.

Network Voice Protocol (NVP)

Network Voice Protocol (NVP)-An older voice protocol that was developed to allow real-time voice transmission over the ARPANET (packet voice) system.

Network-Based Call Signaling (NCS)-NCS is a call signaling protocol that is used to control communication units (such as IP telephones). NCS is used to setup, manage, and terminate multimedia communication sessions in a centralized communications system.

Networking-The connection of geographically separate computers and communication devices using transmission line facilities.

Neural Network-A computer system whose information processing is modeled on the structure of the brain and its neurons. Information processing is parallel, in that many processes take place at the same time. Processes are also asynchronous because each step in a process has no time relationship to steps in other processes. To mimic the brain, some functions are stochastic, or random.

Neutral-(1-charge) A device or object having no electric charge. (2- power system) A conductor in an electric power-distribution system that carries no current when the power load is balanced.

New Matter-When a continuation, or divisional, application is filed which contains descriptions, figures, or other information which was not part of the original pending application, the additional information is deemed "new matter" and the application called a continuation-in-part, or CIP.

When determining patentability over the prior art, new matter is awarded the filing date of the CIP, while matter which is common to the CIP and the original parent application is awarded the priority date of the parent application.

Newton-The standard SI metric unit of force. One Newton is the force that, when applied to a body mass of 1 kg, gives it an acceleration of 1 meter per second per second (or one meter per second squared.)

NEXT-Near End Cross Talk

Next Hop Resolution Protocol (NHRP)-An name resolution protocol designed to allows Internet Protocol (IP) datagrams to route across multiple types of access networks (e.g. ATM, SMDS, and X.25).

NextGen or Next Generation Network (NGN)- Next Generation networks (NGN) refers to the infrastructure service providers will require as they migrate from circuit switched systems to packet switching systems. Nextgen networks are still evolving, but have five core components - media gateways, softswitches, media servers, application servers and session border controllers. Also known as next generation network or NGN.

NF-Noise Factor

NFS-Network File System

NGN-NextGen or Next Generation Network

NGRS-National Geodetic Reference System

NGW-Network Gateway

NHRP-Next Hop Resolution Protocol

NI-Network Interface

Nibble-A 4-bit unit of data (half of a byte).

NIC-Network Interface Card

NID-Network Interface Device

Night Answer-A telephone system feature that redirects in-coming calls during designated times of the day, such as after business hours.

Night Service-The processing state of a telephone system (such as a PBX) during hours of operation when the company is closed or in a different state of business operation. Night service usually provides a different greeting messaging and call routing (transfer) capability. Night service may prompt callers to leave messages instead of being routed to an operator.

This diagram shows how a telephone system can change its basic operation for daytime and nighttime telephone service. In this example, during the day, all the incoming calls are routed to (received by) a receptionist at extension 1001. At night (between 5 pm and 8 am), the calls are automatically redirected to an automated telephony call processing system that is connected to extension 1014. When the automated attendant detects a ring signal, answers the phone (off-hook signal) and plays a pre-recorded messaging informing the caller of options they may choose to direct the call to a specific extension. In this example, the automated call attendant software decodes DTMF tones or limited list of voice commands to determine the routing of the call. The automated call attendant software then determines if the destination choice is within the option list and if the extension is available. If the extension is available, the automated attendant will send a command to the computer telephony board (voice card) that can switch the call to the selected extension. If the extension is not valid or not available, the automated attendant will provide a new voice prompt with updated information and additional options.

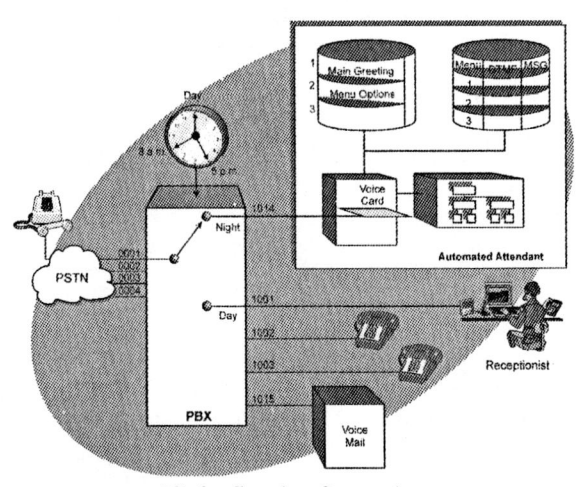

Night Service Operation

This diagram shows that no-return to zero (NRZ) encoding uses the logical level voltage during the entire period of each logical bit. In this example, the data is transmitted at 1 kbps so each logical bit period is 1 msec. During this entire period, the logical level remains at the same voltage associated with the logical level.

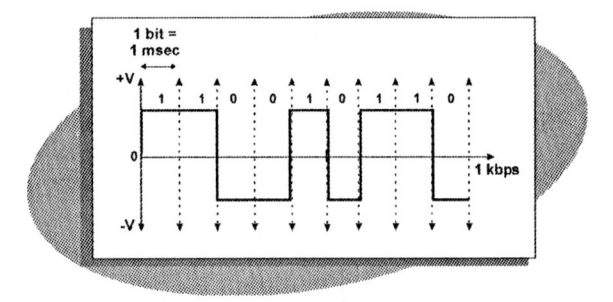

No Return to Zero (NRZ) Operation

NII-National Information Infrastructure
NIS-Network Information Service
NIST-National Institute Of Standards And Technology
NIU-Network Interface Unit
NMC-Network Maintenance Center
NMC-Network Management Center
NMP-Network Management Protocol
NMS-Network Management Station
NMT-Nordic Mobile Telephone
NMT-450-Nordic mobile telephone, 450 MHz. See also: Nordic mobile telephone.
NMT-900-Nordic mobile telephone, 900 MHz, See also Nordic mobile telephone.
NNI-Network Node Interface
NNTP-Network News Transport Protocol
NNX-A term used for qualifying dialing digits. N stands for any number from 2 through 9 and X stands for any number 0 through 9. An example of a valid NNX is 230.
NO-Normally Open
No Charge Traffic-Traffic, such as 611 service request and 911 emergency calls, classified as "no-charge" in a tariff on file with an appropriate regulatory agency.
No Circuit Tone-A low tone, interrupted 120 times per minute (02 5 on and 0.3 5 off) indicating that no trunk is available. This term also is known as a reorder, all-circuits busy, or fast-busy tone.
No Return To Zero (NRZ)-A digital code in which the signal level is low for a 0 bit and high for a 1 bit and does not return to 0 between successive 1 bits.

NOC-Network Operations Center
Nodal Multiplexer-A multiplexer that has the capability of dynamically routing channels onto different communication circuits.
Node-(1-network) In network topology, a terminal of any branch of a network, or a terminal common to two or more branches of a network. (2-ascending) The point where a satellite crosses the plane of the equator when moving north. (3-current) The points at which the current is at minimum in a transmission system in which standing waves are present. (4-descending) The point where a satellite crosses the plane of the equator when moving south. (5-network) A terminal on any branch of a network. (6-switching) The switching points in a switched communications network, including patching and control facilities. (7-transmission line) A point of interconnection on a transmission line. (8-tree structure) A point where subordinate data originates. (9-telephony) A switching office or facility junction. (10-test facility) A remote test facility. (11-voltage) The points at which the voltage is at a minimum in a transmission system in which standing waves are present.
Node B-The name assigned to the universal terrestrial radio access network (UTRAN) radio equipment located at a transmitter cell site. Node B was a name that was temporarily given to the UTRAN

base station during the standard's development and was not changed after the standard was released.

Noise-An undesired signal generated by a physical mechanism that cannot be "turned off" in theory. The most typical example is the electric current produced by the random motion of electrons, a manifestation of temperature related kinetic energy that is zero only when the material involved is cooled to superconducting temperatures. In contrast to noise, a distinct type of undesired signal, called interference, in theory can be "turned off" because it originates from a device like a separate radio transmitter, a faulty electric device, or other artifact. In some cases, writers use the term "noise" to describe all undesired signals, both physical noise and interference together, often inappropriately.

Noise Equivalent Power (NEP)-In optoelectronics, the radiant power, at a given modulation frequency and for a given bandwidth, that produces a signal-to-noise ratio of I at the output of a given detector. in this sense, NEP is the minimum detectable power at the given frequency and for the given bandwidth.

Noise Factor (NF)-The ratio of the noise power measured at the output of a receiver to the noise power that would be present at the output if the thermal noise resulting from the resistive component of the source impedance were the only source of noise in the system.

Noise Figure-A measure of the noise in decibels generated at the input of an amplifier, compared with the noise generated by an impedance method resistor at a specified temperature.

Noise Filter-A network that attenuates noise frequencies.

Noise Floor-The power level of background noise signals.

Noise Generator-A generator of noise signals.

Noise Immunity-The capability of a device to receive and decode signals or information in the presence of noise.

Noise Level-The ratio of noise on a given circuit to a reference noise, expressed in decibels above reference noise for an electrical system or decibels sound pressure level for an acoustical system.

Noise Margin-An assigned minimum signal-to-noise ratio, expressed in decibels, that is required for a specific type of signal to be useful.

Noise Power Ratio (NPR)-The ratio, expressed in decibels, of signal power to intermodulation product power plus residual noise power, measured at the baseband level.

Noise Suppressor-A filter or digital signal processing circuit in a receiver or transmitter that automatically reduces or eliminates noise.

Noise Temperature-The temperature, expressed in Kelvin, at which a resistor will develop a particular noise voltage. The noise temperature of a radio receiver is the value by which the temperature of the resistive component of the source impedance should be increased (if it were the only source of noise in the system) to cause the noise power at the output of the receiver to be the same as in the real system.

Noise Voltage-The RMS component of the detected electric output voltage (current) of a fiber system that is incoherent with the signal radiant power. The noise voltage value usually is determined with the signal power removed.

Noise Weighting-The assignment of a specified amplitude-vs.-frequency characteristic to a noise signal prior to measurement so that the measured value closely approximates the relative effect on a customer using the circuit.

Nominal-The most common value for a component or parameter that falls between the maximum and minimum limits of a tolerance range.

Nominal Value-A specified or intended value.

Non-Associated Mode - SS7 Signaling-The mode where SS7 messages for a signaling relation involving two non-adjacent signaling points are conveyed between those signaling points, over two or more signaling links in tandem passing through one or more signal transfer points.

Non-Blocking-A characteristic of a switch fabric implying that it is capable of handling traffic at the maximum number of circuit switched channels for a circuit-switched fabric, or for a packet switch the maximum frame arrival rate, on all interfaces simultaneously, without requiring some packets to wait for resources to become available before transmission. This term is used with the assumption that no port will be presented more that the line-rate. In older switching technologies, blocking was common. This means that, due to the design of the switch, it was possible for a packet to arrive, destined for a port that currently has no traffic, but the packet was still required to be buffered. Except for the most massive switches on the market, most modern day packet switching equipment is non-blocking. Most small to medium size circuit switches are non-blocking today, but large circuit switches having 10000 lines or ports are typically configured for less than 1 percent probability of blocking rather than being non-blocking.

Non-Disclosure Agreement-A binding contract between parties not to disclose and to keep confidential information shared among the parties from being spread to other parties. Commonly referred to as an NDA, in certain circumstances such an agreement can preserve the novelty of an invention. Documents exchanged under a properly written and executed NDA may not be considered as a publication, or public dissemination of an invention and can preserve the right of the inventor to apply for patent protection.

Non-facilities Based Carrier-Refers to carriers that do not operate their own switches and networks

Non-Obvious-According to 35 USC 103 a) a patent may not be obtained if the differences between the subject matter sought to be patented and the prior art are such that the subject matter as a whole would have been obvious at the time the invention was made to a person of ordinary skill in the art.
Non-Obvious is analogous to Inventive Step.

Non-Payment Churn-The most common form of involuntary churn. Non-payment churn occurs when a customer fails to pay their bills and the telco terminates their service.

Non-Recurring Charge (NRC)-A cost for a facility or product that only occurs one time or is not periodically charged.

Non-Recurring Engineering Costs (NRE)-Costs associated with a product or service that are associated with the development of the product and not associated with the marketing or support of the product sales.

Non-Wireline-Term used to describe the A side cellular carrier. Non wireline refers to the fact that the carrier did not have in place "wired" or landline service originally.

Non-Wireline Carriers-Cellular service providers that are not engaged in the business of providing landline telephone service within the CGSA they are operating in. Non-wireline carriers also are known as A carriers.

Nonce-A number (usually a random number that may be encrypted) that is transferred from a authentication challenging device (such as a service provider) to another device so that the receiving device can process this value with other information and return a response so the challenger can compare the result to confirm the identity of the recipient.

Nonlinear-A device or circuit whose output is not directly proportional to its input.

Nonlinear Distortion-The usually undesirable difference between a signal at the input to a system and at the output caused by the nonlinear functioning of the system.

Nonlinear Scattering-In optoelectronics, the direct conversion of a single photon from one wavelength to one or more other wavelengths.

Nonvolatile-A memory device or system whose stored data is unaffected by the removal of operating power.

Nonvolatile Memory-A form of computer memory that will store information for an indefinite period of time with no power applied. If the storage area cannot be rewritten with new information, the non-volatile memory can also be called read only memory (ROM).

NOR Gate-A logic device that only gives a logic one (1) output if all inputs are zero (0). It is equivalent to an OR gate followed by a NOT gate.

Nordic Mobile Telephone (NMT)-A mobile cellular telephone system that was introduced to Europe in 1981. The NMT system has been deployed on two frequency bands; 450 MHz and 900 MHz. These systems use FM (analog) radio modulation and many of these systems are converting to the GSM (digital) system.

Normal-A line perpendicular to another line or to a surface.

Normal Burst (NB)-A 577 bit burst period that is primarily used to transfer data between the base station and the mobile station GSM systems. Each normal burst can transfer 114 bits of user information (less after error protection is removed).

Normal Business Hours-Those hours during which most similar businesses in the community are open to serve customers. In all cases, "normal business hours" must include some evening hours at least one night per week and/or some weekend hours.

Normal Routing-The routing of a given signaling traffic flow under normal conditions, that is, in the absence of failures.

Normalized Frequency-The ratio between the actual frequency and its nominal value.

Normally Closed (NC)

Normally Closed (NC)-Switch contacts that are closed in their non-operated state, or relay contacts that are closed when the relay is de-energized.

Normally Open (NO)-Switch contacts that are open in their non-operated state, or relay contacts that are open when the relay is de-energized.

North American Cellular Network (NACN)-An interconnection of regional cellular carriers that allow cellular customers to travel anywhere within that network and still have their telephone behave and respond the same way as if they were in their home system. Members must adhere to certain customer service standards and hours among other things.

North American Numbering Plan (NANP)-The NANP is an 11 digit-dialing plan that is used within North America. It contains 5 parts: international code, optional intersystem code (1 +), geographic numbering plan area (NPA), central office code (NXX), and station number (XXXX). The NPA code defines a geographic area for the serving telephone system (such as a city). The NXX defines a particular switch that is located within the telephone system. Finally, the station code identifies a particular line (station) that the switch provides service to.

NOS-Network Operating System

Notch Filter-A circuit designed to attenuate a specific frequency band; also known as a "band-stop" or "band reject" filter. Notch filters are sometimes used to restrict access to video signals that are transmitted through a cable television distribution system.

This diagram shows a notch (band reject) filter that is used to block a specific frequency band (television channel) from a multi-channel input signal. In this example, a television system is broadcasting many television channels. This diagram shows how a notch filter can block a specific channel (such as a pay for subscription channel) from being received by a customer.

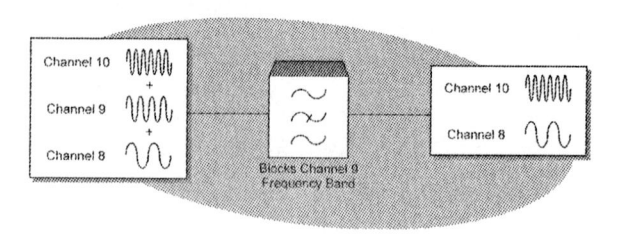

Notch Filter Operation

Notched Noise-A noise signal in which a narrow band of frequencies has been removed.

Notice of Allowance-Notice of Allowance is a formal notification from the USPTO that a patent will be granted. The notice of allowance normally includes a request that the issue fee be paid. Continuation, Continuation-In-Part, and Divisional applications may be filed after receipt of a notice of allowance, but before payment of the issue fee.

Notification Server-A computing device (typically a computer with communications software) that provides notification to users or devices when specific events occur.

Notifications-Notifications are synonymous with SNMP traps (from SNMPv1) or informs (from SNMPv2). Events on network devices or SNMP agents send notifications to the network management system when something of interest occurs on the agent.

Notify Message-A message that is used to provide information to a person or device about an event that has occurred. The event criteria may have been set by sending a Subscribe message along with the parameters of the event (such as exceeding a maximum count or a level that has been exceeded.) The notify message is defined in session initiation protocol (SIP) toolkit.

NPA-Numbering Plan Area

NPA Codes-Interchangeable Numbering Plan Area

NPCS-Narrowband PCS

NPP-Non Premium Percent

NPR-Noise Power Ratio

NPS-Network Planning System

NRC-Non-Recurring Charge
NRE-Non-Recurring Engineering Costs
NRZ-No Return To Zero
NSA-National Security Agency
NSAP-Network Service Access Point
NSC-Non-Standard Facilities Command
NSC-Network Service Center
NSDB-Network And Services Database
nsec-Nanosecond
NSEP Treatment-National Secutity Emergency Preparedness
NSF-Non-Standard Facilities
NSIS-Network Server Interface Specification
NSP-National Signaling Point
NSP-Network Service Part
NSP-Network Service Provider
NSS-Non-Standard Facilities Setup Command
NSS-Network Subsystem
NT-Network Termination
NT1-Network Termination 1
NT2-Network Termination 2
NTCA-National Telephone Cooperative Association
NTE-Network Termination Equipment
NTFS-New Technology File System
NTIA-National Telecommunications and Information Administration
NTIZ-Network Identity And Timezone
NTL-National Transcommunications Limited
NTP-Network Time Protocol
NTSC-National Television System Committee
NTSC Video-The NTSC television system standard was developed in the United States and is used in many parts of the world. The NTSC system uses analog modulation where a sync burst precedes the video information. The NTSC system uses 525 lines of resolution (42 are blanking lines) and has a pixel resolution of approximately 148k to 150k pixels.

This figure demonstrates the operation of the basic NTSC analog television system. The video source is broken into 30 frames per second and converted into multiple lines per frame. Each video line transmission begins with a burst pulse (called a sync pulse) that is followed by a signal that represents color and intensity. The time relative to the starting sync is the position on the line from left to right. Each line is sent until a frame is complete and the next frame can begin. The television receiver decodes the video signal to position and control the intensity of an electronic beam that scans the phosphorus tube ("picture tube") to recreate the display.

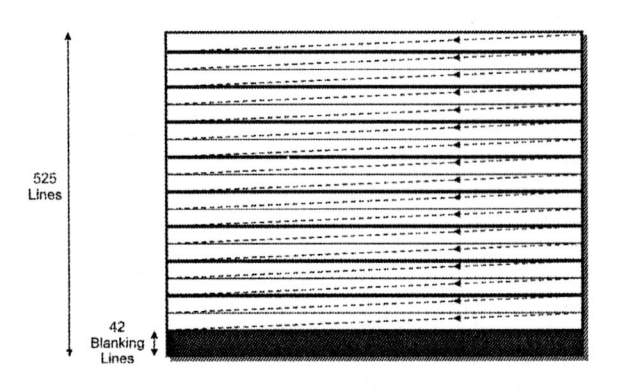

NTSC Video

Null-A zero or minimum amount or position. A binary character that has all the binary digits set to zero. ASCII 0 is a 7 bit character that represents the null (no) value. A null character is used in programming to indicate padding information (filler) or it may be used to indicate an end of a field or block of information (delimiter).

This diagram shows how a null modem cable can be used to connect two data terminal units without the need for a data hub or other network access device. The null modem cable reverses the data and control lines so the transmitter lines from one data communication device are connected to the receiver lines of the other data communications device.

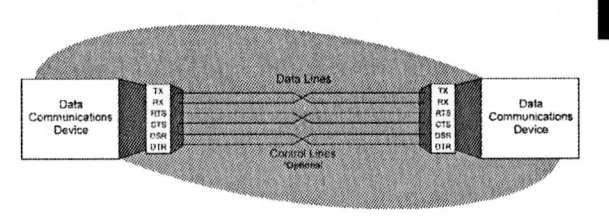

Null Modem Cable

Null Modem Cable-A cable that is configured to cross-connect computers without the need of a modem. A null modem cable reverses the data and control lines (e.g. transmit to receive and receive to transmit).

Number Assignment Module (NAM)-A type of integrated circuit called a PROM that is programmed to contain information specific to a cellular phone, such as its electronic service number (ESN)and the phone number assigned to it. The information contained in its NAM is what identifies the phone to a cell site and mobile telephone switching office (MTSO). It is the memory storage area that contains certain North American Standards profile information.

Number Portability-Number portability involves the ability for a telephone number to be transferred between different service providers. This allows customers to change service providers without having to change telephone numbers. Number portability involves three key elements: local number portability, service portability and geographic portability.

The first part of the telephone number (NPA-NXX) usually identifies a specific geographic area and specific switch where the customer subscribes to telephone service. If a telephone number is assigned to another system (different NXX) in the same geographic area (same NPA), the interconnecting carriers (IXCs) connecting to that system must know which local system to route the calls based on the selected local service providers. In this case, the IXC must look up the local telephone number in a database (called a database dip) prior to delivering the call to the end customer.

Numbering Plan-A numbering plan is a system that identifies communication points within a communications network through the structured use of numbers. The structure of the numbers is divided to indicate specific regions or groups of users. It is important that all users connected to a telephone network agree on a specific numbering plan to be able to identify and route calls from one point to another.

Telephone numbering plans throughout the world and systems vary dramatically. In some countries, it is possible to dial using 5 digits and others require 10 digits. To uniquely identify every device that is connected to public telephone networks, the Consultative Committee for International Telephony And Telegraphy (CCITT) devised a world numbering plan that provides codes for telephone access to each country. These are called country codes. Coupled with the national telephone number assigned to each subscriber in a country, the country code telephone makes that subscribers number unique worldwide. The International Telecommunications Union (ITU) administers the World Numbering Plan standard E.164 and publishes any new standards or modifications to existing standards on the Internet.

Numbering Plan Area (NPA)-A 3-digit code that designates one of the numbering plan areas in the North American Numbering Plan for direct distance dialing. Originally, the format was NO/IX, where N is any digit 2 through 9 and X is any digit. From 1995 on, the acceptable format is NXX.

Numbering Scheme-British English synonym for North American "Numbering Plan."

Numeric-A display, message or readout that contains numerals only, such as paging.

Numeric Paging-A paging application in which information is sent and displayed as numeric characters. Usage is typically to display phone numbers which user call back to get messages or information.

Numerical Aperture-A measure of the acceptance angle for an optical device (e.g., an optical fiber). The numerical aperture is equal to the square root of the difference between the squares of the indexes of refraction for the two media (e.g., air and fiber).

NVOD-Near Video On Demand

NVP-Network Voice Protocol

NXX-A term used for qualifying dialing digits. N stands for any number 2 through 9 and X stands for any number 0 through. 9An example of a valid NXX is 201.

Nyquist Frequency-The lowest sampling frequency that can be used for analog-to-digital conversion of a signal without resulting in significant aliasing. Normally, this frequency is twice the rate of the highest frequency contained in the signal being sampled.

Nyquist Rate-The maximum rate at which data can be transmitted over a limited-bandwidth channel without inter-symbol interference. The Nyquist rate, in bauds, is twice the channel bandwidth in Hertz. The term was named for the American physicist who determined the rate, Harry Nyquist.

O

O/E-Optical To Electronic Conversion
OA-Optical Amplifier
OA&M-Operations Administration And Maintenance
OADM-Optical Add Drop Multiplexer
OAI-Open Application Interface
OB-Outside Broadcast
OBEX-Object Exchange
OBF-Ordering And Billing Forum
Object Code Program-The representation of machine language computer programs in binary form.
Object Exchange (OBEX)-A session-layer protocol for object exchange originally developed by the Infrared Data Association (IrDA) as IrOBEX. Its purpose is to support the exchange of objects in a simple and spontaneous manner over an infrared or Bluetooth wireless link.
Object Identifier (OID)-The object identifiers or OIDs are defined in the Structure of Management Information (SMI) for SNMP (RFC1065). Object identifiers define the structure of all objects defined in a tree format. All names of objects have corresponding numbers. For example, the path to SNMP objects is represented in the following two ways as an OID: iso.org.dod.internet or 1.3.6.1. OIDs can be represented as all text, all numeric, or a combination of both.
Object Linking And Embedding (OLE)-A mechanism by which Microsoft Windows applications can include each other's creations and files. This often refers to a screen-scraping technique that enables the incorporation of a character-based system into a graphical user format.
OC&C-Other Charges & Credits
OC-1-The SONET optical carrier 1, operating at a data rate of 51.84 Mbps.
OC-12-The SONET optical carrier 12, operating at a data rate of 622.08 Mbps.
OC-3-The SONET optical carrier 3, operating at a data rate of 155.52 Mbps.
OC-48-The SONET optical carrier 48, operating at a data rate of 2488.32 Mbps.
OC-n-Optical Carrier Hierarchy
OC1-Optical Carrier 1
OC12-Optical Carrier 12
OC192-Optical Carrier 192
OC3-Optical Carrier 3
OC768-Optical Carrier 768
OC9-Optical Carrier 9
OCA-Outside Collections Agency
OCC-Other Common Carrier
Occupancy-The fraction of the time that a circuit or equipment is in use, expressed as a decimal. Occupancy is the erlangs carried and is equal to the hundred call seconds (CCS) carried divided by 36. It includes both message time and setup time.
OCN-Operating Company Number
OCR-Optical Character Recognition
Octal-A numbering scheme with the base 8.
Octave-Any frequency band in which the highest frequency is twice the lowest frequency.
Octet-A group of eight binary digits, usually operated upon as an entity. See also Byte.
ODBC-Open Database Connectivity
Odd Parity-A data error detection method in which one extra bit, the parity bit, is added to the cede signal for each data character such that the total number of ones in the data, including the parity bit, is an odd number.
ODI-Open Data Link Interface
ODMA-Opportunity Driven Multiple Access
ODS-Operational Data Store
ODSI-Open Directory Services Interface
Odyssey-A MEO satellite system that uses OCDMA technology.
OEM-Original Equipment Manufacturer
OEO-Optical to Electronic to Optical
Of An Earth Satellite-Inclination Of An Orbit
of the ACIF-Cabling Reference Panel
OFDM-Optical Frequency Division Multiplexing
OFDMA-Orthogonal Frequency Division Multiple Access
Off Line (Offline)-(1-general) A condition of devices or subsystems not connected into, not forming a pant of, and not subject to the same controls as an operational system. (2-computer system) A circuit or device that is disconnected from a system, usually a remote computer, and not available for use.
Off Peak-A time period where a telecommunication system usage is lower, typically after normal business hours. Some telecommunications service

providers charge a reduced rate for the use of services during off-peak hours.

Off Premises Extension (OPX)-A call processing feature that allows a call to be forwarded to a telephone at a secondary location that is located off the premises of the phone system that is transferring the telephone call.

Off-Hook-A electrical signal that occurs when a customer typically removes a telephone receiver off of its cradle, thus releasing the hook switch. When the hook switch is released (off-hook), this typically causes a drop in telephone line voltage due to connecting of the local loop telephone wires together. Automatic devices such as a computer modem can also initiate an off-hook signals.

Offline-Off Line

Offset-An intentional difference between the realized value and the nominal value.

Offset Operation-A method of establishing a frequency difference between radio channels that are operating on the same frequency (co-channels) to minimize distortion. Offset operation is commonly used for television and paging systems that use simulcast transmission.

OHCI-Open Host Controller Interface

Ohm-The unit of electrical resistance through which one ampere of current will flow when there is a difference of one volt. The unit is named for the German physicist Georg Simon Ohm (1787-1854).

Ohm's Law-A law that sets forth the relationship between voltage (E), current (I), and resistance (R). The law states that E=IxR.

Ohmic Loss-The power dissipation in a line or circuit caused by electrical resistance.

Ohmmeter-A test instrument used for measuring resistance. Often part of a multimeter.

Ohms Per Volt-A measure of the sensitivity of a voltmeter.

OHX-Foreign Dial Tone

OHY-Second Dial Tone

OI-Operator Interrupt

OID-Object Identifier

OLAP-Online Analytical Processing

OLC-Overload Class

OLE-Object Linking And Embedding

OLT-Optical Line Termination

OMAP-Operation, Maintenance and Administration Part

OMC-Operations And Maintenance Center

OMC-R-Operations and Maintenance Center — Radio

Omnidirectional Antenna-An antenna that transmits its radio signal in all directions equally at a particular azimuth.

On Board Repeater-A radio station that receives and automatically retransmits signals between on-board communication stations.

On Line (Online)-(1-general) A device or system that is energized and operational, and ready to perform useful work. (2-computer system) A circuit or device that is connected to a system, usually a remote computer, and available for use.

On Peak-A time period where a telecommunication system usage is higher, typically during normal business hours. Some telecommunications service providers charge a premium rate for the use of services during peak hours.

On-Demand Streaming-In audio or video streaming, a stream which begins at the time that the client requests it. Usually the client may also pause the stream, skip to a different time in the presentation, or fast-forward or rewind.

On-Hook-A electrical signal that occurs when a customer typically replaces a telephone receiver onto its cradle, thus opening the hook switch. When the hook switch is opened (on-hook), this typically causes a increase in telephone line voltage due to removal of the connection between the telephone wires on the local loop line.

ONA-Open Network Architecture

One Channel Evolution Version Data Only (1XEVDO)-The evolution of existing systems for data only (EVDO) is an upgraded version of the cdma2000 system. The EVDO system uses the same 1.25 MHz radio channel bandwidth as the existing cdma2000 system that provides for multiple voice channels and medium rate data services. The EVDO version changes the modulation technology to allow for a maximum data transmission rate of approximately 2.5 Mbps. The EVDO system has an upgraded packet data transmission control system that is allows for bursty data transmission rather than for more continuous voice data transmission.

One Flat Business Line (1FB)-A telephone line used by a business that is charged a single monthly fee regardless of how many calls that are originated or received during each month.

One-Way Paging-A paging technology whereby the signal is sent from the base station to the paging unit only, without a return verification signal or other 2-way capabilities.

This figure shows a one-way paging system. In this diagram, a high power transmitter broadcasts a paging message to a relatively large geographic area. All pagers that operate on this system listen to all the pages sent, paying close attention for their specific address message. Paging messages are received and processed by a paging center. The paging center receives pages from the local telephone company or it may receive messages from a satellite network. After it receives these messages, they are processed to be sent to the high power paging transmitter by an encoder. The encoder converts the pagers telephone number or identification code entered by the caller to the necessary tones or digital signal to be sent by the paging transmitter.

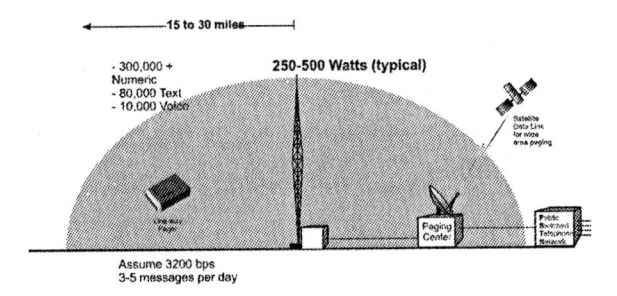

One-Way Paging Operation

Oneshot-A circuit that produces an output signal of fixed duration when an input signal of any duration is applied. A oneshot is also called a monostable multivibrator.
Oneway Trunk-A trunk that can be seized at only one end.
ONI-Operator Number Identification
Online-On Line
ONU-Optical Network Unit
OOO-Optical to Optical to Optical
OOP-Object-Oriented Programming
OP-Outside Plant
Opaque Systems-A optical transmission system that transmit and receive through a transmission medium, multiple optical wavelengths are separated by photonic demultiplexer, and each optical channel is converted into electrical form.
OPC-Originating Point Code
Open-An interruption in the flow of electric current, as caused by a broken wire or connection.

Open Architecture-A design that permits the interconnection of system elements provided by many vendors. The system elements must conform to interface standards.
Open Database Connectivity (ODBC)-An interface for accessing data in a environment of relational and non-relational database management systems. ODBC provides a vendor-neutral way of accessing data in a variety of personal computers, minicomputer and mainframe databases.
Open Host Controller Interface (OHCI)-The Universal Serial Bus (USB) host interface used by Compaq, Microsoft and National Semiconductor.
Open Interface-A connection or access point between two assemblies or systems that is well defined and is readily available to manufacturers or users of the interface. Open interfaces are usually defined to encourage competition as multiple manufacturers can compete to produce products that have open interfaces.
Open Loop Power Control-A process of controlling the transmission power level for the mobile radio using the received power level. Open loop power control is used in systems where rapid power control is required (such as in CDMA systems).
This diagram shows how open loop RF power control can be used in a mobile telephone system to provide approximately the same level of RF signal received by the base station from the mobile telephone regardless of the distance the mobile telephone is located from the base station. This diagram shows that the mobile telephone's coarse (open loop) RF amplifier adjustment is controlled by feedback from its receiver section. The mobile telephone continuously measures the radio signal strength received from the base station to estimate the signal strength loss between the base station and mobile telephone. This diagram shows that as the mobile telephone moves away from the base station, the received signal level decreases. When the received signal is stronger, the mobile telephone reduces its own RF signal output; conversely, when the mobile received signal level is weaker, the mobile telephone increases the amplification of its own RF signal output. The end result is that the signal received at the base station from the mobile telephone remains at about at the same power level regardless of the mobile telephone's distance.

Open Network Architecture (ONA)

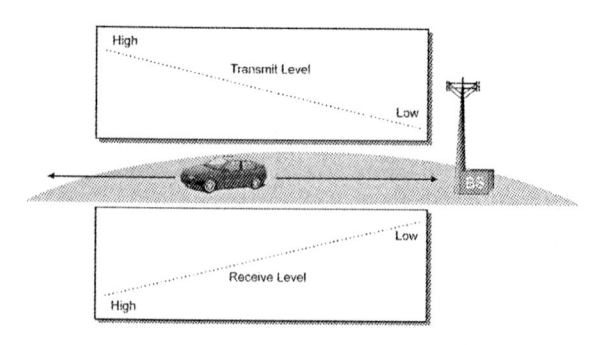

Open Loop Power Control

Open Network Architecture (ONA)-In the context of the FCC's Computer inquiry III, the overall design of a communication carrier's basic network facilities end services to permit all users of the basic network to interconnect to specific basic network functions and interfaces on an unbundled, equal-access basis. Note: the ONA concept consists of three integral components (a) basic service arrangements (BSAs), (b) basic service elements (BSEs), and (c) complementary network services.

Open Numbering Plan-A numbering plan in which local numbers comprise a different quantity of digits, even in the same city, and each area or zone code typically comprises a different quantity of digits. The national telephone numbering plans of many European countries are open plans. For example, there are both 6 digit and 7 digit telephone numbers in the same city in some countries. Some area codes for small towns have more digits than the area codes for larger towns in the same country, etc. The ITU international numbering plan is an open plan, with different national telephone systems being reached via "country codes" having different numbers of digits. The country code for North America is 1, a single digit. The country code for the United Kingdom is 44, a pair of digits. The country code for Israel is 972, comprising three digits. In each case the quantity of digits that must follow the country code is also different for each destination country, and may vary among different cities in the same country. Please note that an assembly of closed numbering plans may comprise an overall open numbering plan! In most local numbering plans the quantity of digits in a number is tied to the leading digits of that number. This is called a "deterministic" open numbering plan. For example, in such a plan, all numbers beginning with the digits 23, 24 or 25 are 5 digits in length, while all numbers beginning with any other two digits are 6 digits in length. When an open local numbering plan is not deterministic, as in some cities in Austria, the originating telephone switch must use a "time out" method to determine when the originator has dialed the last digit. An open numbering plan has the advantage of allowing residents of small towns to dial a minimum quantity of digits actually required to distinguish the small quantity of local telephones in their local dialing area. However, it also increases the complexity of determining accurately when an originator has dialed the final digit of a non-local call. Most systems that handle open numbering plans use the "time out" method. That is, they assume that the originator has completed dialing when an interval of typically 6 seconds elapses without the originator dialing any further digits. Some systems will wait as long as 20 seconds to ensure that no further digits are dialed. This either prolongs the time to set up the call, if the waiting time is very long, or occasionally causes incorrect number connection attempts if the waiting time is too short. In some systems such as in North America, the "time out" method is used for international calls, but an originator dialing an international call can also indicate the end of the dialed digits by using the # key, but this is only possible for a originator who has a touch-tone dial. See also Closed Numbering Plan.

Open Packet Telephony (OPT)-A communication process that uses the call control layer to allow service providers to use open settlement protocol (OSP) to transfer billing records and customer account information. OPT was developed by Cisco.

Open Settlement Protocol (OSP)-A standard protocol that is designed to transfer billing information to allow inter-carrier billing between voice and data communication systems. The OSP format is approved by the European Telecommunications Standards Institute (ETSI). OSP allows communication gateways to transfer call routing and accounting information to clearinghouses for account settlements between carriers (service providers).

Open Shortest Path First (OSPF)-An Internet routing protocol used that provides all the routers within a network domain to know the topology and to use this information when determining the optimal routing (shortest path) through the network. OSPF can also use network loading and bandwidth cost when determining the optical routing path through the network. Routers within an OSPF domain continuously update their stored maps of the network by swapping information with each other.

Open Source Software-Software that includes the original source code from which the product was developed to allow other developers to make changes to the software to meet their specific application needs.

Open Switching Interval (OSI)-A time period that occurs in switching systems (primarily analog switching systems) where circuits or equipments are temporarily disconnected from a line or when other circuits are connected to the line. During OSI periods, other transmission systems (such as custom calling features) may be connected to circuits or equipments.

Open System-A system whose characteristics comply with specified standards and that therefore can be connected to other systems that comply with these same standards.

Open Systems Interconnection (OSI)-The Open Systems Interconnection (OSI) standard layer model was developed by the International Standards Organization (ISO) and the CCITT. The OSI model helps to standardize the inter-connection of computers and data terminals to their applications, regardless of their type or manufacturer. The protocols specify seven layers: physical, link, network, transport, session, presentation, and application. Each layer performs specific functions for data exchange and is independent of the other layers.

This diagram shows the seven layers of the open systems interconnection (OSI) model and how they interact with each other. This example shows how an email application can use the OSI model to allow communication between an email client (user that is checking email) to an email server (computer providing the email information) independent of who controls each layer, provided the interfaces between each layer are specifically defined. This diagram shows that the application layer is the interface to the user that permits the user to request delivery of their email. The application layer presents this request to the transport layer as a data file. The data file is divided up into smaller blocks of data and presented to the session layer. The session layer determines a new session is required (communication link) between the client and the server and this session information is passed on to the transport layer that will oversee the transfer of data during the session. The transport layer sends the destination address of the email server to the network layer. The network layer sends this information to the data link layer that establishes and maintains a data link connection to the network. The data link layer sends information to the physical layer that converts to data signals to either radio, electrical, or optical formats suitable for transmission.

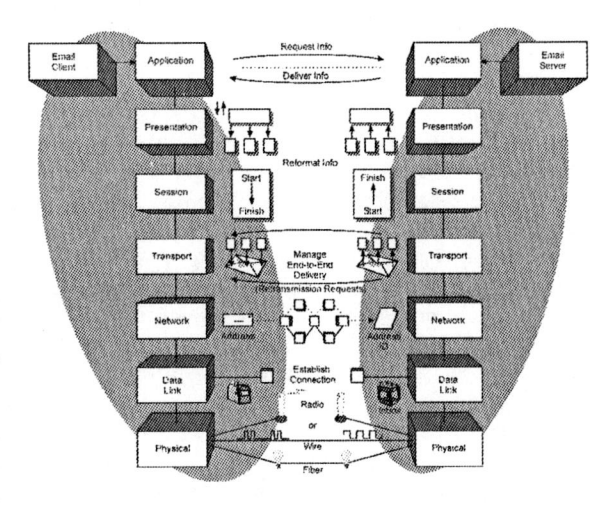

Open System Interconnection (OSI) Protocol Operation

Open Wire-A type of wire installation, now obsolete, in which the electrical conductors need no insulation or sheath for protection from the environment. Open wire is mounted on insulators.

OpenGL-Open Graphics Library

Opening Ticket-An initial work order that is requested by outside plant personnel from the network operations center (NOC) or other maintenance center prior to opening an underground splice closure for either repair or splicing activity. The ticket is closed out at the end of the days activities and a new ticket issued upon request for subsequent work activity.

Operand-Any of the qualities arising out of or resulting from the execution of a computer instruction, a constant, a parameter, the address of any of these quantities, or the next instruction to be executed.

Operating Lifetime-The period of time during which the principal parameters of a component or system remain within a prescribed range.

Operating System (OS)-An operating system is a group of software programs and routines that directs the operation of a computer in its tasks and assists programs in performing their functions. The operating system software is responsible for coordinating and allocating system resources. This includes transferring data to and from memory, processor, and peripheral devices. Software applications use the operating system to gain access to these resources as required.

Operation, Maintenance and Administration Part (OMAP)-The Application Entity that is dedicated to the communications aspects of the Operation, Administration and Maintenance of the Signaling System Network

Operational Data Store (ODS)-A specialized data base system that is created to serve as a storage and holding area for data extracted from operational systems and staged for passive access by other systems and for loading into data warehouses and data marts.

Operational Expenses (OpEx)-The term OpEx is used to define the day-to-day short term expenses paid a telephone company (telco) to support continued business operations (e.g. salaries, rents, commission fees).

Operations-The term denoting the general classifications of services rendered to the public for which separate tariffs are filed, namely exchange, state toll and interstate toll.

Operations Administration And Maintenance (OA&M)-The functions that are necessary to operate, perform administration functions and maintain a communications network.

Operations And Maintenance Center (OMC)-The OMC includes alarms and monitoring equipment to help a network operator diagnose and repair a communications network.

Operations Support System (OSS)-A system that is used to allow a network operator to perform the administrative portions of the business. These functions include customer care, inventory management and billing. Originally, OSS referred to the systems that only supported the operation of the network. Recent definition includes all systems required to support the communications company including network systems, billing, customer care, etc.

Operations System-A general-purpose software system that supports the operations of a telecommunications company. Operations supported include planning, engineering, ordering, inventory tracking, automated designs, provisioning, assignment, installation, maintenance, and testing.

Operator-(1-general) A person who assists customers with the operation or use of a system or service (2-carrier) In telecommunications, this is the company that provides communication services.

Operator Assisted Call-A telephone call made by a customer who dials for an access code for assistance (such as 0 +) or is automatically connected to an operator for assistance in placing person-to-person, collect, coin, third-panty-billed, or credit card calls.

Operator Interrupt (OI)-An operator service whereby the operator may interrupted an ongoing conversation. Sometimes called an Emergency Interrupt (EI)

Operator Number Identification (ONI)-The manual identification of the calling number of a customer dialed toll call and its entry by an operator onto an automatic message accounting tape for billing.

Operator Relay Services-A program to assist those with hearing and/or speech disabilities to communicate over telephone networks through the use of a relay operator who translates written text into speech, and spoken replies into text.

Operator Services-Operator services use an operator to assist in the handling of a processing of a call. These special handling services include collect calling (billing to a called number), third party charging (billing to another phone or calling card), identification of a person who has called (call trace services), call information services (assistance with directory number location), rate information services (call charge rates), or any other service that requires an operator for special call processing services.

Operator Services System-A service system that allows for any special handling of the calls. These special handling services include collect calling (billing to the called number), third party charging (billing to a calling card), identification of the per-

Optical Amplifier (OA)

son called (call trace services), call information services (assistance with directory number location), rate information services (call charge rates), or any other service that requires special call processing services.

Operator Trunks-A term that generally refers to multi-channel communication lines (trunks) that are located between manually operated switchboard positions and local dial central offices.

OpEx-Operational Expenses

Opportunity Driven Multiple Access (ODMA)- A radio protocol that is uses available nodes (radios or routers) in a communication network to help route (relay) information to its destination.

Opposition-According to article 99 of the European Patent Convention, within 9 months of the publication of the mention of the grant of the European Patent, any person may give notice to the European Patent Office of opposition to the European Patent granted. The Opposition procedure is conducted inter partes with the applicant and opponent, or opponents, each filing written reasons for and against the grant of a patent.

Similar procedures exist in most national patent offices, with the exception of the United States where "re-examination" is only remedy.

OPS-Order Processing System

OPSIINE-Operations Process System Intelligent Network Elements System

OPT-Open Packet Telephony

Optical Add Drop Multiplexer (OADM)-In an optical network carrying multiple optical signals (WDM or DWDM), an Optical Add/Drop Multiplexer adds or removes (drops) individual optical signals (also known as lightwaves or wavelengths) from the network. This action provides access to specific wavelengths for other networks or access points.

This photograph shows an ONU device that connects a fiber optic cable line to a electrical data communications line or device. The product is the Alcatel 1640 optical add/drop multiplexer (80/160 channel DWDM System).

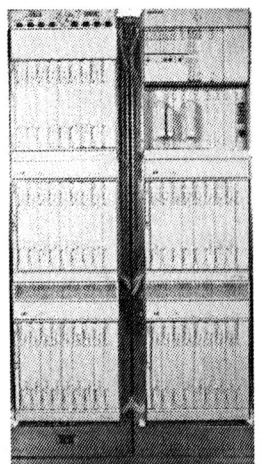

Optical Network Unit (ONU)
Source: Alcatel USA

Optical Amplifier (OA)-A means of amplifying an optical signal through the sensing of energy at particular optical wavelengths and adding optical energy at the same wavelength so the resultant signal is a replica of the input signal at higher (amplified) energy level. The most common optical amplifiers are the Erbium doped fiber amplifier (EDFA), RAMAN, and semiconductor laser amplifier.

This picture shows the model FL8011 series optical signal amplifier. It's design based on Erbium doped fluoride glass fiber that is optimized to produce a spectrally flat (linear) gain profile. This instrument has a gain profile that is uniform (better than 1.5db uniformity) over the entire C-band. This power amplifier series provides low noise amplification up to 21dBm within the wavelength range of 1530nm to 1565nm. Our near flat top gain profile (<1.5dB ripple) is achieved without the use of gain equalization

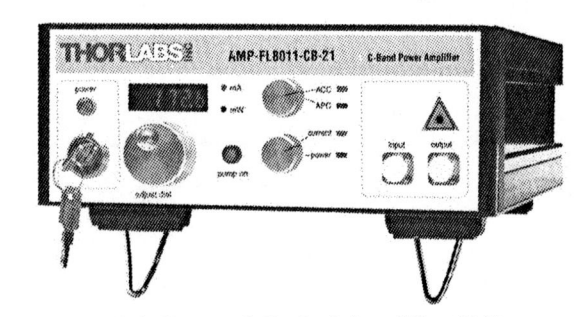

High Powered Optical Amplifier (OA)
Source: Thorlabs Inc.

Optical Attenuator

Optical Attenuator-A device that is used to reduce the intensity (strength) of lightwaves in a fiber optic transmission system when inserted serially into an optical link. Optical attenuators are composed of semitransparent material that absorbs a significant amount of photonic energy within the attenuator.

Optical Axis-The axis of symmetry of an optical fiber or device, typically the direction the light travels through the device. In a fiber, the axis of symmetry is its center, i.e., along its length.

Optical Cable-A cable that contains fibers or bundles of fiber lines that is designed for physical and optical (e.g. optical signal loss at specific wavelengths) specifications that allow it to be used in specific types of optical communication applications (e.g. undersea or in-building.)

Optical Cards-A card similar to a credit card that stores information (such as account codes and account balances) that is stored and retrieved from the card optically.

This diagram shows the optical standards for both SONET and SDH. This table shows that the first common optical level between SONET and SDH is OC3 or STS-1. STS-x and STM-x are the standards that specify the electrical signal characteristics that are input to the respective optical encoding/multiplexing processes.

Optical Carrier (OC) Level	Signal Level	Sonet STS Level	SDH STM Level	DS0 (64 Kbps) Channels
OC-1	51.84 Mbps	STS-1		672
OC-2	103.68 Mbps	STS-2		1,344
OC-3	155.52 Mbps	STS-3	STM-1	2,016
OC-4	207.36 Mbps	STS-4	STM-3	2,688
OC-9	466.56 Mbps	STS-9	STM-3	6,048
OC-12	622.08 Mbps	STS-12	STM-4	8,064
OC-18	933.12 Mbps	STS-18	STM-6	12,096
OC-24	1.24416 Gbps	STS-24	STM-8	16,128
OC-36	1.86624 Gbps	STS-36	STM-12	24,192
OC-48	2.48832 Gbps	STS-48	STM-16	32,256
OC-96	4.976 Gbps	STS-96	STM-32	64,512
OC-192	9.953 Gbps	STS-192	STM-64	129,024
OC-256	13.219 Gbps	STS-256		171,360
OC-768	52.877 Gbps	STS-796		516,096

Optical Carrier (OC) Hierarchy

Optical Carrier 1 (OC1)-Operates at 51.84 Mbps
Optical Carrier 12 (OC12)-Operates at 622.08 Mbps (12 X OC1)
Optical Carrier 192 (OC192)-Operates at 9.95 Gbps
Optical Carrier 3 (OC3)-Operates at 155.52 Mbps (3 X OC1). This rate is the lowest at which Asynchronous Transfer Mode (ATM) is implemented.
Optical Carrier 768 (OC768)-Operates at 39.81 Gbps
Optical Carrier 9 (OC9)-Operates at 466.56 Mbps (9 X OC1)
Optical Carrier Hierarchy (OC-n)-Optical carrier (OC-n) transmission is a hierarchy of optical communication channels and lines that range from 51 Mbps to tens of Gbps (and continues to increase). The "n" is an integer (typically 1, 3, 12, 48, 192, or 768) representing the data rate. Lower level OC structures are combined to produce higher-speed communication lines. There are different structures of OC. The North American standard is called synchronous optical network (SONET) and the European (world standard) is synchronous digital hierarchy (SDH).

Optical Cavity-A cavity is a well-defined space, such as a cylinder or a box. For instance, a segment of optical fiber can be considered a cylindrical cavity and most lasers have a box-shaped cavity. An Optical Cavity is formed when the two ends of the cylinder or box are coated to form mirrors so that light inside the cavity bounces back and forth between the two ends. The purpose of bouncing the light back and forth is to select a specific wavelength, direction of travel and/or other properties of the light. Typically, one or both mirrors are designed to allow a small percentage of the light out, e.g., for a laser beam. A typical use of an optical cavity is in a semiconductor or other laser where the multiple reflections are used to help select the specific photons of interest.

Optical Character Recognition (OCR)-The recognition of printed or handwritten characters by automatic systems, often laser- and photoelectric-based.

Optical Clock-A technique for using optical pulses to distribute timing signals in a computer system. The pulses are carried over optical fiber with virtually none of the timing variations or distortion encountered with the use of metallic wire connections.

Optical Combiner-A device that combines several input optical signals (usually from fibers) into one or several output fibers.

Optical Connector-A connector that has its base permanently attached to the end of a fiber and has a mechanical configuration that allows the fiber to be connected and disconnected from other fiber cables or equipment panels with suitable connectors.

Optical Data Bus-An optical fiber network used to interconnect terminals in which any terminal can communicate with any other terminal.

Optical Demultiplexing-Optical demultiplexing separates individual optical signals (or wavelengths or lightwaves) from each other so that they can be rerouted or processed individually. It is used in WDM and DWDM systems which, by definition, carry multiple optical signals.

Optical Density-A measure of the transmittance of an optical element. The higher the optical density, the lower the transmittance. The optical density value multiplied by 10 is equal to the transmission loss expressed in decibels. For example, an optical density of 0.3 corresponds to a transmission loss of 3 dB.

Optical Detector-Usually a semiconductor device, such as a PIN or avalanche photodiode, that converts light to an electrical signal in fiber optic communications systems. An optical detector also is called an optical receiver.

Optical Disk-A form of data storage using a laser to optically record the data on a disk which is read with a low-power laser pickup. The primary types of optical discs are: read only (RO), write once read many (WORM), erasable/record-able (thermo-magneto-optical TMO) and phase change, (PC).

Optical Fiber-A thin filament of glass (usually smaller than a human hair) that is used to transmit voice, data, or video signals in the form of light energy (typically in pulses).

Optical Fiber Connector-A device whose purpose is to transfer optical power between two optical fibers or bundles, and that is designed to be connected and disconnected repeatedly.

Optical Fiber Coupler-A device whose purpose is to distribute optical power among two or more ports. The term also may be used to describe a device whose purpose is to couple power between a fiber and a source or detector.

Optical Fiber Jacket-The material, often PVC, that covers an optical fiber to protect it from damage and make it easier to handle. A jacket can surround one fiber or a group of them. A fiber may have a plastic buffer layer between it and the jacket.

Optical Fiber Splice-A permanent joint whose purpose is to couple optical power between two fibers.

Optical Filter-A passive or active device that selectively blocks (rejects) or transmits (passes) a range of wavelengths in a fiber optic transmission line.

Optical Frequency Division Multiplexing (OFDM)-A process of using transmitting several high speed communication channels through a single fiber through the use of separate wavelengths (optical frequencies) for each communication channel. OFDM is now commonly called wave division multiplexing (WDM). However, WDM usually refers to optical channels that have very small spacing between them and OFDM refers to multiple optical channels that can have any amount of wavelength spacing between them.

Optical Isolator-A device used to allow light to travel in only one direction through an optical element. They are used to prevent light from reflecting back into lasers or fibers causing damage or signal loss. Optical isolators typically use two polarization filters and a Faraday rotator to accomplish this effect. Light entering in one direction is vertically polarized by the first filter, rotated by 45 degrees, and then passed through the second filter, which is designed to pass linearly polarized light rotated by 45 degrees from the first filter. Light entering from the opposite direction is linearly polarized at an angle of 45 degrees from vertical by the first filter. The Faraday rotator then rotates the angle of polarization another 45 degrees so that the polarization is horizontal. The second filter blocks most of this light because it only passes light polarized in the vertical direction.

Optical Line Termination (OLT)-Optical line termination (OLT) are interfaces in a passive optical network that allow multiple communication channels to be combined to different optical wavelengths for distribution through the PON.

Optical Link-Any optical transmission channel designed to connect two end terminals or to be connected in series with other links. Terminal hardware also may be considered within the bounds of this term.

Optical Loss Test Set-An optical power meter and light source calibrated for use together.

Optical Multiplexer

Optical Multiplexer-The optical multiplexer combines multiple signals onto a common optical medium (such as a fiber). An optical multiplexer is often called an optical add-drop multiplexer ("OADM").

Optical Network Unit (ONU)-ONU's are used to multiplex and demultiplex signals to and from a fiber transmission line. An ONU terminates an optical fiber line and converts the signal to a format suitable for distribution to a customer's equipment. When used for residential use, a single ONU can server 128 to 500 dwellings.

Optical Node-An assembly within an optical network where signals are transferred between optical fibers to other transmission media such as wires or coaxial cable.

This photograph shows the optical node that converts optical signals into multiple electrical communication channels. This optical node is contained in a weather resistant housing that allows it to be installed on or near communication lines.

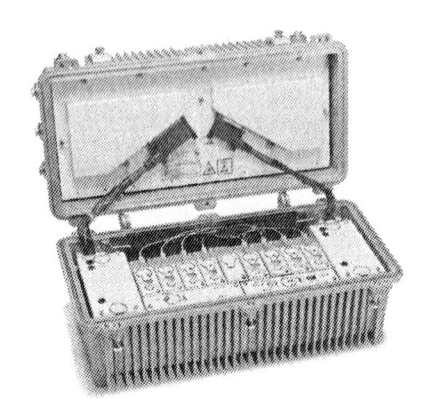

Optical Repeater-A network element that receives and retransmits optical signals after enhancing or regenerating the signal. Optical Repeaters can be all-optical if the signal only needs to be amplified. For more sophisticated regeneration, the optical signal is converted to an electronic signal, in which form the signal-to-noise ration (SNR) can be improved among other processing. See Regeneration and 1R, 2R, and 3R Regeneration.

Optical Source-An optical emitter, usually a semiconductor device (such as a laser or light-emitting diode), that converts an electrical signal into a light pulse for use in a fiber optic communications system. This device also is called an optical transmitter.

Optical Spectrum-The distribution of optical energy as a function of frequency or wavelength.

This diagram shows the different portions of the optical spectrum and their primary characteristics.

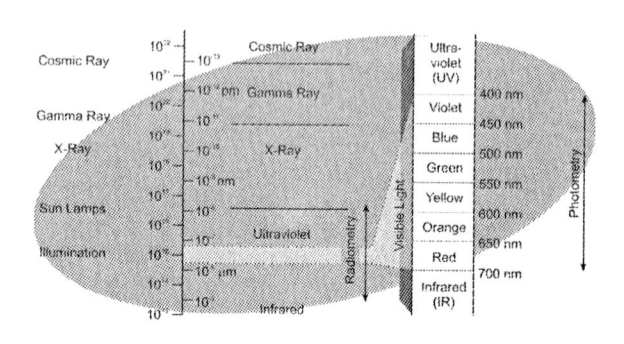

Optical Node
Source: ARRIS

Optical Spectrum Diagram

Optical Path-The path that an optical signal travels from transmitter to receiver. The path can be within optical fiber, through various optical elements, or through free space.

Optical Pigtail-A short length of optical fiber cable (generally less than 3 meters) that serves as a patch cord for coupling a transmission fiber and a component.

Optical Receiver-A device used to convert optical signals into electrical signals.

Optical Switching-The process of directly connecting optical signals between multiple ports or time periods on an optical communication line without the need to convert the optical signals to electrical form.

Optical Time Domain Reflectometer (OTDR)-A test instrument that measures the time for transmission and return of optical signal pulses through a fiber transmission line to determine the distance to specific points in that transmission line (such as a cable break or damaged portion). The OTDR tester operates by sending an optical pulse of energy, timing the response, and analyzing the shape of the pulse to determine the type of change in the transmission line that caused the signal reflection (e.g. un-terminated-terminated cable break).

This picture shows the Tektronix NetTek handheld optical time domain (OTDR) test instrument. This "mini" optical time domain reflectometer (OTDR) and optical power meter modules can be used with a NetTek Analyzer to control and display line condition, power levels, and perform many different measurement tasks. Using different OTDR modules allows for the testing of virtually every prevailing mode, wavelength, and fiber optic network, and both in-band and out-of-band measurements.

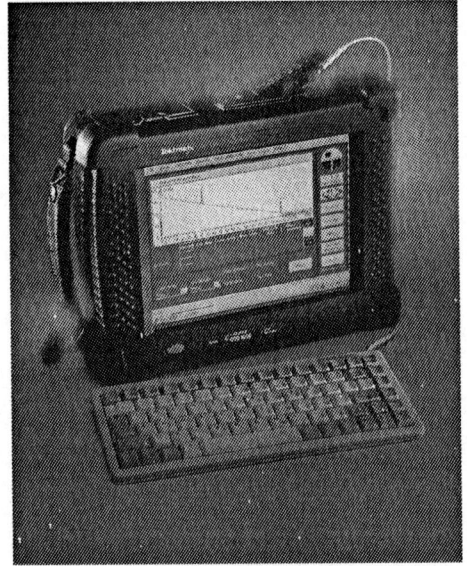

Optical Time Domain Reflectometer (OTDR)
Source: Tektronix, Inc.

Optical To Electronic Conversion (O/E)-The process of changing an optical signal into an electronic one. Optical signals, or lightwaves, are used to transport information rapidly. However, many critical network processing steps can only be performed on electronic signals. For this reason, the optical signals are converted to electronic signals at the endpoints of a transmission and usually several points in between for processing. See also Electronic to Optical Conversion (E/O).

Optical to Electronic to Optical (OEO)-Optical to electronic to optical (OEO) refers to network elements that convert optical signals to electronic ones for signal processing, then convert them back to optical signals for further transport through the optical network. See also OOO.

Optical to Optical to Optical (OOO)-Optical to optical to optical (OOO) is a term used to describe all-optical network elements, such as optical switches, to differentiate them from OEO (Optical to Electronic to Optical) elements.

Optical Transmitter-A device, circuit, or assembly that converts electrical signals into an optical signal.

Optical Transparent-A material that allows photonic energy to the same amount of energy that enters to leave without any loss of energy.

Optical Waveguide-Any physical medium (such as fiber cable) or structure capable of guiding (channeling) optical power from one point to another. An optical fiber is an optical waveguide.

Optical Waveguide Connector-A device whose purpose is to transfer optical power between two optical waveguides or bundles, and that is designed to be connected and disconnected repeatedly

Optical Waveguide Coupler-A device whose purpose is to distribute optical power among two or more ports. This term also may refer to a device whose purpose is to couple optical power between a waveguide and a source or detector.

Optical Waveguide Splice-A permanent joint designed to couple optical power between two waveguides.

Optical Waveguide Termination-A device mounted at the end of an optical waveguide that is intended to prevent signal reflections.

Optically Active Material-A material with the capability to rotate the plane of polarization of light sent through it. Such a material exhibits different refractive indices for left and right circular polarization.

Optically Opaque-A material that decreases the amount of photonic energy from when it enters the medium to when it leaves (a loss of transmitted energy).

Optimize-The process of adjusting for the best output or maximum response from a circuit or system.

Optimized Network-A network in which each trunk group has been sized to operate at its specified economic or service objective when traffic is routed according to a specified plan.

Optimum Layered Pricing Strategy-A pricing plan that allocates different prices to different customers for the same service levels.

Optional Calling Plan-A service offering that gives customers the choice of expanding a local calling area for an additional monthly charge, or selecting a smaller calling area and paying toll charges

Options Message

for all calls outside that area. The plan includes Extended Area Service (EAS).

Options Message-A message that is used to define the parameters of a communication session. The options message is defined in session initiation protocol (SIP).

Optoelectronic Device-A device responsive to electromagnetic radiation (light) in the visible, infrared, or ultraviolet spectral regions, emits or modifies non-coherent or coherent electromagnetic radiation in these same regions, or utilizes such radiation for its internal operation.

Optoelectronic Integrated Circuit-An integrated circuit that contains electronic and optical devices on the same semi-conductor chip.

Optoelectronic Nodes-Devices that receive, convert, and route optical signals.

Optoelectronics-The range of materials and devices that generate, amplify, detect, and control light. Each of these functions requires electric energy to operate and depends on electronic devices to sense and control this energy.

Optoisolator-A coupling device consisting of a light emitter and a photodetector used to couple signals without any electrical connection. Opto-isolators provide voltage and/or noise isolation between input and output, while transferring the desired signal.

OPX-Off Premises Extension

Or Automatic Number Identification-
Centralized Automatic Message Accounting-
Operator

Or Down-Step Up

Oracle NCA-Oracle Network Computing Architecture

ORB-Object Request Broker

Orbit-(1-general) The path, relative to a specified frame of reference, described by the center of mass of a satellite or other object in space, subjected solely to natural forces (mainly gravitational attraction). (2-equatorial) An orbit parallel to a plane through the equator. (3-polar satellite) An orbit in plane running north and south; a satellite in a polar orbit is continually changing its position over the earth. (4-sub-synchronous satellite) A satellite orbit that is at a lower altitude than the 35,880 km requited for a synchronous equatorial orbit (5-synchronous) The path followed by a communication satellite at such a distance above the Earth (about 35,880 km) that the satellite keeps pace with the Earth's rotation, and thus hovers above a particular point on the earth.

Order Handling-Order Handling is the process of entering the orders gathered by the sales organization into the billing and network management systems.

Order Processing System (OPS)-Order processing systems gather information related to orders, process the information into specific orders, and create actionable information that allows the fulfillment of the orders. SIP telephone systems can be integrated with order processing systems to allow interactive control with customers to allow the capturing of order information directly from customers and to assist in fulfillment of the order.

Order Wire-A dedicated voice grade line for communications between maintenance and repair personnel. In digital carrier systems, the order wire is a communication (talk) channel that allows near and far end telephone company (telco) personnel to communicate using telephone sets.

Organizational Learning-Process whereby the entire organization is able to receive feedback from the environment (the market, competition, new technology providers) and dynamically adapt itself accordingly.

Original Equipment Manufacturer (OEM)-The original manufacturer of equipment regardless of who sells the equipment or the name marked or associated with the equipment. The term OEM is sometimes used to refer to companies that use other manufacturers to produce products for them. These companies sell, name, and/or use their distribution system for the product that was produced by the other manufacturer. When a company adds assemblies, software, or documentation to products produced by OEMs. These are referred to as a value-added reseller (VAR). See also Badget.

Originating Number-In telephone number of the device that originated a call. The originating number is often provided as an ANI.

Originating Point Code (OPC)-A identification code that is part of an signaling system (SS7) message that uniquely identifies the originating point of the SS7 message.

Originating Point Code Part-The part of the SS7 message label that uniquely identifies the originating point of the message.

Originating Rate Center-The office in a geographic area that is designated as a rate center for a service originating within that area.

Orthogonal Coding-A method of multiplexing that allows multiple communication channels to exist on the same medium through the use of coded channels. The codes for these channels are chosen so that interference between information elements (symbols) are mutually exclusive.

Orthogonal Variable Spreading Factor (OVSF)-The use of OVSF allows the spreading factor to be changed and remain orthogonal (no interaction between codes) between different spreading codes of different lengths to be maintained.

OS-Operating System

Oscillation-(1-general) A periodic change in a voltage, current or other quantity above and below a mean value. (2-parasitic) An oscillation, usually unwanted, that occurs in a self-resonant element of a device.

Oscillator-(1-general) An electrical device or circuit that converts direct current into a periodically varying output. (2-audio) A circuit used to produce audio frequency alternating current. (3-beat-frequency) A circuit used to generate a signal that is combined with a received radio signal to produce an audio frequency beat signal. (4- crystal-controlled) An oscillator circuit whose frequency is accurately controlled by a quartz crystal. (5-local) An oscillator in a radio receiver whose output is mixed with the received radio signal to produce an in intermediate frequency (IF) signal, which is amplified, then detected. (6-master) A stable oscillator used to control the operating frequency of other oscillators and/or systems. (7- relaxation) An oscillator whose frequency is dependent on the charging time of a capacitor.

Oscilloscope-A test instrument that uses a display, usually a cathode ray tube or liquid crystal display (LCD) to show the instantaneous values and waveforms of a signal that varies with time or some other parameter.

This photograph shows the WavePro 954 test instrument that combines oscilloscope functions along with other test capabilities such as spectrum analysis into an integrated unit. This test instrument is capable of simultaneously analyzing and displaying multiple signal images.

Oscilloscope
Source: LeCroy

OSI-Open Systems Interconnection
OSI-Open Switching Interval
OSI Reference Model-This reference model was created in 1982 by the International Standards Organization (ISO) to standardize communication systems. The model standardized nomenclature across existing protocols and provided guidelines for new protocols using 7 layers. Each successively higher layer builds on the functions of the layers below, as follows:

Application layer 7 The highest level of the model. It defines the manner in which applications interact with the network, including database management, e-mail, and terminal-emulation programs.

Presentation layer 6 Defines the way in which data is formatted, presented, converted, and encoded.

Session layer 5 Coordinates communications and maintains the session for as long as it is needed, performing security, logging, and administrative functions.

Transport layer 4 Defines protocols for structuring messages and supervises the validity of the transmission by performing some error checking.

Network layer 3 Defines protocols for data routing to ensure that the information arrives at the correct destination node.

Data-link layer 2 Validates the integrity of the flow of data from one node to another by synchronizing blocks of data and controlling the flow of data.

Physical layer 1 Defines the mechanism for communicating with the transmission medium and interface hardware.

OSP-Out Side Plant
OSP-Open Settlement Protocol
OSPF-Open Shortest Path First

OSS-Operations Support System
OTAC-Over-The-Air-Control
OTAP-Over-The-Air-Programming
OTAR-Over-The-Air-Rekeying
OTD-Orthogonal Transmit Diversity
OTDR-Optical Time Domain Reflectometer
Other Charges & Credits (OC&C)-Charges and credits which do not fall under any other billing category. An example of OC&C: One time waiver of charge.
Other Common Carrier (OCC)-A pre-divestiture term for a telecommunications common carrier, other than a former Bell operating company, authorized to provide a variety of private line services. This term has been replaced by interexchange carrier. The Federal Communications Commission (FCC) also uses the terms miscellaneous, or specialized common carrier.
Out Of Band Emission-Emission on a frequency or frequencies immediately outside the necessary bandwidth that results from the modulation process, but excluding spurious emissions.
Out of Band Signaling-Signaling, typically related to call processing such as setup, disconnection, (or handover in a cellular system) that is transmitted without using any part of the transmission channel capacity reserved for subscriber traffic. In older frequency division multiplexing equipment, some transmissions of this type utilized a different frequency band than the speech signal, so the name was literally accurate. In modern digital transmission systems, this term is sometimes loosely used for signals that use different digital bits than those reserved for the subscriber traffic. The term is not precisely accurate in this case, because there are not different bands of frequencies involved. The opposite term is "In band signaling"
This figure shows how out-of-band signaling occurs on a telephone network. This diagram shows that a signaling control message is sent on different channels than voice or data information (e.g. voice trunks). This diagram shows that these signaling messages are sent on their own network (packet switching network). Because all the control signals are sent on a common packet network, this is called common channel signaling.

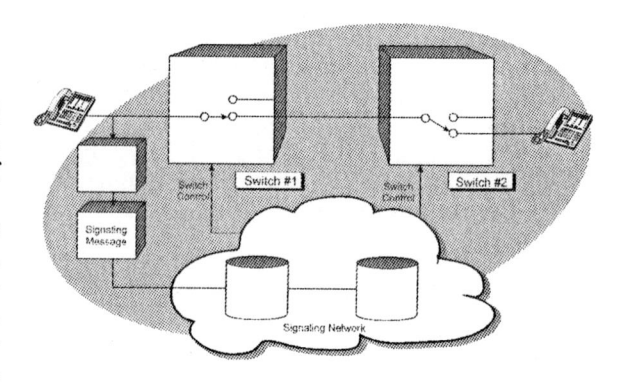

Out of Band Signaling Operation

Out of Slot Signaling-The transmission of signals in a time division multiplexing system in different time slots (or in general in different bit fields) than those reserved for the digital subscriber traffic. A more accurate term than "out of band signaling" when used with digital systems, but actually only precisely accurate when used for time division multiplexing systems.
Out Side Plant (OSP)-All telephone company (telco) facilities that are located from the main distribution frame (MDF) outward toward the subscriber, interoffice or toll facility. This includes all toll, trunk, exchange grade facilities whether copper, fiber, or wireless.
Out-of-Band-Communication-A type of communication that uses frequencies outside the range being used for data or message communications. This name is sometimes questionably applied to digital time division multiplexing systems in which a communication occurs out of the time slot used for subscriber traffic, better called "out-of-slot." Out-of-band communication is generally done for diagnostics or management purpose.
Out-of-Slot-See: out-of-band
OUT-WATS-Outward WATS
Outage-Any disruption of service that persists for more than a specified time period.
Outage Probability-The probability that the outage state will occur within a specified time period. In the absence of specific known causes of outages, the outage probability is the sum of all outage duration's divided by the time period of measurement.

Outage Threshold-A defined value for a supported performance parameter that establishes the minimum operational service performance level for that parameter.

Outbound Call Center-A call center (group of customer service agents) that originate telephone calls from customers. Outbound call centers (telemarketing centers) are often used by companies to solicit new business or to obtain statistical or other business related information.

OutCollects-Outcollects are charges a network operator sends to other telecommunication companies for services they provided to customers that are not registered in the local network (such as completing their calls in local network). Outcollects are call detail records (CDRs) that are sent by service provider A to service provider B for services provided by A to B's customers. An example of an Outcollect is a roaming billing record that is sent to the home service provider of a customer that details the usage of the customer for services that were provided by the visited system.

Outgoing Call Restriction-In telephone call-processing feature that restricts telephone use to specific authorized dialing patterns (typically local phone number).

Outpulsing-The process of transmitting address information over a trunk from one switching environment to another.

Output Power-The radio frequency output power of a transmitter's final radio frequency stage as measured at the output terminal while connected to a load of the impedance recommended by the manufacturer.

Output Stage-The final driving circuit in a piece of electronic equipment.

Outside Broadcast (OB)-A radio or TV program that originates outside the studio. If the program is presented live, the signals must be sent back to the permanent control equipment by temporary links.

Outside Collections Agency (OCA)-An external organization that attempts to collect past due money from delinquent customers. OCAs are generally perform collection services for a fee (commission) that is based on the success of their collection activity.

Outside Plant (OP)-The part of a telephone system that is outside of local exchange company buildings. Included are cables and supporting structures. Microwave towers, antennas, and cable system repeaters are not considered to be outside plant equipment.

Outsourcing-The use of an outside firm to produce products, assemblies or to perform specific business functions that would or could be conducted internally. When a communication carrier contracts to another company to provide teleservices or facilities management is an example of outsourcing.

Ovenized Crystal Oscillator (OXO)-A crystal oscillator that is enclosed within a temperature regulated heater (oven) to maintain a stable frequency despite external temperature variations.

Over Modulation-(1-general) Amplitude or frequency modulation that exceeds 100 percent, resulting in distortion of the transmitted signal and/or excessive bandwidth. (2-digital coder) The condition when a modulating signal level exceeds the dynamic coding range of a digital coder, causing increased quantization distortion in the output.

Over The Air Activation-The ability for a wireless service provider to program or activate service features for a mobile telephone or radio receiver after the product has been purchased.

Over The Air Programming-The ability for a service provider to directly program information stored in a mobile telephone or radio device. This allows over the air activation after the customer initially purchases a mobile telephone.

Overcoupling-A degree of coupling greater than the critical coupling between two resonant circuits. Over-coupling re results in a wide bandwidth circuit with two peaks in the response curve.

Overflow Traffic-That part of an offered traffic load that is not carried.

Overgauging-The installation of a relief cable with pairs of a thicker gauge than prescribed by transmission or signaling requirements. Overgauging is an economical alternative to using two separate cables of different gauges to serve a route that has differing minimum gauge required.

Overhead Bit-Any bit in a digital data stream other than an information bit. Also called a control bit or, simply over-head.

Overhead Information-The digital information transferred across the functional interface separating a user and a telecommunication system (or between functional entities within a telecommunication system) for the purpose of directing or controlling the transfer of user information.

Overhead Messages-System messages that are sent from a system (such as a cellular base station) to receivers (such as mobile telephones) giving the communication device the necessary parameters

Overlap Sending

(such as access codes or initial transmitter power levels) to operate within that system.

Overlap Sending-A sending condition that occurs in an Integrated Services Digital Network (ISDN) when the setup message sent from a user's equipment to a stored-pro gram control system does not contain all of the called-party number information. Certain functions at the terminal will overlap those of the switching system because the system must wait for a complete address.

Overlay-(1-data storage) The technique of repeatedly using the same areas of internal storage during different stages of the execution of a program. (2-radio system) The process of overlaying another radio system technology or capability over an existing (underlying) system.

Overload-In a transmission system, a power greater than the amount the system was designed to carry. In a power system, an overload could cause excessive heating. In a communications system, distortion of a signal could result.

Overload Class (OLC)-A field within the global system messages sent to communication devices such as mobile telephones that indicates if the device is authorized to attempt access or if its overload class has been restricted from accessing the system. The use of OLC allows systems to reduce the number of call access attempts during periods of high system activity.

Overload Control-A process used by the system to control the access attempts initiated by mobiles. Overload Class (OLC) bits sent in the overhead message inhibit operation of groups of mobile telephones.

Overprovisioning-The providing more capacity than is actually required for a given application. Overprovisioning may include reserving excessive switching capacity or using a higher speed communications link. Overprovisioning is sometimes performed to reduce the transmission delay of data transmission.

Overrun-The loss of data that occurs when a receiving device is unable to accept data at the rate at which it is being transmitted.

Overshoot-The first maximum excursion of a pulse beyond the 100% level. Overshoot is the portion of the pulse that exceeds its defined level temporarily before settling to the correct level. Overshoot amplitude is expressed as a percentage of the defined level.

Oversubscription-A situation that occurs when a service provider sells more capacity to end customers than a communications network can provide at a specific time period. This provides a benefit of reduced network equipment and operational (reduced leased line) cost.

Over-subscription is a common practice in communications networks as customers do not continuously use the maximum capacity assigned to them and customers access the networks at different time periods. Unfortunately, over-subscription in telecommunications can cause problems when customers do attempt to access the network at the same time. For example, when customers open their presents at a holiday event (e.g. Christmas) and attempt to access the Internet at the same time.

OVHM-Overhead Message
OVSF-Orthogonal Variable Spreading Factor
OXO-Ovenized Crystal Oscillator

P

P-Peta-

P-clamp-A hardware item that is used to attach either the terminal or station end of aerial dropwire to a J-hook or first attachment.

P-frame-Predicted Frame

PABX-Private Automatic Branch Exchange

PACCH-Packet Associated Control Channel

PACE-Priority Access Control Enabled

Pacing-A method for controlling call flow. In a pacing algorithm, mathematical rules are established to control the rate at which calls are placed by an automatic dialing machine using momentary sampling techniques to forecast the rate of pacing or dialing.

Packet-A small group of digital bits that is routed through a network to their destination. The bit sequence of the packet (field structure) of a packet may be arranged to include the destination address in addition to the data that is being transported and other data such as the packet originator and error protection bits.

Packet Assembler And Disassembler (PAD)- The PAD divides or converts blocks of data (such as data files) to and from small packets of information. In the disassembly process, a PAD usually assigns sequential numbers to the packets as they are created to allow the reassembly PAD to identify the correct sequence of data packets to reproduce the original data signal.

This diagram shows a packet assembler and disassembler (PAD) system operation. This diagram shows that a large file is to be sent over a packet data network. The large file is supplied to the PAD circuit that divides the data file into smaller packets. These packets are sent toward their destination through a data communications network. When they are received, the are reassembled into the original large data file by a packet assembler.

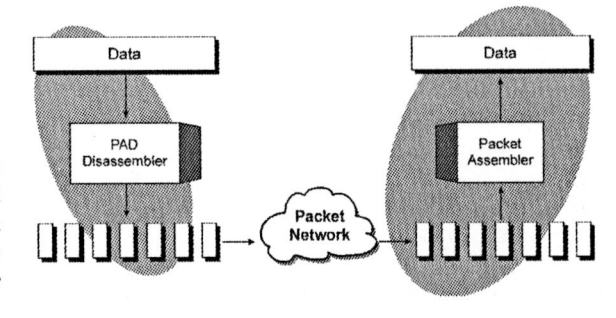

Packet Assembler and Disassembler (PAD) Operation

Packet Buffer-Memory space set aside for storing a packet awaiting transmission or for storing a received packet. The memory may be located in the network interface controller or in the computer to which the controller is connected.

Packet Data-The sending of data through a network in small packets (typically under 100 bytes of information). A packet data system divides large quantities of data into small packets for transmission through a switching network that uses the addresses of the packets to dynamically routes these packets through a switching network to their ultimate destination. When a data block is divided, the packets are given sequence numbers so that a packet assembler/disassembler (PAD) device can recombine the packets to the original data block after they have been transmitted through the network.

Packet Data Convergence Protocol (PDCP)- The packet data protocol that is used in the universal mobile telephone service (UMTS) system. The main function of PDCP is to compress the headers during packet transmission over the radio channel.

Packet Data Transmission Delay-Packet data transmission delay is the accumulated delays that occur for the transmission of a packet from entry into the packet data network (e.g. from entry into a router) to its exit of the system. Common causes of packet data transmission delay include router switching time, alternate routing through other paths, priority queuing delays (other packets may have priority), and retransmission requests.

Packet Filter-The device or circuit that decodes and searches incoming packets to determine and alter its contents or change its routing or contents based on filtering information (e.g. type of service).

Packet Format-A set of rules governing the structure of data control information in a packet. The packet format defines the size and content of the various fields that make up a packet

Packet Forwarding-The process of copying a packet to another node without looking at the destination address.

Packet Header-The front part of a packet that gives the receiving device information such as the message length, the terminal to which the packet is addressed and the packet priority.

Packet Internet Groper (Ping)-A command that is used for data networks that tests for network connectivity by transmitting Internet control message protocol (ICMP) diagnostic packets to a specific node on the network. The ICMP packets for the receiving node to acknowledge that the packet reached the correct destination. If the node responds, this is referred to as a "ping" and it indicates the node is operational.

Packet Loss-The ratio of the number of data packets that have been lost in transmission compared to the total number of packets that have been transmitted.

Packet Mode-The switching of packets of information from different users by statistically multiplexing, or combining, them over shared transmission facilities.

Packet Segmentation-The dividing of a block or packet of data into several segments (pieces). Packet segmentation is often performed to divide a large data packet into smaller data packets so that they can be sent through a network that can only transfer small data packets. When these packets are received at their destination, they are reassembled to their original data packet size. See fragmenting.

Packet Sniffer-A program or process used by to monitor a data stream for a specific pattern such as an address or specific content. Packet sniffers are sometimes used inappropriately to discover passwords or credit card numbers.

Packet Switch-A device in a data transmission network that receives and forwards packets of data. The packet switch receives the packet of data, reads its address, searches in its database for its forwarding address, and sends the packet toward its next destination.

Packet switching is different than circuit switching because circuit switching makes continuous path connections based on a signal's time of arrival (TDM) port of arrival (cross-connect) or frequency of arrival. In a packet switch, each transmission is packetized and individually addressed, much like a letter in the mail. At each post office along the way to the destination, the address is inspected and the letter forwarded to the next closest post office facility. A packet switch works much the same way.

Packet Switched Data-Packet switch data is the transfer of information between two points through the division of the data into small packets. The packets are routed (switched) through the network and reconnected at the other end to recreate the original data. Each data packet contains the address of its destination. This allows each packet to take a different route through the network to reach its destination.

This figure shows data communication using packet switching technology. In this example, a laptop computer is sending a file to a company's remote computer that is connected to a packet data network. The laptop computer data communication software requests the destination address for the packets for the user to connect to the remote computer (202.196.22.45). In this example, the source computer divides the data file into three parts and adds the packet address to each of the 3 data packets. The packets are sent through routers in the packet network that independently determine the best path at the time that will help the packet reach its destination (smart switches). This diagram shows the three packets take 3 different routes to reach their destination. When the 3 packets reach their destination, the remote computer reassembles the data packets into the original data file.

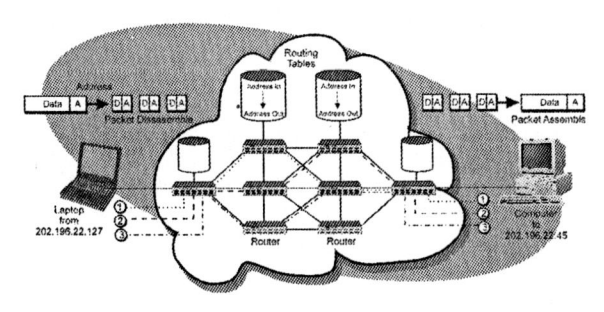

Packet Video

Packet Switched Data Operation

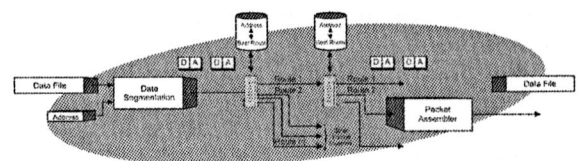

Packet Switching System Types Operation

shows that packets of data arrive at the switch. The routing switch extracts the destination address and possibly the type of message.

Packet Switched Public Data Network (PSPDN)-Public data networks interconnect data communication devices (e.g. computers) with each other through a network that is accessible by many users (the pubic). To allow many different users to communicate with each other, standard communication messages and processes are used. The Internet is an example of a public data network (there are other public data networks) that uses standard Internet protocol (IP) to allow anyone to transfer data from point to point by using data packets. Each transmitted packet in the Internet finds its way through the network switching through nodes (computers). Each node in the Internet forwards received packets to another location (another node) that is closer to its destination. Each node contains routing tables that provide packet-forwarding information.

Packet Switching-A mode of data transmission in which messages are broken into increments, or packets, each of which can be routed separately from a source then reassembled in the proper order at the destination.

This diagram shows two types of packet switching in a communications system. Diagram (a) shows that connection based packet switching sets up a communication circuit prior to transmitting packets that contain data. Diagram (b) shows connectionless packet switching. Connectionless packet switching requires intelligent switching nodes (routers) that can decode the destination address and select the forwarding route based on the results of the lookup in the routing table. This diagram

Packet Switching Exchange (PSE)-Packet switches that are used within the X.25 network.

Packet Video-The sending of video information in packet data format.

The picture below shows a video mail message being viewed on the new Sendo z-100 device. PacketVideo's pvPlayer technology enables the delivery of MPEG-4 videos and images to a variety of handheld devices, including Sendo. Mobile users now have the ability to view full-motion video of family, sports, news, or entertainment, on their handheld devices.

Mobilemedia Phone
Source: PacketVideo and Sendo

Packet-Switched Connection-In the public data wide-area network, the transmission medium is shared by all devices and it is by the address in the packet header that data reaches its destination. Thus, with packet switching the circuit established during call set-up is a virtual circuit. The packet switch directs the packet to the device to which it is addressed.

Packet-Switching Network-A network that uses packet-switching technology to transfer blocks of data. (See also: packet switching and public packet switched network)

PacketCable-The name of the project to define a suite of interoperability specifications to allow for devices within a packetized telephony-over-cable network to function correctly even if provided by many vendors.

Packetized Voice-Packetized voice is the process of converting audio signals into digital packet format, transferring these packets through a packet network, reassembling these packets into their original data form, and then recreating the audio signals.

PACKNEST-Packet Network Simulation Tool

PACS-Personal Access Communications System

PAD-Packet Assembler And Disassembler

PADS-Product Acquisition and Development System

PAGCH-Packet Access Grant Channel

Page-(1-paging) A message that is sent to a receiving device alerting the device of incoming information (2-computer) A measure of computer memory. A memory page can contain up to a given maximum number of bytes. (3-Internet) A chunk of information, such as a document or a file, on the Web. You write them in HTML code. They can contain graphics, audio and video.

Pager-A small radio receiver designed to be carried by a person and to give an audio, visual or tactile indication when activated by the reception of a radio signal containing its specific code. It may also reproduce sounds and/or display messages that were also transmitted. Some pagers also transmit a radio signal acknowledging that a message has been received.

Paging-A method of delivering a message, via a public communications system or radio signal, to a person whose exact whereabouts are unknown by the sender of the message. Users typically carry a small paging receiver that displays a numeric or alphanumeric message displayed on an electronic readout or it could be sent and received as voice message or other data.

Paging And Radiotelephone Service (PARS)-A radio service in which common carriers are authorized to offer and provide paging and radiotelephone service for hire to the general public. This service was formerly titled Public Land Mobile Service.

Paging Channel-(1- paging system) A radio channel in a paging system that sequentially sends paging alert messages along with data such as telephone number digits or alpha messages. (2- cellular and PCS) A logical channel used to transfer call setup pages to a cellular or PCS phone. See also SMS messaging, paging, and access response channel (SPACH).

Paging Encoder-A device that creates (formats) the information that is sent on a paging radio channel to identify a paging device and transfer any information (number of message data) to the device.

Paging Group-Paging messages that are grouped together to allow radio receivers to "wake up" to receive their messages during their group and "sleep" during other groups of messages.

Paging Receiver-A unit capable of receiving the radio signals from a base station for the bearer to hear a page (someone's name or other identifier said in order to find, summon or notify him/her) sent by a radio transmitter.

Paging Service-Transmission of coded radio signals for the purpose of activating specific pagers; such transmissions may include text, voice, or data messages.

This diagram shows how paging systems can provide for one-way and/or two-way paging operation. This diagram shows that paging sytsems typically use high-power transmitter to broadcast paging message to a relatively large geographic area. All pagers that operate on this system listen to all the pages sent, paying close attention for their specific address message. Paging messages are received and processed by a paging center. The paging center receives pages from the local telephone company or it may receive messages from a satellite network. After it receives these messages, they are processed to be sent to the high power paging transmitter by an encoder. The encoder converts the pagers telephone number or identification code entered by the caller to the necessary tones or digital signal to be sent by the paging transmitter. Two-way paging systems allow the paging device to acknowledge and sometimes respond to messages sent by a nearby paging tower. The two-way pager's low power transmitter necessitates many receiving antennas being located close together to receive the low power signal. This Figure shows a high power transmitter (200-500 Watts) that broadcasts a paging message to a relatively large geographic area and several receiving antennas. The reason for having multiple receiving antennas is that the transmit power level of pagers are much lower than the transmit power level of the paging radio tower. The receiving antennas are very sensitive, capable of receiving the signal from pagers transmitting only 1 watt.

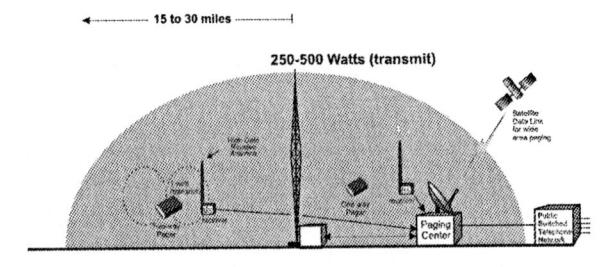

Paging System

Pair-(1-general) The two wires of a transmission circuit. (2-shielded) A pair of wires wrapped with an electrostatic shield to minimize induced interference. (3- twisted) A pair of wires with the conductors twisted together to reduce the effect of inductive interference.

Pair Gain-The effective increase in the capacity of a pair of wires that results when signal concentration, or multiplexing, enables the pair to carry not one, but many, signals simultaneously. A pair gain system can include digital or analog carrier systems, and line concentrator systems. A typical example of pair gain is % subscribers on eight pairs, or a ratio of 12 to 1.

Paired Device-In a Bluetooth system, a device with which a link key has been exchanged, either before connection establishment was requested or during the connecting phase.

Pairing-For a Bluetooth system or device, an initialization procedure whereby two devices communicating for the first time create a common link key that will be used for subsequent authentication. For first-time connection, pairing requires the user to enter a Bluetooth security code or PIN.

Pairs Terminated-The feeder pairs terminated on a main distributing frame.

PAL-Phase Alternating Line Video

PAL-Programmable Array Logic

PAL ID-PAL Identification

PAM-Pulse Amplitude Modulation

PAMR-Public Access Mobile Radio

PAN-Personal Area Network

Panel-A plate, usually vertically mounted, on which components, keys, controls, and/or meters are mounted.

PAP-Printer Access Protocol

PAP-Packet-Level Procedure

PAP-Password Authentication Protocol

Parabolic Antenna-A type of antenna that takes its name from the shape of the structure, described mathematically as a parabola. The function of the parabolic shape is to focus the weak microwave signal hitting the surface of the dish into a single focal point in front of the dish. It is at this point that the feedhorn, or a sub sector that reflects the energy back into the feedhorn, usually is located.

Parabolic Reflector-A curved section of an antenna, resembling a dish, that reflects radio waves, sound waves, or lightwaves and concentrates them into a beam.

Paradigm-An example or pattern that is commonly associated with thought processes or business operations.

Paradigm Shift-A significant change in the thought patterns or business operations that have been previously established. An example of paradigm shift includes sending mail through the Internet instead of sending mail through the postal service.

Parallax-An apparent change in the position of an object because of a change in the viewing position of the observer.

Parallel Circuit-A circuit that operate in parallel with each other.

Parallel Connection-A group of circuit elements connected in such a way that the same voltage appears across each. The current is divided among them in inverse proportion to impedance.

Parallel Data

Parallel Data-The transmission of data bits in groups simultaneously along a cable or collection of circuit board traces referred to as a bus. A typical parallel bus may accommodate transmission of one 8-, 16, or 32-bit byte at a time.

Parallel To Serial Converter-A device that converts parallel input data into a sequenced stream of signal elements (usually data bits).

Parameter-(1-component or circuit) A specific value of some variable as applied to a particular component or circuit. Examples include the resistance of a resistor or the operating frequency of a transmitter. (2-machine language) A variable that identifies and contains information needed to execute a command. (3-variable) A variable that must have a constant value for a specified application.

Parameter Negotiation (PN)-The process of requesting and agreeing on the preferred characteristics for a communication session.

This diagram shows how two data communication devices negotiate for data transmission rates and protocols selection in a data network using the preferences assigned by a user along with the options determined by equipment availability. In this example, data terminal 1 sends a connection request message to data terminal 2. This connection request indicates that the data terminal prefers to use a 56 kbps data transmission rate because it has enough bandwidth. Unfortunately, the data terminal cannot accept the request for 56 kbps because it's access bandwidth is low speed (28 kbps). The receiving data terminal sends back a request to use 28.8 kbps data transmission rate. When the originating data terminal receives this request, it accepts the request because has that data transmission rate and protocol capability available. It then confirms the request and both devices use a data transmission rate of 28.8 kbps.

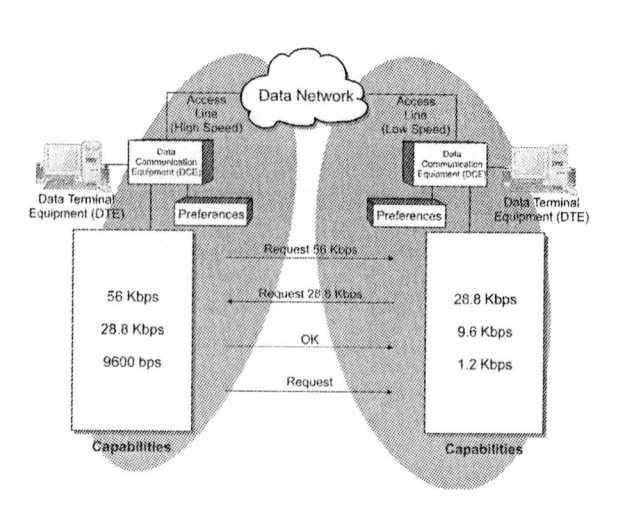

Parameter Negotiation (PN) Operation

Parasitic-An unwanted oscillation, capacitance or inductance in a circuit.

Parity-A method of verifying the accuracy of transmitted or stores data. An extra bit is appended to an array of data as an accuracy check during transmission. This bit is referred to as the parity bit. Parity may be even or odd. In an odd parity system, if the number of ones in the array is even, a one is added as the parity bit to make the total odd. For even parity, if the number of ones in the array is odd, a one is added as the parity bit to make the total even. The receiving computer checks the parity bit and indicates a data error if the number of ones does not add up to the proper even or odd total.

Parity Bit-A check bit indicating that the total number of binary "one" bits in a character or word is odd or even.

Park Mode-In a Bluetooth system, park mode is a long term state that allows the slave device to become inactive so it does not need to participate on the Piconet channel. During park mode, the slave device still remains synchronized to that channel. The parked slave wakes up at regular intervals to listen to the channel in order to resynchronize and to check for broadcast messages.

PARS-Paging And Radiotelephone Service

Partitioning-(1-equipment) The division of a complex entity such as a subsystem into smaller parts that can be dealt with independently, and that will provide the behavior of the entity when the parts are recomposed. (2- PBX) In a private branch exchange (PBX), the provision of separate trunk groups and multiple consoles for separate customer groups for use in shafed4enant-ser-vice systems.

Passband-A range of frequencies that can be passed through a filter or medium essentially unchanged.

Passive-A component, device, circuit, or network that is not powered. This term also may refer to a part of an electric circuit that is not a source of power; examples include resistors and capacitors.

Passive Filter-A filter that does not add energy in the filtering process.

Passive Optical Network (PON)-A passive optical network (PON) combines, routes, and separates optical signals through the use of passive optical filters that separate and combine channels of different optical wavelengths (different colors). The PON distributes and routes signals without the need to convert them to electrical signals for routing through switches.

PON networks are constructed of optical line termination (OLT), optical splitters and optical network units (ONUs). OLTs interface the telephone network to allow multiple channels to be combined to different optical wavelengths for distribution through the PON. Optical splitters are passive devices that redirect optical signals to different locations. ONU's terminate or sample optical signals so they can be converted to electrical signals in a format suitable for distribution to a customer's equipment. When used for residential use, a single ONU can server 128 to 500 dwellings. In 2001, most PON's use ATM cell architecture for their transport between the provider EO or POP and the ONU (in some case even to the user workstation). When ATM protocol is combined with PON system, it is called ATM passive optical network (APON).

This figure shows an ATM passive optical network (APON) system that locates optical network units (ONUs) near residential and business locations. This passive optical network routes different optical signals (different wavelengths) to different areas in the network by using optical splitters instead of switching devices. In this example, the optical distribution system uses ATM protocol to coordinate the PON. ONU interfaces are connected via fiber to an OLT located at the provider's EO or POP. Each ONU multiplexes user channels (between 12 and 40) into an optical frequency spectrum allocated to that ONU. Up 32 ONU's can share access to a single PON using the features of dense wave division multiplexing (DWDM). Some newer PON's use high-density wave division multiplexing (HDWDM). Use of HDWDM increases the number of ONU's per PON from 32 to 64. This diagrams shows that a PON that uses HDWDM can support approximately 2500 residential customers.

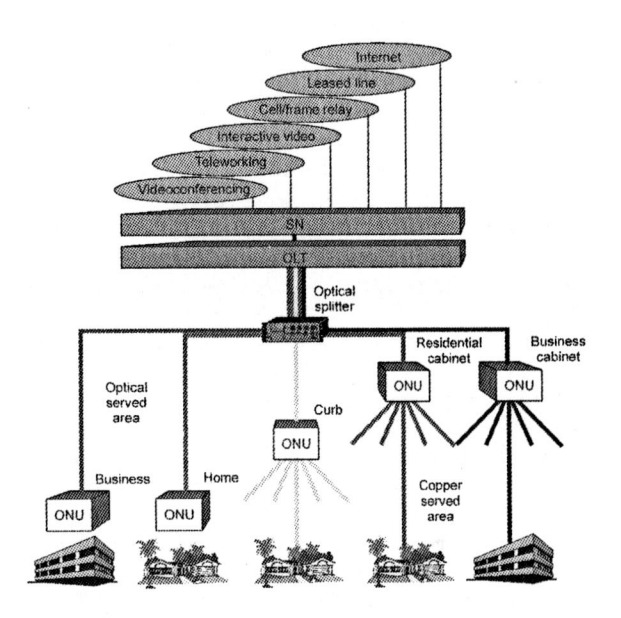

Passive Optical Network (PON) System

Passive Repeater-In a radio communication system (such as microwave links), passive repeaters are antennas that are placed at an intermediate point in a signal path to redirect the signal around terrain or other obstacles.

Passive Sensor-A measuring instrument in the earth exploration-satellite service or in the space research service by means of which information is obtained by reception of radio waves of natural origin.

Passport-PassPort is an enhanced trunking protocol designed for wide area networking. PassPort was developed by Trident Micro Systems, a subsidiary of Trident Datacom Technologies, Inc. PassPort systems are in use throughout the world. PassPort is primarily targeted to business, industrial and commercial SMR applications. The PassPort protocol and NTS system development started in 1994 with the first commercial deployments of the technology in 1997.

Password Authentication Protocol (PAP)

The PassPort protocol is downwardly compatible with analog LTR systems. It supports auto registration and de-registration as radios move between sites. It operates on the NTS networked trunking infrastructure platform developed by Trident Micro Systems. The NTS is a unique distributed networking environment that does not require a central or hub switch with all sites in the network connecting to a common point. The NTS platform provides fast network call set up times. A scaleable switch is located at each RF site. Sites may be connected in any combination of a star, linear, mesh or hub network configuration.

The key attributes for PassPort include wide area dispatch networking, voice mail for interconnect and dispatch, and ESN security. Additionally, NTS is an all digital platform and provides an easy migration path for operators between analog RF and digital RF when such product becomes widely available.

Password Authentication Protocol (PAP)-A security protocol that prompts the user to enter a user name and password before that is transferred to the system before access to services is granted.

Patch-(1-circuit) A temporary circuit rearrangement made by using jacks to bypass faulty circuit components or transmission facilities. (2-software) A temporary software program update that is used to correct software problems between official software release updates.

Patch Panel-An array of switches or connectors that reroute or reconfigure an electrical or optical transmission system.

This figure shows how a patch panel can be used in a data communications network to interconnect computers with data network equipment. This patch panel contains an array of switches and connectors that reroute or reconfigure an electrical or optical transmission system. This patch panel is located within the telecommunications room (a common practice) and it uses drop cables to route the connections to wall jacks located near each workstation. Wall jacks are connected to network interface cards (NICs) by a patch cable.

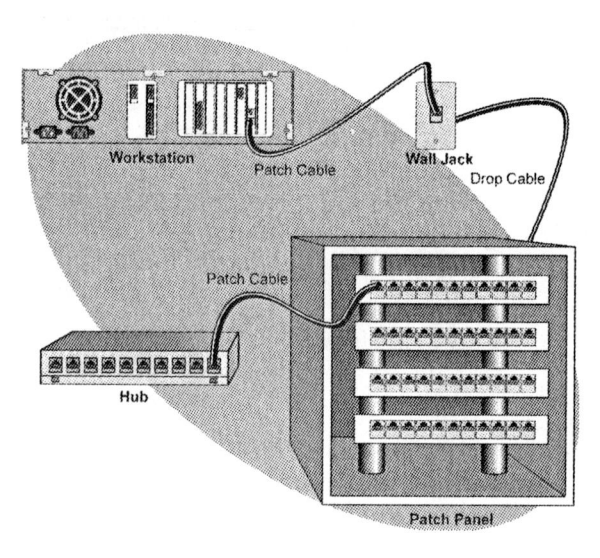

Patch Panel Diagram

Patching-Creating business units with overlapping business functionality in order to encourage competition and excellence in delivery.

Patent-Design Around

Patent-A document granting a monopoly for a limited time to an invention described and claimed in the body of the document.

Patent rights are Intellectual Property Rights which give the owner, or assignee, the right to prevent others from making, using, or selling the invention described and claimed in the patent.

Patent rights must be applied for in the country, or region where they are desired. Most countries have a national patent office, and an increasing number of countries grant patents through a single regional patent office.

Patent-Enablement

Patent Claim-A description of a specific use of technology that is new and described in the patent. There are usually many patent claims that are part of a patent.

Patent Examiner-A patent examiner is a person having a technical degree, employed by a national or regional patent office, and having the responsibility for granting valid patents according to national, and international laws.

Patent Family-The collection of nation and international patents which claim priority to a single common application for patent. A patent family may claim the same invention in many different countries, claim different, but related, inventions in the same country, or as in most cases, comprise a combination of both.

Patent License-A license that is given for the production, use, or sale of products that may use technologies or processes defined by the claims in a patent.

Patent Licensing-The process of obtaining the rights to use technology that is described in one or more patents.

Patent Royalty-A fee paid for the use of technology that is defined in the claims of a patent.

Patent Search-The process of searching for patents.

Path-(1-general) The route a signal travels through a channel, circuit, or network. (2-communications system) In radio and optical communications systems, the route that an electromagnetic wave travels through space. (3- data) A logical connection between the point where a signal is assembled into a data format and the point at which the signal is disassembled. (4-file name) The complete name location for a file or information element in a computing system.

Path Antenna Gain-The change in transmission loss when lossless, isotropic antennas are used at the same locations as the actual antennas.

Path Clearance-In microwave line-of-sight communication, the perpendicular distance from the radio beam axis to obstructions such as trees, buildings, or terrain. For a particular K-factor, the required path clearance usually is expressed as some fraction of the first Fresnel zone radius.

Path Establishment-The process of creating a communication path between two (or more) points. Path establishment may involve a predetermined (permanent) connection or it may involve the dynamic (connection each time) creation of path (switched connection).

Path Establishment Delay-The amount of time it takes to establish a communication path or circuit.

Path Length-The time required for a signal to travel through a piece of equipment or a length of cable. This time also may be referred to as propagation delay.

Path Loss-A decrease in signal energy during transmission from one point to another. The path loss in free space transmission is 20 dB per decade. This means that for each 10 times in distance the signal travels, the energy will decrease by a factor of 100 (99% decrease in energy).

This diagram shows that the amount of signal energy decrease (path loss) in free space as compared to an obstructed path for an omnidirectional (all direction) antenna. The path loss in free space is approximately 20 dB per decade. This diagram shows that for a 10 times increase in distance, the signal energy will decrease by 99%. The path loss through objects will vary by frequency. This diagram shows that the path loss of radio signals through buildings and trees for radio signals below 2 GHz is approximately 40 dB to 60 dB per decade. The lower diagram shows that a 100 Watt signal will be reduced to less than 10 milliWatts as distance increases by a factor of 10.

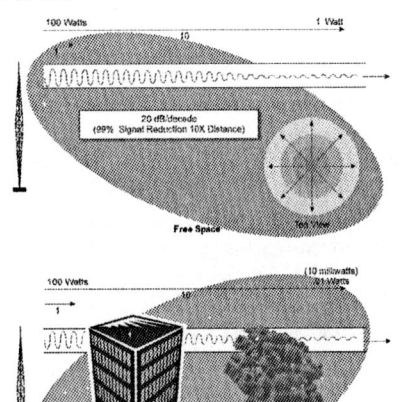

Path Loss Operation

Path Profile-A graphic representation of a propagation path, showing the surface features of the earth, such as trees, buildings, and other features that may cause obstruction or reflection in the vertical plane containing the path.

Path Survey-An assembling of the geographical and environmental data required to design a microwave communication system.

Path Terminating Equipment (PTE)-A communication device that terminates the end of an transmission system.

Pattern-(1-antenna system) A description of how much power is sent in any given direction from a directional antenna system. The pattern often is documented in the form of a graph. (2-video effects) In a production switcher, a variety of geometric shapes that are available for use as wipe transitions, key masks, and other functions.

Pay Per View (PPV)-The providing of television programming such as sports, movies, and other entertainment video that customers view for a fee. PPV services may be ordered by telephone or via an interactive set top box.

Pay Phone-A telephone that requires coins to be inserted or a calling card swiped or inserted before a toll call can be placed. Some pay telephones allow toll free or freephone calls to be originated without the need for coins or calling cards.

Pay-Per-View-A video delivery service that allows customers to select and pay for individual movie or video selections. Pay-per-view services can allow access to movies at any time (video on demand) or may only offer access at pre-scheduled intervals (near video on demand).

Payload-As applied to a data field, a payload is a discrete package of information, also called a packet, which contains a header, user data, and a trailer. In the Bluetooth specification, the format for a data field consists of a payload header that specifies the logical channel, controls the flow on the logical channel, and indicates the length of the payload body. The payload body consists of the user data. The "trailer" consists of a Cyclic Redundancy Check (CRC) code for ensuring the accuracy of the packet.

Payload Pointer-In the Synchronous Optical Network (SONET), the pointer that indicates the location of the beginning of a Synchronous Payload Envelope.

Payload Type Identifier (PTI)-A control code that is located at the beginning of a packet payload (such as in an Asynchronous Transfer Mode (ATM) cell) that indicates the type of data in the payload. The PTI may contain length of the packet, type of information (e.g. voice or data), and additional routing and control information.

PB-Personal Base
PBCCH-Packet Broadcast Control Channel
PBP-Packet Burst Protocol
PBX-Private Branch Exchange
PC-Personal Computer
PC-Point Code
PC-DOS-Personal Computer Disk Operating System
PCA-Password Call Acceptance
PCB-Printed Circuit Board
PCCCH-Packet Common Control Channel
PCF-Physical Control Field
**PCF-802.11 Point Coordination Function
PCH-Paging Channel

PCH-A test project that validated the code division multiple access (CDMA) radio performance that was used for the universal terrestrial radio access network (UTRAN).
PCI-Protocol Capability Indicator
PCI-Peripheral Component Interconnect
PCI local bus-Peripheral Component Interconnect Local Bus
PCL-Percent Common Line Factor
PCM-Pulse Coded Modulation
PCMCIA-Personal Computer Memory Card International Association
PCN-Personal Communications Network
PCP-Post Call Processing
PCS-Personal Communication Services
PCS-Plastic Clad Silica Fiber
PCS Data Card-An adapter card that connects electronic communications equipment (such as a fax machine or computer) to a GSM telephone, typically via a standard PCMCIA type II connector.
PCS Frequency Band-A frequency band auctioned off by the FCC from 1830 MHz to 1990 MHz for mobile voice and data services.
PCSA-Personal Computing System Architecture
PCT Application-PCT stands for the Patent Cooperation Treaty which is an agreement among 179 countries enabling applicants to file patent applications in PCT countries which claim priority to a single application for patent. The PCT is administered by WIPO, the World Intellectual Property Organization a specialized agency of the United Nations.
PDA-Personal Digital Assistant
PDCP-Packet Data Convergence Protocol
PDD-Post Dial Delay
PDF-Portable Document Format
PDFA-Praseodymium Doped Fiber Amplifier
PDH-Plesiochronous Digital Hierarchy
PDM-Pulse Duration Modulation
PDN-Premises Distribution Network
PDN-Public Data Network
PDP-Policy Decision Point
PDP-Plasma Display Panel
PDT-Portable Data Terminal
PDTCH-Packet Data Traffic Channel
PDU-Protocol Data Unit
PE-Partial Echo Field
Peak-(1-signal) The maximum amplitude or value, usually of a voltage or current. In a periodic function, the peak is the instantaneous maximum value, either positive or negative. (2-category of usage) A

category of usage that indicates that the request or usage of service is at its highest level. (3-billing) A rating time period that may assign a different billing rate (generally a higher rate) when customers use services during period of higher network usage.

Peak Amplitude-The maximum amplitude of a periodically repeating quality.

Peak Bandwidth-Expressed in bytes per second, this limits how fast packets may be sent back-to-back from applications. Some intermediate systems can take advantage of this information, so that more efficient resource allocation results.

Peak Envelope Power-The average power supplied to the antenna transmission line by the transmitter during one radio frequency cycle at the highest crest of the modulation envelope, taken under conditions of normal operation.

Peak Hour-The one or two hours of the day where the number of calls to a call center is at its highest level.

Peak Load-The load that results from higher-than-average traffic volume. Peak load usually is expressed as the load during a 1-hour period. It also can be expressed in terms of any of several functions of an observing interval, such as a peak hour during the day.

Peak Power-The power at the highest level in a signal (usually a pulse signal).

Peak Power Output-The maximum output power of a radio transmission burst.

Peak Time-The hours designated as the heaviest usage of a system. In cellular systems, peak time usually is defined as being between 7 a.m. and 7 p.m., Monday through Friday. Usage rates are sometimes higher during peak time than during off-peak time.

Peak Transmit Power-The peak power output as measured over an interval of time equal to the frame rate or transmission burst of the device under all conditions of modulation. Usually this parameter is measured as a conducted emission by direct connection of a calibrated test instrument to the equipment under test.

PEB-PCM Expansion Bus

Peer Entities-Entities in the same layer of an SS7 message but in different systems (nodes) that must exchange information to achieve a common objective.

Peer Protocol-A formal language used by peer entities to exchange information. In the Signaling System 7 protocol, a formal language used by peer entities to exchange user data.

Peer-To-Peer-A system in which two or more network nodes or processes can initiate communications with each other. Peer-to-peer usually describes a network in which all nodes have the capability to share resources with other nodes so that a dedicated server can be implemented but is not required.

Peg Count-The total number of any traffic event that occurs during a given period. An example is the number of times a switching-system component functions during one hour.

PEL-Picture Cell

PEM-Privacy Enhanced Mail

Penetration-(1-skin effect) A measure of the depth of current in a conductor. As the frequency through a conductor is increased, the current tends to travel only on the outer surface (skin) of the conductor. (2-broadcasting) The extent to which the population within a service area can receive a broadcast signal, or the extent to which receivers in a service area are available. Penetration may be measured as a percentage of all households, or of the total population. (3-cable TV) The ratio of the number of subscribers to the total number of households passed by a cable TV system.

Penetration Tap-A device used in Ethernet to connect transceiver to the bus without requiring that the bus be interrupted for the installation of fittings. This is accomplished by a needlelike device that penetrates the insulation on the bus (a co-axial cable) and reaches the center conductor of the co-ax. Also called a Vampire Tap.

Pentode-An electron tube with five electrodes: the cathode, control grid, screen grid, suppressor grid, and plate.

PEP-Policy Enforcement Point

Per Call Block-A feature of calling number identification that allows the originator of a call to disable the display of their calling number. This feature may be used when a caller wishes to place an anonymous complaint.

Per Line Block-A feature of calling number identification that allows the user of the telephone device to disable the display of their calling number when making calls. This feature may be used when a caller wishes to place calls from a private number.

Percent Interstate Usage (PIU)-The percent of interexchange carrier interstate traffic carried on a message trunk or special-service circuit provided by a local exchange carrier. An allocation factor between Inter and Intra-State traffic.

Percent Local Usage (PLU)-The percent of allocation between IntraLATA and Local traffic.

Perceptual Noise Substitution (PNS)-A technique of transform audio compression for efficiently encoding portions of the signal which are perceived as noise. Psychoacoustics suggests that one noise-like signal is perceptually equivalent to another with the same coarse energy spectrum, so reproducing the exact waveform is not critical. Therefore rather than encoding individual frequency components within parts of the spectrum that are considered to be noise-like, PNS encodes a single energy level for a group of adjacent frequencies. Especially when encoding at lower bit rates with quality compromises being made, PNS can improve quality significantly by allowing more bits to be used for encoding the non-noise-like parts of the signal.

Performance Counter-A counter that increases each time a performance threshold has been reached.

Perigee-The point in the orbit of an earth satellite that is at a minimum distance from the center of the earth.

Period-The time required for one complete cycle of a regular, repeating series of events.

Periodic-A recurring function or signal that repeats at regular time intervals.

Peripheral-An auxiliary device, such as a printer, mouse or graphics pad, that is connected to a computer and provides service to it but does not participate in principal computer functions.

Peripheral Bus-A serial communications bus between a master controlling device and peripheral or slave devices. The master sends out commands to remotely control the peripherals.

Peripheral Component Interconnect (PCI)-A standard data communication connection that allows accessory cards to be installed into a personal computer. The PCI specification defines both the electrical and physical (connector) requirements. PCI was introduced by Intel and it allows up to 10 PCI-compliant expansion cards in a PC. The PCI standard has replaced the previous industry standard architecture (ISA) bus.

Permanent Virtual Circuit (PVC)-A PVC is a virtual circuit is manually created for a continuous communication connection.

After a permanent communications circuit is established, a data path (logical connection) is maintained.

This diagram shows how a permanent virtual circuit (PVC) is used to allow the transfer of data through a communications network through a pre-established logical (virtual) path. In this example, a PVC is created by programming routing tables in 4 switches before any data is sent. These routing tables assign data transfer connections between input and output channel on each switch. For example, as data from the sending computer (portable computer) is sent into input channel 3 of the first switch, it is transferred to the output channel 5. This process will repeat for any data that is sent from the sending computer to the destination computer. This example also shows that the PVC path remains active even if the portable computer is disconnected for a period of time.

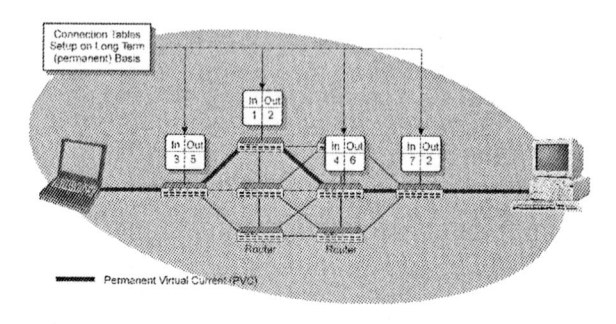

Permanent Virtual Circuit (PVC) Operation

Permeability-The magnetic permeability of a material is the ratio of the magnetic field intensity, usually denoted B (measured in units of volt seconds per square meter, or the equivalent unit tesla), to the magnetic induction, usually denoted H (measured in amperes per meter). For a linear magnetic material, this ratio is a constant at all levels of magnetic field intensity, and is usually denoted by the Greek letter Mu (μ). The unit of magnetic permeability is one henry per meter. The magnetic permeability of vacuum and most non-magnetic materials, usually denoted μo, is symbolically 4 pi · 0.000

000 1 henry per meter, where pi represents 3.14159. The numerical value of μo is 0.000 001 256 henry per meter (1.256 microhenrys per meter).

Permeance-A measure of the ease of establishing a magnetic field. Permanence is the ratio of magnetic flux to the magnemotive force. It is the reciprocal of reluctance.

Permissible Interference-Observed or predicted interference which complies with quantitative interference and sharing criteria contained in these [International Radio] Regulations or in CCIR Recommendations or in special agreements as provided for in these Regulations.

Permittivity-The dielectric permittivity of a material is the ratio of the so-called displacement field, usually denoted D (measured in units of ampere seconds per square meter), to the electric field intensity, usually denoted E (measured in volts per meter). For a linear dielectric material, this ratio is a constant at all levels of electric field intensity, and is usually denoted by the Greek letter epsilon (e). The unit of dielectric permittivity is one farad per meter. The dielectric permeability of vacuum, usually denoted eo, is numerically 0.000 000 000 008 85 farad per meter (8.85 picofarads per meter)

Persistence-The characteristic of a material to continue producing light after the exciting energy source is removed.

Persistence Of Vision-A characteristic of the human eye. The sensory cells of the eye are quickly saturated and require a short period of time to recover after the light source is extinguished. It is this "persistence" that allows the phenomenon of television, which relies on the use of individual image frames that are repeated at a given rate.

Personal Access Communications System (PACS)-Personal access communications system (PACS) is a short range cordless telephone system which is highly integrated with the wired telephone network. PACS was developed with influence from wireless access communications system (WACS) and the Japanese personal handy phone (PHP) system. The PACS system uses a 384 kbps digital radio channel which provides each customer with a 32 kbps ADPCM voice channel.

Personal Area Network (PAN)-A network concept in which all the devices in a person's life communicate and work together, sharing each other's information and services.

Personal Communication Services (PCS)-Refers to the emerging market of wireless communications that is personalized with services selected by the individual. The wireless PCS networks use radio signals as the access point to the network; the wireless network is then tied back into the public switched network for call routing to or from the wireless subscriber to the other party. Many PCS carriers occupy the newer PCS frequencies auctioned by the government.

Personal Communications Network (PCN)-A digital cellular system which is based on the DCS 1800 specification. It is a short-range wireless communications network whose radio frequencies carry personal communications service (PCS) signals between portable handsets and radio base stations that are connected to a public switched network. Typically, service is available from 100 to 2000 feet from a radio base. See also: Personal Communications Service (PCS).

Personal Computer Memory Card International Association (PCMCIA)-A standard physical and electrical interface that is used to connect memory and communication devices to computers, typically laptops. The physical card sizes are similar to the size of a credit card 2.126 inches (51.46 mm) by 3.37 inches (69.2 mm) long. There are 4 different card thickness dimensions: 3.3 (type 1), 5.0 (type 2), 10.5 (type 3), and 16 mm (type 4 - unofficial size). The number of connections (pins) is 68. The first PCMCIA standard was approved in 1991 with the concept of standardizing computer memory cards. Because the interface defined how to address, control, and transfer data in standard formats, the PCMCIA standard was quickly used for other types of communication and storage accessories such as modems, network interface cards (NICs), hard disk storage, and controller adapters. A picture of a PCMCIA local area network (LAN) network interface card (NIC). This card can be inserted into a computer (typically a laptop computer) to provide communication to Ethernet data networks.

Personal Computing System Architecture (PCSA)

PCMCIA LAN Card
Source: Xircom, an Intel company

Personal Digital Assistant (PDA)
Source: Hewlett-Packard Company

Personal Computing System Architecture (PCSA)-A personal computer implementation of DECnet that allows PCs to work in a DECnet environment. DECnet is a trademark of Digital Computer.

Personal Digital Assistant (PDA)-Small computing devices that contain it's own software operating system that allows the user to run software processing applications. Personal digital assistants are often used to organize personal activities and may provide access to communication services (such as web browsing and email).

This photograph shows a Jornada 560 series personal digital assistant (PDA) that includes Robust Microsoft © Pocket PC 2002 Premium Edition software. It has a built-in compact flash type 1 extender slot that allows the addition of memory cards and other optional accessories such as wireless network solutions and HP Pocket camera. This PDA is small enough to fit in your hand and big enough to organize your confusing life.

Personal Firewall-A device or software program that runs on your computer that provides protection from Internet or data network intruders. Firewalls can restrict access types and may monitor for advanced security threats by analyzing certain types of data communication activities. Although firewalls are important for protecting data that is connected to pubic networks, they can be complicated to setup, can cause problems with desired communications and generally can slow down the transfer of data communication.

Personal Handiphone System (PHS)-A low power public cordless telephone system. The PHS system uses a 300 kHz radio carrier signal that is divided using TDMA.

Personal Identification Number (PIN)-A number assigned to an individual subscriber which is used to gain access to specified services, such as credit card calling or prepaid wireless services.

Personal Identification Number (PIN) Code-A unique number that identifies a user of a card or information and allows a call to be completed or a service to be authorized.

Personal Identification Number Blocking (PIN Blocking)-The use of personal identification numbers (PINs) to control access to 976 Information Access Service. In practice, such blocking allows customer access only after a valid PIN is entered.

Personal Information Manager (PIM)-Are like computerized appointment books, and are used on many personal computers, to help a person keep track of appointments, telephone numbers, and other contract information. A telephony enabled PIM could provide point and click telephone dialing, automated conference calling, or automated access to databases and other applications.

Personal Unblocking Key (PUK)-(1 - GSM) A key code that is used to reset the personal identification number (PIN) number that is stored in a GSM subscriber identity module (SIM) card after the previous PIN has been invalidated. The PIN can accidentally become invalid if the user attempts to enter an incorrect PIN number 3 times in a row. If the PUK is entered incorrectly 10 times in a row, the SIM card is completely disabled to the user.

Personalized Ringing-The ability for a user to select and setup the audio sound that is used to indicate an incoming call. Personalized ringing helps individuals to identify their telephone is ringing when there are several telephones located within the same area.

Persons of Population (POPs)-The population within a geographic area that can have access to the service of a communications service provider. This is commonly used for evaluating the worth of a system. The number of pops is multiplied by the price to be paid per pop. One Pop generally equal to one person.

Pervasive Computing-Refers to the emerging trend toward numerous, casually accessible, often invisible computing devices. The computing devices are frequently mobile or embedded in the environment, and connected to an increasingly ubiquitous network structure. The objective is to facilitate computing anywhere it is needed.

PFC-Paging Frame Class

PFC-Page Frame Class

PFM-Pulse Frequency Modulation

PGP-Pretty Good Privacy

Phantom Call-A phone that rings inadvertently without a calling party online.

Phantom Circuit-An electric circuit using center-tapped connections to a transformer coil for the purpose of transmitting an additional electric current signal on the transmission wires that connect to these transformer coils. There are several arrangements of this type, but typically the phantom current flows to the center-tapped connection, where it divides into two parts, each part flowing through half of the turns of the coil but in opposite directions. The current then returns to its source via the two parallel wires that are connected to the two ends of the coil. Because this center-tap current flows in equal but opposite amounts in the two parts of the transformer coil, it does not produce a net signal on the other (secondary) coil of the transformer. This current can exist without causing any effect on a second simultaneous current that flows from one end of the coil to the other, not using the center-tap wire. Phantom circuits are used to transmit direct current for power purposes in some ISDN s-interface customer premises wiring, and also in some proprietary PBX equipment. Phantom circuits were first used in the 19th century on telegraph circuits to transmit an additional telegraph channel over existing wires.

Phase-(1-general) Any distinguishable state in a periodic phenomenon. (2-displacement) The time displacement between two currents or two voltages or between a current and a voltage measured in electrical degrees, where an electrical degree is 1/260 part of a complete cycle. (3-power) The number of separate voltage waves in a commercial alternating-current supply, designated as single-phase, three-phase, and soon.

This figure demonstrates the operation of the basic PAL analog television system. The video source is broken into 30 frames per second and converted into 625 lines per frame. Each video line transmission begins with a burst pulse (called a sync pulse) that is followed by a signal that represents color and intensity. The time relative to the starting sync is the position on the line from left to right. Each line is sent until a frame is complete and the next frame can begin. The television receiver decodes the video signal to position and control the intensity of an electronic beam that scans the phosphorus tube ("picture tube") to recreate the display.

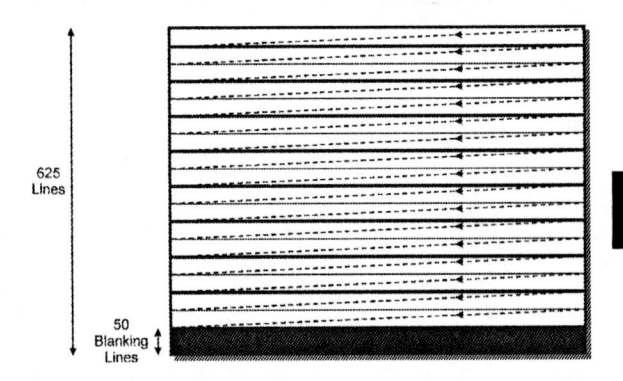

Phased Alternate Each Line (PAL) Video

Phase Alternating Line Video (PAL)

Phase Alternating Line Video (PAL)-A television system that was developed in the 1980's to provide a common television standard in Europe. PAL is now used in many other parts of the world. The PAL system uses 7 or 8 MHz wide radio channels. The PAL system provides 625 lines of resolution (50 are blanking lines).

Phase Constant-The change of phase to which a signal is subject when propagating through a medium, such as a transmission line. Phase constant, which serves as one measure of performance for a medium, is measured in radians per unit length.

Phase Delay-The ratio of the phase shift of a sinusoidal wave traveling through a transmission system to the frequency of that wave.

Phase Distortion-Distortion resulting from the selective phase shifting of the frequency components of a signal as it travels through a transmission medium. (See also: dispersion, phase velocity.)

Phase Inversion-A condition whereby the output of a circuit produces a wave of the same shape and frequency but 180. out of phase with the input.

Phase Jitter-(1-general) The unwanted phase variations of a signal, expressed in degrees. (2-PCM) An abrupt variation in the phase of a pulse code modulated signal caused by impulse noise, cross-talk, or timing changes in a digital bit stream.

Phase Locked-A condition in which signals of the same frequency are in phase at all times.

Phase Locked Loop (PLL)-A circuit that compares the output of a voltage controlled oscillator with a reference signal and uses the resulting difference signal to adjust the oscillator so that it will be locked to the reference in both phase and frequency. Phase-lock loops are used in frequency synthesizers.

This diagram shows how a phase locked loop circuit allows the precise creation of any frequency using a standard precision (crystal controlled) signal source. This example shows that a voltage-controlled oscillator (VCO) is used to provide a radio frequency (RF) signal based on an approximate DC voltage level. This RF signal is sampled and divided by a frequency divide. The frequency divider is programmed from the frequency control panel so it produces an output frequency that is the same as the crystal controlled (precision) frequency reference. This divided signal phase is compared against the reference frequency. As the phase begins to change, it produces a small DC voltage that is used to fine tune the VCO therefore phase locking it to the reference crystal frequency.

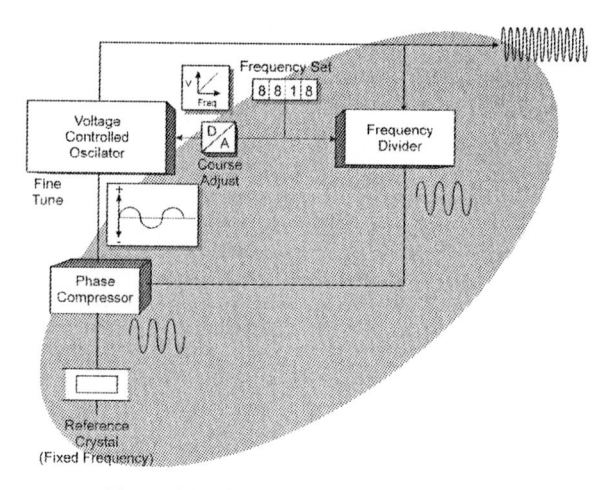

Phased Locked Loop (PLL) Operation

Phase Modulation (PM)-Phase modulation is a modulation process where the phase (relative time shift of the carrier sine wave relative to an unshifted "clock" sine wave of the same frequency) of the carrier signal is modified by the amplitude of the information (e.g., audio or data) signal.

This figure shows a sample of phase modulation (PM). In this diagram, a digital signal (on top) creates a phase modulated carrier signal (on bottom). As the digital signal voltage is increased, the frequency of the radio signal changes briefly so the phase (relative timing) of the transmitted signal advances compared to the unmodulated radio carrier signal. This results in a phase-shifted signal (solid line) compared to an unmodulated reference radio signal (dashed lines). When the voltage of the digital signal is decreased, the frequency changes again so the phase of the transmitted signal retards compared to the unmodulated radio carrier signal.

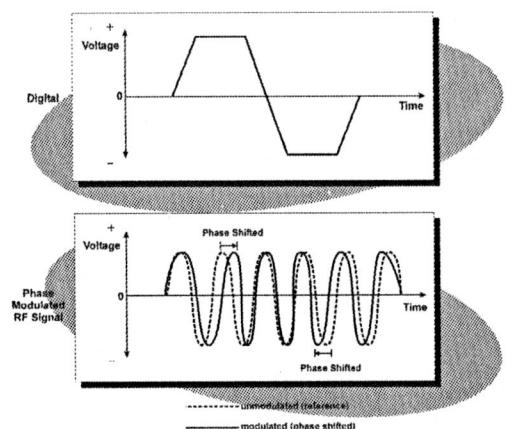

Phase Modulation (PM) Operation

Phase Noise-A noise signal that changes the phase of a desired signal. Phase noise is a measurement of a frequency source instability.

Phase Shift-(1-general) A change in the relative phase of a sinusoidal component of a signal, caused by a transmission medium. (2-reference) The change in phase of a periodic signal with respect to a reference.

Phase Shift Keying (PSK)-A method of modulation used for digital transmission wherein the phase of the carrier is discretely varied in relation to a reference phase, or the phase of the previous signal element, in accordance with the data to be transmitted.

Phase Splitter-A device that produces from a single input wave two or more output waves differing in phase from one another by a specified amount

Phase Velocity-The speed with which a sine wave component of a signal travels on a transmission medium. Distortion results if all of the components do not travel at the same phase velocity.

Phased Array-An array antenna whose directivity pattern is controlled largely by the relative phases of the excitation coefficients of the radiating elements.

Phone Doubler™-The telephone service that provides a customer an indication that an incoming call (usually a voice call) is waiting for them while they are connected to the Internet. The service allows the customer to temporarily hold the Internet connection while they communicate on the other communications channel (answer the voice call). When the customer hangs up the phone, the Internet connection is restored and Internet service resumes.

Phone Patch-A device or circuit that is used to temporarily connect a radio system to a public telephone network.

Phonelet-A phonelet is a small program that runs on a server to provide specific types of functions (such as message waiting light control) for particular models of telephones.

Phoneline Network-A computer network technology that allows standard telephone wiring to be used as network cabling without the need to disconnect standard telephones. The Phoneline Network uses high frequency signals that are above standard telephone and DSL frequency bands. To install a phoneline network, end users install Phoneline NICs in a similar method to adding an Ethernet card. The Phoneline networking system allows computers to be connected to each other without the use of a hub (daisy chain).

Phosphor-A chemical that gives off visible light when bombarded by electrons or other subatomic particles.

Phosphorescence-The characteristic of a material to produce light when excited by a radiant energy source. Minimal heat, or none at all, results from the process. The time between excitation and emission of light is generally much longer than that for fluorescence.

Photo ID-Photo Identification

Photo Transistor-A junction transistor in which the collector current varies with the light focused on the base.

Photocathode-An electrode in an electron tube that emits electrons when bombarded by photons of light.

Photocell-A device that produces an electric output from a visible light input.

Photoconductivity-The property of some nonmetallic materials that causes increased electrical conductivity when they absorb incident photons.

Photocurrent-The current that flows through a photosensitive device, such as a photodiode, as a direct result of exposure to radiant energy.

Photodarlington-A light detector in which a phototransistor is combined in a circuit with a second transistor to amplify its output.

Photodetector-A device used to detect light. In optical networks, a photodetector is typically a semiconductor p-n junction diode designed to generate a voltage or current when it absorbs a photon. The voltage or current is representative of the number of photons that reach the photodetector, so it converts an optical signal to an electronic one that can be processed or measured.

Photodiode-A diode having a current-vs.-voltage characteristic that is dependent on the level of optical power incident on the device. Photodiodes are used for the detection of optical power and for the conversion of optical power to electric power.

Photoelectric Effect-The emission of electrons when certain materials are struck by incident photons. Photoelectric materials are used to fabricate solar cells.

Photoelectromagnetic Effect-The production of a potential difference by virtue of the interaction of a magnetic field with a photoconductive material subjected to incident radiation.

Photogrammetry-A process through which terrain elevation data can be accumulated from using a stereo image or photographs in 3-D.

Photolithography-A technique used in the manufacture of integrated circuits. Complex circuit designs are photographed, and reduced-size masks are used to build the various levels of the device.

Photon-A Photon is a particle of light as an electron is a particle of electrical energy. Photons are the smallest measurable part of light. Each Photon is a packet of energy that resonates at a specific optical frequency.

Photonic Device-An optical device, such as an optical switch, that processes light signals without first converting them to electrical form.

Photonic Switches-Devices that can directly permit or inhibit (switch) the optical transmission between two or more optical connections. Photonic switches eliminate the need for the conversion of optical signals to electrical form to allow switching to different communications lines or ports.

Photonics-The field of science and engineering that deals with photons of light and their utilization.

Photons-Quanta of electromagnetic radiation. Light can be viewed as either waves or as a series of Photons

Photoresist-A class of various materials used in photolithography. The photoresist forms polymers on exposure to light, and these act as a barrier during etching and other processing stages. Positive photoresist acts in the reverse direction. It starts off as a polymer, but on exposure to light, no longer acts as a barrier.

Phototransistor-A transistor that detects light energy and amplifies the resulting electrical signal. Light falling on the base-collector junction generates a current, which is amplified internally.

Photovoltaic Effect-The production of a voltage across a pn junction resulting from the absorption of photon energy. The potential is caused by the internal drift of hole electron pairs; hence, the phenomenon leads to direct conversion of a part of the absorbed energy into a usable voltage.

PHS-Personal Handiphone System

PHY-Physical Layer

Physical Address-A grouping of numbers that identifies a particular piece of computer hardware connected to a local area network or other data communications system. Contrast with logical address.

Physical Address Space-The set of memory locations where information can be stored for program execution. Virtual memory addresses can be mapped, relocated, or translated to produce a final memory address that is sent to hard-ware memory units. The final memory address is the physical address.

Physical Control Field (PCF)-The access control (AC) and frame control (FC) bytes that are in the header portion of a data message of a Token-Ring system.

Physical Layer (PHY)-The physical layer performs the conversion of data to a physical medium (such as copper, radio, or optical) transmission and coordinates the transmission and reception of these physical signals. The physical layer receives data for transmission from an upper layer, such as the Open System Interconnection (OSI) Data Link layer, and converts it into physical format suitable for transmission through a network (such as bursts, slots, frames, and superframes). An upper layer provides the physical layer with the necessary data and control (e.g. maximum packet size) to allow conversion to a format suitable for transmission on a specific network type and transmission line. The physical layer is layer 1 in the OSI protocol layer model.

PIC-Preferred Interexchange Carrier

PIC-Plastic Insulated Cable

PICC-Primary Interexchange Carrier Charge

Pico-A prefix of a unit of measure meaning one-trillionth.

Picocell-A radio coverage area that has a radius of less than few hundred feet (usually less than 100-200 feet).

Picofarad-A measurement of capacitance that is one millionth of a microfarad.

Piconet-A small network of Bluetooth communication devices (e.g. less than 8). Bluetooth is a wireless personal area network (WPAN) communication system standard that allows for wireless data connections to be dynamically added and removed between nearby devices. Multiple Piconets can be linked to each other to form Scatternets.

This diagram shows Bluetooth devices that have created temporary connections. In this diagram, the personal digital assistant (PDA) device is synchronizing (deleting, changing, and adding) addresses with a laptop computer. The laptop computer is also connected to the Internet through a Bluetooth enabled access node. A mobile phone is also synchronizing its phone book listing with the laptop computer. However, because it is out of direct range of communicating with the laptop, it communicates through the access node. The mobile phone is also communicating with a wireless headset.

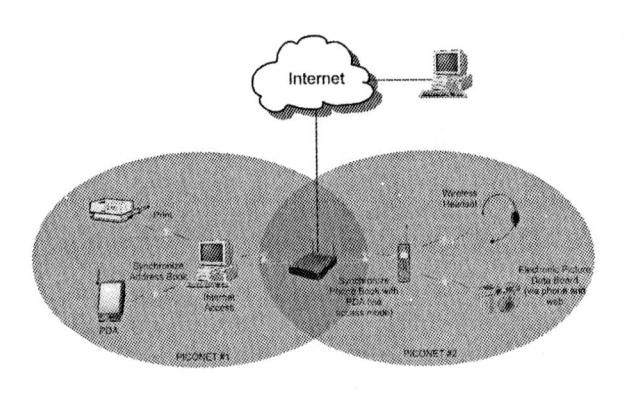

Piconet System

Picosecond (ps)-One trillionth of a second. In one picosecond, light travels about one-third of a millimeter.

PICSIDCPR-Plug-In Inventory Control System Detailed Continuing Property Record

Picture Cell (PEL)-An abbreviation for picture cell (or picture element). A pel is one sample of digital picture information, which can be an individual sample of R, G, B, luminance or chrominance or a collection of such samples if they are co-sited and as a group produce one picture element. Also called a pixel.

Picture Element (Pixel)-(1-video) The smallest distinguishable and resolvable area in a video image. A pixel is a single point on the screen. The term pixel is derived from the words picture element.

Picture Tube-A tube that is used to display an image by the varying of intensity of a scanning electron beam onto phosphors that are deposited on the inner surface of the display portion area of the tube. The beam of electrons is produce by a electron gun. The electron beam is directed to different positions on the tube by horizontal and vertical magnetic fields (the yoke assembly). To create the image, the electron beam is swept across the picture tube line by line horizontally from top to bottom until a complete frame is displayed. The electron beam is momentarily blanked as it is repositioned to across the screen (horizontal blanking) and when it is repositioned to the top of the screen to begin the next frame image (vertical blanking).

This diagram shows how a picture tube directs an electron beam to a screen to produce a video image. This example shows that a gun produces the electron beam (stream of electrons) that is focused and positioned (directed) towards a high voltage screen. The screen has a phosphor coating that emits light when the electrons hit the screen area. The electron beam is directed from side to side (horizontal scan) with the intensity of the electron beam varying so the amount of light (brightness) varies. When the electron beam reaches the end of the sweep, the electron beam is stopped (blanked) during the changing of the focusing (directing) controls. Horizontal lines are produced from top to bottom until the entire picture is created. This diagram shows that the final lines of the video signal are video blanking lines that indicate the next frame of the picture is to begin. This allows the television system to reposition the electron beam to the top of the picture tube.

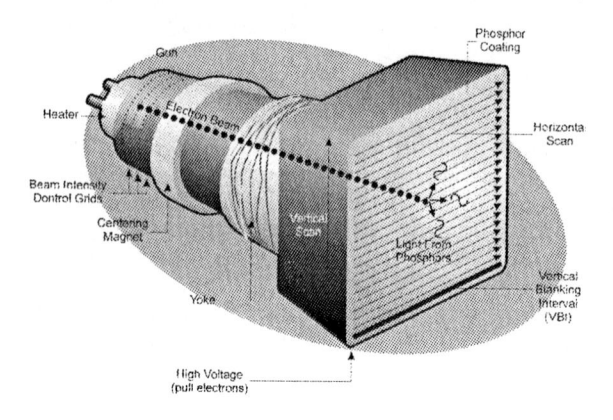

Picture Tube Operation

Piezoelectric Effect-A phenomenon observed in some partially ionic crystals such as certain types of quartz. When such a material is compressed or stretched, a voltage develops across the crystal. If a voltage is applied, the crystal compresses or expands. This effect is the basis for SAW and crystal devices used in many telecommunications applications.

Pigtail-(1-copper pair) A short length of copper-pair cable, typically for connections from a repeater or loading coil case. (2-coax) A short length of coaxial cable extending from a transmitter or receiver and used to make a connection to that equipment. (3-optical) A length of optical fiber with one end terminated with a connector and the other end attached to a light source or detector. The pigtail couples light from a source to a fiber optic cable or from a cable to a detector.

PIL-Percent Interstate

Pilot-(1-general) A signal, usually a single frequency, transmitted over a system for supervisory, control, synchronization, or reference purposes (2- FM) A signal at 19kHz transmitted with the FM broadcast carrier. The FM stereo subcarrier at 38kHz, modulated with a L-R difference signal, is generated from the pilot, leaving the two in-phase. In FM receivers, a pilot detector circuit serves the present of the 19kHz signal, which is used to decode the stereo information.

Pilot Frequency-A single-tone frequency transmitted at a specified level to enable receive side equipment to maintain synchronization and/or proper level control.

Pilot Subcarrier-A subcarrier that serves as a control signal for use in the reception of FM stereophonic sound broadcasts. A subcarrier used in the reception of TV stereophonic aural or other sub-channel broadcasts.

Pilot Tone-An unmodulated tone of a specified frequency appropriate to the transmission system concerned. The pilot tone is transmitted over the system together with the information channels. The pilot performs synchronization and level control functions at the receiver.

Pilot Trunk-The first trunk, or line, of a private branch exchange (PBX) group of trunks, and the trunk that represents the entire PBX group.

PIM-Personal Information Manager

PIM-Protocol Independent Multicast

PIMS-Profit Impact of Market Share

PIN-Personal Identification Number

PIN-Procedure Interrupt Negative

Pin-A terminal on the base of a component, such as an electron tube.

PIN Blocking-Personal Identification Number Blocking

PIN Code-Personal Identification Number

PIN Number-Personal ID Number given a subscriber to access service. Prepaid wireless services often use PIN numbers when using phones with the billing platform built into the handset to access the network.

PIN Photodiode-A diode with a large intrinsic region sandwiched between p and n doped semiconducting regions. Photons absorbed in this region create electron-hole pairs that are then separated by an electric field, thus generating an electric current in a load circuit.

PIN Photodiodes are used as a photodetector in optical networks. It is a semiconductor p-n junction device with an extra layer added to the junction to improve performance. The letters p, i and n refer to p-type doped material, intrinsic material and n-type doped material, respectively. The intrinsic material is electrically neutral while the doped materials have extra charge carriers. The thicker the intrinsic material, the more sensitive it is, but the slower its response time.

Ping-Packet Internet Groper

Ping-Ping is a technique utilized in TCP/IP based networks to test connectivity between network devices. Ping is a common method that is used to measure network availability, to troubleshoot network operation, and to measure transmission parameters. Ping uses Internet Control Message Protocol (ICMP) Echo and Echo Reply packets to determine if an IP destination address is active and how long packets take to travel and return between their source and destination

Ping Pong Handoff-A situation that can occur when cellular system radio coverage areas (cells) are marked as preferred to each other or when the cell parameters are set in a way that destabilizes the handoff process. For example, when two cells in the same geographical area are considered the best candidates for each other, causing the phone to "bounce" between the two cells.

Ping-Pong-A process of time compression multiplexing where transmitters and receivers constantly exchange information.

Pink Noise-A random noise whose frequency spectrum has been contoured to appear flat across the audible spectrum of human auditor perception.

Pinout-The pin configuration for a connector or system cabling.

PINT-PSTN and Interworking

PIP-Procedure Interrupt Positive

PIP-Picture-in-Picture

Pipe-A software interface or a hardware device that acts as an interface or buffer between network applications and devices.

Piracy-(1-transmission) The operation of unauthorized commercial stations, usually in international waters near target regions. Pirates violate flu frequency regulations, copyright laws, and national broadcasting laws. (2-programming) The duplication and sale of program materials, particularly TV programs on videotape, in violation of copyright laws. (3-signals) The unauthorized use of cable TV or satellite signals.

PIU-Percent Interstate Usage

Pixel-Picture Element

PKZip-A file compression program developed by the late Philip Klein, which uses the Lempel-Ziv-Welch algorithm. WinZip is a similar program having a Windows-compatible graphic user interface.

PL-Power Level

PL-Preferred Language

PLA-Programmable Logic Array

Plain Old Telephone Service (POTS)-Plain old telephone service is the providing of basic telephone service without any enhanced features. It is the common term for ordinary residential telephone service. The POTS system uses in-band signaling tones and currents to determine call status (e.g. call request). Because POTS allow for the transfer of audio signals below 3.3 kHz, POTS systems are also used for modems that allow data transmission (called dial up connection). Whenever a new service or feature is described, the author may refer to the previous available package of features and services as POTS, even when the previous package included several very sophisticated capabilities.

Planar Waveguide-A transmission waveguide that is fabricated in a flat material form such as a film.

Plant-A general term applied to all of the physical property and facilities of a telephone company that contribute to the provision of communications services.

Plant Location Record-Records that show the placement of distribution terminals and the location, length, size, date, and nomenclature of cables.

Plant Retired-Plant which has been removed, sold, abandoned, destroyed, or otherwise withdrawn from service.

Plasma-(1-arc) An ionized gas in an arc-discharge tube that provides a conducting path for the discharge. (2-solar) The ionized gas at extremely high temperature found in the sun.

Plasma Display-A type of flat visual display device, usually based on ionization of gas for light emission.

Plastic Film Capacitor-A capacitor whose dielectric is a plastic-based film.

Plastic Insulated Cable (PIC)-Telephone wire or cable insulated with (typically polyethylene) plastic, in contrast to older types of insulation such as compressed paper pulp or woven cotton or other fabric threads.

Plate-(1-electron tube) The anode of an electron tube. (2-battery) An electrode in a storage battery. (3-capacitor) One of the surfaces in a capacitor. (4-chassis) A mounting surface to which equipment may be fastened.

Plating-The application, by electrolysis, of a coating of one metal on the surface of another.

PLC-Power Line Carrier

PLD-Programmable Logic Device

Plenum Cable-A cable constructed of flame retardant material that generates little smoke when exposed to fire, for installation in air ducts.

Plesiochronous-The relationship between two signals whose corresponding significant instants occur at nominally the same rate, any variation in rate being constrained within specified limits. Two signals having the same nominal digit rate, but not stemming from the same clock are usually plesiochronous.

Plesiochronous Digital Hierarchy (PDH)-The original multiplexing hierarchy used in T1/E1 and T3/E3 systems (plesiochronous = almost synchronous). While multiplexing to higher rates, justification and synchronization bits had to be added to the original T1/E1 channels. These bits were discarded when demultiplexing thus creating a very inefficient and inflexible structure.

PLL-Phase Locked Loop

PLMN Code-Public Land Mobile Network

PLMR-Public Land Mobile Radio

PLMS-Public Land Mobile Service

PLU-Percent Local Usage

Plug And Play (PnP)-A compatibility system that simplifies the installation and removal of hardware devices into a personal computer. PnP includes the automatic recognition of hardware installation and removal, activation and deactivation of software drivers, and system accessory management functions.

PLV-Production Level Video

PM-Phase Modulation

PM-Pulse Modulation

Pm!-Pulse Time Modulation

PMD-Physical Media Dependent

PMD-Polarization Mode Dispersion

PMMU-Paged Memory Management Unit

PMOS-P-Channel Metal Oxide Semiconductor

PMR-Public Mobile Radio

PMS-Paging Message Service

PN-Parameter Negotiation

PN Junction-In semiconductors, the interface of positively doped material (P type) and negatively doped material (N type).

PNAP-Private Network Access Point

PNCH-Packet Notification Channel

PNNI-Private Network Node Interface

PNNI-Private Network-to-Network Interface

PNP-Private Numbering Plan

PnP-Plug And Play

PNS-Perceptual Noise Substitution

Pocketphone-A handheld portable cellular telephone typically small enough to be carried in a purse or pocket.

POCSAG-Post Office Code Standard Advisory Group

POH-Power-On Hours

POI-Point Of Interface

POI-Point Of Interconnection

Point Code (PC)-A code that is assigned to each node (signaling point) in an SS7 network. Point codes can be unstructured 14 bit (ITU) point codes or structured 24 bits (ANSI) point codes. The structured ANSI point codes are divided (hierarchical structure) a network identification, network cluster, and network cluster element (the specific device or assembly). Point codes are transferred in the signaling messages that exchanged between signaling end points and they identify the destination and source of the signaling message. The assignment of point codes is managed by a government agency.

Point Of Presence (POP)-A physical location that allows an interexchange carrier (IXC) to connect to a local exchange company (LEC) within a LATA. The point of presence (POP) equipment is usually located in a building that houses switching and/or transmission equipment for the LEC.

Point Of Termination (POT)-The point within a local access and transport area (LATA) at which a local exchange carrier's responsibility for access service ends and an inter-exchange carrier's responsibility begins.

Point to Multipoint Messaging-The process of sending data, text or alphanumeric messages from one communication device to several communication devices.

This figure illustrates how a multipoint messaging transfers a message to a group of users in a cellular (wireless) communication system. Like the point-to-point message service, this diagram shows the message first goes to the MSC message center (step 1) where the message center determines that this message is designated for multiple units. The designated list of message recipients may be pre-arranged (such as a sales staff) or it may be included with the originating message. In either case, the message center stores the message and recipient list (step 2). To complete message delivery, the MSC then searches for each mobile telephone in the list. The cellular system then individually alerts each mobile telephone that a message is coming (steps 3-5). The mobile telephones tune to the message channel (voice or control) and the system sends the message to each unit that the units receive and store. If the transmission is successful, the mobile telephone typically sends an acknowledgment, and the message may be removed from MSC message center. If the message transmission to one or more of the receivers was not successful, the cellular system attempts delivery again later.

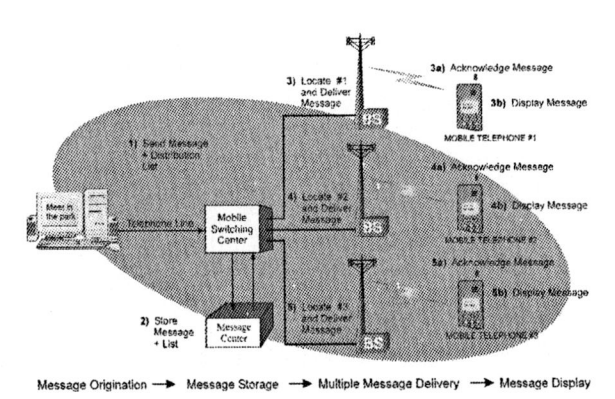

Point-To-Point Link

Point to Multipoint Messaging Operation

Point To Point Messaging-The process of sending data, text or alphanumeric messages from one communication device to one other communication device.

This diagram shows how a point to point message system sends a short message from one source to one receiver. Initially, the message goes to the Mobile Switching Center (step 1) to be routed and stored in the short message service center (SMSC) (step 2.) The cellular system searches for the mobile telephone (step 3) and alerts the mobile telephone that a message is coming. The mobile telephone tunes to the traffic channel where the message will be sent. The system then attempts to send the message (step 4). As the message is being sent, the system waits for acknowledgment messages (step 5) to confirm accurate delivery of each part of the message. If the transmission is successful, the message may be removed from SMSC memory. If unsuccessful, the system attempts delivery again.

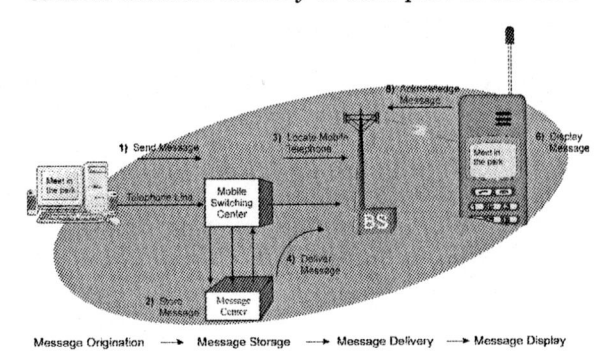

Point-To-Point Communication-The transmission of signals from one specific point to another, as distinguished from broadcast transmission which blankets the general public. Radio relay links are a common type of point-to-point communication.

This figure shows examples of how point-to-point services can be implemented. The first method is a paging system that uses device addressing to uniquely identify a specific receiver of the information. While all the pager devices receive the same radio signal, only the receiver that has the correct paging code will be able to descramble the paging message. The second method uses network routing in the Internet to store and forward data to a specific recipient in the network. When a router in the Internet receives a point-to-point message, it will use the address to lookup the best forwarding path to transfer the information towards its destination. Using point-to-point communications by network routing, only the designated recipient will receive the data.

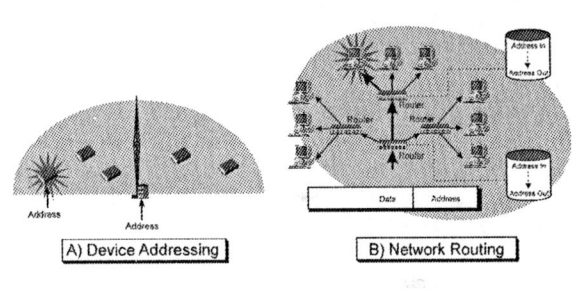

Point to Point Messaging Operation

Point-to-Point Operation

Point-To-Point Link-A connection between two devices or node that uses point-to-point protocol (PPP).

This diagram shows the structure of a point-to-point protocol (PPP) frame. This example shows that the field structure includes a start frame flag, address code for the PPP connection, a control flag (for PPP link), a link protocol identifier, data field (to hold the payload packet), and frame check

Point-To-Point Protocol (PPP)

sequence (FCS) for error detection. The PPP frame is only used to transport the data between points so the data payload includes the user's data along with the destination address of the packet

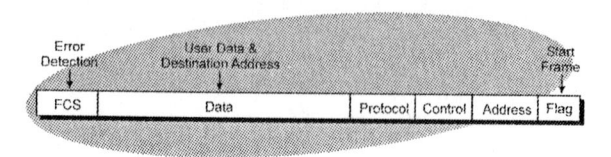

PPP Frame Structure

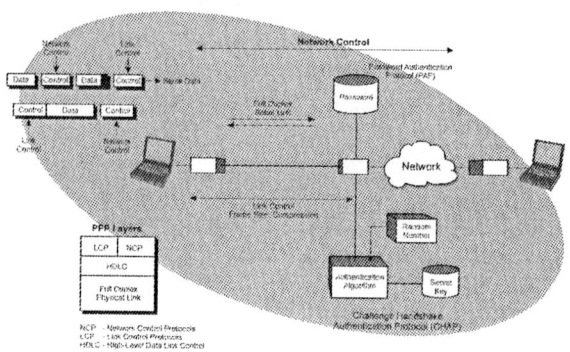

Point to Point Protocol (PPP) Operation

Point-To-Point Protocol (PPP)-A connection oriented protocol that is established between two communication devices that encapsulates Internet protocols such as Transmission Control Protocol and Internet Protocol (TCP/IP). PPP allows end users (end points) to setup a logical connection and transfer data between communication points regardless of the underlying physical connection.

This diagram show how a point-to-point protocol (PPP) connection allows a computer connection to be established, verified, and maintained. This example shows that PPP protocol can be used on any type of access connection. The PPP protocol allows for the separation of the link protocol on the PPP link and the network control protocol that operate above the high level data link control (HDLC) protocol that is used to coordinate the PPP link. The PPP protocol also includes security features such password authenticated protocol (PAP) and a more secure challenge handshake authentication protocol (CHAP).

Point-To-Point Protocol Over Ethernet (PPPoE)-Point-to-point protocol (PPP) data transmission over an Ethernet connection. PPPoE works with a network interface card (NIC) to create a point-to-point communication link between the NIC and a connection point to the Internet. The use of PPPoE allows the setup, management, and disconnection of a communication link between a customer and an Internet service provider (ISP). The PPPoE manages the transport of wide area IP packets that use 32-bit or 128-bit IP addressing over a local area Ethernet system that uses 48-bit Ethernet addressing.

Point-To-Point Tunneling Protocol (PPTP)-Point-to-point tunneling protocol (PPTP) is a virtual private network solution. PPTP software is included with Microsoft windows95, 98, NT, and 2000 and is built into some routers.

PPTP is an easy protocol to setup. However its first revision does not provide for data encryption. This lowers the privacy for data transmission.

Polar-(1-general) Relating to one or more separated points, such as the north and south poles of the earth, or more figuratively two intellectual positions such as saying that, "The organization has become polarized into two groups, one that supports the proposal and another that opposes it." (2-science and technology) A distinction based on direction in space for vector quantities such as force or electric field, or based on positive vs. negative sign for scalar quantities that do not have a space direction.

Polarity-The property of a physical system in which there are opposing characteristics, such as negative and positive charges or north and south magnetic poles.

Polarization Mode Dispersion (PMD)-In some optical fibers, the effective wave speeds for infrared light rays having different polarization directions are different. As a result, wave components whose polarization is in the direction corresponding to a higher wave speed will arrive at the receiver at an earlier time than the wave components whose polarization is the direction corresponding to a lower wave speed. This produces and undesired time broadening of the pulse, which limits the bit rate. The physical cause of the discrepancy in wave speed for different polarizations can be due to one or more of the following: Oval rather than circular core shape. A preferred orientation of oscillation direction of electrons on some atoms in the core that contribute to the absorption and re-radiation of light(See: Dispersion). One of the objectives of fiber manufacturers is to make fibers that have the same wave speed regardless of the polarization orientation of the light, thus preventing PMD.

Polarized-(1-waves) Transverse waves such as radio and light have their electric field in a transverse plane, at right angles to the direction of wave travel. (Sound waves, in contrast, are not transverse waves since the sound pressure vector is longitudinal or parallel to the direction of wave travel.) A polarized transverse wave can be produced (or can be made by appropriate filtering of a non-polarized wave that has some electric field components in the desired direction) so that the direction of the electric field is not merely in the transverse plane, but is always in the same direction within that plane. Polarizing filters transmit only components of polarized waves that have their direction of electric field polarization parallel to the preferred polarization direction of the polarizing filter. The trade name Polaroid describes polarizing optical filters made by the Polaroid Corporation, based on the methods invented by the late Edwin Land. Polaroid instant cameras, another product of this same corporation, have no technological connection with polarized light. (2-batteries) A degradation of the current capacity of an electric battery due to formation of tiny gas bubbles on one of the plates of the battery.

Polarizing Connector-In an optical network, a connector designed to join separate fibers carrying polarized light so that the direction of polarization is the same in both fibers, from input to output.

Pole Attachment-Any attachment by a line (such as a cable television system) to a pole, duct, conduit, or right-of-way owned or controlled by a utility.

Pole Attachment Contracts-Agreements that are government-regulated that affect utility pole owners, joint use contracts, and companies that enter into agreements to lease space on poles to attach facility cabling, wiring, terminals, etc.

Policy Server-A policy server is a communications server (computer with a software application) that coordinates the allocation of network resources based on predetermined policies on the priorities and resources required by communication services and applications within the network. A policy server is used to help manage network operation in the event of loss of resources. The pre-set policies define which communication services are critical (such as voice) and how much resources should be allocated to these critical services at the expense of other communication services (such as web browsing).

Polling-The process of sending a request message (usually periodically) for the purpose of collecting events or information from a network device. The receipt of a polling message by a device starts an information transfer operation. Polling of network devices is regularly performed to gather usage information that is processed by the billing system rating engine. Polling involves the use of a network control system that periodically invites its tributary stations to transmit in any sequence specified by the control station.

Polyethylene-A thermoplastic synthetic polymer used in a highly refined state for cable insulation and for cable jacket material.

Polymer-A long chain molecule, usually a good insulator, made by combining several smaller molecules.

Polystyrene-A plastic frequently used as an insulator in electrical products.

Polyvinyl Chloride (PVC)-A thermoplastic made of polymers that is tough, non-flammable, and water-resistant. PVC commonly is used as an insulator.

PON-Passive Optical Network
POP-Post Office Protocol
POP-Point Of Presence
POPs-Persons of Population

Port

Port-(1-general) A physical or logical connection point between a computer or computer-based machine and other hardware devices. (2-network) A place of access to a device or network where energy may be supplied or withdrawn, or where the device or network variables may be measured. (3-software) The process of moving source code and executable programs from one computing system to another of a different type without substantive changes to the source code.

Port Identifier-A value assigned to a port that uniquely identifies it within a switch. Port Identifiers are used by both the Spanning Tree and Link Aggregation Control Protocols.

Port Mirroring-A process whereby one switch port (the Mirror Port) is configured to reflect the traffic appearing on another one of the switch's ports (the Monitored Port).

Port Number-For internet protocols such as TCP and UDP, when a machine with a particular IP address receives an IP packet, a destination port number in the packet header can be used to determine how to handle the packet. Typically certain applications are expected to process packets sent to specific well-known port numbers for specific protocols. For example, if a machine runs a web server application, that application typically receives all TCP packets directed to port number 80.

Port Sharing-The process of allowing multiple virtual connections (logical channels) share the same port connection in a data network where.

Portability-This means that application software can be dragged across different computing platforms and operating systems. With the TAO spec, developers can write an application that will run on a PC, Alpha Server or Tandem host. SCSA's operating-system-independent APIs give developers a uniform method for supporting multiple operating systems.

Portal-An Internet Web site that acts as an interface between a user and an Information service.

POS-Point of Sale. That moment or place in which the sale occurs.

Position Location System-A system that gathers and presents the geographic position of devices or equipment. The position location information may be gathered by the network sensing the position of the device relative to one or several antennas in a network (such as the Teletrac system) or by the vehicle reporting its location using external position locating devices (such as the Global Positioning System).

Positive Acknowledgement-Used to signify information has been successfully transferred. For example, message signal units (MSUs) that are sent in an SS7 signaling system require positive acknowledgement.

Positive Temperature Coefficient (PTC)-One type of over voltage protection for telephone wiring that uses a component that increases its resistance when their temperature increases. Some resistors made of metal flakes suspended in a polymer plastic have a much greater quantitative positive temperature coefficient than pure metal wire, and are designated by the name PTC. When high current flows due to a fault, PTC resistors in series with the affected lines increase their resistance due to self heating and thus limit the current. See also BORSCHT.

Positron-An elementary particle identical in all respects to an electron but with a positive rather than negative charge. The positron is the antiparticle of the electron.

POSIX-Portable Operating System Interface

POST-Power On Self Test

Post Dial Delay (PDD)-The time period between when a user dials the last digit of a phone number and hears the phone at the other end begin to ring.

Post Dialing-Refers to the ability of a terminal to send dialing information after the outgoing call request setup message is sent.

Post Office Code Standard Advisory Group (POCSAG)-Post Office Code Standard Advisory Group (POCSAG) protocol is a pager addressing format is accepted throughout most parts of the world. The POCSAG standard was developed by the British Post Office in 1978.

The electronic address of the pager is controlled by an industry trade association (the Personal Communications Industry Association) so duplicate addresses will not be programmed into POCSAG pagers. POCSAGs support word messaging as an optional feature. The POCSAG system can operate at 512, 1,200 and 2,400 bps.

Post Office Protocol (POP)-A protocol that is used with simple mail transfer protocol (SMTP) to store and transfer e-mail messages from one system to another.

Post Pickup Delay-The period between when a user lifts a handset off-hook and hears a dial tone.

Post, Telephone And Telegraph (PTT)-A term used for a government agency in many countries that supplies and maintains the infrastructure and provides basic telecommunication services.

418

Postpay Service-A billing or coin telephone service that allows for payment of service after communication has started (e.g. called party answers.) Postpay, now rarely used, is the simplest form of coin service but requires additional operator attention on toll calls.

POT-Point Of Termination

Potentiometer-A 3-terminal variable resistor usually connected as a voltage divider. A potentiometer has a rotating shaft or sliding contact that can vary the value of resistance while the component is connected to a circuit. A potentiometer often is referred to simply as a pot.

POTS-Plain Old Telephone Service

Potting-The process of sealing components under a plastic cover to keep out moisture and other contaminants.

Power-(1-general) Power is the time rate at which energy is generated, consumed, or converted from one form to another. The electrical unit of power is the watt. One watt is the product of one volt of voltage and one ampere of current. In mechanical units, one watt is the product of one Newton of force with one meter per second of velocity. In non-technical discourse, the four words force, energy, power and momentum are often unfortunately used as synonyms for each other. In science and technology, each of these words describes a distinct and separate physical quantity. (2-alternating current) The product of the effective voltage, the effective current, and the cosine of the phase angle between them. (3-apparent) The product of the voltage and the RMS current. (4-available) The maximum power available from a source by a suitable adjustment of the load. (5-average) In a pulsed laser, the energy per pulse (joules) multiplied by the pulse repetition rate (in Hertz). (6- average speech) The total speech energy over a period of time, divided by the length of the period. (7-carrier) The average power supplied to the antenna transmission line by a radio transmitter during one radio frequency cycle under conditions of no modulation. For each class of emission, the condition of no modulation should be specified. (8-direct current) The product of the voltage and the current, or of the resistance and the square of the current. (9-effective radiated, ERP) The product of the power supplied to an antenna and the antenna gain relative to a half-wave dipole. (10-mean) The power supplied to an antenna transmission-mission line by a transmitter during normal operation averaged over a time sufficiently long compared with the period of the lowest frequency encountered in the modulation. A time of 0.1 seconds during which the mean power is greatest normally will be selected. (11-peak) The maximum power emitted by an optical source or a radio frequency transmitter. (12- peak envelope) The average power supplied to the antenna transmission line by a transmitter during one radio frequency cycle at the highest crest of the modulation envelope taken under conditions of normal operation. (13-real) The power in an alternating current circuit that is used in doing work. It is the product of the RMS voltage times the RMS current times the power factor (the cosine of the angle between voltage and current).

Power Class-A classification of the maximum power level or range of power level associated with mobile telephones.

Power Class Of MS-The maximum power rating of a mobile station.

Power Control-(1-RF level) The process of regulating the power level of a transmitter (2-cellular) The process of controlling the power level in a cellular system where the base station receiver monitors the received signal strength of a mobile telephone and control messages are transmitted from the base station to the mobile telephone commanding it to raise and lower its transmitter power level as necessary to maintain a good radio communications link.

Power Control Message-A message that commands a mobile station to increase or decrease its power level.

Power Density-The power in watts per Hertz, or the total power in a band of frequencies divided by the bandwidth in hertz.

Power Dialing-The process of dialing calls in lists, connecting the call to a customer service representative if it is answered by a person, remembering unanswered calls for a later callback, and updating records to indicate the calls that have been completed to avoid repeat calling. Power dialing systems are sometimes linked to customer information databases to allow the customer service representative to see the call recipients account information when the call is connected.

Power Efficiency-The ratio of the emitted power of a source to the electric input power.

Power Factor-The ratio of the true power consumed in an alternating-current circuit to the apparent power (the power actually supplied by the

Power Level (PL)

source). The value of the power factor is the cosine of the phase angle between the voltage and current in the circuit.

Power Level (PL)-(1-measurement) The amount of power produced by a transmitter. (2-cellular classification) A power level setting for a transmitter as commanded by a base station.

Power Line Carrier (PLC)-A carrier wave signal that can be simultaneously transmitted on electrical power lines. A power line carrier signal is above the standard 60 Hz powerline power frequency (50 Hz in Europe).

This diagram shows three types of communication systems that use an electrical power distribution system to simultaneously carry information signals along with electrical power signals. In this example, the high voltage portion of the transmission system is modified to include communication transceivers that can withstand the high voltages while coupling/transferring their information to other receivers that are connected to the high voltage lines. This type of communication could be used to monitor and control power distribution equipment such as relays and transformers. This example also shows a power line distribution system that locates a communication node (radio or fiber hub) near a transformer and provides a data signal to homes connected to a transformer. This system could allow customers to obtain Internet access or digital telephone service by plugging the computer or special telephone into a standard power socket. The diagram also shows how a consumer may use the electrical wiring in their home as a distribution system for data (e.g. Ethernet) communication.

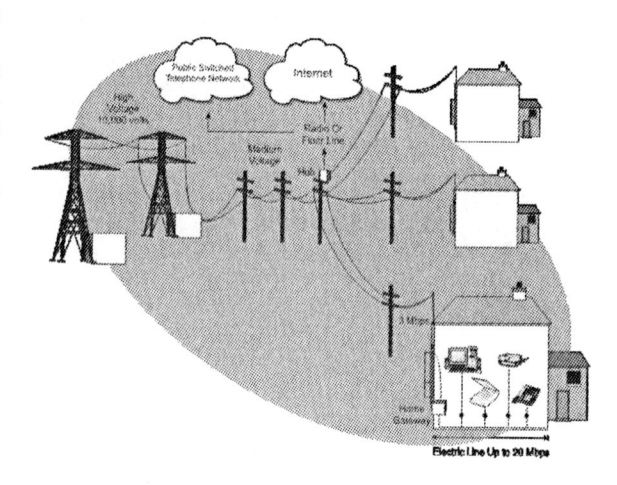

Power Line Data Transmission System

Power Line Carrier Systems-The use of the electrical power distribution system used by an electric power utility to transmit information signals. Power line carrier systems can be used for power system equipment control and monitoring and can be used to transfer other information signals such as commercial voice or data signals.

Power Line Ethernet-A data transmission system uses power lines to transmit data in Ethernet packet format on standard (110-220 VAC) electric power lines.

Power Saving-A mode of operation where electronic circuits are de-energized and energized to conserve battery life. See also sleep mode.

Power Time Template-A time profile requirement of a transmitted burst during any transmission time slot in a communication system. The power time template is used to determine if the transmitter can meet and not exceed specific the power levels necessary to conform to the system specification.

PPC-Pay Per Call
PPCH-Packet Paging Channel
PPDN-Public Packet Data Network
PPM-Pulse-Position Modulation
PPM-Pulse Position Modulation
PPP-Point-To-Point Protocol
PPP Session-The events that occur during a period of time when logical connection exists between two communication devices that are using point to

point protocol (PPP). PPP is a connection oriented protocol that can be established between two communication devices that encapsulates Internet protocols such as Transmission Control Protocol and Internet Protocol (TCP/IP) into the underlying transmission system (such as dialup connections). PPP allows end users (end points) to setup a logical connection and transfer data between communication points regardless of the underlying physical connection.

PPPoA-Point-To-Point Protocol Over Asynchronous Transfer Mode

PPPoE-Point-To-Point Protocol Over Ethernet

PPSN-Public Packet Switched Network

PPTN-Public Packet Telephone Network

PPTP-Point-To-Point Tunneling Protocol

PPV-Pay Per View

PRACH-Packet Random Access Channel

Praseodymium Doped Fiber Amplifier (PDFA)-A device used to amplify signals across a broad range of wavelengths in an optical network. This device is very similar to the Erbium Doped Fiber Amplifier (EDFA). Praseodymium doping causes the metastable electrons to have the right energy to enable stimulated emission of photons in the 1300 nm window often used in optical networks. They are in less common use than EDFAs.

PRBS-Pseudo-Random Bit Stream

PRD-Permanent Reference Documents

Pre-Emphasis-The increase in the amplitude of the high-frequency components in a transmitted signal to overcome noise and attenuation in the transmission system. The relative balance of high- and low-frequency components is restored in a receiver by de-emphasis. Frequency-modulation broadcast stations use pre-emphasis.

Preamble-A unique sequence of bits in synchronous transmission that the receiving device must recognize and lock onto. Once the receiving device locks onto the preamble, it knows where to find the rest of the packets.

Precipitation Attenuation-The loss in radiated electromagnetic energy through absorption, refraction, and scattering as it travels through atmospheric moisture, such as rain, snow, sleet, or hail. This effect is most prominent above frequencies of 8 GHz. Precipitation attenuation also is known as rain attenuation.

Precipitation Static-A type of interference experienced in a receiver during snowstorms, rainstorms, and dust storms. The static is caused by the impact of charged particles against the antenna.

PreConditions Met (COMET)-A process that is used between communication devices that is used to determine if all the preconditions have been met to allow a communication session to begin. The use of COMET prevents the establishment of communication sessions (such as a telephone call through the Internet) that cannot be completed because the communication devices do not have all the capabilities (such as the same type of speech coding) to communicate with each other.

Predatory Acquisition Campaign-Marketing campaign developed to encourage prospects to switch from a competitor.

Predicted Frame (P-frame)-In video compression, a frame encoded with reference to a single past reference frame which is either a P-frame or an I-frame. Using motion compensation, the difference between the pixels of the P-frame and the reference frame is encoded in a way similar to JPEG for still images. Also called P-pictures.

Predictive Dialing-An automated method for making outbound telephone calls in which a pacing algorithm determines the number of calls placed in advance of actual operator availability.

Preferred Interexchange Carrier (PIC)-The assignment or use of an inter-exchange carrier to complete calls from a customer outside their systems calling area. The PIC code is obtained when a customer dials a number that requires inter-exchange carrier (IXC) service. The PIC code is used to route the call to the IXC carriers point of presence (POP) switching center.

Preferred Language (PL)-Provides a telephone service subscriber with the ability to specify the language for network services. This service allows the subscriber to specify service in English, Spanish, French or Portuguese.

Prefix-Any digit dialed before a destination address. Prefixes indicate service options. For example, a prefix of "1 indicates a 10-digit call address in some areas, and a "0" indicates a request for the services of an operator.

PREMIS-Premises Information System

Premises Distribution Network (PDN)-A premises distribution network (PDN) is a short-range network that is located at a customer's facility or even within their personal area. A PDN is used to connect terminals (computers) to other networks and each other. The most common types of PDN are Ethernet, asynchronous transfer mode 25

Preorigination Dialing

(ATM 25), universal serial bus (USB), home packet data network (HomePDN), and FireWire (IEEE-1394).

This diagram shows several forms of PDN. This diagram shows that the data transfer rate varies with the length and type of interconnection cable. This diagram also shows that some PDN technologies are better suited for multimedia applications than others. For example, ATM25 can transfer (multiplex) multiple communication channels with different levels of quality of service (QoS). Other PDN systems are capable of very high-speed data transfer rates (up to 480 Mbps) for very short distances.

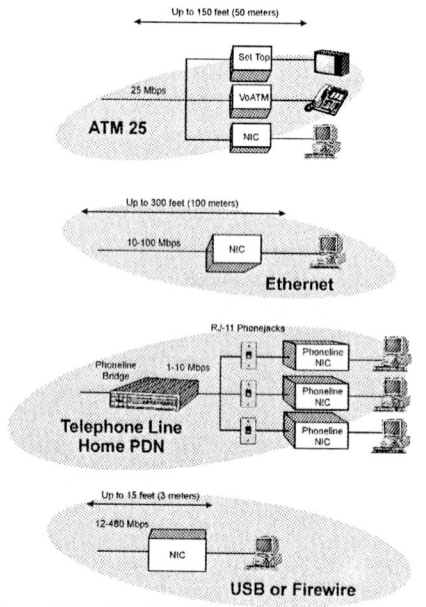

Premises Distribution Networks (PDNs) Systems

Preorigination Dialing-A process where the dialing sequence takes place before the mobile's first communication with the cellular system.

Prepaid Calling Card-A card that is issued by a telecommunications service provider that contains coded identification information that permits the card holder to initiate a call or request and receive an information service. Calling cards contain a number or code contained on a magnetic stripe that uniquely identify the card and authorized services to the system.

Prepaid Service-Service in which future use is prepaid. Prepaid communication services is a process that is used by service providers (carriers) to obtain revenue for services without the risk of bad debt and it eliminates the need and cost for billing operations. Prepaid service is often associated with customers that may be credit challenged or who want more control over bills.

Prepaid Wireless-Wireless connection whereby service is prepaid before usage is accumulated. Typical users hear an announcement prior to the call noting how many minutes or dollars or units they have remaining on account.

Prepay Service-Coin telephone service in which an initial coin deposit is required before a connection is established on chargeable calls. Prepay service is provided either by coin-first or dial-tone-first coin service.

Presence Entity-Presentity

Presence Server-The Presence Server is a server that is responsible for receiving, storing and distributing presence information received from User Agents. In SIP the Presence Service may be co-located with other SIP servers.

Presence User Agent (PUA)-A SIP User Agent which support presence subscriptions and notifications and interacts with software known as the presentity. The presentity is responsible for providing presence information to the Presence Service.

Presentation Layer-Layer 6 of the OSI model. This layer responds to service requests from the Application Layer and issues service requests to the Session Layer. The Presentation Layer relieves the Application Layer of concern regarding syntactical differences in data representation within the end-user systems. Note: an example of a presentation service would be the conversion of an EBCDIC-coded text file to an ASCII-Coded file.

Presentity (Presence Entity)-The presentity provides presence information to a presence service (server), which can distribute that information to any user that has subscribed for it.

Pressure Plug-An 8 " poured and hardened composite plug installed around cable conductors to provide back pressure for maintaining acceptable underground air pressurization systems. They are installed to encircle the conductors while maintaining sheath grounding integrity and without degradation of voice/data quality of service.

Pressure Transducer-A monitoring device that is located on aerial, buried underground cable plant to measure air pressurization and signals alarm conditions to a central system when thresholds are compromised. Pressure transducers convert pneumatic pressure readings to resistance readings (in ohms) and are read by the alarm monitoring system and OSP personnel.

Pretty Good Privacy (PGP)-A public-key encryption and certificate program that is used to provide enhanced security for data communication. It was originally written by Phil Zimmermann and it uses Diffie-Hellman public-key algorithms.

PRF-Pulse Repetition Frequency

PRI-Primary Rate Interface

PRI Gateway-Primary Rate Interface Gateway

PRI Offload-Primary Rate Interface Offload

PRI-EOM-Procedure Interrupt-End Of Message

PRI-EOP-Procedure Interrupt-End Of Procedures

PRI-MPS-Procedure Interrupt-Multipage Signal

Price Elasticity -Economic indicator used to define how sensitive consumers will be to changes in prices. A high elasticity means that customers will change their demand drastically with small changes in price.

Pricing Models -Mathematical models developed to help finance determine what prices should be charged to what customers for different services.

Primary Colors-Colors that combine to produce a full range of colors within the limits of a system. non-primary colors are mixtures of two or more primary colors. In television, the primary colors are specific sets of red, green, and blue.

Primary Interexchange Carrier Charge (PICC)-A recurring fee that is added to the bill of the local telephone service customer that charges them for access to inter-exchange carrier (IXC) services. This fee is charged in addition to other fees or tariffs (such as a percentage of billed long distance usage) that are paid by the IXC to the local exchange carrier (LEC). The purpose of the PICC charge is to help recover the cost of providing local loop access service in a marketplace that has declining long distance revenue charges.

Primary Rate Interface (PRI)-An standard high-speed data communications interface that is used in the ISDN system. This interface provides a standard data rates for T1 1.544 Mbps and E1 2.048 Mbps. The interface can be divided into combinations of 384 kbps (H) channels, 64 kbps (B) channels and includes at least one 64 kbps (D) control channel.

Primary Rate Interface Gateway (PRI Gateway)-PRI gateway is a "line-side" access gateway that converts PSTN PRI circuits for interconnection to other networks such as IP PBX or IP Centrex systems. It translates between H.323 and H.320 protocols and converts media, including audio, video and data, from circuit-switched packets to IP packets. The trunk gateway is considered to be more efficient for larger POPs. PRI gateway and trunk gateway are 2 different types of PSTN voice or media gateways. The only difference is in the type of facilities used for interconnection to the PSTN.

Primary Rate Interface Offload (PRI Offload)-Primary Rate Interface offload is the diversion of dial-up modem data traffic away from the PSTN, and on to an IP network. This was an early driver of VoIP adoption in the mid-1990s, providing relief to PSTN network congestion brought about by the advent of dial-up Internet. Also known as Internet offload.

Primary Ring-The primary path used in a communication network ring. A primary ring is used in the fiber distributed data interface (FDDI) system.

Prior Art-The term prior art means different things in different countries, but generally refers to information in the public domain which affects the patentability of an invention.

For European Patent Applications, prior art includes any public display, offer for sale, or use of the invention prior to the filing date of the application. Publications, books, other patents which were published and publicly available prior to the filing date of the application are also considered as prior art.

Prioritization-Frame and packet prioritization assigns different priority codes to packet that are transmitted through a communication network. This allows some frames or packets to receive a higher transmission priority for time sensitive data communications (such as packetized voice).

Priority Call Operation-A system access control process of allow users with a higher priority level than other users to gain access to communication system resources. Priority call operation may allow high priority users to automatically disconnect lower priority users so they may gain access to the system.

Priority Date-In most cases, the priority date is equivalent to the filing date of the first application for patent in a family of patents. Foreign applica-

Privacy

tions claiming priority to a United States application for patent can only claim the filing date of the US application for patent even if the applicant is entitled to another US priority date due to an earlier Date of Invention.

Privacy-(1-encryption) A term used with regard to encryption to indicate a level of protection that is minimal, corresponding to a moderate amount of effort on the part of the eavesdropper to understand the private communication, but not so good as the better levels designated by the words "secret" or "secure."

(2-channel separation) An electrical capability to prevent other extensions on a multi-extension analog telephone line or Key Telephone System from connecting during a conversation, typically via diodes or other devices actuated by the decrease in subscriber loop voltage when one set is off hook. Also the name of an equivalent capability using an electronic key system.

Private Automatic Branch Exchange (PABX)- A telephone switch that is generally located on a customer premise. Often referred to as a PBX, CBX, EPABX. This provides for the transmission of calls internally as well as to and from the public network.

Private Branch Exchange (PBX)-PBX systems are private local telephone systems that are used to provide telephone service within a building or group of buildings in a small geographic area. PBX systems contain small switches and advanced call processing features such as speed dialing, call transfer, and voice mail. PBX systems connect local telephones ("stations") with each other and to the public switched telephone network (PSTN).

This diagram shows a private branch exchange (PBX) system. This diagram shows a PBX with telephone sets, voice mail system, and trunk connections to PSTN. The PBX switches calls between telephone sets and also provides them switched access to the PSTN. The voice mail depends on the PBX to switch all calls needing access to it along with the appropriate information to process the call.

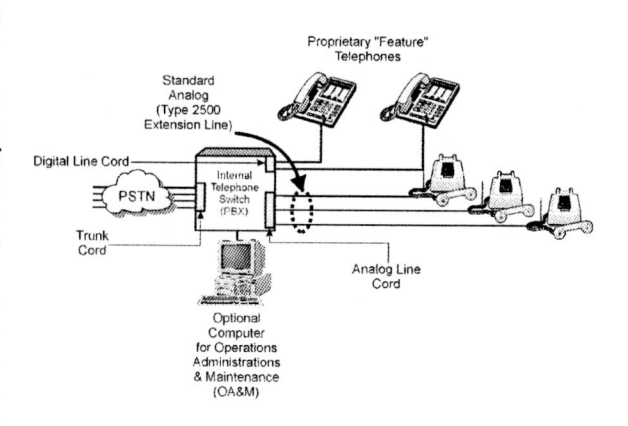

Private Branch Exchange (PBX) System

This photograph shows the NEC NEAX ® 2000 Internet Protocol Server (IPS), the first IP telephony system that supports 100 % peer to peer connectivity and offers all the system reliability and features associated with traditional NEC TDM PBXs. The NEAX 2000 IPS is designed to provide telephony solutions for the mid-sized business and enterprise branch office market.

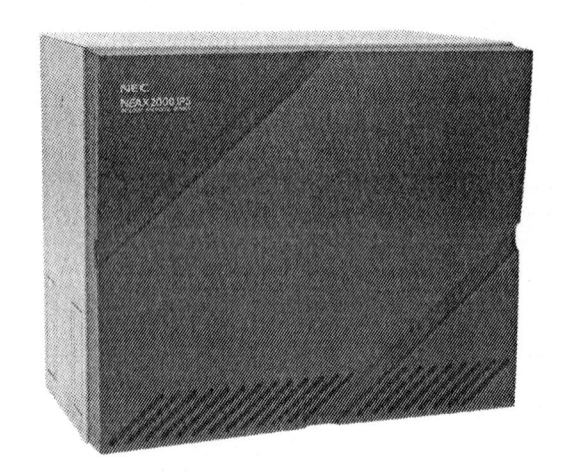

Private Branch Exchange (PBX) Switch
Source: NEC America, Inc.

Private Carrier-An entity licensed in the private services and authorized to provide communications service to other private services on a commercial basis.

Private Internet Address-An Internet address that can be transferred by routers on a private data network.

Private Key-A key used to decrypt a message encrypted with the owner's public key or to sign a message from the key's owner. Sometimes called a secret key.

Private Line-A dedicated communications circuit that is leased by a customer from a telephone service provider for voice, data, or video services. While a private line may be connected through a switching facility, the connection resources are constantly dedicated to the customer who is leasing the line.

Private Network-A network designed for the exclusive use of one customer. Often, such a network is nationwide in scope and serves large corporations or government agencies.

Private Network Access Point (PNAP)-An key private physical (layer 2) interconnection point that interconnects regional Internet systems and subsystems.

Private Numbering Plan (PNP)-A feature that enables subscribers to call defined private-network extensions using an abbreviated dialing pattern. When a PNP subscriber dials a private-network extension, the cellular network translates the dialed extension to a number in the North American numbering plan (NANP) and routes the call.

Private Telephone System-Private telephone systems are independent telephone systems that are owned or leased by a company or individual. Private telephone networks include key telephone systems (KTS), private branch exchange (PBX) and computer telephone integration (CTI).

This diagram shows the basic types of private telephone systems; key telephone systems (KTS), private branch exchange (PBX), and computer telephony integration (CTI) systems. The most simplistic private telephone network is a single telephone attached to a business line. Key systems allow each telephone in a business to answer and originate calls on several business lines. A PBX system allows many extensions within a business to call each other and the PSTN. Computer telephony (CT) systems are communication networks that merge computer intelligence with telecommunications devices and technologies.

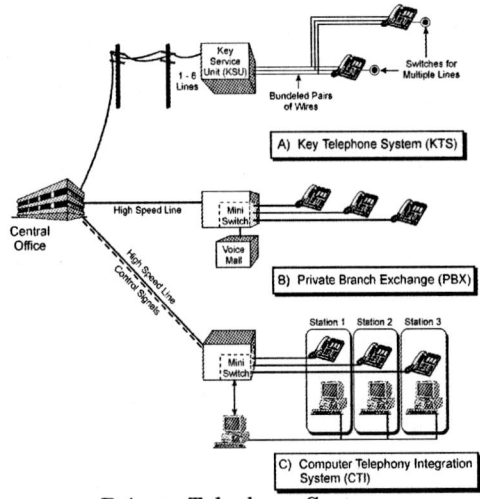

Private Telephone Systems

Privatization-The process of selling or reallocated government owned assets or resources to companies or investors. Privatization in the telecommunications industry often includes the issuance or re-allocation of licenses to provide a telecommunications services.

PRM-Private Mobile Radio

PRN-Provide Roaming Number

Probability Block-IS-136 specifies the order in which a phone must scan one of 16 divisions of the 800 MHz spectrum, each containing approximately 26 cellular channels when searching for a DCCH.

Probe-(1-test) A test prod used to check components for the presence of signals. (2-cavity) A wire loop inserted in a cavity for coupling energy to an external circuit. (3-communication) A process of sending a message or alert signal to another device or group of devices that discovers if there is information to be gathered.

Proc Amp-Processing Amplifier

Processing Delay-The time required for processing information. Processing delay for coding and decoding of voice or data information is usually measured in msec.

Processing Power-The number of computations that a computer, microprocessor, or digital signal processor can complete in a fixed time interval. May be measured in millions of instructions per second

(MIPS) or millions of floating point operations (MFlops.) Typical low-cost digital signal processing (DSP) chips provide 10-20 MFlops.

Processor-A circuit or device that systematically executes specific operations on digital data, as directed by a program stored in memory.

Product Acquisition and Development System (PADS)-A process that is used for the identification and coordination of product development activities. The PADS system is divided into 5 phases; concept, feasibility, planning, development, and introduction.

This diagram shows a basic product development process to help ensure the successful development of communication products and services. This example shows that this product development process evolves through a series of gates (called toll gates) that are used to determine if a product should continue in the normal development process or eliminated (discarded) so additional resources can be allocated to other products. In this example, the product development cycle begins with product ideas (concept), business evaluation (feasibility), resource scheduling (planning), technology design and production testing (development), and market introduction and distribution support (introduction).

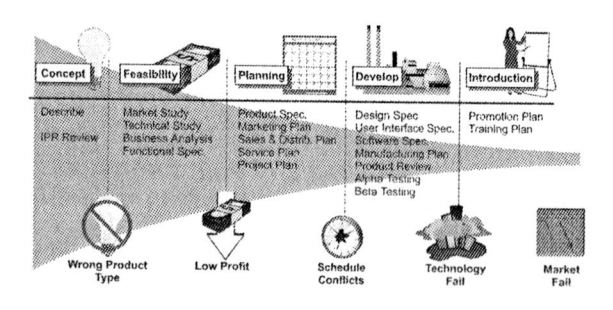

Product Acquisition and Development System (PADS) Operation

Product Marking-According to 35 U.S.C. § 287(a) a patent holder is required to give notice that an item is patented. This may be accomplished by physically marking the item, or accompanying packaging, with the US patent number, or alternatively with the phrase "Patent Pending" where an application has been applied for, but not yet granted.

The purpose of marking is to provide constructive notice to the world of the patent, and is an important issue in many patent infringement suits. Defendants may be able to avoid some damages if patentee did not properly mark the article.

Production House-A facility that typically does everything to generate final video productions except shooting the original videotape. Services typically include editing raw master tapes, modifying recorded material, and 'creating new effects. Projects include advertising, training, promotion, music videos, and TV shows.

Profile-A particular implementation or instantiation of a more general protocol. Many protocols are extremely general and allow one to specify a restricted set of messages and their actions for a particular purpose. Such a set is known as a profile. For example, NCS is a profile of MGCP with a few extensions.

Profit Impact of Market Share (PIMS) -Series of studies conducted in the 1970's that allegedly "proved" that the companies
with the largest market share were the most profitable. (These findings have since
been disproved in many cases).

Program-A sequence of instructions used to tell a computer how to receive, process, store, and transfer information.

Program Language-C

Programmable Array Logic (PAL)-A programmable device that provides discrete logic functions and are commonly used to link integrated circuits and components within a device or system together.

Programmable Logic Array (PLA)-A general-purpose integrated circuit logic device containing an array of gates that can be programmed to perform various functions.

Programmable Logic Device (PLD)-An IC containing a large number of logic gates whose interconnections are programmable for specific applications.

Programmable Read Only Memory (PROM)-A ROM device that can be programmed by the equipment manufacturer.

Programmer-A person who prepares sequences of instructions for a computer.

Programming-The process of developing an assembly of instructions for a computer to enable it to carry out a particular job.

Programming Language-C+ Or C++

Programming Language-A language designed to be under-stood by a computer. A high-level programming language is converted into the required machine code by a program called a compiler.

PROM-Programmable Read Only Memory

Promiscuous Mode-A mode of operation of a network interface that allows it to attempt the reception of all data regardless of the destination address of the incoming data.

Prompt-A cue to help the operator choose the next action.

Propagation-The process of transfer of a radio signal (electromagnetic signal) or acoustic signal (sound) from one point to another point.

Propagation Constant-A measure of signal propagation through a transmission line or other medium through which a wave passes. Propagation constant consists of the attenuation constant in decibels or nepers per unit length and the phase constant in radians per unit length.

Propagation Delay-The delay in the reception of a signal caused by its travel from the transmitter to the receiver. Propagation delay can be quite lengthy when signals travel great distances, such as over satellite circuits.

Propagation Path Obstruction-A physical material or assembly that effects the travel (path) of a radio signal (loss) other than the effect of reflecting the signal. Obstructed path loss from an omnidirectional antenna (all directions) for 900 MHz mobile communications through buildings and foliage is approximately 40 dB per decade. That means for each 10 times increase in distance, the amount of energy decreases by 99.99% (10,000 times less). A signal of 1 Watt will be reduced to 0.0001 Watts (0.1 mWatts).

Propagation Time-The time required for a signal to travel between points on a transmission path.

Prorating-The process of fractionalizing charges for a partial period. In order to determine the number of days for which to charge, a "multiplier" is used (See "Multiplier").

Prosecution-Prosecution refers to the examination and exchange of communications between the patent office and the applicant, or the applicant's agent or attorney.

Protected Field-That portion of a packet that is to be protected by some mechanism, typically either encryption or a cyclic redundancy check.

Protector-A device or circuit that prevents damage to lines or equipment by conducting dangerously high voltages or currents to ground. Protector types include spark gaps, semiconductors, varistors, and gas tubes.

Protector Frame-Protector frames are usually part of a main distributing frame (MDF) and it serves as the termination mounting assembly for local loop cables. The protector frame includes devices that will isolate loops and/or electrically ground communication equipment to help protect the equipment when undesired voltages (e.g. high voltage spikes) are detected (such as lightning strikes.)

Protocol-Protocols are the languages, processes, and procedures that perform functions used to send control messages and coordinate the transfer of data. Protocols define the format, timing, sequence, and error checking used on a network or computing system. While there are several different protocol languages used for Internet telephone service, the underlying processes (setup and disconnection of calls) are fundamentally the same.

Protocol Adaptation-The process of adapting one protocol to another protocol. This may involve syntax changes (text format and command name changes,) timing relationships, and other functional processes.

This figure shows an example of how protocols from different communication systems can be adapted to work with each other. This diagram shows a data file is being sent from a local computer to a destination computer that has a specific Internet protocol (IP) address. The local computer divides the data file into small IP packets. Each IP packet has its destination IP address. The local computer sends these IP packets over a universal serial bus (USB). The USB identifies that the packets must be routed through a digital subscriber line (DSL) modem. The network address of the DSL modem is added to the USB packet to allow the packet to be transferred to the DSL modem. The original IP packets (IP address and data) are transferred in the data portion of the USB packet (called encapsulation). When the DSL modem receives the USB packet, it extracts the IP packet(s) and determines that this must be routed to a destination ISP. The DSL system creates a logical channel through the switches to the destination ISP and each path between switches is assigned a logical routing address. Packets sent by the DSL modem continually follow this path to the destination ISP. Because the DSL modem communicates through the system

Protocol Analyzer

using small asynchronous transfer mode (ATM) packets, the DSL modem divides up the IP data packets into very small 48 byte packets. Each of these packets has their own address code that indicates the logical channel through the DSL network where these packets are routed. When the packets are received at the remote ISP location, these packets are reassembled into their original IP packet form (address and data). Since the remote data storage device is connected through an Ethernet local area network (LAN), the LAN determines the media access control (MAC) address of the destination computer using the received IP address. The LAN then encapsulates each IP packet an Ethernet frame and uses the destination MAC address. The remote computer then receives the IP packets and can confirm their reception to the sending computer. This diagram shows that each of these address translations, packet division, re-assembly, and flow control steps are required and those protocol adaptations take time. As a result, the data transmission delay increases as the number of protocol conversions increase.

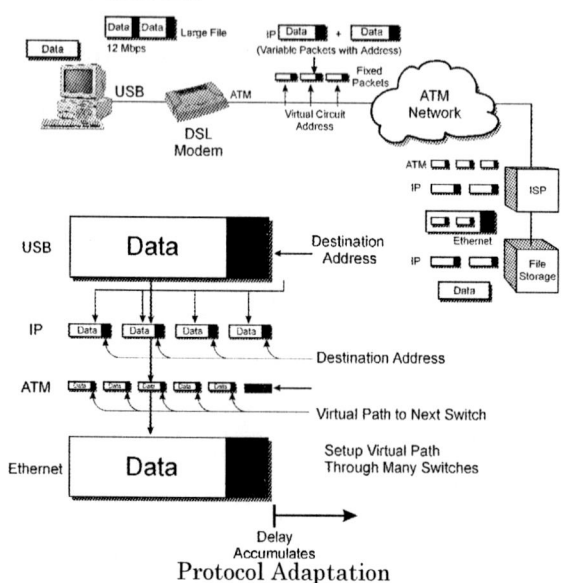

Protocol Adaptation

Protocol Analyzer-A test instrument that is designed to monitor a network and provide analysis of the communication taking place on the network. This allows a technician to monitor a network and provides information for problem determination and resolution.

Most modern day protocol analyzers are aware of all commonly used, industry standard protocols. More advanced protocol analyzers sit "in-line" between two devices, without the devices being aware that the analyzer is present. Other less sophisticated protocol analyzers can be created using standard PCs with network interface cards in "promiscuous mode", whereby the copy all packets that appear on the network, irregardless of destination address.

Protocol Conversion-Protocol conversion involves the translation of the protocols of one system to those of another to enable different types of equipment, such as data terminals and computers, to communicate. This is done by an inter-working function (IWF). An IWF system (such as a data bridge) adapts the communications between two different types of networks. Protocol conversion may be used to interconnect circuit switched or packet switched networks.

Protocol Converter-A device for translating the protocols of one terminal or system to those of another, enabling equipment with different formats and procedures to intercommunicate.

Protocol Data Unit (PDU)-The package of data that contains header and data protocol information that is used to communicate with a specific layer in a software stack.

Protocol Interworking-The ability of different protocols to work with each other. An example of protocol interworking is how well all of the commands and messages in VoIP protocols (such as H.323, SIP, or MEGACO) interwork with commands and messages for telephone protocols (such as SS7 and E&M).

Protocol Layers-A hierarchical model of network functions. The divisions of the hierarchy are referred to as layers or levels, with each layer performing a specific task In addition, each protocol obtains services from the protocol layer below it and performs services to the protocol layer above it.

Protocol Multiplexing-(1-general) The identification and sharing of multiple protocols on a single logical channel. (2-Bluetooth) A function performed at the Logical Link Control and Adaptation Protocol (L2CAP) layer. L2CAP must support protocol multiplexing because the baseband protocol does not support a "type" field to identify the higher-layer protocol being multiplexed above it. L2CAP must therefore be able to distinguish between upper-layer protocols such as the Service Discovery

Protocol (SDP), Radio Frequency Communication (RFCOMM), and the Telephony Control Specification (TCS).

Protocol Stack-A hierarchical structure of information processing functions that are logically separated into layers that theoretically only interact with higher and lower functional layers. The use of a protocol stack allows software programs and devices to be independently created (such as from different manufacturers) that only provide parts of the overall operation. Protocol stacks can be proprietary (owned exclusively by a company or group of companies) or protocol stacks can be created as an industry standard.

Protocol Suite-A hierarchical set of related protocols.

Protocols-(1-rules) Protocols are a precise set of rules, timing, and a syntax that govern the accurate transfer of information between devices or software applications. Key protocols in data transmission networks include access protocols, handshaking, line discipline, and session protocols. (2- connection) A procedure for connecting to a communications system to establish, carry out, and terminate communications.

Provisional Application for Patent-An application for patent filed in the USPTO under 35 U.S.C. §111(b) without a formal patent claim, oath or declaration, or any information disclosure (prior art) statement. A provisional application for patent is a faster and lower cost means to provides a means to establish an early effective filing date as compared to filing an ordinary application for patent. A provisional application for patent, commonly called a provisional application, has a pendency lasting 12 months from the date the provisional application is filed which cannot be extended. An applicant who files a provisional application must file a corresponding ordinary application for patent (a non-provisional application) during the 12-month pendency period of the provisional application in order to benefit from the earlier filing of the provisional application

Provisional Response-A temporary (interim) signaling response that is transmitted from a receiver of a command message that indicates the command has been received is currently being processed. Provisional responses are used to extend the amount of time a sender is willing to wait for a response so the sender does not retransmit the command message or terminate the communication session.

Provisioning-A process within a company that allows for establishment of new accounts, activation and termination of features within these accounts, and coordinating and dispatching the resources necessary to fill those service orders. Provisioning is also known as customer care.

Provisioning Server-A server that coordinates the activation setup, authorization of features, and elimination of users from a communications system.

Proximity Effect-A non-uniform current distribution in a conductor, caused by current flow in a nearby conductor.

Proxy Server-Proxy servers are computing devices (typically a server) that interface between data processing devices (e.g. computers) and other devices within a communications network. These devices may be located on the same local area network or an external network (e.g. the Internet). A proxy server usually has access to at least two communication interfaces. One interface communicates with a device requesting services (e.g. a client) and a device that is being requested for a service (the server).

This figure shows how a SIP proxy server is responsible for establishing communication connections between devices within a specific domain. In this example, when the proxy server receives a request message from a User Agent (UA) inviting another party to join a session, the proxy server forwards this invitation onwards (it acts as a proxy) to the designated User Agent. If the designated User Agent is unavailable, the proxy server may direct the connection request to the second User Agent or the connection request may be forwarded via one (or more) proxy servers.

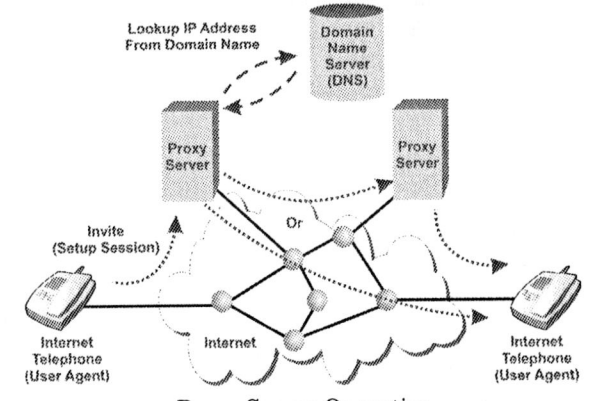

Proxy Server Operation

ps-Picosecond
PSA-Partitioned Service Area
PSAP-Public Safety Answering Point
PSC-Public Service Commission
PSDN-Public Switched Digital Network
PSDS-Public Switched Digital Service
PSE-Professional Server Expert
PSE-Packet Switching Exchange
Pseudo Ternary Line Coding-A method of encoding a serial bit stream which is a combination of binary bit inversion followed by AMI line coding. The result is that a binary zero is mapped into a positive or negative voltage pulse, the polarity being controlled by a simple memory device so that the polarity of each pulse is the opposite of the last pulse. A binary 1 is mapped into an interval of zero volt output, or no pulse. Pseudo Ternarly line coding is used in the DS-1 system and for S-interface wiring in ISDN.
Pseudo-Random Bit Stream (PRBS)-A sample data signal that is used for telecommunications testing.
PSID-Private System Identity
PSK-Phase Shift Keying
PSM-Protocol Service Multiplexer
PSPDN-Packet Switched Public Data Network
PSS-Professional Server Specialist
PSTN-Public Switched Telephone Network
PSTN and Interworking (PINT)-A protocol that allows certain telephony services to be invoked from an IP network, these services might include click-to-dial and click-to-fax, where an Internet user may select a link and the protocol allows a call to be completed on the PSTN. The PINT protocol is a SIP enhancement.
PSTN Fallback-The ability to use the public switched telephone network as a backup connection in the event that another network (such as a data network or the Internet) are not able to provide communication services. PSTN fallback is commonly used in voice over data communication systems.
PSTN Gateway-A PSTN gateway is a communications device or that transforms data that is received from one network (such as the Internet or DSL network) into a format that can be used by the PSTN network. The PSTN gateway may be a simple device that performs simple call completion and adaptation of digital audio into a compatible signals for the PSTN. Or, it may be a more complex device that is capable of advanced services such as conference calling, call waiting, call forward and other PSTN like services. The PSTN gateway must create signaling protocols and compensate for timing differences between a end users computer and the public switched telephone network (PSTN).
Psychoacoustic Distortion-A mathematical model, based on experiments done with human listening test subjects, of the amount of perceptible noise or distortion introduced into an audio signal by a process such as a codec. Current psychoacoustic distortion formulas are based on absolute hearing threshholds (a sound might be too quiet to hear at all) and masking threshholds (one sound may not be audible due to another sound nearby in frequency and time). An audio encoder can use a psychoacoustic distortion model to choose the best among alternative encodings for a signal.
PTC-Positive Temperature Coefficient
PTE-Path Terminating Equipment
PTI-Payload Type Identifier
PTN-Public Telephone Network
PTO-Patent and Trademark Office
PTP-Peak-To-Peak
PTR-Percent Traffic Routing
PTT-Push To Talk
PTT-Post, Telephone And Telegraph
PUA-Presence User Agent
Public Cells-Cells in the public cellular system that provides the same basic cellular service to all customers.
Public Internet Address-An Internet address that can be recognized and transferred (routed) on the public Internet.
Public key-A cryptographic key that is made public for the purpose of encrypting messages to the key's owner or validating signed messages from him or her.
Public Land Mobile Network (PLMN) Code-(1-GSM) A unique code assigned to each GSM network operator. The code is made up of a mobile country code (MCC) and a mobile network code (MNC).
Public Mobile Radio (PMR)-A land mobile radio system that is accessible to the public.
Public Network-A network operated by common carriers, local and long-distance, that is made available for private users and the public.
Public Packet Data Network (PPDN)-A packet data network available for customers.
Public Packet Switched Network (PPSN)-An exchange access capability that provides packet-switched data transport for terminals, value-added networks, host computers and interexchange carriers.

Public Packet Telephone Network (PPTN)-A public telecommunications network that uses packet based communication as its core communication technology.

Public Safety Answering Point (PSAP)-An agency that receives and processes emergency calls. The PSAP usually receives the calling number identification information that can be used to determine the location of the caller.

Public Switched Digital Network (PSDN)-A high-speed data network that uses a public switched telephone system to link users over greater distances than can be provided by local area networks.

Public Switched Telephone Network (PSTN)- Public switched telephone networks are communication systems that are available for public to allow users to interconnect communication devices. Public telephone networks within countries and regions are standard integrated systems of transmission and switching facilities, signaling processors, and associated operations support systems that allow communication devices to communicate with each other when they operate.

This diagram shows a basic overview of the Public Switched Telephone Network (PSTN) as deployed in a typical metropolitan area. PSTN customers connect to the end-office (EO) for telecommunications services. The EO processes the customer service request locally or passes it off to the appropriate end or tandem office. As Different levels of switches interconnect the parts of the PSTN system, lower-level switches are used to connect end-users (telephones) directly to other end-users in a specific geographic area. Higher-level switches are used to interconnect lower level switches.

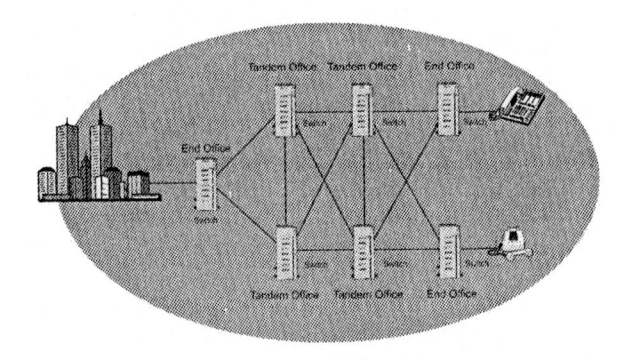

Public Switched Telephone Network (PSTN)

Public Telephone-A telephone provided by a telephone company through which an end user may originate interstate or foreign telecommunications for which he pays with coins or by credit card, collect or third party billing procedures.

Public Telephone Network (PTN)-An unrestricted dialing telephone network that is available for public use. The network is an integrated system of transmission and switching facilities, signaling processors, and associated operations support systems that is shared by customers. PTN is the common term used for public telephone networks outside North America. Inside North America, the public telephone network is called the public switched telephone network. In the United States, the PTN is referred to as the Public Switched Telephone Network (PSTN).

Public Utilities Commission (PUC)-A state regulatory body that sets rates and rules for local exchange carriers and interexchange carriers that provide long distance (interLATA) service within a state boundary. In some states, it is called a public service commission (PSC), board of public utilities (BPU), or corporation commission (CC).

Publication Date-With some exceptions, applications for patent are published and made public 18 months after the date of filing. The publication date is important in that it is the date on which pending patents can be used as prior art against non-US applications for patent.

According to US law, pending US applications for patent may be considered as prior art as of the date of filing against a later US application for patent.

Publisher Database-A database that contains billing and usage information about a subscriber to a service. This usually includes authorized features, billing records, and other related data that may be regularly updated (read-write). The original subscriber database that provides information to the publisher database is often read-only.

PUC-Public Utilities Commission

PUK-Personal Unblocking Key

Pull Box-A rectangular or square steel box installed inline within a buildings conduit run. They facilitate the initial installation and subsequent maintenance of indoor telephone company (telco) plant cabling. Pull boxes are suitable for cables of 2" OD or less and are installed in rigid steel conduit runs over 30 meters in length. They also must be installed if the conduit run has more than two 90 degree bends. The pull box length must be at least 12 times the diameter of the largest conduit.

Pulse-A signal of short duration characterized by a rapid change in amplitude. A pulse signal is often characterized by a rise time from 10% to 90% of the signal height, its fall time from 90% to 10%, and it's pulse duration between the beginning and end of the pulse.

This diagram shows a series of pulse signals and typical characteristics associated with pulse signals. In this diagram, these pulse signals are characterized by a rise time from 10% to 90% of the signal height, their fall time from 90% to 10%, and the pulse duration between the beginning and end of the pulse. This diagram also shows how a pulse signal may include signal ringing. Signal ringing is caused by a rapid signal level change in a system that does not have effective impedance matching.

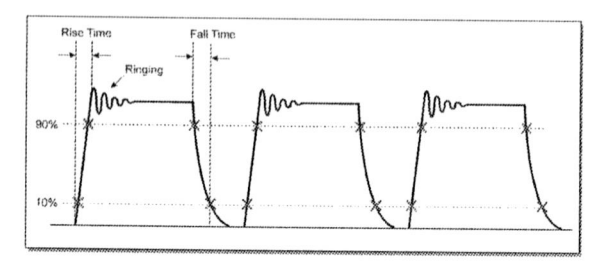

Pulse Wave Diagram

Pulse Amplitude Modulation (PAM)-A form of modulation in which the amplitude of the pulse carrier is altered in accordance with some characteristic of the modulating signal.

Pulse Coded Modulation (PCM)-That form of modulation in which the modulating signal is sequentially sampled, quantized, and coded into a binary form for transmission over a digital link.

Pulse Dialing-A dialing process that uses momentary pulses to transfer a telephone number from a telephone (rotary telephone) to a telephone system.

Pulse Frequency Modulation (PFM)-Modulation of a carrier signal by a repetitive pulse whose repetition frequency is varied to represent a modulating signal. Rectangular waves cause an abrupt shift of frequency; triangular pulses cause the carrier to shift frequency gradually toward its maximum.

Pulse Interval-The time between the start of one pulse and the start of the next

Pulse Position Modulation (PPM)-A method of conveying information by transmitting a series of extremely brief pulses of equal amplitude and duration but varying time occurrence within each time window of a sequence of consecutive equal-duration time windows or time slots. To encode a telephone-quality voice signal, each time window is typically about 100 microseconds in duration. For example, the position (time relative to the beginning of the time window) of a pulse within the window is typically proportional to the sampled voltage of an analog speech waveform that is represented by the PPM signal. A PPM waveform is approximately the time derivative of a corresponding pulse width modulation (PWM) waveform.

Pulse Rate-The number of pulses that are transmitted during a time interval, usually measured in pulses per second (pps).

Pulse Repetition Period-The time interval between the repetitive pulse signals.

Pulse Train-A series of pulses having similar characteristics.

Pulsed Laser-A laser that emits light in pulses rather than continuously.

Pump-The energy source used to drive electrons up to a high energy state called the pump level. From the pump level, the electrons lose energy and reach the metastable state needed for stimulated emission. Lasers and fiber amplifiers must be pumped before stimulated emission can take place.

Pump Laser-A laser used to prime a fiber amplifier so that stimulated emission can take place. The laser pumps the energy of electrons in the fiber up to a high energy state called the pump state. They then decay to the metastable state where they can interact with photons to generate new photons of the same wavelength.

Punchdown Block-A device used to connect one group of wires to another. This is also referred to as a terminating block, a connecting block, and a cross-connect block.

Purge and Ventilate-to clear and refresh the air in a closed, controlled environment (generally underground) for prescribed times and volumes based on the area's square footage.

Push Technology-The ability of a communication user to push content or control to another communication user. Push technology is commonly used in web seminar (webinar) systems where an instructor can push presentations to others who are participating in a webinar session.

Push To Talk (PTT)-(1-system) The process of initiating transmission through the use of a push-to-talk button. (2-switch) A control switch (usually part of a handheld microphone) that must be pushed before the user can transmit.

Pushbutton Dialing-The use of keys or pushbuttons instead of a rotary dial to generate the sequence of digits that establishes a circuit connection. Touch tones are the most common form of pushbutton dialing.

PVC-Permanent Virtual Circuit
PVC-Polyvinyl Chloride
PVR-Personal Video Recorder
PWM-Pulse Width Modulation

Q

Q-Quality Factor

Q.931-A telecom call processing signaling protocol that is used in telephone communication systems. The Q.931 protocol defines the messages and formats are control messages that are created by the end communication device. Some of the common information contained in Q.931 messages include call setup and tear down messages, called and calling party telephone numbers, and other access control signaling messages.

This diagram shows the basic structure of a Q.931 signaling message. The examples in this diagram show that Q.931 messages are used to provide information to setup, maintain, and tear down communication connections. Although Q.931 is ISDN (digital) related, the signaling message structure is also used for some analog communication systems.

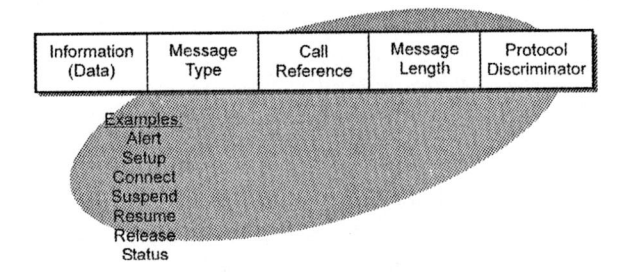

Q 931 Message Structure

QA-Quality Assurance
QAM-Quadrature Amplitude Modulation
QBE-Query By Example
QBone-Quality Of Service Backbone
QC-Quality Control
QCIF-H.261 Quarter Common Intermediate Format
QoS-Quality Of Service
QPSK-Quadrature Phase Shift Keying
QSIG-QSIG is a peer-to-peer signaling standard that enables call setup between voice-enabled equipment made by different manufacturers. QSIG is also known as Private Signaling System No. 1 (PSS1) and falls under the auspices of both the European Telecommunications Standards Institute (ETSI) and the International Standards Organization (ISO). In addition to QSIG's ability to set up a basic call, supplementary QSIG services specify how calling features, such as line forwarding, call transfer, call forwarding, and many others, can work across different vendors' platforms. QSIG grew out of the Integrated Services Digital Network (ISDN) standard called Q.931.

Quadraphonic-Surround-sound reproduction involving the recording and playback of four channels of sound.

Quadrature-A state of alternating current signals separated by one-quarter of a cycle (90 degrees).

Quadrature Amplitude Modulation (QAM)-QAM is a combination of amplitude modulation (changing the amplitude or voltage of a sine wave to convey information) together with phase modulation. There are several ways to build a QAM modulator. In one process, two modulating signals are derived by special pre-processing from the information bit stream. Two replicas of the carrier frequency sine wave are generated; one is a direct replica and the other is delayed by a quarter of a cycle (90 degrees). Each of the two different derived modulating signals are then used to amplitude modulate one of the two replica carrier sinewaves, respectively. The resultant two modulated signals can be added together. The result is a sine wave having a constant unchanging frequency, but having an amplitude and phase that both vary to convey the information. At the detector or decoder the original information bit stream can be reconstructed. QAM conveys a higher information bit rate (bits per second) than a BPSK or QPSK signal of the same bandwidth, but is also more affected by interference and noise as well.

This diagram shows that amplitude and phase modulation (QAM) can be combined to form an efficient modulation system. In this example, one digital signal changes the phase and another digital signal changes the amplitude. In some commercial systems, a single digital signal is used to change both the phase and the amplitude of the RF signal. This allows a much higher data transfer rate as compared to a single modulation type.

Quadrature Distortion

Amplitude Modulation (QAM) Operation

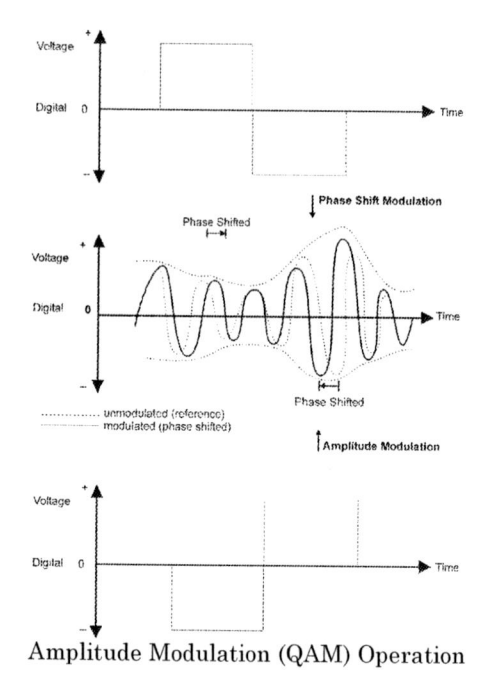

Quadrature Distortion-A distortion resulting from the asymmetry of sidebands used in vestigial sideband TV transmission. Quadrature distortion appears when envelope detection is used, but can be eliminated by use of a synchronous demodulator.

Quadrature Modulation-Modulation of two carrier components 90° apart in phase by separate modulating functions.

Quadrature Phase Shift Keying (QPSK)-A type of modulation that uses 4 different phase shifts to represent the digital information signal. These shifts are typically +/- 45 and +/- 135 degrees. See also Binary Phase Shift Keying - BPSK.

Quality Assurance (QA)-All those activities, including surveillance, inspection, control, and documentation, aimed at ensuring that a given product will meet its performance specifications.

Quality Control (QC)-A function whereby management exercises control over the quality of raw material or intermediate products to prevent the production of defective devices or systems.

Quality Factor (Q)-(1-inductor) A figure of merit that defines how dose a coil comes to functioning as a pure inductor. High Q describes an inductor with little energy loss resulting from resistance. Q is determined by dividing the inductive reactance of a device by its resistance. (2-capacitor). A figure of merit that defines how close a capacitor comes to functioning as a pure capacitance. High Q describes a capacitor with little energy loss resulting from resistance. Q is determined by dividing the capacitive reactance of a device by its resistance.

Quality Of Service (QoS)-One or more measurements of desired performance and priorities of a communications system. QoS measures may include service availability, maximum bit error rate (BER), minimum committed bit rate (CBR) and other measurements that are used to ensure quality communications service.

Quality Of Service Backbone (QBone)-A high speed backbone network used for Internet2 that allows for different levels of Quality of Service.

Quantization-The process of representing a value with a less precise value. In analog-to-digital conversion, a continuous analog value is represented by one of a finite number of quantized values. In lossy signal compression (such as an A-law encoding) one digital value is represented by another one which is usually not precisely the same. Except in lucky cases where the quantized value is exactly the same as the original, quantization introduces error (or noise).

Quantization Error-The difference between the actual value of the analog signal at the sample time and the resulting digital word value. The greater the resolution of the analog-to-digital converter, the lower the quantization error.

Quantization Noise-Noise (or distortion) of a quantized signal that results from the process of coding a continuous analog signal into a finite number of digital samples. Quantization noise is reduced by increasing the number of samples. This term also is known as quantization distortion.

Quantizing-The process of sampling an analog waveform to convert its voltage levels into digital data.

Quantizing Distortion-The distortion resulting from the quantization process.

Quantum-A discrete package as opposed to a continuously variable process. In the electronic field, the quantum of electromagnetic radiation is the photon, but this unit is so small that, for all large scale purposes, it may be considered that energy is carried by continuously varying electromagnetic fields obeying classical laws.

Quartz-A crystalline mineral that, when electrically excited, vibrates with a stable period. Quartz typically is used as the frequency determining element in oscillators and filters.

Quasar-A quasi-stellar radio source. A quasar is a star-like body near the limit of the presently observable universe that emits radio and visible light radiation.

Quasi-Associated Mode-In the Signaling System 7 protocol, a non-associated mode of signaling in which the route for each signaling message is determined, for each signaling message, by information contained in this message (namely in its routing label) and is fixed in normal operation.

Query By Example (QBE)-A database front-end routine that requests the user to supply an example of the type of data to be retrieved.

Queuing-A process of delaying or sequencing messages. Queuing involves receiving requests for service, prioritizing these requests, storing them in appropriate order and transferring the messages when the facilities (channels) are available to send them.

Queuing systems may change the order of messages or services to be provided based on priority access. For example, communication requests from a public safety official may be given priority over a communication request from a consumer.

Queuing System-A method of trunked radio communication in which users attempting to obtain a channel are placed in a queue, and the user at the front of the queue is allocated the first channel that becomes free.

Quiescent-An inactive device, signal, or system.

QUIL-Quad-In-Line

R

R-170A-A document prepared by the Electronics Industries Association describing recommended practices for NTSC color TV signals in the United States. The changes from RS-170 involve pulse timing parameters.

R-Spec-Resource Flowspec

R/C Transmitter-A transmitter that operates or is intended to operate at a station authorized in the R/C service. R/C may stand for Radio Control or Remote Control.

R/N-Received/Not Received

RAA-Rural Allocation Area

RAB-Random Access Burst

RACE-Research In Advanced Communications In Europe

RACES-Radio Amateur Civil Emergency Service

Raceway-A covered trough or channel for internal wiring and cabling.

RACH-Random Access Channel

Rack-An equipment rack, usually measuring 19 inches (48.26 cm) wide at the front mounting rails.

RACON-Radar Beacon

Racon Station-A radionavigation land station that uses a racon. A racon (radar beacon) is a transmitter-receiver associated with a fixed navigational mark, which when triggered by a radar, automatically returns a distinctive signal which can appear on the display of the triggering radar, providing range, bearing and identification information.

Rad-The unit used to measure the absorption of ionizing radiation.

RAD-Rapid Application Development

Radar-Radio Detection And Ranging

Radar Beacon (RACON)-A transmitter-receiver associated with a fixed navigational mark which, when triggered by radar, automatically returns a distinctive signal which can appear on the display of the triggering radar, providing range, bearing and identification information.
A receiver transmitter which, when triggered by a radar, automatically returns a distinctive signal which can appear on the display of the triggering radar, providing range, bearing and identification information.

Radiate-The process of emitting electromagnetic energy.

Radiation-(1-verb) The emission and propagation of electromagnetic energy in the form of electromagnetic waves. (2-noun) Radiation also is called radiant energy.

Radiation Dose-The term radiation dose and the associated units described in this entry are used only for ionizing radiation. The rad is a unit of absorbed dose. One rad is equal to an absorbed dose of 100 ergs/gram. (1 erg = $6.24E^{11}$ eV) The SI unit of absorbed dose is the Gray (Gy). 1 Gy = 1 joule/kilogram = 100 rad.
An exposure of 1 R results in an absorbed dose of 0.87 rad. The unit of radiation exposure in air is the roentgen (R). It is defined as that quantity of gamma or x-radiation causing ionization in air equal to $2.58E^{-4}$ coulombs per kilogram of air. Exposure applies only to absorption of gammas and x-rays in air.

Radiation Pattern-(1-radio frequency) The magnitude of the relative electric field strength radiated from an antenna in a given plane as a function of the angle from a given reference direction. (2 - fiber optic) The output radiation of an optical waveguide, specified as a function of angle or distance from the waveguide axis. The near field radiation pattern is specified as a function of angle. The near field radiation pattern is specified as a function of distance from the waveguide axis. The radiation pattern is a function of the length of waveguide measured, the manner in which the waveguide is excited, and the wavelength.

Radiator-(1-radio) Any part of an antenna that radiates electromagnetic waves. (2-heat) Devices that radiate heat rather than electromagnetic waves.

Radio-(1) Radio is the transfer of information signals through the air by the by means of electromagnetic waves. These electromagnetic waves repeat their cycle in a frequency range of approximately 150 kHz to 300GHz. (2) The electronic equipment used to receive or transmit radio (electromagnetic waves).

Radio Access-A method or technology used to coordinate access to radio channels or portions of radio channels.

Radio Beam

Radio Beam-A radiation pattern from a directional antenna such that the energy of the transmitted electromagnetic wave is confined to a small angle in at least one dimension.

Radio Broadcasting-The transmission by program material (audio or video) that can be simultaneously received by receivers that are capable of receiving and decoding the radio signal to recover the original audio or video signal.

Radio Carrier-(1-radio signal) A radio wave at a specific frequency. (2- broadcast company) A company that broadcasts information services using radio transmission.

Radio Channel-A communications channel that uses radio waves to transfer information.

This diagram shows how a radio carrier wave is modulated by a user's information signal to produce a radio channel signal. In this example, the user data is multiplexed (time shared) with a control information signal. The combined data signal (user data and control data) is supplied to the radio frequency (RF) modulator along with the radio carrier signal. This diagram shows that the modulation process (changing) of the radio carrier signal causes the frequency of the carrier to change and the energy of the carrier wave to be distributed within a predefined frequency band (channel bandwidth). This example shows that some of the modulated signal energy does fall out of the prescribed channel bandwidth. This is typical for most systems and radio system specifications (and/or government transmission requirements) usually exist to ensure the levels transmitted outside the prescribed frequency band do not interfere with other systems.

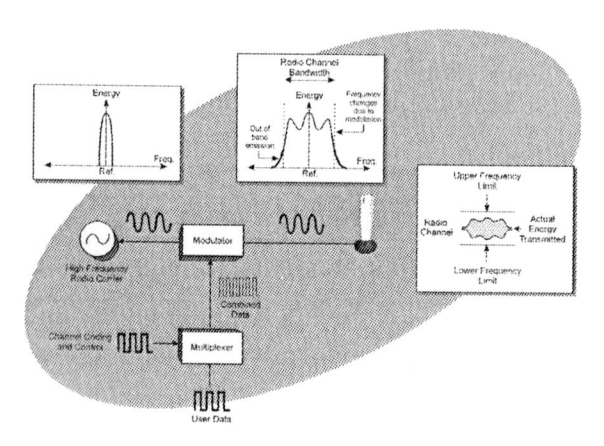

Radio Channel Operation

Radio Communications-The process of telecommunication by means of radio waves.

Radio Coverage-The geographic area that receives a radio signal above a specified minimum level.

Radio Data System (RDS)-A transmission system that combines a low-speed data signal with existing FM radio broadcast signals. In the United States, the RDS carrier signal of 57 kHz is transmitted along with an FM modulated audio broadcast signal inside the 200 kHz RF radio broadcast channel. The RDS carrier signal transfers data at 1187.5 kbps.

Radio Detection And Ranging (Radar)-A system that determines the distance and direction of objects through the transmission and reflection of electromagnetic energy.

Radio Direction Finder (RDF)-A radio receiver with a steer-able antenna that can determine the direction of arrival of a radio signal.

Radio Frequency (RF)-Those frequencies of the electromagnetic spectrum normally associated with radio wave propagation. RF sometimes is defined as transmission at any frequency at which coherent electromagnetic energy radiation is possible, usually above 150kHz.

Radio Frequency Communication Port (RFCOMM)-A communication protocol used in the Bluetooth system that uses packet messages to emulates operation of a serial communication port. It is capable of simulating the software and hardware operations of RS-232 (EIA/TIA-232-E) serial ports and is based on ETSI TS 07.10.

Radio Frequency Interference (RFI)-Undesired signals at radio frequencies that interfere with radio reception or the proper operation of equipment. When they interfere with television reception, such signals are called television interference (TVI). Many electronic devices that are not intended to produce radio frequency power, including radio and television receivers, computers, and peripherals, can interfere with other signals in the radio-frequency range by producing incidental electromagnetic radiation. The amount of such incidental radiation is regulated by FCC Regulations Part 15 in the United States.

Radio Landline-A circuit that connects a cellular switching office to a cell site or to a public switched network. It also denotes any wireline circuit from a control station to remote transmitters or receivers.

Radio Link- A radio frequency communications channel between a fixed cellular radio site and mobile units. Also, a radio system established between two points.

Radio Navigation-The process of radio location intended for the determination of position or direction, or for obstruction warning in navigation.

Radio Network Controller (RNC)-An assembly in a mobile radio network that controls the transmitter radio station equipment (base station).

Radio Port (RP)-A radio base station (node) that links mobile radio customers with the wireless network.

Radio Propagation-Radio propagation is the process of transferring a radio signal (electromagnetic signal) from one point to another point. Radio propagation may involve a direct wave (space wave) or a wave that travels along the surface (a surface wave).

Radio Propagation Prediction-Radio propagation charts that indicate the future the transmission characteristics for frequencies used for radio communication services. The charts predict radio transmission performance using criteria including variations in ionospheric levels during day to night changes, geographic factors, sunspot activity, and seasonal changes.

Radio Recognition-In military communications, the determination by radio means of the friendly or unfriendly character of an aircraft or ship.

Radio Relay System-A point-to-point radio transmission system in where radio signals are received, amplified, and retransmitted by one or more transmission radios or systems.

Radio Sub-System (RSS)-The combination of radio base stations and their controllers in a GSM system.

Radio System-A system that is composed of mobile or fixed communication devices. The types of radio systems include broadcast (one to many), multicast (one to several) or point-to-point (one-to-one). When a radio system allows communication devices to select from or be assigned to one of several radio channels, this is called trunking. The radio access control procedure allows the mobile to access alternate radio channels in the system in the event that the channel it has requested is busy.

Radio Tower-Radio towers are poles, guided towers or free standing constructed grids that raise one or more antennas to a height that increases the range of a transmitted signal. Radio towers typically vary in height from about 20 feet to more than 300 feet.

Radio Waves-Electromagnetic waves at frequencies lower than 3000 GHz (3 Tera Hertz - THz) that travel in free space (commonly air). Above 3 THz, the energy is considered optical signals.

Radiovision-A system that combined television broadcasting with radio audio broadcasting to provide for early interactive television programming.

RADIUS-Remote Access Dial In User Server

Radome-A cover that protects an antenna from the extremes of climate while allowing electromagnetic signals (radio waves) to pass through without attenuation. The Radome is usually constructed from plastic or fiberglass.

This photograph show a Radome antenna cover that protects the underlying antenna from the elements of weather. This prevents weather damage such as corrosion and the effects of hail and ice.

Radome Antenna Cover
Source: Wade Antenna Ltd.

RADSL-Rate Adaptive Digital Subscriber Line
RAI-Resource Availability Indicator
RAID-Redundant Array of Inexpensive Disks
Rain Attenuation-The attenuation of a radio signal traveling through rain or moisture bearing clouds.

Rake Receiver

Rake Receiver-A receiver has that has the ability to receive and decode multiple (mulitpath) signals to improve the quality of the received signal. This adaptive equalizer compensates for multipath radio transmission or geometric dispersion. The name "rake" is due to the appearance of several different pulses separated in delay time and displayed on an oscilloscope. They resemble the prongs of a rake. Although the rake receiver is not an entire receiver but only an equalizer, it has a distinct name due to its historical origin in connection with radar.

This diagram shows how a rake receiver can add a multipath signal to the direct signal to improve the performance of a receiver in a spread spectrum system. This diagram shows a radio channel that has two code spreading sequences. The shaded codes are time delayed because the original signal was reflected and received a few microseconds later. The original signal is decoded by mask #1. The mask is shifted in time until it matches the delayed signal. The output of each decoded channel is combined to produce a better quality signal.

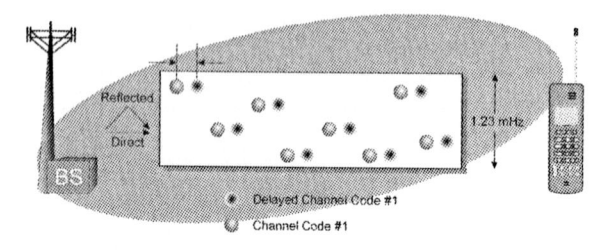

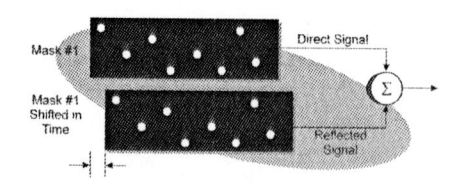

Rake Reception Operation

RAM-Random Access Memory

Raman Amplification-A method for amplifying the amplitude of infrared light in optical fibers by means of interaction with sound waves. Named for the east Indian physicist who first discovered the Raman effect, upon which this method is based, in the 1930s.

Raman Scattering-A nonlinear effect in optical fiber that has been used for signal amplification. Raman scattering occurs when a photon impinges on a material and excites a mechanical vibration or phonon. Two photons are then emitted, one with the energy of the original photon minus the energy absorbed by the phonon, the other with the original energy plus the phonon energy.

RAN-Radio Access Network

RANAP-Radio Access Network Application Protocol

RAND-1. A random number sent from wireless systems to be used as part of the subscription validation (authentication) process.

2. Stands for Reasonable and non-Discriminatory. Many standardization bodies, such as ETSI, require that members who have patented inventions which are incorporated into published standards agree to make licenses available to anyone who wants to use the standard on fair, reasonable, and non-discriminatory terms.

Random Access Channel (RACH)-A random access channel (RACH) is a shared bandwidth resource used by digital cellular phones to access the system.

Random Access Memory (RAM)-A data storage device, often an integrated circuit, from which a word of data can be removed without regard for the time it was stored or its location. Internal semiconductor RAM, static or dynamic, is volatile in nature and does not retain the data stored within if power is removed from the device.

Random Early Discard (RED)-The RED mechanism was proposed by Sally Floyd and Van Jacobson in the early 1990s to address network congestion in a responsive rather than reactive manner. Underlying the RED mechanism is the premise that most traffic runs on data transport implementations that are sensitive to loss and will temporarily slow down when some of their traffic is dropped. TCP, which responds appropriately and robustly to traffic drop by slowing down its traffic transmission, effectively allows RED's traffic-drop behavior to work as a congestion-avoidance signaling mechanism. RED aims to control the average queue size by indicating to the end hosts when they should temporarily slow down transmission of packets. RED takes advantage of TCP's congestion control mechanism. By randomly dropping packets prior to periods of high congestion, RED tells the packet source to decrease its transmission rate. Assuming the packet source is using TCP, it will decrease its transmission rate until all the packets reach their destination, indicating that the congestion is cleared. You can use RED as a way to cause TCP to slow down transmission of packets. TCP not only pauses, but it also restarts quickly and adapts its transmission rate to the rate that the network can support.

Random Errors-Errors in a digital signal that occur in such a way that each error can be considered statistically independent from any other error.

Random Noise-Electromagnetic signals that originate in transient electrical disturbances and have random time and amplitude patterns. Random noise generally is undesirable, but it is often generated for testing purposes.

Random Number-A number formed by a set of digits in which each successive digit is equally likely to be any of the digits in a specified set.

Random Separation-A method by which facilities of two or more utilities are placed with no deliberate separation of the facilities.

RAO-Revenue Accounting Office

RAR-Read After Read

RARP-Reverse Address Resolution Protocol

RAS-Registration, Admission, Status

RAS-Remote Access Server

Raster-A predetermined pattern of scanning the screen of a CRT. Raster also can refer to the illuminated area reduced by scanning lines on a CRT when no video is present.

Rate-Message Unit

Rate-The price charged a customer for a particular service, as specified and defined in a tariff.

Rate Adaptive Digital Subscriber Line (RADSL)-A hybrid analog and digital subscriber line technology that allows the data transmission rate to dynamically change dependent on the electrical transmission characteristics and/or the settings provided by the DSL service provider. RADSL features have been incorporated into standard ADSL technology.

Rate Band-A rate band is comprised of the combination pair of originating and terminating numbers where the "banded rating" amount (rate) is fixed within the same band. International calling is an example of banded rating (e.g. all calls from anywhere in the US to anywhere in France will carry the same rate.)

Rate Center-As defined by rate map coordinates, the point within an exchange area that is used as the primary basis for determining toll rates. rate centers also can be used for determining selected local rates. A rate center also is called a rate point.

Rate Class-A rate classification determined by the type of message, station, or customer, and the rate in effect at the originating rate center at the time a telephone call begins.

Rate Plan-The structure of service fees that a user will pay to use telecommunications services. Rate plans are typically divided into monthly fees and usage fees.

Rating-The function within the billing system which assigns a rate (cost parameter) to an event. Parameters typically used in this function: Originating Number/Address, Terminating Number/Address, Date, Rate Period, Call Type, Jurisdiction.

Rayleigh Fading-The signal strength fading resulting from multipath that results from phase interference between multipath signals and is approximated by Rayleigh distribution.

RBDS-Radio Broadcast Data System

RBOC-Regional Bell Operating Company

RBP-Radio Burst Protocol

RC-Recurring Charge

RCA-Radio Corporation Of America

RCA Connector-A connector developed by the Radio Corporation of America that is commonly used to connect low to medium frequency electrical signals.

RCC-Radio Common Carrier

RCDD-Registered Communications Distribution Designer

RCF-Remote Call Forwarding

RCF-Read Control Filler

RCS-Reduced Cluster Size

RDC-Regional Data Center

RDF-Radio Direction Finder

RDF-Resource Descriptor Framework

RDP-Remote Desktop Protocol

RDS-Radio Data System

RDT-Remote Digital Terminal

RDTC-Reverse Digital Traffic Channel

Re-Examination-The process by which any person may challenge the validity of a US Patent. Essentially, re-examination is a request that the USPTO re-open the examination of a granted US patent. Not to be confused with opposition, there are two types of US Re-Examination: ex parte and inter partes. Ex parte means that a person who files a request for re-examination may not participate in the subsequent examination. Thus, ex parte re-examination is nearly identical to the original examination process. Since 1999, the US Re-Examination process has been expanded to allow the requestor to participate in some stages of the process. Re-Examination is also often used by a patentee who may wish to strengthen his patent by re-examining his own patent over prior art which may not have been known during the original prosecution.

Re-Invite-A message that is used to inform a person or device that the communication session parameters of the original communication session invite request have changed. The re-invite message usually contains information about the new communication parameters such as the bandwidth and media types. The re-invite message is defined in session initiation protocol (SIP).

Re-Issue-The process by which substantive errors in the claims of a US Patent may be corrected.

Within two years after the grant of a US Patent, the patentee may request a re-issue in order to broaden, or narrow, his claims if it can be shown that the patentee is entitled to a different scope of patent protection. Often re-issue is used to broaden the scope of claims in order to encompass an infringing product, or to narrow the scope of claims to avoid prior art. After two years, patent claims may only be narrowed, but this may be done at any time until the patent expires, or lapses.

REA-Rural Electrification Administration

Reachability-A measurement used in network management that indicates the capability of successfully pinging or successfully communicating with a device in the network from another device in the network.

Reactance-The part of the impedance of a network resulting from inductance or capacitance. The reactance of a component varies with the frequency of the applied signal. Reactance represents an opposition to the flow of alternating current.

Read Only Memory (ROM)-A memory circuit or device in which any address can be read from, but not written to, after initial programming. The ROM is an asynchronous device with an access time dictated by internal circuit time delays. Semiconductor ROM storage is nonvolatile and retains data when power is removed.

Readout-A visual display of the output of a device or system.

Ready Access Terminal-A ready access (RA) terminal is a non-protected terminal that is spliced into a cable and still allows full access to the entire cable count (all conductors).

Real Time Billing-Real time billing involves the authorizing, gathering, rating, and posting of account information either at the time of service request or within a short time afterwards (may be several minutes). Real time billing is primarily used for prepaid services such as calling cards or prepaid wireless.

This figure shows a real time prepaid billing system. In this example, the customer initiates a call to a prepaid switching gateway. The gateway gathers the account information by either prompting the user to enter information or by gathering information from the incoming call (e.g. prepaid wireless telephone number). The gateway sends the account information (dialed digits and account number) to the real time rating system. The real time rating system identifies the correct rate table (e.g. peak time or off peak time) and inquires the account determine the balance of the account. Using the rate information and balance available, the real time rating system determines the maximum available time for the call duration. This information is sent back to the gateway and the gateway completes (connects) the call. During the call progress, the gateway maintains a timer so the caller cannot exceed the maximum amount of time. After the call

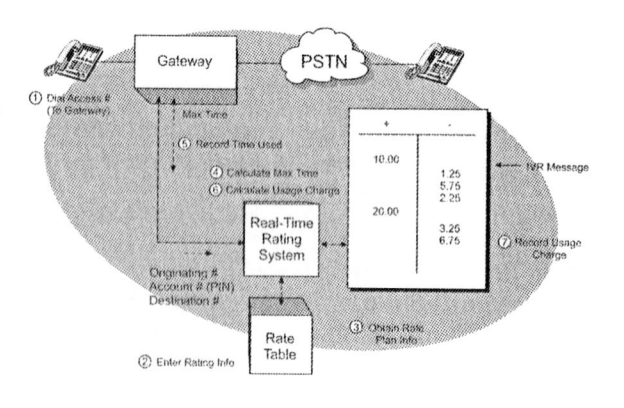

Real Time Billing Operation

is complete (either caller hangs up), the gateway sends a message to the real time rating system that contains the actual amount of time that is used. The real time rating system uses the time and rate information to calculate the actual charge for the call. The system then updates the account balance (decreases by the charge for the call).

Real Time Billing, ITSP-This figure shows how real time billing for Internet telephone service commonly allows the customer to display their billing records immediately after they are created (in real time). This example shows how the ITSP keeps track of each call as it helps to connect the call and to process advanced features. It uses the call setup and termination information to adjust your bill and in this example, these charges can be displayed immediately through an Internet web page.

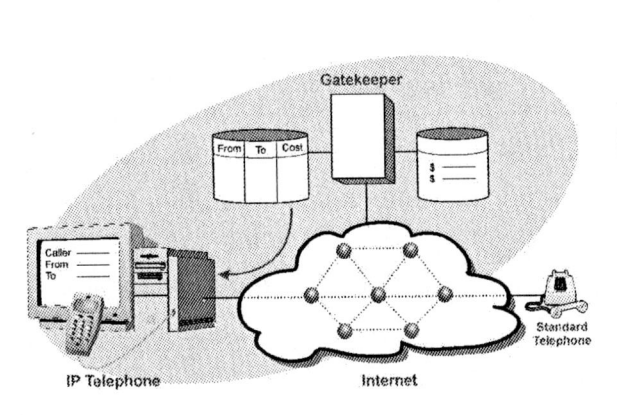

audio signal) be converted to digital form (digital audio) prior to transmission. This digital signal is divided into small packets. The RTP protocol is a high-level protocol and each packet of data each of the transmitted packets starts with an IP header that contains the destination address of the packet. An additional flow control protocol header is added (usually UDP protocol header) to identify the specific port the data will be routed to at it's destination. The RTP system then adds a third header (the RTP control header). The RTP system uses a precise clock to add time stamp information to each packet along with other signal recreation control information. Because the packets may have different types of compression and their recreation time can dramatically vary, the RTP protocol header uses the time stamp and other information to decode the and recreate the data packet.

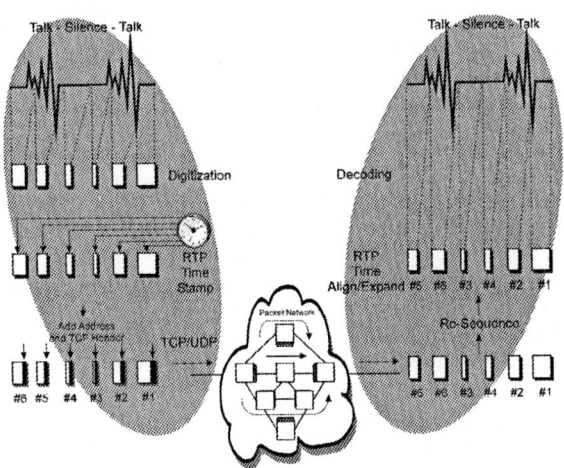

ITSP Real Time Billing Operation

Real Time Transport Protocol (RTP) Operation

Real Time Transport Protocol (RTP)-RTP is a packet based communication protocol that adds timing and sequence information to each packet to allow the reassembly of packets to reproduce real time audio and video information. RTP is defined in RFC 1889. RTP is the transport used in VoIP environments.

This diagram shows how real time transmission control protocol (RTP) operates to send real time data through a packet network that may have variable transmission delays. This diagram shows that an RTP system requires that a real time signal (e.g.

Real-Time Transport Control Protocol (RTCP)-A signaling control protocol that allows for the controlling and reporting on the flow of real-time data that is transmitted in a packet network. The RTCP protocol sends messages in addition to the communication data to monitor the transmission quality within a packet network. RTCP is defined in RFC 1889.

Realtime Streaming Protocol (RTSP)-A Internet protocol that is used for continuous (streaming) audio and video sessions. RTSP is capa-

Reassembly

ble of bi-directional (two-way) sessions. RTSP is defined in RFC 2326.

Reassembly-The process of reconstructing a packet from its fragments.

Reboot (Reset)-To restart a an electronic assembly or computer.

Rebroadcast-The broadcast of material picked up directly (off-the air) from another broadcasting station.

RECC-Reverse Control Channel

Received Signal Strength Indicator (RSSI)-A signal level indicator, usually on a mobile radio, that regularly displays the approximate level of a received signal. The RSSI indicator allows a user to determine if the radio signal strength in that area is sufficient to initiate or complete a call.

Receiver-(1-general) A device that receives a transmitted signal and makes it perceptible to a human user or converts it into some other useful form. (1- telephone handset) Older name for the earphone of a telephone handset, or used as an adjective for signals associated with that earphone. Also the entire handset (microphone and earphone, et al). (2-radio) Radio receiver, a device that responds to a modulated radio signal by producing a replica of the original modulating information. (3-tone receiver) A device in a telephone system responsive to in-band telephone tone signals such as MF, SF or DTMF signals, interpreting the significance of such tone signals by producing the corresponding digital representation in a form useable by the destination telephone switch.

Receiver Off-Hook Tone-A tone that is provided when the phone is off-hook.

Reception-The act of receiving, listening to, or watching information carrying signals.

Recharge-(1-battery) To store energy in a battery. (2-calling card) A process of adding additional time to a calling card or prepaid wireless account using either a credit card or debit card.

Reciprocal Compensation-The 1996 Telecommunications Act mandated that local telecommunications companies exchange revenue for the cost of terminating calls that originated on the wireline network. Previously, only wireless companies were obligated to pay compensation for calls originated on their networks but terminated on the wireline network.

Recombination-Combination of an electron and a hole in a semiconductor that releases energy, sometimes leading to light emission.

Recommended Standard (RS)-A set of standards documents that are managed by the electronic industries association (EIA) that specify the format of electronic signals and protocols.

Recon-Reconnect

Reconnect (Recon)-In outside plant construction, the moving of subscriber dropwires onto new cable facilities as engineered on a work order. This often relieves congestion at poles and terminals in addition to improving quality of service for customers.

Record Route-The accumulation of routing information that is added to packets of data as they transfer through a network.

Rectify-The process of converting alternating current into direct current.

Recurring Charge (RC)-A predetermined charge associated with a product or service that is assessed on a regular interval I.e. monthly, quarterly, annually.

RED-Random Early Discard

Red, Green, Blue, (RGB)-The three additive primary colors used in video processing, often referring to the three unencoded outputs of a color camera. The sequence of GBR indicates the mechanical sequence of the connectors in the SMFTE standard. Although GBR is the preferred name, the term RGB commonly is used to denote the same signals.

Redirection Server-A redirection server assists in the establishment of communication sessions by providing alternative locations where the designated recipient can be found. The redirect server does not initiate any action. It only provides information back to the requesting device as to the potential locations of the designated recipient.

Reduced Instruction Set Computer (RISC)-A generation of microprocessors that implement instructions that are reduced from a complex instruction set computer (CISC) format. These instructions are typically executed directly on the hardware state machine, rather than (microcoded) as in CISC architectures. Therefore, many RISC instructions can execute in a single clock cycle. However, it may take multiple RISC instructions to complete the same function as one CISC instruction.

Reduction to Practice-In the United States, where patents are awarded on a First to Invent basis, the act of invention is not complete until a working model of the invention is built and operated. Alternatively, the filing of a US Application for Patent is considered as completing the act of invention.The construction of a working model is known as an actual Reduction to Practice. Filing a US Application for Patent is known as a constructive Reduction to Practice. The Date of Conception of an invention cannot be used unless the inventor works diligently towards the Reduction to Practice. If the Reduction to Practice is an actual, diligence must also be shown from this date up to the filing date of a US Application for Patent.

Redundancy-A system design that includes additional equipment for the backup of key systems or components in the event of an equipment or system failure. While redundancy improves the overall reliability of a system, it also increases the number of equipment assemblies that are contained within a network. Redundancy usually increases cost.

Redundancy Plan-A network structure plan that defines the alternate equipment configurations and routes that are used when specific failures occur.

Redundant Array of Inexpensive Disks (RAID)-Redundant array of independent disks (RAID) is a computer information storage system that uses multiple independent disk storage devices (hard disks). RAID systems were first defined in 1988 and they were originally called redundant array of inexpensive disks. RAID systems can use standard hard disk drive interfaces such as SCSI or Integrated Drive Electronics (IDE).

RAID systems can be configured in various ways to ensure data integrity and to provide the ability to remove and replace disks while the system continues to operate in the event of equipment failure ("hot-swap".) The different types of configurations are called RAID levels. There were six original levels (RAID 0 through RAID 5). Several manufacturers have combined the RAID levels to produce new unique levels above RAID 5. For RAID 0, the data is distributed (striped) over several drives but there is no redundant storage of data. RAID 1 combines two hard disks of equal storage capacity that simultaneously store the same information (mirror). RAID 2 allows data to be corrected on one drive from another drive by interleaving (distributing) the information across multiple drives and using a data protection formula (algorithm) that relates the information. RAID 3 stores information on several disk drives where only one drive is used for parity bits that are used for error detection and correction. RAID 4 systems store data in multiple drives without the need to mirror information at the bit level. RAID 5 writes data and parity information on the same disk but to different sectors to increase the data transfer performance.

Redundant Server-The inclusion of a second communication server that allows the automatic transfer of service (reconfigure) in the even a failure occurs on part or all of the other server. In a redundant server, an equipment failure should cause no loss of data or information.

Redundant System-The duplicate configuration of equipment in communication systems that allows some of the equipment to automatically reconfigure in the even a failure occurs on part or all of the duplicated system. In a redundant system, a failure should cause no loss of data or information.

Reed Solomon Code (RSC)-A 2-dimensional error correction code that derives the correction information by applying two different mathematical formulas on the same data. Two correction words, P and Q, are generated from this process. Simply stated, the P word is a binary sum, and the Q word is a binary product of the individual bits in the data word. As a result of this powerful cross correlation of data, Reed Solomon codes are particularly effective in correcting large scale burst errors.

Refarming-A process of moving or eliminating radio users from a frequency band to allow new users to use that frequency band.

Refer Message-A message that is used to request the transfer of a communication session from one device to another device. The refer message is sent from one device (the device that desires to be transferred) to a new device that is not part of the original communication session. If the new device accepts the invitation request (it is willing to accept the transfer request), the transfer device informs the device that it is communicating with of the desire to transfer the session to a new device along with the parameters (such as the new device network address) that will be used by the existing device to connect to the new device. After the two devices have successfully established a new communication session, the transfer device is no longer part of the original communication session. The refer message is defined in session initiation protocol (SIP) toolkit.

Reference Frequency-A standard fixed frequency from which other frequencies can be derived for operations or testing purposes. The reference frequency is distributed between offices in a telecommunications system for synchronization purposes.

Reference Level-A measured or objective value used as a starting or comparison point for other measurements. The term usually infers a power level, such as one milliwatt. The international standard test power for telephone circuit alignment, signal-to-noise, and impairment measurements is 40 dBm0.

Reference Voltage-A voltage used for control or comparison purposes.

Reflecting Satellite-A satellite intended to reflect radiocommunication signals.

Reflection-When a traveling sound wave, radio wave or light wave encounters certain types of changes in the properties of the medium through which the wave travels, a reflected wave will be generated. A reflected wave is generated if the wave impedance is different just before and just after the region of interest. Wave impedance of a sound wave is the ratio of the sound pressure to the acoustic fluid velocity in the wave. Wave impedance of an electromagnetic wave is the ratio of the transverse electric field intensity to the transverse magnetic induction intensity in the same polarization direction. A different acoustic wave impedance is the result of a different mass density and/or a stiffer or less stiff compressibility of the acoustic medium. A different electromagnetic wave impedance is the result of a different dielectric permittivity (usually) and/or a different magnetic permeability of the medium. These differences may in turn be the result of a higher or lower mass density of the medium for the case of a compressible fluid. Since many of these same properties also affect the wave speed, the reflection properties can sometimes be described in terms of the change in wave speed instead of being described in terms of a change in wave impedance. When the wave impedance on both sides of the region of interest are exactly the same value, there is no reflected wave power. The fraction of the wave amplitude that is reflected can be described by a reflection coefficient, which approaches 1 in absolute magnitude (that is, 100% reflection) in the two extreme cases where the ratio of the wave impedance before and after the region of interest approaches either zero or infinity. Full 100% reflection can also theoretically occur from a perfectly conducting surface for an electromagnetic wave, or from a perfectly rigid reflecting surface for an acoustic wave.

Reflection Coefficient-The ratio between the amplitude of a reflected wave and the amplitude of the incident wave.

Reflection Factor-In transmission-line theory, the ratio of two currents delivered to a load: the first current when the characteristic impedance of a line is not matched to a load and the second when there is a proper match. if there is a mismatch, the signal is reflected from the load back to the source. See: characteristic impedance, echo.

Reflection Loss-(1-general) The apparent loss of signal strength caused by an impedance mismatch in a transmission line or circuit The loss results from the reflection of part of the signal back toward the source from the point of the impedance discontinuity. The greater the mismatch, the greater the loss. (2 -optical) The total loss from reflections at the junction between two optical components. (3- transmission line) The ratio in decibels between the incident and the reflected wave at any discontinuity or impedance mismatch.

Reflector-The metal elements placed behind the active element of an antenna to make it directive.

ReFLEX-ReFLEX is a two-way version of the FLEX paging technology. In addition to providing flexible paging data transmission rates, ReFLEX can provide a reverse data channel at 9,600 bps. The protocol is designed for the most recent Narrowband PCS frequencies for two-way messaging services. ReFLEX-25 has a bandwidth of 25 kHz and a forward data rate of 12,800 bps. ReFLEX-50 has a bandwidth of 50 kHz and a forward data rate of 25,600 bps. ReFLEX service has been available from several paging carriers since 1995.

ReFLEX-25 can be set up as a simulcast system or frequency re-use system. When set up as a simulcast system, the capacity of the system is limited to the single channel that is re-broadcast by each tower. If the system approaches its maximum capacity, frequency re-use is possible to increase system capacity.

ReFLEX first sends out a broadcast locating signal and the intended pager must respond or the page/message will not be sent. When the intended pager receives the page/message, ReFLEX requires an acknowledgment indication back from the pager.

ReFLEX 25-A narrowband two-way text messaging PCS technology developed by Motorola.

ReFLEX 50-A two-way paging system technology implemented by Mobile Telecommunication Technologies Inc.

Refraction-The gradual bending or changing of propagation direction of a sound wave, radio wave, or lightwave as it passes from a medium of one density to a medium of another density. The bending of the wave results from a change in the speed of signal propagation in the different types of materials.

Refractive Index-The ratio of the speed of light in a vacuum to the speed of light in a material; abbreviated n.

Reframing-The process of recovering or relocating the framing pulse in a bit stream after a mis-frame.

Regeneration-Digital signal regeneration is the process of reception and restoration of a digital pulse or lightwave signal to its original form after its amplitude, waveform, or timing have been degraded during transmission. The resultant signal is virtually free of noise or distortion.

This figure shows the process of digital signal regeneration. This example shows how an original digital signal (a) has a noise signal (b) added to it that produces a combined digital signal with noise (c). The regeneration process detects maximum and minimum expected values (threshold points) and recreates the original digital signal (d)

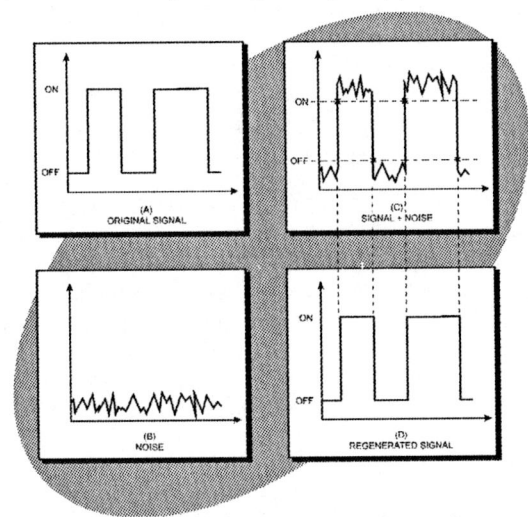

Digital Signal Regeneration Operation

Regenerative Repeater-A repeater that receives, amplifies, reshapes, and retransmits digital signals.

Regenerator-A receiver and transmitter combination used in fiber optic systems to reconstruct signals for digital transmission. The receiver converts incoming optical pulses to electrical signals, decides whether the pulses are is or Os, regenerates the electrical pulses, and converts them to an optical signal for further transmission on the fiber.

Regional Bell Operating Company (RBOC)-A United States telephone company that is one of the seven telephone companies that were created as a result from the division of AT&T in 1983. RBOCs are also known as the Baby Bells. The RBOCs were Ameritech, Bell Atlantic, BellSouth, Nynex, Southwestern Bell Corporation, Pacific Telesis, and US West.

Regional Data Center (RDC)-A regional data center (RDC) is a collocation facility that allows a carrier interconnection point-of-presence (POP) within a rate center to provide information services.

Regional Enterprise Network (REN)-A Regional Enterprise Network (REN) is a subset of the Enterprise Network (EN) that interconnects corporate offices located within the same city.

Register Message-A message that is used to inform a service provider or communication system that a user is connected to the system. The register message usually contains information about the connection of the device including the address (or temporary address) that the device is using. The register message is defined in session initiation protocol (SIP).

Registered Communications Distribution Designer (RCDD)-A certification for professionals that setup network communication systems such as twisted pair and optical networks. Because RCDD is managed by BICSI, the RCDD certification is often called a BICSI (pronounced "bik-see") certification.

Registrar Server-A registrar is a server that accepts registrations from users and places these registrations, (which is essentially location information), in a database known as a Location Service. The process of registration associates a user with a particular location, (IP address); this association is known as a 'binding' in SIP. When there is an incoming session for a user within a domain, the proxy server will interrogate the Location Server to determine the route for the signaling messages.

Registration-(1- FCC) A legally required procedure whereby vendors must submit their telephone equipment for testing and certification before it can be directly connected to a public telephone network. (2-wireless system notification) A process where a mobile radio transmits information to a wireless system that informs it that it is available and operating in the system. This allows the system to send paging alerts and command messages to the mobile radio. (3-VoIP) The process of a Internet telephone terminal registering with a gatekeeper.

Registration Identification

Registration Identification-The process where the mobile accesses the cellular network to inform the system it is in its operating area.

Registration, Admission, Status (RAS)-Service authorization is the processes associated with maintaining information about the location and status of users within a network. In a packet voice network, RAS is used by gatekeepers to control the gateways and media (audio and video) flows.

Registry-A database of information that contains computer's hardware and software information that is used by it's operating system to manage communication between hardware and software applications. Information store in the registry is usually updated automatically as configuration information changes through the addition, removal, or modification of hardware and software controlled by the system. The registry database holds information that was commonly stored in initialization files (.INI) files.

Regression Analysis-A forecasting method used where the factor to be predicted may be expressed as a function of one or more variables. The variables themselves may have to be forecasted.

Regulation-(1- electrical quantity) A process of adjusting the parameters of some signal or system (such as circuit gain). (2 - government) Rules established by a government agency that are designed to maintain the service public communications systems.

Regulator-(1-gas pressure) A control device that acts as an interface between gas in a high-pressure tank and the low-pressure gas in a cable pressurization system. (2-voltage) A device that maintains its output voltage at a constant level. (3-government) An agency or department of a government that establishes, changes, or enforces rules and regulations.

Relative Antenna Gain-The ratio of the radiation intensity of an antenna in a given direction to the radiation intensity of a reference antenna in the same direction, with the same power input to both antennas. If the reference antenna is a lossless half-wavelength dipole, the gain is expressed in decibels relative to the dipole antenna (dBd).

Relative Envelope Delay-The difference in envelope delay at various frequencies when compared with a reference frequency that is chosen as having no delay.

Relay-(1 - general) A device by which current flowing in one circuit causes contacts to operate that control the flow of current in another circuit. (2 - general) The process of re-transmitting signals through an intermediate system. A relay point is a place in a network where a signal is "repeated" after being modified or regenerated to remove impairments or other communication degrading effects, as in a telegraph or digital repeater device, or in a telecommunications relay service (TRS) for the deaf. (3- hermetically sealed) A relay in which the contacts are sealed in an airtight glass or metal enclosure. (4-latching) A relay that latches into its operated position and is held there without the need for a holding current. Release is accomplished by energizing a release winding. (5 - mercury-wetted) A relay with contacts in sealed en-closures constantly wetted by mercury to provide a low-resistance contact. (6 - multi-contact) A relay in which a large number of spring sets are operated simultaneously. (7 - overload) A relay that operates only when current in a circuit exceeds a specified value. (8 - reed) A relay in which ferromagnetic reeds are sealed in small glass tubes and surrounded by operating coils. (9- solid-state) A device in which a current or voltage in one circuit controls the switching of another circuit, but involves no mechanical movement, armatures, moving contacts, or reeds. (10 - thermal) A heat operated relay that normally depends on the bending of a bimetallic strip for contact closure. (11 - radio system) An intermediate station on a multi-hop radio system.

Release-A signal that indicates that a line or circuit has been released from use.

Reliability-The ability of a network or equipment to perform within its normal operating parameters to provide a specific quality level of service. Reliability can be measured as a minimum performance rating over a specified interval of time. These parameters include bit error rate, minimum data transfer capacity or mean time between equipment failures (MTBF).

RemDev-Remote Device

Remote Access Dial In User Server (RADIUS)- A network device that receives identification information from a potential user of a network service, authenticates the identity of the user, validates the authorization to use the requested service and creates event information for accounting purposes. RADIUS is specified in RFC's 2138 and 2139,

RADIUS is a client/server protocol that uses UDP.

Remote Call Forwarding (RCF)-A service offering in which calls to a given directory number are redirected to another number, often in another city. Also, a service that enables a customer to invoke call forwarding whether the customer is at home or away. Control normally is limited to direct local access to a line from which forwarding is to be authorized. Remote control is invoked by calling a shared number at a home office, entering a PIN (personal identification number), then processing as if local control applied.

Remote Console-The computer access device that allows an operator to communicate with or control a network from a remote location (not directly connected to the network). A remote console commonly consists of a display monitor, keyboard, data modem or other interconnection line.

Remote Control-A system used to control a device from a distance.

Remote Digital Terminal (RDT)-The RDT provides an interface between a high speed digital transmission line (e.g. DS1) and the customer's access line. The RDT can dynamically assign time slots from a high speed line to customer access lines. Because customer access lines are not used at the same time, an RDT that interfaces to a DS1 line usually provides service to 96 customer access lines. The RDT is divided into three major parts; digital transmission facility interface, common system interface and line interface. The digital transmission interface terminates the high speed line and coordinates the signaling. The common system interface performs the multiplexing/de-multiplexing, signaling insertion and extraction. The line interface contains digital to analog conversions (if the access line is analog) or digital formatting (if the line is digital).

This picture shows a remote digital terminal (RDT) device that combines multiple local loop communication channels over a high speed digital line (96 telephone lines to a 24 channel communication line). This RDT conforms to GR-303 specification.

Remote Digital Terminal (RDT)
Source: Zhone Technologies

Remote Monitoring Specification (RMON)-Protocols that allow network management software to configure, poll and trap events on network elements.

Remote network MONitoring (RMON)-Remote network MONitoring or RMON is in essence just another SNMP MIB, but allows an Network Management Station (NMS) to request that an agent do many types of functions. There are 9 functions or groups associated with RMON as originally defined in RFC1757: Ethernet statistics, history statistics, alarm and threshold setup, MAC-layer host statistics, busiest hosts, MAC-layer conversation pair statistics, filters, capture packets, and event and alarm notification. This is also known as RMONv1. RFC2819 now obsoletes the original RFC1757.

Remote Shell (RSH)-In Unix, an application that allows running a command on a remote machine, usually over a TCP connection.

REN-Regional Enterprise Network

Reorder Tone-A tone applied 120 times per minute indicating that all switching paths or trunks are busy. The reorder tone also is called a channel busy or fast busy tone.

Repeat Dialing-A service that automatically dials the last dialed number a repeated number of times or until an event occurs.

Repeater

Repeater-(1-general) A device or circuit that is located between transmitting and receiving devices to improve the quality the signal that is delivered between them. A repeater obtains some or all of the signal from the transmitter, amplifies and may adjust (change a frequency) or filter the signal, and retransmits the signal to the receiver(s). (2-LAN) In a local area network, a device which operates at layer 1 (physical layer) of the OSI reference model. This device does not inspect packets, but instead regenerates all input signals on its output(s). Repeaters were common in shared-media Ethernet based on IEEE 802.3 10-Base-2 and 10-Base-5 protocols. In recent years, the need for repeaters has been greatly deminished as new physical layer transmission technologies have provided better transmission capabilities.

This picture shows a ICOM Desktop Repeater that can receive a radio signal from a nearby mobile radio or radio tower and amplifies (repeats) the signal for transmission to another area that does not have a strong radio signal (usually in a shadow from the nearby tower). This diagram shows that this repeater can be programmed to receive signals on one frequency and retransmit the signals on a different frequency.

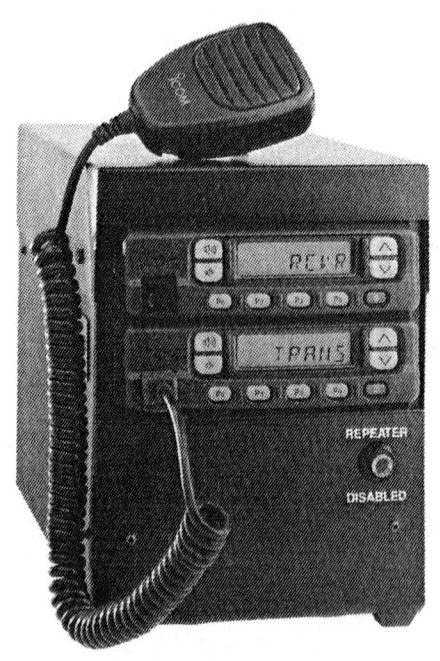

Repeater
Source: ICOM America, Inc.

Reporting System-A system that provides information about specific events (such as call or service usage). The reporting system usually provides details on the cause of the event (user identification), type of event (telephone call type), and amount of event (minutes or amount of data transferred).

Request For Budgetary Proposal (RFBP)-A statement of requirements and technical specifications designed to provide information on the costs of products or services from suppliers.

Request For Comments (RFC)-A requirements or draft standard document created by a standards body that solicits comments from manufacturers, carriers and industry experts to finalize the standard. When used by the Internet engineering task force (IETF) every major Internet Protocol is specified first by an RFC. There are many RFC documents available and they are a significant method used to define Internet protocols and technical standards.

Request For Information (RFI)-A formal statement of requirements for a technical system's acquisition that is submitted to vendors for information on the feasibility of designed concepts.

Request For Proposal (RFP)-A formal statement of requirements and technical specifications that constitute user needs when actually procuring a system and seeking bid information for the evaluation and selection of services or products.

Request For Quotation (RFQ)-A document that is prepared by a manufacturer, carrier, or other company that defines the needs of a company and solicits a response for quotation for equipment and services that fill the requirements needs as defined in the RFQ.

Request To Send (RTS)-A control signal on a communication line (such as an RS-232 serial bus) that indicates the device is requesting to send communication information.

Rerouting-Use of a short-term alternate transmission path for selected traffic items. Rerouting can be permanent or, for a nonrecurring situation, temporary.

Research In Advanced Communications In Europe (RACE)-A cooperative research program started in Europe commissioned to develop the technology for Broadband Integrated Services Digital Network (B-ISDN) systems. The RACE members are reviewing a Mobile Broadband System operating in the 60 GHz bands for mobile service applications in the approximate range of 2-100 Mb/s.

Reseller-A reseller buys network services in bulk from an existing carrier for resale to the public or other customers. The reseller provides sales and support services to the customer and the customer usually pays the reseller for the communication services it receives.

Reset-The act of restoring a device to its default or original state. Reset also may refer to restoring a counter or logic device to a known state, often a zero output.

Reset-Reboot

Residential Gateway-Residential gateways facilitate the circuit-to-packet conversion from analog endpoints in the home out to an IP network. Typical devices include IADs, cable modems, DSL modems, and wireless broadband units.

Residential System Identity (RSID)-An identity code that is used to differentiate residential systems from public system in a cellular or PCS system. One or more RSIDs can be used to create single location access points or neighborhood residential systems by broadcasting the appropriate RSIDs. When a mobile phone is operating in its designated residential system, the phone typically displays "At Home" and therefore may receive special services (such as reduced billing, and so on). A primary use of RSIDs is to allow cellular or PCS phones to also be used as a cordless phone.

Residual Interconnection Charge (RIC)-An additional telecommunication charge that allows a telecommunications carrier recover revenue losses (shortfalls) that were created by the implementation of the FCC's Local Transport Restructure.

Resilient Packet Ring-An emerging network architecture and technology designed to meet the requirements of a packet-based metropolitan area network. Standardized by the IEEE 802.17 committee, RPR is designed to effectively manage a shared metro fiber ring by transporting packets. Unlike SONET, which uses TDM, RPR is designed to interoperate with packet-based networks such as Ethernet, while reducing the non-deterministic nature of metro Ethernet.

Resistance-(1-general) The opposition of a material to the flow of electric current. Resistance is equal to the voltage drop through a given material divided by the current flow through it. The standard unit of resistance is the ohm, named for the German physicist Georg Simon Ohm (1787-1854). (2 - antenna) The impedance of an antenna at its operating frequency. (3 - contact) The resistance measured across a pair of contacts. Mercury-wetted contacts normally will have zero contact resistance. Good, dry contacts have resistance measured in milliohms. (4 - effective) At all alternating frequencies, the total resistance, including dc resistance and resistance from the skin effect, eddy currents, hysteresis, and dielectric losses. (5 - insulation) The resistance offered to the flow of dc through insulation. (6 - leakage) The resistance of a path, normally to

Resistive Load-A load in which the sine wave voltage is in phase with the current.

Resistor-(1 - general) A device whose primary function is to introduce resistance into an electric circuit. (2 - carbon) A small cylindrical resistor made of carbon. (3- current limiting) A resistor, inserted in a circuit for safety reasons, that limits the value of the current flow under a fault condition. (4- non-inductive) A resistor wound in such a way as to minimize inductance.

Resistor Color Code-The colored markings on a resistor that indicate the value and tolerance of the device.

Resolution-(1 - display) The number of pixels per unit of area. A display with a finer grid contains more pixels, and therefore has a higher resolution, capable of reproducing more detail in an image. (2 - digital) The number of bits into which an analog signal has been converted. The greater the number of bits, the more accurate the digital representation. (3-horizontal) The amount of resolvable detail in the horizontal direction in a picture. Horizontal resolution usually is expressed as the number of distinct vertical lines, alternately black and white, that can be seen in three quarters of the width of the picture. This information usually is derived by observation of the vertical wedge of a test pattern. A picture that is sharp, clear, and detailed has good (high) resolution. If the picture is soft, blurred, and does not show small details distinctly, it has poor (low) resolution. Horizontal resolution depends upon the high frequency amplitude and phase response of the pickup equipment, the transmission medium, and the picture monitor, as well as the size of the scanning spots. (4 - instrument) The smallest increment that can be acted upon, or read from a measuring instrument. (5- vertical) The amount of resolvable detail in the vertical direction in a picture. Vertical resolution usually is expressed as the number of distinct horizontal lines, alternately black and white, that can be seen in a test pattern. Vertical resolution is fixed primarily by the number of horizontal scanning lines per frame. Beyond this, it depends on the size and shape of the scanning spot of the pickup equipment and picture monitor.

Resonance-(1 - general) A tuned condition conducive to oscillation, when the reactance resulting from capacitance in a circuit is equal in value to the reactance resulting from inductance. (2-parallel) The condition in a circuit with capacitance and inductance in parallel, when the frequency is such that the current entering the circuit from outside is in phase with the voltage across the parallel circuit. (3 - series) The condition in a circuit with capacitance and inductance in series, when the frequency is such that the current through the circuit is in phase with the voltage across the circuit.

Resonant Frequency-The frequency at which the inductive reactance and capacitive reactance of a circuit are equal.

Resonator-A resonant cavity.

Resource Availability-The amount of available system capacity to provide services to additional users. Resource availability may be measured by amount of data transfer (Mbps), number of users, or number of calls in a specific time period.

Resource Availability Indicator (RAI)-A message that indicates the ability of a gateway to provide more communication services. A RAI message is defined in the H.323 packet voice system.

Resource Descriptor Framework (RDF)-A W3C standard framework for the description and exchange of metadata between computer systems.

Resource Reservation Protocol (RSVP)-RSVP is a quality of service (QoS) protocol that is used to control the amount of bandwidth dedicated to packet flow for specific communication sessions in a packet data network. RSVP is primarily used in real-time communication sessions (such as voice over packet).

Response-(1-message) A reply to a query. (2 - audio) The fidelity with which equipment reproduces an audio signal. (3 - frequency) The gain or loss of a system over its specified frequency band. (4 - spurious) The response of tuned equipment to an undesired signal at a frequency to which the equipment is not tuned. (5 - transient) The response of an amplifier or other circuit to high-frequency signals, such as those represented by a square wave test signal.

Response Time-The elapsed time between the generation of an inquiry and the receipt of a reply in a communications system. The response time includes the transmission time, processing time, time for searching records and files to obtain relevant data, and transmission time back to the inquirer. In a data system, it is the elapsed time between the end of transmission of an inquiry message and the beginning of the receipt of a response message, measured at the inquiry originating station.

Response Time Reporter (RTR)-RTR is a feature set found on Cisco routers that allows them to send ICMP pings and other types of packets that devices may respond to. RTR can be configured with thresholds for event and alarm notifications, similar to RMON.

Responsivity-A measure of the sensitivity of a photosensor. Responsivity is the ratio of the output current or voltage to the input flux in watts or lumens. When responsivity is indicated at a particular wavelength (in amperes/watt), it denotes the spectral response of the device.

Restoration-The repair or returning to service of one or more telecommunication services that have experienced a service outage or are unusable for any reason, including a damaged or impaired telecommunications facility. Such repair or returning to service may be done by patching, rerouting, substitution of component parts or pathways, and other means, as determined necessary by a service vendor.

Restricted Digital Information-Typically a 64 kb/s digital bit stream that is restricted to require that at least one out of each eight consecutive bits must be a binary 1. This restriction is the result of using a digital transmission link that has a maximum zeros density limitation. See also Unrestricted Digital Information.

Retail Channel-Distribution channel that makes use of retail stores to provide goods and services to customers.

Retention-Retention is the process of actively or passively preventing customers from leaving and switching to other service providers or products. A measurement of retention is churn rate.

Retrace-The return of the electron beam in a CRT to the starting point after scanning. To retrace, the beam typically is turned off. All the sync information is placed in this invisible portion of the video signal. Retrace may refer to beam movement after each horizontal line or after each vertical scan (field).

Retrace-Flyback

Retransmission-A method of network error control in which hosts receiving messages acknowledge the receipt of correct messages and either do not acknowledge, or acknowledge in the negative, the receipt of incorrect messages. The lack of acknowledgment, or receipt of negative acknowledgment, is an indication to the sending host that it should transmit the failed message again.

Retrial-Any subsequent attempt by a customer, operator, or switching system to complete a call within a measurement period.

Retrieval Services-A service that allows authorized users of the service to retrieve information from an information center.

Return Call-A feature that allows a telephone user to program the telephone system to return a call (usually to the originating telephone) when a specific telephone number that is busy becomes available for call delivery.

Return Loss-The difference, usually expressed in decibels (dB) compared between the incident voltage or current on a transmission line (forward signal) and the reflected current or voltage as measured at a particular point (e.g. termination point or impedance discontinuity).

Return on Investment (ROI)-A financial measurement that compares the profit with the original investment. ROI is to evaluate the impact of an investment on the telephone company's profitability or operational efficiency. Return on Investment is reported in terms of dollars spent compared to benefits gained.

Return on Marketing Activity (RoMA)-A financial measurement that is utilized to evaluate the effectiveness of a marketing activity. Return on Marketing Activity is reported in terms of dollars spent / market share realized.

Return To Zero (RZ)-A data format in which the logic level for a data 1 is a 1 during the time the data clock is high, but returns to 0 during the time the data clock is low. For a data 0, the logic level is 0 for both high and low states of the data clock.

This diagram shows that return to zero (RZ) encoding uses the logical level voltage during half of the bit period of each logical bit. In this example, the data is transmitted at 1 kbps so each logical bit period is 1 msec. During this entire period, the logical level remains at the same voltage associated with the logical level for 1/2 a bit period (0.5 msec).

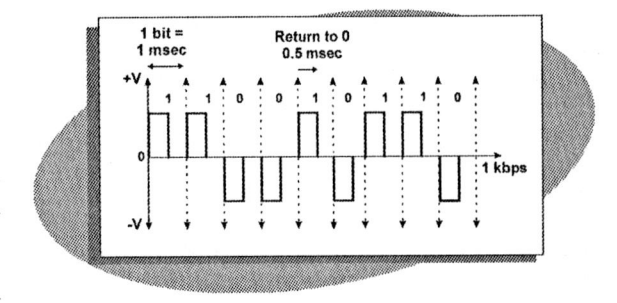

Return to Zero (RZ) Operation

Reuse-The sharing of identical radio channels in two or more cells of a cellular network. Sufficient cell separation is required to obtain adequate signal to noise ratio.

Revenue Accounting Office (RAO)-An accounting office that handles billing and related data processing activities for telecommunications services provided to customers.

Reverberation-The persistence of sound resulting from repeated echoes, as in a large hall, alter the sound source has stopped.

Reverse Address Resolution Protocol (RARP)-An Internet protocol that controls the translation of a data link control (DLC) address to an Internet protocol (IP) address. RARP allows a computer to obtain an IP address from a server when only the hardware address is known.

Reverse Control Channel (RECC)-The FSK modulated channel that completes the control channel signaling transmission path from the subscriber unit to the Base Station.

Reverse Digital Traffic Channel (RDTC)-A digital channel from a cellular phone to a cell used to transport user information and signaling. There are two separate control channels associated with the RDTC: the fast associated control channel (FACCH) and the slow associated control channel (SACCH).Also see FACCH, SACCH.

RF-Radio Frequency

RF Amplifier-Radio Frequency Amplifier

RF Channel-An RF channel is a communication link that use radio signals to transfer information between two points.

RF Energy-Radio Frequency

RF Power Control-A process of controlling the power level of a mobile radio. This is typically the result of a wireless system sensing the received signal strength level and sending power control messages that adjust the output power that increases or decreases its power level.

This diagram shows how the radio signal power level output of a mobile telephone is adjusted by commands received from the base station to reduce the average transmitted power from the mobile telephone. This lower power reduces interference to nearby cell sites. As the mobile telephone moves closer to the cell site, less power is required from the mobile telephone and it is commanded to reduce its transmitter output power level. The base station transmitter power level can also be reduced although the base station RF output power is not typically reduced. While the maximum output power varies for different classes of mobile telephones, typically they have the same minimum power level.

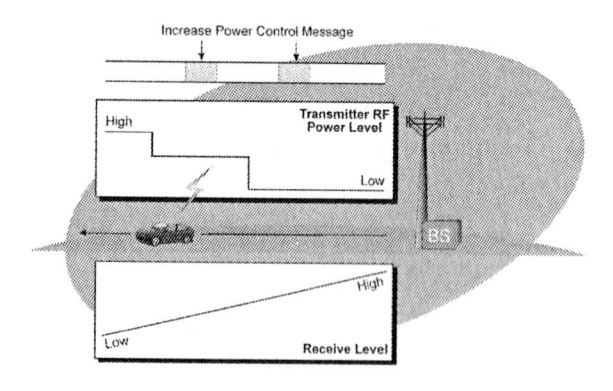

RF Power Control

RF Power Meter-A test instrument that can measure the power contained in a RF signal within specific frequency ranges.

RFBP-Request For Budgetary Proposal
RFC-Request For Comments
RFCH-Radio Frequency Channel
RFCOMM-Radio Frequency Communication Port
RFI-Radio Frequency Interference
RFI-Request For Information
RFN-Reduced Frame Number
RFP-Request For Proposal
RFQ-Request For Quotation
RFS-Remote File System
RGB-Red, Green, Blue,
RGB System-The basic parallel component set (red, green, and blue) in which a signal is used for each primary color. The same signals also may be called GBR in recognition of the mechanical sequence of connections in the SMFTE interconnect standard.
RHCs-Bell Regional Holding Companies
Rheostat-A 2-terminal variable resistor, usually constructed with a sliding or rotating shall that can be used to vary the resistance value of the device.
Rhombic Antenna-A horizontal directional high-frequency band antenna constructed as a diamond shaped array of radiating wires, fed at one apex and terminated at the opposite apex with an impedance, usually a resistor. When unterminated, the antenna is bi-directional.
Ribbon Cable-Flat cable with multiple parallel conductors that have been individually insulated.
RIC-Residual Interconnection Charge
RIFF-Resource Interchange File Format
Ring-(1-wire) One of the two subscriber loop/line wires. It is connected to the so-called ring conductor on a manual switchboard plug, hence its name. In North America, red insulation color is used to identify it. Corresponds to the European "B" subscriber wire. (2-audio) Audible alerting signal at subscriber set indicating incoming telephone call. May be a ringing bell or other sound.
Ring Generator-A component of virtually all phone systems, ranging from large offices to small key systems, that supplies the power to ring the bells inside phones, typically 90 volts AC at 20 Hz.
Ring Network-A network where each communication device (typically computers) are connected to a neighboring computer and the interconnected computers form a ring where the last computer in the string is connected to the first computer in the network. The data transmission process in a ring network involves computers passing all network data from its neighboring computer to the next neighboring computer. When a computer in the ring network receives data, it looks for data information designated for its address and removes (does not retransmit) the data.

This figure shows a ring network. In this diagram, each data communication device receives information from one computer in the ring and sends (and possibly forwards received information) information to another computer in the ring.

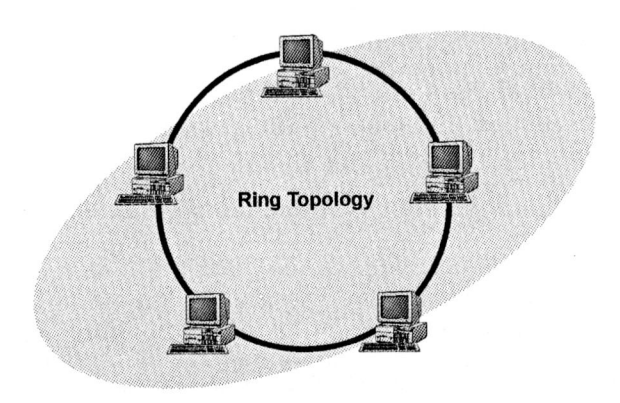

Ring Topology System

Ring Operation

Ring Side-The "hot" conductor in a 2-wire telephone circuit or loop, referenced at -48 volts DC.

Ring Signal-An alternating high voltage signal that is sent to a telephone set to alert the user (usually by creating a ring sound) of an incoming call. In North America, the ring cadence is usually 2 seconds on and 4 seconds off. The ring signal may also include messages that operate at low voltage on a different frequency that includes caller identification information.

This diagram shows that the ring signal is a high voltage alternating current signal that alerts the user of an incoming call. This diagram shows that this ring signal include a caller identification messages and the data is transferred by FSK modulation between the AC ring periods.

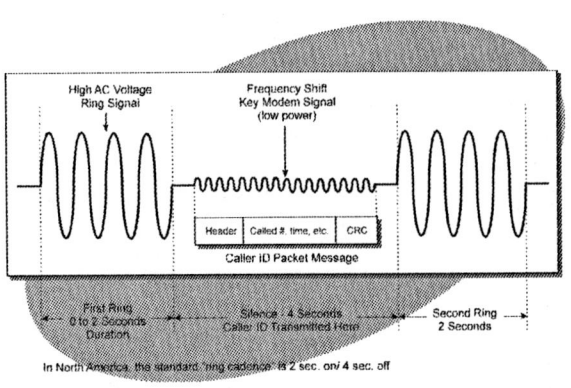

Ring Tone-The tone that is used to announce an incoming telecommunications call. There may several different types of ring tones and some telecommunications devices allows the user to program in their own unique selection for a ring tone.

Ring Topology-A design for a local area network (LAN) formed by placing stations along a closed loop, connected by sections of a medium, such as optical fiber or copper wire. Data circulates around the ring from station to station. Some networks, such as Token Ring Networks, utilize a logical ring topology while being physically wired in a star topology. The drawback to a physically wired ring topology is the inability to quickly "wire around" a failing station. Star wiring removes this drawback. (See also: bus topology, star topology, tree topology.)

Ringback-An indication (usually audible) that the phone at the far end is ringing. Ringing may be either local or remote.

Ringdown Circuit-A tie-line facility of connected telephones in which picking up a phone automatically rings another phone at a different location. This technique is often used in network control centers and high-volume operating environments where communications are often repetitive.

Ringer-A bell, or an electronic equivalent, that responds to an alternating current signal to produce a ringing, or audible alerting sound, from a telephone set.

Ringing-(1- telephone) An alternating-current signal that rings a bell or other sounder in a telephone set to alert a subscriber to an incoming call. A ring-

ing voltage is applied to a customer loop and, through ringers or ring-up circuitry, provides an audible and/or visual signal to the called person.
Ringing Generator-A circuit or device in a phone system that generates the ringer signal. The ringer signal voltage usually ranges from 90 to 115 VAC (typically 105) with a frequency of 30 Hertz.
RIP-Routing Information Protocol
Ripple-An ac voltage superimposed on the output of a dc power supply, usually resulting from imperfect filtering.
RISC-Reduced Instruction Set Computer
RISC-Reduced Instruction Set Computing
Rise Time-The amount of time that it takes a signal to rise from a low level to its high level (e.g. peak) value. Rise time is usually measured when the signal increases between 10% to 90% of maximum output.
Riser Cable-A high-strength cable intended for use in vertical shafts between floors in a building.
RIUT-Rate Of Information Transfer
RJ-Registered Jacks
RJ-11-A modular connector that has 2 to 6 conductors that is commonly used to interconnect end-user telephone equipment.
RJ-12-A six-wire or three-pair connection using the same connector as RJ-11.
RJ-14-A jack that looks and is exactly like the standard RJ-11 that you see on every single line telephone. Where as the RJ-11 defines one line — with the two center, red and green, conductors being tip and ring, the RJ-14 defines two phone lines. One of the lines is the "normal" RJ-11 line — the red and green center conductors.
RJ-45-A standard 8 wire modular connector. RJ-45 connectors are commonly used in telephone and data communication systems.
RLAC-Radio Link Access Control
RLP-Radio Link Protocol
RLV-Re-Usable Launch Vehicle
RMAS-Remote Memory Administration System
RMON-Remote Monitoring Specification
RMON-Remote network MONitoring
RMONv2-RMONv2 is an extension to the original Remote network MONintoring (RMON) standard as defined in RFC1757. RMONv2 introduces 9 more groups and functions, mostly concentrating on higher layer protocol statistics: protocols supported by the device, statistics per protocol, MAC to higher-layer protocol mapping, higher-layer host statistics, higher-layer conversation pairs, application-layer host statistics, application-layer conversation pairs, history statistics for user-defined criteria, and configuration of RMON features. RMONv2 is defined in RFC2021.
RMS-Root-Mean-Square
RNC-Radio Network Controller
RNS-Radio Network Sub-Systems
Roamer-A subscriber operating in a cellular system other than its registered home system.
Roamer Validation-Roamer validation is the verification of a wireless telephone's identity using registered subscriber information. Validation is necessary to limit fraudulent use of cellular service. The two types of roamer validation are "post-call" and "pre-call". Post-call validation occurs after a call is complete, and pre-call validation occurs before granting access to the system.
Roaming-The capability to move from one wireless carrier's system coverage area to another carrier's coverage area and obtain service. While it is desirable to roam without loosing functionality of the phone, some wireless systems offer advanced features (such as high speed data) that other systems may not have installed. This may limit the operation of advanced features.
Robbed-Bit Signaling-Robbed-Bit Signaling is used to convey supervisory (idle vs. busy channel) status in some installations of the DS-1 (T-1) digital multiplexing carrier for pulse code modulation (PCM), used in North America and Japan. First a synchronized sequence of either 12 frames or 24 frames (called extended super frame - ESF) is established by means of a specific binary sequence of 0 and 1 signals in the framing bit of each frame. Then the eighth bit of every channel of the sixth and twelfth frames (and also the 18th and 24th frames in ESF installations) is preempted, and typically replaced with a bit whose value (0 or 1) indicates the idle vs. busy (on-hook vs. off-hook)status of that channel. In some types of signaling (example, E&M tie line signaling) these bits have other meanings as well. Robbed bit signaling is not necessary when common channel signaling or associated bit signaling is used.
Robot (Bot)-A software application that works with the world wide web that automatically locates and gathers information within Web sites that meet specific criteria. Bots can be used to create databases of Web sites of specific categories.
ROI-Return on Investment
Roll Off-The increase of loss in a frequency selective device as the bandwidth limits are approached.

Rollover Lines-Telephone lines that are placed in a hunt sequence where calls will move to the next available line when those higher in the hunt group are found busy. This also refers to the movement of inbound call traffic from one telephone switch or ACD queue to another based on current busy conditions or programmed instructions within the telephone switch.

ROM-Read Only Memory

RoMA-Return on Marketing Activity

Rotary Dial-A process of dialing using a spring-loaded mechanical switch that produces pulses as it rotates through 10 positions (1 through 9, and 0). As the rotary dial turns, a switch briefly interrupts the loop current. The number of pulses per rotation is counted to determine the number dialed. A time pause between rotary dials is used to determine when additional digits are dialed or when the caller is finished dialing.

This diagram shows how a mechanical rotary switch can be used to gather dialed digit information. In this example, a spring-loaded rotary dial is turned to produces pulses that represent the numbers on the dial. As the rotary dial turns, a contact switch briefly interrupts the loop current. The line card in the telephone system counts number of pulses to determine the number dialed. This diagram shows that after a specified time pause occurs after a series of pulses, the counter resets and is ready to count another dialed digit.

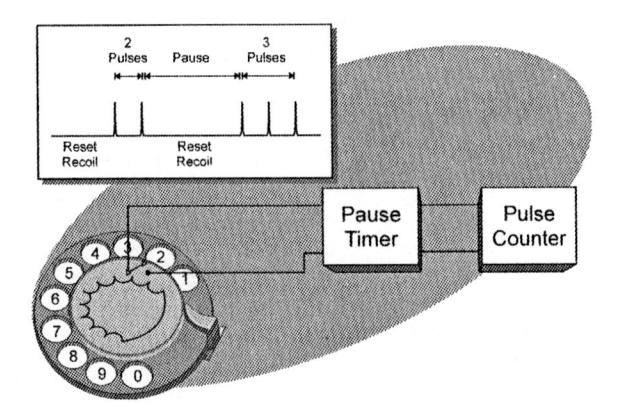

Rotary Dial Operation

Rotor-The rotating part of an electric generator or motor.

Round Trip Delay-The time required for a wave to travel from a signal source and return from either the distant end of a circuit or an impedance mismatch within the circuit.

Route Flapping-The continual changing of a network connection path that results from a intermittent congestion or loss of circuit connection that indicates to the current router connection path that there is a loss in connection or that a better connection path exists. This causes the packet routing path to continually change. These different paths can result in significant variance in transmission delay times (excessive jitter). Router flapping is overcome by newer IPV6 protocols and reservation protocols.

Route Flutter-The variance of packet delay due to the continual changing of the routing paths in a communication network. See route flapping.

Route Set Congestion Control-A procedure included in the SS7 signaling that is used to update the congestion status of a signaling route in a given signaling point.

Router-A router is a device that directs (routes) content (data, voice or video) from one path to another on a network. Routers base their switching information on one or more parameters contained in the packet of content. These parameters may include availability of a transmission path or communications channel, destination address contained within a packet, maximum allowable amount of transmission delay a packet can accept, along with other key parameters.

Routers forward data packets between multiple interfaces based on the network layer. Most modern day routers support one or more of the following protocols: Internet Protocol (IP), Novell IPX or AppleTalk. Routing occurs at layer 3 of the OSI reference model and can be used to limit the broadcast domain of a bridged network.

This diagram shows how a router is used to receive and forward data packets toward their destination. This diagram shows that a router receives packets from other routers, examines the destination address of the packet, and uses its stored routing tables to determine which router is the best choice for forwarding the packet towards its destination.

Routine

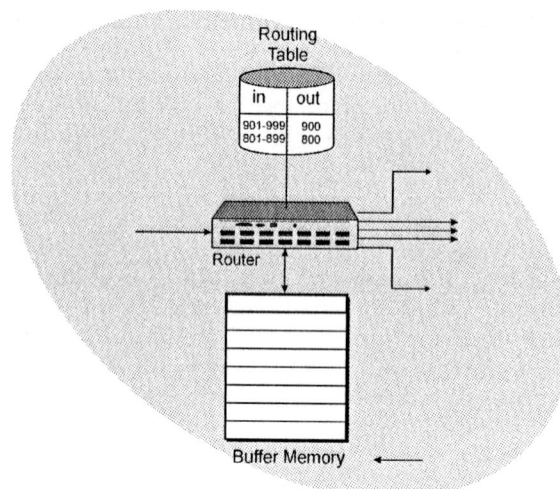

Router Operation

This is a picture of Cisco router, model ONS 15454.

Router
Source: Cisco Systems, Inc.

Routine-A group of instructions for carrying out a specific processing procedure. Routine usually refers to part of a larger program. Routine and subroutine tine have essentially the same meaning, but a subroutine could be interpreted as a self contained routine nested within another routine or program.

Routing-(1-telecommunications traffic) The physical path, circuit group, and switching systems through which telecommunications traffic flows. (2-communications path) The selection of a communications path. Traffic routing pertains to the path between switching offices, regardless of the physical paths of the connecting circuits. Circuit routing pertains to the alternative physical paths used in connecting switching offices. (3-packet network) The process of reviewing the destination address against a routing table and algorithm and the forwarding to a path that will help it reach its destination.

This figure shows a how a router can dynamically forward packets toward their destination. This diagram shows that a router contains a routing table (database) that dynamically changes. This diagram shows a router with address 100 is connected to two other routers with addresses 800 and 900. Each of these routers periodically exchanges information allowing them to build routing tables that allow them to forward packets they receive. This diagram shows that when router 100 receives a packet for a device number 952, it will forward the packet to router 900. Router 900 will then receive that packet and forward it on to another router that will help that packet reach its destination.

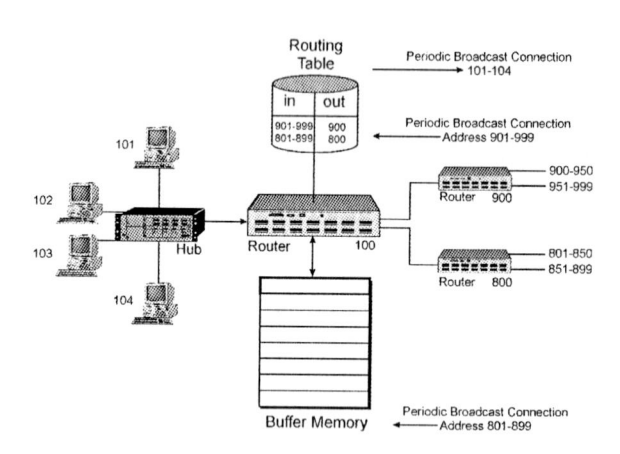

Data Network Router Operation

Routing Information Field-In Token Ring Networks, the optional field at the end of the MAC header, before the user data which contains explicit route information used by source routing bridges.

Routing Information Protocol (RIP)-A Internet routing protocol that uses a list of reachable networks to calculate the degree of difficulty that may be involved in reaching a specific, usually by determining the lowest hop count. RIP has been succeeded by the routing protocol Open Shortest Path First (OSPF).

Routing Label-A route identifying part of a data message that is used for message routing in the signaling network. The routing label of an SS7 message includes the destination point code (DPC), the originating point code (OPC) and the signaling link selection field.

Routing Loop-The process of passing a data packet from a router to other routers that return the packet back to the original sending router. This may occur because routing tables are dynamically changing as routers continually learn of connections to new routers and this results in new paths for forwarding packets. For packets that get caught in a routing loop, the time to live (TTL) field in the packet header will eventually be depleted (decreases each time a routing hop occurs) so the packet will eventually be discarded.

Routing Tables-The database table that is located within a router that is used to determine the forwarding path (route) for incoming packets.

Royalty-When patents were issued by English Queens and Kings, it was common to provide the monarch with a small (or large) payment in return for the grant. In modern times, a royalty is a payment to and author or composer for each copy of a work sold, or performed, or a payment to an inventor for items sold under license to a patent

Royalty Free-Royalty free means that a patent is licensed without cost to the user. Royalty free licensing terms are mandatory for patents incorporated into Open Source software such as Linux, and W3C standards.

RP-Radio Port

RPC-Remote Procedure Call

RPE-Remote Peripheral Equipment

RPG-Report Program Generator

RS-Recommended Standard

RS-232-Serial communications standard RS-232 is an electronics industry association (EIA) standard protocol that is used to transfer information in asynchronous (unscheduled) form. The RS-232 specification defines optional physical (connector), electrical (signal levels), and software data formats. RS-232 serial connectors are a common connector that is used on the back of computers to connect to modems and other external devices. RS-232 communication uses universal asynchronous receiver and transmitter (UART) technology and adds communication negotiation features to the serial data communication session.

This diagram shows how an RS-232 communication system transfers information in serial form. To control the flow of data, the RS-232 system uses control lines to indicate the status of transmitting and receiving devices (e.g. ready or not ready). The control lines include a ring indicator to indicate a modem connection is requested (not used for this direct connections), carrier detect (CD) lines to sense a modem signal, data set ready (DSR) and data terminal ready (DTR) to indicate the data communication device is ready to operating. The control lines also include clear to send (CTS) and request to send (RTS) as a method of flow control. This example shows that the data can be transmitted in two directions on the transmit data (TD) and receive data (RD) lines.

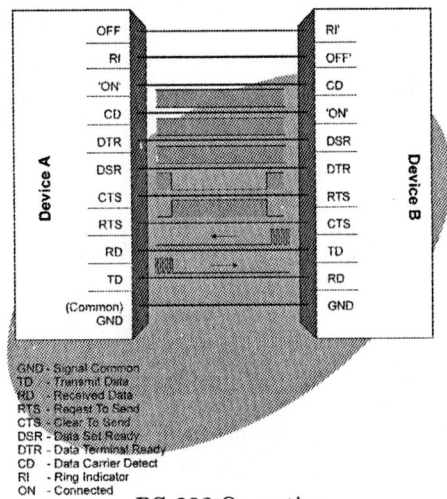

RS-232 Operation

RS-232-C-A technical specification published by the Electronic Industries Association (EIA) that establishes electrical interface requirements between computers, modems, terminals, and communications lines. Also known as EIA/TIA-232.

At the time of original issue, the EIA was known as the Radio Manufacturers Association, and its standards were then designated "Radio Standards" or RS and then the identifying number. The standard defines the specific electrical and functional characteristics used in asynchronous transmissions between a computer (data terminal equipment, or DTE) and a peripheral device (data communications equipment, or DCE). The suffix C denotes the third revision of that standard. RS-232-C is sub-

RS-449

stantially compatible with the CCITT V.24 and V.28 standards, as well as ISO IS2110. (The original version of RS-232, now obsolete, used contact closures for some signals. The modern version uses different voltage levels to convey all control information.)

RS-232-C is almost always implemented using a 25-pin or 9-pin DB type connector, although the standard does not specify the mechanical shape or form of the connector. The accompanying illustration shows the pinouts used in a DB-25 male connector. It is used for serial communications between a computer and a peripheral device, such as a printer, modem, or mouse. The maximum cable limit of 15.25 meters (50 feet) can be extended by using high-quality cable, line drivers to boost the signal, or short-haul modems. A related standard, RS-449, uses separate balanced 2-wire circuits (instead of using a "one wire" circuit for each control pin, with a common return wire shared by several circuits) and can operate over longer cables, but is not as widely available. Peripheral device (data communications equipment, or DCE). RS is the abbreviation for recommended standard, and the C denotes the third revision of that standard. RS-232-C is compatible with the CCITT V.24 and V.28 standards, as well as ISO IS2110.

RS-232-C uses a 25-pin or 9-pin DB connector. The accompanying illustration shows the pinouts used in a DB-25 male connector. It is used for serial communications between a computer and a peripheral device, such as a printer, modem, or mouse. The maximum cable limit of 15.25 meters (50 feet) can be extended by using high-quality cable, line drivers to boost the signal, or short-haul modems.

RS-449-An EIA specification on general purpose interfaces of data terminal equipment (DTE) and data-circuit-terminating equipment (DCE) that employs serial binary data interchanges. An RS-449 connection may use a 37-pin or 9-pin DB connector.

RSA-Reliable Service Area
RSA-Rural Statistical Area
RSA-Rural Service Area
RSA-Rivest, Shamir, Adleman
RSB-Record Storage Buffer
RSC-Reed Solomon Code
RSH-Remote Shell
RSID-Residential System Identity
RSL Or RXLEV-Received Signal Level
RSS-Radio Sub-System
RSSI-Received Signal Strength Indicator
RSVP-Resource Reservation Protocol
RTA-Remote Trunk Arrangement
RTC-Runtime Control
RTC-Rural Telephone Company
RTCP-Real-Time Transport Control Protocol
RTL-Resistor Transistor Logic
RTN-Retain Negative
RTP-Real Time Transport Protocol
RTR-Response Time Reporter
RTS-Request To Send
RTS-Ready to Send
RTSP-Realtime Streaming Protocol
RU-Rack Unit
Rural Service Area (RSA)-A cellular service area in rural (small population) regions.
Rural Statistical Area (RSA)-A geographic area designated by the FCC for service to be provided for by cellular carriers that falls outside the Metropolitan Statistical Area (MSA) regions. There are over 400 RSAs in the United States.
RVC-Reverse Voice Channel
Rx-Receiver.
RXQUAL-Received Signal Quality
RZ-Return To Zero

S

S-Supervisory Function Bit
s-Second
S-siemens
S Format-Supervisory
SA-Signaling Agent
SA-Security Association
SA-Source Address
SAA-Systems Application Architecture
Sabin-A unit of sound absorption that is the equivalent of one square foot of surface of perfectly absorptive material.
SABME Frame-A set, asynchronous balanced mode extended frame. The SABME frame is used in the Integrated Services Digital Network (ISDN) and the Public Packet-Switched Network (PPSN) to place a data link into modular 28 multiple frame acknowledged operation.
SAC-Service Access Code
SAC-Single Attached Concentrator
SACCH-Slow Associated Control Channel
Safe Action Area-Safe Title Area
SAID-Security Association Identifier
Sales-(1-operations) The operational department responsible for identifying potential new customers, negotiating offers, and closing contracts. (2-Revenue) The amount of products or services that are sold or committed to be sold.
Sales Plan-A plan that is used to define and manage the sales process for products or services. The sales plan usually includes objectives (sales targets) and it may identify the salespeople or representatives and their responsibilities along with their territories and commission structures.
This diagram shows how a sales process can be managed. In this example, the sales process is divided key steps that can be defined and managed. The prospecting step is used to identify new customers or expanded needs for existing customers. The qualification step is used to determine how many of the prospects are qualified (real candidates) for the product or service. The interest assessment step is used to determine how motivated the prospect is to take action to satisfy their need for the product or service. The fact-finding stage is used to determine who are the decision makers and what steps are necessary to complete the sale. The close involves the consolidation of the previous steps (coordination of motivated decision makers) so purchases can result. The progress between each step can be tracked and optimized. Included in the chart is an example activity time sheet showing that step may require different levels of time commitment and that the allocation of resources (time for each salesperson) is usually distributed along each step.

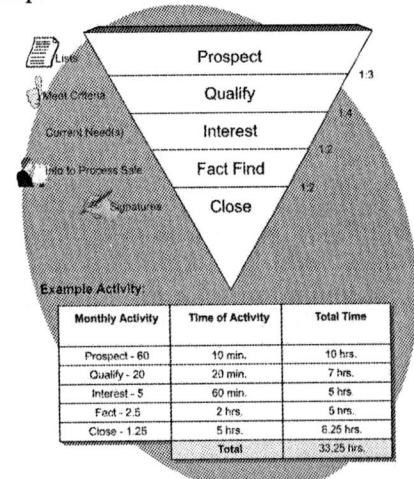

Sales Management Diagram

Sales Program-A series of related sales campaigns targeting the same objectives.
Salutation-A procedure for looking up, discovering, and accessing services, and information. This architecture defines abstractions for devices, applications, and services; a capabilities exchange protocol; a service request protocol; standardized protocols for common services; and application programming interfaces (APIs) for information access and session management. See also salutation Architecture.
SAM-Security Accounts Manager
Sampling Rate-The rate at which signals in an individual channel are sampled for subsequent modulation, coding, quantization, or any combination of these functions. The process of taking samples of an electronic signal at equal time intervals to typically convert the level into digital information. frequency usually is specified as the number of samples per unit time.

SAN-Storage Area Network
SAP-Service Advertising Protocol
SAP-Secondary Audio Program
SAP-Service Access Point
SAP-Session Announcement Protocol
SAPI-Service Address Pint Indicator
SAPI-Speech API
SAS-Single Attached Station
SAT-Supervisory Audio Tone
SATAN-Security Administrator Tool For Analyzing Networks

Satellite-(1-general) An object that revolves around another about of greater mass (such as the earth) and has a motion that is determined by the force of attraction (gravity) of the larger object. (2-communications) A space vehicle that orbits the earth which contains one or more radio transponders that receive and retransmit signals to and from the earth. (3-equipment) A piece of equipment or system that operates at a remote location from a central control system.

Satellite Acquisition-The process of lining up a ground station with a telecommunications satellite so that the antenna locks on and tracks the satellite.

Satellite Availability-The probability that a satellite will be on station and available for a particular task.

Satellite Communications-The use of orbiting satellites to relay communications signals from one station to many others.

Satellite Downlink-A microwave radio link from a satellite to a ground station on earth.

Satellite Earth Station-A complete ground station designed to control a satellite system and to interconnect this system with a terrestrial system.

Satellite Relay-An active satellite repeater that relays signals between two earth terminals.

Satellite System-A network of satellites and associated ground stations that transmit telephone, audio, video, and data signals between terrestrial points.

This diagram shows the different types of satellite communication systems. The GEO satellite system is primarily used for television broadcast services, as their satellites appear stationary above the Earth. MEO and LEO systems are used for mobile communications as they are located much closer to the Earth. However, these satellites continuously move relative to the surface of the Earth.

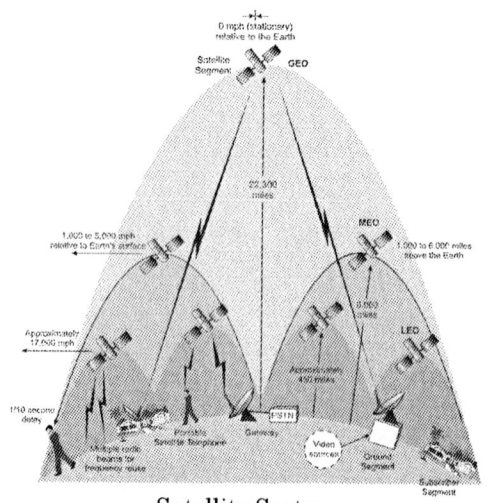

Satellite System

Satellite Telephone-A wireless telephone that uses mobile satellite service to send voice and data.

Satellite Transponder-A combination receiver and transmitter that receives a signal from an Earth station (uplink) and retransmits it to ground receiving stations (downlink).

Satellite Uplink-The transmission equipment used to modulate, amplify, and radiate a signal to an orbiting satellite.

Saturation-(1- chroma, color) The intensity of the colors in the active picture; the voltage levels of the colors. Saturation relates to the degree by which the eye perceives a color as departing from a gray or white scale of the same brightness. A nth percent saturated color contains no white; that is, adding white reduces saturation. In NTSC and PAL video signals, the color saturation at any particular instant in the picture is conveyed by the corresponding instantaneous amplitude of the active video subcarrier. (2- amplifier circuit) The point on the operational curve of an amplifier at which an increase in input amplitude will no longer result in an increase in amplitude at the output. (3- inductor or transformer) A condition in which the ferromagnetic core material of an inductor or transformer is subjected to strong magnetic induction so that its internal magnetic field has reached its maximum possible value of physical magnetization. At the atomic level, all of the magnetic domains in the core material are now oriented in the same space direction, and all of the elementary intrinsic electron spins in the outer electron shells of the atoms in the core are oriented in the same space direction.

SAW-Surface Acoustic Wave
SAW Device-Surface Acoustic Wave
SBC-Session Border Controller
SC-Subcarrier
Sc/H Phase-Subcarrier To Horizontal Phase
SCADA-Supervisory Control And Data Acquisition
Scalability-The ability of a system to an increase in the number of users or amount of services it can provide without significant changes to the hardware or technology used.
Scalar Analyzer-A test instrument that measures the characteristics of a transmission signal that includes return loss (reflected signals), voltage standing wave ratio (VSWR), and insertion loss.
Scanner-A device used to input graphic images in digital form. A fax machine's scanner determines the brightness of a document's pixels for transmission
Scanning-The process of searching through multiple radio channels.
Scanning Receiver-A receiver that is capable of searching through multiple radio channels.
Scatter-The diffusion of the direction of propagation of a wave into multiple paths, caused by discontinuities in the transmission medium.
Scattering-(1 - radio) The diffusion of radio waves when they encounter air masses or ionized layers in the troposphere or ionosphere. The new waves so produced have random direction and polarization. (2 - acoustics) The irregular diffuse dispersion of sound in many directions, caused by reflection, refraction, or diffraction. (3 - optoelectronics) The deflection of light from the path it would follow if the refractive index were uniform or gradually graded. Scattering is caused primarily by microscopic or submicroscopic fluctuations in the refractive index. Scattering is the principal cause of the attenuation of an optical waveguide.
(1 - ionosphere) The elliptical patch in the ionized E-layer by day and F-layer by night that is illuminated by a beam from a transmitting antenna and toward which a receiving antenna is directed. (2 - troposphere) The elliptical patch about 10 km above ground in which air turbulence causes radio wave refraction.
SCC [Field]-SAT Color Code
SCCP-Signaling Connection Control Part
SCCP-Skinny Client Control Protocol
SCCP Routing-An SS7 function based on the called party address information, which evaluates and translates the information, checks the addressee availability, and the need for coupling of connection sections.
SCCP User Adaptation (SUA)-SCCP User Adaptation layer is a protocol for the transport of any SS7 SCCP user Signaling (e.g. TCAP, RANAP or RNSAP messages) over IP Between two signaling endpoints.

This diagram shows the basic operation of the SCCP User Adaptation (SUA) protocol. This diagram shows that the SUA protocol allows the upper layer signaling applications to directly communicate with the IP based SCTP signaling transport protocol. This diagram shows that the SCTP protocol maintains the connection control through the network while the SUA protocol maps the communication channel from the upper layer application to the transport channel over the IP network

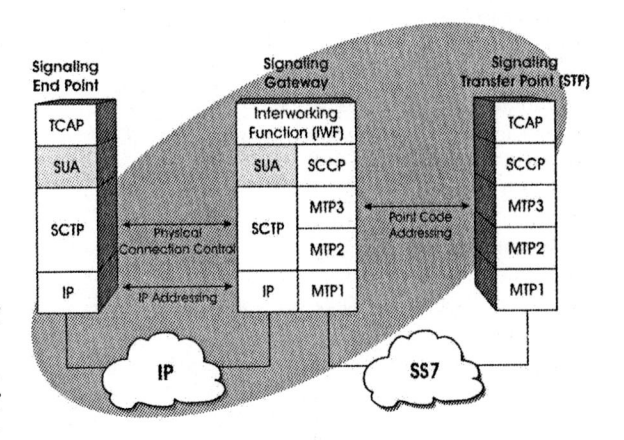

SCCP User Adaptation Protocol (SUA) Operation

SCE-Service Creation Environment
SCF-Shared Channel Feedback
SCH-Synchronization Channel
Scheduled Conferencing-A conference bridge that allows callers to become members of a conference group during a predetermined time period.
Scheduled Transmission-A feature allowing the user to schedule a fax transmission at a specific data or time in the future. Key benefits are convenience and cost savings. Scheduling jobs at a period of low telephone rates can have immediate considerable savings.
Schematic-A diagram of a circuit or flow of a system.

Scintillation-In radio propagation, a random fluctuation of the received field about its mean value. The deviations usually are relatively small. The effects of this phenomenon become more significant as frequency increases.

SCN-Subchannel Number

SCP-Service Control Point

SCR-Same Cell Reuse

SCR Mode-Same Cell Reuse

Scrambler-A device that transposes or inverts signals, or otherwise encodes a message at the transmitter to make it unintelligible at a receiver that is not equipped with a descrambling device.

Scrambling-A process of altering or changing an electrical signal (often the distortion of a video signal) to prevent interpretation of the signals by users that can receive the signal but are unauthorized to receive the signal. Scrambling involves the changing of a signal according to a known process so that the received signal can reverse the process to decode the signal back into its original (or close to original) form.

Scratch Pad-A read/write random access memory space used for the temporary storage of data; it is the working area of a memory unit.

Screen Based Telephony-The use of computer screens to provide a telephone user with information about a call in progress and/or the status of their telephone calls. Screen based telephony may allow the telephone user to dial, answer, and control their telephone calls.

Screen Name-An identifying name that is used for instant messaging systems to identify a instant messaging person or device. The instant messaging system provides the relationship between a screen name and an email or IP address. The use of a screen name allows a person to avoid providing an email address (for privacy) and it allows the use of dynamically changing IP addresses (for systems that dynamically assign IP addresses when the user logs onto the system).

Screen Pop-The automatic display of a screen on a computer monitor that is automatically triggered by an event such as an incoming call or customer selected feature request.

This figure shows that because VoIP telephone systems can share the same type of data network, the telephone system can be more easily integrated with the company's information system. In this example, a customer service representative (CSR) is receiving a call from John Doe. The screen pop shows that John Doe has already purchased a book. The CSR can use the account information from John Doe to help him find additional products to purchase.

Customer Service Representative (CSR)

Screen Pop Operation

SCSA-Signal Computing System Architecture

SCSI-Small Computer Systems Interface

SCTP-Stream Control Transport Protocol

SCTP Packet-Stream control transmission protocol (SCTP) packets contain a common header and variable length blocks (chunks) of data. The SCTP packet structure is designed to offer the benefits of connection-oriented data flow (sequential) with the variable packet size and the use of Internet protocol (IP) addressing.

This figure shows that the SCTP packet structure includes a common header format along with chunks of data. The header includes the source and destination port numbers that are associated with the specific IP addresses to uniquely identify the packet for a specific communication application. The verification tag uniquely identifies (validates) the sender of the packet. Each packet has a checksum to ensure all the data has been reliably sent. The data is sent in chunks to allow the near real-time streaming (continuous one-way delivery) of information.

Second Generation (2G)

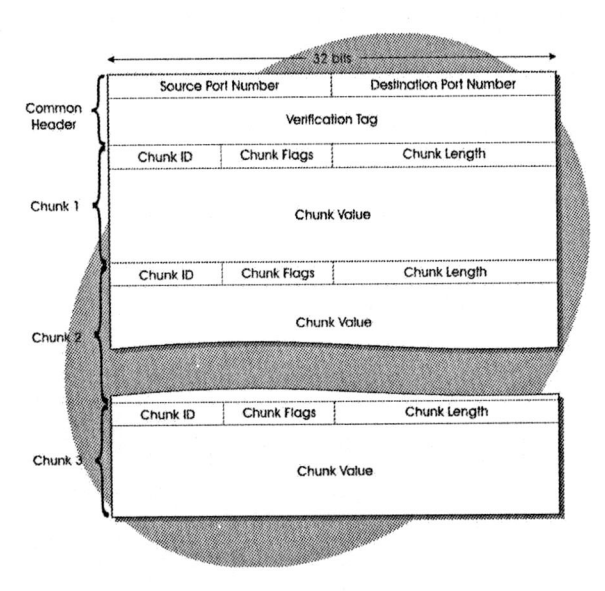

SCTP Packet Structure

SDAP-Service Discovery Application Profile
SDCC-Supplementary Digital Color Code
SDCCH-Stand-Alone Dedicated Control Channel
SDD-Space Division Duplex
SDH-Synchronous Digital Hierarchy
SDI-Serial Data Interface
SDK-Software Development Kit
SDL-System [or State-machine] Description Language
SDLC-Synchronous Data Link Control
SDM-Statistical Division Multiplexing
SDMA-Spatial Division Multiple Access
SDN-Software Defined Network
SDP-Session Description Protocol
SDP-Service Discovery Protocol
SDPng-Session Description Protocol Next Generation
SDSL-Symmetrical Digital Subscriber Line
SDU-Service Data Unit
Sealing Current-A method that is used to decrease the effects of corrosion is to continuously run current through the copper wire pair. This "sealing current" is a small amount of direct current that is passed through a copper wire to reduce the corrosion effects of the splice points. The sealing current effectively maintains conductivity of mechanical splices that are not soldered. The direct current effectively punches holes the corrosive oxide film that forms on the mechanical splices.
SECABS-Small Exchange Carrier Access Billing
SECAM-Sequential Couleur Avec MeMoire
Second (s)-1- time unit) Unit of time. It is defined as 9,192,631,770 cycles of the radiation associated with the transition between the two hyperfine levels of the ground state of the cesium-133 atom. Originally historically defined as 1/ 31 557 600 of a year. (2- order) The item in a sequence following the first item.
Second And A Half Generation (2 1/2G)-A term commonly used to describe one or more interim technologies that are used to help a specific application or industry transition from one capability to a much more advanced capability. In cellular telecommunications, 2 1/2 generation systems used improved digital radio technology to increase their data transmission rates and new packet based technology to increase the system efficiency for data users.
Second Generation (2G)-A term commonly used to describe the second technology used in a specific application or industry. In cellular telecommunications, second generation systems used digital radio technology with advanced messaging and data capabilities. For second-generation cordless telephones, second generation of products used multiple channels (using analog FM technology) radios. This photograph shows an image of a 2nd generation mobile phone, model SPH-i300. This telephone has a wireless phone in combination with a personal digital assistant (PDA). This device utilizes the advanced messaging and data features of second generation mobile communications technology.

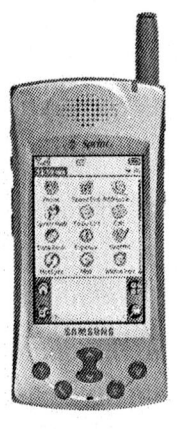

2nd Generation Wireless Telephone
Source: Samsung Telecommunications America

Secondary-The output winding of a transformer.

Secondary Audio Program (SAP)-The second audio program channel in the BTSC stereo TV system. The SAP channel often is used for second-language or other specialized programming.

Secondary Dedicated Control Channels-A second set of nationwide allocated control channels that may be used by a digital cell site switch.

Secondary Ring-The secondary path used in a communication network ring. A secondary ring circulates data in the opposite direction as the primary ring. It is used in the fiber distributed data interface (FDDI) system.

Secondary Route-The circuit to be used when the primary route is congested. In manual and semi-automatic operations, a secondary route also may be used when transmission on the primary route is not of sufficiently good quality, or if traffic is to be handled outside the normal hours of service on the primary route.

Secondary Voltage-The operating voltage on the load side of a transformer.

Secondary Winding-The output winding of a transformer. The secondary receives power by induction from current flowing in the primary winding.

Section Terminating Equipment (STE)-A device or assembly that terminates a section in an optical system. STE network equipment terminates the physical and section layers of a Synchronous Optical Network (SONET).

Section Throw-An aerial or underground engineering order involving the splicing of new cabling in parallel with live working cable. This is conducted by half-tapping the new cable into the existing live cable and then trimming out the old, defective section. Cable counts remain the same in the new working cable.

Sector-(1- radio) One of the radio transmission portions of a sectorized cell site. See sectorization. (2- computing) A logical divisions of storage disk's tracks. Each track is typically divided into several sectors.

Sector Antenna-A directional antenna that produces coverage of one or more sectors in a cellular cell site system.

Sectorization-The dividing of transmitter radio coverage areas into smaller coverage areas by using directional antennas. Most cellular systems initially use cell sites that transmit in omni-directional 360-degree coverage patterns. To increase the system capacity, the antenna system is changed so 3 or 6 focused antenna patterns are created.

This diagram shows how an omni-directional cell site radio coverage area has been divided (sectorized) into three smaller areas (sectors). This diagram shows that frequency number 2 now can be reused in an adjacent cell site due to the directional properties of the antennas (directional antennas have reduced gain in the opposite direction.)

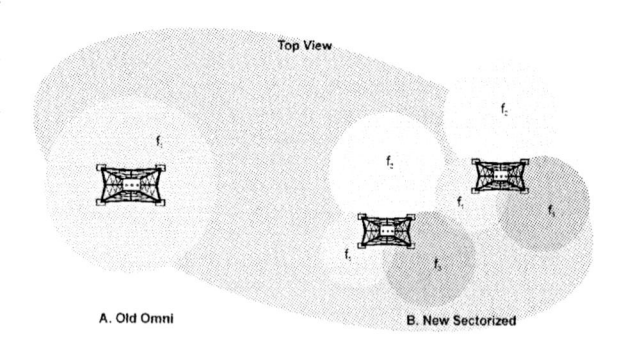

Cell Site Sectorization

Secure Real Time Protocol (SRTP)-An enhanced version of real time protocol (RTP) that provides increased security (e.g. confidentiality and message authentication).

Secure Shell (SSH)-A secure alternative to RSH, SSH allows remote command execution and remote terminal sessions. SSH creates an encrypted and authenticated channel between hosts for all communication. SSH may also be used to securely tunnel TCP traffic between two hosts' networks.

Secure Sockets Layer (SSL)-SSL version 2 provides security by allowing applications to encrypt data that goes from a client, such as Web browser, to a matching server (encrypting your data means converting it to a secret code) SSL version 3 allows the server to authenticate that the client is who it says it is.

Security-(1-General) Security management of a network involves identity validation (authentication), service authorization, and information privacy protection. Authentication processes identifies the device or person that is requesting the use of the telecommunications device or network services.

Selective Call Acceptance

Authorization is the process of determining what services devices are customers are permitted to use. Privacy or encryption services are used to help ensure that the information transmitted or received is not available to unauthorized recipients. (2-FCAPS) Security is one of the five functions defined in the FCAPS model for network management. Security is responsible for protecting the network against hackers, unauthorized users, and physical or electronic sabotage.

Security Proxy-A proxy server that receives and filters information designated for a user it is serving. The security proxy restricts information based on rules it has received from the network operator and/or user of the information. A security proxy is also called a firewall.

Security System-Security systems are monitoring and alerting systems that are configured to provide surveillance and information recording for protection from burglary, fire, environmental hazards, and other types of losses.

SEF-Source Explicit Forwarding

Segmentation-(1-network) The dividing of a network into multiple networks. (2-data segmentation) The division of data blocks or packets into smaller segments or packets. Segmentation is also known as fragmentation as in the data and Internet community. (3-database) In database management, the process of separating filed information or records into groups (segments) that allow for more rapid information access. (4-software) The dividing of a program into functional parts (segments). These segments can usually be changed or updated independently making the software program easier to develop and change.

This diagram shows how a large data packet may be divided into several smaller segments to allow transmission through networks that have a smaller packet size than packet of data traveling through the network. In this example, a packet that is 1200 bytes enters into a newtork that can only transport packets with a maximum size of 400 bytes. If the packet header indicates it is acceptable to divide the packet, each smaller packet will include a the destination address plus additional information that indicates which part of the packet segment it represents. When the packet reaches its destination, the smaller fragments will be reassembled into the original data packet. Some time-sensitive systems (such as real-time voice systems) do not want packets to be divided.

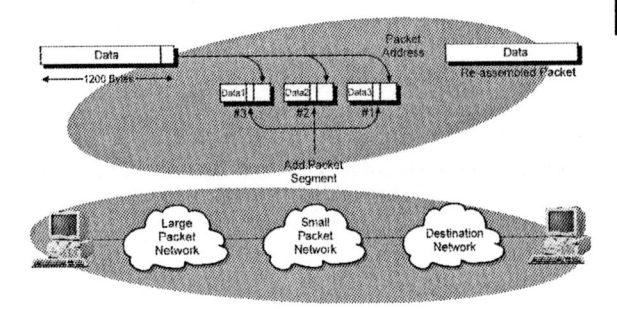

Packet Segmentation

Selective Call Acceptance-A service feature that only delivers calls to their dialed destination if they are on a previously specified selective call acceptance telephone number list. Calls that are received by other numbers are provided with a pre-recorded announcement that states the number is not accepting their call or the call may be routed to an alternate directory number.

This diagram shows how selective call acceptance can be used to only deliver calls from a specific list of callers. This diagram shows that regional managers from a service center can call from numbers that are pre-defined (their office numbers) and that when their calls are received, the will be connected to the specified telephone or extension. Call that are received from numbers that are not on the selective call acceptance list are transferred to an automated message unit that plays a pre-recorded announcement that states the number is not accepting their call.

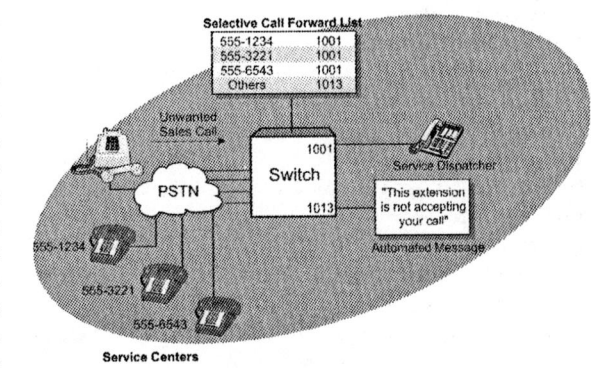

Selective Call Acceptance Operation

Selective Call Forwarding

Selective Call Forwarding-A service feature that forwards calls to one (or multiple) telephone numbers dependent on the incoming call forwarding criteria. Selective call forwarding can be used to redirect calls of a specific type (such as fax calls) to a pre-designated number (such as an office fax machine.)

This diagram shows how selective call forwarding can be used to deliver calls to alternate number based on a specific criteria type. This diagram shows a selective call forwarding service that routes fax calls to different telephone number or extension after it detects the call is a fax call. After the system detects that the incoming call is a fax (by the fax tones), the switch call processing software transfers the call to the destination number that is connected to a fax machine.

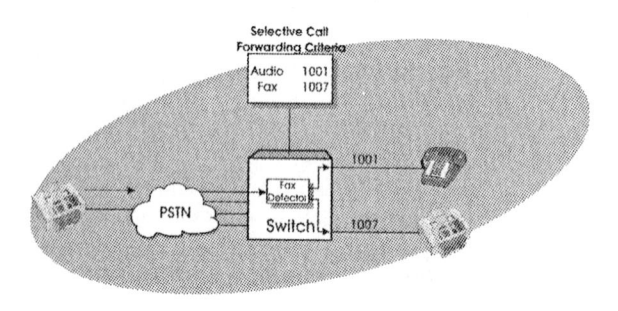

Selective Call Forwarding Operation

Selective Call Rejection-A service feature that restricts the delivery of calls to their dialed destination if they are on a previously specified call rejection telephone number list. Selective call rejection is used to block calls from undesired callers such as prank callers or harassing bill collectors. Calls that are received by numbers on the call rejection list are provided with a pre-recorded announcement that states the number is not accepting their call or the call may be routed to an alternate directory number.

Selective Combiner-A circuit or device that dynamically selects the stronger of two or more signals (usually from different antennas) so that the output signal is selected from the received signal that has the higher signal strength.

Selective Ringing-A technique for ringing the telephone of only one customer on a multiparty line.

Selectivity-The capability of equipment to select and operate upon a signal of a particular frequency despite the presence of other signals at frequencies close to that to which the equipment is tuned.

Self Healing Ring-A transmission system that is designed to cover automatically from cable breaks or equipment failures, using fiber optics or radio facilities in a closed loop.

Semantic Web-The next generation World Wide Web characterized by the ability to search for information based upon metadata content.

Semi-Private Cells (System)-Cells that provide the same basic cellular service to all customers, and also provide special services to a group or groups of private customers.

Semi-Residential Cells (System)-Cells that provide the same basic cellular service to all customers, and also provide special services to a group of groups of residential customers.

Semiconductor-A material whose resistivity is between that of a conductor and an insulator, and can sometimes be changed by light, an electric field, or a magnetic field. Current flow can result from movement of negative electrons or from the transfer of positive holes. Semiconductors are used in a variety of solid-state devices, including transistors, integrated circuits, and light emitting diodes. Silicon is the most commonly used semiconductor material.

Semiconductor Device-A device, such as a transistor, that utilizes the properties of a semiconductor.

Semiconductor Laser-An injection laser in which lasing action occurs at the junction of n-type and p-type semiconductor materials.

Semitransparent-A material that absorbs an amount of photonic energy from when it enters the medium to when it leaves.

Sensitivity-(1 - general) The capability of a circuit or device to respond to a minimal signal input level and still pro duce an acceptable level of reliable operation. (2-optoelectronics) An imprecise synonym for responsivity. (3-radio receiver) The minimum input signal required to produce a specified output waveform having a specified signal-to-noise ratio. (4 - voltmeter) The current required to produce a full-scale deflection.

Sensor-A detection device that is sensitive to changes in level or state.

Sequential Access-Serial

Sequential Couleur Avec MeMoire (SECAM)- SECAM is a color TV system that provides 625 lines per frame and 50 fields per second. This system was developed by France and the former U.S.S.R., and the former Eastern Block Countries, and in some Middle East Countries. In order to transmit color the information is transmitted sequentially on alternate lines as a FM signal.

Serial-An arrangement whereby one element of data is linked to the next so that progress must proceed from the first element through the next, then the next, and so on.

Serial Data Bus-A communication line or system that transfers information sequentially, bit by bit. The serial data bus (such as an RS-232 system) may use additional lines to help coordinate the transmission (e.g. ready, wait, or error).

This diagram shows the process used to transmit data in serial form. In this diagram, parallel data is converted to serial form by a parallel to serial data converter. Prior to sending any bits, a framing circuit inserts a pre-established sequence (frame indicator) to identify the start of a frame of information. This diagram shows that each frame of data contains groups of bits that represent each logical channel. Data arrives at the serial transmitter in groups of bits or bytes where each group represents a portion of a logical channel. The relationship between the frame of data and bits assigned to logical channels is called the mapping. The framing circuit will usually append some bits to the end of the frame that are used for error checking.

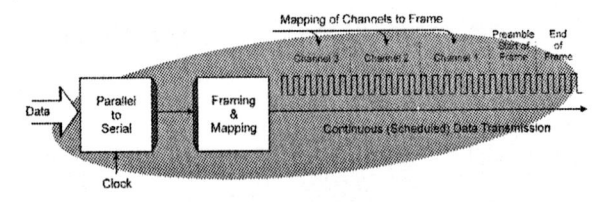

Serial Data Transmission

Serial Line Interface Protocol (SLIP)-A simple protocol that was developed to allow computers to connect to the Internet using a serial connection (such as a dialup connection). The SLIP protocol is being replaced by the point-to-point protocol (PPP).

Serial To Parallel Converter-A device or assembly that accepts information in time sequence and groups the information into groups of digits or elements where each group is transmitted at the same time (in parallel).

Serial Transmission-A sequential rather than simultaneous parallel transmission of data bits.

Server-A computer that can receive, process, and respond to an end user's (client's) request for information or information processing.

Service-(1 - telecommunications) The provision of telecommunications to customers by a common carrier, administration, or private operating agency using voice, data, and/or video technologies. (2- performance) The overall quality of telephone system performance, sometimes stated in terms of blocking or delay.

Service Access Code (SAC)-The 3-digit codes in the NPA (N 0/1 X) format which are used as the first three digits of a 10-digit address in a North American Numbering Plan dialing sequence. Although NPA codes are normally used for the purpose of identifying specific geographical areas, certain of these NPA codes have been allocated to identifying generic services or to provide access capability, and these are known as SACs. The common trait, which is in contrast to an NPA code, is that SACs are non-geographic.

Service Access Point (SAP)-(1-ISDN) In the Integrated Services Digital Network (ISDN), the logical point at which a data link layer provides services to the network layer. (2-LAN network) The part of an IEEE 802 frame that identifies the network frame protocol type.

Service Area-The region covered by a given cellular carrier. It is also a landline term which means a geographical area in which local exchange carriers provide local exchange service to end users as well as network access to interexchange carriers.

Service Bureau-A company that owns its own network or switching facilities and provides turnkey services to other companies.

Service Control Point (SCP)-The signaling control point (SCP) is a computer database that receives information request messages from the SS7 network and returns information that is necessary for the completion of calls or services. The SCP usually receives requests for a service switching point (SSP) via signaling transfer points (STPs) that determine that additional information is necessary to complete the call (such as an 800 toll free/freephone destination number lookup).

Service Control Point 800-A service control point database system that provides call processing information required for 800 number inquiries.

Service Creation Environment (SCE)-A development toolkit that allows the creation of services for advanced intelligent network (AIN) that is used as part of the signaling system 7 (SS7) network.

Service Data Unit (SDU)-The SDU consists of the data originated at the source program and the cumulative headers and trailers appended to the data by the higher layers.

Service Discovery-The ability to discover the capability of connecting devices or hosts.

Service Discovery Application Profile (SDAP)- In a Bluetooth system, SDAP defines the procedures whereby an application in a Bluetooth device can discover services registered in order Bluetooth devices and retrieve information pertinent to the implementation of these services.

Service Discovery Protocol (SDP)-Service discovery protocol (SDP) is the communication messaging protocol used by a communication system (such as the Bluetooth system) to allow devices to discover the services and capabilities of other devices.

Service Information Octet (SIO)-Eight bits contained in an SS7 message signal unit (MSU), comprising the service indicator and the sub-service field.

Service Level Agreement (SLA)-An agreement between a customer and a service provider that defines the services provided by the carrier and the performance requirements of the customer. The SLA usually includes fees and discounts for the services based on the actual performance level received by the customer.

Service Life-The period of time that equipment is or can be expected to be in active use.

Service Location Protocol (SLP)-A protocol that is used to help users discover servers that can provide them with specific services. SLP is defined in RFC 2608.

Service Management System (SMS)-A computer system that administers service between service developers and signal control point databases in the SS7 network. The SMS system supports the development of intelligent database services. The system contains routing instructions and other call processing information.

Service Multiplexing-The multiplexing of different types of services such as voice, data and video onto one type of communications channel. Service multiplexing allows one piece of equipment and/or communication channel to provide services that have different delay, bandwidth and other quality of service (QoS) requirements.

Service Number Portability (SNP)-Service number portability allows a customer to take their telephone number to a different type of service provider. Service number portability involves determination of the type of service provider (e.g., wireless or wired) who is responsible for completing the call using the telephone number (e.g. area code and NXX.) Service number portability may differ from local number portability as the interconnection and call processing for different types of service providers may vary.

Service Order-A record that describes a customer request to establish, change, or terminate a service. The service order contains all information required to meet a customer's needs.

Service Provider-An generic name given to a company or organization that provides telecommunications service to customers (subscribers). See also Network Provider and Reseller.

Service Provisioning-Service provisioning is the process of an authorized agent or process that processes and submits the necessary information to enable the activation of a service. This typically includes: transmission, wiring, and equipment configuration.

Service Quality Churn -Voluntary churn that occurs when customers are dissatisfied with their service quality,
and change carriers to get better quality.

Service Record-A database record in a communication system that contains a description of a service and the attributes that are necessary (e.g. protocols) to use that service.

Service Switching Point (SSP)-In an Intelligent Network (IN), a stored-program controlled switching system that has the functional capability to differentiate intelligent network calls and interact with service control points (SCPs). SCP databases are accessed by the SSP in providing database query oriented services such as the 800 Data Base Service and Alternate Billing Services. (See also: Intelligent Network. SSP is an IN term for the Class 4/5 Switch that have SS7 capabilities. The SSP has an open interface to the IN for switching signaling, control and handoff.

Servicemark-A unique symbol, word, name, picture, or design, or combination thereof, used by firms to identify their own services and distinguish them from the services sold by others.

Servicemarks are Intellectual Property Rights which give the owner the right to prevent others from using a similar mark which could create confusion among consumers as to the source of the services. Servicemark protection must be registered in the country, or region, where it is desired. In most countries and regions, servicemarks are administered by the same government agency which administers trademarks.

Services URL-An address that is preprogrammed into or used by devices (such as an IP Telephone) where information is kept regarding services that may be used by the device. The services URL may be associated with a menu button or an icon on the display of an IP telephone.

Serving Area-(1 - central office) A geographic area served by a central office or exchange. (2 - outside plant) In outside plant, a distribution area that connects with a feeder route through a serving area interface.

Serving General Packet Radio Service Support Node (SGSN)-A switching node in a wireless network that coordinates the operation of packet radios that are operating within its service coverage range. The SGSN operates in a similar process of a MSC and a VLR, except the SGSN performs packet switching instead of circuit switching. The SGSN registers and maintains a list of active packet data radios in its network and coordinates the packet transfer between the mobile radios. The GGSN

Servlet-A servlet is a small program (written in a language such as Java) that runs on a server (such as a web server).

Servo-An electromechanical system for relaying positional or angular information.

Session-The time a process is active for the operation of a software program or logical connection between two communications devices. In communications systems, the session involves the establishment of a logical channel with configuration transmission parameters, operation of higher level applications, and termination of the session when the application is complete. During a session, many processes or message transmissions may occur.

Session Announcement Protocol (SAP)-An Internet protocol that distributes information about multicast sessions. SAP is defined in RFC 2974.

Session Border Controller (SBC)-A session border controller (SBC) is an interface to a network firewall that facilitates the secure hand-off of voice packets from one IP network to another IP network. In an enterprise network, the SBC "controls" the communications "session" as it crosses the "border" from the LAN to IP. Conventional firewalls support the secure traversal of data streams, but for IP networks, SBCs are needed to facilitate secure, real time, multimedia communication. SBCs are a recent, but important component in today's nextgen network infrastructure.

Session Description Protocol (SDP)-A text based protocol that is used throughout to provide high-level definitions of connections and media streams. The SDP protocol is used with session initiated protocol (SIP). The SDP protocol is used in the PacketCable system. SDP is defined in RFC 2327.

Session Description Protocol Next Generation (SDPng)-An enhanced version of SDP that allows high-level definitions of connections and media streams independent of the media control protocols that are used (such as H.248 or MGCP). SDPng protocol includes feature descriptions that are necessary by these different types of media control systems.

Session Initiation Protocol (SIP)-SIP is an application layer protocol that uses text format messages to setup, manage, and terminate multimedia communication sessions. SIP is a simplified version of the ITU H.323 packet multimedia system. SIP is defined in RFC 2543.

Session Initiation Protocol Project Investigation (SIPPING)-An IETF working group that was formed to help the SIP working group to develop proposed standards to draft standards.

Session Invitation Protocol Version 1 (SIPv1)-The initial SIP text based protocol that was developed to provide multimedia services using Internet protocol (IP) networks. SIPv1 was first submitted to the IETF in February 1996.

Session Layer-The session layer protocol coordinates the information transmission between endpoints during a communication session. The session layer receives requests for transmission from an application layer and converts it into network addressable data formats that can be transferred through a network or transmission line. The session layer usually establishes a communication session,

Session Level

coordinates the overall control of the session (such as handling retransmission and restart requests), and termination procedures. The location of the session layer within the protocol stacks varies dependent on the protocol. The session layer is layer 5 in the open system interconnection (OSI) protocol layer model.

Session Level-The fifth level in the ISO layered model. The session level receives the services of protocols located in the transport level and below, it performs services for the presentation and application levels above it.

Session Transfer-The process of transferring a communication session from one device to another device. The process used by a session transfer is the sending of an invite message from one device (the device that desires to be transferred) to a new device that is not part of the original communication session. If the new device accepts the invitation request (it is willing to accept the transfer request), the transfer device informs the device that it is communicating with of the desire to transfer the session to a new device along with the parameters (such as the new device network address) that will be used by the existing device to connect to the new device. After the two devices have successfully established a new communication session, the transfer device is no longer part of the original communication session.

Set Top Box-An electronic device that adapts a communications medium to a format that is accessible by the end user. Set top boxes are commonly located in a customers home to allow the reception of video signals on a television or computer.

This figure shows a block diagram of a video set-top box. This device converts incoming RF channels into a lower frequency that can be provided to a television. This diagram shows an analog setup box usually involves the conversion of the incoming RF channel frequency to channel 3 or channel 4. The conversion of a digital cable channel is more complex as it requires demodulation, decoding (data decompression), and RF modulation back to a form suitable for a television. Optionally, a digital decryption section and/or video descrambling section may be included.

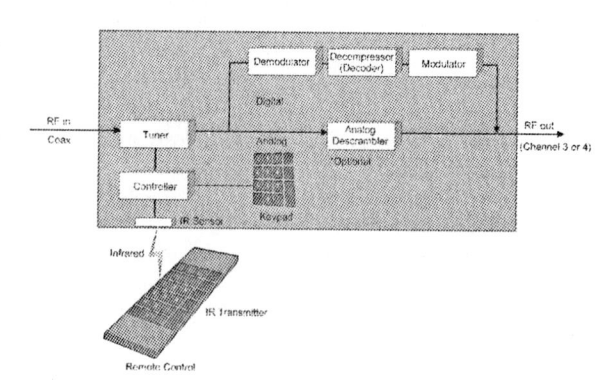

Set Top Box Operation

This picture shows cable set top box that adapts cable television signals into a format suitable for a television set. Scientific-Atlantic's Explorer 8000 digital set-top box has an integrated hard drive so you can automatically record your favorite programming. The Explorer 8000 set-top also delivers access to new interactive TV services, including video on demand, email on TV, Web browsing on TV and pizza on demand.

Set Top Box
Source: Scientific-Atlanta, Inc.

Set-Request-The set-request SNMP operation is virtually identical to the get operation, with the exception that the manager fills in the desired values for the MIB objects and the agent is requested to change the value of these objects. A read-write community string is usually required to issue a set-request operation.

Settlement Code-A code that identifies a billing message type and the preferred settlement procedures. Some sample settlement codes include interstate, intrastate, domestic, and overseas.

Settlements Procedures-A process for distributing revenues among carriers in proportion to services provided or assets used.

SF-Single Frequency Signaling
SFC-Switch Fabric Controller
SFD-Start Of Frame Delimiter
SFID-Service Flow Identifier
SFP-Sequenced Packet Protocol
SFT-System Fault Tolerance
SG-Signaling Gateway
SGCP-Skinny Gateway Control Protocol
SGCP-Simple Gateway Control Protocol
SGML-Standard Generalized Markup Language
SGSN-Serving General Packet Radio Service Support Node
SGSN-Serving GPRS Support Node
SHA-1-Secure Hash Algorithm Number 1

Shadow Market Affinity -Market strength of a company that is unaccounted for in the current financial statements because the good will, awareness, familiarity and preference have been established in the consumers mind during previous marketing cycles.

Shadow Network Capacity-An amount of network capacity that the network service provider has that can be utilized to produce revenue that is unaccounted for in the current financial statements. Shadow network capacity exists because it may not be readily measurable or it has already been depreciated and amortized.

Shared Bandwidth-A characteristic of a communications channel in which the available capacity is shared among all of the attached stations.

Shared Bus-A type of switch fabric that uses a common communications channel (e.g., a backplane) as the mechanism for frame exchange among the switch ports.

Shared Media-A physical communications medium that supports the connection of multiple devices, each of which may transmit and/or receive information across the common communications channel. Shared media systems require some method for the attached stations to arbitrate for the use of the common channel. See also Dedicated media and MAC algorithm.

Shared Secret Data (SSD)-Secret data that is stored in a mobile phone in the IS-136 and IS-95 cellular standards. and in a cellular or PCS network that is used to create key codes that are transferred through the network to validate the identify of the mobile phone. The SSD information is created by an authentication A-key.

Shared Tenant Service-The sharing of centralized equipment, facilities, and telecommunication services by occupants in a building or office complex.

This figure shows the basic operation of a shared wireless data system network. The wireless data device scans radio channels (step 1) for a free one. After it has found and locked onto a free radio channel, it will begin to transmit data (step 2). If the radio channel is interrupted by another activity (such as a voice signal), the wireless data device will re-tune to the next available wireless data radio channel and continue to send data. This technique treats the radio channels that have voice activity (voice sessions) as noise. This is an example of the frequency hopping technique described earlier. The important point here is that voice is treated as the superior application. If voice has the channel, the wireless data system looks elsewhere for an unoccupied radio channel. If the channel is quiet, the data hops on. If data is active and another voice session comes up, the data hops again. This is pos-

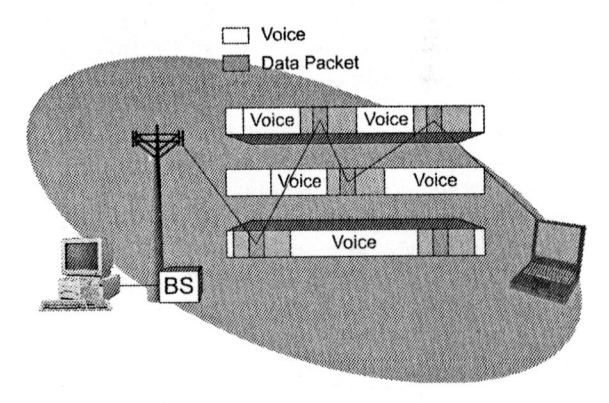

Shared Voice and Data Operation

Shared Wireless Access Protocol (SWAP)

sible since the data is operating in packet mode. The packets can be reassembled at the receiving end no matter which air channel is used.

Shared Wireless Access Protocol (SWAP)-A common industry specification that allows wireless communication between devices in the home. The first SWAP specification was created in 1998.

Sheath-The outer covering of a cable, the main function of which is to provide protection for the insulated conductors that makeup up the cable.

Sheath Miles-The number of route miles installed (excluding pending installations) along a telecommunications cable path multiplied by the number of wires or fibers existing within cabling along the same path.

SHF-Super High Frequency

Shield-A metal covering used to restrict the transfer of electrostatic and electromagnetic signals (radio signals) to or from equipment or electronic circuits. Shields are used to eliminate unwanted signals might otherwise produce interference to the normal operation of systems are circuits.

Shielded Enclosure-A screened or solid metallic housing that isolates an area from external electromagnetic fields or prevents fields from escaping such an enclosure.

Shielded Pair-A pair of wires enclosed in an electrostatic shield to prevent interference.

Shielded Twisted Pair (STP)-Wire that includes twisted pairs and a shield that surrounds the twisted pairs of wires. STP uses a metal shield (foil) that surrounds the twisted pair wires to help protect against unwanted electromagnetic interference.

Shielding-The process of enclosing a wire or circuit with a grounded metallic structure, so that electrical signals outside the structure cannot reach the wire, and so that signals on the wire cannot reach beyond the structure, except to the apparatus to which the wire connects.

Shift Register-A circuit that stores a stream of digital data that is changing at a fixed rate. Data enters the shift register one bit at a time, shifting the previous bit toward the output. When the register is full, the next bit to enter causes a bit to shift out from the circuit.

SHL Station-Cable Television Relay Service Studio To headend Link

Short Circuit-A very low-impedance path between two or more conductors. The voltage output of a circuit terminated in a short circuit is zero or negligible, but the current is at a maximum, potentially capable of causing damage to the circuit. This also is called a short.

Short Message Service (SMS)-A messaging service that typically transfers small amounts of text (several hundred characters). Short messaging services can be broadcast without acknowledgement (e.g. traffic reports) or sent point-to-point (paging or email). Most digital cellular systems have SMS services. Short messaging for mobile telephones may include: numeric pages (dialed in by a caller), messages that are entered by a live operator via keyboard, an automatic message service that sends a predefined message when an event occurs (such as a fire alarm or system equipment failure), network operator announcements to customers, to and from other message capable devices in the system, from the Internet, advertisers or other information providers.

Short Message Service Center (SMSC)-The facility that processes and routes short messages to telecommunications devices (such as a mobile telephone.)

Shortened Burst-A shortened transmit burst used by the mobile when initial transmit occurs in a large diameter cell where timing information has not been established. This is required to overcome propagation delays which may cause burst collisions (overlapping received bursts).

Shortwave (SW)-A frequency within the limits of 3 MHz and 30 MHz (wavelength within the limits of 100m and 10m).

Shot Noise-The noise developed in a vacuum tube or photoconductor resulting from the random number and velocity of emitted charge carriers. The effect is noticeable as errors in a fiber optic system, in the sound output of a radio receiver, or in the picture of a TV set. Shot noise also is called Schottky noise.

SI-MKS System of Units, see: Systém International
SI-Systéme International
SID-System Identity
SID-System Identification
SID-Sudden Ionospheric Disturbance
SID-Service Identifier
SID-Silence Insertion Description
SID-Security Identifier

Side Band-The frequency band on either the upper or lower side of the carrier frequency band where the frequencies produced by the modulation process take place. Various transmission techniques make use of one or both of these side bands when sending data.

Side Lobe-The spurious radiation of an antenna that causes signal leakage in an undesired direction. Excessive side lobes reduce antenna efficiency and can cause interference to other communication systems.

Sidetone-The hybrid coil directional coupler used in an analog telephone set is not ideal, a user hears his or her own voice appearing in his or her own earphone to some extent. This is called sidetone, and most users are accustomed to hearing a little sidetone in a correctly working telephone. Because of this, some artificial sidetone is intentionally generated in some four-wire telephone or radio systems to give the use the usual perception of a small amount of sidetone. Sidetone is very undesirable for modems and certain other systems, and echo cancellers must be used to minimize sidetone in such situations.

SIF-Signaling Information Field

SIG-Special Interest Group

Sign a Message-The process of adding a digital signature to a message. The digital signature is calculated from the contents of the message using a private key and appended or embedded within the message. The signing of a message allows a recipient to check the validity of file or data by decoding the signature to verify the identity of the sender.

Signal-(1 - electrical quantity) An electrical, optical, or other indicator, impulse, or fluctuating electric quantity that conveys information for messages, such as voice, data, or video. (2 - alarm) An acoustic or visual device that attracts attention by lighting up or emitting a sound. (3- data) The information transferred over a communications system by electrical or optical means. (4 - message) A message communicated by electrical, optical or other means. (See also: carrier.)(5- telecommunications network) An electrical, optical, or other indicator, impulse, or fluctuating electrical quantity used for network control, such as call routing or network management, or for the internal operation of network elements, such as timing or control of switching systems.

Signal Computing System Architecture (SCSA)-A standard for the telecom/voice processing/computing industry announced by Dialogic in early Spring 1993. SCSA integrates with other standards used with PC-based switching systems.

Signal Conditioning-The processing of a signal so as to make it compatible with a given device, including pulse shaping, pulse clipping, and other modifications.

Signal Converter-A device in which the input and output signals are formed according to the same code, but not according to the same type of electrical modulation.

Signal Distributor-A circuit in an electronic switching system that transmits control signals from a central point to other circuits.

Signal Egress-The emission of a portion of a transmitted signal from a transmission line. The emission is usually unwanted and may cause interference to neighboring cables or electronic circuits.

Signal Fading-An reduction of signal strength that is typically the result of combined radio signals that subtract from each other. Radio signals subtract from each other when one signal is delayed relative to the other by 1/2 signal cycle. This can be a result of the direct radio signal reception combined with a reflected signal that has to travel a longer path. See also: Multipath

Signal Generator-A test instrument that can be adjusted to provide a test signal at some desired frequency, voltage, modulation, and/or waveform.

Signal Ingress-The absorption of radio signal energy from an external source into a communications circuit or communications link. Signal ingress may occur when electrical signals from sources such as radio or lightning spikes occur.

This diagram shows a source of signal ingress from a nearby radio tower that may occur in a transmission system. This diagram shows that a high power AM radio transmission tower that is located near a telephone line couples some of its energy onto the

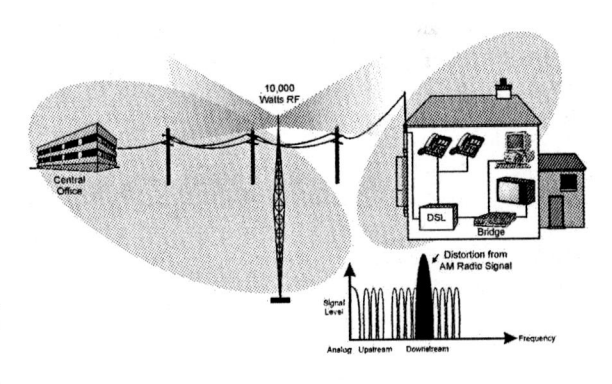

Signal Ingress Operation

telephone line. This interference signal (radio ingress) usually reduces the data transmission capacity of a digital subscriber line (DSL).

Signal Path-Bus

Signal Processing-The process of modifying a signal from one form to another.

Signal Regeneration-The restoration, to the extent that is practical, of a signal to an original predetermined configuration or position in time or space.

Signal Ringing-An undesired oscillatory signal is imposed on another signal as a result of a rapid signal change (such as a pulse) or signal processing element mismatches.

Signal Sampling-The process of taking samples of a particular characteristic of a signal, usually at equal time intervals.

Signal To Interference Ratio (SIR)-The ratio of a desired signal as compared to the total interference that is received with the signal.

Signal Unit-(1- signaling link) Data bits that can be transferred as a group to convey information on a signaling link. (2 - SS7) In the Signaling System 7 protocol, the smallest defined group of bits on the signaling channel used for the transfer of signal information.

Signal Unit (SU)-A group of bits forming a separately transferable entity used to convey information on an SS7 signaling link.

Signal-To-Noise And Distortion (SINAD)-A test measurement that determines the quality of a signal (audio output) when a communications receiver demodulates a low-level RF signal. SINAD compares the audio signal output with a low-level signal input to the audio output level with no signal input (noise only). SINAD is commonly used to test the performance of analog radio receivers. For example, when an analog FM AMPS radio is operating properly with an adequate radio signal strength (RSSI above -116 dBm), the SINAD should be 12 dB or more.

Signal-To-Noise Ratio (SNR)-The ratio of information-carrying signal power to the noise power in a system. In a phone call, the signal would be your voice, while the noise could be static, clicking from relays, or any other sound that is not voice. Analogous effects occur in optical networks. SNR is often used as one measure of signal quality.

Signaling-The process of transferring control information such as address, call supervision, or other connection information between communication equipment and other equipment or systems.

Signaling Connection Control Part (SCCP)- The functional part of a common channel signaling system that provides additional functions to the MTP to cater for both connectionless as well as connection-oriented network service and to achieve an OSI compatible network service.

Signaling Data Link-The data link in an SS7 Network that provides an interface to signaling terminals and is made up of digital transmission channels and digital switches or their terminating equipment.

Signaling Gateway (SG)-A signaling gateway (SG) is used to interface a signaling control system (e.g. such as SS7) and a network device (e.g. a transfer point, database, or other type of signaling system). The signaling gateway may convert message formats, translate addresses, and allow different signaling protocols to interact.

Signaling Information-(1-telecommunications) The information content of a signal that pertains to supervisory control and management of calls. (2- SS7) All information transferred over Signaling System 7 using its protocol.

Signaling Information Field (SIF)-The bits of a message signal unit (MSU) which carry information particular to a certain user transaction and always contain a label.

Signaling Link-A communication path that carries common channel signaling messages between two adjacent signaling nodes.

Signaling Link Management Functions- Functions that control and take actions when required to preserve the integrity of locally connected signaling links.

Signaling Link Selection Code (SLS)-An link identifier that is used within SS7 system.

Signaling Link Selection Field-A message field of the SS7 routing label used by the message routing function to perform load sharing among different signaling links or link sets.

Signaling Message-In the Signaling System 7 protocol, signaling information pertaining to a call, management transaction, or other event that is transferred as an entity.

Signaling Message Discrimination-In the Signaling System 7 protocol, the process that decides if each incoming message, whether the signaling point is a destination point or if it should act as a signal transfer point (STP) for that message.

Signaling Point (SP)-In the Signaling System 7 protocol, a node in a signaling network that originates and receives signaling messages, transfers signaling messages from one signaling link to another, or both.

Signaling Protocol-A protocol that is used to coordinate the operation of devices and services within a system or network.

Signaling Route-In the SS7 protocol, a pre-determined path described by a succession of signaling points toward a specific destination.

Signaling System-A system that receives, processes and sends control information. Signaling may be in-band (replaces voice or data information) or out-of-band (is sent separately from voice or data information).

Signaling System 6 (SS6)-An internationally standardized protocol for medium-speed data communication between intelligent nodes.

Signaling System 7 (SS7)-The signaling system #7 (SS7) is an international standard network signaling protocol that allows common channel (independent) signaling for call-establishment, billing, routing, and information-exchange between nodes in the public switched telephone network (PSTN). SS7 system protocols are optimized for telephone system control connections and they are only directly accessible to telephone network operators.

Common channel signaling (CCS) is a separate signaling system that separates content of telephone calls from the information used to set up the call (signaling information). When call-processing information is separated from the communication channel, it is called "out-of-band" signaling. This signaling method uses one of the channels on a multi-channel network for the control, accounting, and management of traffic on all of the channels of the network.

An SS7 network is composed of service switching points (SSPs), signaling transfer points (STPs), and service control points (SCPs). The SSP gathers the analog signaling information from the local line in the network (end point) and converts the information into an SS7 message. These messages are transferred into the SS7 network to STPs that transfer the packet closer to its destination. When special processing of the message is required (such as rerouting a call to a call forwarding number), the STP routes the message to a SCP. The SCP is a database that can use the incoming message to determine other numbers and features that are associated with this particular call. In the SS7 protocol, an address, such as customer-dialed digits, does not contain explicit information to enable routing in a signaling network. It then will require the signaling connection control part (SCCP) translation function. This is a process in the SS7 system that uses a routing tables to convert an address (usually a telephone number) into the actual destination address (forwarding telephone number) or into the address of a service control point (database) that contains the customer data needed to process a call. Intelligence in the network can be distributed to databases and information processing points throughout the network because the network uses common channel signaling A set of service development tools has been developed to allow companies to offer advanced intelligent network (AIN) services.

This diagram shows the basic structure of the SS7 control signaling system. This diagram shows that a customer's telephone is connected to a local switch end office (EO). The service switching point (SSP) is part of the EO and it converts dialed digits and other signaling indicators (e.g. off-hook answer) to SS7 signaling messages. The SS7 network routes the control packet to its destination using its own signal transfer point (STP) data packet switches using separate interconnection lines. In some cases, when additional services are provided, service control point (SCP) databases are used to process requests for advanced telephone services. This diagram also shows that the connections used for signaling are different than the voice connections. This diagram shows that there are multiple redundant links between switches, switching points, and network databases to help ensure the reliability of the telephone network. The links between points in the SS7 system have different functions and message structures. Access links (A-links) are used for access control between EOs and SCPs. Bridge links (B-links), cross links (C-links), and diagonal links (D-links) interconnect STPs. Extended links (E-links) are optionally used to provide backup connections from an EO to the SS7 network. Fully associated links (F-links) share (associate with) the connection between EOs.

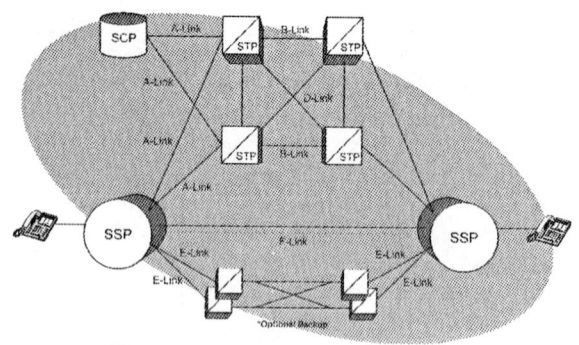

Signaling Tone

Signaling System 7 (SS7) Network

This photo shows the EAGLE 5 Signaling Application System (SAS) that provides for fast, protocol agnostic applications in the signaling system 7 (SS7) network. The EAGLE 5 system provides the capabilities of a signal transfer point (STP), signaling server, service control point (SCP) and application server. This system also can provide for SS7 over Asynchronoous Transfer Mode (ATM); Internet Protocol (IP); Universal Mobile Telephone System (UMTS); and General Packet Radio Services (GPRS). It operates as a high-capacity, fault-tolerant packet switch, delivering reliable global signaling and real-time, transaction-oriented applications. The EAGLE 5 supports all of the requirements of a next-generation signaling server in the emerging UMTS/3G network architectures. It functions as a signaling gateway, enabling the interworking of legacy and packet networks

Signaling Tone-A tone that is used to indicate a status change or to transfer a signaling message on a communication system. The signaling tone is mixed with or replaces an audio signal in a communication system. Signaling tones are used on analog mobile communication system between the mobile station and the base station to indicate event changes such as handoff, the end of a call, or a special service request (e.g. hookflash).

Signaling Transfer Point (STP)-A signaling switch used in the SS7 common channel signaling network. These transfer points are used to route signaling messages (packets) to other signaling transfer points or network parts.

Signaling Transport (SIGTRAN)-A set of standard that were defined by the Internet engineering task force (IETF) that contain a set of protocols that are suitable to provide signaling control messages (such as SS7 message) over in Internet Protocol (IP) network.

This diagram shows that the Sigtran protocol stack is composed of the packet transport layer (IP), common signaling transport layer, and adaptation protocol layer. This protocol stack allows the Sigtran system to transport signaling control messages on a connectionless IP based system. The IP communication channel is managed by the SCTP connection-oriented protocol layer to allow for sequential and secure transport. The adaptation layers convert the protocols (e.g. IP addressing to Point Code addressing) between the Sigtran system to the SS7 system.

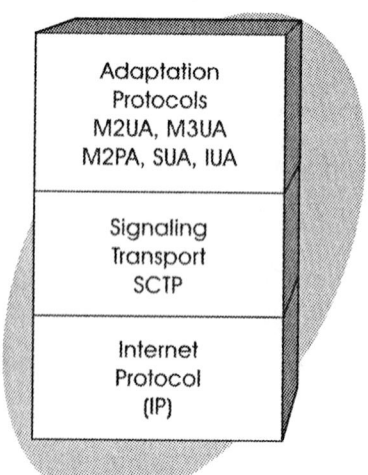

Signaling System 7 Network Node
Source: Tekelec

Sigtran Protocol Stack

Signaling Unit Error Rate Monitor (SUERM)- An error rate monitor that is used in the SS7 system that is used to estimate the error rates associated with a signaling link.

Signed Response (SRES)-(1- GSM) The calculated response value of the GSM authentication process.

SIGTRAN-Signaling Transport

Silence Insertion Description (SID)-A frame of information that describes the silence period of audio. The SID information requires much less bandwidth than voice (typically 1 kbps).

Silence Suppression-A technique that is used in speech compression devices (such as speech codecs like G.729) that sends a reduced (minimal) number of bits during periods when the speaker is silent.

Silent Churn-The process where customers change their usage patterns in preparation for conversion (disconnection) from one carrier (service provider) to another. Silent churn can involve the complete or partial disconnection of service(s) by the customer. Silent churn is usually indicated by a reduction in usage of a particular service.

Silicon-The chemical element used as the basis of most semiconductors. Silicon is a dark gray, hard crystalline solid, the second most abundant element in the Earth's crust. It is widely used in the manufacture of fiber optic waveguides.

Silicon On Sapphire (SOS)-An integrated circuit manufacturing technique in which sapphire substrate is used as the base on which a thin layer of silicon is deposited. Complementary metal oxide semiconductor (CMOS) circuits then are fabricated in this silicon layer. The insulation and low capacitance provided by the sapphire, after the silicon layer has been partially etched away, enable the CMOSSOS chips to operate at a faster speed than conventional CMOS chips.

SIM-Subscriber Identity Module

SIM Card-Subscriber Identity Module (SIM) card.

SIM Cards-Multi-Function Subscriber Identity Module

SIMPLE-SIP for Instant Messaging and Presence Leveraging Extensions

Simple Gateway Control Protocol (SGCP)- Simple Gateway Control Protocol (SGCP) is a protocol used with SGCI for controlling Voice over IP Gateways from external call control elements.

Simple Mail Transfer Protocol (SMTP)-The protocol that provides a simple e-mail service and is responsible for moving e-mail messages from one e-mail server to another. SMTP provides a direct end-to-end mail delivery, rather than a store-and-forward protocol. The e-mail servers run either Post Office Protocol (POP) or Internet Mail Access Protocol (IMAP) to distribute e-mail messages to users. SCTP is defined in RFC 821.

Simple Message Desk Interface (SMDI)- Simplified message desk interface (SMDI) defines a way for a phone system to provide voice-mail systems with the information needed to intelligently process incoming calls. Each time the phone system routes a call, it sends an EIA/TIA-232 message to the voice-mail system that tells it the line it is using, the type of call it is forwarding, and information about the source and destination of the call.

Simple Network Management Protocol (SNMP)-Simple Network Management Protocol (SNMP) is a standard protocol used to communicate management information between the network management stations (NMS) and the agents (ex. routers, switches, network devices) in the network elements. By conforming to this protocol, equipment assemblies that are produced by different manufacturers can be managed by a single program. SNMP protocol is widely used via Internet protocol (IP) and operates over UDP well-known ports of 161 and 162. SNMP was originally defined in RFC1098 and is now obsolete and updated by RFC1157.

Simple Network Management Protocol version 1 (SNMPv1)-The Simple Network Management Protocol (SNMP) version 1 is the "original" standard protocol used to communicate management information between the network management stations (NMS) and the agents (ex. routers, switches, network devices) in the network elements. By conforming to this protocol, equipment assemblies that are produced by different manufacturers can be managed by a single program. SNMPv1 is defined in RFC1157. See also SNMP.

Simple Network Management Protocol version 2 (SNMPv2)-SNMPv2 Builds upon the SNMPv1 (Simple Network Management Protocol) protocol definition as defined in RFC1157. SNMPv2 adds the functionality of security and increased performance. It added a party-based security mechanism, the get-bulk operation, and the inform operation. SNMPv2 also introduced the use of 64-bit counters instead of the 32-bit used in SNMPv1. SNMPv2 is defined in RFCs 1441, 1445-1449, 1901, 1905, 1906, and 1908. SNMPv2 is sometimes

Simple Network Management Protocol version 3 (SNMPv3)

referred to as SNMPv2p, where "p" designates support for the party-based security mechanism. SNMPv2 never really became a standard due to dissention in the standards committee about SNMPv2 security. SNMPv2c came out briefly to the same community-based security (denoted by the "c") used in SNMPv1.

Simple Network Management Protocol version 3 (SNMPv3)-Defined in RFCs 2571 through 2575. SNMP defines a protocol designed to allow a network operator to manage the individual devices on his network. SNMPv3 specifically builds on SNMPv2 or SNMPv2c by adding in a user-based security model or USM. It supports secure authorization of the user sending the packet, as well as the privacy or encryption of the packet.

Simple Object Access Protocol (SOAP)-An XML-based protocol that that can be used for simple one-way messaging and for performing Remote Procedure Call (RPC) request-response communication sessions. SOAP is not part of any particular transmission protocol group however it is commonly implemented on web based (HTTP) systems. Using the SOAP message structure, it is possible for clients and servers to use any language that can create and understand SOAP messages. Conformance to SOAP allows companies to independently develop modules for applications (building block) and applications can effectively interoperate with each other through a data network (such as the Internet).

Simplex-Simplex communication allows the transmission of information between users, but only one direction at a time on the same channel or frequency. The common use of Simplex systems is traditional television or audio broadcast radio systems that transmit a signal from a single transmitter to many receivers.

This diagram shows the communication process between two people using alternating (simplex) transmission and reception. This diagram shows that person 1 turns on (keys-on) the transmitter and begins to transmit audio. When person 1 transmits, their receiver is off or disconnected from the antenna. Person 2 hears the communication from person number one. When person 2 determines that person 1 has finished talking, person 2 turns on (keys-on) their transmitter and begins to talk on the same radio channel frequency. When person 2 transmits, their receiver is off or disconnected from the antenna. This diagram shows that simplex communications does allow two-way conversation. However, only one person can talk at any specific time.

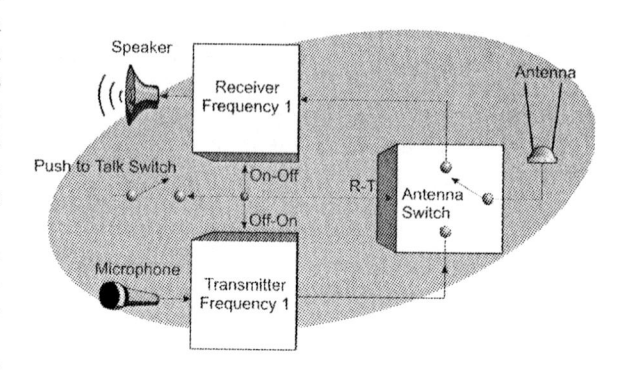

Simplex Operation

Simulated Annealing -The process of integrating "noise," conflict, and discord into the operational business model in order to encourage change and flexibility within the organization.

Simulation-A mathematical model that employs physical and mathematical quantities to portray a real-life situation and movement conforming natural phenomena.

Simulcast Transmission-Simulcast transmission is a process of transmitting a radio signal on the same frequency (or frequency this is very close) from multiple locations to allow the radio coverage area from adjacent radio transmitters to overlap. This overlap of radio coverage helps to ensure the radio signal more evenly covers a geographic area (preventing dead spots). Simulcast is extensively used in radio paging systems. Because the transmission time from each of the transmitted signals may not be the same (one of the transmitter towers may be closer to the mobile radio than the other), the information sent by adjacent transmitters may be synchronized in time so that it arrives at the same approximate amplitude and phase as the other transmitter. Otherwise, this may cause radio signal distortion that could cause transmission errors. This is especially important in radio transmission systems that operate at high data rates (e.g. 9600 bps compared to 1200 bps).

This figure shows a paging system that uses simulcast transmission. In this example, the same paging message is sent to two paging transmitters. As can be seen in this example, the challenge with simulcast paging is as the pager is closer to one tower then the other, the transmit delay time can cause the signals to not directly overlap. Because radio signals travel so quickly, this delay is minor. However, it can result in some dead spots due to signal adding or subtracting. This diagram also shows that the ability to simulcast also depends on the distance and data rate of each of the transmitted signals to the radio (paging) receiver.

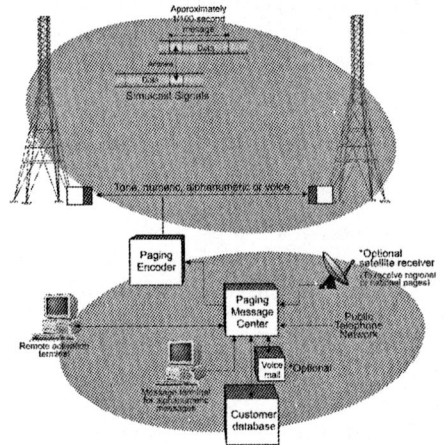

Simulcast Transmission

SINAD-Signal-To-Noise And Distortion

Sine Wave-A signal wave whose characteristics repeat (oscillate) where the signal wave constantly varies as the mathematical sine of a linear function of time. Also called a sinusoidal ware.

This diagram shows that a sine wave continually changes during its cycle. This example shows that a sine wave can be characterized by its cycle (time between one complete cycle), its peak value, and its peak-to-peak value.

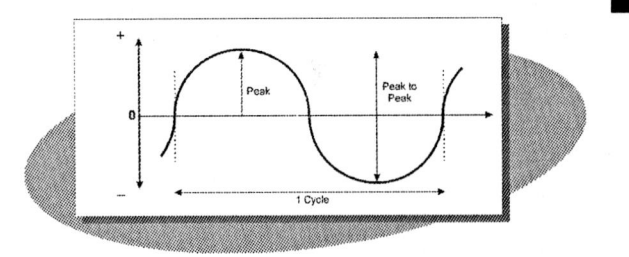

Sine Wave Diagram

Singing-A continuous whistle or howl on a circuit, caused when repeater gain exceeds circuit losses and an imbalance exists at a 2-wire-to-4-wire conversion point.

Single Attached Concentrator (SAC)-A data communication device that is used in a FDDI network that is used to extract and insert information to the fiber ring to a group of data terminals (or other communication devices).

Single Attached Station (SAS)-A data communication device that is used in a FDDI network that is used to extract and insert information to the fiber ring to a single data terminal (or other data communication devices).

Single Frequency Signaling (SF)-A type of inband telephone inter-switch signaling using a single frequency (2600 Hz in North America) to indicate that a particular voice channel is idle or available for use. In some installations, the tone signal was also turned on and off rapidly with the same timing as rotary dial pulsing to convey the dialed number digits. It has been supplanted almost completely because of susceptibility to fraud and because of the better versatility of later signaling systems, first MF and then SS7.

Single Mode Fiber-An optical fiber designed to carry only one mode of a given wavelength of light. This effect is achieved by making the diameter of the fiber core less than about 10 micrometers (0.4 thousandths of an inch). This type of fiber can support higher data rates over longer distances because it avoids modal dispersion. See also Multimode Fiber.

This diagram shows how single mode fiber transmission uses a very small (10 micron) fiber strand to only allow a specific narrow wavelength of light

Single Sideband (SSB)

to pass through the fiber. Because the optical channel is so small, the signals that make it through the fiber primarily travel directly through the center of the fiber. This results in very small group delay and as a result, very small amounts of signal dispersion

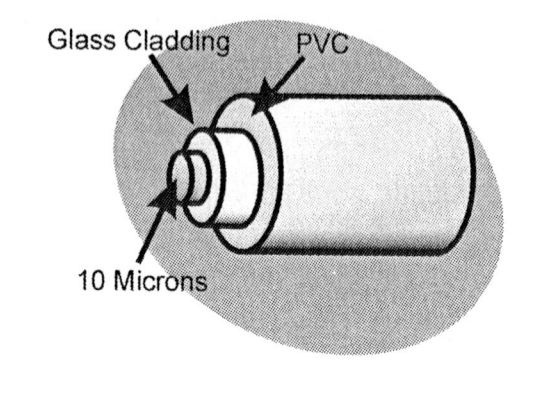

Single Mode Fiber Operation

Single Sideband (SSB)-A specialized method of analog modulation (AM) in which the bandwidth of the modulated radio signal is not wider than the baseband bandwidth of the audio signal being modulated onto the carrier. SSB is used in many analog FDM telephone multiplexing systems (such as N carrier), and in some military and amateur radio systems. To produce SSB, one of the two side bands of spectral power produced via ordinary AM is removed by either a frequency filter or via phasing and cancellation of two different signals derived from the baseband signal. In most SSB transmitters, the carrier frequency itself is not transmitted in order to reduce the total amount of transmitted power, and an equivalent constant amplitude sine wave artificial carrier signal is inserted at the receiver by means of a local oscillator.

SINR-Signal-To-Interference-Plus-Noise-Ratio
SIO-Service Information Octet
SIP-Sequenced Packet Protocol
SIP-Session Initiation Protocol

SIP Extension-A set of commands or protocols that are used to extend the capabilities SIP protocol. SIP extensions are commonly used to rapidly extend the capabilities of an existing application or protocol without changing the underlying application or protocol.

SIP for Instant Messaging and Presence Leveraging Extensions (SIMPLE)-A IETF working group that was formed in 2001 to address the numerous submissions to expand the capabilities of the session initiation protocol (SIP).

SIP Suite-Session Initiated Protocol Suite

SIP Toolkit-A group of software programs that assists designers or developers to create products or services that use session initiated protocol (SIP).

Siphon-The first aerial section of cable leaving a manhole, rising vertically up a pole, and inclusive of the first splice closure. Siphon synonymous with riser.

SIPPING-Session Initiation Protocol Project Investigation

SIPv1-Session Invitation Protocol Version 1

SIR-Signal To Interference Ratio

Sit Tones-Standard Information Tones

SIU-Subscriber Interface Unit

Skew-(1- digital effects) An effect in which the picture is slanted along its horizontal or vertical axis. (2 - videotape machine) A curve at the top of the displayed picture resulting from improper VTR tape tension.

Skills Based Call Routing-The routing of an incoming call via an automatic call distribution (ACD) system to an extension that has a specific skill set (e.g. salesperson) based on information gathered by the caller (e.g. telephone DTMF key presses).

Skin Effect-An effective increase in the resistance of a conductor caused by alternating current flow, which tends to travel near to the surface of a wire, thus reducing the total cross section of the conductor that will actually carry the current flow.

Skinny Client Control Protocol (SCCP)-Skinny protocol is used to establish, control and clear audio calls. Skinny protocol was developed by Cisco to provide for real-time calls and conferencing using the standard suite of Internet Protocols (IP). Skinny protocols permit Cisco IP phones to co-exist in an environment of other VoIP protocols such as H.323. The skinny protocol is more efficient than other VoIP protocols that allows it to operate in less memory size, using less CPU processor power, and it has lower complexity.

Skinny Gateway Control Protocol (SGCP)-A control protocol that was developed by Cisco Systems to provide communication signaling between the IP telephones and Cisco's CallManager product. It runs over TCP utilizing ports 2000-2002. MGCP, H.323, and SIP are similar protocols.

Skip-A phenomenon produced by ionized layers in the atmosphere that results in radio signals being received at greater distances from the transmitting point than would normally be expected. Skip varies depending on the time of day and the day of the year.

Skip Distance-The minimum distance between a transmitting station and the point of return to the earth of a transmitted wave reflected from the ionosphere.

Skip Zone-A ring-shaped region within the transmission range wherein signals from a transmitter are not received. The skip zone is the area between the farthest points reached by the ground wave and the nearest points at which the reflected sky waves come back to earth.

SKU-Stock Keeping Unit

Skywave-The propagation path of a radio wave that travels through the ionosphere level of the atmosphere that reflects or refracts back to the earth.

SLA-Service Level Agreement

Slack-An extra (excess) length of wire or cable between two mounting poles or attachment points.

Slamming-Slamming is the unauthorized transfer of customer's preferred service provider to a different service provider.

Slave-A component or machine in a system that does not art independently, but only under the control of another component (the master) or machine.

SLC-Subscriber Loop Carrier

SLC-Subscriber Line Charge

SLED-Single Large Expensive Disk

Sleep Mode-A feature where a radio receiver or other electronic circuits and a mobile radio product are deactivated or put into a low power consumption mode (such as back lighting off) to save battery energy.

Sleeve-(1-covering) A tubular covering designed to protect a splice, connection, or cable. (2-control lead) A control lead used in some patch cords, switchboards, and switching systems to represent the busy/idle status of the associated channel by means of the presence/absence of a voltage on the sleeve wire. It is connected to the sleeve conductor on a manual switchboard plug, hence its name. Corresponds to the European "C" wire. Only electromechanical switches use a sleeve wire. It is not used in electronic switches.

Slew Rate-The maximum rate of change of the output voltage of an amplifier operated within its linear region.

SLIC-Subscriber Loop Interface Circuit

Sliding Window-A technique used to provide flow and/or error control whereby the sender is allowed to transmit only that information within a specified window of frames or bytes. The window is shifted (slid) upon receipt of proper data acknowledgements from the receiver.

Slip-In a synchronous digital transmission system, the advance or delay of one or more signal elements from their expected time of arrival, resulting when transmitting and receiving terminals are not synchronized. Slip causes the loss or repetition of signal elements.

SLIP-Serial Line Interface Protocol

Slope-(1 - transmission line) The rate of change, with respect to frequency, of transmission line attenuation over a given frequency spectrum. (2- telecommunications) A comparison of transmission losses at various frequencies in the voice band compared to the loss at 1004Hz. If the transmission loss is measured at 404 Hz and 2804 Hz relative to loss at 1004 Hz, it is known as three-tone slope.

Slot-(1- frequency) A narrow band of frequencies. (2 - digital system) The time period for a signal element or sample of a voice channel. (3 - satellite) A term used to express the longitudinal angular position geosynchronous orbit into which a communications satellite is parked. Above the United States, communications satellites typically are positioned in slots that are spaced 3' to 5, apart.

Slow Associated Control Channel (SACCH)-In digital cellular radio, a SACCH is used to continuously transmit certain call processing and control signals at a low bit rate. It does not detract from the digital subscriber traffic like the FACCH does, since it uses pre-reserved bits in the digital transmission separate and distinct from those used for digital subscriber traffic. It is therefore sometimes called "out of band" transmission. In full-rate GSM systems, the SACCH data is transmitted in the same time slot that would otherwise be used for digital subscriber traffic. During a scheduled sequence of

Slow Frequency Hopping

26 transmission frames, 24 of these carry digital subscriber traffic, one carries SACCH data, and one is not used. In IS-136, 12 bits (not in the bit field used for digital subscriber traffic) are reserved in each and every time slot for transmission of SACCH bits.

Slow Frequency Hopping-A process of changing the radio frequencies of a communications on a regular basis (pattern). The duration of transmission on a single frequency is typically much longer than the amount of time it takes to send several bits of digital information.

Slow Scan Video-A method of transmitting pictures over a voice grade system by scanning an image once every several seconds, rather than every few milliseconds as in a conventional TV system.

SLP-Service Location Protocol

SLS-Signaling Link Selection Code

Slug Tuning-The adjustment of a resonant frequency by changing the inductance of a coil by moving a ferrite slug into or out of the coil.

SMA-Standard Microwave Adapter Connector

SMAE-System Management Application Entity

Small Computer Systems Interface (SCSI)-A general-purpose parallel interface for interconnection of one or more computers and peripherals. A maximum of eight devices may be connected to one bus. If a device is a disk controller, it may control up to eight disks. In general, the interface uses a 50 conductor cable and connector and permits a daisy-chain arrangement from one device to the next. SCSI is pronounced scuzzy.

SMAP-System Management Application Process

Smart Antenna-An antenna technology that uses active components to allow the forming or selection of specific antenna patterns. Smart antennas may have multibeam capability that allows for the reuse of the same frequency in the same radio coverage area.

Smart Card-A portable credit card size device that can store and process information. When the card is inserted into a smart card socket, electrical pads on the card connect it to transfer Information between the electronic device and the card. Smart cards are used with devices such as mobile phones or bank card machines. Smart cards can be subscriber identity modules (SIM) for the GSM system, however not all SIM cards are smart cards. See also: Subscriber Identity Module (SIM).

Smart Phone-An end user telephone device that combines computer functionality with telephone services. Smart phones often have enhanced display and keypad options along with advanced telephone call process feature controls. Some smart phones provide access to the Internet and they may reformat information such as telephone directories and personal schedules in a format through which the user can easily navigate and interact with.

This photograph shows an Internet Protocol (IP) screen phone that provides for advanced telephone services and allows access to the Internet. This type of screen phone is used in hotels and executive offices because they provide easy access to information without the need for a separate computer.

IP ScreenPhone (Smart Phone)
Source: Avaya Inc.

Smart Switch-Smart switches are interconnection switching systems that can dynamically change its connection data rates and services that they provide. Smart switches are used to provide for voice, data, video, and advanced call processing (call feature) services.

This diagram shows how a smart switch can efficiently route information more directly to computers connected within its network. A smart switch builds a routing table based on a device's medium access control (MAC) address. As the data is addressed to a destination computer, the smart switch dynamically maintains a list of the computer addresses that are connected to each of its port's. After it has updated its table, packets are routed directly towards the port that the MAC address table has identified for that address.

Smart Switch Operation

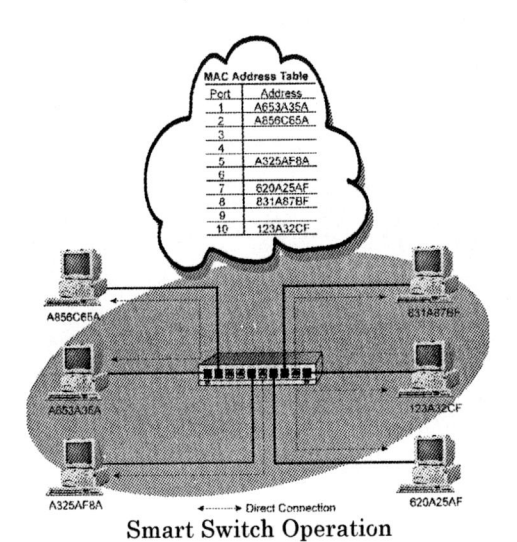

Smart Telephone (SmartPhone)-Smart phones are telephone device that includes intelligent call processing to provide enhanced audio and display capability. Some of the advanced features offered by smart phones are communication list management, enhanced caller information displays, and information storage services.
Smart Terminal-An interface device (terminal) that has both idenpendent (local) computing capability and the ability to communicate with other devices or systems.
SmartPhone-Smart Telephone
SMB-Server Message Block
SMDI-Simple Message Desk Interface
SMDI-Centrex Standard Message Desk Interface
SMDR-Station Message Detail Recording
SMDS-Switched Multimegabit Data Service
SMDS Frame Structure-Switched Multimegabit Data Service
SMDS Switch-A switch that is used in a switched multi-megabit data service (SMDS) system.
SMF-Single Mode Fiber
SMG-Special Mobile Group
SMI-Structure of Management Information
SMM-System Management Mode
SMP-SCSA Message Protocol
SMP-Symmetric Multiprocessing
SMPTE-Society Of Motion Picture And Television Engineers
SMR-Specialized Mobile Radio
SMRS-Specialized Mobile Radio Service
SMS-Storage Management Services
SMS-Service Management System

SMS-Short Message Service
SMS Messaging, Paging, And Access Response Channel (SPACH)-A control channel on a digital cellular system that is divided into a messaging, paging and access response (coordination) channels. The SMS channel (transfers point-to-point messaging (designated to a single recipient), paging messages (calls to be received) and access response channel (radio channel assignment after a service request).
SMS Point-To-Point Messaging-A short message service (SMS) that transfers messages from one point to a specific destination point.
SMS-PP-Short Message Service Point-To-Point
SMSA-Standard Metropolitan Statistical Area
SMSC-Short Message Service Center
SMSCB-Short Message Service Call Broadcast
SMSCH-SMS Channel
SMT-Station Management Task
SMT-Surface Mount Technology
SMTP-Simple Mail Transfer Protocol
SNA-System Network Architecture
SNA-Systems Network Architecture
Snark-A 3 headed beast!
SNDCP-Subnetwork Dependent Convergence Protocol
Sneakernet-The process of manually delivery information (such as letters) through a network of information users by walking between the distribution points.
Snell's Law-A law that calculates how much a lightwave is refracted as it crosses between two materials that have different indices of refraction.
SNET-Southern New England Telecommunications Corporation
SNI-Standard Network Interface
SNI-Subscriber Network Interface
SNMP-Simple Network Management Protocol
SNMP Trap-An SNMP trap is a method for an SNMP agent, like a network device, to asynchronously deliver event notifications to a network management station (NMS) or trap receiver. SNMP traps are defined in MIB definitions. SNMP traps are sent on a best-effort basis and without any method to verify whether they were received by the trap receiver.
SNMP Views-SNMP views is a method for allowing network managers to restrict what MIBs and community-strings can be accessed by what users. SNMP views can be applied, depending on the vendor implementation, to SNMPv1 community

strings, SNMPv2 party, or an SNMPv3 user. In SNMPv3 views is renamed to View-based Access Control Models (VACMs).

SNMPv1-Simple Network Management Protocol version 1

SNMPv2-Simple Network Management Protocol version 2

SNMPv3-Simple Network Management Protocol version 3

Snow-A form of noise picked up by a TV receiver, characterized by alternate dark and white dots randomly appearing on the screen.

SNP-Service Number Portability

SNR-Signal-To-Noise Ratio

SNS-Single Number Service

Snubber-An electronic circuit that is used to suppress high-frequency noise signals.

SOAC-Service Order Analysis And Control System

SOAP-Simple Object Access Protocol

SOC-System Operator Code

Social/Psychological Churn -Type of voluntary churn that occurs when customers change carriers in order to enhance their image or self esteem. Often teenagers will churn for these reasons.

Socket-A socket connection is the combination of a data communication device address and the logical channel (port) number.

Socks-Socks is a protocol uses sockets and a proxy to identify and track communication connections. Socks provides a secure channel between two TCP/IP systems, usually between a web server and an internal corporate intranet.

Soft Capacity Limit-A system subscriber serving capacity limit which allows more users to receive service at a less than optimal quality.

This figure shows that a soft capacity system allows for the gradual decay of voice quality in a communication system when additional users are added in a system. The limit of over capacity in a soft capacity system is determined when voice quality falls below the allowable minimum (usually determined by an acceptable bit error rate). To achieve increased capacity, users in the system are provided lower bit rates to allow more subscribers to be added to the system. Assigning lower bit rates to users as service demand increases trades off voice quality for voice quality (creating a soft capacity limit).

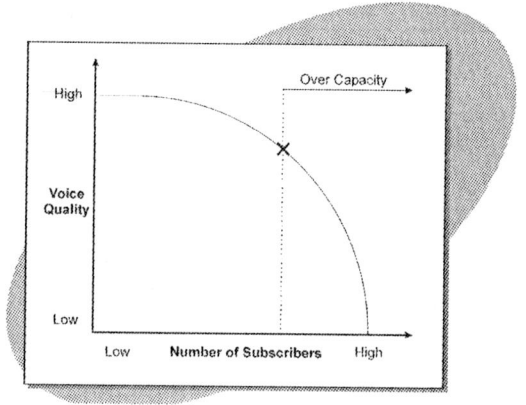

Soft Capacity Operation

Soft Console-An access point to an information system that is defined by software that allows users to monitor and/or interact with a system. The use of a soft console allows the controllable features of a console to change based on the authorization and console program software.

Soft Handoff-This is a handoff where the connection with the initial base station is continued during the connection with one or several new base stations.

Soft Keys-Soft keys are buttons on the keypad of an electronic device that have the ability to redefine their functions. Soft keys are typically located adjacent to a display that provides a description of the key function. This allows an electronic assembly to reduce the number of keys which is especially important for portable handheld telephones. Figure 8.x shows a sample of soft keys.

Soft Knobs, Soft Controls-Soft Panel

Soft Switch (Softswitch)-Softswitches are call control processing devices that can receive call requests for users and assign connections directly between communication devices. Soft switches only setup the connections, they do not actually transfer the call data.

Softswitches were developed to replace existing end office (EO) switches that have limited interconnection capabilities and to transfer the communication path connections from dedicated high-capacity lines to other more efficient packet networks (such as packet data on the Internet). This allows a single softswitch to operate anywhere without the need to be connected to high-capacity trunk connections.

Softer Handoff-A soft handoff process (communication with multiple base stations during handoff) where the connections are made with multiple transmission systems on the same base station.

Softphone-A software application that allows a computer to be used to dial numbers, make phone calls, and perform other functions that a traditional phone would be capable of doing. A softphone combines a software program that operates on a computer along with a signal processing card (a sound card) to provide telephone services. A firmphone relies on its own circuitry for the telephone signal processing so that a it does not burden the computer's operating system when processing the call. More importantly, if the operating system is busy servicing other software applications (such as a word processor), it does not degrade the quality of the telephone call.

Softswitch-Soft Switch

Software-The programs, instructions and procedures that enable a computer to perform designated tasks.

Software-Loop

software defined radio This figure shows a radio receiver that uses software to control all parts of the radio signal demodulation. In this example, the signal from the antenna is amplified and filtered and applied to a high-speed analog to digital converter. This digitized signal represents the entire frequency band that the radio can use. The software installed in the radio analyzes the rapidly changing digital signals and decodes the specific frequency channel and modulation type that is used for this radio. To change channels or modulation types, only the software needs to be changed in the digital radio.

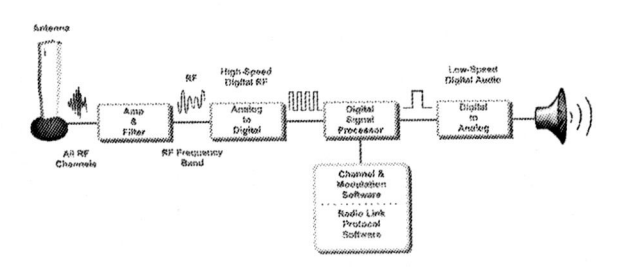

:Software Defined Radio

Software Development Kit (SDK)-A combination of software development tools that allow a company or person to develop software applications. The kit usually includes code editors, compilers, debuggers, utility libraries, along with technical information that instructs the developer on how to use the kit.

Software Extensions-A set of commands or protocols that are used to extend the capabilities of an existing application or protocol. Software extensions are commonly used to rapidly extend the capabilities of an existing application or protocol without changing the underlying application or protocol.

SOH-Start-Of-Heading Character

SOHO-Small Office/Home Office

Solder-(1-general) A material (usually a mixture of tin and lead) that can be melted and flowed across metals to provide an electrical connection between them. (2-rosin core) A solder that includes a rosin flux material to assist in the cleaning and flowing of solder. (3-silver) A solder material that includes a silver alloy to provide a stronger, more electrically conductive bond than solder that is composed of tin and lead.

Soldering-The process of joining metals by fusing them by means of a molten metal with a relatively low melting point.

Soldering Iron-A wedge shaped piece of copper fitted with a long shank and a heat isolated handle, used to melt solder. Usually heated electrically by a resistive element, the soldering iron also can be heated by an external source, such as a torch, especially when used for heavy soldering jobs, such as the repair of lead cable sheaths. When fitted with a pistol grip, the device is called a soldering gun.

Solenoid-An electromagnetic coil fitted with an iron core that moves when the coil is energized. Often, the core is used as an actuator when a short mechanical movement is desired, as in an electric lock.

Solid-A single-wire conductor that has a contiguous core as contrasted with a stranded or braided type of wire.

Solid State Laser-A laser whose active medium is a semiconductor junction.

Soliton-A optical pulse that naturally retains its original shape as it travels along an optical fiber. A soliton pulse has such a high amplitude that it causes changes in the local physical properties of the transmission medium (for example, optical fiber) as the wave passes by. It is known as a "non-linear" wave because it is not described by the usual linear wave equation.

SOM-System Object Model

Sone-The subjective unit of loudness; one sone is the apparent loudness of a 1 kHz tone 40 dB above the threshold of audibility, or 0.0002 microbar.

SONET-Synchronous Optical Network

SOP-Standard Operating Procedure

SOR-Standard Offset Reference

SOS-Silicon On Sapphire

Sound-The periodic or random oscillations in pressure, particle displacement, or particle velocity in a medium, such as air. The oscillations normally occur within an audible frequency range.

Sound Analyzer-An instrument consisting of a microphone, amplifier, and wave analyzer used to measure the amplitude and frequency of the components of a complex sound.

Sound Pressure Level (SPL)-A measure of acoustic wave force. SPL is the force a sound can exert against an object, measured in decibels.

Source Address (SA)-That address of the device the originally transmitted a message.

Source Explicit Forwarding (SEF)-A network security feature that allows transmissions only from specified stations to be forwarded by bridges.

Source File-A tile on disk which as programming language statements for a specific assembler or compiler. From this data, the assembler or compiler creates an object file which is in machine code or language.

Source Identification-A brief message, keyed into video, that defines the originator or point of origin of the signal.

Source Quench-A process of flow control used in a packet data network (such as the Internet) to indicate that a router or switching system is receiving packets faster than they can be processed. This allows the sender of the packets to slow down packet data transmission so the buffer of the receiver does not become full.

SP-Signaling Point

SPA-Software Publishers Association

Space-(1-communication) Space represents a logical value of 0. Space was defined from the open circuit condition in a teletypewriter system that actuates a printer function. Space is the opposite of a mark. (2-universe) The continuous 3-dimensional expanse outside the Earth's atmosphere.

Space Diversity-A method of transmission or reception, or both, employed to minimize the effects of fading by the simultaneous use of two or more antennas spaced a number of wavelengths apart.

Space Link-The transmission link from an ground station through a communications satellite to other ground stations.

Spacecraft-A man-made vehicle designed to go beyond the major part of the earth's atmosphere.

SPACH-SMS Messaging, Paging, And Access Response Channel

SPACH Channel-A SMS messaging, paging, and access response channel.

Spanning Tree-A loop-free topology used to ensure that frames are neither replicated nor resequenced when bridged among stations in a catenet.

SPARC-Scalable Performance Architecture

Spare Wire-A spare pair placed in a cable for use when a regular pair develops a fault.

Spark Gap-A gap between two electrodes designed to produce a spark under given conditions.

Spatial Division Multiple Access (SDMA)-Spatial division multiple access (SDMA) is a system access technology that allows a single transmitter location to provide multiple communication channels by dividing the radio coverage into focused radio beams that reuse the same frequency. To allow multiple access, each mobile radio is assigned to a focused radio beam. These radio beams may dynamically change with the location of the mobile radio. SDMA technology has been successfully used in satellite communications for several years.

Spatial Interpolation-An interpolation process performed across a static video frame. This technique is used to create texturing and filtering effects, such as softening of an image. Spatial interpolation is used extensively in digital video effects systems to ensure clean anti-aliased images.

SPC-Signal Processing Component

SPC-Stored Program Control

SPCAS-SPC Allocation Service

SPDT-Single Pole Double Throw

SPE-Signal Processing Element

Speakerphone-An audio terminal, consisting of a transmitter and loudspeaker unit, that is part of a telephone for hands-free conversations.

SPEC Benchmarks-Systems Performance Evaluation Cooperative benchmarks

Special Interest Group (SIG)-A group that works to help develop and promote information about a specific technology, product, or service. A SIG is usually part of or related to an industry association.

Special Mobile Group (GSM)-Created by CEPT to establish a digital cellular standard to be used in Europe. This system, while theoretically similar to IS-54 is not compatible with the U.S. GSM system.

Specialized Mobile Radio (SMR)-Specialized mobile radio (SMR) systems allow communication between mobile radios and one or more radio towers and typically provides advanced services not possible with traditional LMR systems. SMR systems always use radio channel pairs, whether or not the subscriber units operate half-duplex (requiring push-to-talk) or full duplex. Although talk-around (direct mobile to mobile communications) is possible, it is not the usual mode of operation. Most SMR operators are licensed to provide service in the 800-900 MHz frequency range. (See also ESMR).

This figure shows a typical two-way SMR radio system. In this diagram, a single radio channel is shared by several mobile units that operate within the radio coverage limits of a high power radio base station. To help differentiate between different groups of users, a separate squelch tone signal is mixed in with audio signals for specific groups of users (for example, a taxi company and a flower delivery service). This squelch tone is ordinarily a frequency that is below the audio frequencies so the users cannot hear the tone.

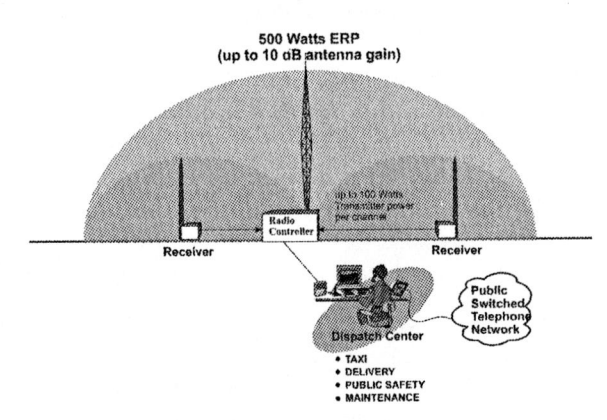

Specialized Mobile Radio (SMR) System

Specification-A document intended primarily for use in procurement, which clearly describes the essential technical requirements for items, materials, or services, including the procedures by which it will be determined that the requirements have been met.

Spectral Efficiency-A measurement characterizing a particular modulation and coding method that describes how much information can be transferred in a given bandwidth. This is often given as bits per second per Hertz. Modulation and coding methods that have high spectral efficiency also typically are very sensitive to small amounts of noise and interference, and often have low geographic spectral efficiency. (See also Geographic Spectral Efficiency)

Spectral Linewidth-The spread of the wavelength values from an optical source away from a single narrow spectral line or wavelength.

Spectrum-(1-general) A range or distribution of physical characteristics in a given system, such as the color spectrum and the electromagnetic spectrum. (2-frequency) A continuous band of frequencies within which waves have some common characteristics. (3 - audio) The range of sound frequencies that can be detected by normal human hearing, ranging from about 20 Hz to 16 kHz - 20 kHz. Most people are not cognizant of hearing above about 17kHz, but harmonics in this area are known to be effective in music. (4-microwave) The range of radio frequencies not strictly defined but usually taken as between I GHz and 100GHz. (5-radio) The electromagnetic frequencies used for radio communication ranging from extremely low frequency (ELF) at about 200 Hz to tremendously high frequency (THF) at 3000GHz.

Spectrum Analyzer-A test instrument that presents a graphic display of signals over a selected frequency bandwidth. Amplitude in the vertical axis and frequency in the horizontal axis.

Spectrum Management-The regulation of spectrum use to achieve maximum efficiency while minimizing destructive interference from other authorized stations.

Spectrum Signature-The pattern of radio signal frequencies, amplitudes, and phases that characterize the output of a particular device, and tend to distinguish it from other devices.

Speech Coder-Speech coding (also called voice coding) is a data compression device that characterizes and compresses digital speech information. A speech coder is also called a Coder/Decoder (CoDec).

Speech Coding

Speech Coding-IS-641

Speech Coding-A process of analyzing and compressing a digitized audio signal, transmitting that compressed digital signal to another point, and decoding the compressed signal to recreate the original (or approximate of the original) signal.

This diagram shows the basic digital speech compression process. The first step is to periodically sample the analog voice signal (5 - 20 msec) into pulse code modulated (PCM) digital form (usually 64 kbps). This digital signal is analyzed and characterized (e.g. volume, pitch) using a speech coder. The speech compression analysis usually removes redundancy in the digital signal (such as silence periods) and attempts to ignore patterns that are not characteristic of the human voice. In this example, this speech compression processes use pre-stored code book tables that allow the speech coder to transmit abbreviated codes that represent larger (probable) digital speech patterns. The result is a digital signal that represents the voice content, not a waveform. The end result is a compressed digital audio signal that is 8-13 kbps instead of the 64 kbps PCM digitized voice.

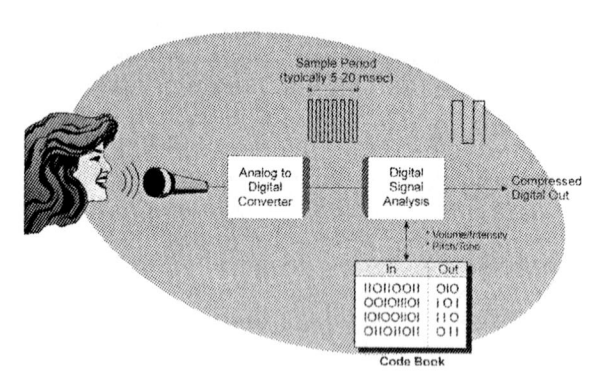

Speech Coding Process

Speech Recognition-Voice recognition is the ability of a machine to recognize your particular voice. This contrasts with speech recognition, which is different. It is the ability of a machine to understand human speech — yours and everyone else's. Voice recognition needs training. Speaker independent recognition does not require training.

Speech Scrambler-A device in which speech signals are converted into an unintelligible form before transmission and are restored to an intelligible form at reception, used for security reasons.

Speed-(1-velocity) Synonym for velocity, in some cases a synonym for the absolute value of a velocity (that is, a positive quantity regardless of whether the velocity is positive or negative). Speed is measured in meters per second (or km/s, miles/s, etc.) (2-data rate) Often loosely used when the term "bit rate" would be more precise. When a higher bit rate is used to transmit a quantity of data, the complete transmission is finished in a briefer time, thus leading to the imprecise description of this result as being "faster" or "higher speed." In fact, the wave speed of the signal through the transmission path occurs at the same speed for both low bit rates and high bit rates.

Speed Dialing Quick Keys-The ability of a telecommunications device to use abbreviated identifiers (typically numbers) to dial lengthy numbers. Speed dialing numbers are often tagged with the name associated with the phone number. For example, to dial 1-919-557-2260, the user may scroll through letter options to find "APDG." APDG may be associated with speed dial number 1.

Spherical Antenna-A type of satellite receiving antenna that permits more than one satellite to be accessed at any given time. A spherical antenna has a broader angle of acceptance than a parabolic antenna. See also Luneburg Lens.

SPI-Service Provider Interface

SPI-Security Parameter Index

SPID-Service Profile Identifier

Spiff-The providing of marketing incentives from one company directly to sales representatives of another company (usually a retail sales company). Spiffs are often used to focus sales representatives on demonstrating and giving preference to products or services from a specific manufacturer.

Spike-A high energy, short duration pulse superimposed on an otherwise regular waveform.

SPL-Sound Pressure Level

Splice-The process of joining together two entities permanently, to provide an electrical or optical path from one wire or waveguide to another.

Split 323-A technology developed by Dialpad, which splits the H.323 protocol into two parts: call control and data streaming. The call setup and control process, which is the heavy work, is handled on the server level. Then the voice streaming, which is the light task, can be done by a small software program on a client computer. The split architecture design makes it easier for individual users to use the full functionality of H.323 with an easier download and setup process.

Splitter-A circuit, device or component that divides a complex input signal into several outputs. A splitter may simply divide the signal energy power from a single source to 2 or more outputs or the splitter may separate frequency components to different output ports.

Splitting Ratio-The ratio of the power emerging from the output ports of a coupler.

SPM-Service Provider Messages

SPNP-Support Of Private Numbering Plan

Spontaneous Emission-A radiation emitted when the internal energy of a quantum mechanical system drops from an excited level to a lower level without regard to the simultaneous presence of similar radiation.

Spooler-A computer program that queues input for later output. For example, a print spooler can accept files at a high transfer rate, then send them to a printer at whatever rate that printer can handle.

Sporadic-An event occurring at random and infrequent intervals.

Spot Beam-A narrow and focused downlink transmission that allows a satellite to use different frequencies, or to reuse the same frequencies in other downlinks.

SPP-Signal Processing Platform

SPP-Serial Port Profile

Spread Spectrum-Spread spectrum transmission is a process of spreading information signals (typically digital signals) so the frequency bandwidth of the radio channel is much larger than the original information bandwidth.

To create a spread spectrum signal, the information signal is multiplied by a spreading code. The multiplication process produces several bits of information for each bit of information signal. This makes the transmitted signal over a wide frequency band. To receive the signal, the spread spectrum receiver must use the same spreading code to recover the original information signal.

This technique changes the frequency components of a narrowband signal so they are spread over a relatively wide frequency band. The resulting signal resembles electrical noise (called white noise).

Spread spectrum transmission is a technique that has been used to achieve very high signal security and privacy, and to enable the use of a common frequency band by many users. This is because the transmitted signal energy is relatively low compared to the natural background noise of received signals. Because the signal is spread over such as wide band, other narrowband signals (such as a high power radio jamming signal) have little effect on the overall transmission of the spread spectrum signal. This makes the spread spectrum signal very hard to find and very hard to interfere with.

SPST-Single Pole Single Throw

Spurious Emissions-Unwanted electromagnetic signal emissions from an equipment or assembly. Spurious emissions may be caused by devices such as microprocessors, oscillators, clock assemblies, electrical switching equipment, and other signal processing equipment. Unwanted spurious emissions may interfere with equipment that operate on frequencies where spurious emissions may occur.

This diagram shows a transmitter and some of the sources of spurious emissions from the equipment. This diagram shows that signal energy may leak from a local oscillator, computer clock, microprocessor, or any other signal processing or transmission device.

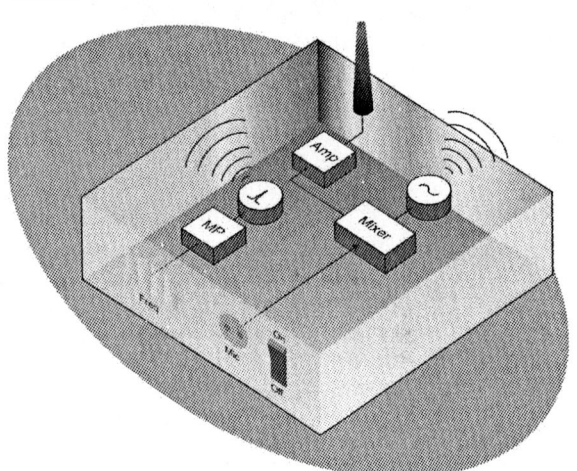

Spurious Emissions Operation

SPX-Sequenced Packet eXchange

SQL-Structured Query Language

SQL Server-Structured Query Language Server

Square Wave-A signal wave whose characteristics repeat (oscillate) of the square wave period. The square wave has a peak maximum and minimum amplitudes remain constant in value for some part

Squelch

of the oscillation period. The graphic representation of such a wave resembles a square or rectangle.

Squelch-Squelch is a process where the audio output of a radio receiver is controlled (enabled) by the reception of an incoming RF signal that is above a predetermined level. Squelch allows a radio user to avoid listening to noise or interference signals of distant radio transmissions that occur on the same frequency.

Squelch Circuit-A circuit that reduces background noise (typically to zero) in the absence of desired input signals.

SRBP-Signaling Radio Burst Protocol

Sreaming Services-Streaming services are audio and video services delivered to customers where the content stays on the server. The delivery of information in streaming form has two main advantages: 1-The customer doesn't have to wait until the entire file downloads to view or listen to the content, 2- The content is kept on the server helps to reduce the copyright fears among content owners.

SRES-Signed Response

SRLP-Signaling Radio Link Protocol

SRNC-Serving Radio Network Controller

SRTP-Secure Real Time Protocol

SrvDscApp-Service Discovery Application

SS-Supplementary Service

SS6-Signaling System 6

SS7-Signaling System 7

SS7 Emergency Changeover-A special changeover procedure used when the normal one cannot be accomplished. For example in case of some failures in the signaling terminal equipment or in case of inaccessibility between the two involved signaling points.

SS7 Message Transfer Part (MTP) level 3-MTP Level 3 provides message routing between signaling points in the SS7 network. MTP Level 3 re-routes traffic away from failed links and signaling points and controls traffic when congestion occurs. MTP Level 3 is equivalent to the OSI Network Layer.

SS7 Protocol Stack-The hardware and software functions of the SS7 protocol are divided into functional abstractions called "levels". These levels map loosely to the Open Systems Interconnect (OSI) 7-layer model defined by the International Standards Organization (ISO). Together these levels are called the SS7 Protocol Stack.

This diagram shows the SS7 protocol stack and how it relates to the OSI 7-Layer Reference Model. The bottom half of the SS7 protocol consists of the Message Transfer Part (MTP). There are three levels to the MTP: Level 1 corresponds to the OSI Layer 1 (Physical Layer): Level 2 corresponds to OSI Layer 2 (Data Link Layer): and, Level 3 corresponds to the bottom of OSI Layer 3 (Network Layer). The upper half of the SS7 protocol consists of several parts. The SS7 Signaling Connection Control Part (SCCP) corresponds to the top of OSI Layer 3. The ISDN-User Part (ISDN-UP) maps onto OSI layer 3 as well, and, in addition, it maps onto Layer 4 (Transport Layer), Layer 5 (Session Layer), Layer 6 (Presentation Layer), and Layer 7 (Application Layer). The Transaction Capabilities Application Part (TCAP), the Application Service Elements (ASE), and the Operations, Maintenance and Administration Part (OMAP) of the SS7 protocol all map onto OSI Layer 7 as well.

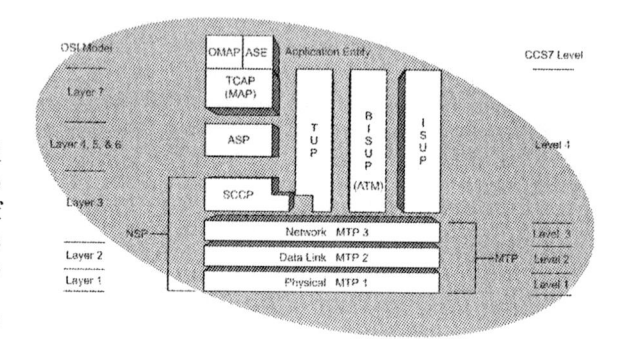

SS7 Protocol Stack

SSA-Serial Storage Architecture
SSAP-Source Service Access Point
SSB-Single Sideband
SSC-Special Services Center
SSD-Shared Secret Data
SSH-Secure Shell
SSI-Server Side Include
SSL-Secure Sockets Layer

SSP-Service Switching Point
SSRC-Synchronization Source
STA-Spanning Tree Algorithm

Stability-(1 - general) The capability of a device or circuit to remain stable in frequency, power level, and/or other specified parameters. (2- operating value) The maintenance of a constant frequency or other value during a specified time interval. Variations from an initial value typically are called drift if the change is relatively slow; they are called jitter or noise if the change is relatively fast (3- oscillation) The freedom of a circuit from unwanted oscillation.

Stack-(1 - interrupt) An area of memory needed to hold information about the status of a computer at the instant an interrupt occurs, so the computer can continue processing after the interrupt has been handled. (2 - memory) A storage area for handling the accessing sequence of nested subroutines. In a UFO (last-in-first-out) stack, the last instruction pushed onto the stack is the first that may be removed or popped from the stack. Some interpreted languages permit limited manipulation of information on the stack. Others use multiple stacks for retention of "dictionaries" of special instruction features.

Stack Pointer-A counter or register in a computer system used to track the storage and retrieval of each byte stored in a system stack.

Stand-Alone Dedicated Control Channel (SDCCH)-A low data rate control channel that is primarily used to setup calls. The low speed channel maintains a call connection until the call is answered. At that time, the call is transferred to a standard traffic channel. The SDCCH channel is used to conserve system capacity.

Standard (STD)-A description of the processes, protocols and specific parameters that are used by a system or particular device. Standards may be developed by industry associations or government agencies.

Standard Microwave Adapter Connector (SMA)-A screw type connector that is specifically designed to have low insertion loss when connecting microwave cables.

Standard Network Interface (SNI)-A device that provides a connection between customer premises equipment and a public switched telephone network. This interface isolates the two systems to the systems from each other (example: protection from lightning strikes).

Standard Operating Procedure (SOP)-A procedure (or set of procedures) that are used (referred to) to ensure the same repeated step by step actions are followed each time an action is required. SOPs are often used to ensure the same quality of service or to ensure the reliable configuration and testing of equipment.

Standards-Documents that describe an agreed-upon way of doing things such that independent groups or companies can design and build hardware, firmware, software, or combinations there of and have them interwork with similar products designed and built by others.

Standards Committee-(1-general) Committees that are commissioned to developed industry standards or operating procedures. Standards committees allow technical experts from several manufacturers (often from competing manufacturers) with a defined objectives and generally have a commitment to unbiased development activities. (2-humor) An excuse to travel to interesting places at nice times of the year to drink beer and discuss arcane technical issues with 400 geeks.

Standby Time-The amount of time a mobile telephone or radio can listen for incoming messages between battery recharges or replacement of its disposable batteries.

Standing Wave-A pattern caused by two waves of the same frequency moving in opposite directions along a transmission-mission line. Where voltages add together, the pattern forms a voltage antinode; where they subtract, it forms a voltage node. The nodes and antinodes are stationary in a standing wave.

Standing Wave Ratio (SWR)-The ratio of the maximum to the minimum value of signal energy that is a result of the adding of a forward transmission signal to a reflected transmission signal.

Star Change-The splicing of a working cable and a working telephone line where the line will not tone out (indicate a path) or identify itself. When this occurs, a new spare facility (line) is requested from engineering or estimate assigning. The old line is then transferred to the new facility (cable pair).

Star Coupler-A passive optical coupler whose purpose is to distribute optical power from one port to all other ports, or to a set of all other ports.

This figure shows a star topology network. Each node in this network is connected directly to a central node. Each station must communicate with or

Star Topology

Star Topology-A network connection scheme in which each station is connected directly to a central node. Some networks, such as ATM are both a logical and physical star topology. 100-Base-T Ethernet however is a star-wired, logical bus network, while Token Ring is a star-wired, logical ring network. (See also: bus topology, ring topology, tree topology.)

This figure shows a star topology network. Each node in this network is connected directly to a central node. Each station must communicate with or through the central node (usually a hub) to reach other stations in the network.

Star Topology (Centralized)

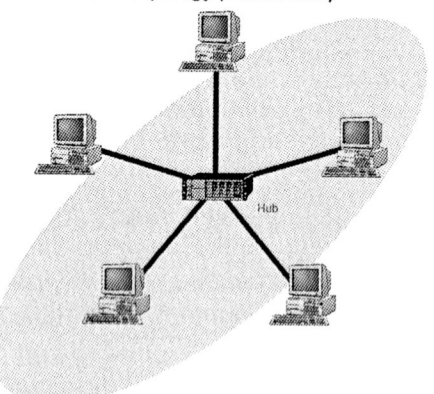

Star Topology System

Start Bit-In asynchronous transmission, the first element in each character that prepares the receiving device to recognize the incoming information elements.

Stateful Proxy-A proxy server that remembers the call state information during a specific communication session transaction (such as a call transfer). All the call state messages associated with a specific communication state must pass through the stateful proxy. The stateful proxy is defined in the session initiation protocol (SIP) standard.

Stateless Proxy-A proxy server that does not remember any call state information during a specific communication session. Each call state messages associated with a specific communication state is completely processed by the stateless proxy and no information is stored about the event or call status. Stateless proxies are used in the center of a communication system to assist in the connection of communication sessions. The stateless proxy is defined in the session initiation protocol (SIP) standard.

Statement-A single unit of program command in a high-level language.

Static Buffering-The adding of supplemental air pressure to aerial or underground cable plant by using air or pressurized gas (nitrogen). Static buffering systems use bottled nitrogen that is strapped to a telephone pole (safely strapped) and is connected via hose to the cable through the hook holes in a manhole cover. This buffering procedure protects splices or other sheath openings from water ingress thereby preventing subsequent damage to conductors.

Static Fatigue-The decrease in wire or fiber strength with time when the cable suffers stress, including exposure to humidity and high temperature.

Static IP Addressing-An process of assigning a fixed Internet Protocol (IP) address to a computer or data network device. Use of a static IP address allows other computers to initiate data transmission (such as a video conference call) to a specific recipient.

Station Address-A grouping of numbers that uniquely identifies a station in a local area network. Data transmitted from the station will bear this grouping of numbers as a source address, and data destined for this station will bear this grouping of numbers as a destination address.

Station Hunting-A telephone call-handling feature that causes a transferred call to "hunt" through a predetermined group of telephones until it finds an idle station.

Station Management Task (SMT)-The portion of a data communication device that is part of a data network that coordinates the access to and from the data communication transmission lines.

Station Message Detail Recording (SMDR)-The telephone switch generated record of all calls and those that are received by the PBX/ACD system. For call accounting, a feature of a centrex, private branch exchange (PBX), or electronic-key system that generates a chronological listing of every call leaving the system.

Statistical Model-Tool for testing different hypotheses through the use of statistical evaluation techniques. For example, a marketing manager would like to know how many customers will leave if the price is raised by 20%. A statistical model could be developed to predict the most likely customer reaction.

Statistical Multiplexer-A time-division multiplexing switch that dynamically assigns channel bandwidth as each connected device needs it to transmit data. Typically, a statistical multiplexer enables dozens of terminals to share a leased line and a pair of modems, thus saving line, modem, and dial-up costs.

Statistics Message-A formatted message containing call data collected and compiled to reflect cellular telephone system usage, billing information and subscriber activity.

Status-The present operational condition of a device or system.

Status Code-A code that indicates the status of a message or communiation session. An example of a status code is a message that indicates that a communication session is active.

STD-Standard

STE-Section Terminating Equipment

Steady State-A condition in which circuit values remain essentially constant, occurring after all initial transients or fluctuating conditions have passed.

Steerable Antenna-An antenna that has a steerable transmission pattern. Steerable antennas may use mechanical or electrical systems to adjust the beam pattern.

Step By Step Switching System (SXS)-A switching system that automatically selects the correct line or path for a telephone number through a switch through the use of step-by-step switches that are progressively assigned to each digit of the dialed telephone number. The SXS system operated using the dialed pulses created when a customer dialed each number using a rotary dial. The output of each switch is cascaded to the next switch in an SXS system.

Step Index Fiber-An optical fiber with an abrupt transition between the index of refraction of the core and that of the cladding. Both single mode and multimode fibers can be made with a step index profile. The profile is designed for total internal reflection within the fiber.

Stereo-The generation and reproduction of a 2-channel sound source.

Stimulated Emission-Emission of multiple photons from a material or device due to the presence of another photon of the same energy. In the type of emission observed in a light bulb and most other light sources, the photon created by the emission process either leaves the device or may be absorbed, possibly leading to the creation of another photon. In the case of stimulated emission, the device is 'primed' with high energy electrons that can interact with a photon so that each can release a similar photon. Each new photon as well as the original one can stimulate the release of additional photons. The result is very bright light in which the photons all have the same energy, which means they also have the same wavelength, or color.

STL-Studio-Transmitter Link

Stock Keeping Unit (SKU)-The SKU of a particular item is the package size used to store or dispense it. Each different package size has a distinct identification number, often represented by a bar coded number on a label. For example, one SKU for electrical resistors is a package of 10 resistors. Another SKU is a roll comprising 1000 resistors held by disposable tapes so that the roll can be used in an automatic component inserting assembly machine.

Stop Bit-In asynchronous transmission, the last transmitted element in each character, which permits the receiver to come to an idle condition before accepting another character.

Stop Signal-In start-stop transmission, one or more bits that terminate each character transmitted.

Storage-(1 - nonvolatile) A storage medium in which information can be retained despite the absence of power, and which becomes available again as soon as power is restored ~ stored. (2- parallel) A storage medium in which all bits, characters, or words are equally available, access time not being dependent upon the order in which they were stored. (3 - serial) A storage medium in which access time varies according to the order in which information was stored. Storage can be serial by word, by character, or by bit. (4-volatile) A storage medium in which information is lost whenever power is removed from the store.

Storage Area Network (SAN)-A network that is used primarily in a data center or server room. The "storage area" is defined by a cluster of servers and storage devices that share common resources and users. Occasionally, the term may be used for a wide area or metro area network that is used for data center redundancy and disaster protection. SANs typically have very low latency, high throughput and offer assured delivery of block I/O. The most common SAN implementation is Fibre

Storage Capacity

Channel, however this is not the only alternative. Recently there have been efforts, such as iSCSI to implement storage networks with Ethernet and IP (Internet Protocol) infrastructure.

Storage Capacity-The quantity of information that can be retained in a memory system, usually measured in MegaBytes, Gigabytes, or TeraBytes.

Stored Program Control (SPC)-A stored program control system uses a computer program (software instructions) to control the operation of a communication system. Prior to the use of SPC, mechanical switches controlled the interconnection of communications networks.

STP-Signaling Transfer Point

STP-Shielded Twisted Pair

Straight Splice-A engineered splice method where the entire incoming cable count is spliced through. For example cable 10 pairs 1-600 is "pushed-through" or spliced straight for further distribution or feeder activity.

Stratosphere-The atmospheric layer above the troposphere, the outer limit of which is approximately 90 km above the surface of the earth.

Stratum Clock-One of a hierarchy of digital system clocks that provide long-term accuracy, short-term stability and pull-in range to synchronized digital networks.

Stream Control Transport Protocol (SCTP)-A protocol that is used to coordinate the sending of signaling information over real time communication sessions. SCTP is defined in RFC 2960.

Streaming-A method that provides a continuous stream of information that is commonly used for the delivery of audio and video content with minimal delay (e.g. real-time). Streaming signals are usually compressed and error protected to allow the receiver to buffer, decompress, and time sequence information before it is displayed in its original format.

Streaming Protocol-The network protocol used for delivering real-time streaming media. The true streaming protocols currently used by the major software vendors for their streaming systems are proprietary, requiring exclusive use of their software components at both server and client end. HTTP pseudostreaming, while not proprietary, is not a real-time streaming protocol strictly speaking, because it is based on TCP which is not real-time.

Strengths, Weakness, Opportunities, and Threats (SWOT)-The analysis of strengths, weaknesses, opportunities and threats that compares the strategic position of a company in a given industry or situation.

Structure of Management Information (SMI)-SMI is used for defining how SNMP objects and the MIBs in which they are defined are structured. SMI was first introduced in SNMPv1 or in RFC1065. SMI defines that the structure of all objects is a tree, starting at or rooted at ISO. Object identifiers (OIDs) are designated as this SMI structure.

Structured Query Language (SQL)-A widely used computer programming language for manipulating database information.

STS-Synchronous Transport Electrical Signal

STS-1-Synchronous Transport Signal Level 1

STSNc-Concatenated Synchronous Transport Signal Level N

STX-Start-Of-Text Character

Stylus-(1 - audio) The part of a phonograph pickup that transfers the sound vibrations, as pressed into the sides of the record groove, to the phono cartridge or transducer. In the transducer, the vibrations are converted to electrical signals. (2 - computer) A small pointer used like a pencil for inputting information (drawing) on a computer graphics system.

SU-Signal Unit

SUA-SCCP User Adaptation

Sub Band Signaling-A signaling method that uses a frequency band that is located within the communication channel but outside the normal communication channel (e.g. audio band) bandwidth to transfer signaling control messages.

A unique signaling feature used by some radio systems is the sub-band digital audio signaling. In many radio broadcast and mobile communication systems, an audio bandpass filter blocks the audio channel's lower range. It is possible to combine a sub-band digital signaling channel (a low speed digital signal) using the lower frequency range (below 300 Hz). This figure shows how sub-band digital and audio signals are combined with standard audio.

Sub-band Signaling

Sub Rate Multiplexing

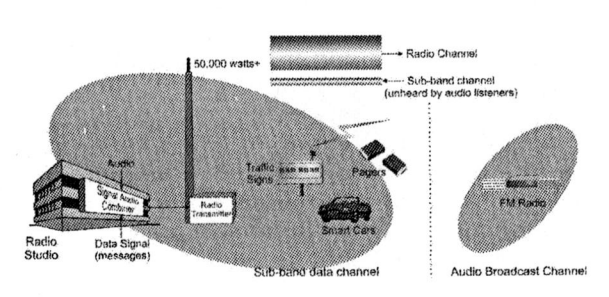

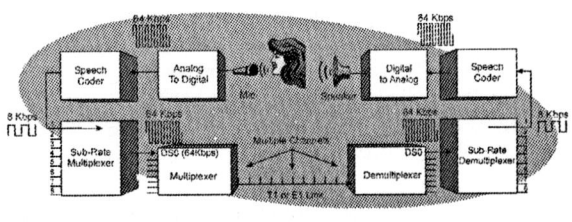

Sub Rate Multiplexing-A process that divides a single transmission path to sub channels that operate at lower (sub rate) data transfer rates.

This figure shows how eight compressed voice signals can be sub rate multiplexed to share a single DS0 (64 kbps) channel. In this system, an analog voice signal is converted to a PCM digital signal at 64 kbps. A speech coder analyses and compresses the voice signal to 8 kbps. This compressed digitized voice channel is time shared (time multiplexed) on a single 64 kbps DS0 channel with eight other 8 kbps compressed voice channels. Each of the 64 kbps DS0 channel is then time multiplexed onto a leased line (24 for T1 and 30 for E1 lines). The process is reversed on the receiving end where the T1 or E1 channels are split back into DS0 channels. Then each DS0 channel is further de-multiplexed again to produce eight 8 kbps compressed voice signals. This digital signals are then converted back into their the 64 kbps signal by a speech decoder and a digital to analog converter then changes the digital signal back into its original analog form.

Subassembly-A functional unit of a system.
Subcarrier (SC)-(1 - general) A carrier applied as modulation on another carrier, or on an intermediate sub-carrier. (2-color video) In NTSC or PAL video, a continuous sine wave that constitutes a portion of the color video signal. The sub-carrier is phase modulated to carry picture hue information and amplitude modulated to carry color saturation information. The NTSC subcarrier frequency is 3.579545 MHz, and the PAL-I frequency is 4A3361875 MHz. A sample of the sub-carrier, called color burst, is included in the video signal during horizontal blanking. Color burst serves as a phase reference against which the modulated sub-carrier is compared to decode the color information.
Submarine Cable-A cable designed to be laid underwater.
Subnet-In an Internet Protocol (IP) network, a group of devices which share a common IP address prefix, in other words, they have the same high-order bit values. The subnet mask is used to identify which bits are used for the network portion of the IP address and which bits are used for the host address portion. See also Classless Inter-Domain Routing (CIDR).
Subnet Address-The portion (sub portion) of an IP address that identifies the portion of a host host which the IP address belongs.
Subnet Mask— The assignment of a number that is used to separate sections of a network by using a specific sequence of numbers in a network address. When used with the internet, a subnet mask is 32 bits.

Subnetting

Subnetting-A process of configuring a network to route specific groups of network addresses to specific local domains (subnets) within the network. To create subnets, a binary subnet mask is used to mask unwanted addresses from entering into the subnet.

Subroutine-A program that carries out a particular function and can be called from another program. A subroutine needs to be placed only once in memory; it can be called by the main program on multiple occasions.

Subscribe-The process of indicating to a communication server or other network service provider that the user is requesting services to be provided in the future and where those services can be delivered.

Subscribe Message-A message that is used to request services from a communication server. The process indicates to a communication server that the user is requesting services and the specific events that may occur to cause the service to be provided. The subscribe message is defined in session initiation protocol (SIP) toolkit.

A SIP message that is sent by an entity to subscribe to resource or call state information. The Subscribe message can be used for example to request presence information relating to another user, changes in the state of that user would be indicated by the Notify message.

Subscriber-The end user of telecommunications services.

Subscriber Database-An informational and relational database which includes subscriber information, including usage patterns, billing records, personal information and other related data. This base is often "mined" for information which helps identify potential churn candidates as well as useful marketing information.

Subscriber Fraud-Fraud perpetrated by the end-user in which false user ID information was used to obtain service.

Subscriber Identity Module (SIM)-The subscriber identity module (SIM) is a small "information" card that contains service subscription identity and personal information. This information includes a phone number, billing identification information and a small amount of user specific data (such as feature preferences and short messages). This information is stored in the card rather than programming this information into the phone itself. This intelligent card, either credit card-sized (ISO format), or the size of a postage-stamp (Plug-In format), can be inserted into any SIM ready wireless telephone.

This figure shows a block diagram of a SIM. This diagram shows that SIM cards have 8 electrical contacts. This allows for power to be applied to the electronic circuits inside the card and for data to be sent to and from the card. The card contains a microprocessor that is used to store and retrieve data. Identification information is stored in the cards protected memory that is not accessible by the customer. Additional memory is included to allow features or other information such as short messages to be stored on the card.

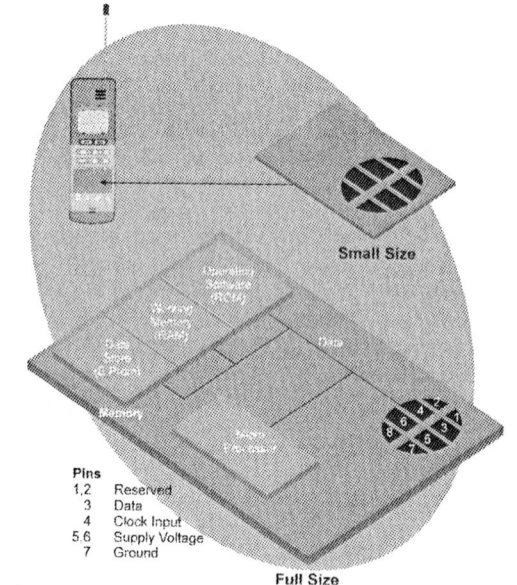

Subscriber Identity Module (SIM) Block Diagram

Subscriber Interface Unit (SIU)-An electronic assembly that is used to convert a digital signal (such as from a network interface unit) to a suitable format for a customers user (such as a television signal).

Subscriber Line-A telephone line that connects from a switching office to a customer's wired phone.

Subscriber Line Charge-A monthly flat-rate charge that recovers a portion of local loop costs paid by an end user. The charge is the result of a Federal Communications Commission (FCC) effort to eliminate unreasonable discrimination and undue preferences among rates for interstate service, to promote efficient use of a local network, to prevent uneconomic bypass, and preserve universal service.

Subscriber Loop Carrier (SLC)-Subscriber loop carrier works in conjunction with digital carrier (e.g. T-carrier or E-carrier) systems to increase the circuit capacity of a distribution plant without adding additional lines. SLC involves adding equipment to end offices (EOs) and remote plant locations that are connected by digital cable pairs, fiber optic cable, or digital radio media. The equipment at the end office (EO) and remote terminals allows more efficient use of cable pairs. The systems that use PAM and PCM technology can generally serve up to 96 subscribers from 10 cable pairs.

Subscriber Loop Interface Circuit (SLIC)-Electronic version of the two to four-wire hybrid interface that supplies an analog signal from a line card to a subscriber's phone or network terminal equipment. It provides what is known as the BORSCHT functions in telephony (Battery Feed, Overvoltage Protection, Ringing, Signaling, Coding, Hybrid, and Test).

Subscriber Number-The phone number that enables a user to reach a wireless subscriber within the same local network or numbering area. This term is a synonym for directory number.

Subscriber Tables-A collection of tables containing standard as well as optional subscriber features and other information, including dialog plans and other data required to complete or deny service to or from a mobile phone.

SUERM-Signaling Unit Error Rate Monitor

Super High Frequency (SHF)-The frequencies ranging from 3GHz to 30GHz; wavelengths are 10cm to 1 cm.

Super Node-A super node is used in a radio carrier or wireless distribution system that is used to provide a connection and gathering point for several smaller sites that are backhauled for interfacing the main network provider. Super nodes may be mounted on a main rooftop or tower-mounted collector site.

Superconductivity-The reduction to zero resistance that occurs in some metals when they are cooled to temperatures close to absolute zero (-273 Degrees).

Superframe-For the IS-136 TDMA system, a superframe is a DCCH burst structure made up of sixteen 40ms TDMA frames equivalent to 32 consecutive TDMA blocks at full-rate, creating a sequence of 32 DCCH carrying bursts spread through 96 TDMA bursts. Each DCCH burst in the superframe is designated for either broadcast, paging, SMS messaging, or access response information. The superframe structure is continuously repeated on the DCCH channel. In the GSM cellular system, a superframe is a sequence of approximately 6.21 seconds duration comprising 51 sequences each called a mutiframe of the 26-frame type (or the GSM superframe may equally well be described as 26 multframe sequences of the 51-frame type). In the North American T-1 digital multiplexing system, the historical superframe has a duration of 12 ordinary frames (with at total duration of 1.5 milliseconds) and the extended superframe has twice this duration (24 frames or 3 milliseconds). In the European E-1 primary rate 2.048 Mb/s digital multiplexing system, a superframe of 16 frames duration (2 milliseconds total) is used only when channel associated bit signaling is in use (not with SS7 common channel signaling). The purpose of superframe synchronization is to allow transmission of certain predefined information types (usually call processing data) during certain pre-defined parts of the superframe time interval without further identification.

Supergroup-A bandwidth allocation in frequency division multiplexed systems that provides for 60 voice-bandwidth channels between 312 kHz and 552 kHz.

Superheterodyne Receiver-A radio receiver in which all signals are converted to a common frequency for which the intermediate stages of the receiver have been optimized for tuning and filtering. Signals are converted by mixing them with the output of a local oscillator whose output is varied in accordance with the frequency of the received signals to maintain the desired intermediate frequency.

Superluminescent Diode-A solid-state light-producing device with a more narrow angle of emission than that of a light-emitting diode (LED), but a wider emission angle than that of a laser.

Superradiance-The amplification of spontaneously emitted radiation in a gain medium, characterized by moderate line narrowing and moderate directionality. This process generally is distinguished from lasing action by the absence of positive feedback, hence the absence of well defined modes of oscillation.

Supervision-(1-control signals) The act of monitoring a line or trunk for on-hook or off-hook supervisory signals. (2-call status) Signaling that indicates the status of a call or the readiness of an item of equipment to respond to an attempt or release a connection.

Supervision vs. Signaling-Historically, the word "supervision" denoted only the information regarding the busy vs. idle status of a telephone channel, but not the dialed digits or other information. The term "signaling" historically refereed to the more inclusive set of information used in call processing, such as dialed digits as well as busy vs. idle status. In recent years, the two terms have both been used for both meanings in many published documents, and the reader must infer with care exactly what the writer means in confusing cases.

Supervisory Audio Tone (SAT)-A tone that is combined with an information signal on a communication channel to provide an indication of connectivity and signal changes in the condition of the channel. SAT tones are used on analog cellular systems. For the AMPS system, one of several tones (at approximately 6 kHz) are combined with a voice signal during radio transmission to and from a mobile radio and a base station. The audio frequency of SAT tones are within the range normally audible to the human ear, but low-pass filters are used in the cell phone to attenuate audio power above about 3.5 kHz, so SAT tones are not heard by the user. To the extent possible, different SAT tones are used in each of those cells that have the same frequency plan assignments, to attempt to recognize and distinguish false reception of a radio signal from another cell having the same set of assigned carrier frequencies. Unfortunately with only 3 distinct SAT tone frequencies, this is not always feasible. In digital cellular systems, there are an adequate number of distinct DVCC code values in IS-136 systems (or distinct synchronization and training bit sequences for GSM systems) used to distinguish radio signals that use the same carrier.

Supervisory Control And Data Acquisition (SCADA)-The process of monitoring or controlling devices.

Supervisory Signals-Signals used to indicate or control the status or operating states of circuits that establish a connection. A supervisory signal indicates that a particular state in a call has been reached and can signify the need for additional action.

Supplementary Service (SS)-Supplementary services provide a network user with capabilities beyond those of elementary call control. Supplementary services enrich the basic services functions and are not specific to a telephone or system features. Often, the subscriber (user) can specify some of the operations of supplementary services (such as call forwarding). Supplementary services may be defined or installed in systems before complete testing or industry consensus can be reached.

Supply Analytics -Collection of data mining models utilized to help evaluate the impact that spending will
have on the customer utility generated by customer relationship management and service delivery operations.

Suppressor Grid-The fifth grid of a pentode electron tube, which provides screening between plate and screen grid.

Surcharge-An additional charge for a service that is in addition to the basic charge. Examples of surcharges include additional charges for using pay telephone or toll free access lines.

Surface Acoustic Wave (SAW)-A mechanical wave that travels on the surface of a piezoelectric material. The wave is created by a transducer that couples electrical to acoustic energy (mechanical vibrations) to the piezoelectric. A second transducer (pattern of metal fingers) converts the acoustic energy back into an electrical form. Depending on the shape of the transducers and the properties of the material through which the SAW travels, the signal characteristics can be changed (such as the passing or rejecting of particular frequency bands). Other designs are used to delay signals to facilitate processing.

This figure shows a transmission system that uses surface acoustic wave (SAW) technology. This device is a radio frequency filter that uses SAW technology. The radio signal is applied to an IDT. The IDT converts the electrical signal to an acoustic (mechanical) wave that moves across the surface of the SAW filter substrate. When the acoustic wave reaches the destination IDT, it is converted back into an electrical signal.

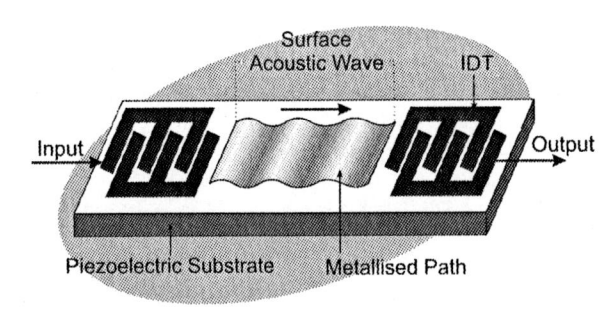

Surface Acoustic Wave (SAW) Filter Operation

Surface Acoustic Wave (SAW) Device-A device exploiting Surface Acoustic Wave phenomena via interdigitated transducers on a piezoelectric substrate. SAW devices are used for bandpass and band reject filter applications, primarily in wireless infrastructure and handsets. They are also used to make delay lines for radar and synchronization applications. Both filters and delay lines have been used in voltage-controlled oscillators to provide high performance frequency sources for optical network applications.

Surface Emitting Diode-An LED that emits light from its flat surface rather than from its side.

Surface Emitting Laser-A semiconductor laser that emits light from the wafer surface. A laser that directs light upward off its surface. In contrast, an edge-emitting laser directs light horizontally. Many surface-emitting lasers can be densely packed onto a singe chip for such applications as fiber optic communications.

Surface Leakage-A leakage current from line to ground over the face of an insulator supporting an open wire route.

Surface Mount-A method of mounting sub-miniature integrated circuits and other components directly on the surface of a printed circuit board. Surface mount construction permits greater component density on boards, making the electronic equipment smaller.

Surface Mount Technology (SMT)-A electronics manufacturing process that uses components that have flat leads that are mounted direct on top of the printed circuit (PC) board. This eliminates the requirement of drilling holes in the PC board, allows the use of smaller components (allowing more parts on the PC board), simplifies the assembly process, and allow components to be mounted on both sides of the PC board.

Surface Mounted Assembly-A circuit board or assembly that uses surface mounted technology (SMT) components that are mounted onto the surface of a printed circuit board. SMT components can be more easily handled by assembly machines, are generally smaller in size, and do not require holes to be drilled through the printed circuit board.

Surface Wave-A transmitted signal wave that is guided along the surface between two points.

Surge-A rapid rise in current or voltage, usually followed by a fall back to a normal value.

This figure shows a surge protector that provides protection for power overage. This surge protector includes an induction coil that restricts rapid current changes and a fuse that will open in the event of a very large and rapid voltage change. Voltage transients can be caused by power surges and lightning pulses. Surge protectors may not protect against direct lightning strikes or extreme voltage changes.

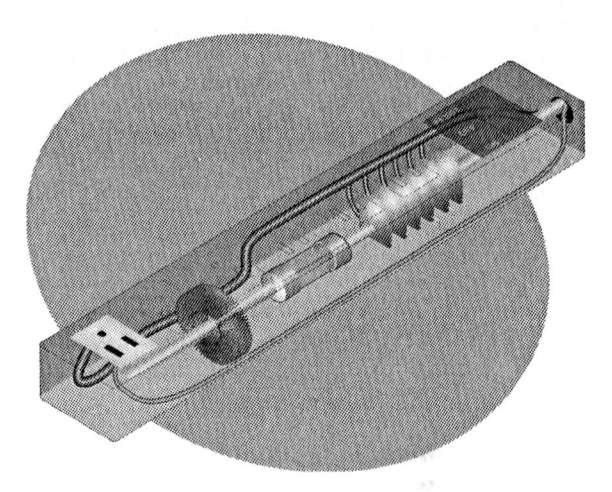

Surge Protector System

Surge Suppressor-A filter designed to protect computers and other electrical equipment from brief bursts of high voltage or surges.

Surround-Sound reproduction that surrounds the listener with sound, as in quadraphonic recording and reproduction.

Survivability-The capability of a communications network to continue to provide service after major damage to any part of the system.

SVC-Switched Virtual Circuit
SVID-System V Interface Definition
SW-Shortwave

SWAP-Shared Wireless Access Protocol
SWC-Serving Wire Center
Sweep-To vary the frequency of a signal over a specified bandwidth.
Sweep Circuit-A generator that produces a periodic deflection of an electron beam on a CRT.
Sweep Generator-A test oscillator, the frequency of which is constantly varied over a specified bandwidth.
Sweetening-The process of electronically improving the quality of an audio or video signal, such as by adding sound effects, laugh tracks, and/or special effects.
Switch-A network device (typically a computer) that is capable of connecting communication paths to other communication paths. Early switches used mechanical levers (cross-bars) to interconnect lines. Most switches use a time slot interchange memory matrix to dynamically connect different communications paths through software control. See also: Mobile Switching Center (MSC).
Switch Fabric-A device which interconnects multiple communications stations. Switch fabrics are responsible for making connections between sources and destinations, usually with specific quality of service and speed requirements. All LAN and WAN switches use a switch fabric of some type. The three major classes of switch fabric are Shared Memory, Shared Bus, and Crossbar.
Switch Hook Flash-A signaling technique whereby the signal is originated at the passive end of a subscriber loop by momentarily depressing the switch hook. This causes an interruption in loop current for an interval of typically 1 to 1.5 seconds. A shorter interruption is ignored or may be considered as equivalent to a wink or dialing the digit 1 with a rotary dial. A longer interruption typically causes a disconnection. See also Cradle Switch Flash.
Switch Port-A resource that allows a Group to communicate with an Other Group. All Groups implicitly posses a Switch Port as a secondary resource, but in order to use it, the application must explicitly connect the Switch Ports of two Groups.
Switchboard-A manually operated switching system, now rarely used, that connects a limited number of telephones within a building or provides operator services.
Switchboard Operator-A telephone system employee who provides information to customers and assists with the connection of telephone calls. The first switchboard operators connected calls by using cables with plugs to connect calls between connection points on their switchboard panel.
Switched Access Transport Services-Transmission of switched voice or data telecom traffic along dedicated facilities between LEC central offices and long distance carrier POPs.
Switched Circuit-A circuit that may be temporarily established at the request of one or more of the connected stations.
Switched LATA Access-The providing of connection lines through switches within a LATA area. This is called Switched Access Service.
Switched Multimegabit Data Service (SMDS)-A high-speed data transmission service that is often used in metropolitan areas that allows for the dynamic creation and disconnection of virtual circuits through the network. It is based on the 802.6 standard and may use T1 and T3 circuits to provide Ethernet, Token Ring, and Fiber Distributed Data Interface (FDDI) interconnection services.
Switched Virtual Circuit (SVC)-A switched virtual circuit is an automatically and temporarily created virtual connection that is used for a communication session.
This diagram shows how a switched virtual circuit (SVC) uses an address provided by the user to establish a logical (virtual) path through a communications network. In this example, a SVC is created by using the destination address to determine the required programming of routing tables in switches within the data network. These routing tables assign data transfer connections between input and output channel on each switch. For example, as data from the sending computer (portable

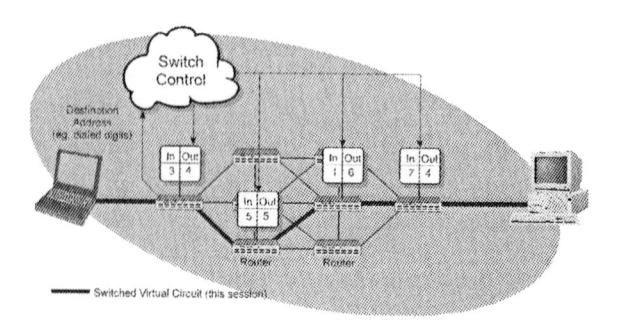

Switched Virtual Circuit (SVC) Operation

computer) is sent into input channel 3 of the first switch, it is transferred to the output channel 4. This process will repeat for any data that is sent from the sending computer to the destination computer during the switched connection.

Switchhook-The switch of a telephone set that closes and opens a customer local loop circuit. It is used to indicate control (supervisory) signals from the user. The switchhook is usually operated by lifting the handset or returning it to its cradle (holder). A switchhook is also called a Hookswitch.

Switching-A switching system connects two (or more) points together within a network or communication devices. These connections can be physically connected (mechanical switch) or connected logically (through software). The first telephone systems performed the switching of calls by human operators. The operators interconnected telephone lines by manually connecting cables at switchboards.

Switching systems have evolved from manual switchboard systems (wires and plugs) to logical (digital) switches. The earlier types of manual switchboard systems were changed to automatic switching systems to eliminate the need for operators to setup every call. The first types of automatic switching systems used crossbar switches. Crossbar switches used mechanical arms to physically connect to wires (or busses) together. This has progressed to time slot interchange (TSI) switches. TSI switches logically interconnect communication lines through the temporary storage of data in memory time slots.

Switching Center-A location where either toll or local telecommunication traffic is switched or connected from one line or circuit to another. Also called a switching office.

Switching Hub-A data hub (broadcast device) that also has capability to directly route (switch) packets to specific computer or device.

Switching Office-A switching center within a building (central office).

Switching System-Switching systems are assemblies of equipment that setup, maintain, and disconnect connections between multiple communication lines. Switching systems are classified by their network types and the methods used to control the switches. The term "switch" is sometimes used as a short name for switching system. Public telephone switching systems have many switches within their network. A typical switch can handle up to 10,000 communication lines each.

SWOT-Strengths, Weakness, Opportunities, and Threats

SWR-Standing Wave Ratio

SXS-Step By Step Switching System

Symbol-A recognizable electrical state associated with a signal element. In binary transmission, a signal element is represented as one of two possible states or symbols. In ternary transmission, a signal element is represented as one of three possible states or symbols. In quaternary transmission, the signal element represents one of four possible states or symbols.

Symmetrical-A type of Asynchronous Connectionless Link (ACL) that offers the same data rate in both the send and receive directions. For symmetrical connections, the Bluetooth specification specifies a maximum data rate of 433.9 Kbps in both directions, send and receive. See also Asymmetrical.

Symmetrical Compression-A compression system that re quires equal processing capability for compression and decompression of data. This form of compression is used in applications where both compression and decompression will be used frequently. Examples include still image database, still image transmission, color fax, video production, video mail, videophone, and videoconference. (See also: asymmetrical compression.)

Symmetrical Digital Subscriber Line (SDSL)-An all-digital transmission technology that is used on a single pair of copper wires that can deliver near T1 or E1 data transmission speeds. SDSL is a symmetrical service that ranges from 160 kbps to 2.3 Mbps and can reach to 18000 feet from the central switching office.

SYNC-Synchronization

Sync Pulse-Framing pulses that are added to a signal (such as a video signal) to keep the entire video process synchronized in time.

Sync Pulse Generator, SPG-Sync Generator

Synchronization-(1- general) The process of adjusting the corresponding significant instants of signals, such as the zero-crossings, to make them synchronous. The term synchronization often is abbreviated as sync. (2 - digital) An arrangement for operating digital switching and transmission systems at a common (synchronized) clock rate to prevent the loss of portions of a bit stream during transmission. The required synchronization pulses commonly are referred to as clock pulses or clock signals. (3 - video) The pulses and timing signals that lock the electron beam of the picture monitor in step, both horizontally and vertically, with the electron beam of the pickup tube. The color sync signal (in the NTSC system) is known as color burst.

Synchronization Channel (SCH)-A logical channel that provides a mobile station with a timing reference to assist in the demodulation of the radio channel. This timing reference is a unique training sequence of digital information.

Synchronization Word-A unique sequence of bits or information symbols that is used to provide a timing reference that is relative to other information that will follow.

Synchronized Television (SyncTV)-A video program delivery application that simultaneously transmits hypertext markup language (HTML) data that is synchronized with television programming. Synchronized television allows the simultaneous display of a video program along with additional information or graphics that may be provided by advertisers or other information providers.

Synchronizing Pulse Generator-A device or circuit that creates (generates) synchronization signals (pulses).

Synchronous-A system or signal that involves the transfer of information in a predefined serial time sequence.

Synchronous Data Link Control (SDLC)-A bit oriented synchronous communication protocol that organizes information into sequenced frames (groups of bits) of data. SDLC is similar to the high-level data link control (HDLC) protocol defined by the International Organization for Standardization (ISO).

Synchronous Detection-A demodulation process in which the original signal is recovered by multiplying the modulated signal by the output of a synchronous oscillator locked to the carrier.

Synchronous Digital Hierarchy (SDH)-A digital transmission format that is used in optical (fiber) networks to transport high-speed data signals. SDH uses standard data transfer rates and defined frame structures formats in a synchronous (sequential) format. SDH is similar to SONET.

Synchronous Modem-A modem that is able to transmit timing information in addition to data. The modem must be synchronized with the associated data terminal equipment by timing signals. A synchronous modem sometimes is referred to as an Isochronous modem.

Synchronous Network-A network in which the data communication lines are synchronized to each other or to a common clock signal that allows the exact determination of groups of bits (frames or fields) that are defined within the transmission of digital information.

Synchronous Operation-A process that is used to coordinate the timing of data transmission between communication devices. This allows the communication devices to know the specific timing relationship for information that is transported by the communication system.

Synchronous Optical Network (SONET)-A digital transmission format that is used in optical (fiber) networks to transport high-speed data signals. SONET uses standard data transfer rates and defined frame structures formats in a synchronous (sequential) format.

Synchronous Orbit-An orbit in which a satellite has an orbital angular velocity synchronized with the rotational angular velocity of the earth, and thus remains directly above a fixed point on the surface. See also Geosynchronous (or Geostationary) Earth Orbit.

Synchronous Satellite-A satellite in a synchronous orbit.

Synchronous Serial Data Transmission-The sequential transmission of data, bit by bit, of that is related to a reference clock signal and a specific channel mapping format.

Synchronous Transmission-(1 - general) A mode of transmission during which sending and receiving terminals operate at precisely the same frequency. Timing or synchronization is maintained by synchronizing pulses rather than by start/stop pulses, as in asynchronous transmission. (2-phase relationship) A mode of transmission in which the data transmitter and receiver are maintained in a uniform phase relationship. (See also: asynchronous transmission.)

Synchronous Transport Signal Level 1 (STS-1)-A basic building block of the SONET transport hierarchy. STS-1 provides an electrical signal line rate of 51.84 Mbps.

SyncTV-Synchronized Television

Syntax-The relationships among characters or groups of characters, independent of their meanings or the manner of their interpretation and use.

Synthesizer-A device used to modify an external or internal audio signal by a preset controlled process.

Syslog-The syslog protocol was first defined as part of the UNIX operating system to log messages with the operating system (OS). Syslogs allow a computer or device to deliver messages to another computer. Syslog messages have a particular format that associates a facility, and a severity or priority with a message. Syslog servers are widely used in data networks to log information about network devices.

Syslog Facility Code-Syslog facility code is a component of the syslog protocol and function. There are 24 unique syslog facilities defined. The facility code allows syslog to group messages from different sources and take action based on this facility or group. See also syslog priority levels.

Syslog Priority Levels-Syslog priority levels operate with the syslog protocol or function. There are 8 priority levels defined: Emergencies (0), Alerts (1), Critical (2), Errors (3), Warnings (4), Notifications (5), Informational (6), and Debugging (7). 0 is the highest and 7 is the lowest. Network devices can be configured to send syslog messages based on a user-defined priority setting. Debugging is very verbose, while emergencies indicate systems are unusable.

System-Semi-Residential Cells

System-Trunked

system-Autonomous Cells

System-Residential Cells

System-Semi-Private Cells

System-An assemblage of all facilities, united by some form of regular interaction or interdependence, required to accomplish a comprehensive function or functions.

System [or State-machine] Description Language (SDL)-A set of graphic symbols defined by ITU-T standard Z.100. SDL is an enhancement of traditional computer programming flow charts that includes a graphic representation of real time programming steps such as setting or clearing a hardware clock or timer, and responding to an interrupt signal. SDL is related to the programming language CHILL.

System Administrator-A person who oversees the operational functionality of a computer or related telephone equipment. Also included in this individual's responsibilities are introducing new user Ids and organizing phone numbers, commonly referred to as moves, adds, or changes (MAC).

System Identity (SID)-Home System Identification which uses a 5 digit number for assignment of the local cellular system. The SID is an identity structure that allows cellular phones to distinguish between public, private, semi-private, and domestic base stations. As in IS-54B, the SID represents an international identification and a system number identifying the service area and cellular band (A or B cellular).

System Management Application Entity (SMAE)-An SS7 software function that provides communications functions that support applications that are part of system manage application process (SMAP). There may be several SMAE functions for each SMAP process.

System Management Application Process (SMAP)-A SS7 software function or process that monitors, controls, and coordinates resources for application layer protocols.

System Network Architecture (SNA)-A communications system protocol developed by IBM that allows for control and data communication via different types of network communication equipment.

System Operator Code (SOC)-An IS-136 system identity code that allows a phone to recognize cell sites belonging to a specific cellular operator.

System Or Station Speed Dialing-A system or device that allows for automatic dialing of telephone numbers.

Système International (SI)-The SI version of the metric system of units is compatible with practical electric units such as the volt, ampere, and watt. The basic mass unit is the kilogram, the basic length unit is the meter, and the basic time unit is the second. The basic unit of electric current is the ampere, and the basic unit of electric charge is the coulomb or ampere second (see: charge). It is also called the Georgi system, after the name of the Italian physicist who first formulated the appropriate mechanical units for its use. The older version of the metric system, sometimes called the cgs system, is based on the centimeter, gram and second.

T

T-tesla
T-Tera
T Carrier-Trunk Carrier
T Coupler-A signal coupler that has three ports.
T-1-T1 Carrier
T-Carrier-A system operating at one of the standard levels in the North American digital hierarchy.
Each digital signaling level supports several 64 kbps (DS0) channels. t-carrier was initially used in North America and now is used throughout several parts of the world. The different digital signaling levels include;
- DS-1, 1.544 Mbps with 24 channels
- DS-1C, 3.152 Mbps with 48 channels
- DS-2, 6.132 Mbps with 96 channels
- DS-3, 44.37 Mbps with 672 channels
- DS-4, 274.176 Mbps with 4032 channels
T-Spec-Traffic Flowspec
T.38-A standard protocol used for the transmission of fax information over IP data networks. Created by the ITU-T, it defines real-time fax for IP-enabled fax devices and fax gateways.
T1 Carrier (T-1)-A North American and Japanese digital primary rate telephone multiplexing system that combines 24 channels of digitally coded speech or other subscriber data, at 64 kb/s for each such channel, with an 8 kb/s synchronization bit stream (the framing bits or F bits) into a 1.544 Mb/s bit stream. T-1 was historically a trade name of one manufacturer, although it is now often used for compatible products regardless of manufacturer. The synonym DS-1 is often used in standards and other documents.
TA-Terminal Adapter
TA-Transaction Capabilities
TACS-Total Access Communications System
TADIG-Transferred Account Data Interchange Group
TAF-Terminal Adaptation Function
Talk Around-The ability of a mobile radio (typically LMR) to directly communicate with another mobile radio without the need to communicate with a network.
Talk Time-The amount of time a mobile telephone or transmitting device can continuously transmit between battery recharges or the replacement of its disposable batteries.

Tandem-(1 - general) The connection of the output terminals of one network, circuit, or link directly to the input terminals of another. (2-message network) A switching system that establishes trunk-trunk connections but has no subscriber lines connected to it. Tandem types include local tandems, LATA tandems, and access tandems.
Tandem Compression-The connection of one coder/decoder (such as 32 kbps ADPCM) through another coder/decoder (such as 8 kbps G.729) when a call is routed through interconnecting tandem switches or routers. The cascading of coders (such as speech coders used in mobile or VoIP communication systems) usually results in significant degradation of voice quality.
Tandem Free Operation (TFO)-Tandem free operation involves the direct connection of switching centers in mobile communication system without the need to decompress and re-compress (transcoding) speech information. TFO overcomes the challenges of cascading the speech coding process. Each time speech information is compressed and decompressed, some audio distortion occurs and time delay is added.
The logical function for coordinating TFO is the transcoder rate adaptation unit (TRAU). The TRAU is usually located in the MSC (it is possible to put the TRAU in the base station) and it negotiates the ability of the MSC to use TFO with another MSC. The TRAU is also responsible for disabling TFO if the call is transferred to another MSC or system that is not capable of TFO.
Tandem Office-A switching system that is used to interconnect end offices with each other.
Tandem Switching System-A switching system in a tandem office that handles trunk-trunk traffic. Local tandems switch calls from one end office to another within the same area. Access tandems switch calls to and from an interexchange carrier.
TAO-Telephony Application Object Framework
TAP-Test Access Port
Tap-(1 - circuit) A branch or intermediate circuit in a communications system. (2- optical) A device for extracting a portion of the optical signal from a fiber. (3- telephony) Short for wire tap.(4 In a local area network an electrical connection permitting signals to be transmitted onto or received from a bus.

TAP-Transferred Account Procedures
TAPI-Telephony Application Programming Interface
Tapped Delay Line-(1 - analog technology) A string of inductors and capacitors that simulates a transmission line and delays the passage of a signal. The line is tapped at points along its length to enable sampling of the signal after various amounts of delay. (2 - digital technology) An application of a shift register, an integral part of an echo canceller, and an important component in many spread-spectrum systems.
Target Market-A segment of a potential market for a product or service that has common characteristics. These characteristics are often used to focus (target) groups of customers that have specific wants or needs for a product or service.
Tariff-The published rates and conditions for provision of a regulated communications service by a common carrier. Tariffs include a schedule of rates, prices, and regulations for services that have been approved by national or international regulatory bodies (e.g. FCC, OFTEL, etc.).
TASI-Time Assigned Speech Interpolation
TBC-Time Base Corrector
TCAP-Transaction Capabilities Application Part
TCF-Training Check Frame
TCH/F-Full Rate TCH
TCH/F-Traffic Channel Full Rate
TCH/H-Half Rate Data TCH
TCL-Tool Command Language
TCM-Trellis Coded Modulation
TCO-Total Cost Of Ownership
Tcommerce-Television Commerce
TCP-Transmission Control Protocol
TCP Header-Transmission Control Protocol
TCP Ports-A logical assignment of a computer connection (such as in a TCP/IP system) that allows a communication session between data communication devices. A port is part of a socket connection (channel identifier and port number.) Port numbers are commonly assigned based on specific protocols or applications. Some of the more common port assignments for Internet based TCP communications include port 21 for file transfer protocol (FTP), port 23 for Telnet, port 25 for Simple Mail Transfer Protocol (SMTP), and port 80 for Hypertext Transfer Protocol (HTP).
TCP/IP-Transmission Control Protocol And Internet Protocol

TCP/IP Stack-The software that allows a computer to communicate through TCP/IP. Stack refers to the fact that five layers of protocols operate on a TCP/IP network.
TCS-Telephony Control Service
TCXO-Temperature-Compensated Crystal Oscillator
TDD-Time Division Duplex
TDD-Teletypewriter Device for the Deaf
TDD-Teleprinter Device for the Deaf
TDD-Telecommunications Device For The Deaf
TDM-Time Division Multiplexing
TDMA-IS-54
TDMA-IS-136
TDMA-IS-54C
TDMA-Time Division Multiple Access
TDMoIP-Time Division Multiplexing over Internet Protocol
TDR-Time Domain Reflectometer
TE-Terminal Equipment
Technician-A skilled craftsman who is capable of diagnosing, servicing, and repairing electronic or electrical assemblies (such as a PBX system or telephone device).
Technological Churn-Type of Voluntary, Deliberate churn, where the customer changes carriers because
the new carrier has newer and better technological options. (i.e. customers switching
to digital from analog)
TEK-Traffic Encryption Key
TeLANophy-The use of local access network (LAN) systems as a communication line for telephone systems.
TeLANophy-LAN Telephony
Telco-Telephone Company
Telco Product Market Planning (TPMP)-Discipline employed by telco product managers in order to determine the potential
market for a new product. TPMP is utilized to help decide which products to invest in.
Telco Product Portfolio Management (TPPM)-Formal discipline utilized to help product managers to organize and manage large groups of products for maximum efficiency. TPPM is used to help:
1. Size product market potential
2. New product capacity planning
3. Network capacity planning
4. Operational support profile definition
5. Set sales targets and marketing objectives

Telcordia-Successor to Bellcore.

Telecom Branding -The process of adding to the value of a telecommunications organization, product line or offer through the creation of "psychological" value. Branding causes prospects to perceive special value in the branded item, encouraging loyalty, acquisition and sometimes higher billing rates.

Telecommunication-The transmission and reception of audio, video, data, and other intelligence by wire, radio, light, and other electronic or electromagnetic system.

Telecommunication Closet-A room or space that is dedicated for telecommunication equipment and cable connection points.

Telecommunication Regulation-The regulation of telecommunications systems and services are developed to help or improve the ability of citizens in a country to reliably communicate with each other at reasonable cost. Telecom rules and regulations are usually imposed by a government agency. These rules are designed to maintain the quality and cost of public utility services.

In the United States, the Telecommunications Act of 1996 was created to allow competition into the telecommunications industry. It provides a national framework for the deregulation of the local exchange market, a deregulation that was already taking place on a state-by-state basis through the actions of state regulatory commissions. Its summary impact on the local exchange market is to require current LECs to remove all barriers to the competition (e.g., interconnect, white and/or yellow pages access, co-location, and wholesaling of facilities restrictions) in return for LEC access to the long distance market.

Telephone companies in the United States are regulated by the government, but not owned by the government. For most European countries and many other countries, local telephone service is provided by government owned post telephone and telegraph (PTT) operators. In some European countries, the post (mail) network has been separated from the operation of telephone and telegraph networks. In some countries, the telephone and telegraph systems have become privatized, and are no longer owned by the government.

Telecommunication Resellers Association (TRA)-A telecom trade association, TRA recently merged with the National Wireless Resellers Association (NWRA) and promotes the expansion of telecom services through lobbying, legal efforts and other forms of support.

Telecommunication Services-
Telecommunications services are the underlying communications processes that provide information for telecommunications applications. It is common to use the word services in place of applications, especially when the service is very similar to the application. Examples of communication applications include voice mail, email, and web browsing. Telephone services include voice, data, and video transmission. Voice services can be categorized into quality of service and voice privacy. Data services use either circuit-switched (continuous connection) data or packet-switched (dynamically routed) data. Video transmission is the transport of video (multiple images) that may be accompanied by other signals (such as audio or closed-caption text).

Telecommunications-The transmission, between or among points specified by the user, of information of the user's choosing (including voice, data, image, graphics, and video), without change in the form or content of the information.

Telecommunications Act Of 1996-The U.S. Telecommunications Act of 1996 provides a national framework for the deregulation of the local exchange market, a deregulation that was already taking place on a state-by-state basis through the actions of state regulatory commissions. Its summary impact on the local exchange market is to require current LECs to remove all barriers to the competition (e.g. interconnect, white and/or yellow pages access, co-location, & wholesaling of facilities restrictions) in return for LEC access to the long distance market.

Telecommunications Device For The Deaf (TDD)-A small communications terminals with a keyboard and visual display that connects to a telephone circuit to relay writ-ten messages to and from hearing- and/or speech-impaired persons. An acoustic coupler is used to send audio tones from a TDD through the handset of a conventional telephone instrument.

Telecommunications Industry Association (TIA)-Telecommunications Industry Association. This telecom industry trade association represents the manufacturers of telecom equipment.

Telecommunications Operational Model (TOM)-An industry standard template describing the major operational components that make up a telecommunications organization.

Telecommunications Profitability Formula-A modified business profitability formula that reflects the unique nature of the telecommunications business model where Profit = (Revenue - (CapEx + OpEx))

Telecommunications Relay Service for the Deaf (TRS)-Each state and province in North America operates one or more telecommunications relay service centers for the deaf. Human communications assistants (CAs) at the TRS center provide translation between ordinary speaking and hearing telephone users at one end of a conversation, and deaf or speech impaired users at the other end. The deaf or speech impaired persons use a teletypewriter to communicate with the CA. In the year 2000 the US FCC designated 711 as the universal telephone access number for TRS centers.

Telecommunications Technology Association (TTA)-An association in Korea that oversees the creation of telecommunications standards.

Telecommunications Value Chain-An operational model that describes the core functions that are required to deliver telecommunications service to the end customer. The links in the telecommunications value chain include Market Acquisition, Market Research, Network Planning, Network Build, Network Maintenance, Service Order Processing, Provisioning, Activation, Sales, Marketing, Advertising, Billing, Customer Service, Credit Processing.

Telecommuting-The process of an employee that is conducting business related activities at a remote location (usually at a home) through the use of telecommunications services and equipment. Telecommuting allows employees to work at home without the need to commute to the office and reduces the need for the business to maintain office space for workers.

Telecomputing-The use of remotely located computers and databases for computing and access to computer-based services, such as home shopping, banking, and electronic mail.

Teleconferencing-A process of conducting a meeting between two or more people through the use of telecommunications circuits and equipment. Teleconferencing usually involves sharing video and/or audio communications.

Telecopier-European term for fax.

Teledapt-Trade name used in Canada for modular telephone cord connectors.

Teledesic-Teledesic is a "Big LEO" satellite system that is proposed to provide high speed data services (broadband) to consumers. The Teledesic system is a $9 billion project proposed by Bill Gates and Craig McCaw. Teledesic, is planned to become available in the year 2001.

Telegraphy-A form of telecommunications, which is concerned with the process of providing transmission and reproduction at a distance of text material or fixed images. The transmission of such information may be physical transmission facilities or over the air using some form of signaling protocol.

Telemanagement-A technique involving the management of the telephone and telecommunication expenses of an organization. This also refers to third-party software packages directed at monitoring inbound and outbound calling on the PBX switch with the capability to format detailed reports by departments and per-minute charges.

Telemarketing-Telemarketing is the process of conducting marketing and sales programs using telecommunication systems. Telemarketing call centers generally combine advanced call processing systems (e.g. computer telephony automatic call distribution) with customer order processing systems.

Telemedicine-Processes that assist with health care service that employ communications services and equipment. Examples of telemedicine include delivery of medical images, remote access to medical records, remote monitoring of heath care equipment and distant monitoring of biological functions such as heart rate and blood pressure.

Telemetry-The transfer of measurement information to a monitoring system through the use of wire, optical fiber, or radio transmission. The term telemetry is often used with the gathering of information.

Telephone Answering Service-A service company that answers telephone calls and takes messages for subscribers who are away from their homes or offices.

Telephone Company-Telephone companies (also known as service providers or carriers) provide communication services to the general public. They are usually are regulated by the government and in some countries, may be partly or wholly owned by the government. For most European countries and many other countries, local telephone service is provided by government owned posts, telephone and telegraph (PTT) operators. In some European countries, the post (mail) network has been separated from the operation of telephone and telegraph networks. In some countries, the telephone and telegraph systems have become privatized, and are no longer owned by the government.

Telephone Company (Telco)-A contraction of telephone company, generally signifying an operating telephone company

Telephone Eye-A system that combined television broadcasting with telephone service to provide for early interactive television programming.

Telephone Network For The Deaf (TND)-A network for relaying the written messages of hearing- and/or speech impaired persons. (See also: operator relay services, telecommunications device for the deaf.)

Telephone Numbering Plan-A numbering plan is a system that identifies communication points within a communications network through the structured use of numbers. The structure of the numbers is divided to indicate specific regions or groups of users. It is important that all users connected to a telephone network agree on a specific numbering plan to be able to identify and route calls from one point to another. See also Dialing Plan, open numbering plan, closed numbering plan.

Telephone Routing over Internet Protocol (TRIP)-A protocol that allows for the dynamic assignment of call routes through the advertising of the availability of destination devices (such as telephones) and for providing information relatives the available routes and preferences for these routes to reach the destination device(s).

Telephone Set-The terminal equipment at a subscriber's premises used for voice telephone service. The instrument includes a microphone and earphone, called a transmitter and a receiver, as well as a switchhook, keypad or dial, ringing device, and associated circuitry. The set also is called a telephone instrument.

This block diagram shows a standard plain old telephone service (POTS) telephone set (also known as a 2500 series phone). This telephone continuously monitors the voltage on the telephone line to determine if an incoming ring signal (high voltage tone) is present. When the ring signal is received, the telephone alerts the user through an audio tone (on the ringer). After the customer has picked up the phone, the hook switch is connected. This reduces the line connection resistance (through the hybrid)

and this results in a drop in line voltage (typically from 48 VDC to a few volts). This change in voltage is sensed by the telephone switching system and the call is connected. When the customer hangs up the phone, the hook switch is opened increasing the resistance to the line connection. This results in an increase in the line voltage. The increased line voltage is then sensed by the telephone switching system and the call is disconnected

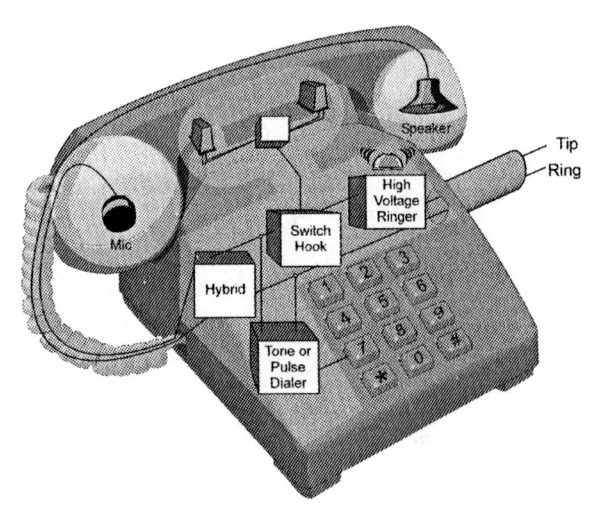

Plain Old Telephone Service (POTS) Telephone

Telephone Station-Telephone stations are telephone instruments that are connected to a telephone network for the purpose of telephony. When these telephone stations are used as part of the private network, they are often identified by the type of system they are connected with. For example, telephone stations that are connected to key systems are called Key Telephones.

Telephone stations can vary from simple POTS telephones (sometimes called 2500 telephones) to complex Internet telephones (IP Telephones). Telephone stations usually receive their power from the telephone line (loop current) but may receive their power from an external source (such as the PBX switching system).

This diagram shows the difference between standard analog telephone stations and more advanced PBX stations. This diagram shows that analog telephones receive their power directly from the telephone line and digital PBX telephones require a

Telephone User Part (TUP)

control section that gets its power from the PBX system. Analog telephones also use in-band signaling to sense commands (e.g., ring signals) and to send commands (e.g., send dialed digits). Digital telephones use out-of-band signaling on separate communication lines to transfer their control information (e.g., calling number identification).

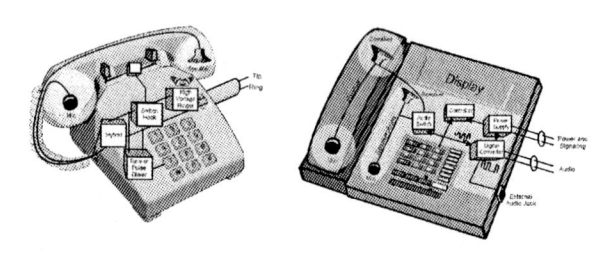

Telephone Station Functional

Telephone User Part (TUP)-The User Part specified in ISDN for telephone services.

Telephony-The use of electrical, optical, and/or radio signals to transmit sound to remote locations. Generally, the term means interactive communications over a distance. Often, telephony relates to a telecommunications infrastructure designed and built by private or government-operated telephone companies.

Telephony and Internet Protocols Harmonization over Networks (TIPHON)-A ETSI initiated project for multimedia communications that combines and coordinates multiple data compression and communication standards to allow audio, picture, and video transmission between users on packet switched networks. Information about TIPHON can be found at www.etsi.org/tiphon.

Telephony Application Programming Interface (TAPI)-An industry standard that defines the application interface between computers and telecommunications devices. TAPI was introduced in 1993 as the result of joint development by Microsoft and Intel. The standard supports connections by individual computers as well as LAN connections serving many computers. Within each connection type, TAPI defines standards for simple call control and for manipulating call content.

Telephony Call Dispatcher-A server feature on an AVVID system that allows users to receive and forward calls to other users.

Telephony Control Service (TCS)-In a Bluetooth system, TCS defines the call control signaling for the establishment of speech and data calls between Bluetooth devices. In addition, it defines mobility management procedures for handling groups of Bluetooth TCS devices. The Bluetooth SIG has defined the set of AT commands by which a mobile phone and modem can be controlled in the multiple usage models.

Telephony Over Internet Protocol (ToIP)-A process of providing telephony services using Internet protocol (IP).

Telephony Services Application Programming Interface (TSAPI)-TSAPI is a software communication standard developed primarily by the companies Lucent and Novell to allow PBX or Centrex systems to communicate through the use of NetWare communications software.

Teleprinter Device for the Deaf (TDD)-See teletypewriter device for the deaf.

Teleservices-Teleservices are telecommunication services that provide added processing or functionality to the transfer of information between users. Teleservices are categorized by their high level (application) characteristics, the low level attributes of the bearer service(s) that are used as part of the teleservice, and other general attributes. High-level attributes include: application type (for example voice or messaging) and operation of the application. The low-level description includes a list of the bearer services required to allow the teleservice to operate with their data transfer rate(s) and types. Other general attributes might specify a minimum quality level for the teleservice or other special condition. The categories of teleservices available include voice (speech), short messaging, facsimile, and group voice.

This diagram shows a typical teleservice in a mobile communication system. In this diagram, a telephone user wishes to send a fax to a recipient who is traveling. The designated recipient has setup a fax forwarding service where the delivery of incoming faxes can be instructed. The sender is given the recipient's fax number. When the sender dials the number, the call is routed through the telephone network to the fax forwarding service provider (step 1). When the incoming call is detected, the fax forwarding service receives the fax into a fax mailbox (step 2). Later that day, the recipient of the fax forwarding service calls in and enters a fax forwarding number (step 3). The fax forwarding service then checks the fax mailbox and automatically sends all the waiting faxes to the new number (possibly a hotel fax number) that has been updated by the recipient (step 4). Because this service involves both the transport and processing of information, it is categorized as a teleservice.

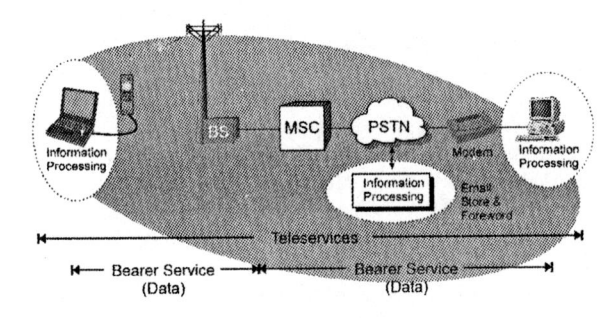

Teleservice Operation

Teletext-(1-television) A service that transfers data information along with a standard television signal to allow the simultaneous display of text and video on the television. Teletext information is usually encoded into the video blanking interval (VBI) and decoded by the receiver in the television or cable converter box.

Teletrac-A position location system that uses signal sensing from a Teletrac transceiver at multiple radio towers to determine the location of a vehicle or other equipment.

Teletype-Trade name for teletypewriters made by the Teletype Corporation. Widely used informally as a synonym for teletypewriter.

Teletypewriter (TTY)-A text communications device comprising an alphanumeric keyboard and a display (or paper printer) that connect to an appropriate telecommunications channel. Older teletypewriters, and most teletypewriters used by the deaf, use the Baudot-Murray (ITU alphabet No. 2) 5-bit character code, while newer teletypewriters use ASCII 8-bit character code (similar to ITU alphabet No. 5).

Teletypewriter Device for the Deaf (TDD)-A teletypewriter using a special modem to support text communication for deaf people via the public telephone network.

Teletypewriter Exchange Service (TWX)-TWX service was originally set up in North America by the Bell Telephone system in the 1950s as a competitor to the pre-existing telex service operated by Western Union Telegraph Co. TWX service uses modems at each teletypewriter and transmits text via the public switched telephone network. TWX teletypewriters also use ASCII code at a higher character rate than the slower Baudot-Murray code telex machines. Eventually several interconnection and code conversion switches were installed to permit interconnection between TWX and telex services, and the TWX system was purchased by Western Union in the 1970s.

Television-Television broadcasting is the transmission of video and audio to a geographic area that is intended for general reception by the public, funded by commercials or government agencies. Television broadcasters transmit at high power levels from several hundred foot high towers. A high power television broadcast station can reach over 50 miles. Commercial television service started in the United States in 1946. By 1949, television receiver sales were exceeding 10,000 per month.

This figure shows a television broadcast system. This television system consists of a television production studio, a high-power transmitter, a communications link between the studio and the transmitter, and network feeds for programming. The production studio controls and mixes the sources of information including videotapes, video studio, computer created images (such as captions), and other video sources. A high-power transmitter broadcasts a single television channel. The television studio is connected to the transmitter by a high bandwidth communications link that can pass video and control signals. This communications link may be a wired (coax) line or a microwave link. Many television stations receive their video source from a television network. This allows a single video source to be relayed to many television transmitters.

Television Commerce (Tcommerce)

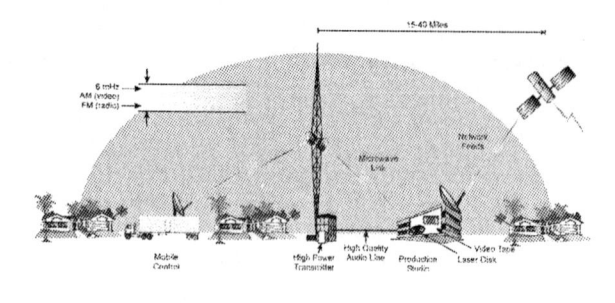

Television Broadcast System

Television Commerce (Tcommerce)-A shopping medium that uses a television network to present products and process orders.

Television Receive Only (TVRO)-A ground station used for the reception of TV signals transmitted from satellites. A typical station consists of a parabolic dish antenna, pre-amplifier, down-converter, tunable receiver, and video monitor.

Television Receiver-A device that receives a broadcast TV signal and converts it into a picture accompanied by its associated sound.

Television Signal-A radio signal that includes both the video component and the audio component of the transmitted program.

Teleworker-A worker who performs their job at remote location through the use of a telecommunications connection. An example of a teleworker is a customer service representative who answers customer service inquiry messages from their home.

Telex-An international teletypewriter exchange service for communications among its subscribers. Operation is typically at a rate of 50 bauds, via Baudot-Murray 5-bit character codes.

Telnet-The standard TCP/IP remote login protocol specified in RFC-854. Using Telnet, a user can work from a personal computer as if it were a terminal attached to another machine by a hardwired line.

TEM-Transverse Electromagnetic Mode

Temperature Compensation-An equipment design criterion such that the effect of a temperature change in one component of the system is offset by the effect of the same temperature change in one or more other components of the system, thereby keeping the overall response substantially independent of temperature.

Temporary Boot-A splice closure that is temporarily installed over a cable using rubberized coverings, cements and bindings to protect conductors during a construction or maintenance project. They are replaced with permanent plastic or lead closures at the completion of repairs or splicing and acceptance testing.

Temporary Location Directory Number (TLDN)-A temporary location directory number (TLDN) is a temporary identification number that is used to route calls from a home system and a visited communication system. TLDN numbers are commonly used in mobile communication systems where mobile telephones regularly operate in other (visited) systems. The TLDN is usually assigned for each call delivery request received from the home system. When the call is received into the visited system, it is mapped (translated) to the current resources (e.g. cell site and mobile number) that are currently being used in the visited system.

Temporary Mobile Station Identity (TMSI)-A 20-bit or 24-bit number representing a temporary mobile telephones identity. It is assigned to a phone by the system at initial registration to provide enhanced paging capacity in the air interface.

Tera (T)-A prefix that means 10 to the 12th power 1,000,000,000,000. Commonly referred to as one trillion in the American numbering system, and one million million in the British numbering system.

TeraByte-A quantity equal to one trillion bytes (actually 1,099,511,627,776 bytes.)

TeraHertz (THz)-A unit equal to one trillion of a Hertz (1,000,000,000,000 Hertz).

Term Plan-A product/service that offers the customer a special discount in return for committing to a certain length of time. Penalties are typically assessed for early termination.

Terminal-(1 - circuit) A point at which a circuit element may be directly connected to one or more other elements. (2 - communications) Any type of equipment at the end of a communications circuit. User terminals include telephone sets, teletypewriters, and computing equipment. Carrier terminals include modulation, demodulation, and multiplexing equipment used to transmit, combine, and separate communications channels in a transmission system. (3- computer) An input/output device connected to a processor or a computer in order to communicate with it and control processing. An intelligent terminal has some local computing power and an associated data store. (4 - loop plant) The hardware that facilitates the connection and removal of drop or service wires to and from cable pairs. Examples include distribution terminals and cross-connection terminals. (See also: central office terminal.)(5 - post) A binding past, tag, or lug to which an external circuit may be readily connected.

Terminal Emulation-A microcomputer or personal computer mimics, pretends to be (I.e. emulates) a data terminal. It does this with special printed circuit boards inserted into its motherboard and/or special software.

Terminal Equipment (TE)-Equipment that is located at the end of one or more communication lines that send or receive signals for services. Terminal equipment can be wired or wireless devices.

Terminals-Devices that typically provide the interface between the telecommunications system and the user. Terminals may be fixed (stationary) or mobile (portable).

Terminating Point Master File (TPMF)-Also known as "V&H file", it contains a correlation between an NPA-NXX and the name of the locale where the serving switch resides. This "Place Name" appears in the detail section for each call. This file also contains the Vertical & Horizontal coordinates of the NPA-NXX; this is used by the rating engine for distance-based rating.

Termination-(1 - circuit) An impedance source connected to the end of a circuit under test to simulate an actual connection. (2 - device) A device used to terminate a transmission line or coax. To accurately send a signal through a transmission ~-mission line, there must be an impedance at the end that matches the impedance of the source and of the line itself. otherwise, amplitude errors and reflections will result. Video systems typically use 75 ohm hardware, and RF systems typically use 50 ohm hardware. (3- switching network) The points on a switching net-work to which a trunk or a line can be attached.

Termination Charges-Fees paid by telephone operators to other access providers for terminating (routing calls to their destination) through the other access networks. An example of this is the payment by wireless carriers to local exchange carriers (local telcos) for the termination of mobile telephone calls into the wired public telephone network.

Terminator-An electrical device that can be attached to the end of a cable to simulate the attachment of an indefinite amount of additional cable. If this device has the same impedance as the cable, signals arriving at the end of the cable and encountering this device will not experience an impedance discontinuity, and signal reflections will not be created.

Terrestrial Link-A direct, overland radio circuit not relying on satellite transmission. This term also may refer to the carrier facilities, usually microwave radio or coaxial cable, between a ground station and a facility terminating office.

Terrestrial Service-Maritime Mobile

tesla (T)-Unit of magnetic flux density, equal to one volt second per square meter, or to one Weber per square meter. Named for Serbian-American electrical engineer Nikola Tesla (1856-1943). One tesla is 10 000 gauss. (A gauss is the older cgs unit of magnetic flux density.)

Test Antenna-An antenna of known performance characteristics used in determining the transmission characteristics of equipment and associated propagation paths.

Test Line-A central office test facility, including testing equipment, circuits, and communication channels, used for the maintenance of trunks. A test line can range from a simple passive termination and tone generations to complex electronic equipment for signal and transmission testing. A test line also can be referred to as a test termination.

Test Point-A post or terminal on a circuit board to which test equipment can be connected to check a parameter of the circuit.

Test Set-Any instrument designed to perform tests.

Test Signal-An electronic signal with defined standard characteristics used to check the capability of a circuit.

Test Signal Generator-A device used to generate test signals.

TETRA-Trunked Enhanced Terestrial Radio

TETRA POL-Trunked Enhanced Terestrial Radio POL

Tetrode-A 4 element electron tube consisting of a cathode, control grid, screen grid, and plate.

Text Encoding-The use strings of characters to define (encode) control messages.

Text Messages (Textual Messages)-A message that contains strings of characters (text) to define meaning of the message. The use of text messages results in easy to read messages at the expense of using more data bits for each message.

Text Messaging-See Short Message Service.

Text Paging-Text messages which are sent via operator or computer to a Text Pager. A text pager is commonly called an Alpha pager.

Text To Speech (TTS)-The conversion of text (ASCII) information into synthetic speech output. This technology is used in voice processing applications that require the production of broad, unrelated and unpredictable vocabularies, e.g., products in a catalog, names and addresses, etc.

Textual Messages-Text Messages

Texture Mapping-The ability of a digital effects system to create textured surfaces that can be applied to shapes.

TF-Transport Format

TFCI-Transport Format Combination Indicator

TFCS-Transport Format Combination Set

TFI-Transport Format Indicator

TFO-Tandem Free Operation

TFT-Thin Film Transistor

TFTP-Trivial File Transfer Protocol

TFTP Server-A server (computer) that is commonly used to supply supplies essential startup or configuration information (boot information) to data communication devices (such as IP telephones).

TGCP-Trunking Gateway Control Protocol

TGID-Talk-Group Identifier

TGL-Transmission Gap Lengths

TGS-Ticket Granting Server

TGW-Trunking Gateway

THD-Total Harmonic Distortion

Thermal Noise-A noise resulting from the movement of electrons in conductors and semiconductors, caused by thermal effects.

Thermionic-A method of device operation pertaining to the emission of ions or electrons by a hot material.

Thermistor-A resistive device whose value varies with temperature. Thermistors are used to measure temperature and as sensing elements in protective circuits and alarm systems.

Thermocouple-A junction of two dissimilar metals that, when heated on one side, generates a direct current. Thermocouples often are used as temperature-sensing elements in protective circuits, alarm systems, and test instruments.

Thermoelectric Cooling-A method of cooling in which an electric current is passed through two dissimilar metals joined at two points. Heat is liberated at one junction and absorbed at the other.

Thermoelectric Junction-A junction between dissimilar metals that are capable of producing electricity when heated.

Thermoelectric Power-Power generated by thermoelectricity. Many isolated radio repeater stations are powered by thermoelectric generators using propane, butane, or natural gas as fuel. The gas burns to heat one end of a thermo pile while the other end is kept cool.

Thermomagnetic-The effect of temperature on the magnetization of a material, or the heating effect of a magnetic change.

Thermoplastic-A resin that can be softened repeatedly when heated.

Thermostat-A device used to control temperature. Basic types are activated by the heat-induced bending of a bimetallic strip.

THF-Tremendously High Frequency

Thick Film-A film pattern made by applying conductive and insulating materials to a ceramic substrate by a silk-screen process. Thick films can be used to form conductors, resistors, capacitors, and other components.

Thicknet-A network cabling scheme using twin axial cable.

Thin Ethernet-A coaxial cable (approximately 0.2-inch or RG58A/U50-OHM) that uses a smaller diameter coaxial cable than the standard thick Ethernet. Thin Ethernet systems tend to have transceivers on the network interface card rather than in external boxes. Thin Ethernet is also referred to as Thin Net and Thin Wire.

Thin Film Filter-A device that separates one wavelength from others by passing light through a series of thin films of specific thickness and index of refraction. The change in index of refraction causes light to be reflected. The thickness of the films (as well as other factors, such as the angle of incidence on the filter surface) determines the wavelength of light that will be reflected due to interference. Other wavelengths are not affected significantly by the multiple layers. These filters are also known as interference filters.

Thin-Film Circuit-A circuit in which components are manufactured by means of a applying a thin film of material, usually only a few micrometers thick. This film may be deposited on a ceramic or glass base.

Third Generation (3G)-A term commonly used to describe the third generation of technology used in a specific application or industry. In cellular telecommunications, third generation systems used wideband digital radio technology as compared to 2nd generation narrowband digital radio. For third generation cordless telephones, products used multiple digital radio channels and new registration processes allowed some 3rd generation cordless phones to roam into other public places.

This diagram shows a 3rd generation broadband wireless system. This system uses two 5 MHz wide radio channels to provide for simultaneous (duplex) transmission between the end-user and other telecommunication networks. There are different channels used for end- user to the system (called the "uplink") and from the system to the end-user (called the "downlink"). This diagram shows that 3G networks interconnect with the public switched telephone network and the Internet. While the radio channel is divided into separate codes, different protocols are used on the radio channels to give high priority for voice information and high-integrity to the transmission of data information.

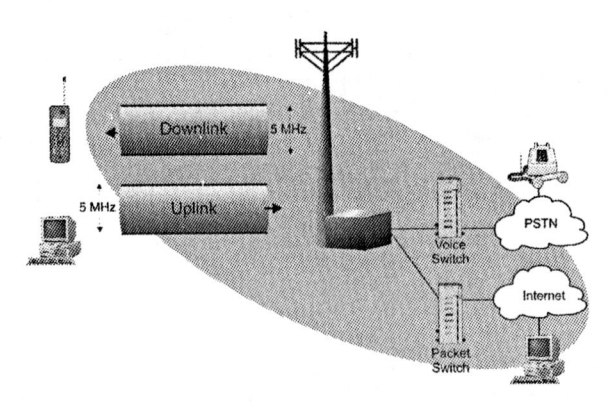

Third Generation (3G) Wireless

This photograph shows the SPH-X2500 model third generation mobile telephone. This advanced communication device can access mobile communication systems using the CDMA2000 third generation wireless technology. Third generation technology allows this device to send and receive data at speeds up to 144 kbps. In addition to wireless access, this device includes a 256 color LCD display, phonebook, organizer, messaging, games, and Bluetooth transmission to connect to a wireless headset and other computing devices.

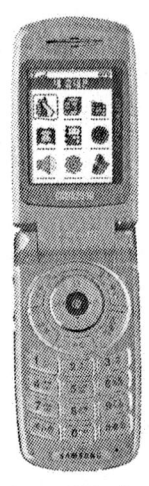

3G Mobile Phone Wireless Terminals
Source: Samsung Electronics Co., Ltd.

Third Party Billing-The billing of customers by one telephone company ("the Biller") on behalf of another Telco (the "Third Party").

ThM-Transverse Electromagnetic Mode

Three Way Calling (3WC)-A service that provides a telephone service customer the capability of adding another party (third party) to an established two-party call. During a three-way call, all three parties may communicate at the same time. However, to prevent annoying audio feedback, sophisticated audio volume control is typically offered that reduces the amplified audio signal level for parties that are not talking.

Three-Way Handshake-A series of three sets of messages that are exchanged between communication devices operating on a system to establish communications links for further transmissions of voice or data. Three-way handshake is used in the session initiation protocol (SIP) system to ensure call control messages (such as invite) can be confirmed in an unmanaged packet data transmission system (such as the Internet).

Threshold-(1-maximum) The minimum signal value that can be detected by a system. (2-FM improvement) The level at which signal peaks entering a FM receiver equal the peaks of internally generated thermal noise power. (3-noise) The radio frequency input level in a radio receiver at which the signal power equals the internally generated thermal noise power. (4-trigger) The value necessary to activate a device. (5-network Management) Threshold in network management terms applies mainly to agent-based thresholds that allow network devices to directly generate events when something interesting happens on the network. Thresholds are usually set with SNMP or the RMON MIB.

Threshold Alarm-An alarm indication that informs operators that a predetermined threshold that has been exceeded. An example of a threshold alarm is switch capacity.

Threshold Current-In telecommunications, the minimum forward current at which a laser is in a lasing state at a specified temperature.

Threshold Of Audibility-The minimum audio sound power intensity that the average human ear can hear, is usually taken to be 0.000 000 000 001 watts per square meter. This is the nominal zero dB level for acoustic dB values.

Threshold Value-The input voltage at which a change in output state is triggered in a logic device.

Throughput-(1 - telephone system) The number of telephone call attempts successfully completed each second. (2- data communication) The number of bits, characters, or blocks of data that a system working at maximum speed can process during a specified period of time. (3- ISDN/PPSN) In the Integrated Services Digital Network (ISDN) and the Public Packet-Switched Network (PPSN), the average information rate of a particular virtual circuit.

Throughput Delay-The total time in ms between the initiation of a voice or data signal, i.e.. push-to-talk, until the reception and identification of the identical signal at the received output speaker or other device.

THz-TeraHertz

TIA-Telecommunications Industry Association

TIA/EIA-568-A-An EIA standard for commercial building wiring that covers; the medium, the topology of the medium, terminations and connections, and general admission. This involves planning and installing of structured wiring systems, abiding to cable specifications (both performance and installation requirements), physical star topology design, and cabling divisions (horizontal and backbone cabling).

TIE Line-Terminal Interface Equipment Line

TIFF-Tagged Image File Format

Tightly Buffered Fiber-A fiber that is held rigidly inside its jacket through the use of a buffer. The fiber typically has a soft plastic buffer surrounded by a hard plastic layer.

Time Assigned Speech Interpolation (TASI)-TASI is a process that dynamically assigns communication channels to transfer information that is based in speech activity. The TASI system senses activities of speech signals and availability of communication channels in a system and dynamically transmits information signals on available communications channels. Because speech conversation is composed of pauses and alternating directions of communications (usually one person speaks at a time), the use of TASI increases the efficiency of a communications system of approximately 2:1. For example, a communications circuit that has 96 voice channels (64 kbps DS0 channels) can use TASI to provide service approximately 192 calls.

Time Delay-The amount of time for a signal to travel from one point to another point.

Time Division-The separation in the time domain of a number of transmission channels between two points.

Time Division Duplex (TDD)-Time division duplex (TDD) is a process of allowing two way communications between two devices by time sharing. When using TDD, one device transmits (device 1), the other device listens (device 2) for a short period of time. After the transmission is complete, the devices reverse their role so device 1 becomes a receiver and device 2 becomes a transmitter. The process continually repeats itself so data appears to flow in both directions simultaneously.

Time Division Multiple Access (TDMA)-Time division multiple access (TDMA) is a process of sharing a single radio channel by dividing the channel into time slots that are shared between simultaneous users of the radio channel. When a mobile radio communicates with a TDMA system, it is assigned a specific time position on the radio channel. By allow several users to use different time positions (time slots) on a single radio channel, TDMA systems increase their ability to serve multiple users with a limited number of radio channels.

Time Slot Interchange Switching (TSI)

Time Division Multiplexing (TDM)-A method for sending two or more signals over a common transmission path by assigning the path sequentially to each signal, each assignment being for a discrete time interval. All channels of a time-division multiplex system use the same portion of the transmission links' frequency spectrum - but not at the same time. Each channel is sampled in a regular sequence by a multiplexer

This figure shows how a single carrier channel is time-sliced into three communication channels. Transceiver number 1 is communicating on time slot number 1 and mobile radio number 2 is communicating on time slot number 3. Each frame on this communication system has three time slots.

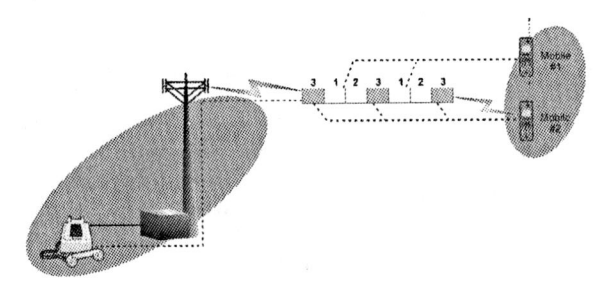

Time Division Multiplexing (TDM) Operation

Time Division Multiplexing over Internet Protocol (TDMoIP)-The providing of time based (time division) communication circuits through the Internet. TDMoIP allows for T1 and E1 communication lines to be transferred by a managed or unmanaged packet network (such as the public Internet).

Time Domain Reflectometer (TDR)-A test instrument used to identify and locate discontinuities or breaks in a transmission line. At a discontinuity, the impedance of the line is irregular and causes reflections to occur. The instrument measures the distance to a discontinuity by sending a voltage pulse through a line and measuring the time it takes for the signal to reach the point of and reflect from the discontinuity.

Time Out-A specified period of time allowed to elapse in a system before a specified event takes place, unless another specified event occurs first.

Time Sharing Option (TSO)-Time Sharing Option (TSO) is software that provides interactive communications for IBM's MVS operating system. It allows a user or programmer to launch an application from a terminal and interactively work with it. The TSO counterpart in VM is called CMS. Contrast with JES, which provides batch communications for MVS.

Time Slot (TS)-The smallest division of a communications channel that are assigned to particular users in the system. Time slots can be combined for a single user to increase the total data transfer rate available to that user. In some systems, time slots are assigned dynamically on an as-needed basis.

Time Slot Interchange Switching (TSI)-A process of connecting incoming and outgoing digital lines together through the use of temporary memory locations. A computer controls the assignment of these temporary locations so that a portion of an incoming line can be stored in temporary memory and retrieved for insertion to an outgoing line.

This diagram shows a TSI switching system. This diagram shows a simplified matrix switching system. Each input line (port) is connected to a multiplexer. The multiplexer places data from each port in time sequence (time slot) on a communications line (e.g., a T1 or E1 line). This time multiplexed signal is supplied to a matrix switching assembly. The matrix switching assembly core has two memory parts: a section that holds the pulse coded modulation (PCM) data and Control Memory - CRAM that holds switching addresses data. The time slots (voice channels) from the incoming multiplexed sent through switch S1 to be sequentially stored in the PCM data memory. The data is later retrieved by switch S2 and placed on a specific time slot on an outgoing line. The outgoing multiplexed line is supplied to a de-multiplexer so each time slot is routed to an output port.

Time Stamp (Timestamp)

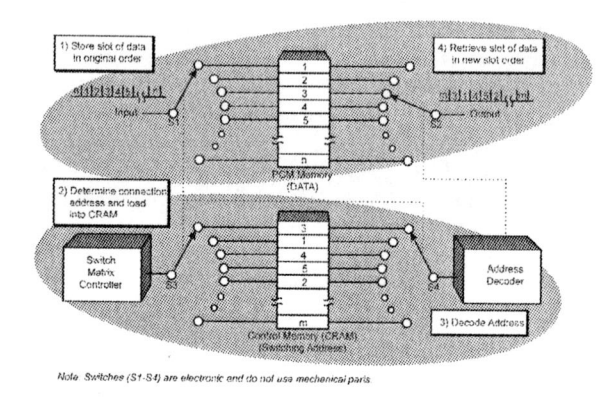

Time Slot Interchange (TSI) Switching System

Time Stamp (Timestamp)-The insertion of information that indicates a time that an event occurred into a record a block of data. Time stamps are used to indicate the time a file was created or when an error or network failure has occurred in a system.

Time To Live (TTL)-A field within a data packet that is used to limit the maximum number of routing or switching points a packet may pass through during transmission in a data network. The TTL counter is decreased as it progresses through each router or switching point in the network. If the TTL counter reaches 0, the packet can be discarded. The use of TTL ensures packets will not be transmitted in an infinite loop.

This diagram shows a packet data network uses a time to live (TTL) control field to avoid the potential for routing packets through long travel paths or infinite loops. This diagram shows that a data packet enters into a packet network contains its destination address, time to live field, and data payload. The TTL field is initially set by the sending computer and its value may vary dependent on the type of service (e.g. real time voice compared to file transfer.) As the packets are routed through the packet network, each router (packet switching device) forwards the packet towards the destination it believes will send the packet towards its destination. Each time the packet passes through a router, the TTL field value is decreased. This diagram shows that the routers in this diagram accidentally send the packet into an infinite loop due to a broken line between routers that force packets to be rerouted through alternate paths. Eventually the routers will adjust their packet forwarding tables. However, in this case the packet travels through too many routers and the TTL field expires and the packet is discarded.

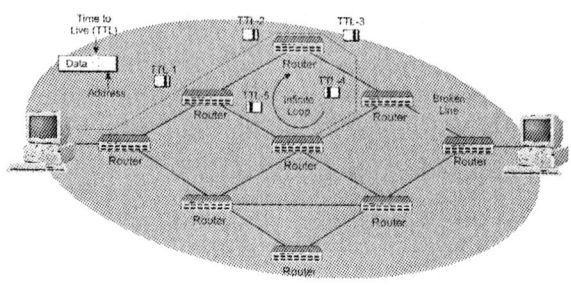

Time to Live (TTL) Operation

Timed-Sc/H Phased
Timed-Release Disconnect-A disconnect process that results as a result of exceeding a pre-determined maximum time interval for a process or service.
Timeslot-The smallest time division dedicated to a mobile station in the GSM system. Each time slot is a 577 microsecond time period.
Timestamp-Time Stamp
Timestamp Synchronization-The process of using time information embedded in multiple sources of data (such as an audio stream and a video stream) to present them together in synchronization. An example of a streaming protocol with embedded synchronization timestamps is RTP.
Timing Pulse-A pulse that is used to synchronize signals.
Timing Recovery-The process of determining the appropriate sampling times for a data stream by deriving a clock signal from that bit stream.
TIMS-Transmission Impairment Measurement System
Tinned-An element covered with a thin layer of metallic tin to inhibit corrosion and facilitate soldering.
Tip & Ring-An old fashioned way of saying "plus" and "minus," or ground and positive in electrical circuits. Tip and Ring are telephony terms. They derive their names from the operator's switchboard

plug, and the ring wire was connected to the slip ring around the jack. A third conductor on some jacks was called the sleeve. That's it. Nothing more sinister. Nothing more interesting.

Tip Wire-One of the two subscriber loop/line wires. It is connected to the tip conductor on a manual switchboard plug, hence its name. In North America, green insulation color is used to identify it. Corresponds to the European "A" subscriber wire.

TIPHON-Telephony and Internet Protocols Harmonization over Networks

TISOC-Telecom Industry Services Operating Center

TK-Toolkit

TLDN-Temporary Location Directory Number

TLLI-Temporary Logical Link Identity

TLS-Transport Layer Security

TLS-Transparent LAN Services

TLV-Type-Length-Value

TM-Phone Doubler

TM-Transverse Magnetic

TM Services-CLASS

TMSI-Temporary Mobile Station Identity

TN-Timeslot Number

TND-Telephone Network For The Deaf

TNDS!IX-Total Network Data System/Trunking

TO-Tandem Office Switches

ToIP-Telephony Over Internet Protocol

Token-The code passed among nodes (typically computers) in network in a particular sequence for each node. This sequence indicates which note has permission to transmit data next.

Token Bucket Size-The size of the token bucket (i.e. buffer) in bytes. If the bucket is full, then applications must either wait or discard data. For best-effort service, the application gets a bucket as big as possible. For guaranteed service, the maximum buffer space will be available to the application at the time of the request. See also Token Rate.

Token Bus-In local area networking, a topology in which a network node transmits its data only when it receives an electronic symbol, or token, that is transmitted continuously from node to node around a logical data bus called a token-ring. This process is called token passing.

Token Passing-A method of controlling the use of a communications channel, especially a ring network A token packet is circulated from node to node when there is no live traffic. Possession of the token gives a node access to the network for transmission of data.

Token Reservation-A mechanism that allows stations to arbitrate for the future use of a token at a given priority level.

Token Ring-Token ring is a LAN system that passes a token to each computer connected to the network. Holding of the token permits the computer to transmit data. The token ring specification is IEEE 802.5 and token ring data transmission speeds include 4 Mbps, 16 Mbps, 100 Mbps and 1 Gbps.

Token ring networks are non-contention based systems, as each device connected via the token ring network must receive and hold a token before it can transmit. This ensures only one device will transmit data at any given time. Token ring systems provide an efficient control system when many devices are interconnected with each other. This is the reason token ring systems will not see data traffic degradation when many new stations are added, compared to collision-based (non-switched) Ethernet systems which degrade exponentially as new stations are added. Passing tokens does add a small overhead (additional control messages) and can slightly increase a packet's transmission time if the transmitting station must wait for a free token to arrive in efficient control system when many computers are interconnected with each other. This is the reason token ring systems will not see data traffic degradation when many new users are added compared to Ethernet systems. However, passing tokens does add overhead (additional control messages) that reduces the overall data transmission bandwidth of the system.

Token Ring Frame Structure-The structure of data packet that is used in the Token ring data network. The data packet frame length is variable from between 18 bytes and approximately 18,000 bytes. The maximum packet size is limited to 10 milliseconds. The fields within the data packet include a start delimiter for synchronization, access control, frame control, 48 bit destination address, 48 bit source address, optional routing information, optional data field, error detection field and ending delimiter. There are two types of frames used in the Token ring system: media access control (MAC) frames and logic link control (LLC) frames. MAC frames are used to control access to the ring and report errors, while LLC frames are used to transport user data.

Token Ring Network

This figure shows the structure of a token ring packet. This example shows that the fields within the data packet include a start delimiter for synchronization, access control, 48 bit destination address, 48 bit source address, control fields, data packet, and error detection. The data field has a variable length packet and there is no specified maximum frame length.

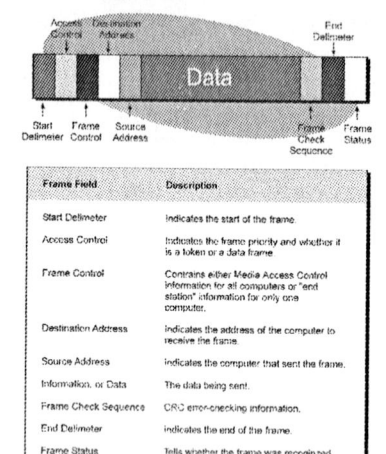

Token Ring Frame Structure

Token Ring Network-IBM's implementation of the IEEE 802.5 token-ring standard. This protocol implements a token-passing mechanism at rates of 4 or 16Mbps on shielded or unshielded twisted pair cabling. As many as 250 stations may be connected on a single ring. Multiple rings may be connected with bridges to create virtually unlimited sized networks. Token Ring Networks utilize a star-wired topology with a physical ring created by the wiring hubs, also called Multistation Access Units (MAU). MAUs can be chained together to produce large rings. On a token-ring network, one station is elected to be the controlling network interface card, known as the Active Monitor. The Active Monitor is responsible for providing the synchronization clock for all stations on the ring, as well an ensuring that a token is always circulating on the ring. If the Active Monitor does not receive a token every 10 milliseconds, it will begin a ring recovery protocol to ensure that there are no faults in the ring and produce a new token. When a station has data to transmit, it captures the token, sets its Access Control Field to busy, and then adds a MAC header and user data. All other nodes continuously receive data from the ring and search for valid frames and their own destination MAC address. If a frame is destined for a station, it will copy the packet to a buffer while repeating the frame to the next station. This enables the transmission of multicast frames. For unicast frames, the receiver will set an Address Recognized (AR) bit in the Ending Delimiter to inform the sender that the frame was seen by the receiver. In addition to the AR bit, a Frame Copied (FC) bit is set by the receiver if it has buffer space available to receive the entire frame. The sender is responsible for removing any packet it transmits. If, due to a sender error, a frame is seen by the Active Monitor more than once, it will destroy the frame and begin the ring recover protocol.

This diagram shows a typical token ring LAN. This diagram shows that token ring networks are non-contention based systems, as each computer connected via the token ring network must have received and hold a token before it can transmit. This ensures computers will not transmit data at the same time. Token ring systems provide an efficient control system when many computers are interconnected with each other. This is the reason token ring systems will not see data traffic degradation when many new users are added compared to Ethernet systems. However, passing tokens does add overhead (additional control messages) that reduces the overall data transmission bandwidth of the system.

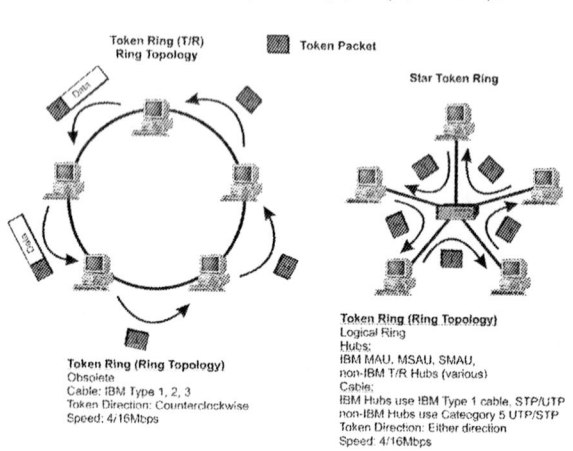

Token Ring System

Toll-A charge for telecommunications service that results when a call is routed beyond a local calling area. Due to telecommunications deregulation, toll charge boundaries have changed and in some cases, have been eliminated throughout entire country areas so all calls are charged the same rate even when calling out of their local service area.

Toll Bypass-The routing of calls or communication sessions around any other networks facilities to avoid toll charges. An example of toll bypass is the use of voice over internet protocol (VoIP) services that allows customers to bypass the public switched telephone network (PSTN) switches in order to utilize the packet network (ex. IP data network) for long-distance (also known as toll) voice calls. This technique is used in conjunction with the H.323 protocol.

Toll Call-A call made beyond a subscriber's local calling area. The charges for such calls vary, but may take into account the distance, duration, day of week, and time of day.

Toll Fraud-The theft of long-distance service including hacking or using stolen credit cards, computers, and 800 numbers to access a switch and determine a method by which other calls can be placed.

Toll Free-A service that allows callers to dial and telephone number without being charged for the call. The toll free call is billed to the receiver of the call. In the United States, toll free calls are preceded by an 1-800, 1-888 or 1-877 exchange. In Europe and other parts of the world, toll free calls are called Freephone and typically begin with 0-800.

Toll Quality-A quality of service (QoS) that is acceptable for voice communication on a telephone system. Because the measurement of voice quality is subjective, it is measured by a mean opinion score (MOS). The MOS is number that is determined by a panel of listeners who subjectively rate the quality of audio on various samples. The rating level varies from 1 (bad) to 5 (excellent). Good quality telephone service (called "toll quality") has a MOS level of 4.0.

TollGate-A milestone step in a product development process that is used as an evaluation point to determine if further product development steps will be taken. There are often tollgates between concept, feasibility, planning, development, and introduction phases. Tollgates may require specific types of documents such as product descriptions, marketing evaluations, and intellectual property reviews.

TOM-Telecommunications Operational Model

Tone Clock-A circuit or assembly that provides timing signals that are used by tone generators to create tones such as call progress, ringing, and signaling tones.

Tone Controlled Squelch System-Tone controlled squelch systems mute the audio of a radio receiver unless the incoming radio signal contains a specific tone. Tone controlled squelch allows a radio user to avoid listening to noise or interference signals of other radio transmissions that occur on the same frequency.

This figure shows a tone controlled squelch system. In this diagram, the radio receiver only un-mutes (connects) the audio signal when an incoming RF signal level contains a particular squelch tone. This permits to receiver to block out conversations from other nearby users that do not have the same squelch tone.

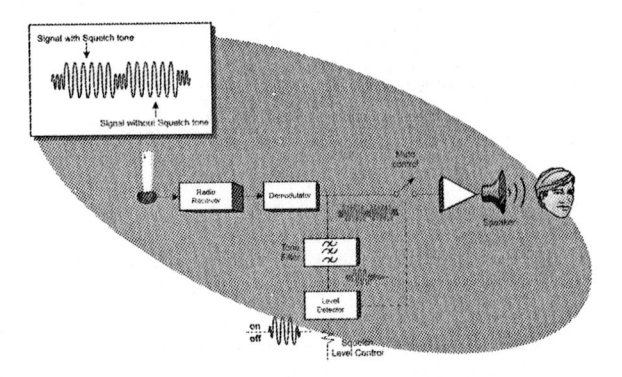

Tone Controlled Squelch System

Tone Only Paging-A paging system that only sends a tone or beep.

Tone Paging-Paging service whereby the user is notified of a message via a tone. This tone usually is designated to mean a callback to a single location is requested.

This figure shows a tone pager typical rate plan offered in 2003 in New Zealand. This rate plan shows that the monthly fee for tone paging varies based on the geographic coverage area. The rate for wide area national coverage is $15 per month and that this rate decreases to $8.50 per month for a

Tone Receiver

smaller metro area coverage zone. This plan also shows that the user of the pager also pays charge of 17 cents per page and that the calling party also pays a per message rate of 10 cents per page.

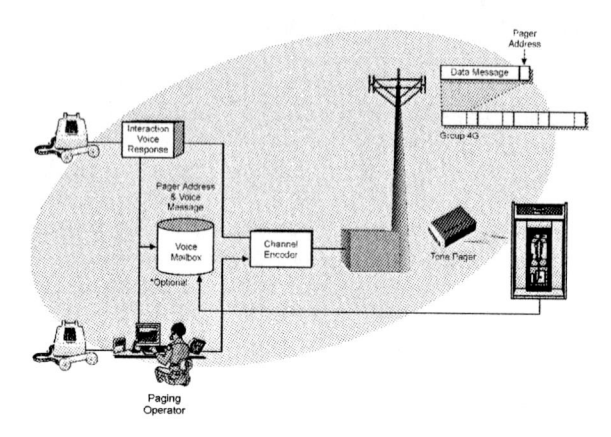

Tone Paging Operation

Tone Receiver-A circuit or program that can receive and decode tone information into their original information (message) form. Examples of tone receives include DTMF and MF tone receivers.

Tone Signaling-The process of transmitting supervisory, address, and alerting signals over a telephone circuit by means of voice frequency tones.

Tool Command Language (TCL)-A scripting language that is commonly used by communication service providers to customize their interactive voice response (IVR) systems. TCL uses relatively simple (and modifiable) syntax and can be used as a separate program or combined (embedded) in other programs. TCL is open source language so its' use is free. See http://www.tcl.tk for more information.

Toolkit-A software program or system that assists designers or developers to create products or services.

Toolkit (TK)-A graphic user extension language for tool command language (TCL). TK allows for the creation of graphic user interfaces for software applications.

TOP-Technical And Office Protocols

Topology-The topology of a network is it's physical or logical interconnection layout. It is the connection map of a network. The physical and logical toplogy does not have to be the same. The physical topology describes the connection of cables to equipment. The Logical topology is the data communication paths that messages take to move between locations on the network.

TOS-Type Of Service

Total Access Communications System (TACS)-An analog system in use in England and several other countries. It is an enhanced version of AMPS (analog cellular) that primarily operates on the 900 MHz frequency range. Some TACS systems are converting to GSM systems.

Total Cost Of Ownership (TCO)-A term used to quantify the real cost of a particular system or service. TCO encompasses the direct costs of hardware and software along with the costs for operation and maintenance.

Total Cross-Connect Automation-Rather than using wire or fiber optic jumpers to link telecommunications facilities and equipment total cross connect automation uses electronic cross-connect systems to perform this function.

Total Internal Reflection-The mechanism by which light is contained in an optical fiber. Optical fiber has a core that typically has a large index of refraction surrounded by cladding with a low index. The index of each part is chosen so that virtually all the light that reaches the interface between the core and the cladding is reflected back into the core. If the difference in index is not sufficient, light will be refracted into the cladding, resulting in a loss of signal.

Touch-Tone Signaling-A signaling system that uses the combination of two tones to represent digits that are transferred in a communication system. Also known as dual-tone multi-frequency (DTMF) signaling.

Tower-A tall structure that supports antennas or antenna wires. If made of metal, the tower itself can serve as the antenna radiating element.

TP4-ISO 8073-DAD2

TPMF-Terminating Point Master File

TPMP-Telco Product Market Planning

TPPM-Telco Product Portfolio Management

TR Switch-Transmit Receive Switch

TRA-Telecommunication Resellers Association

Trace-The pattern on an oscilloscope screen when displaying a signal.

Traceroute-A utility used on TCP/IP networks to trace the route that IP datagrams take between one network device (ex. server, router) and another system on the network. A traceroute also tells you how long each hop takes, it can be a useful tool in identifying system or network trouble spots. Traceroute utilizes the time to live (TTL) field in the IP packet and typically runs over the UDP protocol (best effort). Traceroute is sometimes indicated also by the syntax "tracert".

Trade Secret-Information, data, documents, formulas, or anything which has commercial value and which is kept and maintained as confidential.

Trademark-A unique symbol, word, name, picture, or design, or combination thereof used by firms to identify their own goods and distinguish them from the goods made or sold by others.

Trademark rights are Intellectual Property Rights which give the owner the right to prevent others from using a similar mark which could create confusion among consumers as to the source of the goods.

Trademark protection must be registered in the country, or region, where it is desired. In most countries and regions, patents and trademarks are administered by the same government agency.

Traffic-The amount of data transferred over a communications link or number of messages processed by a communication server over a specified period of time. An example of traffic measurement is centum call seconds (CCS). A single CCS equals 100 seconds of a call on the same communication circuit.

Traffic Channel-The combination of voice and data signals existing within a communication channel.

Traffic Channel Full Rate (TCH/F)-(1- GSM) A logical channel that dedicates one slot per frame for a communication channel between a user and the cellular system.

Training Sequence-A sequence of data bits that are previously known to the sender and receiver of the data bits. This allows the receiver to adjusts its reception process by using the known sequence.

Transaction Capabilities Application Part (TCAP)-The portion of the SS7 protocol which is responsible for information transfer between two or more nodes in the signaling network. The application layer of the Transaction Capabilities protocol consists of transaction capabilities that manage remote operations.

Transceiver-The combination of a radio transmitter and receiver into one radio device. A portable cellular phone is a transceiver.

This diagram shows a block diagram of a mobile radio transceiver. In this diagram, sound is converted to an electrical signal by a microphone. The audio signal is processed (filtered and adjusted) and is sent to a modulator. The modulator creates a modulated RF signal using the audio signal. The modulated signal is supplied to an RF amplifier that increases the level of the RF signal and supplies it to the antenna for radio transmission. This mobile radio simultaneously receives another RF signal on a different frequency to allow the listening of the other person while talking. The received RF signal is then boosted by the receiver to a level acceptable for the demodulator assembly. The demodulator extracts the audio signal and the audio signal is amplified so it can create sound from the speaker.

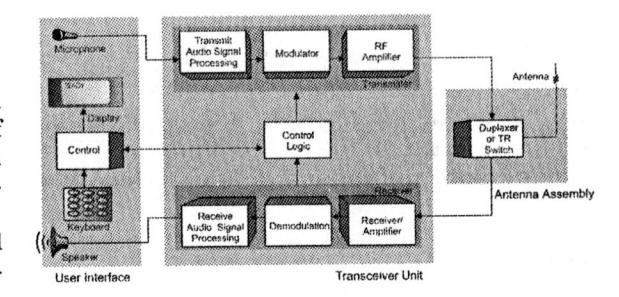

Mobile Radio Functional

Transcoder-A device that enables differently coded transmission systems to be interconnected with little or no loss in functionality.

Transcoding-The conversion of digital signals from one coding format to another. An example of transcoding is the conversion of u-Law encoded PCM to A-Law encoded PCM signals.

This diagram shows the basic transcoding process between mu-Law PCM coding and A-Law PCM coding. This diagram shows that a telephone that uses A-LAW PCM speech coding in North America system is communicating with a telephone in Europe that is using u-LAW PCM speech coding. This diagram shows that the transcoding system must identifies the type of PCM audio used by each system and the location of the transcoding gateway function. The PCM transcoder converts the A-LAW PCM signal to u-LAW PCM.

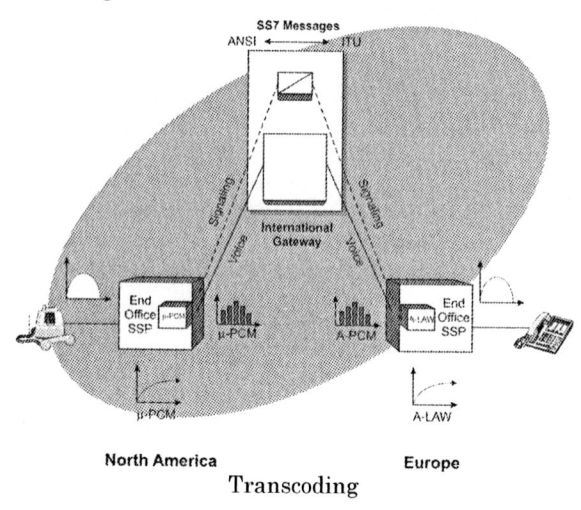

Transcoding

Transducer-Any device or substance that converts energy from one form to another. The transducer on a SAW device converts voltage to mechanical vibrations (phonons) and vice versa. A microphone converts mechanical vibrations (sound) to voltages. A photodiode converts light to electrical energy. A laser or an LED converts electrical energy to light.

Transferred Account Data Interchange Group (TADIG)-A group within the GSM Association that is responsible for defining data inter-change procedures for roaming.

Transferred Account Procedures (TAP)- Transferred accounting process (TAP) is a standard billing format that is primarily used for global system for mobile (GSM) cellular and personal communications systems (PCS). As of 2001, the versions of TAP, TAP II, TAP II+, NAIG TAP II, and TAP 3. Each successive version of TAP provided for enhanced features.

Due to the global nature of 3G wireless and GSM, the TAP billing standard provides solutions for multi-lingual and multiple exchange rate issues. TAP3 was released in 2000 as a significant revision of TAP2. TAP3 has changed from the fixed record size used in TAP2 to variable record size and TAP3 offers billing information for many new types of services such as billing for short messaging and other information services. The TAP standard is managed by the GSM association at www.GSMmobile.com.

Transformer-(1-electrical) An electrical component typically comprising two or more coils of insulated wire surrounding a common core space. The core in some transformers may be only air, but most transformers use a core of ferromagnetic material such as iron or ferrite. Transformers are used in both electric power and telecommunications applications. When an alternating current power source is connected to one coil of a transformer, a voltage having the same waveform but a different (larger or smaller) voltage amplitude is produced via electromagnetic induction at the second coil terminals. The ratio of the two voltages is typically the same as the ratio of the number of turns of wire in each coil, respectively. Thus a transformer having a primary coil (connected to the power source) comprising 100 turns and a secondary coil comprising 200 turns will produce a voltage at the terminals of the secondary coil that is double the source voltage. The current at the secondary coil in this example will be one half of the primary coil current, so the power output is not larger than the power input. Because the voltage to current ratio is four times larger at the secondary coil terminals compared to the primary coil terminals, we describe this transformer as a device that changes the impedance level of the secondary circuit compared to the primary circuit by a factor of four.

Transformers with the same number of turns in the primary and secondary coils are extensively used in telecommunications to couple the alternating current component (such as speech waveform) from the primary to the secondary, but not the direct (constant) component of the current.

Certain devices that do not use coils of wire also have the capability of producing higher or lower voltages as well. For example, a resonant quarter wavelength transmission line stub, driven by a low voltage source of resonant frequency near the short circuit end, produces a higher output voltage than the input voltage at the open circuit end.

(2-toy) In non-telecom usage, there is a completely unrelated child's toy called a transformer that has pieces that can be moved about so that the toy resembles a robot in one configuration and a truck or rocket, for example, in another.

Transformer Ratio-The ratio of the number of turns in the secondary winding of a transformer to the number of turns in the primary winding, also known as the turns ratio.

Transient-A sudden variance of current or voltage from a steady-state value. A transient normally results from changes in load or effects related to switching action.

Translate-(1- radio frequency) The process of converting a signal or band of signals to a different frequency spectrum. (2-data) The process of converting information from one language or code to another.

Translation-The conversion of information from one form to another or the conversion of all or part of a telephone address destination code to routing instructions or routing digits.

Translational Bridge-A transparent bridge that interconnects LANs that use different frame formats (e.g., an Ethernet-to-FDDI bridge). A translational bridge must map the salient fields between the dissimilar frame formats used on its ports.

Transmission-(1 - general) The transfer of electric power, signals, or intelligence from one location to another by wire, fiber optic, or radio means. (2- beam) The use of a directional radio antenna to concentrate radio power into a small angle. (3- parallel) The simultaneous transmission of a number of signals through one or more systems. (4 - radio) The transmission of electromagnetic radiation at radio frequencies. (5 - serial) The transmission of sequential signals over a given system. (6 - stereophonic) The use of various methods to transmit two separate, but related, channels of audio information in order to convey the stereo effect

Transmission Channel-The assembly of circuits, components, and media necessary to transport a message produced by a source and deliver its information content to the input of a sink.

Transmission Control Protocol (TCP)-A session layer protocol that coordinates the transmission, reception, and retransmission of packets in a data network to ensure reliable (confirmed) communication. The TCP protocol coordinates the division of data information into packets, adds sequence and flow control information to the packets, and coordinates the confirmation and retransmission of packets that are lost during a communication session. TCP utilizes Internet Protocol (IP) as the network layer protocol.

This diagram shows how transaction control protocol (TCP) operates to reliably send data through a packet network. This diagram shows that the TCP system receives the data from a specific communication port (port number). The TCP system then packetizes (divides) the sender's data into smaller packets of data (maximum 1500 bytes). Each of these packets starts with an IP header that contains the destination address of the packet. The TCP system then adds a second header (the TCP control header) that includes a sequence number along with other flow control information. The packets are sent through the system where they may be received at different time periods. The sequence numbers can be used to reorder the packets. The TCP protocol also includes a window size that indicates to the receiving device how many packets it can receive before it must acknowledge their receipt. This window defines how much data the sending device must keep in temporary memory to enable the retransmission of a packet in the event that a packet is lost in transmission. If a packet is lost, the receiving device requests the transmitting device to re-send the packet with a specific sequence number.

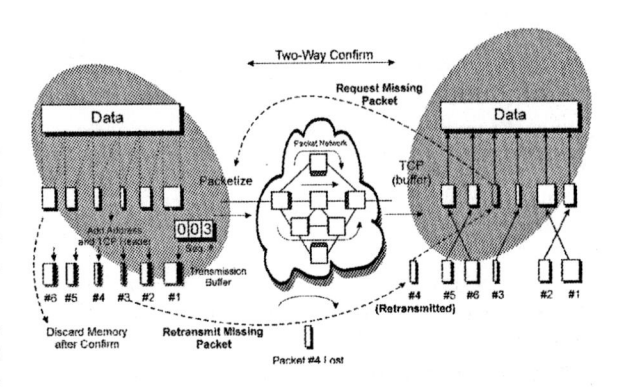

Transmission Control Protocol (TCP) Operation

Transmission Control Protocol (TCP) Header-A control header that is added to the data portion of an Internet protocol datagram packet. The TCP header holds sequencing information that is used to coordinate the reliable transfer of information during a communication session.

This diagram shows the TCP header field structure for version 4 IP. The TCP header is session level protocol and is it is located after the network IP addressing portion of each IP datagram packet. The TCP control structure is used to ensure reliable communication using IP addressing. The TCP packet header starts with port numbers that indicates the logical channel where the data originated from

Transmission Control Protocol And Internet Protocol (TCP/IP)

the sending device and where it is to be routed (e.g. which software application) by the receiving communication device. The header includes sequence numbers and acknowledgement sequence numbers to track the sending and confirmation of each packet (or groups of packets) that have a TCP header. The TCP header includes a field that defines the confirmation window size. The window size indicates the maximum number of TCP packets that can be received before the receiving device must confirm their reception.

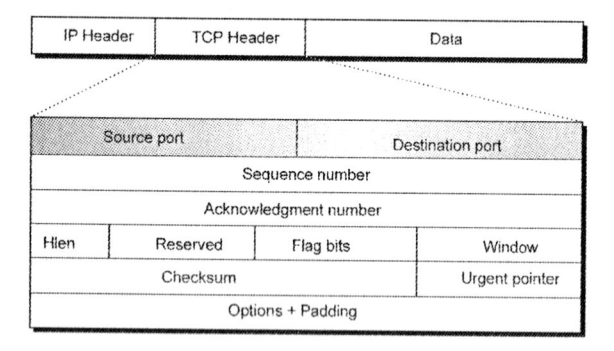

IP Header	TCP Header	Data

Source port	Destination port
Sequence number	
Acknowledgment number	

Hlen	Reserved	Flag bits	Window
Checksum			Urgent pointer
Options + Padding			

Transmission Control Protocol Version 4 (TCP) Header Structure

Transmission Control Protocol And Internet Protocol (TCP/IP)-TCP/IP is standard set (suite) of protocols that defines how the Internet messages are transferred reliably through a data network. The Transmission Control Protocol (TCP) portion ensures message delivery between two points and the Internet Protocol (IP) defines the routing and addressing of packets.

Transmission Delay-Transmission delay is the time that required for transmission of a signal or packet of data from entry into a transmission system (e.g. transmission line or network) to its exit of the system. Common causes of transmission delay include transmission time through a transmission line (less than the speed of light), channel coding delays, switching delays, queuing delays waiting for available transmission channel time slots, and channel decoding delays.

Transmission Line-(1-general) Any transmission medium, including free space. (2- transmitter) A circuit that connects a transmitter to a load over a distance. (3-wave) Any circuit whose dimension is large compared with the wavelength of the signals passing through it.

Transmission Loss-The ratio, usually measured in decibels, of the power of a signal at a point along a transmission path to the power of the same signal at a more distant point along the same path. This value often is used as a measure of the quality of the transmission medium for conveying signals. Changes in power level normally are expressed in decibels by calculating 10 times the logarithm base 10) of the ratio of the two powers.

Transmission System-Transmission systems interconnect communication devices (end nodes) by guiding signal energy in a particular direction or directions through a transmission medium such as copper, air, or glass. A transmission system will have at least one transmitting device, a transmission medium, and a receiving device. The transmitting communication devices is capable of converting information into form electrical, electromagnetic wave (radio), or optical signals that allows the information to be transferred through the medium. The receiving communication device converts the transmitted signal into another form that can be used by the device or other devices that are connected to it. Transmission systems can be unidirectional (one direction) or they can be bi-directional (two directions).

Transmit Delay-The time delay between when a signal is first originated to when it is first received at its destination. Also called transmission time or propagation delay for radio signals.

Transmit Flow Control-A means of adjusting the rate at which data may be transmitted from one terminal in a data communication system so that it is equal to the rate at which the data can be received by another terminal in the same system.

Transmit Receive Switch (TR Switch)-A switch that provides a connection from an antenna to either a transmitter or receiver. The use of a TR switch avoids the possibility of the high power transmitter being connected to a sensitive receiver. This diagram shows how a TR switch connects an antenna to either a transmitter or receiver. This TR switch is controlled by the transmitter on ("key on") switch. When the transmitter is on, the TR switch moves to connect the antenna to the transmitter. When the transmitter is off, the TR switch connects the antenna to the receiver allowing the user to hear the channel when they are not talking.

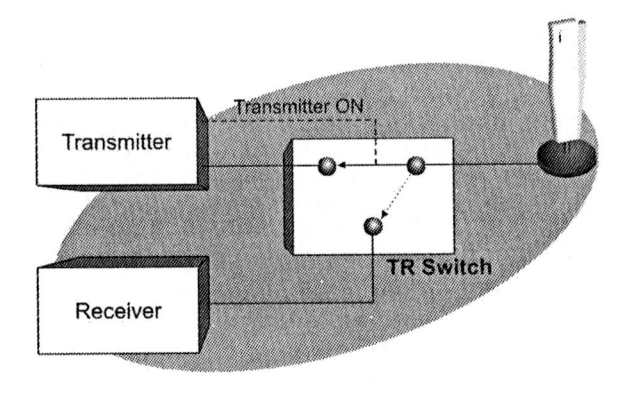

TR Switch Operation

Transmittance-A measure of the capability of a material to permit light signals to pass through.

Transmitter-(1-telephone) Older term for the microphone in a telephone handset, and used as an adjective to identify signals associated with that microphone. (2-radio) Device for transmitting a radio signal, or the location thereof.

Transmitter (XMTR)-A device or system that receives and information signal and converts it to a form that is suitable for transmission.

Transparency-A property of a communications medium (such as a coaxial cable or optical fiber) to carry a signal without altering or other-wise affecting the photonic or electrical characteristics of the signal.

Transparent Bridge-A local area network bridge which relies solely on MAC address learning for all forwarding decisions. Each transparent bridge inspects each packets source MAC address to associate each MAC address with a destination port. When a packet arrives its source MAC address is inspected for forwarding table maintenance, while its destination MAC address is inspected for to determine the egress port. Contrast this to a source route bridge, which relies on explicit information in each packet to determine the path he packet will take. Transparent bridges are generally more difficult to implement than source route bridges, but allow end stations to be on bridged LAN segments without being aware of that fact. Thus they are called transparent bridges. Source route networks require each end station to maintain not only destination MAC address information, but also path route information. However, source route bridges are easier to implement and allow complex networks to be built with multiple active parallel paths for better traffic distribution and reliability.

Transparent LAN Services (TLS)-Transparent LAN Services (TLS) are provided by ILECs, CLECs or other service providers to allow end customer to leverage their existing LAN infrastructure in a Wide Area Network. To implement a TLS, the service provider will typically provide the end user with termination equipment, sometimes in the form of an edge router, which presents the end user with a standard Ethernet or other LAN interface.

The customer utilizes the service providers wide area network to connect multiple enterprise sites, but manages it as a private LAN. The details of the underlying WAN are hidden from the customer, thus reducing the amount of management personnel required to maintain the network.

Transparent Systems-An optical transmission system that does not require conversion to electrical form for switching or routing through a network.

Transponder-(1-relay) A combination receiver and transmitter, frequently part of a communications satellite, that receives a signal from an uplink station and retransmits it to ground receiving stations. (2-response system) An electronic circuit that receives a signal from an interrogating station, such as a ground-based radar unit, and transmits an appropriate response.

Transport Layer-A protocol layer that responds to transmission service requests from upper layers (such as the session layer) and creates service requests to the lower network communication layers. The primary purpose of the Transport Layer is to provide the transfer of data between end users.

Transport Overhead-In the Synchronous Optical Network (SONET), the overhead added to the Synchronous Payload Envelope of a Synchronous Transport Signal for transport purposes. Transport overhead consists of line and section overhead.

Transportable-A full 3-watt cellular mobile telephone, sometimes switchable to .3 watts, complete with battery pack and antenna, facilitating portable operation.

TRAU-Transcoder And Rate Adaptation Unit

Tray-Electronics Frame

Treatment-The handling of customers or vendors for a specific purposes such as recovering overdue balances. When special treatment involves collection activities, it is called "dunning".

Treble-The highest audible sound frequencies, between approximately 2000 to 20,000 cycles per second.

Tree Topology-In a local area network, a configuration that resembles the distribution system of a tree trunk and its branches.

Trellis Coding-A method of forward error correction used in certain high-speed modems where each signal element is assigned a coded binary value representing that element's phase and amplitude. It allows the receiving modem to determine, based on the value of the preceding signal, whether or not a given signal element is received in error.

Tremendously High Frequency (THF)-The frequency band from 300GHz to 3000GHz.

TRF-Tuned Radio Frequency

Trigger-The process of initiating an action as a result of a specific event occurring or a parameter associated with an event (such as exceeding a voltage level).

Trigger Point-An event, incident, or development in an outside plant network that initiates a step or stimulates a reaction. An example of a trigger paint is the exhaustion of existing facilities or structures.

Triggers-Statically armed detection points.

Trim-The process of making fine adjustments to a circuit or a circuit element.

Trimmer-A small, mechanically adjustable component connected in parallel or in series with a major component so that the net value of the two can be finely adjusted for tuning purposes.

Triode-A 3 element electron tube, consisting of a cathode, control grid, and plate.

TRIP-Telephone Routing over Internet Protocol

Triple Play-Triple play refers to providing of three main services such as data, voice, and video on one network. For cable MSOs, this usually means building out the next generation network to DOCSIS 2.0 specifications, for Carriers this often means building out fibre or VDSL (very fast DSL) networks. Usually it is the larger MSOs and Telecom Carriers that roll out triple play services, and the advantage is that they can sign customers to a bundle of three services, thereby increasing revenue and customer loyalty.

Trivial File Transfer Protocol (TFTP)-Trivial file transfer protocol (TFTP) is a protocol that is used to transferring files between devices in a data communication network. TFTP is a simplified version of file transfer protocol (FTP) and it is commonly used in devices to allow for the transfer of setup and configuration information. TFTP is defined in RFC 1350.

Tropopause-The upper boundary of the troposphere above which the temperature either increases slightly or remains constant.

Troposphere-The layer of the earth's atmosphere, between the surface and the stratosphere, in which about 80 percent of the total mass of atmospheric air is concentrated and in which temperature normally decreases with altitude.

Tropospheric Wave-A radio wave propagated by reflection from a point of abrupt change in the dielectric constant or dielectric gradient in the troposphere.

Trouble-A failure or fault affecting the service provided by a system.

Trouble Ticket-A corrective maintenance work order used by equipment and facilities maintenance personnel.

Troubleshoot-The process of investigating, localizing and (if possible), correcting a fault.

TRS-Telecommunications Relay Service for the Deaf

TRUA-Transcoder Unit

Truncation-The removal of lower significant bits on a digital system, possibly leading to errors or artifacts.

Trunk-A communication path that connects two network elements (such as switching systems, networks or data devices). Trunks are usually shared by many users. Trunks may be classified by the type of equipment they connect. For example, a PBX trunk connects a PBX system to the public switched telephone network (PSTN). Trunks carry conversations for different subscribers at different times. The name comes from a trunk of a tree, via the prior usage for railroad trunk lines.

Trunk Carrier (T Carrier)-Trunk carrier (T Carrier) are a hierarchy of digital communication lines that are used for multiple channel transmission lines that range from 1.544 Mbps to 565 Mbps. T carrier is the actual transmission channel as opposed to the digital signal (DS) channel that is the digital format into the channel transmitter. Tx has been used to represent the digital transmission standards where the "x" denotes which service is under discussion.

Trunk Group-A number of trunks that can be used interchangeably between two switching systems.

Trunk Occupancy-The percentage of some time period, usually an hour, during which a trunk is in use. Trunk occupancy also may be expressed as the carried hundred call seconds per hour per trunk.

Trunk Side Connection-Trunk side connections are used to interconnect telephone network switching systems to each other. Trunk side connections are usually high capacity lines.

Trunk Signaling-In interoffice signaling, the use of trunks carrying voice or data traffic to also transfer signals between switching systems.

Trunk Usage-The percentage of usage of a trunk compared to its maximum capacity. Trunk usage may be measured at peak time or it may be an average measurement.

Trunk-Side Connection-The connection of a transmission path to the trunk side of a local switching system.

Trunked Enhanced Terestrial Radio (TETRA)-TETRA is a digital land mobile radio system that was formerly call Trans European Trunked Radio. The TETRA is being developed by the European Telecommunications Standards Institute (ETSI) to create a more efficient and flexible communication services from both private and public-access mobile radio users.

TETRA is capable of sending and receiving short data messages simultaneously with an ongoing speech call. It effectively supports voice groups and has capacity for over 16 million identities per network (over 16 thousand networks per country). TETRA permits direct mode operation (talk around) that permits direct communication between mobile radios without the network. TETRA includes a priority feature to help guarantee access to the network by emergency users. The system allows independent allocation of uplinks and downlink to increases system efficiency. The signaling protocol supports sleep modes that increases the battery life in mobile radios.

The TETRA system is fully digital and allows for mixed voice and data communication. It is specified in open standards. The TETRA system allows up to 4 users to share each 25 kHz channel. It allows inter-working with other communication networks via standard interfaces. TETRA is capable of call handoff between cells and it has integrated security (user/network authentication, air-interface encryption, end-to-end encryption).

Trunked Radio-A radio base station system in which each radio channel carries conversations for different subscribers at different times. Trunked radio systems may utilize reuse of the same radio carrier frequencies in different base stations, like cellular radio systems. Unlike cellular radio systems, the mobile radios used in trunked radio systems do not have the ability to change their operating frequency in response to commands from the base stations and thus cannot handover a connection in progress.

Trunked Radio System-Trunked radio is a mobile communication system that allows mobile radios (called trunked radios) to access more than one of the available radio channels in that system. The radio access control procedure allows the mobile to access alternate radio channels in the system in the event that the channel it has requested is busy.

This figure shows a typical trunked SMR system. In this example, there are several available radio channels. Mobile radios that operate in this system that wish to communicate may search for an available radio channel by looking for identification tones. Optionally, some trunked radio systems use dedicated radio channels to coordinate access to radio channels.

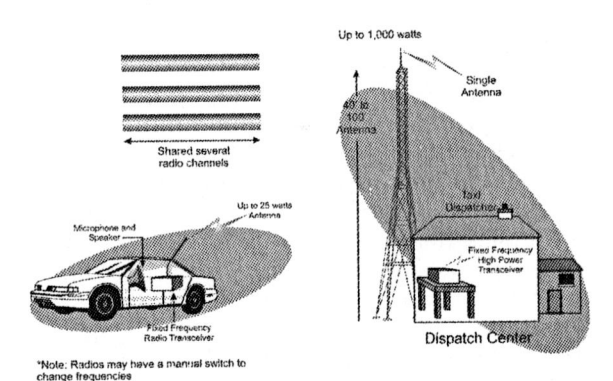

Trunked Radio System

Trunked System

Trunked System-Trunked systems are two-way radio networks that use an automatic (dynamic) process to select an unused radio channel (trunk) from a pool of available channels when communication service is requested. The use of dynamic channel assignment (trunk assignment) increases the probability that a mobile radio transceiver can gain access to the system when there are several available channels.

This picture shows a mobile Land Mobile Radio (LMR); model Maxon Enduro SM-6000, that is used by public safety or other private personnel to communicate with each other and to dispatchers.

Land Mobile Radio (LMR)
Source: Topaz3

Trunking-An infrastructure dependent technique where communications resources, comprised of more than one logical channel (trunk) are shared amongst system users by means of an automatic resource allocation management technique based upon statistical queuing theory and resident in the systems fixed infrastructure. Typically usage requests follow a Poisson arrival process (statistical probability distribution) and the resource allocator assigns communications resources in response to requests from system users. As demand for service exceeds system capability at that time, service must be increasingly denied immediate access. This action is termed "blocking", with the blocked service request being queued for a later service response. The offered grade of service of the system is inversely proportional to the probability of blocking (e.g. lower probability of blocking offers a higher grade of service potential).

The dynamic resource allocation methodology of trunking results in the establishment of functional channels defining resource availability by means of dynamically allocating logical channels both to particular subscribers and for specific functions. These functional channels can be used for the conveyance of payload information, system control or a combination thereof. This results in the development of three (3) specific types of functional channels, these are: control, digital voice, and digital data.

Dedicated Control Trunking: Refers to a logical channel resource which supports only control function channel type (e.g.. resource management signaling, requests for service, etc.) between the fixed trunking system infrastructure and the subscriber and associated units.

Composite Control Trunking: Refers to a logical channel resource which supports control functional channel type as well as digital voice and/or digital data functional channel types (e.g. teleservice related payloads).

Trunking Efficiency Ratio-A ratio of the measured system trunk usage efficiency to the engineered trunk occupancy. The amount of trunking efficiency can be computed for different trunk groups, network types, and common network usage characteristics (e.g. average duration of a call).

Trunking Gateway-Trunking gateways are voice gateways that provide bearer connectivity between the PSTN and the Voice-over-Packet network. Typically, T1 or T3 trunks provide the physical interface to the PSTN. The voice gateway converts voice bearer traffic from TDM voice to packetized voice (VoIP) and vice versa.

Trunk gateway is considered a "trunk-side" gateway, requiring IMTs (inter-machine trunks) for interconnection into the PSTN to carry voice traffic and "A"-link for interconnection to the SS7 network to carry signalling. In this case, a separate IP-to-SS7 signalling gateway is required to translate and mediate signals between the IP network and the SS7 network. Media Gateways also perform voice compression using CODECs; common CODECs for VoIP applications are: G.711 (64kbps), G.729a/b (8kbps), and G.723 (5 - 6 kbps).

Trunks-Groups of wires or fiber optic communication lines that are used to interconnect communications devices. Trunks usually have many physical and/or logical communication channels.

Trunks In Service-The number of trunks in a group that are in use or available to carry calls. The number of trunks in service may differ than the actual number of communication lines due to line failure or alternative assignment/uses of wire circuits.

Trunks Required-The number of trunks that are required between switches to provide a specific grade of service such as maximum percentage of blocked calls.

Trusted Device-(1-general) A device that is previously known or suspected to only communicate information that will not alter or damage equipment of stored data. Trusted devices are usually allowed privilege levels that could allow data manipulation and or deletion. (2-Bluetooth) A device using Bluetooth wireless technology that has been previously authenticated and allowed to access another Bluetooth device based on its link-level key. See also Untrusted Device.

TS-Time Slot

TSAPI-Telephony Services Application Programming Interface

TSAT-T1 Small Aperture Terminal

TSI-Time Slot Interchange Switching

TSI-Transmitting Subscriber Information

TSO-Time Sharing Option

TSP-Telephony Service Provider

TSPI-Telephony Service Provider Interface

TSR-Terminate And Stay Resident

TSR-Telephone Service Representative

TSS-Telecommunication Standardized Sector

TTA-Telecommunications Technology Association

TTI-Transmission Time Interval

TTI-Transmit Terminal Identification

TTL-Transistor Transistor Logic

TTL-Time To Live

TTS-Transaction Tracking System

TTS-Text To Speech

TTY-Teletypewriter

Tube-(1 - electron) An evacuated or gas-filled tube enclosed in a glass or metal case in which the electrodes are maintained at different voltages, giving rise to a controlled flow of electrons from the cathode to the anode. (2- cathode ray, CRT) An electron beam tube used for the display of changing electrical phenomena, generally similar to a Th picture tube. (3 - cold-cathode) An electron tube whose cathode emits electrons without the need for a heating filament. (4 - gas) A gas-filled electron tube in which the gas plays an essential role in operation of the tube. (5 - mercury vapor) A tube filled with mercury vapor at low pressure, used as a rectifying device. (6 - metal) An electron tube enclosed in a metal case. (7 - traveling wave, TWT) A wideband microwave amplifier in which a stream of electrons interacts with a guided electromagnetic wave moving substantially in synchronism with the electron stream, resulting in a net transfer of energy from the electron stream to the wave. (8- velocity-modulated) An electron tube in which the velocity of the electron stream is changing continually, as in a klystron. In British English the word "valve" is used instead of "tube.

Tune-The process of adjusting the frequency of a device or circuit, such as for resonance or for maximum response to an input signal.

Tuned Radio Frequency Amplifier-An amplifier that operates at a specific (tuned) radio frequency.

Tuner-(1 - radio) The radio frequency and intermediate frequency pants of a radio receiver that produce a low-level audio output signal. (2 - waveguide) A device that permits adjustment of the impedance of a waveguide.

Tuning-(1 - general) The process of adjusting a given frequency; in particular, to adjust for resonance or for maximum response to a particular incoming signal. (2-broad) The tuning of a circuit that is not sharply resonant. (3- electron) The adjustment of the frequency of a vice by changing electrode potential. (4 - ganged) The simultaneous tuning of several circuits by a single control, often involving varying capacitors mounted on a common shaft. (5 - permeability) The tuning to resonance of a circuit by moving a core (usually ferrite) into or out of a coil to change its inductance. (6 - sharp) A tuning technique that is sharply resonant over a narrow frequency bandwidth. (7 - staggered) A method of obtaining a broader effective bandwidth by tuning each individual IF stage to a slightly different frequency.

Tuning Core-A ferrite or powdered iron core that can be moved relative to a coil to change its inductance.

Tuning Screw-An impedance adjusting rod that tunes by penetrating a cavity or waveguide.

Tunnel-A logical path through a network. Data that enters into a tunnel end will appear at the tunnel exit unchanged regardless of the network type that it is being transmitted through. Network tunneling is the process that creates a secure communication path through the use of a virtual connection within a network link through the encryption of data that is transmitted between the virtual connection points.

TUP-Telephone User Part

Turnkey-The development of a product or system where the product or system is ready for use when it is delivered to a customer.

Turnover-(1 - equipment) The point at which the installation of central office equipment is complete, and a telephone company accepts the equipment from an installation agent or supplier. (2 - wiring) In a cable pair, the reversal of the tip and ring connections.

Turns Ratio-In a transformer, the ratio of the number of turns on the secondary to the number of turns on the primary. (See also: transformer.)

TVI-Television Interference

TVM-Time-Varying Media

TVRO-Television Receive Only

TVUI-Telephony Voice User Interface

Tweaking-The process of minor adjusting an electronic assembly circuit to optimize its performance or to bring the system to within required specifications.

Tweeter-A loudspeaker designed to efficiently handle high (3 kHz to 20 kHz) audio frequencies.

Twisted Pair-A pair of insulated copper wires used in transmission circuits to provide bi-directional communications. The wires are twisted about one another to minimize electrical coupling with other circuits. Paired cable is made up of a few, to several thousand, twisted pairs.

Two Binary, One Quarternary (2B1Q)-A line code (modulation and signaling structure) that transmits two binary bits of data at one time using a multi-level code. This code was initially used by the Integrated Services Digital Network (ISDN) standard for basic rate service (2 64 kbps + 1 16 kbps channels). 2B1Q is used by some DSL technologies including IDSL, HDSL and SDSL.

Two Wire Circuit-A circuit that carries information signals in both directions over a single pair.

Two-Way Paging-Two-way paging is a process where paging messages (signals) are sent from a radio tower to a pager that has the capability to return a delivery verification signal. Two-way paging systems usually have the ability to also receive messages that are initiated from the paging device to the system. Because two-way paging systems can determine the location of the pagers, it can send messages directly to the paging transmitter where the two-way pager has been located. This avoids the need to send paging messages to many paging transmitters to larger geographic regions thus dramatically increasing the number of pagers it can serve with a limited number of radio channels or bandwidth.

Two-way paging systems allow the paging device to acknowledge and sometimes respond to messages sent by a nearby paging tower. The two-way pager's low power transmitter necessitates many receiving antennas being located close together to receive the low power signal. This Figure shows a high power transmitter (200-500 Watts) that broadcasts a paging message to a relatively large geographic area and several receiving antennas. The reason for having multiple receiving antennas is that the transmit power level of pagers are much lower than the transmit power level of the paging radio tower. The receiving antennas are very sensitive, capable of receiving the signal from pagers transmitting only 1 watt.

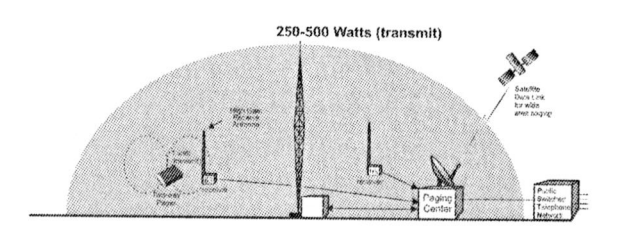

Two-Way Paging

Two-Way Radio System-Two-way mobile radio systems allow communication on a dedicated radio channel between two or more mobile radios. Traditionally, two-way radio systems have served public safety and industrial applications. These systems typically only allow push-to-talk service (one-way at a time) to allow sharing of a single radio channels frequency. Two-way radio systems are available in a wide range of frequency bands.

This figure shows a typical analog two-way radio system. In this diagram, a mobile radio has an FM transmitter and FM receiver bundled together. Because the mobile radio is used for "push-to-talk" service, the antenna does not need to be connected to both the transmitter and receiver at the same time. When the user presses the push to talk button, the antenna is connected to the transmitter. When the button is released, the antenna is connected to the receiver. A squelch circuit is connected to the receiver to allow the receiver audio to reach the speaker only when the receiver level or tone code is of sufficient level to ensure a good received signal. When the level is below that threshold (normally set by the user via a squelch knob), the speaker is disconnected and no sound (such as noise) can be heard. The mobile radio communicates with a radio tower that has a high power base station. The base station usually is connected to a control console that allows a dispatcher to communicate or to patch the audio channel to another location (such as a telephone line connection).

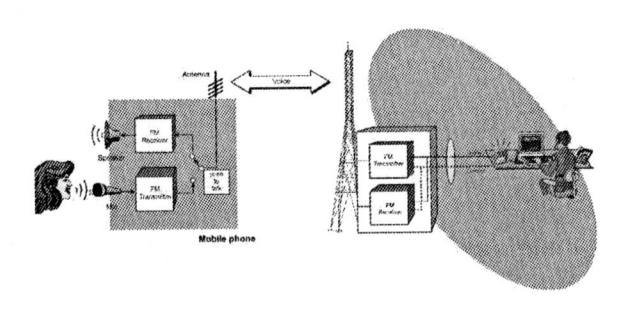

Two-Way Radio
Source: Motorola

Two-Way Radio System

This picture shows a two-way handheld mobile radio that can receive and reply to one another within 500 feet of each other. These two way radios are often used by nature lovers and professional that desire to routinely coordinate (communicate) their activities.

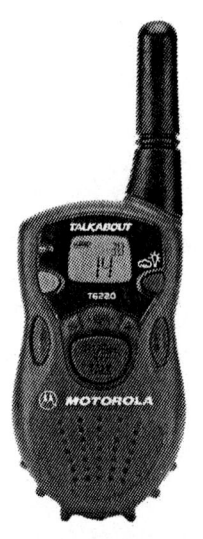

Two-Wire To Four-Wire Conversion-The connection of a two-wire cable facility to a four-wire transmission facility with a separate path in each direction. This conversion uses a hybrid to separate the two directions of transmission on the two-wire side.

TWX-Teletypewriter Exchange Service
TXD-Transmit Data
Type 1-Data Connector
Type 1 Connection-Type 1 connections are trunk-side connections to an end office. The end office uses a trunk-side signaling protocol in conjunction with a feature known as Trunk With Line Treatment (TWLT). This connection was originally described in technical advisory 76 published by AT&T in 1981. This interconnection was developed because dial line and DID connections did not provide enough signaling information to allow the connection of public telephone networks to other types of networks (such as wireless and PBX networks). The switch must be equipped to provide TWLT, or its equivalent to offer Type 1 service. As a result, type 1 is not universally available. The TWLT feature allows the end office to combine some line-side and trunk-side features. For example, while trunk-side signaling protocols are used, the calls are recorded for billing purposes as if they were made by a line-side connection.

Type 2A Connection

Type 1 connections are usually used as 2-way trunks. Two-way trunks are always 4-wire circuits, meaning they have separate transmit and receive paths, and almost always use MF address pulsing and supervision. The address pulsing normally uses wink-start control. One-way Type 1 connections can be provided on a 2-wire basis using E&M supervision or reverse battery like the DID connection.

Type 2A Connection-Type 2A connections are true trunk-side connections that employ trunk-side signaling protocols. Typically, they are two-way connections that are 4-wire circuits using E&M supervision with multifrequency (MF) address pulsing. The address pulsing is almost always under wink-start control. Type 2A connections allow the other public or private telephone network switching systems to connect to the PSTN and operate like any other EO.

Type 2A connections may restrict calls to specific NXX (exchange) codes and access to operator services (phone number directories, emergency calls, freephone/toll free) may not be permitted. For some interconnections, additional connections (such as a type 1) may be used to supplement the type 2A connection to allow access to other operator or network services.

Type 2B Connection-Type 2B connections are high usage trunk groups that are used between EOs within the same system. The type 2B connection can be used in conjunction with the Type 2A. When a type 2B is used, the first choice of routing is through a Type 2B with overflow through the type 2A. Because the type 2b connection is used for high usage connections, it can access only valid NXX codes of the EO providing that it is connected to. Type 2B connections are almost always 4-wire, two-way connections that use E&M supervision and multifrequency (MF) address pulsing.

Type 2C Connection-Type 2C connections were developed to allow direct connection to public safety centers (E911) via a tandem or local tandem switch. This interconnection type must provide additional such as the return phone number (complicated on mobile telephone systems) and the location of the caller. This information is passed on to a public safety answering point (PSAP).

Type 2D Connection-Type 2D interconnection lines allow direct connection from a operator services system (OSS) switch. The OSS switch is special tandem that contains additional call processing capabilities that enables operator services special directory assistance services. The type 2D connection also forwards the automatic number identification information to allow proper billing records to be created. Type 2D connection will normally use trunks employing E&M signaling with wink start, and multifrequency (MF) address pulsing.

Type A USB Connector-A USB connector that is used to connect to the USB host (upstream) device.

Type Approval-A approval from the a regulatory agency (such as the FCC) that identifies the equipment that is manufactured has passed tests certifying it meets the minimum requirements for that type of electronic or radio equipment. Most electronic devices must meet several regulatory specification requirements to receive type approval.

Type B USB Connector-A USB connector that is used to connect to the USB data downstream) device.

Type Of Service (TOS)-The field within the IP datagram header that indicates the type of packet so the system can vary the type of service (typically the priority of routing) that is performed on the IP data packet. The TOS field as specified in RFC 760 has and RFC 2475 defines a application of TOS as used in DiffServ networks.

Type S-Type S connections transfer signaling messages that are associated with other interconnection types (out-of-band signaling). The type S is a data link (e.g. 56 kbps) that is used to connect the signaling interfaces between switches. Type S connections permit additional features to be supported by the network such as finding and using call forwarding telephone numbers. Because type S connections cost money, some smaller public telephone networks do not offer or use type S connections.

U

u -Occasional typographic substitute for Greek letter μ.

μ-Law Encoding-The type of non-linear digital voice coding (digital signal companding) that is commonly used in the Americas and other parts of the world. The U Law (pronounced Mu Law) coding process is used to compress the 13 bit sampling of a digitized audio signal into the equivalent of an 8 bit sample. It does this by assigning a non-binary (non-linear) value to each of the binary bits. Another non-linear voice coding system is the A Law coding system that is used in Europe and other parts of the world.

UA-User Agent
UAC-User Agent Client
UART-Universal Asynchronous Receiver And Transmitter
UAS-User Agent Server
UAT-User Acceptance Testing
UBR-Unspecified Bit Rate
UCD-Upstream Channel Descriptor
UCD-Uniform Call Distribution
UCH-User Channel
UDLC-Universal Data Link Control
UDLC-Universal Digital Loop Carrier
UDP-User Datagram Protocol
UDP-Uniform Dial Plan
UDP Header-User Datagram Protocol
UE-User Equipment
UGS-Unsolicited Grant Service
UHCI-Universal Host Controller Interface
UHF-Ultra High Frequency
UI-User Interface
UI-Unnumbered Information
UIC-Union Internationale Des Chemins De Fer
UL Approval-Underwriters Laboratories
ULF-Ultra Low Frequency
Ultra High Frequency (UHF)-The frequency range from 300 MHz to 3000 MHz.
Ultra Low Frequency (ULF)-The frequency band from 300 Hz to 3000 Hz.
Ultrasonic-An acoustical signal of a frequency higher than can be heard by a human (above 20 kHz).
Ultraviolet-Electromagnetic waves invisible to the human eye, with wavelengths about 10-400 nm.

Um Interface-The RF channel between the cell site and the cellular mobile unit. The label commonly given to the Common Air Interface reference point in the General System Model.
UMB-Upper Memory Block
UMTS-Universal Mobile Telecommunications System
UMTS Frequency Bands-UMTS frequency bands assigned for 3G wireless systems. Additional UMTS frequency may be added in the future to allow cost effective systems capacity expansion. Some of the UMTS frequency bands have already been assigned for other services such as PCS systems in the United States.

This figure shows the different frequency bands that have been identified for UMTS use. It is likely that additional frequency bands will be added in the future to allow increased competition and cost effective systems capacity expansion. Some countries already use frequency bands that have been designated as UMTS. This includes part of the personal communication service (PCS) systems in the United States. These bands may be converted to UMTS systems.

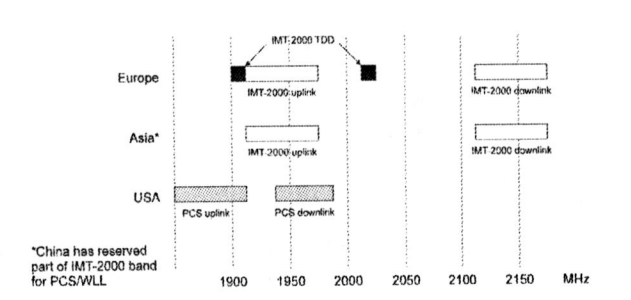

Universal Mobile Telecommunications System (UMTS) Frequency Bands

UMTS Subscriber Identity Module (USIM)

UMTS Subscriber Identity Module (USIM)-A smartcard that holds a subscribers identity information, authentication and voice privacy algorithms, and some customer feature preference information for user equipment (UE) in the universal mobile telephone system (UMTS).

UMUX-Universal Data Stream Multiplexer

Unavailability-A measure of the degree to which a system, subsystem, or piece of equipment is not operable and not in a committable state at the start of a mission when the mission is called for at a random point in time.

Unbalanced-A condition of circuits or lines in which the impedance of one side of the terminal differs from that of the other side of the same terminal. Examples include a cable pair in which the impedance of the ring and tip differ, and an amplifier in which one of the input terminals is connected to ground. (See also: single ended, balanced.)

Unbalanced Line-A transmission line in which the magnitudes of the voltages on the two conductors are not equal with respect to ground. A coaxial cable is an example of an unbalanced line.

Unbundled-A term describing services and programs that are sold separately by the manufacturers of telephone a computer-related equipment.

Unbundling Services-Unbundling is the process of separating portions of a telecommunication network that are owned or operated by a service provider. Unbundling is a common term used to describe the separation of standard telephone equipment and services to allow competing telephone service providers to gain fair access to parts of incumbent telephone company systems. An example of an unbundled service is for the incumbent phone company to lease access to the copper wire line that connects an end user to the local telephone company. The competing company may install high-speed data modems (such as ADSL) on the copper line to enhancing the value of the telecommunications service.

UNC-Universal Naming Convention

Uncertainty-An expression of the magnitude of a possible deviation of a measured value from the true value. Frequently, it is possible to distinguish two components: the systematic uncertainty and the random uncertainty. The random uncertainty is expressed by the standard deviation or by a multiple of the standard deviation. The systematic uncertainty generally is estimated on the basis of the parameter characteristics.

Unchannelized Carrier-A communication line (carrier) that allows the user to have unrestricted access to the entire data transmission capacity (after the line control overhead is removed) of the communication bearer circuit. Unchannelized carriers are sometimes called unstructured carriers.

Underground-A type of construction in which cables are pulled through conduits buried in the ground. Access to such cables and conduits is provided by utility access holes placed at distances of every 500 to 1000 feet.

Underlying Carrier-A common carrier that provides services or facilities to other common carriers.

Undervoltage Protection-The automatic disconnection by a circuit breaker of loads from a power source when the incoming voltage is too low for safe and reliable operation.

UNE-Unbundled Network Element

UNE-P-Unbundled Network Elements Platform

Unexposed Cable-Generally refers to underground toll cabling that is does not have aerial risers or terminations other than straight splices. Outside plant engineering can design/install non-protected apparatus cases (e.g. 466 or 818 type).

UNI-User Network Interface

Unicast-Within the Internet, a unicast is delivery of data to only one client, that is, the opposite of multicast. In particular, unicast is used to describe a streaming connection from a server to a single client.

Unicast Address-An address used to route data to a specific device that is connected to a communications network. The Unicast address is also called a Physical or Hardware address.

Unicode-Universal Code

unicom-Aeronautical Advisory Station

Unidirectional-A signal or current flowing in one direction only.

Unidirectional Carrier-A telephone company (telco) carrier system that utilizing 2 cable runs.

Unified Messaging-Unified messaging allows you to store, manage, and transfer different forms of messages from a variety of access devices. Unified messages include audio (voice messages), electronic mail (email), data messages (such as fax or files), and video (video mail). Unified messaging provides you with access to these multiple types of messages using standard telephones, (text to audio), Internet web pages (playing back voice messages), and other devices such as fax machines and mobile telephones.

This diagram shows that the key parts of a unified messaging system are access devices, storage system, and unified messaging services (programs). This diagram shows that there are several ways to access and receive information from the unified messaging system including pagers, wireless telephones, standard (wired) telephones, email, faxes, and web pages. The core of the unified messaging system is a digital message storage and retrieval system. This system can provide various services and in this example, it includes voice mail, fax mail, e-mail, text-to-audio, operator-to-audio, and directory information services. The digital storage system is accessed via a media converter processing assembly. This processor is capable of converting audio to data, data to audio, and transforming data from one format (such as a email) to data of another format (such as a fax page). The system operation is controlled by the feature programs and requests from the access devices (such as play message).

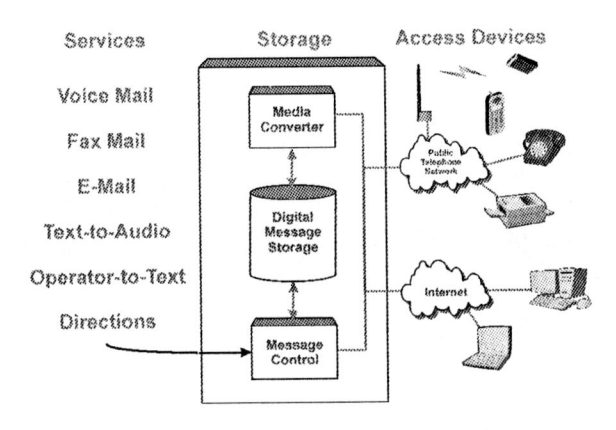

Unified Messaging System

Uniform Dial Plan (UDP)-The dialing plan (digit sequences) that are used by all standard users within a telephone system. The use of UDP in private telephone systems (such as a PBX) allows callers to dial telephone extensions (other users within the private system) using a predefined dialing codes (such a 4 digit or 5 digit extension codes).

Uniform Resource Identifier (URI)-A set of characters in an HTTP message header that identifies a resource (such as a image or media file) that exists on the Internet.

Uniform Resource Name (URN)-A naming process defined by the Internet Engineering Task Force (IETF) that is used as an identifying name for Internet resources. The URN has no relation to where the resources are located located within the Internet.

Uniform Service Ordering Code (USOC)-A code comprising a few letters and/or numbers used to order services and/or equipment in the telephone industry. Established in the 1960s to provide a short and uniform description for the same service and equipment provided by various different telephone service providers and manufacturers, Some specific USOC codes have become well known to the public, such as a RJ11 socket for a telephone set connection to the central office wiring.

Unimodem-Universal Modem Driver

Uninterruptible Power Supply (UPS)-A battery backup system designed to provide continuous power in the event of a commercial power failure or fluctuation. A UPS system is particularly important for network servers, bridges, and gateways.

Unipolar-A transistor formed from a single type of semiconductor material, n-type or p-type, as employed in a field effect transistor.

Unit Data-An SS7 message that contains information for ISuP and TCAP services.

Universal ADSL Working Group (UWAG)-A working group that was established to assist in the standardization of ADSL equipment for the consumer marketplace. The UWAG was setup primarily by telecommunications service providers and computer equipment manufacturers.

Universal Asynchronous Receiver And Transmitter (UART)-A UART is an assembly or integrated circuit that coordinates communication on a communication line using serial data communications. UARTS receive data from a microprocessor or other data source, divides the information into groups of data bits, adds error detection bits and control information, and then controls the serial transmission of the data on a transmission line. This diagram shows how a universal asynchronous receiver and transmitter (UART) can randomly send blocks of data that are related to multiple channels (logical channels) of data over a serial data bus. In this example, the UART converts data

Universal Connector

from a parallel format into a send data format, transmits the data at a speed determined by the UART serial data bus clock, frames the blocks of data into different sized data packets, multiplexes control and data packets by using logical channel identifiers, and delimits (identifies) the start and stop points of data by specific length start and stop transmission bits.

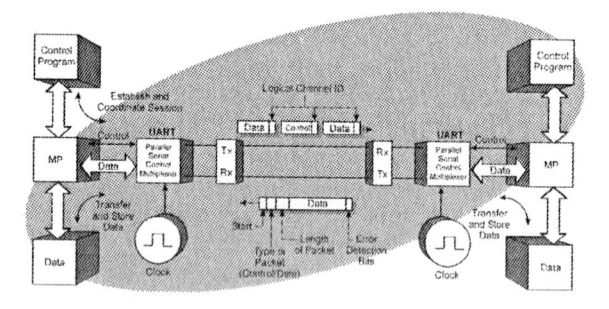

Additionally, since the "special" wiring configuration is in an external jumper it is easily changed as technology shifts. Typical RJ-45 jacks wired in the wall are wired with all 4 twisted pairs when only 2 are used in most available networks.

Universal Asynchronous Receiver And Transmitter (UART) Operation

Universal Connector
Source: InCon of Fuquay Varina Corp. ("InCon")

Universal Connector-A connector that can be used for voice, data, or video services. Universal connectors use industry standard connection types to allow end users to connect a variety of device types to the connector.

This figure shows a Multi-Function Communication Faceplate that supports 3 wire types through one hole in the wall. Power, Coaxial Cable and twisted pair communication wire (phone/digital). The InCon Tri-Plate attaches directly to a standard decorator style A/C receptacle that is mounted conventionally in a standard single size electrical outlet box. The InCon cable that is used for these universal connectors consists of one RG-6 Coaxial Cable that is thinly webbed to a 4 pair Category-5 bundle of Twisted Pair communication wire. For a Local Area Network (LAN) a cable is made up with the LAN (RJ-45) computer jack at one end and 2 phone type (RJ-11) jacks at the other. This leaves the other 2 RJ-11 jacks and their respective twisted pairs for use for phone or other devices.

Universal Host Controller Interface (UHCI)- The Universal Serial Bus (USB) host interface used by Intel.

Universal Mobile Telecommunications System (UMTS)-A Universal Mobile Telecommunications System (UMTS) that offers personal telecommunications services that uses the combination of wireless and fixed systems to provide seamless telecommunications services to its users. It is expected that UMTS will allow on-demand transmission capacities of up to 2 Mb/s in some of its radio locations. It should be compatible with broadband ISDN services.

Universal Resource Locator (URL)-A standardized addressing process used to identify resources that are connected to the Internet. The URL is a text string that defines the location of a resource (such as an address of a web site the Internet), as well as the protocol to be used to access the resource.

Universal Serial Bus (USB)-An industry standard data communication interface that is installed on personal computers. The USB was designed to replace the older UART data communications port.

There are two standards for USB. Version 1.1 that permits data transmission speeds up to 12 Mbps and up to 127 devices can share a single USB port. In 2001, USB version 2.0 was released that increases the data transmission rate to 480 Mbps.

This diagram shows how a universal serial bus (USB) system interconnects devices in a personal distribution network (PDN). This example shows that a USB system uses a host controller interface (HCI) to coordinate the access to all other devices that it is attached to. As each device is added, the host controller registers the device (called device enumeration) and coordinates all communication to and from the devices. This diagram also shows that there are two types of connectors used in the system to ensure that a host device is not accidentally connected to another host device. A hub is used to allow the connection of additional devices (up to 127 can be attached to one host system). The USB system allows for the supply of power through the USB cable (5 volts) or an external power supply can be used.

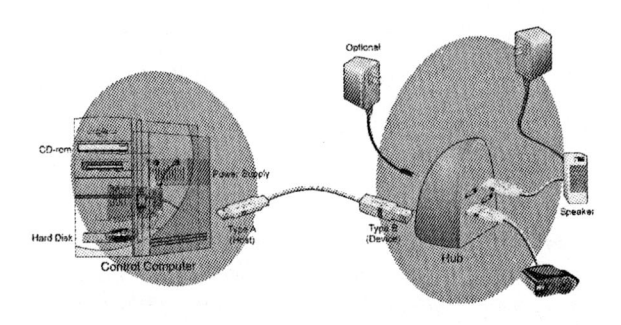

Universal Serial Bus (USB) System

Universal Service-The objective set by many state regulatory agencies and the Federal Communications Commission to keep telephone services affordable for as many customers as possible.

Universal Service Fund–The Universal Service Fund is a collection of money from existing communication service providers that is used to develop communication systems that ensure that areas of the U.S. that are economically or geographically challenged have the funding to receive the same telecommunications capabilities every other part of the U.S. Since the early 1900's, carriers have been faced with difficult financial burdens of providing telephone access to everyone "universally". For example, the cost/benefit analysis of wiring rural areas for telephone service has always deterred carriers from dedicating equal attention to areas that are not as dense in population.

Universal Synchronous Asynchronous Receiver Transmitter (USART)-An integrated circuit provided in many data communications devices that converts data in parallel form from a processor into serial form for transmission.

Universal Synchronous Receiver And Transmitter (USRT or USART)-An electronic module that combines the transmitting and receiving circuitry needed for synchronous communications over a serial line.

Universal Terrestrial Radio Access (UTRA)-A worldwide standard for the 3rd generation wireless communications system developed by the 3rd generation partnership project (3GPP).

UNIX-A computer operating system originally developed and deployed by the Bell Telephone laboratories and now an industry standard. UNIX is a registered trademark mark of UNIX System Laboratories.

Unknown Device-In a Bluetooth system, a Bluetooth device that is currently not connected with a local device and the local device has not paired with it in the past. No information about the device is stored, either by address, link key, or other information. An unknown device is also called a new device.

Unlicensed Frequency Band-A frequency band that can be used by any product or person provided the transmission conforms to transmission characteristics defined by the appropriate regulatory agency.

Unlicensed Spectrum – A frequency band that is used for low-power devices that do not require a user license from regulatory agencies (such as the FCC or other agencies). These low power bandwidths have been commonly used for consumer electronics with low RF requirements such as electronic garage door openers, remote control toys, and car alarm devices.

UNMA-Unified Network Management Architecture

Unmanaged Connection-A communication connection that does not provide guaranteed performance. Examples of unmanaged connections include residential DSL and cable modem lines.

Unmodulated

Unmodulated-A carrier signal that is not modulated by an information carrying signal.

Unshielded-Wiring not protected from electromagnetic and radio frequency interference by a conductive braid or foil.

Unshielded Twisted Pair (UTP)-The transmission line for DSL systems is typically unshielded twisted pair (UTP). UTP consists of pairs of copper wires twisted around each other and covered by plastic insulation. The twisting of the copper wire pair reduces the effects of interference as each wire receives approximately the same level of interference (balanced) thereby effectively canceling the interference. UTP is by far the most popular cabling used for local access lines and computer LANs (such as 10BaseT and 100BaseT).

Unspecified Bit Rate (UBR)-Unspecified bit rate (UBR) is a category of telecommunications service that provide an unspecified data transmission rate of service to end user applications. Applications that use UBR services do not require real-time interactivity nor do they require a minimum data transfer rate. UBR applications may not require the pre-establishment of connections. An example of a UBR application is Internet web browsing.

Unstructured Supplementary Service Data (USSD)-(1- GSM) A service that allows a mobile telephone to send messages to different parts of the GSM network without conforming to specific message structures. USSD allows for flexibility to deploy new network services.

Unterminated-A device or system that is not terminated.

Untrusted Device-In a Bluetooth system, a device that is unknown to another Bluetooth device. An untrusted device may require authentication based on some type of user interaction before access is granted.

Unwanted Emission-A spurious or out-of-band emission.

Up Converter-A frequency translation device in which the frequency of the output signal is greater than that of the input signal. Such devices are commonly found in microwave radio and satellite systems.

Up-Sell Campaign-Marketing campaign with the objective of encouraging customers to purchase more of a product that they are already buying.

Uplink-(1- Satellite) The earth-to-satellite microwave link and related components such as earth station transmitting equipment. The satellite contains an uplink receiver. Various uplink components in the earth station are involved with the processing and transmission of the signal to the satellite. (2- cellular systems) The radio link between the mobile station and the base station.

Upper Sideband (USB)-The higher of the two bands of frequencies produced by amplitude modulation. The upper sideband is equal to the sum of the carrier and the modulating signal frequencies. It can be transmitted with or without a full-level carrier and retains all information impressed during the modulation process.

UPS-Uninterruptible Power Supply

Upstream-(1 - general) A device or system placed ahead of other devices or systems in a signal path. (2-network) The direction opposite the direction of distribution of network timing signals. (See also: down-stream.) (3- video keyer) A term that describes the location of keyers in a mix/effects level or in the overall switcher architecture. (4- video switcher) A term relating the priority of the video signals as they are combined through a production switcher.

Uptime-The uninterrupted period of time that network or computer resources are accessible and available to a user.

URC-Uniform Resource Characteristic
URI-Uniform Resource Indicator
URI-Uniform Resource Identifier
URL-Universal Resource Locator
URN-Uniform Resource Name
US TDMA-IS-136

Usable Field Strength-The minimum value of the field strength necessary to permit a desired quality of reception under specified receiving conditions in the presence of noise and interference, either in an existing situation or as determined by agreements or frequency plans. For fluctuating interference or noise, the percentage of time during which the required quality must be ensured is specified.

Usage-A measurement of the load carried by a server or group of servers, usually expressed in hundred call seconds (CCS) per hour or erlangs.

Usage Stimulation Campaign-Marketing campaign with the objective of encouraging customer to increase the amount of service they using.

USART-Universal Synchronous Asynchronous Receiver Transmitter
USB-Universal Serial Bus
USB-Upper Sideband

USB Bulk Endpoint-The terminating logical channel (or address) in a data communication system that receives large blocks of data. The bulk endpoint is defined as part of the Universal Serial Bus (USB) system.

USB Bus-A serial data communication line that allows multiple devices to share a data communication channel.

USB Control Endpoint-The terminating logical channel (or address) in a data communication system that sends and receives signaling messages. The control endpoint is defined as part of the Universal Serial Bus (USB) system.

USB Enumeration-USB enumeration is the process of discovering the communication parameters and configuration states of a data device that has been connected to a computer. When the computer (the host) discovers a new device, it determines what type of data transfers it anticipates transferring. Enumeration is a process used in the many computing systems.

USB Power-Power can be applied to USB devices from the USB host or from an external power source. When power is applied from the USB host, the maximum power that can be used is 500 mA.

USB Version 2.0-A high speed version of the USB bus that was released in mid 2001. This specification increases the data transmission rate to a maximum of 480 Mbps.

usec-Microsecond

User-A person, company, or group that uses the services of a system for the transfer of voice, data information or other purposes.

User Acceptance Testing (UAT)-A set of tests that are performed by or for a potential user (often a buyer) of a piece of equipment or system that is supposed to ensure the equipment or system will meet the functional requirements of the user.

User Agent (UA)-End user devices in a SIP system are called user agents (UA). The UA is a conversion device that adapts signals from a data network into a format that is suitable for users. Examples of user agents include dedicated IP telephones (hardphones), analog telephone adapters (ATAs), or software (softphones) that operate on a computer that has multimedia (audio) capabilities.

User Agent Client (UAC)-User agent clients are the requesting user part of the session initiation protocol (SIP) communication session.

User Agent Server (UAS)-User agent servers are the request processing part of the session initiation protocol (SIP) session. The user agent server receives requests from a user agent client (UAC) and generates responses.

User Authentication-A security strategy that verifies the identity of a person requesting access to a network, system, and/or computer.

User Channel (UCH)-The raw data portion of a channel that is available to the user. In the Synchronous Optical Network, a channel allocated to a network provider for data communication, such as is used in maintenance activities and proprietary remote notification of alarms external to the span equipment.

User Datagram Protocol (UDP)-UDP is a high-level communication protocol that coordinates the one-way transmission of data in a packet data network. The UDP protocol coordinates the division of files or blocks of data information into packets and adds sequence information to the packets that are transmitted during a communication session using Internet protocol (IP) addressing. This allows the receiving end to receive and re-sequence the packets to recreate the original data file or block of data that was transmitted. UDP adds a small amount of overhead (control data) to each packet relative to other high-level protocols such as TCP. However, UDP does not provide any guarantees to data delivery through the network. UDP protocol is defined in request for comments 768 (RFC 768).

This diagram shows how user datagram protocol (UDP) operates to efficiently send data through a packet network. This diagram shows that the UDP system first packetizes (divides) the sender's data into smaller packets of data (maximum 1500 bytes). Each of these packets starts with an IP header that contains the destination address of the packet. The UDP system then adds a second header (the UDP control header) that includes a destination port. The packets are sent through the system where they may be received or lost in transmission. Because the UDP protocol does not contain any guarantee of delivery, it is up to the user on how to handle lost packets of data.

User Datagram Protocol (UDP) Header

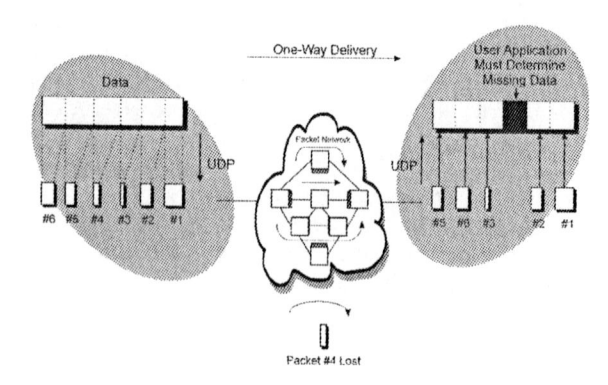

User Datagram Protocol (UDP) Operation

User Datagram Protocol (UDP) Header-A control header that is added to the data portion of an Internet protocol datagram packet. The UDP header holds a limited amount of control data and it is used to coordinate the transfer of information during a communication session that does not need to retransmit (confirm) information.

This diagram shows the UDP header field structure for version 4 IP. The UDP header is session level protocol and is it is located after the network IP addressing portion of each IP datagram packet. The UDP header structure is simplified used to allow efficient unacknowledged communication using IP addressing. The UDP packet header contains with port numbers that indicates the logical channel where the data originated from the sending device and where it is to be routed (e.g. which software application) by the receiving communication device.

UDP Packet Version 4 Header Structure

User Equipment (UE)-A mobile radio telephone operating within a universal mobile telephone system (UMTS). This includes hand held units, transceivers installed in vehicles and fixed wireless units.

User Group-A group of users of a specific product or software system that share information. User groups may have newsletters and chat rooms to help gather and distribute information relative to a product or service.

User Interface (UI)-The portion of equipment or operating system that allows the equipment to interface with the user. Also called the man machine interface.

User Network Interface (UNI)-The interface between an end user and a telecommunications network. A UNI could be a industry standard set of protocol rules and data transmission specifications or may be a proprietary protocol.

User Part-A functional part of a common channel signaling system that transfers signaling messages via the message transfer part. Different types of user parts exist (e.g. for telephone and data services), each of which is specified to a particular use of the signaling system.

User Plane-The portion of a network system that is involved in the transfer of the users media (voice, data, or video).

User-Based Security Model (USM)-The security model was originally defined in RFC 2274 for version 3 of SNMP. It is obsoleted by RFC2574. It supports secure authorization of the user sending a packet, as well as the privacy or encryption of the packet.

User-To-User Signaling Supplementary Service (UUS)-Services that are provided or operated between users that are not part of standard telecommunication services of the system connecting the users.

USIM-UMTS Subscriber Identity Module
USM-User-Based Security Model
USO-Universal Service Order
USOA-Uniform System Of Accounts
USOAR-Uniform System Of Accounts Rewrite
USOC-Uniform Service Ordering Code
USOC-Uniform Service Order Code
USPTO-United States Patent and Trademark Office, a branch of the United States Department of Commerce
USRT-Universal Synchronous Receiver And Transmitter

USRT or USART-Universal Synchronous Receiver And Transmitter
USSD-Unstructured Supplementary Service Data
USTA-United States Telephone Assodation
USTSA-United States Telephone Suppliers Association
UTC-Coordinated Universal Time
Utility Patent-Utility patents are all other patents than design patents and plant patents. In general, utility patents protect the way an article is made and used and design patents protect the way the article looks. Utility patents usually have a longer duration than design patents and maintenance fees are typically required.
Utilization-(1-Facilities) The use of telecommunications facilities and equipment, expressed as a percentage of working to working plus spare facilities. (2-Performance Management) Utilization as used in performance management for network management measures the use of a particular resource over time. The measure is usually expressed in the form of a percentage in which the usage of a resource is compared with its maximum operational capacity, like bandwidth utilization on a network link.
Utilization Factor-The ratio of maximum demand for service to the rated capacity of a system that must provide that service.
UTP-Unshielded Twisted Pair
UTRA-Universal Terrestrial Radio Access
UTRAN-UMTS Terrestrial Radio Access Network
Uu Interface-The radio interface between mobile equipment (ME) and the base station (node B) in a universal mobile telephone service (UMTS) network.
uucp-Unix-to-Unix copy
UUS-User-To-User Signaling Supplementary Service
UUT-Unit Under Test
UWAG-Universal ADSL Working Group
UWB-Ultra Wide Band

V

V-volt

V.17-ITU recommendation for simplex modulation technique for use in extended Group 3 facsimile applications only. Provides 7,200, 9,600, 12,000, and 14,400bps trellis-coded modulation.

V.21-ITU recommendation for 300 bps duplex modems for use on the switched telephone network. V.21 modulation is used in a half-duplex mode for Group 3 fax negotiation and control procedures.

V.22-ITU recommendation for 1,200 bps duplex modems for use on the switched telephone network and leased lines.

V.22bis-ITU recommendation for 2,400 bps duplex modems for use on the switched telephone networks. V.22bis also provides for 1,200 bps operation for V.22 compatibility.

V.24-A CCITT standard with definitions for interchanged circuits between data terminal equipment (DTE) and data communications equipment.

V.25-The automatic calling or answering equipment on a switched telephone network that includes the disabling of echo suppression on manually established calls.

V.27ter-ITU recommendation for 2.4/4.8-kbps/s modem for use on the switched telephone network. Half-duplex only. It defines the modulation scheme for Group 3 facsimile for image transfer at 2,400 and 4,800 bps.

V.29-ITU recommendation for 9,600 bps modem for use on point-to-point leased circuits. This is the modulation technique used in Group 3 fax for image transfer at 7,200 and 9,600 bps. V.29 uses a carrier frequency of 1,700 Hz which is varied in both phase and amplitude. V.29 can be full duplex on four-wire leased circuits, or half duplex on two-wire and dial-up circuits.

V.32-ITU recommendation for 9,600 bps two-wire full duplex modem operating on regular dial-up lines or two-wire leased lines. V.32 also provides fallback operation at 4,800 bps.

V.32bis-ITU recommendation for full-duplex transmission on two-wire leased and dial-up lines at 4,800, 7,200, 9,600, 12,000, and 14,400 bps. Provides backward compatibility with V.32. It includes a rapid change renegotiation feature for quick and smooth rate changes when line conditions change.

V.33-ITU recommendation for 14.4 kbps and 1.2 kbps modem for use on four-wire leased lines.

V.34-A modem standard for correcting errors and compressing data at 28,800 bits per second. (28.8 Kbps).

V.34 VBIS-A modem standard for correcting errors and compressing data at 34,400 bits per second (34.4 Kbps).

V.35-CCITT's standard for trunk interfaces between network access devices and packet-network-defined signaling for data rates greater than 19.2 Kbps.

V.42-ITU recommendation, primarily concerned with error correction and compression for modems. The protocol is designed to detect errors in transmission and recover with a retransmission.

V.90-A CCITT (ITU) standard for modems that is used for 56K modems. The V.90 system uses an asymmetric connection to provide speeds of up to 56Kbps downstream and up to 33.6Kbps upstream.

VA-Volt Ampere

VACMS-View-based Access Control Models

Vacuum Evaporation-A manufacturing technique in which material to be deposited in a thin layer on another material is heated in a vacuum in the presence of the base material, which remains cool. Atoms evaporate from the heated solid and condense on the base material in a thin film of readily controllable thickness.

Vacuum Fluorescent Display (VFD)-A light-emitting triode utilizing fluorescent phosphors that can be used in alphanumeric display panels.

Vacuum Relay-A relay whose contacts are enclosed in an evacuated space, usually to provide reliable long-term operation.

Vacuum Tube-An electron tube. The most common vacuum tubes include the diode, triode, tetrode, and pentode.

VAD-Voice Activity Detector

Valence Band-The energy band for electrons bound to individual atoms. In metals, such electrons easily can be transferred to the conduction band of the atom to facilitate electric conduction.

Value Added Reseller (VAR)-A company or organization that adds assemblies, software, or documentation to products produced by another manufacturer or service provider so they may be sold in

Value Added Services (VAS)

their sales and distribution system. VARs may modify a standard product (such as a laptop computer) and modify for use in a specific industry (called a vertical application.)

Value Added Services (VAS)-Services that provides benefits to a customer that are not part of the standard telecommunications services associated with a basic communication service. VAS services include voice mail, information services and content delivery.

Services offered by prepaid provider (e.g., voice mail, fax store and forward, interactive voice response, and information services) in addition to calling time.

Value Added Tax (VAT)-A tax that is added on to the value of the product or service.

Value Proposition-A statement, made by a business person, which describes how the business will make use of the information delivered to improve operational efficiency, profitability, or market strength.

Vampire Tap-A type of connector that connects one cable segment to another without the need to cut and splice the cable. The vampire tap uses a needle to pierce the cable insulation so it can make a connection to the wire inside the cable. See also Penetration Tap.

This diagram shows a vampire tap connector is used to connect one cable segment to another. Within the vampire tap, a blade and needle are used to pierces the cable insulation to make a connection to the wire within. Vampire taps are typically used on a Thicknet Ethernet system and connect directly to the network backbone.

VAN-Value Added Network
VAP-Value-Added Process
Vaporware-A sarcastic name for a product that has been announced but is not available and has not been produced. Vaporware products often do not get released.
VAR-Value Added Reseller
Varactor Diode-A semiconductor device whose capacitance is a function of the applied voltage. A varactor diode, also called a variable reactance diode or simply a varactor, often is used to tune the operating frequency of a radio circuit.
Variable Bandwidth-A communication system that allows for a variable data transmission rate or changes in communication channel frequency bandwidth dependent on the need of the end user applications and/or the ability of the system to provided the desired data transmission or frequency bandwidth. Because variable bandwidth systems help match the system resources used to the actual data transmission needs of the end customer (e.g. reduce the bandwidth when the user has nothing to send or say), variable bandwidth systems are more efficient that constant bandwidth systems.
Variable Bit Rate (VBR)-A category of telecommunications service that provide an variable data transmission rate of service to end user applications. Applications that use VBR services usually require some real-time interactivity with bursts of data transmission. An example of a VBR application is videoconferencing.
Variable Gain Amplifier-An amplifier whose gain can be controlled by an external signal source.
Variable Rate Speech Coding-A speech compression process that offers multiple speech coding rates. The use of variable compression rates allows a lower bit rate (higher compression rates) coding process to be used when system capacity is limited and more users need to be added to the system.

This diagram shows a speech compression process that can vary based on speech activity. This example shows an analog speech signal that is sampled in short 10-20 msec intervals. The speech coder produces data rates in the range of approximately 8 kbps to 1 kbps. As the speech activity decreases, the bit rate decreases (speech compression increases). The speech coder bit rate is controlled by commands received by the communication network.

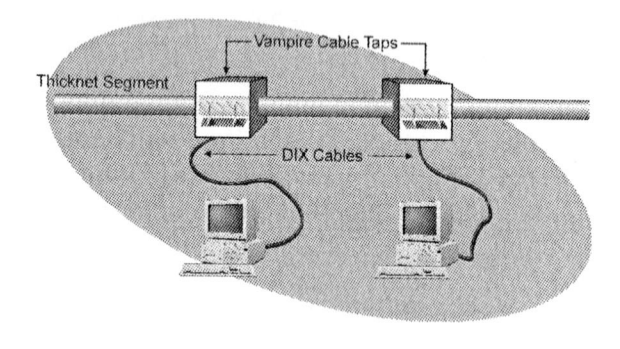

Vampire Tap Diagram

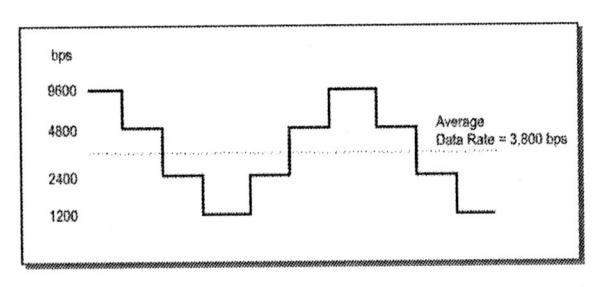

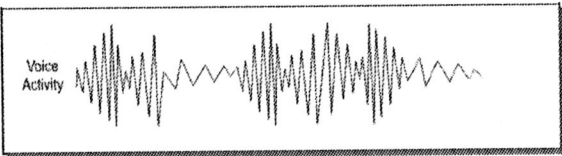

Variable Rate Speech Coding Operation

Varicap-A diode used as a variable capacitor.
Varistor-A semiconductor device whose resistance is a function of the applied voltage. The resistance of a varistor decreases nonlinearly as the applied voltage is increased.
VAS-Value Added Services
VAT-Value Added Tax
VB-Microsoft Visual Basic
VBA-Visual Basic For Applications
VBI-Video Blanking Interval
VBR-Variable Bit Rate
VBS-Voice Broadcast Service
VC-Virtual Circuit
vCalendar-A format for calendar and scheduling information. The vCalendar specification was created by the Versit consortium and is now managed by the Internet Mail Consortium (IMC).
vCard-A format for personal information such as would appear on a business card. The vCard specification was created by Versit consortium and is now managed by the Internet Mail Consortium (IMC).
VCC-Virtual Channel Connection
VCI-Virtual Channel Identifier
VCO-Voltage Controlled Oscillator
VCOMM-Virtual Communications
VCSEL-Vertical Cavity Surface Emitting Laser
VCXO-Voltage Controlled Crystal Oscillator
VDA-Video Distribution Amplifier
VDRV-Variable Data Rate Video
VDSL-Very High Bit Rate Digital Subscriber Line
VDT-Video Display Terminal

Vector Sum Excited Linear Predictive Coding (VSELP)-A speech analysis and compression technology by that uses digital software processing which linear audio (voice) samples are collected, analyzed, and then compressed using an encoding algorithm. VSELP technology is the algorithm used to code and decode speech information in IS-54B and IS-136 TDMA cellular environments.
Version Number-A number that identifies a particular software or hardware product that uses the same name as other products. These products usually undergo revisions or updates and the version number often relates to the date of release. The version number is usually assigned by the manufacturer or developer of the product. that often includes numbers before and after a decimal point; the higher the number, the more recent the release. The version number is important as features and operation of a product or software program may vary between different versions. The ability to determine the specific version number of a product may allow more reliable interaction between programs and products. Version number 1.0 often indicates an initial version of a product or software.
Vertical Application-An application that is designed or used for a specialized industry application or profession. An example of a vertical application is a software programmed to provide wireless meter reading services to utility companies.
Vertical Cavity Surface Emitting Laser (VCSEL)-A type of semiconductor laser in which the cavity is perpendicular to the p-n junction. In most lasers, the cavity is parallel to the junction that is the source of the electron-hole pairs that recombine to form photons. The advantages to the vertical arrangement are that they are relatively easy to manufacture and package, can be modulated directly at higher rates than can LEDs and use less power than other types of lasers. Their light is not as bright as other lasers and is usually of shorter wavelengths around 800 nm, so they tend to be used only for short haul applications.
Vertical Interval Test Signal (VITS)-A signal that may be included during the vertical blanking interval to permit on the air testing and adjustment of video circuit functions.
Vertical Retrace-The return of the electron beam from the bottom to the top of the raster after completion of each field.

Vertical Serrations, Serration Pulses- Serrations

Vertical Services-Products or services that customers can add, at an additional charge, to enhance their basic exchange service. (See also: custom calling services, enhanced services.)

Vertical Sync Pulse-The synchronizing pulse at the end of each field that signals the start of vertical retrace.

Very High Bit Rate Digital Subscriber Line (VDSL)-A communication system that transfers both analog and digital information on a copper wire pair. The analog information can be a standard POTS or ISDN signal and the typical downstream digital transmission rate (data rate to the end user) can vary from 13 Mbps to 52 Mbps downstream and the maximum upstream digital transmission rate (from the customer to the network) can be 26 Mbps. The data transmission rate varies depending on distance, line distortion and settings from the VDSL service provider. The maximum practical distance limitation for VDSL transmission is approximately 4,500 feet (~1,500 meters). However, to achieve 52 Mbps, the maximum transmission length is approximately 1,000 feet (~300 meters).

This diagram shows a VDSL system is commonly used with a fiber distribution network that reaches a neighborhood or small group of buildings. The fiber terminates in an optical network unit (ONU). The ONU converts the optical signal into an electrical signal that can be used by the VDSL modem in the DSLAM. The DSL modem signal is supplied to a splitter that combines the analog and digital signal to copper access line. The splitter is actually attached to the last few hundred feet of the copper access line. The figure shows that the analog POTS signal from the local telephone company may still travel thousands of feet back to the central office. At the customers' premises, the VDSL signal arrives to a splitter that separates the analog signal from the high-speed digital VDSL signal. Because VDSL has a much higher data transfer rate, the CPE may include a digital video set top box that allows for digital television.

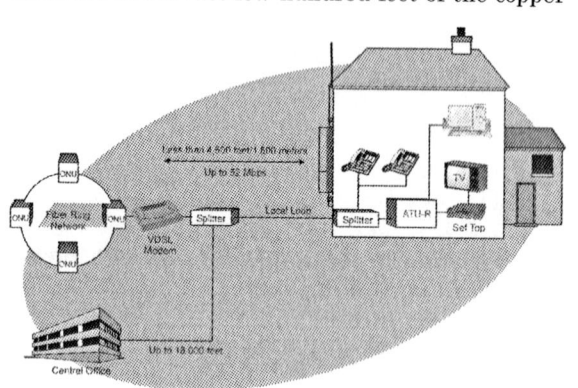

Very High Frequency (VHF)-The frequency band from 30 MHz to 300 MHz (wavelengths 10 m to 1 m).

Very Low Frequency (VLF)-The radio frequency band from 3 kHz to 30 kHz.

This diagram shows how a very small aperture (VSAT) satellite communication system can be used to efficiently distribute data messages to many receivers in a large geographic region. In this example, a computer at a corporate headquarters building sends a data messages to the VSAT communication system via a landline data connection. The VSAT system receives the data message (a corporate price change in this example) and transfers this data message through an Earth station satellite transmitter to the satellite. The satellite receives this data message and retransmits it to the large radio coverage area. Each of the company's retail stores that have the VSAT antenna (a small antenna dish) and receiver decode the message and forward the information to their local computer system which updates the prices.

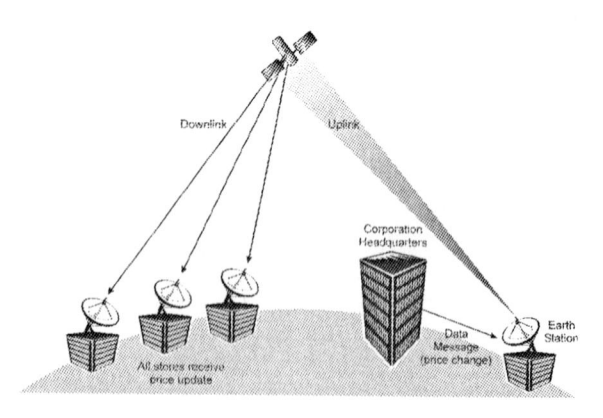

Very High Bit Rate Digital Subscriber Line (VDSL) System

Very Small Aperature Terminal (VSAT) Operation

Very Small Aperture Terminal (VSAT)-A Very Small Aperture Terminal (VSAT) consists of a small, dish-shaped antenna and associated electronics which allow satellite access to a geosynchronous, communications satellite. A VSAT system is an entire network which includes the central hub, the remote sites, and the network software to run the system. VSAT utilizes geosynchronous satellites located 22,500 miles above the equator, as the communication backbone. The satellite connects the VSAT locations to the central hub facility which routes messages to the appropriate destination.
VESA-Video Electronics Standards Association
Vestigial Sideband-A form of transmission in which one sideband is significantly attenuated. The carrier and the other sideband are transmitted without attenuation.
VFD-Vacuum Fluorescent Display
VFO-Variable Frequency Oscillator
VGA-Variable graphics Array
VGCS-Voice Group Call Service
VHE-Virtual Home Environment
VHF-Very High Frequency
VIA-D-Voice Interface Access-Disabled System
Vibration Testing-A testing procedure in which subsystems are mounted on a test base that vibrates, thereby revealing any faults resulting from badly soldered joints or other poor mechanical design features.
Video-An electrical signal that carries TV picture information.
This figure demonstrates the operation of the basic NTSC analog television system. The video source is broken into 30 frames per second and converted into multiple lines per frame. Each video line transmission begins with a burst pulse (called a sync pulse) that is followed by a signal that represents color and intensity. The time relative to the starting sync is the position on the line from left to right. Each line is sent until a frame is complete and the next frame can begin. The television receiver decodes the video signal to position and control the intensity of an electronic beam that scans the phosphorus tube ("picture tube") to recreate the display.

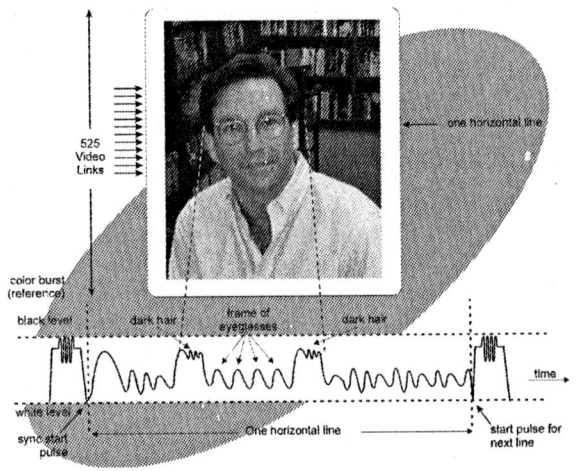

Analog Video

Video Amplifier-An amplifier designed to operate over the band of frequencies used for TV signals.
Video Blanking Interval (VBI)-Video blanking interval is a number of video lines that are transmitted in addition to the video display lines of a television signal to allow the television to blank the picture tube while the electron beam is repositioned from the bottom of the picture tube to the top of the picture tube. For NTSC signals, the VBI is 21 lines of the 525 transmitted video display lines. Because the VBI signal is not used for video display, some systems (such as Teletext or closed caption) use the VBI signals
Video Broadcasting-The process of transmitting video images to a plurality of receivers. The broadcasting medium may be via radio waves or wired systems such as CATV or the Internet.
Video Camera-A device that converts images (light signals) into electrical video (multiple frame) signals.
This diagram shows how a video camera uses a cathode ray tube (CRT) to convert light energy into a video signal. This diagram shows that the CRT includes a photosensitive plate that receives an optical signal through a lens and also receives energy from an electron beam signal. When there is light on the plate and the electron beam hits the plate in a specific spot, a small amount of current flows from the CRT tube. The video signal genera-

Video Coding

tor controls the horizontal and vertical position of the electron beam through the deflection coils. Because the video generator knows the exact position of the beam, it can create a composite video signal that represents the intensity (amplitude) and position (timing) of the image (light).

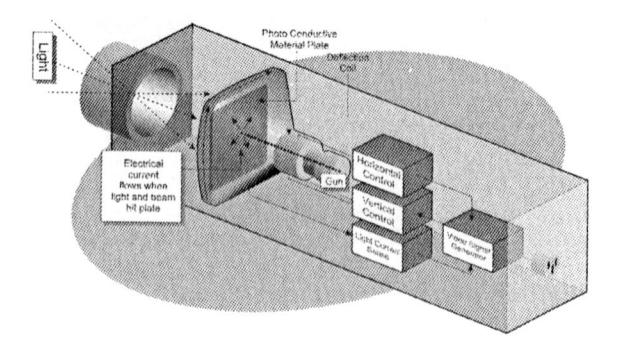

Video Camera Operation

Video Coding-A coding algorithm that converts video signals into streaming data signals. Some of the common compression video compression technologies include H.261 and H.263.

Video Communication-Video communication is the transmission and reception of video (multiple images) and other signals that can be represented by the frequency band used for video signal transmission. Telecommunications systems can transfer video signals in analog or digital form.

This figure shows the basic process used for video signal transmission. In this example, a television camera converts an image and audio sounds to electrical signals. The video signal is created by a camera scanning the viewing area line by line. At the beginning of each line scan, the camera create a synchronization pulse and the image (light level) is created by varying the electrical signal level after the synchronization pulse. The audio signal is created by using a microphone. These video and audio electrical signals are combined to form a composite video electrical signal. The composite video signal (baseband) modulates the radio transmitter frequency (broadband) signal. This low level radio signal is amplified to a very high power level for transmission. A video receiver (typically a television) receives the radio signal and many others from its antenna. It's receiver selects the correct radio signal by using a variable frequency filter (television channel selector) that demodulates the incoming radio signal to create the original video and audio electrical signals. The video signal is connected to a display device (typically a picture tube) and the audio signal is connected to the speaker.

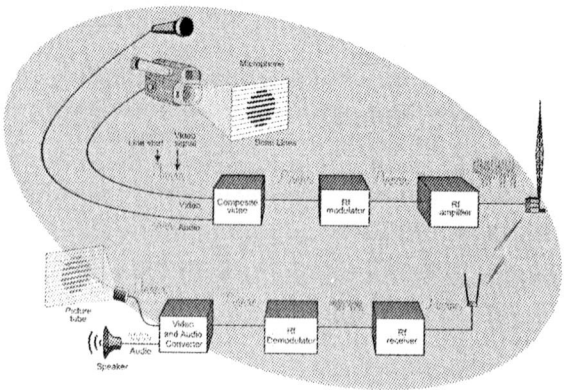

Video Transmission Operation

This photo shows a Sony MICROMV(tm) Handycam ® Camcorder. This camcorder was the worlds smallest and lightest video camera available in 2002. This innovative camcorder has a memory

Camcorder with Wireless Link
Source: Sony Electronics

554   www.ITMag.Com

Video Conferencing

chip, Internet access, along with wireless personal area network Bluetooth © technology that allows the video signal to be linked to recorders and viewers without wires.

Video Compression-The reduction, by digital coding techniques, in the number of digital bits required to represent a video signal. When compressed, a video signal can be transmitted on circuits with relatively slow data rates, such as the 64 kbps B channel of an Integrated Services Digital Network (ISDN).

Video Conferencing-A process of conducting a face-to-face meeting between two or more people in different locations through the use of telecommunications circuits and equipment that allows video and audio communications. Video conferencing usually requires real-time two-way transmission of audio and video communications between two or more locations. Transmitted video images may be in the form of full TV-quality images or freeze frame still images, where the picture is repainted every few seconds.

This figure shows the basic operation of sending video over an Internet connection. This diagram shows a computer with video conferencing capability that calls a destination computer. Computer #1 initiates a video conference call to computer #2 using the address 223.45.178.90. When computer #2 receives a data message from computer #1, a message is displayed on the monitor and an audio tone (ring alert) occurs. If the user on computer #2 wants to receive the call, they select the answer option (via the mouse or keyboard) that is generated by the software. Computer #1 then initiates a data connection with computer #2. The video conferencing software and data processing software in the computers (e.g., USB data bus and sound card) convert the analog audio signal from the microphone and digital video signal into a digital form that can be transmitted via the data link between the computers.

This photograph shows the InnoMedia IP VideoPhone (all-in-one video conferencing system), an Internet protocol (IP) desktop video conferencing telephone that integrates video images with key telephone features like speakerphone, mute, last number redial and an on-screen phone book into a system that is as easy to use as a standard phone. It uses 64 kbps to 192 kbps IP connections. It has a built-in 4" LCD screen and high quality CCD camera. It can interoperate with industry standard H.323 gateways and gatekeepers and supports multi-party video conferencing with external MCU. It can also be remotely upgraded and configured through a web browser.

Video Conferencing

Video Conferencing
Source: InnoMedia Incorporated

Video Dial Tone-An access and transport service for carrying full-motion video in much the same way as a dial-up call is carried on a conventional voice network.

Video Display-A computer output device that presents data to the user in the form of an image, including text and/or graphics.

Video Display Terminal (VDT)-A computer terminal equipped with a keyboard and an electronic readout, such as a cathode ray tube or liquid crystal display. Video display terminals often are used to connect remote locations to a distant host computer.

Video Mail (VMail)-A process of recording and sending short video messages (typically 1-2 minute video clips) in digital form via an electronic mail (email). Video mail messages may be sent as an attachment to standard Email addresses.

Video Monitor-A high utility TV set (without RF circuits) that accepts video baseband inputs directly from a TV camera, videotape recorder, or other TV source.

Video On Demand (VOD)-A service that provides end users to interactively request and receive video services. These video services be from previously stored media (entertainment movies or education videos) or have a live connection (news events in real time).

This figure shows a video on demand (VOD) system. This diagram shows that multiple video players are available and these video players can be access by the end customer through the set-top box. When the customer browses through the available selection list, they can select the media to play.

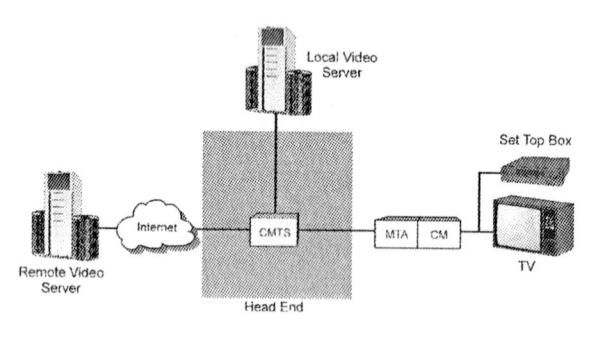

Video On Demand (VOD) Operation

Video Signal-An electrical signal that includes all of the intensity and position information related to a sequence of images. The video signal includes a horizontal sync pulse that indicates the start of an image sweep across the screen. The video signal then includes a composite signal that indicates the intensity of the image at each position along sweep. The video signal also includes a vertical blanking pulse or signal to indicate the end of a frame and to allow the image sweep to be repositioned at the top of the screen.

Video Stream-A process of delivering a continuous stream of digital video information that is commonly with minimal delay (e.g. real-time). Video streaming signals are usually compressed and error protected to allow the receiver to buffer, decompress, and time sequence information before it is displayed in its original format.

Video Streaming-A real-time system for delivering video, usually along with synchronized accompanying audio, typically over the internet. Upon request, a server system will deliver a stream of video and audio (both compressed) to a client. The client will receive the data stream and (after a short buffering delay) decode the video and audio and play them in synchronization to a user.

Video Tape Recorder (VTR)-A device that permits audio and video signals to be recorded onto magnetic tape.

Video Telephony-A telephone service that allows customers to hear and see another telephone user or video source. Video telephony applications include video on demand (VOD) movies, distance learning (remote education), telemedicine, teleconferencing and other applications that can benefit from the combination of video and audio signals.

Video Transmission-Video transmission is the transport of video (multiple images) that may be accompanied by other signals (such as audio or closed caption text).

Videoconferencing-Conducting conferences via a video or multimedia telecommunications system.

Videophone-A videophone is a communication device that can capture and display video information in addition to audio information. A videophone converts multiple forms of media; audio and video into a single transmission format (such as Internet Protocol). The use of videophones with an Internet telephone service allows the video portion of the communications session to share the data connection.

Videotape-A magnetic tape used for recording video programs.

Videotex-An interactive electronic information retrieval systems that allows users at home or office to select and return information to computer centers or data banks.

View-based Access Control Models (VACMS)- VACMs is a method for allowing network managers to restrict what MIBs and community-strings can be accessed by what users. VACMs is synonymous with SNMP views.

VIM-Vendor Independent Messaging

VIR-Vertical Interval Reference

Virtual-A facility or arrangement that gives the effect of being a dedicated facility but, in fact, is relatively shared.

Virtual Call Center-A call center where calls are answered and originated, typically between a company and a customer that uses customer service representatives (CSRs) that can be located at different places via virtual connections. Virtual call centers can assist customers with requests for service activation and help with product features and services as if they were located in a company office. Virtual call centers may use many virtual connections to connect CSRs with customers.

Virtual Channel Connection (VCC)-A logical connection between two end stations in an ATM network.

Virtual Channel Identifier (VCI)-(1-general) The identification of a logical channel on a virtual path. (2-ATM) In an Asynchronous Transfer Mode (ATM) cell header, a 16-bit field used to identify virtual channels between users or between users and networks.

Virtual Circuit (VC)-A logical connection between two communication ports in one or more communication network. Virtual circuits are used to temporarily connect data terminals to host computers. Because virtual circuits logically connect communication ports together, a single network switching system may be used to provide for many virtual circuits.

Virtual Desktop-A desktop workplace for the employee that consists primarily of computing devices. The virtual desktop devices usually include a computer, printer, and a telephone.

Virtual Extension-A communication extension that is created through the use of system programming rather than through the installation of physical equipment.

Virtual Home Environment (VHE)-A concept that a network supporting mobile users should provide them the same computing environment on the road that they have in their home or corporate computing environment.

Virtual Keypad-A software program that operates on a computer or other interactive display device to provide a user with the ability to enter keypad or keyboard information.

Virtual LAN-A number of devices (a subset) that are linked to each other within a larger network by logical channels to allow each device to communicate with other devices in the virtual network using these logical channels. Virtual networks often appear as a separate network to the users of the network. An example of a virtual network is the connection of computers in a city to computers in another city via logical channels (and encrypted channels for security) through the Internet. This allows the computers in one city to access the computers in the other city as if they were connected as a separate network.

Virtual Local Area Network (VLAN)-A local area network (LAN) that is logically (virtually) setup through one or more data networks that independently managed. VLAN connections are setup to allow data to safely and privately pass over other types of data networks (such as the Internet).

Virtual Memory-A memory management operating system technique that allows programs or data to exceed the physical size of the main internal directly accessed memory. Program or data segments or pages are swapped from disk storage as needed. The swapping is invisible to the programmer.

Virtual Pairs-Virtual pairs are multiple circuits that share a common physical communication channel (called "pair gain" systems).

Virtual Path Connection (VPC)-(1-general) An identifier of a physical channel an a logical channel that is used as a connection path between two points. (2-ATM) In Asynchronous Transfer Mode (ATM), a set of logical Virtual Channel Connections (VCCs) between two end stations. All channels in a specific VPC connect the same two end stations.

Virtual Path Identifier (VPI)-(1-general) An identifier of a physical channel or portion of a physical channel that is used as a connection path between two points. (2-ATM) In an Asynchronous Transfer Mode (ATM) cell header, an 8-bit field

Virtual PBX (vPBX)

used to identify virtual paths between users or between users and networks.

Virtual PBX (vPBX)-A virtual PBX offers business users the ability to make and receive calls through the company's PBX system using telephones that can be connected to any of the company's PBX systems at locations that have the ability to connect to a PBX access port (such as an Internet connection).

Virtual Phone-A software program that operates on a computer to provide telephone service.

Virtual Private Network (VPN)-Secure private communication path(s) through one or more data network that is dedicated between two or more points. VPN connections allow data to safely and privately pass over public networks (such as the Internet). The data traveling between two points is encrypted for privacy.

This figure shows the operation of a virtual private network (VPN). This diagram shows that the virtual private network is constructed of network access points that are under the control of the network operator. These network access points usually encrypt the data entering into the network to provide secure private communication path(s) through the network. These secure VPN connections allow a company to safely and privately pass over public networks (such as the Internet). A VPN management system is used to program the access points (e.g. IXC switch) for key parameters (e.g. data rates and QoS.) While this diagram shows virtual paths, the connections may actually pass through one or more switches have been set so a reserved amount of bandwidth is assigned so the end user can reliably receive a Quality of Service (QoS) characteristics that allows the connections to appear as dedicated lines.

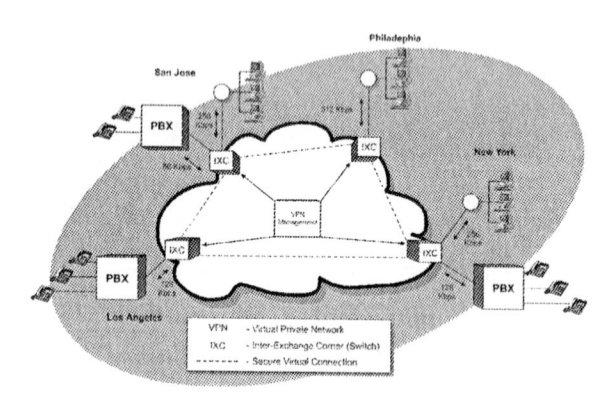

Virtual Reality Modeling Language (VRML)-A text based language that is used to allow the creation of three-dimensional viewpoints, primarily for use with Web browsing. VRML was created by Mark Pesce and Tony Parisi in 1994 and is a subset of Silicon Graphics' Inventor File Format,

Virtual Storage-An auxiliary storage mapped into real ad-dresses so that a computer user views it as an addressable memory store.

Virtual Tributary (VT)-In the Synchronous Optical Network (SONET), a structure designed for the transport and switching of Synchronous Transport Signal Level I (STS-1) payloads, typically in units of 1.544 Mbps each.

Virus-A software program spread by automatic copying from disks or computer networks and intended to interrupt or destroy the functioning of a computer.

Visible Light-Electromagnetic radiation visible to the human eye at wavelengths of 400-700 nm.

Visitor Location Register (VLR)-The database part of a wireless network (typically cellular or UMTS) that holds the subscription and other information about visiting subscribers that are authorized to use the wireless network.

Visual J++-Visual Java Development Environment

Visual Voice Mail-An application displaying and controlling voice messages on a desktop computer. Usually associated with unified messaging.

VITC-Vertical Interval Time Code

Viterbi Decoder-An algorithm for maximum-likelihood decoding of a convolutionally encoded data sequence, given a limited amount of memory.

Vitreous Silica-A glass consisting of almost pure silicon dioxide.

VITS-Vertical Interval Test Signal
VLAN-Virtual Local Area Network
VLF-Very Low Frequency
VLM-Virtual Loadable Module
VLR-Visitor Location Register
VLSI-Very Large Scale Integration
VM-Voice Mail
VMAC-Voice Mobile Attenuation Code
VMail-Video Mail
VMLA-Virtual Mobile Location Area

Virtual Private Network (VPN)

VMS-Voice Mail System
VMSC-Visited MSC
VoATM-Voice Over ATM
VoBB-Voice over Broadband
VoCable-Voice Over Cable
Vocal-Vovida Open Communications Application Library
VoCoder-Voice Coder
VOD-Video On Demand
VoDSL-Voice Over DSL
VoFR-Voice Over Frame Relay

Voice Activity Detector (VAD)-An electronic circuit that senses the activity (or absence) of voice communication signals. VAD is often used to inhibit a transmission signal during periods of voice inactivity or as a control source to allow digital speech interpolation (DSI) or time assigned speech interpolation (TASI).

Voice Band-The frequency spectrum, from approximately 300Hz to 3400Hz, that is considered adequate for speech transmission.

Voice Broadcast Service (VBS)-A voice communications service that allows a single voice conversation or message to be transmitted to a geographic coverage area to be received by subscribers that are capable of identifying and receiving the voice communications.

Voice Channel-In a cellular telephone system, a channel on which voice or data communication occurs, and on which brief digital messages may be sent from a cell to a mobile unit or from a mobile unit to a cell site.

Voice Circuit-A circuit for the interchange of human speech. Normally, the standard band provided is 300 Hz to 3400 Hz, but narrower bands also provide commercially acceptable circuits in some circumstances.

Voice Coder (VoCoder)-Voice coding is digital compression device that consists of a speech analyzer that converts analog speech into its component parts digital signals and speech synthesizer for the recreation of audio signals from the component parts. Voice coders are only capable of compressing and decompressing voice audio signals.

Voice Communication-Voice communication is the transmission and reception of audio and other signals that can be represented by the frequency band used for voice signal transmission. Telephone systems transfer voice signals in a variety of forms through by wire, radio, light, and other electronic or electromagnetic systems. These forms include analog and digital voice signals. Options for voice communications include different voice quality of service levels and voice privacy options.

Voice Compression-Refers to the process of electronically modifying a 64 Kbps PCM voice channel to obtain a channel of 32 Kbps or less for the purpose of increased efficiency in transmission.

Voice Dialing-A process that uses the callers voice to dial a call. Voice dialing involves the activation of the voice dialing feature (either by pressing a key or by saying a key word), saying words in the vocabulary of the voice dialing processor, and providing feedback to the user (usually by audio messages) of the status of the voice dialing process. Voice dialing can be a system (network provided) or device (stored in the telephone device) feature.

There are two basic forms of voice dialing; speaker independent and speaker dependent. Speaker independent voice dialing allows any user to initiate voice commands from a predefined menu of commands. Speaker dependent voice dialing requires the user to store voice commands so these voice commands can be activated by the user and others are unlikely to match the speaker dependent voice commands. Speaker dependent voice recognition allows a user to program specific names into the telephone or network voice recognition system.

This diagram shows different types of dialing using voice commands. In this example, both the telephone set and telephone network have voice dialing

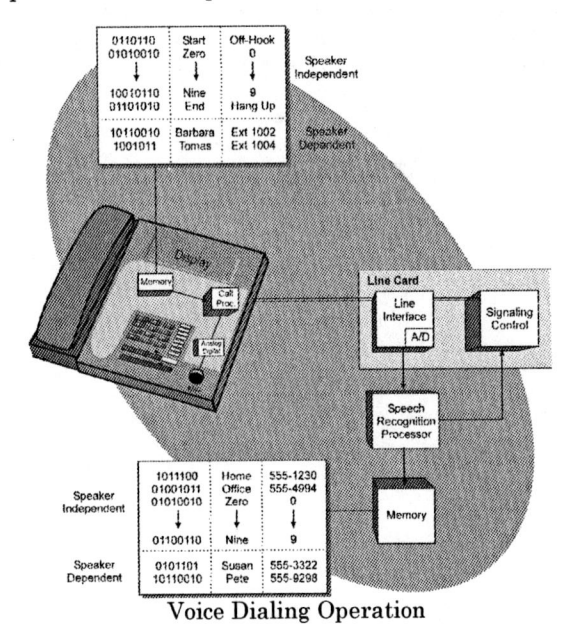

Voice Dialing Operation

Voice Digitization

control capability. When the telephone is used for voice control, the voice from the user is converted to digital form by and analog to digital converter. After the audio is converted to digital form, it is analyzed for patterns and matched to previously stored voice control digital sound patterns. This example shows that the telephone set has some speaker independent patterns (such as start and digits) that have been previously stored. It also shows that this telephone also has a speaker dependent memory storage area that allows the user to store specific names. When these specific names are spoken, the telephone set will retrieve the pre-stored telephone numbers or extensions.

This diagram shows similar voice dialing capabilities that are located in a telephone network. This network voice control system has more accurate voice processing capability than the telephone set and each voice control module can service many line cards as users only use voice control for brief periods.

Voice Digitization-This figure shows how an analog signal is converted to a digital signal. An acoustic (sound) signal is first converted to an audio electrical signal (continuously varying signal) by a microphone. This signal is sent through an audio band-pass filter that only allows frequency ranges within the desired audio band (removes unwanted noise and other non-audio frequency components). The audio signal is then sampled every 125 microseconds (8,000 times per second) and converted into 8 digital bits. The digital bits represent the amplitude of the input analog signal.

This figure shows how an analog signal is converted to a digital signal. An acoustic (sound) signal is first converted to an audio electrical signal (continuously varying signal) by a microphone. This signal is sent through an audio band-pass filter that only allows frequency ranges within the desired audio band (removes unwanted noise and other non-audio frequency components). The audio signal is then sampled every 125 microseconds (8,000 times per second) and converted into 8 digital bits. The digital bits represent the amplitude of the input analog signal.

Voice Gateway-A voice gateway is a communications device or assembly that transforms audio that is received from a telephone device or telecommunications system (e.g. PBX) into a format that can be used by a different network. A voice gateway usually has more intelligence (processing function) than a bridge as it can select the voice compression coder and adjust the protocols and timing between two dissimilar computer systems or voice over data networks.

This diagram shows the functional structure of a voice gateway device. This diagram shows that this voice gateway interfaces between a public telephone network to a packet data network. Input signals from the public telephone network pass through a line card to adapt the information for use within the voice gateway. This line card separates (extracts) and combines (inserts) control signals from the input line from the audio signal. If the audio signal is in analog form, the voice gateway converts the audio signal to digital form using an analog to digital converter. The digital audio signal is then passed through a data compression (speech

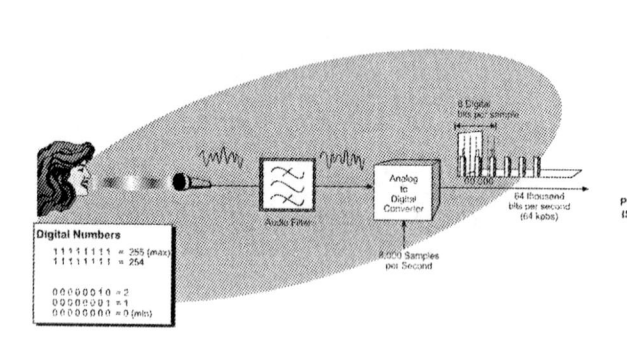

Voice Digitization

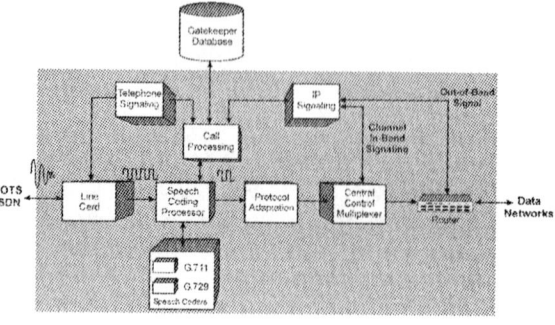

Voice Gateway Operation

coding) device so the data rate is reduced for more efficient communication. This diagram shows that there are several speech coder options to select from. The selection of the speech coder is negotiated on call setup based on preferences and communication capability of both voice gateways. After the speech signal is compressed, the digital signal is formatted for the protocol that is used for data communication (e.g. IP packet or Ethernet packet). This call processing section of the voice gateway may insert control commands (in-band signaling) to allow this gateway to directly communicate with the remote gateway. These digital signals are sent through a data access device (e.g. router shown here) so it can travel through the data communication network. The overall operation of the voice gateway is controlled by the call processing section. The call processing section receives and inserts signaling control messages from the input (telephone line) and output (data port). The call processing section may use separate communication channels (out-of-band) to coördinate call setup and disconnection.

Voice Group Call Service (VGCS)-A voice communications service that allows a single voice conversation to be simultaneously received by a predefined group of service subscribers.

Voice Mail (VM)-A service that provides a telephone customer with an electronic storage mailbox that can answer and store incoming voice messages. Voice mail systems use interactive voice response (IVR) technology to prompt callers and customers through the options available from voice mailbox systems. Voice mail systems offer advanced features not available from standard answering machines including message forwarding to other mailboxes, time of day recording and routing, special announcements and other features.

This diagram shows how a voice mail system provides electronic storage mailboxes to users within the telephone system. In this example, the voice mailbox system connects to a switching system through 2 extensions (ports) on the switching system (other voice mail systems may have many more ports). To access the voice mail system, users may select the voice mailbox system extension (usually programmed into a button on a telephone set that says voice mail). In this example, when a user dials into the telephone system to reach extension 1001, the line is busy. The system has been setup to forward calls to extension 1015 (the voice mail system) when extension 1001 is busy. To help ensure the voice mail system is accessible, if extension 1015 is busy, the call will be forwarded to extension 1016. When the call has entered the voice mail system, the interactive voice response (IVR) system will prompt the caller or user to enter information using touchtone or voice commands. This will allow callers or users to either store or retrieve messages from the digital message storage area (e.g. a computer hard disk drive).

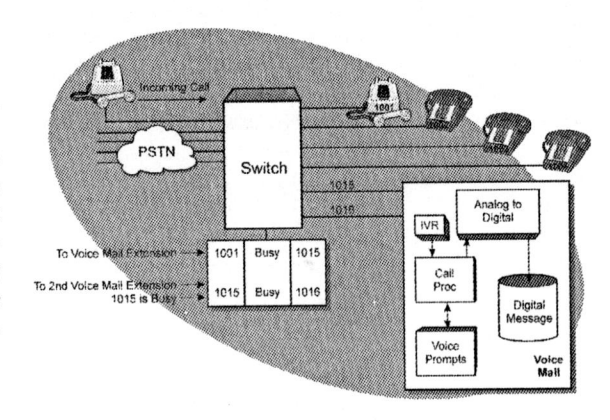

Voice Mail Operation

Voice Mail System (VMS)-A telecommunications system that allows a subscriber to receive and play back messages from a remote location (such as a PBX telephone or mobile phone).

Voice Mailbox-A portion of memory, usually located on a computer hard disk, that stores and plays

Voice Messaging

audio messages. The audio messages are often in compressed digital audio format.

Voice Messaging-A storage and retrieval system for voice messages. Commonly called "voice mail."

Voice Mobile Attenuation Code (VMAC)-A field in the extended address word in a cellular system handoff message that instructs the mobile telephone to the level it must adjust its initial transmitter power level when assigned to a voice or traffic channel.

Voice On the Net (VON)-The process of sending voice over a data network (such as sending voice over the Internet).

Voice Over ATM (VoATM)-A hybrid communication system that allows voice communications to be transmitted over digital ATM channels. If analog voice signals are used, they are first converted to a digital format by an Integrated Access Device (IAD).

Voice over Broadband (VoBB)-The process of sending voice over digital broadband connections.

Voice Over Cable (VoCable)-Voice over Cable solution is a complete Voice over Internet Protocol (VoIP) packet based broadband solution that supports DOCSIS and the PacketCable 1.0 specification.

This diagram shows how a cable television can offer telephony services. In this example, the cable television system has been modified to offer telephone service by adding voice gateways to the cable network's head-end cable modem termination system (CMTS) system and multimedia terminal adapters (MTAs) at the residence or business. The voice gateway connects and converts signals from the public telephone network into data signals that can be transported on the cable modem system. The CMTS system uses a portion of the cable modem signal (data channel) to communicate with the MTA. The MTA converts the telephony data signal to its analog audio component for connection to standard telephones. MTAs are sometimes called integrated access devices (IADs).

Voice Over Data Networks-A process of sending digitized voice signals over data networks (such as the Internet).

Voice Over DSL (VoDSL)-Sending voice over a digital subscriber line system (VoDSL) is a process that sends audio band (also called "voice band") signals (e.g. voice, fax or voice band modem) via a digital channel on a digital subscriber line (DSL) system. VoDSL requires conversion from analog signals to a digital format and involves the formatting of digital audio signals into frames and time slots so they can be combined onto a digital (DSL) channel. To communicate to other users, VoDSL requires one or more communication device that are capable of sending and receiving with the DSL network and conversion of a digital channel back into its analog voice band signal. This can be as simple as a computer with a sound card, a DSL modem and VoDSL software or as complex as a companies telephone network with an integrated access device (IAD). Optionally, some DSL systems have a PSTN gateway that can convert digital audio on a DSL system into telephone signals that can be sent through the public switched telephone network.

This diagram shows the different methods that can be used by a digital subscriber line (DSL) system to provide telephony services. In this example, the

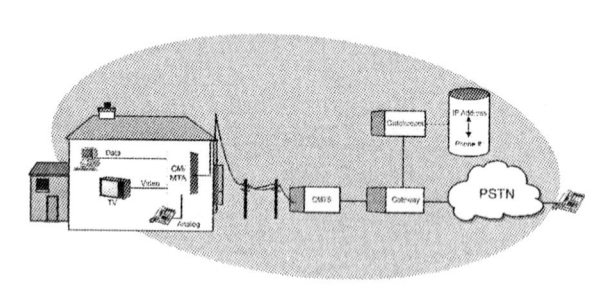

Voice over Cable Television Operation

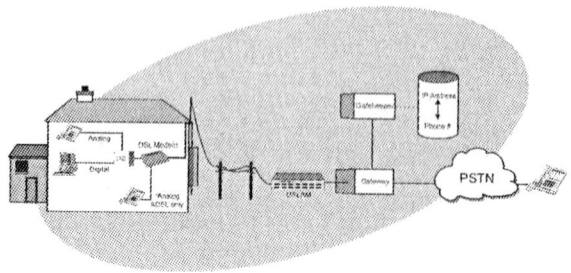

Voice over DSL (VoDSL) Operation

DSL system can offer telephone service by either using an analog channel or a digital channel. When the analog channel is used, the lower frequency portion of the telephone line is used to transfer a standard, low frequency telephone signal. If the digital channel is used to offer telephone service, the DSL system uses a voice gateway to connect a portion of the data channel on the DSL network to the public telephone system. The voice gateway connects and converts signals from the DSL data network to the public telephone network. The voice gateway also communicates with an integrated access device (IAD) that is installed at the residence or business. The IAD converts the telephony data signal to its analog audio component for connection to standard telephones.

Voice Over Frame Relay (VoFR)-A process of sending digitized voice signals over frame relay data networks.

Voice Over Internet Protocol (VoIP)-A process of sending voice telephone signals over the Internet or other data network. If the telephone signal is in analog form (voice or fax, the signal is first converted to a digital form. Packet routing information is then added to the digital voice signal so it can be routed through the Internet or data network.

This diagram shows how an Internet network (public or private) can be used to provide telephone service. In this example, a calling telephone or multimedia capable computer dials a telephone number. This telephone number is provided to a voice gateway. The voice gateway decodes the dialed digits and determines the destination address (IP address) of the gateway that can service the dialed telephone number. The remote gateway signals the caller of an incoming call (rings the phone or alerts a multimedia computer). When the user answers the call, a message is sent between the gateways and a virtual path can be created between the gateways. This virtual path takes the audio, converts it to digital form, compresses and packetizes the information, adds the destination gateway address to each packet, routes the packets through the Internet to the destination gateway, and converts the digital audio back to its original analog form.

Voice Paging-Refers to paging service whereby the transmission of information is in the form of actual voice data. Messages can be stored and different volume controls are selected when receiving pages, including an ear piece which allows privacy. This diagram shows how a voice paging system receives voice messages from callers and forwards these messages on to a voice pager. In this example, a caller dials a paging access number. This number either connects the caller to an interactive voice response unit or an operator that can direct the caller to a voice mailbox associated with the voice pager. After the caller's message is stored in the voice mailbox, it will be placed in the queue for the voice mail system. When the message reaches the top of the queue (available time to send), it will be encoded (formatted) to a form suitable for transmission on a radio channel. In this example, the message is sent as part of group 4. Sending the messages in groups allows the pager to sleep during

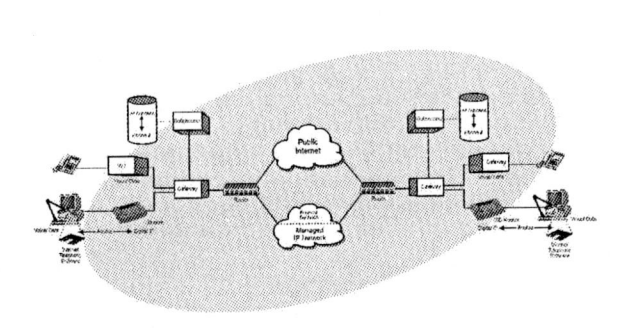

Voice over the Internet (VoIP) System

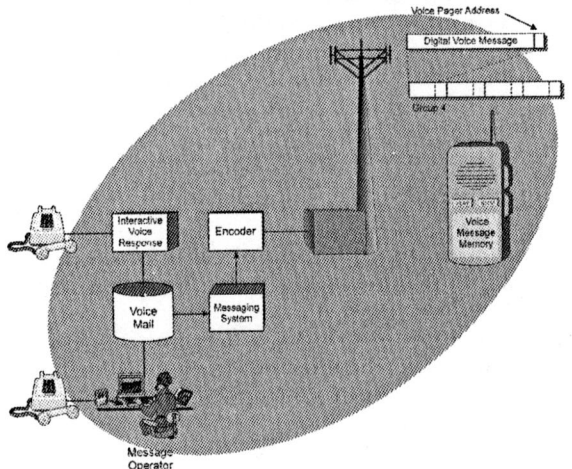

Voice Paging Operation

transmission of pages from other groups that are not intended to reach the voice pager. The voice message includes the voice pager address along with the voice message in digital form. During the reception of the message, it is stored into the voice message memory area so the voice pager can play the message one or many times after it is received.

Voice Privacy-Voice privacy is a process that is used to prevent the unauthorized listening of communications by other people. Voice privacy involves coding or encrypting of the voice signal with a key so only authorized users with the correct key and decryption program can listen to the communication information.

Voice Quality-Voice quality is a measurement of the level of audio quality, often expressed in mean opinion score (MOS). The MOS is number that is determined by a panel of listeners who subjectively rate the quality of audio on various samples. The rating level varies from 1 (bad) to 5 (excellent). Good quality telephone service (called toll quality) has a MOS level of 4.0.

Voice Recognition-A computer-based technology that analyzes audio signals (typically spoken words) converts them into digital signals for other processing (e.g. voice dialing).

Voice Response Unit (VRU)-A equipment that provides a caller with audio messages in response to their touch-tone(tm) key presses or voice commands . VRU are part of interactive voice response (IVR) systems.

Voice Service-Voice service is a type of communication service where two or more people can transfer information in the voice frequency band (not necessarily voice signals) through a communication network. SIP based voice service involves the setup of communication sessions between two (or more) users that allows for the real time (or near real time) transfer of voice type signals between users.

Voice Synthesis-Computer-generated sounds that simulate a human voice.

VoIP-Voice Over Internet Protocol

VoIP Management-A set of procedures, equipment, and operations that keep a voice over Internet protocol (VoIP) communication network operating near maximum efficiency. VoIP management includes system operations (configuration management), tracking and maintaining service availability (uptime), ensuring communications quality (audio and visual quality), and accurately accounting for the usage of the system (cost allocation or billing).

Volatile-A term applied to information held in a memory store that depends on power being continuously available.

Volatile Memory-A type of read/write memory whose content is irretrievably lost when the operating power is removed.

volt (V)-The standard unit of electromotive force equal to the potential difference between two points in a conductor that is carrying a constant current of one ampere when the power dissipated between the two points is equal to one watt. One volt is equivalent to the potential difference across a resistance of one ohm when one ampere is flowing through it. The volt is named for the Italian physicist Alessandro Volta. Also called voltage or "electromotive force." Called electric "tension" in some languages.)

Volt Ohm Milliammeter (VOM)-A general-purpose multi-range test meter used to measure voltage, resistance, and current. Also known as a multimeter.

Voltage Controlled Crystal Oscillator (VCXO)-A device whose output frequency is determined by an input control voltage and is referenced to a crystal oscillator..

Voltage Controlled Oscillator (VCO)-An oscillator circuit which has an output frequency that changes proportionally with a input voltage.

Voltage Drop-A decrease in electrical potential resulting from current flow through a resistance.

Voltage Regulator-A circuit used for controlling and maintaining a voltage at a constant level.

Voltage Standing Wave Ratio (VSWR)-A ratio of maximum to minimum voltage in the standing wave pattern that appears along a transmission line that is due to the adding of the forward and reverse traveling waves. Because the amount of VSWR results from the amount of reflected signal, VSWR can be used as a measure of impedance mismatch between the transmission line and its load.

Voltmeter-A test instrument that used to measure differences or level of electrical potential.

Volume-(1-sound) The loudness (intensity) of a sound. (2-data) A certain portion of data, together with its data carrier, that can be handled conveniently as a unit. (3-graphics) A three-dimensional array of raster data. (4-general) The amount of a cubic space measure (cubic meters, cubic inches, etc.) contained in a given three-dimensional region of space.

Voluntary Churn -A disconnection of service that occurs when a customer decides to drop a service from an existing carrier and initiate the service with another carrier.

VOM-Volt Ohm Milliammeter

VON-Voice On the Net

Voting Receivers-A wireless communication system that uses multiple radio receivers to receive the strongest radio signal possible from mobile radios operating in the system.

This figure shows a system that uses voting receivers to allow low power mobile radios to effectively communicate in a system that has a high power transmitter. In this system, there is one transmitter that that operates a base station that has at 100 Watts transmitter power. As the mobile radio moves throughout the transmitter site radio coverage area, the mobile radio continues to receive the signal from the same base station transmitter. However, because each mobile radio in the system can only transmit at five Watts, the system selects a receiver in the system that has the strongest received signal from the mobile radio. As the mobile radio moves from voting receiver #1 to voting receiver #2, the voting system will eventually select voting receiver #2 as the best choice to receive communication from the mobile radio.

munications account or device.

Vovida Open Communications Application Library (Vocal)-Software applications that are used by developers to create telephone systems that use Internet protocols.

Vox-A voice operated relay circuit that permits the equivalent of push-to-talk operation of a transmitter by the operator.

vPBX-Virtual PBX

VPC-Virtual Path Connection

VPI-Virtual Path Identifier

VPLMN-Visited PLMN

VPN-Virtual Private Network

VPN-Vitual Phone Network

VRAM-Video RAM

VRML-Virtual Reality Modeling Language

VRU-Voice Response Unit

VS&F-Voice Store And Forward

VSA-Vendor-Specific Attribute

VSAT-Very Small Aperture Terminal

VSELP-Vector Sum Excited Linear Predictive Coding

VSWR-Voltage Standing Wave Ratio

VT-Virtual Tributary

VT SPE-VT Syncluonous Payload Envelope

VTAM-Virtual Telecommunications Access Method

VTR-Video Tape Recorder

VU-Volume Unit

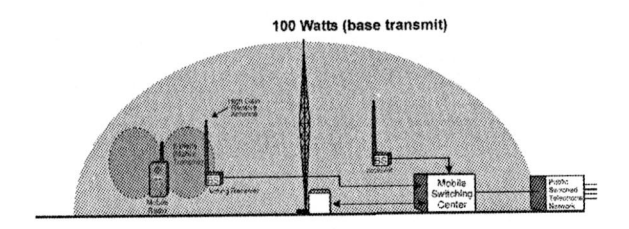

System with Voting Receivers

Voucher-A card, printed brochure, or electronic record that authorizes a customer to activate or recharge (add usage time) to their prepaid telecom-

W

W-watt
W3C-World Wide Web Consortium
WABI-Windows Application Binary Interface
WAE-Wireless Application Environment
Wafer-A thin slice of semiconductor material on which integrated circuits are built.
WAIS-Wide Area Information Service
Wake-A process of re-activating electronic circuits (such as radio receivers) after they have been in "sleep" mode so they can perform normal operations.
Walkie-Tallkie-A mode of voice communication in which one subscriber talks and other subscribers listen on the same talk group.
Wallet Share Improvement-Wallet Share Improvement is the process of actively and passively doing something that result in the customer either using more services or buying more products than before.
Wallet Share, Acquisition, and Retention (WAR)-Three core marketing objectives used by telephone companies to provide higher value to customers, acquire new customers, and keep customers from reducing or disconnecting services.
Walsh Code Administration-The process of assigning codes in a code division multiple access (CDMA) system to minimize interference between the codes. For the CDMA2000 system, Walsh codes are different traffic channel and have variable length that vary from hundred-twenty eight chips to four chips in length for each traffic channel, dependent on the data transmission rate.
WAN-Wide Area Network
WAP-Wireless Application Provider
WAP-Wireless Access Protocol
WAR-Wallet Share, Acquisition, and Retention
WARC-World Administrative Radio Conference
Warm Start-The process of rebooting a computer system without turning the power off.
Watchdog Process-A device or system that continually monitors specific functions of devices or systems (usually mission critical systems) to ensure they continue to operate within predetermined limits.
Watchdog Timer-A hardware timer that upon counting down to zero resets the central processing unit and therefore brings about a reset of the system. Software within the system must set the timer back to its starting point often enough that it does not reach the zero count. As long as the software functions normally it will continue to "pet" the watchdog and prevent a system restart. Software that gets stuck or that crashes will be reset once the watchdog counts down.
Watermark-An imperceptible signal hidden in another signal, such as audio or an image, which carries information. Watermarking is related to the general field of steganography, or information hiding. Ideally a watermark would not be destroyed (that is, the signal altered so that the hidden information could no longer be determined) by any imperceptible processing of the overall signal, for example high-quality lossy compression, slight equalization, or digital-to-analog-to-digital conversion. Sophisticated techniques for successfully destroying watermarks make that ideal difficult to achieve.
WATS-Wide Area Telecommunications Service
watt (W)-Unit of electric power, equal to one joule per second. One watt is produced (or consumed) when a current of 1 ampere flows through a voltage difference of 1 volt. The watt is named after the Scottish inventor James Watt (1736-1819).
WAV-Wave Audio
Waveform Coder-Waveform coding consists of a analog to digital converter and data compression circuit that converts analog waveform signal into digital signals that represent the waveform shapes. Waveform coders are capable of compressing and decompressing voice audio and other complex signals.
Waveform quality or RHO-Waveform Quality or RHO is a correlation measurement within the code division multiple access (CDMA) spread spectrum system.
Wavefront-A continuous surface that is a locus of points having the same phase at a given instant. A wavefront is a surface at right angles to rays that proceed from the wave source. The surface passes through those parts of the wave that are in the

Waveguide

same phase and travel in the same direction. For parallel rays, the wavefront is planar; for rays that radiate from a point, the wavefront is spherical.

Waveguide-Generally, a rectangular or circular pipe that constrains the propagation of an acoustic or electromagnetic wave along a path between two locations. The dimensions of a waveguide determine the frequencies for optimum transmission.

Waveguide Couplers-A coupler in which light is transferred between planar waveguides.

Waveguide Slug Tuner-A dielectric slug used for fine tuning an RF waveguide. Tuning is accomplished by varying the penetration of the slug into the waveguide.

Wavelength-(1-general) The length of a wave through one full cycle, e.g., from a starting point of zero amplitude through maximum amplitude, then minimum and back to zero. In optical networks, light of about 800 nanometers (nm), 1300 nm, or 1550 nm is typically used. These wavelengths are in the infrared portion of the electromagnetic spectrum and are not visible to the naked eye. (2-light) Each beam of light that carries information in an optical network has a specific wavelength (e.g., 1548 nm). They are sometimes referred to as wavelengths rather than beams or lightwaves.

This diagram shows that the wavelength of a signal is the distance that the signal travels over one cycle of the signal. In this example, two signals are transmitted in free space; 300 MHz and 3GHz. The wavelength in free space at 300 MHz is 1 meter and the wavelength at 3 GHz is 1/10th meter. This example also shows that the wavelength changes dependent on the transmission medium, such as coaxial cable as shown in this example. Because the signal wave travels slower in other mediums, the wavelength is shorter. This is why antennas are physically a little shorter than their ideal electrical requirements.

Wavelength Division Multiplexing (WDM)-A process of transmitting several distinct communication channels through a single optical fiber via the use of a distinct separate infrared wavelength (optical frequency or "color") for each communication channel. Each such channel may be further subdivided into several logical channels via time division multiplexing or other methods.

This diagram shows how a wave division multiplexing over fiber operates. This diagram shows that there are several lasers operating at different optical wavelengths (different colors/frequencies). Each laser converts an electrical signal into a pulsed light signal. These optical signals (optical carriers) are combined by an optical multiplexer (lens) for transmission through the optical fiber. At the receiving end, the different optical carriers are separated by an optical demultiplexer (lens) and each optical carrier is sent to a photo-detector. The photo-detector converts the optical signal back into its original electrical form.

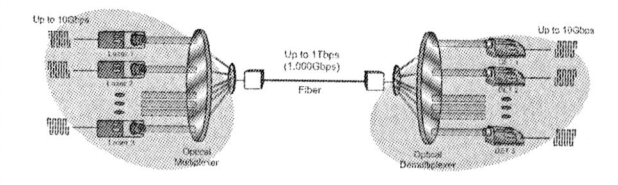

Wavelength Operation

Wavelength Division Multiplexing (WDM) System

Wavemeter-A device for measuring the wavelength of a radio or optical signal.
wb-weber
WBEM-Web-Based Enterprise Management
WCDMA-Wideband Code Division Multiple Access
WDM-Wavelength Division Multiplexing
WDMA-Wavelength Division Multiple Access
Web Browser-Software that used to graphically view information on Web servers. Web browsers request, receive, and reformat Information receives from web servers.

Web Hosting-The providing of web program application services (html or file transfer), allocation of information storage space, and interconnection to the Internet for customers.

Web Seminar (Webinar)-A seminar or instruction session that uses the Internet Web as a real time presentation format along with audio channels (via web or telephone) that allow participants to listen and possibly interact with the session. Webinars allow people to participate in information or training sessions from anywhere that has Internet and audio access.

Web Server-Web servers are computer systems that are used provide access to data that is stored and retrieved by commands in Hypertext Transfer Protocol (HTTP). HTTP is a protocol that is used to request and coordinate that transfer of documents between a web server and a web client (user of information). The typical use of web servers is to allow web browsers (graphic interfaces for users) to request and process information through the Internet.

WebCam
Source: Logitech

Web Television (WebTV)-A set-top box (cable converter) that provides the user with the ability to use and display Internet services on a standard television.

WebCam-A Webcam is a PC video camera that captures and posts live images to a Website. These images are refreshed every few seconds. Webcams can be used for video email and video conferencing and video instant messaging. This photograph shows the QuickCam Pro 3000 that can capture and send images to the Internet. This WebCam acts as a personal computer peripheral accessory that takes continuous still pictures and posts them to a website.

Webcast-The live presentation of information in a continuous (streaming) format delivered through the Internet web. A webcast might be associated with other web pages or other web-browser-based content in addition to the live stream.

weber (wb)-Unit of magnetic flux, equal to one volt second. Named for 19th century German physicist Wilhelm E. Weber (1804-1891).

Webinar-Web Seminar

WebTV-Web Television

Weight-The force exerted on an object due to gravitational attraction to the planet it is on or near. In the SI metric system, this force is measured in newtons (N). In contrast to weight, which depends on the properties of the planet you are on or near, the mass of an object is the same regardless of which planet you are on or near. For example, a 1 kg object has a mass of 1 kg everywhere in space. But it has a weight of 9.8 N on the surface of the earth (more precisely, the weight is slightly smaller at the equator than at the poles, due to the effect of the earth's rotation), and the same 1 kg object has a weight of only 4.8 N on the moon's surface, and 25.9 N on Jupiter's surface. Despite this difference, the words weight and mass are loosely used as synonyms in everyday speech.

Weighted Fair Queuing (WFQ)-A priority scheduling method that gives preferential treatment traffic that has a higher priority (such as real-time traffic) while allocating some bandwidth for lower priority traffic so these applications continue to operate under heavy traffic conditions.

Weighted Random Early Discard (WRED)-WRED combines the capabilities of the RED algorithm with IP Precedence to provide for preferential traffic handling of higher priority packets. WRED can selectively discard lower priority traffic when an interface/port begins to get congested and provide differentiated performance characteristics for different classes of service.

WER-Word Error Rate

Wet Cell-A type of battery (or cell section within a battery) that uses a liquid electrolyte.

Wet Circuit-An analog trunked cable pair that carries speech or data signals, plus a direct current. The dc component is used for signaling and supervision, for talk battery feed, or for sealing current.

Wetting Current-A continuous current ("sealing current") that is applied to a communication line to minimize the effects of oxidation on splices and junctions that could cause poor communication.

WFM-Wired For Management

WFOM-Wait For Overhead Message

WFQ-Weighted Fair Queuing

What you See is What You Get (WYSIWYG)-An expression that is used for a computer system that displays information in the same style in which it will be printed. It is pronounced "wizzy-wig."

White Noise-A mathematical idealization of random noise. white noise has an equal amount of energy over a wide frequency spectrum.

Whiteboard-A device that can capture images or hand drawn text so they can be transferred to a video conferencing system. Whiteboards allow video conferencing users to place share documents, images and/or hand written diagrams with one (or more) video conference attendees.

Wi-Fi-Wireless Fidelity

Wide Area Network (WAN)-A communications network serving geographically separate areas. A WAN can be established by linking together two or more metropolitan area networks, which enables data terminals in one city to access data resources in another city or country.

This figure shows that a WAN is usually composed of several different data networks. Different types of communication lines such as leased lines, packet data systems, or fiber transmission lines can interconnect these networks.

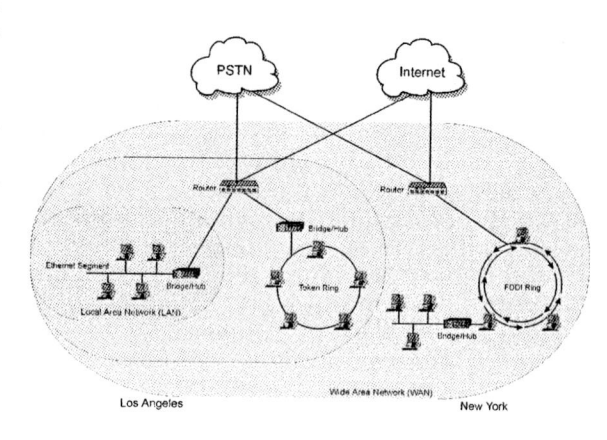

Wide Area Network (WAN) Systems

Wide Area Telecommunications Service (WATS)-A service that allows companies to be billed using different rate (such as flat rate) structures for calls within their wide area telephone system (WATS) system area.

Wideband Code Division Multiple Access (WCDMA)-A 3rd generation digital cellular system that uses radio channels have a wider bandwidth than 2nd generation digital cellular systems such as GSM or IS-95 CDMA.

This figure shows a wideband code division multiple access (WCDMA) system. The WCDMA system uses paired 5 MHz channels (FDD) to provide 3rd generation wireless broadband data transmission. Multiple WCDMA physical channels can co-exist in the same frequency band (even on the same cell site) by using different spreading codes (4-256). The spreading code is used to create several chips per bit of information. These codes are chosen to be orthogonal (non-interfering) with each other. The code chip rate is 3.84 Mcps and each coded channel is divided into frames of 10 msec each. The WCDMA system can dynamically change its spreading codes to provide bandwidth on demand (BoD) services.

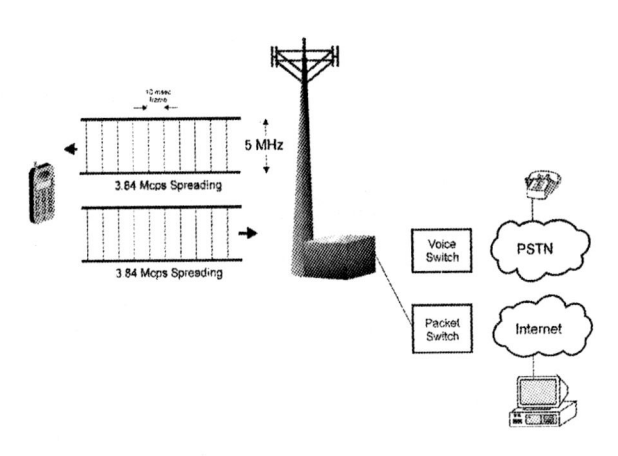

Wideband Code Division Multiple Access (WCDMA) System

WIN-Wireless Intelligent Network
Win32 API-Win32 Application Programming Interface [API]
Wind Loading-The pressure placed upon an antenna structure by the wind.
Wind Velocity Rating-The maximum wind velocity that an antenna assembly can withstand without physical damage.
Winding-(1 - general) A coil of wire used to form an inductor. (2 - noninductive) A winding with specified resistance but negligible inductance, usually made by winding two wires at the same time and using one for each direction of current flow so that the inductive effects of the two cancel out. (3- primary') A transformer winding that receives an input signal from a source, thereby creating magnetic flux in the core. This flux induces current in a secondary winding. (4- secondary) A transformer winding that feeds an output. The secondary receives input by electromagnetic induction from the current flowing in the primary winding.
Window-(1-transmission buffer) An indication of the amount of time or data that should not be exceeded waiting for successful reception of information. If this amount is exceeded, retransmission may occur. (2-video) Video containing information or allowing information entry from a keyboard, time code generator, or other device. A window dub is a copy of a videotape with time code numbers keyed into the picture. (3-graphics) A diagonal area of the display screen specified by the input of two probes where the probes show the lower left and upper right coordinates of the area desired.
Window Size-In the X.25 and X.75 packet-switching protocols, the number of unconfirmed frames or packets that are transmitted across a connection before additional frames or packets can be sent. Window size affects the control of data transmission and reception on the user and network sides of an X.25/X.75 connection.
Wink-(1-control signal) A telephone line signal that is a single supervisory pulse. When caused by the passive end of a subscriber loop, it is usually transmitted as an off-hook signal followed by an on-hook signal where the off-hook signal is of a very short specified duration compared to the on-hook signal. When caused by the powered end of a subscriber loop, it is usually transmitted as a short removal of central office battery voltage from the loop wires. When used with robbed bit or associated bit signaling on a digital multiplexed channel, the robbed bit or associated bit changes binary value for a short time interval. The duration of any form of a wink signal is typically 200 milliseconds. (2-indicator light) A rapid flashing cadence of an indicator light, typically used to indicate which line is ringing with an incoming call on a multi-line telephone set. Winking cadence is typically 5 winks per second, faster than flashing cadence.
Wink Release-A control signaling process used by most end office telephone switches, when the person or device at the other end hangs up, your local central office will disconnect the central office battery for a short time (typically 200 milliseconds). That brief removal of end office battery voltage is called a wink release. Such a signal can be used to alert a data device that the device at the other end has hung up. (Remember it can't tell by just listening — like you and I.) If a key telephone system or a recorded answering machine has a line on hold, the wink release is supposed to cause the key telephone system or the answering machine to hang up.
Wink Start-A control signaling process that momentarily changes (winks) that is used to introduce or indicate the beginning of other following signaling information. Typically the following signal is an inband signal such as MF or DTMF/TouchTone digits. See Wink Operation.
WINS-Windows Internet Naming Service

WIPO-World Intellectual Property Organization, a specialized agency of the United Nations

Wire Protocol-A specification that describes the methods (information flow) and signaling (control) attributes by which a communication signal is placed upon the medium that will carry it from point A to point B in a communication network.

Wire Speed-The bandwidth of a particular transmission or networking system. For an example, the wire speed of 10BaseT Ethernet is 10 Mbps. When data is said to run at wire speed or at "wire rate," it implies there is little or no software overhead associated with the transmission and that the data travels at the maximum speed of the hardware.

Wire Wrapping-Terminating wires on tags by firmly wrapping the wire around a sharp cornered tag that bites through the insulator to the conductor.

Wire-Speed-See line-rate.

Wireless-Devices that transmit over wireless networks rather than over telephone lines. Historically, at various times during the 20th century, this had specialized meanings that have come and gone. For many years, British English used the word "wireless" while North American English used "Radio" instead. In the past, the word "wireless" was occasionally used to describe the transmission via radio of Morse code, but not voice. Today the term wireless is used primarily for cellular systems, and secondarily also for other short range radio systems used directly by end users, such as for example 802.11b short range data transmission.

Wireless Access Protocol (WAP)-A standard protocol specification that allows advanced messaging and information services to be delivered to wireless devices independent of which wireless technology they use.

Wireless Application Environment (WAE)-A software application that utilizes wireless communication. Wireless communication often has limited of costly access to bandwidth.

Wireless Application Provider (WAP)-A 3rd party company that creates wireless business or consumer applications, and provides them to customers through an existing wireless carriers network.

Wireless Broadband-The transfer of high-speed data communications via a wireless connection. Wireless broadband often refers to data transmission rates of 1 Mbps or higher.

Wireless Cable-"Wireless Cable" is a term given to land based (terrestrial) wireless distribution systems that utilize microwave frequencies to deliver video, data and/or voice signals to end-users. There are two basic types of wireless cable systems, Multichannel Multipoint distribution Service (MMDS) and Local Multichannel Distribution Service (LMDS).

Wireless Data-A system or the transmission of digital information through a wireless network such as wireless packet data systems or cellular mobile communications. Wireless data systems are specifically designed to reliably transfer information (data) between a sender and receiver. The term wireless data can apply to mobile or fixed devices and the transmission may be in the form of radio or optical (e.g. infrared systems) communication systems. Wireless data transmission can be sent over dedicated wireless data communication systems (such as Reflexion, Ardis, or Mobitex) or the transmission may share a common channel for voice and data (such as on GSM or 3G cellular systems).

This figure shows the three key types of wireless data networks. This diagram shows a wireless LAN system that has multiple access nodes. These access nodes operate as gateways between the data communication devices (e.g., mobile computer) and the

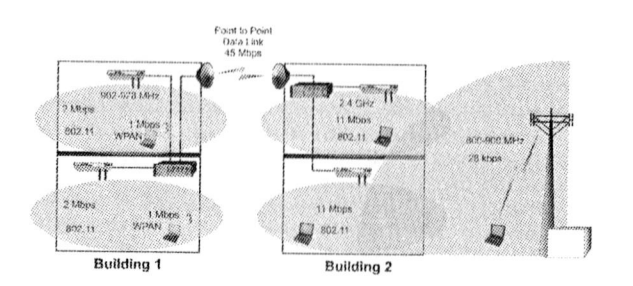

Wireless Data Networks

572

Wireless Email

data network hub. Building 1 uses an older 801.11 wireless LAN system that operates from 902-928 MHz at 2 Mbps. Building 2 uses a newer 802.11 wireless LAN system that operates at 2.4 GHz providing up to 11 Mbps data transfer rate. This diagram also shows a microwave data link that provides a 45 Mbps interconnection between campus buildings. Finally, a user who is operating in a remote area outside the core campus is using the wide area mobile system to transfer data files (at a data transfer rate below 28 kbps).

This is a picture of a BlackBerry RIM model 957 two-way wireless data device that provides wireless access to email and the Internet. This device is used by mobile professionals to gain remote access to company information systems and email services. This product is a totally integrated package that includes server software, desktop tools, nationwide airtime and choice of wireless handheld to provide easy access to corporate email on the go.

This is a picture of a BlackBerry RIM model 957 two-way wireless data device that provides wireless access to email and the Internet. This device is used by mobile professionals to gain remote access to company information systems and email services. This product is a totally integrated package that includes server software, desktop tools, nationwide airtime and choice of wireless handheld to provide easy access to corporate email on the go.

This figure shows a typical wireless data service. In this example, a computer is sending a file. As the computer sends digital information (1's and 0's), their baseband electrical signal (high or low voltage) transfers the information by modifying (modulating) the radio transmitter frequency (broadband) signal. This low level radio signal is boosted by an RF amplifier and converted to electromagnetic wave by the antenna. Another wireless data device receives the radio signal and many others from its antenna. It's receiver selects the correct radio signal by using a frequency filter and compares the modified incoming radio signal to an unchanged frequency to remove the audio electrical signal (called demodulation). The audio electrical signal is converted to the original digital by a modem and routed to another data device.

Wireless Data
Source: Research in Motion Ltd.

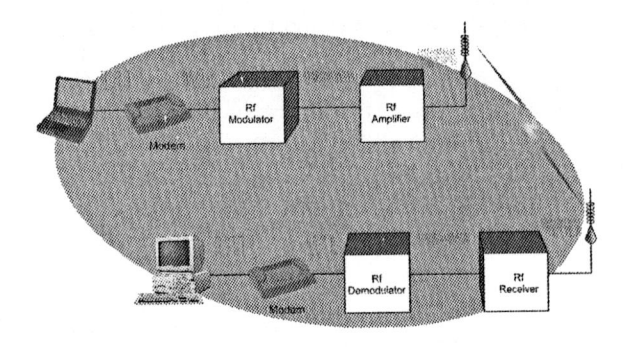

Wireless Data Communication

Wireless Email-e-mail that can be downloaded through a wireless network via a wireless modem.

www.ITExpo.Com

573

Wireless Ethernet

Wireless Ethernet-A wireless version of packet based Ethernet system. Wireless Ethernet systems typically use the ISM frequency band in the 2.4 GHz or 5.7 GHz range. Because bandwidth is limited, wireless Ethernet systems are limited to 1 Mbps to 54 Mbps compared to the 10 Mbps to 10 Gbps for wired Ethernet systems.

Wireless Fidelity (Wi-Fi)-Another name for the 802.11 wireless LAN system.

Wireless LAN Interoperability Forum (WLIF)- A forum established in 1996 to oversee the development of specifications that allow WLAN products from different manufacturers to inter-operate. The first standard produced by WLIF was the WLIF OpenAir(TM) standard.

Wireless Local Area Network (WLAN)-A Wireless Local Area Network (WLAN) allows computers and workstations to communicate with each other using radio propagation as the transmission medium. The wireless LAN can be connected to an existing wired LAN as an extension, or can form the basis of a new network. While adaptable to both indoor and outdoor environments, wireless LANs are especially suited to indoor locations such as office buildings, manufacturing floors, hospitals and universities.

This figure shows products that are typically used in a WLAN system. This WLAN system includes radio access ports and extension ports. The extension ports shown in the figure are PCMCIA cards that plug into a laptop computer. These extension ports communicate via radio-to-radio access ports. The radio access ports convert the WLAN radio signal back into computer network signals (such as Ethernet or token ring).

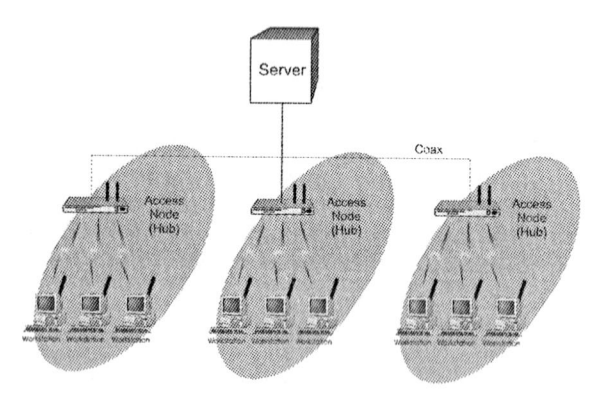

This photograph shows a portable wireless LAN diagnostic device. This device can test wireless LAN's networks for interference levels and and find and isolate unauthorized users.

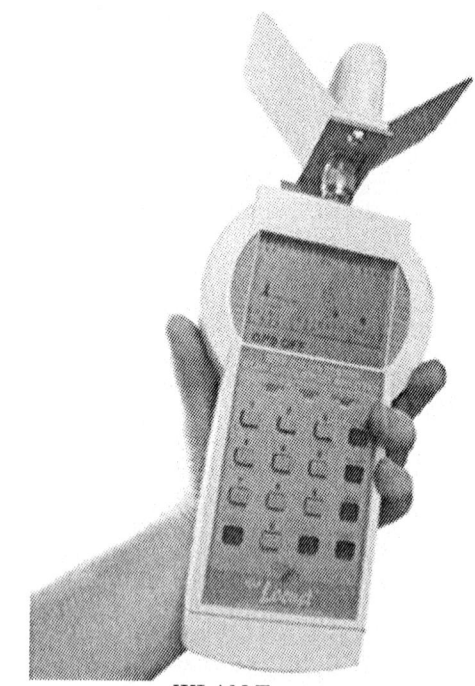

WLAN Tester
Source: Berkeley Varitronics Systems, Inc.

Wireless Local Loop (WLL)-The providing of local telephone service via radio transmission. Wireless local loop systems often use a radio conversion device located at the home or business to create a dialtone signal to allow the use of standard telephones.

This diagram shows a wireless local loop system. In this diagram, a central office switch is connected via a fiberoptic cable to radio transmitters located in a residential neighborhoods. Each house that desires to have dial tone service from the WLL service provider has a radio receiver mounted outside

Wireless Local Area Network (WLAN) System

with a dial tone converter box. The dial tone converter box changes the radio signal into the dial tone that can be used in standard telephone devices such as answering machines and fax machines. It is also possible for the customer to have one or more wireless (cordless) telephones to use in the house and to use around the residential area where the WLL transmitters are located.

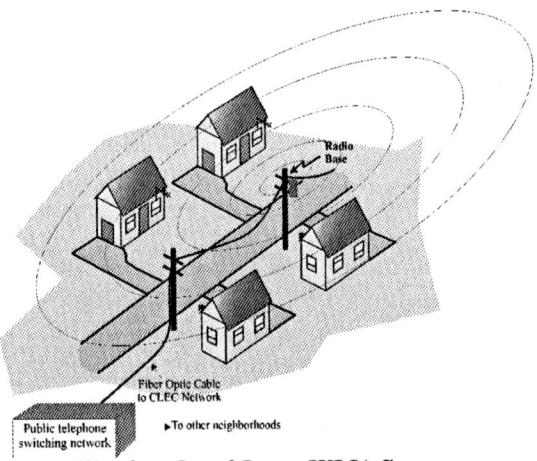

Wireless Local Loop (WLL) System

Wireless Metropolitan-Area Network (WMAN)-WMANs are usually private wireless packet radio networks often that cover an urban or city geographic area. They are commonly used for law-enforcement, utility, or public safety applications.

Wireless Network-Wireless networks are primarily designed to transfer voice and or data from one point to one or more other points, (multipoint). Many networks make use of some wireless technologies as a transport medium even though we do not consider them to be wireless networks. Examples of wireless networks include cellular, personal communication service, (PCS), paging, wireless data, satellite, and broadcast radio and television. Wireless network is a term commonly used for wireless local area network (WLAN).

This figure shows the basic types of wireless networks ant that these networks vary from broadcast (one-way) systems to complex switching two-way systems. This diagram shows a private land mobile radio system, television broadcast system, paging system, mobile telephone system, and satellite communication system. Although all wireless networks can transmit information from one point to another, different types of networks better suited to provide

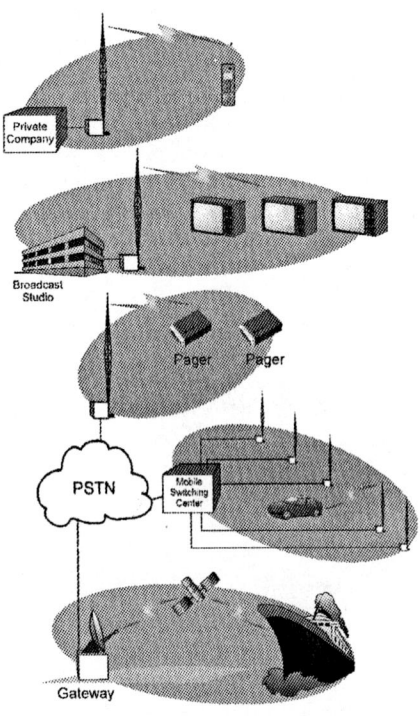

Wireless Network

specific types of services (e.g., paging compared to television broadcasting).

Wireless Network Interface Card (WNIC)-A WNIC is a device that adapts wireless communication signal and protocol to a data bus or data interface in a computer. The WNIC is installed in a wireless LAN system and a computer data bus (such as a PCMCIA or PCI socket). The WNIC is can be connected through a PC expansion board connector and operating system. Software in the computer is installed and setup to recognize the WNIC card.

Wireless Office Telephone System (WOTS)-A wireless telephone system that is used in a business environment that has similar features to a private branch exchange (PBX) wired office telephone system.

Wireless Packet Data-Wireless data transmission technology that transmits data in small packets (up to approximately 100 characters each).

Wireless Private Branch Exchange (WPBX

Wireless Private Branch Exchange (WPBX)-A WPBX offers business users the ability to make and receive calls through the company's PBX system using cordless telephones anywhere on a company's premises that has a radio port (wireless access node).

This diagram shows a sample WPBX radio system. A WPBX system typically has a switching system that is located at the company. The WPBX switch interfaces a PSTN communication line and multiple radio base stations. Radio base stations communicate with wireless office telephones that can move throughout the system. A control terminal is used

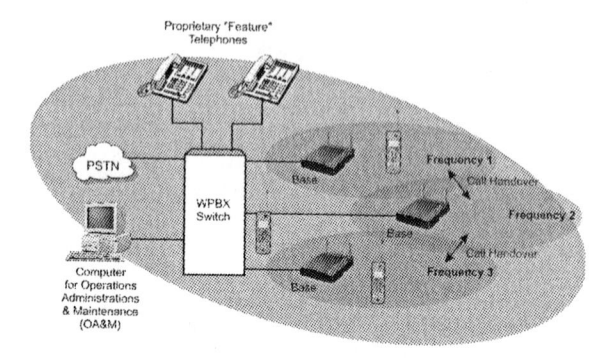

Wireless Private Branch Exchange (WBPX) System

to configure and update the WPBX with information about the wireless office telephones and how they can be connected to the PSTN.

Wireless Resellers Association (WRA)-National Wireless Resellers Association, which has now merged and become part of the Telecommunications Resellers Association. See TRA.

Wireless Session Layer (WSL)-The wireless session layer (WSL) is the management layer of a communication session that is linked by a wireless channel.

Wireless Session Protocol (WSP)-Establishes a relationship between a client application and the WAP server. This session is relatively long lived and able to survive service interruptions. The WSP uses the services of the Wireless Transport Protocol (WTP) for reliable transport to the destination proxy/gateway. See also Home RF.

Wireless Soft Switch-Wireless soft switches are interconnection switching systems that can dynamically change its connection data rates and protocols types by software control to provide for wireless voice, data, and video services. Wireless soft switches were developed to replace existing mobile switching center (MSC) switches that have limited interconnection capabilities. Soft switches are packet based and can simulate multiple protocols such as Internet protocol and asynchronous transfer mode (ATM). This allows for multiple types and simultaneous services to each customer with varying levels of quality of service (QoS).

This figure shows the TELOS Technology, Sonata SE Wireless Softswitch that is used in next generation wireless networks. It provides call processing and other functions that allow voice and multimedia calls to be made over a data network. Like a standard softswitch, it provides control of the media gateways and signaling gateways used to connect to legacy circuit-switched networks. In addition, connects with the radio access network and mobility

Carrier Class Wireless Softswitch
Source: TELOS Technology

network used in wireless systems. Is functionally equivalent to the MSC in a circuit-switched wireless network.

Wireless Telephony Application Interface (WTAI)-A specification that defines the standard telephony specific extensions for WAP devices with WML and WML Script interfaces to such items as call control features, address book and phone book services.

Wireless Transport Layer Security (WTLS)-An optional component of the wireless Application Protocol (WAP) stack that provides a secure data pipe for the session between client and server. WTLS is derived from the Secure Sockets Layer (SSL) specification. As such, it performs the same authentication and encryption services as SSL. See also Secure Socket Layer.

Wireless User Group (WUG)-For a Bluetooth system that has telephony services, a group of terminals restricted to communicating only with each other under the control of a master. The WUG master controls or restricts access to call requests from non-group members. This is a feature supported by the Cordless Telephony Profile.

Wireless Wide-Area Networks (WWAN)-WWANs are wireless data transmission systems that cover large geographic area using cellular or public packet radio systems. These systems are typically limited to data transfer rates below 20 kbps.

Wireline-Telecommunications services provided by wireline common carriers, such as telephone companies. This term also refers to the use of copper wire for transmission of signals rather than radio links.

Wireline Carriers-Cellular service providers that are also engaged in the business of landline telephone service. Band B is allocated for these service providers. Some wireline carriers have been authorized to provide service in band A in other regions.

Wireline Cellular Carriers-Conventional wireline telephone companies that also provide cellular phone service. See also B Band.

WLAN-Wireless Local Area Network

WLIF-Wireless LAN Interoperability Forum

WLL-Wireless Local Loop

WMAN-Wireless Metropolitan-Area Network

WMLscript-A scripting language for use with wireless access protocol (WAP) devices similar to JavaScipt.

WNIC-Wireless Network Interface Card

Wobulator-A device that creates a sweeping frequency signal through the use of a variable capacitor in a tuned frequency generator circuit where one or more plates in the capacitor vibrates (wobble) resulting in a change of frequency.

Woofer-A loudspeaker designed to respond efficiently to lower sound frequencies.

Word-In data communications, a character string, binary element string, or bit string that is considered as an entity.

WORD-Work Order Record And Details

Word - Symbol String-(1- natural language) A string of alphabetic or alphanumeric characters separated from its surrounding text by spaces or other delimiters. (2- computer memory) A group of bits in memory or elsewhere that are copied from place to place or otherwise processed together. Historically, different numbers of binary bits comprised a "word" in different computers. Among others, 4, 6, 8, 12, 18, 60 or 64 bit "words" have been used in different historical computers. Today, the term is typically used to describe 32 bits (four bytes).

Word Error Rate (WER)-The ratio of words received in error with respect to the total number of words sent.

Work Area Protection-The implementation of safe work area measures to ensure outside plant personnel safety that the area is free from contaminated air, combustibles, overhead and underfoot hazards, moving traffic, and foreign electrical voltage.

Work Order-A detailed drawing or print that indicates the addition, removal, or rearrangement of outside plant, also called a work print.

Working Telephone Number (WTN)-A unique number that identifies a specific (telephone) line that is in service. Typically, this is the number recorded by the switch and used in the Billing System to identify the customer who will be billed. A calling card number, or an authorization code may sometimes also be treated as WTNs.

Workstation-A computer that is attached to the network. A workstation has the capability of processing information in addition to requesting and sending information through a network.

World Administrative Radio Conference (WARC)-An international meeting coordinated by the ITU at which countries determine which frequencies will be allocated for what services. Each radio transmitter is assigned an identifying alphanumeric identifier ("call letters") by the licensing authority of the national government. Handsets and portable radio units of cellular and trunked radio systems are deemed to be sub-users of the license identification of their home base service provider. The WARC assigns specific initial call letters or numbers for use by each country, and has done so since 1912. Several examples are obviously based on the name of the country: A for America (USA) although originally assigned to Germany

World Numbering Plan

(from its Latin name Allemania), C for Canada, D for Germany (due to its German-language name Deutschland), F for France, G for Great Britain, I for Italy, J for Japan, R for Russia, and so forth. Nations with large populations also have additional initial letters chosen for various interesting historical reasons: The USA also uses K, N and W, which were chosen to allow certain pre-WARC arbitrarily-chosen station names to continue to be used by the US military, and still fall within the WARC allocations; namely N for NAVY — the official fleet radio station of the US Navy, and W for WAR— the original name of the official US Army radio station (then part of the US War Department, now the US Department of Defense). Originally the letters W and K were used as the initial letters for all continental US broadcast stations regardless of geography, but since 1923 the initial letter K was assigned exclusively for new broadcast stations west of the Mississippi river, and W for those to the east. Legend has it that the letter V was used by the then British Empire in memory of Queen Victoria, and was assigned in combination with various other succeeding letters to various outposts of the then British Empire. The result is the continued use of V today in the independent nations of the former British Commonwealth such as: Australia (VZA), Canada (VE), India (VTA) and so forth.

World Numbering Plan-The numbering plan that assigns each telephone customer in the world a unique telephone. This number that consists of a country code followed by a national number.

World Wide Web (WWW)-A service that resides on computers that are connected to the Internet that allows end users to access data that is stored on the computers using standard interface software (browsers). The WWW (commonly called the "web") is associated with customers that use web browsers (graphic display software) to public users to find, acquire and transfer information.

WORM-Write Once Read Many
WOSA-Windows Open Architecture
WOTS-Wireless Office Telephone System
Wow And Flutter-A common expression relating to the stability of a tape transport. Wow refers to low-frequency variations in pitch; flutter refers to high-frequency variations in pitch caused by variations in the tape speed of the machine.
WPBX-Wireless Private Branch Exchange
WRA-Wireless Resellers Association
WRED-Weighted Random Early Discard

Write-Off-A financial transaction which records as a loss to a company billed services or fees that cannot be collected, either due to customer non-payment, or due to an inability to bill for service or usage.
WS-Work Station
WSL-Wireless Session Layer
WSP-Wireless Session Protocol
WTAI-Wireless Telephony Application Interface
WTLS-Wireless Transport Layer Security
WTN-Working Telephone Number
WTP0-Wireless Transport Protocol
WUG-Wireless User Group
WWAN-Wireless Wide-Area Networks
WWW-World Wide Web
WYSIWYG-What you See is What You Get
WZ1-World Zone 1

X

X.25-An international standard for communications with a packet data switching network. The X,25 standard specifies the protocol between the data device (such as a computer) and the network (such as a public packet data network).

This diagram shows a X.25 packet data system that is used to transfer banking (cash flow machine) information through several different X.25 systems. This diagram shows bank teller machine in Rome is connected to a bank processing system in London through a virtual path. This path is created through each packet switching exchange (PSE) the X.25 networks before any data is sent. This example shows that each switching point in the X.25 networks validate the transfer of each packet to the next node or switching point. This ensures that data reliably passes through each packet node to reach its previously established destination.

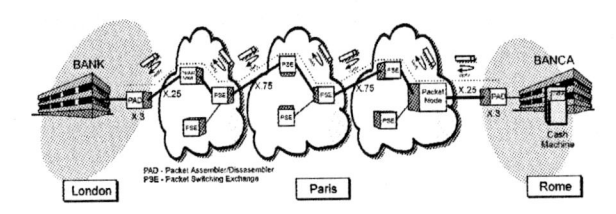

X.25 System

X.75-An international standard for communications between X.25 packet data switches. The X.75 standard specifies the protocol that is used to transfer packets between the network switches. The X.75 standard was developed by CCITT and it is sometimes referred to as the X.25 gateway.

XBar-Crossbar Switches
XCSE-Xylan Certified Switching Expert
XDR-External Data Representation
xDSL-A set of large-scale high bandwidth data technologies that can use standard twisted-pair copper wire to deliver high speed digital services (up to 52 Mbps).
xDSL Splitter-A circuit, device or component that divides a DSL signal into separate voice and data outputs. A DSL splitter is typically used for ADSL and VDSL systems.

The DSL splitter separates the existing telephone signal from the high speed data signal. In the United States, the standard telephone signal (POTS) frequency band extends up to 8 kHz. In Europe, standard telephone signals include additional high frequency components such as 12 kHz billing increment impulses that extend up to 12 kHz. When the DSL splitter is used to allow ISDN signals, the frequency band for the ISDN signal extends up to 80 kHz (120 kHz for ISDN in Germany).

XML-Extensible Markup Language
XMS-Extended Memory Specification
XMTR-Transmitter
XNS-Xerox Network Services
XOR-Exclusive Or Operation
xTalk-Crosstalk

Y

Y-Luminance Signal

YAG-Yttrium-Aluminum Garnet

Yagi Antenna-A directional antenna consisting of a dipole with parasitic directors in front and reflectors behind the active element. This antenna is sometimes called a Yagi-Uda antenna. The actual inventor was Mr. M. Uda. His technical paper was first translated from Japanese into English by F. Yagi, who did not give the author's name when it was first published in English.

Yield Strength-The magnitude of mechanical stress at which a material will begin to deform. Beyond the yield strength point, extension is no longer proportional to stress, and rupture is possible.

YIG-Yttrium-Iron Garnet

YIG Filter-A filter with a yttrium-iron garnet crystal positioned in a magnetic field. Tuning control is obtained by varying the direct current through the solenoid producing the magnetic field.

Yoke-A material that is used to interconnect and focus magnetic cores. A yoke is commonly used in picture tubes to control magnetic fields that adjust the direction of an electron beam.

Yttrium-Iron Garnet (YIG)-A crystalline material used in microwave devices.

YUV-A color space with components Y (luminance), U and V (the color difference components). A color value represented in RGB space can be converted to YUV by a simple linear formula. Image compression (such as JPEG) and video compression (such as MPEG) work on pixel colors in YUV values, often subsampling the U and V components to a lower spatial resolution because the eye is less sensitive to errors in U and V than in Y. See also RGB and CMYK.

Z

Z-The symbol for impedance, measured in ohms.
ZAK-Zero Administration Kit
ZAWS-Zero Administration For Windows
ZBTSI-Zero Byte Time Slot Interchange
Zener Breakdown-In a semiconductor junction, a sudden nondestructive current "breakdown" that results when the electric field in the barrier region is sufficiently high to cause avalanche current multiplication that greatly increases the number of mobile carriers. See Avalanche.
Zener Diode-A type of semiconductor diode, usually fabricated of silicon, in which reverse voltage breakdown results from the avalanche breakdown phenomenon. Named for American physicist Clarence Zener. The silicon in a Zener diode is typically intentionally made with extremely high doping levels so the negative voltage breakdown will occur at a low voltage such as 3 to 10 volts. Zener diodes are also designed with good heat dissipating capacity to prevent overheating due to their high power dissipation. In this breakdown mode, a Zener diode provides a constant voltage that is useful as an accurate voltage comparison or reference, or in a voltage regulator circuit.
Zero Byte Time Slot Interchange (ZBTSI)-A technique for achieving clear channel capability on DS-1 facilities. This is a method of pre-processing the binary data in a DS-1 bit stream so that AMI line coding can be used but there is no zeroes density limitation on the bit stream. ZBTSI permits the DS-1 link to carry any desired bit stream in each channel, even a stream consisting of all binary zeroes. This capability is sometimes called unrestricted or clear channel. In ZBTSI technology, four complete frames of DS-1 bits are stored in memory before transmitting. This comprises 96 time slots (of 8 bits each) and four framing bits. A masking pattern is combined with the bits in all 96 time slots. Any time slots that contain 8 binary zeros (if present) are noted by generating a list of their locations (a "pointer list") relative to the beginning of the group of time slots. Entries in this list are binary numbers representing values from 1 to 96. The data in memory is then "collapsed" by taking out the data from the time slots that contain 8 zeros, and moving all other time slot data towards the high number end of the list. This leaves us with just as many "empty" time slots at the first few time slot locations as the number of original time slots that had eight zeros, but which were removed. The numbers from the pointer list are then put into these "empty" locations. The last pointer has the binary value corresponding to 128 added to it to indicate that it is the last pointer. Blocks of four frames that have no time slots found with 8 zeros are distinguished from blocks that had all the collapsing and pointer setting just described, by means of setting one of the message bits of ESF to 1, but setting it to zero otherwise. The four frames of DS-1 data are then transmitted using AMI line coding, and the entire process just described is reversed step by step. ZBTSI allows test equipment that treats a BPV as a fault to be used without modification, but it introduces more time delay into the transmission of DS-1 signals and is more complex than B8ZS, its "competitor" used by most of the RBOCs. ZBTSI was invented by VeriLink and was installed in the 1980s and 1990s by two of the RBOCs.
Zero Code Suppression-in digital systems, a process of inserting a digital one into a signal to break up long strings of zeros that might cause a loss of timing in line repeaters. (See also: bipolar with eight zeros substitution, zero-byte time slot interchange.)
Zero Suppression-in telecommunications, techniques that limit the number of consecutive data zeros that may be transmitted.
Zero-Slot LAN-A local area network (LAN) that uses existing serial or parallel ports on each network computer rather than a using a network interface card (NIC) that is installed in to the computer's expansion bus. Because the computer's serial or parallel output port has limited maximum data transmission capability, zero-slot LANs are usually communicate much slower than networks that use NICs with high speed LAN technology.
Zip Drive-A removable storage device storing 100 MB to 250 MB of data on 3.5-inch ZIP disks.
Zmodem-A file transfer protocol that is designed to transfer large amounts of data transfers with a limited number of errors. Zmodem is similar to

Xmodem and Ymodem. However, Zmodem includes a feature called checkpoint restart that allows the transfer of data to temporarily interrupted and restarted at the point of interruption.

Zombie-A process that is not used (dead) and has not yet been deleted from the process table in a Unix or other operating systems.

Zone Bandwidth-The amount of bandwidth available (or assigned) in a particular area (zone) of a communication network (such as a data network).

Zone Paging-The process of sending page messages to a particular group of paging transmitters to restrict the paging service to a specific area (zone).

Zone Usage Measurement-A process of calculating toll charges for calls that completed between zones.

Internet Telephone Related Magazines

Broadband Week Magazine
8878 South Barrons Blvd.
P.O. Box 266008
Highlands Ranch, CO 80126-6008
www.broadbandweek.com

Business Leader
3801 Wake Forest Road, Suite 102
Raleigh, NC 27609
919 872-7077
www.businessleader.com

Call Center Magazine
600 Community Drive
Manhasset, NY 11030
United States
215 355-2886
www.commweb.com

Communications Convergence
600 Community Drive
Manhasset, NY 11030
516 562-5000
www.commweb.com

CommVerge
275 Washington St.
Newton, MA 02458
617 964-3030
www.e-insite.net

EE Times.com
600 Community Drive
Manhasset, NY 11030
650 513-4306
www.cmp.com/pubinfo?pubid=20

Global Telephony
9800 Metcalf Ave.
Overland Park, KS 66212
913 341-1300
www.primediabusiness.com

InformationWeek
600 Community Drive
Manhasset, NY 11030
516 562-5000
www.cmp.com

Intele-Card News
523 N. Sam Houston Parkway, East, Suite 300
Houston, TX 77060
281 272-2744
www.intelecard.com

Internet Industry Magazine
101 West 23rd Street, Suite 2286
New York, NY 10011
212 977-3800
www.internetindustry.com

Internet Telephony
One Technology Plaza
Norwalk, CT 06854
United States
800 243-6002
www.tmcnet.com/it/

Lightwave
98 Spit Brook Rd.
Nashua, NH 03062-5737
United States
603 891-0123
www.pennwell.com

Network Computing
600 Community Drive
Manhasset, NY 11030
516 562-5000
www.cmp.com

Internet Telephony Related Magazines

Network Magazine
600 Community Drive
Manhasset, NY 11030
516 562-5000
www.commweb.com

New Telephony
927 18th Street, Suite A,
Santa Monica, CA 90403
310 453-1231
www.newtelephony.com

Phone+
P.O. BOX 40079
Phoenix, AZ 85067-0079
480 990 1101
www.phoneplusmag.com

Portable Design
98 Spit Brook Rd.
Nashua, NH 03062
603 891-0123
www.portable-design.com

Telephony
9800 Metcalf Ave.
Overland Park, KS 66212
913 341-1300
www.primediabusiness.com

WDM Solutions
98 Spit Brook Rd.
Nashua, NH 03062
603 891-0123
www.pennwell.com

Wireless Review
P.O. Box 12901
Overland Park, KS 66212
303 741-8702
www.intertec.com

Wireless Week
P.O. Box 266008
Highlands Ranch, CO 80163
303 470-4800
www.wirelessweek.com

X-Change
3300 N. Central Ave., Suite 2500
Phoenix, AZ 85267-85012
480 990-1100
www.vpico.com

Associations

ADSL Forum
39355 California St. Suite 200
Fremont, CA, 94538
510-608-5905
510-608-5917(F)
www.adsl.com

Advanced Television Systems Committee (ATSC)
1750 K Street, N.W., Suite 1200
Washington, DC, 20006
202-872-9160
202-872-9161(F)
www.atsc.org

Alliance for Telecommunications Industry Solutions (ATIS)
1200 G St. NW, Suite 500
Washington, DC, 20005
202.628.6380
202-393-5453(F)
www.atis.org

American Mobile Telecommunications Association (AMTA)
1150 18th St. NW, Suite 250
Washington, DC, 20036
202-331-7773
202-331-9062(F)
www.amtausa.org

American National Standards Institute (ANSI)
1819 L St. NW
Washington, DC, 20036
202-293-8020
202-293-9287(F)
www.ansi.org

American Registry for Internet Numbers (ARIN)
3635 Concorde Pkwy., ste. 200
Chantilly, VA, 20151-1130
703.227.9840
703.227.0676(F)
www.arin.net

American Teleservices Association (ATA)
3815 River Crossing Parkway, Suite 20
Indianapolis, IN, 46240
317.816.9336
www.ataconnect.org

Association for Local Telecommunications Services (ALTS)
888 17th Street, NW, 12th Floor
Washington, DC, 20006
202.969.ALTS
202.969.ALT1(F)
www.alts.org

Association of Communications Enterprises (ASCENT)
1401 K Street, NW, Suite 600
Washington, DC, 20005
(202) 835-9898
(202) 835-9893(F)
www.ascent.org

Association of TeleServices International (ATSI)
12 Academy Avenue
Atkinson, NH, 03811
(603) 362-9489
(603) 362-9486(F)
www.atsi.org

ATM Forum
Presidio of San Francisco, PO Box 29920
San Francisco, CA, 94129
415-561-6275
415-561-6120(F)
www.atmforum.com

Bluetooth Special Interest Group
www.bluetooth.com

British Standards Institution
British Standards House 389 Chiswick High Road
London, , W4 4AL
England UK
44 (0)20 8996-9000
44 (0)20 8996-7400(F)
www.bsi-global.com

Associations

Building Industry Consulting Service International (BICSI)
8610 Hidden River Parkway
Tampa, FL, 33637
813.979.1991
813.971.4311(F)
www.bicsi.org

Business Technology Association
12411 Wornall Road, Suite 200
Kansas City, MO, 64145
816.941.3100
816.941.4838(F)
www.bta.org

Cable Television Laboratories, Inc (CableLabs)
400 Centennial Pkwy.
Louisville, CO, 80027-1266
303-661-9100
303-661-9199(F)
www.cablelabs.com

California Broadband Users' Group
P.O. Box 27901-391
San Francisco, CA, 94127
415.241.9943
415.753.6942(F)
www.ciug.org or www.isdnworld.com

California Cable & Telecommunications Association (CCTA)
4341 Piedmont Ave. (P.O. Box 11080)
Oakland, CA, 94611
(510) 428-2225
510-428-0151(F)
www.calcable.org

Canadian Standards Association (CSA)
5060 Spectrum Way, Ste 100.
Mississauga, Ontario, l4W 5N6
CANADA
416-747-4044
416-747-2510(F)
www.csa.ca

Canadian Wireless Telecommunications Association (CWTA)
130 Albert Street, Suite 1110
Ottawa, ON, K1P 5G4
Canada
613-233-4888
613-233-2032(F)
www.cwta.ca

CDMA Development Group (CDG)
575 Anton Blvd., Ste. 560
Costa Mesa, CA, 92626
1-888-800-CDMA or 1-714-545-5211
714-545-4601(F)
www.cdg.org

Cellular Telecommunications Internet Association (CTIA)
1250 Connecticut Ave NW, Suite 800
Washington, DC, 20036
202-785-0081
202-785-0721(F)
www.wow-com.com

CommerceNet
10050 N. Wolfe Rd. Ste. SW2-255
Cupertino, CA, 95014
408-446-1260
408-446-1268(F)
www.commercenet.net

Communications Fraud Control Association (CFCA)
3030 North Central Avenue, Suite 707
Phoenix, AZ, 85012
602.265.2322 (CFCA)
602.265.1015(F)
www.cfca.org

Competitive Telecommunications Association (CompTel)
1900 M Street, N.W., Suite 800
Washington, DC, 20036
202.296.6650
202.296.7585(F)
www.comptel.org

Associations

Competitive Telephone Carriers of New York, Inc.
1 Columbia Place
Albany, NY, 14
518-434-8112
518-434-3232(F)

Computer and Communications Industry Association (CCIA)
666 11th St. NW
Washington, DC, 20001
202-783-0070
202-783-0534(F)
www.ccianet.org

Defense Advanced Research Projects Agency (DARPA)
3701 Fairfax Drive
Arlington, VA, 22203-1714
(703) 526-6630
(703) 528-1943(F)
www.darpa.mil

Electronic Industries Association (EIA)
2500 Wilson Blvd.
Arlington, VA, 22201
703-907-7500
703-907-7501(F)
www.eia.org

European Telecommunications Standards Institute (ETSI)
650, route des Lucioles
Sophia-Antipolis, Cedex, 06921
France
33 (0)492944311
33 (0)492385299(F)
www.etsi.fr

Federal Communications Commision (FCC)
445 12 St. SW
Washington, DC, 20554
888-CALL-FCC
202-418-0232(F)
www.fcc.gov

Home Phoneline Networking Alliance (HomePNA)
Bishop Ranch 2, 2694 Bishop Drive, Suite 105
San Ramon, CA, 94583
925-277-8110
925-277-8111(F)
www.homepna.org

Indiana Telecommunications Association (ITA)
54 Monument Circle, Suite 200
Indianapolis, IN, 46204
317-635-1272
317-635-0285(F)
www.itainfo.org

Industrial Telecommunications Association (ITA)
1110 North Glebe Rd. Suite 500
Arlington, VA, 22201-5720
703-528-5115
703-524-1074(F)
www.ita-relay.com

Infared Data Association (IrDA)
P.O. Box 3883
Walnut Creek, CA, 94598
www.irda.org

Information Technology Association of America (ITAA)
1401 Wilson Blvd. Ste. 1100
Arlington, VA, 22209
703-522-5055
703-525-2279(F)
www.itaa.org

Information Technology Industry Council
1250 I Street NW, Suite 200
Washington, DC, 20005
202.737.8888
202.638.4922(F)
www.itic.org

Institute of Electrical and Electronics Engineers, Inc. (IEEE)
1828 L Street, N.W., Suite 1202
Washington, DC, 20036-5104
202-785-0017
202-785-0835(F)
www.ieee.org

Associations

Insulated Cable Engineers Association (ICEA)
P.O. Box 1568
Carrolton, GA, 30117
508-394-4424
www.icea.net

International Multimedia Teleconferencing Consortium, Inc.
Bishop Ranch 2, 2694 Bishop Drive, Suite 275
San Ramon, CA, 94583
925-275-6600
925-275-6691(F)
www.imtc.org

International Municipal Signal Association (IMSA)
165 East Union Street (PO Box 539)
Newark, NY, 14513-0539
315-331-2182 1-800-723-IMSA
315-331-8205(F)
www.IMSAsafety.org

International Telecommunications Union (ITU)
Place des Nations
Geneva 20, Geneva, CH-1211
Switzerland
+41 22 730 51 11
+41 22 733 7256(F)
www.itu.int

International VoIP Council
202-326-1743
www.voipcouncil.org

International Wireless Telecommunications Association (IWTA)
1150 18th St. NW, Suite 250
Washington, DC, 20036
202-331-7773
202-331-9062(F)
www.iwta.org

Internet Mail Consortium
127 Segre Place
Santa Cruz, CA, 95060
831-426-9827
831-426-7301(F)
www.imc.org

InterNIC
P.O. Box 1656
Herndon, VA, 22070
www.internic.net

Mid-America Cable Telecommunications Association (Mid-America)
P.O. Box 3306
Lawrence, KS, 66046
785-841-9241
785-841-4975(F)
www.midamericacable.com

Minnesota Telephone Association
30 East 7th Street
St. Paul, MN, 55101
651-291-7311
651-291-2795(F)
www.mnta.org

National Association of Broadcasters (NAB)
1717 N Street, NW
Washington, DC, 20036-2891
202-429-5300
202-429-4199(F)
www.nab.org

National Association of Radio and Telecommunications Engineers (NARTE)
P.O. Box 678
Medway, MA, 02053
508-533-8333
508-533-3815(F)
www.narte.org

National Association of Regulatory Utility Commissioners (NARUC)
1101 Vermont Avenue NW, Suite 200
Washington, DC, 20005
202-898-2200
202-898-2213(F)
www.naruc.org

National Association of State Telecommunications Directors (NASTD)
PO Box 11910 or 2760 Research Park Dr.
Lexington, KY, 40578-1910
859-244-8186
859-244-8001(F)
www.nastd.org

National Cable TV Association (NCTA)
1724 Massachusetts Ave., N.W.
Washington, DC, 20036
202-775-3669
202-775-3692(F)
www.ncta.com

National Emergency Number Association (NTCA)
422 Beecher Rd.
Columbus, OH, 43230
800-332-3911 or (614) 741-2080
(614) 933-0911(F)
www.nena9-1-1.org

National Exchange Carrier Association (NECA)
80 South Jefferson Rd.
Whippany, NJ, 07981-8597
973-884-8000 or 800-228-8597
973-884-8469(F)
www.neca.org

National Fire Protection Association (NFPA)
1 Batterymarch Park
Quincey, MA, 02169
617-770-3000
617-770-0700(F)
www.nfpa.org

National Institute of Standards and Technology (NIST)
100 Bureau DR
Gaithersburg, MD, 20899
301-975-2000
www.nist.gov

National Technical Information Service (NTIS)
U.S. Department of Commerce
Springfield, VA, 22161
703-605-6000
703-321-8547(F)
www.ntis.gov

National Telecommunications and Information Administration (NTIA)
U.S. Department of Commerce 1401 Constitution Ave. NW
Washington, DC, 20230
202-482-7002
www.ntia.doc.gov

National Telephone Co-op Association (NTCA)
4121 Wilson Blvd., Tenth Floor
Arlington, VA, 22203
703-351-2000
703-351-2001(F)
www.ntca.org

Network Professional Association (NPA)
195 South C St.
Tustin, CA, 92780
714-573-4780
714-669-9341(F)
www.npanet.org

Office of the Federal Register (OFR)
National Archives & Record Administration,
700 Pennsvainia Ave. NW
Washington, DC, 20408
866-325-7208
202-523-6866(F)
www.nara.gov/fedreg

OFTEL
50 Ludgate Hill
London, EC4M 7JJ
England (UK)
44-020-7634-8700
44-020-7634-8845(F)
www.oftel.gov.uk

Optical Storage Technology Association (OSTA)
19925 Stevens Blvd.
Cupertino, CA, 95014
408-253-3695
408-253-9938(F)
www.osta.org

Organization for Promotion and Advancement of Small Telecom Companies (OPASTCO)
21 Dupont Circle, NW, Suite 700
Washington, DC, 20036
202-659-5990
202-659-4619(F)
www.opastco.org

Associations

Pacific Telecommunications Council (PTC)
2454 S. Beretania ST., Suite 302
Honolulu, HI, 96826
808-941-3789
808-944-4874(F)
www.ptc.org

PCI Industrial Computer Manufacturers Group (PICMG)
401 Edgewater Place, Suite 600
Wakefield, MA, 01880
781-246-9318
781-224-1239(F)
www.picmg.org

Personal Communications Industy Association (PCIA)
500 Montgomery St., Suite 700
Alexandria, VA, 22314-1561
703-739-0300
703-836-1608(F)
www.pcia.com

Portable Computer and Communications Association (PCCA)
P.O. Box 2460
Boulder Creek, CA, 95006
541-490-5140
419-831-4799(F)
www.pcca.org

Rural Cellular Association (RCA)
701 Brazos, Suite 320
Austin, TX, 78701
800-722-1872
512-472-1071(F)
www.rca-usa.org

Satellite Broadcasting & Communications Association (SBCA)
225 Reinekers Lane, Suite 600
Alexandria, VA, 22314
703-549-6990
703-549-7640(F)
www.sbca.com

Satellite Industry Association (SIA)
225 Reinekers Lane, Suite 600
Alexandria, VA, 22314
703-549-8697
703-549-9188(F)
www.sia.org

Small Business in Telecommunications (SBT)
1331 H St., NW Suite 500
Washington, DC, 20005
202-347-4511
202-347-8607(F)
www.sbthome.org

Society of Cable Telecommunications Engineers Inc. (SCTE)
140 Phillips Road
Exton, PA, 19341-1318
(800) 542-5040 or (610) 363-6888
610-363-5898(F)
www.scte.org

Society of Motion Pictures & Television Engineers (SMPTE)
595 W. Hartsdale Avenue
White Plains, NY, 10607-1824
914-761-1100
914-761-3115(F)
www.smpte.org

Telecommunications Industry Association (TIA)
2500 Wilson Blvd, Suite 300
Arlington, VA, 22201
703-907-7700
703-907-7727(F)
www.tiaonline.org

The Association for Telecommunications Professionals in Higher Education (ACUTA)
152 W. Zandale Drive, Suite 200
Lexington, KY, 40503
606-278-3338
606-278-3268(F)
www.acuta.org

The Computing Technology Industry Association (CompTIA)
1815 S. Myers Rd.
Oakbrook Terrace, IL, 60181
630-268-1818
630-268-1834(F)
www.comptia.org

The Consumer Electronics Association (CEA)
2500 Wilson Blvd.
Arlington, VA, 22201
703-907-7600
703-907-7675(F)
www.ce.org

The Electronic Frontier Foundation (EFF)
454 Sohotwell St.
San Francisco, CA, 94110-4832
415.436.9333
415.436.9993(F)
www.eff.org

The Open Group
44 Montgomery St. Ste. 960
San Francisco, CA, 94104
415-374-8280
415-374-8293(F)
www.opengroup.org

United States Telecom Association (USTA)
1401 H St. NW, Suite 600
Washington, DC, 20005-2164
202-326-7300
202-326-7333(F)
www.usta.org

United States Telecommunications Training Institute (USTTI)
1150 Connecticut Avenue, NW Suite 702
Washington, DC, 20036
USA
202.785.7373
202.785.1930(F)
www.ustti.org

United Telecom Counsel (UTC)
1901 Pennsylvania Avenue, NW 5th Floor
Washington, DC, 20006
202-872-0030
202-872-1331(F)
www.utc.org

Universal Wireless Communications Consortium (UWCC)
8302 159th Pl. NE
Redmond, WA, 98052
425-580-5031
www.uwcc.org

Wall Street Telecommunications Association (WSTA)
241 Maple Ave.
Red Bank, NJ, 07701
732-530-8808
731-530-0020(F)
www.wsta.org

Wireless Communications Association International (WCA)
1140 Connecticut Ave, NW Suite 810
Washington, DC, 20036
202-452-7823
202-452-0041(F)
www.wcai.com

Wireless Dealers Association (WDA)
9746 Tappenbeck Dr.
Houston, TX, 77055
800 624-6918 or 713 467-0077
800-820-2284(F)
www.wirelessdealers.com

Wireless Industry Association (WIA)
9746 Tappenbeck Drive
Houston, TX, 77055
800-624-6918 or 713-467-0077
800-820-2284(F)
www.wirelessindustry.com

Wireless LAN Alliance (WLANA)
P.O. Box 9097
San Jose, CA, 95157
650-352-4709
650-649-2305(F)
www.wlana.com

International Engineering Consortium (IEC)
549 West Randolph Street
Suite 600
Chicago, IL 60661-2208 USA
1-312-559-4100
1-312-559-4111(F)
www.iec.org

Associations

European Association for Standardizing Information and Communication Systems (ECMA)
114 Rue du Rhône
CH-1204 Geneva, Switzerland
http://www.ecma.ch/

European Telecommunications Standards Institute (ETSI)

650, route des Lucioles
06921 Sophia-Antipolis Cedex
FRANCE
33 (0)4 92 94 42 00
33 (0)4 93 65 47 16(F)

MIT Internet & Telecoms Convergence Consortium (MIT-ITC)
E40-234, 1 Amherst St.
Cambridge, MA 02139 USA
617-253-4138
617-253-7326(F)
http://itel.mit.edu/

Institute of Electrical and Electronics Engineers (IEEE)
1828 L Street, N.W., Suite 1202
Washington, D.C. 20036-5104
1 202-785-0017
1 202-785-0835(F)
www.ieee.org

Multiservice Switching Forum (MSF)
39355 California Street #307
Fremont, CA 94538
510-608-5922
510-608-5917(F)
http://www.msforum.org

International Multimedia Teleconferencing Consortium (IMTC)
Bishop Ranch 2
2694 Bishop Drive, Suite 275
San Ramon, CA 94583
1 925-275-6600
1 925-275-6691(F)
http://www.imtc.org/

Telecommunications Industry Association (TIA)
2500 Wilson Blvd., Suite 300
Arlington, VA 22201   USA
 703-907-7700
 703-907-7727
 703-907-7776(F)
www.tiaonline.org

International Telecommunications Union (ITU)
ITU - Place des Nations
CH-1211 Geneva 20
Switzerland
4122 730 5115
4122 730 5595(F)
http://www.itu.int

TeleManagement Forum (TMF)
1201 Mt. Kemble Ave.
Morristown, NJ 07960-6628
1 973-425-1900
1 973-425-1515(F)
www.tmforum.org

The Internet Engineering Task Force
www.ietf.org

United States Internet Service Provider Association (USIPSA)
1330 Connecticut Avenue, NW
Washington, DC 20036
202-862-3816
202-261-0604(F)
http://www.cix.org

Association of Internet Professionals
4790 Irvine Boulevard, Suite 105-283
Irvine, CA 92620
866-AIP-9700
1-501-423-2248
http://www.association.org

International Internet Marketing Association
PO Box 4018
Vancouver Main
349 West Georgia Street
Vancouver, BC
V6B 3Z4
www.iimaonline.org

Associations

WA Internet Association
250 St Georges Terrace
Perth WA 6000
www.ix.waia.asn.au

Internet Service Providers' Consortium (ISP/C) OR Forum
1301 Shiloh Road, Suite 720
PO Box 1086
Kennesaw, GA 30144-8086
866-533-6990
678-819-1028(F)
www.ispc.org or http://www.ispf.com/

Internet SOCiety (ISOC)
1775 Wiehle Ave., Suite 102
Reston, VA 20190
703 326 9880
703 326 9881(F)
http://www.isoc.org/isoc/

Alliance for Global Internet Services (AGIS)
725 East 175 North
Lindon, UT 84042
Tel: 801-796-9311
http://agis.org

US Internet Industry Association (USIIA)
815 Connecticut Avenue, NW
Suite 620,
Washington, DC 20006
or
5810 Kingstowne Center Drive
Suite 120, PMB 212
Alexandria, VA 22315-5711
703-924-0006
703-924-4203(F)
http://www.usiia.org

Frame Relay Forum
39355 California St., Suite 307,
Fremont, CA 94538
510.608.5920
510.608.5917(F)
www.frforum.com/

CableLabs (Cable Television Laboratories, Inc.)
400 Centennial Parkway
Louisville, CO 80027-1266
303-661-9100
303-661-9199(F)
www.cablelabs.com

DSL Forum
39355 California Street, Suite 307
Fremont, CA 94538
510-608-5905
510-608-5917(F)
www.dslforum.org/

United States Telephone Associations (USTA)
1401 H Street, N.W., Suite 600
Washington, DC 20005-2164
202-326-7300
202-326-7333(F)
www.usta.org

ATM Forum
Presidio of San Francisco
P.O. Box 29920 (mail)
572B Ruger Street (surface)
San Francisco, CA 94129-0920
415-561-6275
415-561-6120(F)
www.atmforum.com

10 Gigabit Ethernet Alliance
1300 Bristol Street North, Suite 160
Newport Beach, CA 92660
949-250-7155
949-250-7159(F)

Association Management Solutions (AMS)
39355 California Street, Suite 307
Fremont, CA 94538
510-608-5900
510-608-5917(F)
www.amsl.com

ACCU
1330 Trinity Dr.
Menlo Park, CA 94025
650-233-9082
www.accu.org

Associations

Internet Telephone Service Providers

4ecalls.com
23 Spruce Avenue, Stillorgran Industrial Park,
Blackrock, Co Dublin,
Ireland
35 3 1 2137981
www.4ecalls.com

Allen-Martin Inc.
217 East 16th Avenue
Grand Rapids, MI49546
251 968-3828
www.allen-martin-inc.com/

Alpha Telecom
391 Richmond Rd.
Twickenham, MiddlesexTW1 2EF
United Kingdom
44 020 7892-3660
www.alphatelecom.com

Anyuser
18000 Studebaker Road, Suite 295
Cerritos, CA90703
562 865-9730
www.anyuser.com/

AudioCodes Inc.
2890 Zanker Road, Suite 200,
San Jose, CA95134
408 577-0488
www.audiocodes.com/

Band-X
61 Broadway, Suite 2220
New York, NY10006
646 835 4900
www.band-x.com/en/

Belgacom Corporate Customers Division
Boulevard du Roi Albert II, 27
Brussels, 1030
Belgium
322 202 62 30
www.belgacom.be/carrier/

Bestnetcall
5075 Cascade Rd. SE, Suite K
Grand Rapids, MI49546
www.bestnetcall.com/index.html

Call2 Limited
69 Soke Rd.
Silchester, ReadingRG7 2PB
United Kingdom
44 118 970-2001
www.call2.com

Callserve Communications Ltd.
2 Harbour Exchange Square
London, E14 9GE
United Kingdom
44 020 7517 7100
www.callserve.net/homepage.asp

CLEAR2PHONE (Clear Internet Phone, LLC)
75 South Broadway, Suite 455
White Plains, NY10161
www.clear2phone.com/

Comease.com
Blk 11 Kallang Place #06-10/12
Singapore, 339155
65 6297 8122
www.comease.com/

Internet Telephone Service Providers (ITSPs)

Connecting Teams
707 780-3000
www.connectingteams.com/

Deltathree
75 Broad Street, 31st Floor
New York, NY10004
212 500-4850
corp.deltathree.com/

Dialpad Communications, Inc.
430 N. McCarthy Blvd, Suite 200
Milpitas, CA95035
408 635-1000
www.dialpad.com/

Ekofon Headquarters
Avenida de las Palmas 555 Piso 8,
Colonia Lomas, Distrito Federal,11000-
Mexico
555 227-7360
www.ekofon.com

Ekofon USA
23679 Calabassas Rd, 758
Calabassas, CA91302
818 337-7374
www.ekofon.com

Equant- U.S. offices
12490 Sunrise Valley Drive, MS
VARESF0257
Reston, VA20196
866 849-4185
www.equant.com

FastWeb
Via Broletto, 5
Milano, 20121
Italy
02/45451
www.fastweb.it

GeoPhones.com / GeoPortals.com
20 Exchange Place, 36th Floor
New York, NY10005
800 700-4099
phones.geoportals.com/

Go2Call
1603 Orrington St. Suite 1190,
Evanston, IL60201
847 864 4123
www.go2call.com

GulfPines Communications
200 North 40th Avenue, Suite One (PO Box 15236)
Hattiesburg, MS39404
601 599-1000
www.gulfpines.com

I-Link
13751 S. Wadsworth Park Drive Suite 200
Draper, UT84020
801 576-5000
www.i-link.com/

iBasis
20 Second Avenue
Burlington, MA01803
781 505-7500
www.ibasis.net/

Innomedia Inc.
90 Rio Robles, Suite 100
San Jose, CA95134
408 432-5424
www.innomedia.com

Internet Telephone Service Providers (ITSPs)

Interoute Americas
6th Floor, 13010 Morris Road
Alpharetta, GA30004
770 576 1970
www.interoute.com

IPCB.net / IPClearingBoard, Inc.
100 Pine Street, Suite 2820,
San Francisco, CA94111
415 362-2360
www.ipcb.net/

Iscom
90 William St., Suite 1202,
New York, NY10038
212 324-1100
www.iscom.net

ITXC
750 College Road East,
Princeton, NJ08540,
609 750-3333
www.itxc.com/

MediaRing.com Inc.
99 West Tasman Drive,
San Jose, CA95134
408 383 9222
www.mediaring.com

MiBridge
446 Highway 35, Building C,
Eatontown, NJ07724
732 544-2322
www.mibridge.com

Net2Phone
171 Main Street
Hackensack, NJ07601
USA
973 412-2882
www.Net2phone.com

NetNumber, Inc.
Wannalancit Technology Center, 650 Suffolk St.
Suite 307, Lowell, MA01854
978 848-2820
www.enum.com

Network Communications International Corp
606 E. Magrill
Longview, TX75601
903 757-4455
www.ncic.com/

NEXCell Telecom Co., Ltd.
401 4Th Fl, Sokchon City B/D, 66-7,
Bangi-Dong, Songpa-Gu,Seoul138-828
Korea
82 2 417 1974
www.nexcell.net

Paramax, Inc.
4 Corporate Plaza Suite: 215
Newport Beach, CA92660
800 525-2324
www.paramax.net/

PCCall.com
One Silicon Alley Plaza, 90 William St.,
Ste.1202
New York, NY10038
212 324-1100
www.PCCall.com

www.ITExpo.Com

Internet Telephone Service Providers (ITSPs)

Phoneserve
2 Harbour Exchange Square London,
E14 9GE
United Kingdom
44 0 20 7517 7100
www.phoneserve.com/

Quicknet Technologies, Inc.
520 Townsend Street
San Francisco, CA94103
415 864-5225
www.quicknet.net/

RingTime
Van Eeghenstraat 80-82 1071 GK,
Amsterdam
The Netherlands
31 20-305 2525
www.ringtime.net/

SingTel
1350 Old Bayshore Highway, Suite #320
Burlingame, CA90410
650 558 3950
business.singtel.com/

Super Technologies, Inc.
914 Scenic Hwy,
Pensacola, FL32503
850 433-8555
www.virtualphoneline.com/

visitalk.com
8936 N CENTRAL AVE,
PHOENIX, AZ85020
602 850-3360
www.visitalk.com/

Voicenet Communications
17 Richard Road
Ivyland, PA18974
215 674-9290
www.voicenet.com

Vonage
2147 Route 27
Edison, NJ08817
732 528-2627
www.Vonage.com

XchangePoint
3rd Floor, 17-18 Henrietta Street,
Covent Garden, LondonWC2E 8QH
United Kingdom
44 020 7395 6020
www.xchangepoint.net

INTERNET TELEPHONY® CONFERENCE & EXPO

TMC

The #1 Global Event Focused on Voice, Video, Fax, and Data Convergence

www.itexpo.com

Launched in 1998.

Get Your FREE Subscription to the VoIP Industry's First and Leading Monthly Magazine.

www.itmag.com

eNews
in your inbox

FREE, laser-focused eNewsletters delivered right to your inbox.

Receive important information in time to take advantage of it.

Get reliable news and analysis from the publishers of *Customer Inter@ction Solutions*® and *Internet Telephony*® magazines

- Internet Telephony
- Communications Solutions
- Next-Gen Service Providers
- Communications Reseller
- Communications Developer
- Next-Gen Wireless
- New! SIP eNews
- The Softswitch Review
- VoiceXML/SALT
- Plus many more!

enews.tmcnet.com

Training
Onsite, Public Courses, Web Seminars

Do you want to get more information or have an expert help you to understand how to use your data networks or the Internet to reduce your telecommunications costs?

Consider using Althos to educate you or your staff on the implementation and technologies used to connect telephones through data networks. Althos offers onsite courses, public courses, and real-time web seminar training. If you want instruction from experts who have setup Internet telephone systems, consider Althos. Althos can customize training to cover your key subject areas. Althos also has standard courses including:

VoIP for Executives

What a management team should know about the costs and risks of sending voice through data networks.

VoIP for Operation Managers

The operation details that operation managers need to know about selecting, implementing, and managing systems that send voice through data networks.

VoIP for Telecom Professionals

The technical information for telecommunication professionals who need to understand competing technologies and how telecommunications systems are adapting to meet the changes.

Althos Training, 404 Wake Chapel Road, Fuquay NC 27526 USA
1-919-557-2260 1-800-227-9681 Fax 1-919-557-2261 WWW.Althos.com
Email: Info@Althos.com

Typical Training Costs

On-Site Training

Althos on-site training cost ranges from $2,200 to $3,600 per day of instruction plus expenses dependent on the length of the course and the type of content (labs, exercise materials, and instructor skill level). Althos does not charge for instructor travel time.

If Althos must incur travel expenses in conjunction with the project (this is typical for an on-site training session or presentation), travel expenses will be reimbursed on the basis of actual cost. The client prior to commitment will approve all travel expenses.

Online Training (Web Seminars)

Althos provides some courses and executive briefings in the form of online web seminars. Althos web training seminars allow two-way audio with all the participants along with presentation materials. The typical cost of web seminars range from about $85 for a 1-hour open enrollment executive briefing to approximately $350 per day for standard course instruction.

Open Enrollment

Althos periodically offers open enrollment to allow individuals to attend courses. The typical cost for individual enrollment ranges from $1,100 to $1,800 per student dependent on the location and type of course. Open enrollment courses include meals and materials (books and workbooks).

Custom Course Development

Althos can customize our courses to meet your specific training need or we can research and use your materials to create a new course. Custom course development fees range from $50 to $200 per presentation slide (graphics + descriptive text).

About Althos

Althos publishing produces books and educational materials for consumers and businesses that help them discover, select, and implement alternative communication technologies and systems. Althos training provides course development and instruction services that help companies transition their communications systems to reduce telecommunication costs and offer enhanced revenue producing services. Althos research performs competitive analysis, customer and vender lists, technology analysis, patent analysis, market analysis, business case analysis, product analysis and expert witness services.

Printed in the United States
39761LVS00003B/3-6